Student Solutions Manual and Study Guide

for

Poole's

Linear Algebra
A Modern Introduction

Second Edition

Robert Rogers
Bay State College

Australia • Brazil • Canada • Mexico • Singapore • Spain • United Kingdom • United States

© 2006 Thomson Brooks/Cole, a part of The Thomson Corporation. Thomson, the Star logo, and Brooks/Cole are trademarks used herein under license.

ALL RIGHTS RESERVED. No part of this work covered by the copyright hereon may be reproduced or used in any form or by any means—graphic, electronic, or mechanical, including photocopying, recording, taping, Web distribution, information storage and retrieval systems, or in any other manner—without the written permission of the publisher.

Printed in the United States of America
2 3 4 5 6 7 09 08 07 06 05

Printer: EPAC Technologies, Inc.

0-534-99858-5

Cover image: Royalty Free/Photodisc/Getty Images

Thomson Higher Education
10 Davis Drive
Belmont, CA 94002-3098
USA

For more information about our products, contact us at:
Thomson Learning Academic Resource Center
1-800-423-0563

For permission to use material from this text or product, submit a request online at
http://www.thomsonrights.com.
Any additional questions about permissions can be submitted by email to **thomsonrights@thomson.com**.

Contents

1 Vectors   **1**
    1.1 The Geometry and Algebra of Vectors 3
    1.2 Length and Angle: The Dot Product 11
          Exploration: Vectors and Geometry 25
    1.3 Lines and Planes 27
          Exploration: The Cross Product 45
    1.4 Code Vectors and Modular Arithmetic 47
          Chapter 1 Review 53

2 Systems of Linear Equations   **61**
    2.1 Introduction to Systems of Linear Equations 63
          Exploration: Lies My Computer Told Me 69
    2.2 Direct Methods for Solving Linear Systems 71
          Exploration: Partial Pivoting 87
          Exploration: An Introduction to the Analysis of Algorithms 89
    2.3 Spanning Sets and Linear Independence 91
    2.4 Applications 109
    2.5 Iterative Methods for Solving Linear Systems 121
          Chapter 2 Review 127

3 Matrices   **135**
    3.1 Matrix Operations 137
    3.2 Matrix Algebra 143
    3.3 The Inverse of a Matrix 155
    3.4 The LU Factorization 163
    3.5 Subspaces, Basis, Dimension, and Rank 177
    3.6 Introduction to Linear Transformations 195
    3.7 Applications 209
          Chapter 3 Review 221

4 Eigenvalues and Eigenvectors   **231**
    4.1 Introduction to Eigenvalues and Eigenvectors 233
    4.2 Determinants 247
          Exploration: Geometric Applications of Determinants 269

4.3 Eigenvalues and Eigenvectors of $n \times n$ Matrices 275
4.4 Similarity and Diagonalization 287
4.5 Iterative Methods for Computing Eigenvalues 299
4.6 Applications and the Perron-Frobenius Theorem 313
    Chapter 4 Review 327

5 Orthogonality   337
5.1 Orthogonality in $\mathbb{R}^n$ 339
5.2 Orthogonal Complements and Projections 349
5.3 The Gram-Schmidt Process and the QR Factorization 355
    Exploration: The Modified QR Factorization 359
    Exploration: Approximating Eigenvalues with the QR Algorithm 361
5.4 Orthogonal Diagonalization of Symmetric Matrices 363
5.5 Applications 369
    Chapter 5 Review 381

6 Vector Spaces   395
6.1 Vector Spaces and Subspaces 397
6.2 Linear Independence, Basis, and Dimension 403
    Exploration: Magic Squares 413
6.3 Change of Basis 415
6.4 Linear Transformations 423
6.5 The Kernel and Range of a Linear Transformation 429
6.6 The Matrix of a Linear Transformation 439
    Exploration: Tilings, Lattices, and Crystallographic Restriction 449
6.7 Applications 451
    Chapter 6 Review 457

7 Distance and Approximation   471
7.1 Inner Product Spaces 473
    Exploration: Vectors and Matrices with Complex Entries 481
    Exploration: Geometric Inequalities and Optimization Problems 485
7.2 Norms and Distance Functions 489
7.3 Least Squares Approximation 497
7.4 The Singular Value Decomposition 505
7.5 Applications 515
    Chapter 7 Review 521
    Appendix I: Key Definitions and Concepts 535
    Appendix II: Theorems 559

Chapter 1

Vectors

1.1 The Geometry and Algebra of Vectors

In this study guide we will explore the thinking behind the exercises.
By exploring the purpose of the exercises we hope to better understand the material.

Q: What is our big goal for Section 1.1?
A: To gain a basic understanding of what vectors are and how they work.

Let's develop our vocabulary so we know what's being talked about.
Consider the vocabulary that appears on page 3:

head	B of $\overrightarrow{AB}$ also called the **terminal point** of vector $\overrightarrow{AB}$
initial point	A of $\overrightarrow{AB}$ also called the **tail** of vector $\overrightarrow{AB}$
position vector	vector with tail at the origin O, i.e. $\overrightarrow{OA}$
tail	A of $\overrightarrow{AB}$ also called the **initial point** of vector $\overrightarrow{AB}$
terminal point	B of $\overrightarrow{AB}$ also called the **head** of vector $\overrightarrow{AB}$
vector	*directed* line segment with *length* and *direction*

Pay close attention to details: $(1, 2, 3)$ is a point, but $[1, 2, 3]$ is a vector.

Contrast new ideas with old ideas we already understand.
For example, points (which we know well) vs. vectors (which are new to us):

point P	uppercase, normal font	*vs.*	vector **p**	lowercase, bold font
point P	$(1, 2, 3)$ parentheses	*vs.*	vector **p**	$[1, 2, 3]$ brackets
point P	definite location	*vs.*	vector **p**	can move to different locations
point P	no length or direction	*vs.*	vector **p**	definite length and direction

Q: How can we tell if our understanding is developing the way that it should?
A: Consider a concrete example like adding vectors numerically and graphically.

Compare the displacement that takes place to the addition of components.
What do we notice? Does the picture make sense and match the algebra?

We should pay particular attention to details that the author spends a lot of time on.
For example, on page 4 the author spends a whole paragraph on the **zero** vector. Why?

Contrast the **zero** vector in $\mathbb{R}^3$ $[0,0,0]$ with the origin O $(0,0,0)$:

- **0** length zero *vs.* **0** no length
- **0** anywhere in $\mathbb{R}^3$ *vs.* **0** always at the intersection of the axes
- **0** points in any direction *vs.* **0** no direction

Afterwards, we should ask ourselves questions that test our understanding.
For example, following up on the **zero** vector, we might ask:

Q: Are there any other vectors of length zero other than the **zero** vector?
A: No. To have zero length every component has to be zero.
   That, by definition, is the zero vector.

Q: What direction does the **zero** vector point in?
A: Every direction. That is how it manages to be orthogonal to every vector in $\mathbb{R}^n$.

Q: What displacement is caused by the **zero** vector?
A: No displacement at all.

1.1 The Geometry and Algebra of Vectors

Key Definitions

Definitions are given to develop the material. They tell us what is important.
For example, we define two vectors to be equal when they have the same length and direction.
That tells us that length and direction are the defining characteristics of a vector.

v = **w**	p.3	1.1	... if and only if *corresponding* components are equal
	p.3	1.1	in symbols: $\mathbf{v} = \mathbf{w} \Leftrightarrow v_i = u_i$ for all i
	p.3	1.1	vectors are equal if they have the same length and direction
	p.3	1.1	in symbols: $\|\mathbf{v}\| = \|\mathbf{w}\|$ and $d_\mathbf{v} = d_\mathbf{w} \Rightarrow \mathbf{v} = \mathbf{w}$
u + **v**	p.5	1.1	$\mathbf{u} + \mathbf{v} = [u_1 + v_1, u_2 + v_2]$ **vector addition**
$\mathbf{u} = \sum c_i \mathbf{v}_i$	p.12	1.1	the c_i are scalars (**linear combination**)
u − **v**	p.8	1.1	$\mathbf{u} - \mathbf{v} = \mathbf{u} + (-\mathbf{v})$ **vector subtraction**
u ∥ **v**	p.8	1.1	$\mathbf{u} \| \mathbf{v} \Leftrightarrow \mathbf{v} = c\mathbf{u}$ if and only if scalar multiples (**parallel**)

Symbolism

Symbols are extremely useful, but beware of gliding over them too quickly.
They pack sophisticated ideas into tiny packages.

v	p.3	1.1	a vector is denoted by a single, boldface, lowercase letter
∥	p.8	1.1	parallel (word is used not symbol)
⇒	p.4	1.1	implies (**if** vectors have the same direction, **then** they are parallel)
⇔	p.4	1.1	if and only if (this phrase is used on p. 4 of Section 1.1)
$\sum c_i \mathbf{v}_i$	p.12	1.1	shorthand for $c_1\mathbf{v}_1 + c_2\mathbf{v}_2 + \cdots + c_n\mathbf{v}_n$
$\overrightarrow{AB}$	p.3	1.1	an overhead arrow (vector $\overrightarrow{AB}$, differs from line segment $\overline{AB}$)

Theorems

Spend enough time with theorems to understand the ideas and reasoning they contain.
At first, we will imitate that reasoning to solve problems and create our own proofs.
As the class progresses, we will be asked to devise our own arguments.

Thm 1.1	p.10	1.1	Algebraic Properties of Vectors in $\mathbb{R}^n$ ($\mathbf{u} + \mathbf{v} = \mathbf{v} + \mathbf{u}$ ...)

Key Definitions and Concepts

In future sections, there will a summary page for key definitions and concepts, and theorems.

components	p.3	1.1	individual coordinates of a vector like 3, 2 of $[3,2]$
head	p.3	1.1	B of $\overrightarrow{AB}$ also called the **terminal point**
initial point	p.3	1.1	A of $\overrightarrow{AB}$ also called the **tail** of vector $\overrightarrow{AB}$
linear combination	p.12	1.1	$\mathbf{u} = c_1 \mathbf{v}_1 + c_2 \mathbf{v}_2 + \cdots + c_n \mathbf{v}_n$ where the c_i are scalars
ordered	p.3	1.1	$[3,2]$ *vs.* $[2,3]$: these are *not* the same vector
parallel	p.8	1.1	if and only if scalar multiples of each other
position vector	p.3	1.1	vector with tail at the origin O, i.e. $\overrightarrow{OA}$
scalar	p.8	1.1	a real number c (that is, c is *not* a vector)
standard position	p.4	1.1	vector with its tail at the origin O, i.e. $\overrightarrow{OA}$
tail	p.3	1.1	A of $\overrightarrow{AB}$ also called the **initial point** of vector $\overrightarrow{AB}$
terminal point	p.3	1.1	B of $\overrightarrow{AB}$ also called the **head** of vector $\overrightarrow{AB}$
vector	p.3	1.1	*directed* line segment with specified *length* and *direction*
vector addition	p.5	1.1	$\mathbf{u} + \mathbf{v} = [u_1 + v_1, u_2 + v_2]$
zero vector	p.4	1.1	$\mathbf{0}$, *all* components are 0, so length is 0

Theorems

In the summary, we will list only the central result of the theorem.

For the complete statement of the theorem, refer to the text.

Thm 1.1　p.10　1.1　Algebraic Properties of Vectors in $\mathbb{R}^n$

1.1 The Geometry and Algebra of Vectors

Exercises

Solutions to odd-numbered exercises from Section 1.1

1.
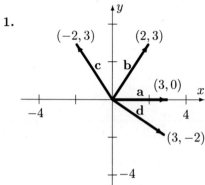

3. See Figures 1.14 and 1.15.

5. (a) $\overrightarrow{AB} = [4-1, 2-(-1)] = [3, 3]$.
(b) $[2, 1]$ (c) $\left[-\frac{3}{2}, \frac{3}{2}\right]$ (d) $\left[-\frac{1}{6}, \frac{1}{6}\right]$.

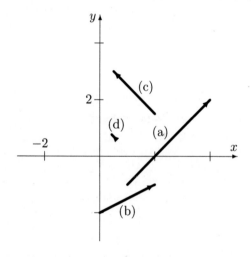
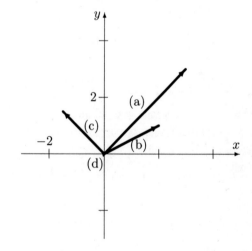

7. $\mathbf{a} + \mathbf{b} = \begin{bmatrix} 3 \\ 0 \end{bmatrix} + \begin{bmatrix} 2 \\ 3 \end{bmatrix} = \begin{bmatrix} 3+2 \\ 0+3 \end{bmatrix} = \begin{bmatrix} 5 \\ 3 \end{bmatrix}$

9. $\mathbf{d} - \mathbf{c} = \begin{bmatrix} 3 \\ -2 \end{bmatrix} - \begin{bmatrix} -2 \\ 3 \end{bmatrix} = \begin{bmatrix} 5 \\ -5 \end{bmatrix}$.

11. $2\mathbf{a} + 3\mathbf{c} = 2\,[0, 2, 0] + 3\,[1, -2, 1] = [2 \cdot 0, 2 \cdot 2, 2 \cdot 0] + [3 \cdot 1, 3(-2), 3 \cdot 1] = [3, -2, 3]$.

13. $\mathbf{u} = [\cos 60°, \sin 60°] = \left[\frac{1}{2}, \frac{\sqrt{3}}{2}\right]$, $\mathbf{v} = [\cos 210°, \sin 210°] = \left[-\frac{\sqrt{3}}{2}, -\frac{1}{2}\right] \Rightarrow$ (implies)
$\mathbf{u} + \mathbf{v} = \left[\frac{1}{2} - \frac{\sqrt{3}}{2}, \frac{\sqrt{3}}{2} - \frac{1}{2}\right]$, $\mathbf{u} - \mathbf{v} = \left[\frac{1}{2} + \frac{\sqrt{3}}{2}, \frac{\sqrt{3}}{2} + \frac{1}{2}\right]$.

15. $2(\mathbf{a} - 3\mathbf{b}) + 3(2\mathbf{b} + \mathbf{a}) \stackrel{\substack{\text{property e.} \\ \text{distributivity}}}{=} (2\mathbf{a} - 6\mathbf{b}) + (6\mathbf{b} + 3\mathbf{a}) \stackrel{\substack{\text{property b.} \\ \text{associativity}}}{=} (2\mathbf{a} + 3\mathbf{a}) + (-6\mathbf{b} + 6\mathbf{b}) = 5\mathbf{a}$.

17. $\mathbf{x} - \mathbf{a} = 2\,(\mathbf{x} - 2\mathbf{a}) = 2\mathbf{x} - 4\mathbf{a} \Rightarrow \mathbf{x} - 2\mathbf{x} = \mathbf{a} - 4\mathbf{a} \Rightarrow -\mathbf{x} = -3\mathbf{a} \Rightarrow \mathbf{x} = 3\mathbf{a}$.

19.

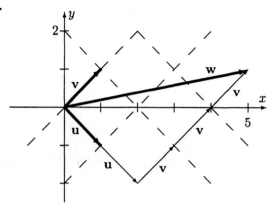

21. See Exercise 19.

23. Property (d) states that $\mathbf{u} + (-\mathbf{u}) = \mathbf{0}$. The first diagram below shows $\mathbf{u}$ along with $-\mathbf{u}$. Then, as the diagonal of the parallelogram, the resultant vector is $\mathbf{0}$.

Property (e) states $c(\mathbf{u} + \mathbf{v}) = c\mathbf{u} + c\mathbf{v}$. The second figure illustrates this.

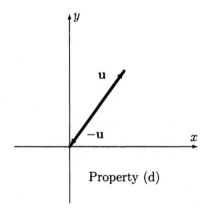

Property (d)

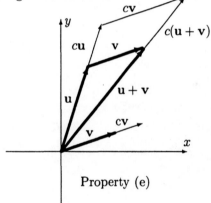

Property (e)

1.2 Length and Angle: The Dot Product

Q: What is our big goal for Section 1.2?
A: To peer into the details of the length and angle of a vector.

We will develop precise definitions for these essential characteristics of a vector.
Note how naturally this flows from Section 1.1 where we got a basic feel for vectors.
Appreciating the organization of the text will help us to follow it and understand it.

Key Definitions and Concepts

angle	p.21	1.2	$\cos\theta = \frac{\mathbf{u}\cdot\mathbf{v}}{\|\mathbf{u}\|\|\mathbf{v}\|}$
distance	p.20	1.2	$d(\mathbf{u},\mathbf{v}) = \|\mathbf{u}-\mathbf{v}\|$
dot product	p.15	1.2	$\mathbf{u}\cdot\mathbf{v} = u_1 v_1 + u_2 v_2 + \ldots + u_n v_n$
length	p.17	1.2	$\|\mathbf{v}\| = \sqrt{\mathbf{v}\cdot\mathbf{v}}$ (Means the same thing as *norm*)
if and only if	p.17	1.2	worth repeating: signals a *double implication*
inner product	p.15	1.2	generalized notion of the dot product (Ch. 7)
norm	p.17	1.2	$\|\mathbf{v}\| = \sqrt{\mathbf{v}\cdot\mathbf{v}}$ (using the dot product)
normalizing	p.18	1.2	a unit vector in same direction ($\mathbf{u} = \frac{1}{\|\mathbf{v}\|}\mathbf{v}$)
orthogonal	p.23	1.2	$\mathbf{u}$ and $\mathbf{v}$ are orthogonal if $\mathbf{u}\cdot\mathbf{v} = 0$
projection	p.24	1.2	$\mathrm{proj}_{\mathbf{u}}(\mathbf{v}) = \left(\frac{\mathbf{u}\cdot\mathbf{v}}{\mathbf{u}\cdot\mathbf{u}}\right)\mathbf{u}$
scalar product	p.15	1.2	another name for dot product (result is a scalar)
standard unit vectors	p.19	1.2	$\mathbf{e}_i$ with a 1 in the *i*th component and 0s elsewhere
unit vector	p.18	1.2	vector of length 1
zero vector	p.23	1.2	$\mathbf{0}$. Note: $\mathbf{0}\cdot\mathbf{v} = 0$ for *every* vector $\mathbf{v}$ in $\mathbf{R}^n$

Theorems

Theorems with names will be listed by name at the end of the Chapter as well.

Cauchy-Schwarz	p.19	1.2	Thm 1.4: $\|\mathbf{u}\cdot\mathbf{v}\| \leq \|\mathbf{u}\|\|\mathbf{v}\|$
Pythagoras	p.23	1.2	Thm 1.6: $\|\mathbf{u}+\mathbf{v}\|^2 = \|\mathbf{u}\|^2 + \|\mathbf{v}\|^2 \Leftrightarrow \mathbf{u}\cdot\mathbf{v} = 0$
Triangle Inequality	p.19	1.2	Thm 1.5: $\|\mathbf{u}+\mathbf{v}\| \leq \|\mathbf{u}\| + \|\mathbf{v}\|$
Thm 1.2	p.16	1.2	Properties of the dot product ($\mathbf{u}\cdot\mathbf{v} = \mathbf{v}\cdot\mathbf{u}$ ...)
Thm 1.3	p.17	1.2	Properties of the norm ($\|\mathbf{v}\| = 0 \Leftrightarrow \mathbf{v} = \mathbf{0}$ ...)
Thm 1.4	p.19	1.2	Cauchy-Schwarz: $\|\mathbf{u}\cdot\mathbf{v}\| \leq \|\mathbf{u}\|\|\mathbf{v}\|$
Thm 1.5	p.19	1.2	Triangle Inequality: $\|\mathbf{u}+\mathbf{v}\| \leq \|\mathbf{u}\| + \|\mathbf{v}\|$
Thm 1.6	p.23	1.2	Pythagoras: $\|\mathbf{u}+\mathbf{v}\|^2 = \|\mathbf{u}\|^2 + \|\mathbf{v}\|^2 \Leftrightarrow \mathbf{u}\cdot\mathbf{v} = 0$

Discussion of Definitions

Q: Why does Section 2.2 start with the definition of the dot product?

A: The dot product is the key to both the length and direction of a vector.

Q: What is the length of a vector?

A: The square root of the squares of its components.

Compare this to the distance formula: $\sqrt{(x_2 - x_1)^2 + (y_2 - y_1)^2}$.

For a vector $\mathbf{v}$ in $\mathbb{R}^2$, we have $\|\mathbf{v}\| = \sqrt{v_1^2 + v_2^2}$.

That is the distance from the origin to the terminal point of the vector.

Q: What is the norm of a vector?

A: Its length. The word *norm* is much more common when discussing vectors because it includes the general notion of *inner products* (see Ch. 7).

Q: How is the distance between two vectors $\mathbf{u}$ and $\mathbf{v}$ defined?

A: As the length (norm) of $\mathbf{u} - \mathbf{v}$: $d(\mathbf{u}, \mathbf{v}) = \|\mathbf{u} - \mathbf{v}\|$.

Discussion of Theorems

Q: What does Cauchy-Schwarz, $|\mathbf{u} \cdot \mathbf{v}| \leq \|\mathbf{u}\| \|\mathbf{v}\|$, say?

A: The absolute value of the dot product is less than or equal to the product of the norms.

Look at this when the vectors have only 1 component and it is positive.

The left hand side (LHS) is $u \cdot v$. The right hand side (RHS) is $\sqrt{u \cdot u}\sqrt{v \cdot v}$.

So we have $u \cdot v \leq \sqrt{u \cdot u}\sqrt{v \cdot v} = u \cdot v$.

This gives us some sense of why this result is true. It is subtle.

Note that the discussion and proof of Cauchy-Schwarz is postponed to Ch. 7.

What does Cauchy-Schwarz look like in $\mathbb{R}^2$?

$$|u_1 v_1 + u_2 v_2| \leq \sqrt{(u_1 u_1 + u_2 u_2)}\sqrt{(v_1 v_1 + v_2 v_2)}.$$

Try to show Cauchy-Schwarz is true in this case. Hint: Square both sides.

Q: What does Pythagoras, $\|\mathbf{u} + \mathbf{v}\|^2 = \|\mathbf{u}\|^2 + \|\mathbf{v}\|^2 \Leftrightarrow \mathbf{u} \cdot \mathbf{v} = 0$, say?

A: The square of the sum of two vectors equals the sum of the square of each vector if and only if those vectors are orthogonal to each other.

Q: What does the Triangle Inequality, $\|\mathbf{u} + \mathbf{v}\| \leq \|\mathbf{u}\| + \|\mathbf{v}\|$, say?

A: The length of the sum of two vectors is less than or equal to the sum of their lengths.

1.2 Length and Angle: The Dot Product

Additional Questions based on the Exercises

23. Q: If two vectors have all positive components, at what type of angle do they meet?

A: An acute angle. Let's look at the definition of $\cos\theta$ to see why that is true.

Recall that $\cos\theta > 0 \Rightarrow \theta$ is acute.

Since $\cos\theta = \frac{\mathbf{u}\cdot\mathbf{v}}{\|\mathbf{u}\|\|\mathbf{v}\|}$, we see that the denominator is always positive.

When all the components of $\mathbf{u}$ and $\mathbf{v}$ are positive, the numerator is positive as well. This is clear because the dot product is $\mathbf{u}\cdot\mathbf{v} = u_1v_1 + u_2v_2 + \ldots + u_nv_n$. So we are multiplying all positive numbers and adding them together.

Note: For the following two questions, we are considering vectors in standard position.

Q: If two vectors lie in the first quadrant, at what type of angle do they meet?

A: An acute angle since they will be squeezed between the x-axis and y-axis.

Q: Can we create a similar argument for vectors with all negative components?

A: Yes, since those vectors lie in the third quadrant. Prove this using $\cos\theta$.

29. Q: Does stretching the vectors $\mathbf{u}$ and $\mathbf{v}$ affect the angle at which they meet?

A: No. Let's show that using the definition of the angle between two vectors.

Let $\mathbf{x} = c\mathbf{u}$, $\mathbf{y} = d\mathbf{v}$ where $c, d > 0$, and apply the definition of $\cos\theta$.

angle between $\mathbf{x}$ and $\mathbf{y}$ = $\frac{\mathbf{x}\cdot\mathbf{y}}{\|\mathbf{x}\|\|\mathbf{y}\|} = \frac{c\mathbf{u}\cdot d\mathbf{v}}{\|c\mathbf{u}\|\|d\mathbf{v}\|} = \frac{\mathbf{u}\cdot\mathbf{v}}{\|\mathbf{u}\|\|\mathbf{v}\|}$ = angle between $\mathbf{u}$ and $\mathbf{v}$

Q: Do we have to adjust our answer if c or d is negative?

A: Yes, because $\|c\mathbf{u}\| = |c|\|\mathbf{u}\|$. How would that affect the string of equalities above?

Q: When equality holds in Cauchy-Schwarz, at what angle do the two vectors meet?

A: When equality holds in Cauchy-Schwarz, we have $|\mathbf{u}\cdot\mathbf{v}| = \|\mathbf{u}\|\|\mathbf{v}\|$.

What does that tell us about the value of $\cos\theta$? It tells us that $\cos\theta = \pm 1$. How?

Since $|\mathbf{u}\cdot\mathbf{v}| = \|\mathbf{u}\|\|\mathbf{v}\|$, we have $|\cos\theta| = \frac{|\mathbf{u}\cdot\mathbf{v}|}{\|\mathbf{u}\|\|\mathbf{v}\|} = 1$.

Thus $\mathbf{u}$ and $\mathbf{v}$ meet at $0°$ or $180°$. So, in fact, $\mathbf{u}$ and $\mathbf{v}$ must be parallel.

45. Q: How does the solution relate to the slope of a line, perpendicular lines, and parallel lines?

Q: Can we use this exercise to show that the axes in $\mathbb{R}^2$ are perpendicular?
A: The vector $[1,0]$ lies in the same direction as the x-axis. What about the y-axis?
The vector $[0,1]$ lies in the same direction as the y-axis.
Since $[1,0] \cdot [0,1] = 0$, the vectors are orthogonal and the axes are perpendicular.

46. Q: How does the formula for $\cos\theta$ help us make the idea of two vectors pointing in the same direction concrete?
A: Pointing in same direction is equivalent to $\mathbf{u} \cdot \mathbf{v} = \|\mathbf{u}\|\|\mathbf{v}\|$.

Q: How could we use absolute value to generalize the question and result of this exercise?

49. The solution here is a good example of how symbols can be great and a trap at the same time.

Q: What does the key fact, $\|-(\mathbf{v}-\mathbf{u})\| = \|\mathbf{v}-\mathbf{u}\|$, of this proof say?
A: The length of a vector is not affected when it points in the opposite direction.

Q: This fact, $\|-(\mathbf{v}-\mathbf{u})\| = \|\mathbf{v}-\mathbf{u}\|$, is an example of what property?
A: The property of the norm that says $\|c\mathbf{v}\| = |c|\|\mathbf{v}\|$.

Q: How might we state the conclusion of this exercise, $d(\mathbf{u},\mathbf{v}) = d(\mathbf{v},\mathbf{u})$, intuitively?
A: The distance between two vectors is independent of the order in which they are considered.

Note: Vectors are like points in this way. Does that make sense? That is ...

Q: Do we need to consider the direction a vector lies in when determining its distance from another vector?
A: Indirectly, perhaps, but directly we use the components of a vector exactly like we use the coordinates of a point in the distance formula.

1.2 Length and Angle: The Dot Product

Solutions to odd-numbered exercises from Section 1.2

1. Following Example 1.8, $\mathbf{u} \cdot \mathbf{v} = \begin{bmatrix} -1 \\ 2 \end{bmatrix} \cdot \begin{bmatrix} 3 \\ 1 \end{bmatrix} = (-1) \cdot 3 + 2 \cdot 1 = -3 + 2 = -1.$

3. $\mathbf{u} \cdot \mathbf{v} = \begin{bmatrix} 1 \\ 2 \\ 3 \end{bmatrix} \cdot \begin{bmatrix} 2 \\ 3 \\ 1 \end{bmatrix} = 1 \cdot 2 + 2 \cdot 3 + 3 \cdot 1 = 2 + 6 + 3 = 11.$

5. $\mathbf{u} \cdot \mathbf{v} = \begin{bmatrix} 1 \\ \sqrt{2} \\ \sqrt{3} \\ 0 \end{bmatrix} \cdot \begin{bmatrix} 4 \\ -\sqrt{2} \\ 0 \\ -5 \end{bmatrix} = 1 \cdot 4 + (\sqrt{2}) \cdot (-\sqrt{2}) + \sqrt{3} \cdot 0 + 0 \cdot (-5) = 4 - 2 = 2.$

7. In the remarks prior to Example 1.11, we note that finding a unit vector $\mathbf{v}$ in the same direction as a given vector $\mathbf{u}$ is called *normalizing* the vector $\mathbf{u}$.

    Therefore, we proceed as in Example 1.12:

    $\|\mathbf{u}\| = \sqrt{(-1)^2 + 2^2} = \sqrt{5}$, so a unit vector $\mathbf{v}$ in the same direction as $\mathbf{u}$ is

    $\mathbf{v} = (1/\|\mathbf{u}\|)\mathbf{u} = (1/\sqrt{5}) \begin{bmatrix} -1 \\ 2 \end{bmatrix} = \begin{bmatrix} -1/\sqrt{5} \\ 2/\sqrt{5} \end{bmatrix}.$

9. Following Example 1.12, we have:

    $\|\mathbf{u}\| = \sqrt{1^2 + 2^2 + 3^2} = \sqrt{14}$, so a unit vector $\mathbf{v}$ in the same direction as $\mathbf{u}$ is

    $\mathbf{v} = (1/\|\mathbf{u}\|)\mathbf{u} = (1/\sqrt{14}) \begin{bmatrix} 1 \\ 2 \\ 3 \end{bmatrix} = \begin{bmatrix} 1/\sqrt{14} \\ 2/\sqrt{14} \\ 3/\sqrt{14} \end{bmatrix}.$

11. $\|\mathbf{u}\| = \sqrt{1^2 + (\sqrt{2})^2 + (\sqrt{3})^2 + 0^2} = \sqrt{6}$, so a unit vector in the direction of $\mathbf{u}$ is

    $\mathbf{v} = (1/\|\mathbf{u}\|)\mathbf{u} = (1/\sqrt{6}) \begin{bmatrix} 1 \\ \sqrt{2} \\ \sqrt{3} \\ 0 \end{bmatrix} = \begin{bmatrix} 1/\sqrt{6} \\ \sqrt{2}/\sqrt{6} \\ \sqrt{3}/\sqrt{6} \\ 0/\sqrt{6} \end{bmatrix} = \begin{bmatrix} 1/\sqrt{6} \\ 1/\sqrt{3} \\ 1/\sqrt{2} \\ 0 \end{bmatrix} = \begin{bmatrix} \sqrt{6}/6 \\ \sqrt{3}/3 \\ \sqrt{2}/2 \\ 0 \end{bmatrix}.$

13. Following Example 1.13, we compute: $\mathbf{u} - \mathbf{v} = \begin{bmatrix} -1 \\ 2 \end{bmatrix} - \begin{bmatrix} 3 \\ 1 \end{bmatrix} = \begin{bmatrix} -4 \\ 1 \end{bmatrix}$, so

    $d(\mathbf{u}, \mathbf{v}) = \|\mathbf{u} - \mathbf{v}\| = -\sqrt{(-4)^2 + 1^2} = \sqrt{17}.$

15. Following Example 1.13, we compute: $\mathbf{u} - \mathbf{v} = \begin{bmatrix} 1 \\ 2 \\ 3 \end{bmatrix} - \begin{bmatrix} 2 \\ 3 \\ 1 \end{bmatrix} = \begin{bmatrix} -1 \\ -1 \\ 2 \end{bmatrix}$, so

    $d(\mathbf{u}, \mathbf{v}) = \|\mathbf{u} - \mathbf{v}\| = -\sqrt{(-1)^2 + (-1)^2 + 2^2} = \sqrt{6}.$

17. (a) $\mathbf{u} \cdot \mathbf{v}$ is a real number, so $\|\mathbf{u} \cdot \mathbf{v}\|$ is the norm of a number, which is not defined.
    (b) $\mathbf{u} \cdot \mathbf{v}$ is a scalar, while $\mathbf{w}$ is a vector.
      Thus, $\mathbf{u} \cdot \mathbf{v} + \mathbf{w}$ adds a scalar to a vector, which is not a defined operation.
    (c) $\mathbf{u}$ is a vector, while $\mathbf{v} \cdot \mathbf{w}$ is a scalar.
      Thus, $\mathbf{u} \cdot (\mathbf{v} \cdot \mathbf{w})$ is the dot product of a vector and a scalar, which is not defined.
    (d) $c \cdot (\mathbf{u} + \mathbf{v})$ is the dot product of a scalar and a vector, which is not defined.

19. From trigonometry, we have:
    $\cos \theta > 0 \Rightarrow \theta$ is acute, $\cos \theta < 0 \Rightarrow \theta$ is obtuse, and $\cos \theta = 0 \Rightarrow \theta$ is right.
    From $\cos \theta = \dfrac{\mathbf{u} \cdot \mathbf{v}}{\|\mathbf{u}\| \|\mathbf{v}\|}$, we see $\mathbf{u} \cdot \mathbf{v}$ determines the sign of $\cos \theta$. Why?
    Therefore, as in Example 1.14, we calculate:
    $\mathbf{u} \cdot \mathbf{v} = 2 \cdot 1 + (-1) \cdot (-2) + 1 \cdot (-1) = 4 > 0 \Rightarrow \cos \theta > 0 \Rightarrow \theta$ is acute.

21. Following the first step in Example 1.14, we calculate:
    $\mathbf{u} \cdot \mathbf{v} = (0.9) \cdot (-4.5) + (2.1) \cdot (2.6) + (1.2) \cdot (-0.8) = 0.45 \Rightarrow \cos \theta > 0 \Rightarrow \theta$ is acute.

23. Since $\mathbf{u} \cdot \mathbf{v}$ is obviously > 0, we have $\cos \theta > 0$ which implies θ is acute.
    Note: $\mathbf{u} \cdot \mathbf{v}$ is > 0 because the components of both $\mathbf{u}$ and $\mathbf{v}$ are positive.

25. As in Example 1.14, we begin by calculating $\mathbf{u} \cdot \mathbf{v}$ (because if $\mathbf{u} \cdot \mathbf{v} = 0$ we're done. Why?):
    $\mathbf{u} \cdot \mathbf{v} = 2 \cdot 1 + (-1) \cdot (-2) + 1 \cdot (-1) = 2 + 2 - 1 = 3$,
    $\|\mathbf{u}\| = \sqrt{2^2 + (-1)^2 + 1^2} = \sqrt{6}$, and $\|\mathbf{v}\| = \sqrt{1^2 + (-2)^2 + (-1)^2} = \sqrt{6}$.
    Therefore, $\cos \theta = \dfrac{\mathbf{u} \cdot \mathbf{v}}{\|\mathbf{u}\| \|\mathbf{v}\|} = \dfrac{3}{\sqrt{6}\sqrt{6}} = \dfrac{1}{2}$, so $\theta = \cos^{-1}\left(\dfrac{1}{2}\right) = \dfrac{\pi}{3}$ radians or $60°$.

27. Following Example 1.14, we calculate:
    $\mathbf{u} \cdot \mathbf{v} = (0.9) \cdot (-4.5) + (2.1) \cdot (2.6) + (1.2) \cdot (-0.8) = 0.45$,
    $\|\mathbf{u}\| = \sqrt{(0.9)^2 + (2.1)^2 + (1.2)^2} = \sqrt{6.66}$, and
    $\|\mathbf{v}\| = \sqrt{(-4.5)^2 + (2.6)^2 + (-0.8)^2} = \sqrt{27.65}$.
    Therefore, $\cos \theta = \dfrac{\mathbf{u} \cdot \mathbf{v}}{\|\mathbf{u}\| \|\mathbf{v}\|} = \dfrac{0.45}{\sqrt{6.66}\sqrt{27.65}} = \dfrac{0.45}{\sqrt{182.817}}$,
    so $\theta = \cos^{-1}\left(\dfrac{0.45}{\sqrt{182.817}}\right) \approx 1.5375$ radians or $88.09°$.
    Note: To minimize error, we do not approximate until the last step.
    Since $\dfrac{0.45}{\sqrt{182.817}} \approx 0.0332816$ is a positive number close to zero,
    we should expect θ to be close to but less than $90°$. Why?

1.2 Length and Angle: The Dot Product

29. Following Example 1.14, we calculate:
$$\mathbf{u} \cdot \mathbf{v} = 1 \cdot 5 + 2 \cdot 6 + 3 \cdot \ +3 \cdot 7 + 4 \cdot 8 = 70,$$
$$\|\mathbf{u}\| = \sqrt{1^2 + 2^2 + 3^2 + 4^2} = \sqrt{30}, \text{ and}$$
$$\|\mathbf{v}\| = \sqrt{5^2 + 6^2 + 7^2 + 8^2} = \sqrt{174}.$$
Therefore, $\cos \theta = \dfrac{\mathbf{u} \cdot \mathbf{v}}{\|\mathbf{u}\| \|\mathbf{v}\|} = \dfrac{70}{\sqrt{30}\sqrt{174}} = \dfrac{35}{3\sqrt{145}},$
so $\theta = \cos^{-1}\left(\dfrac{35}{3\sqrt{145}}\right) \approx 0.2502$ radians or $14.34°$.

Note: To minimize error, we do not approximate until the last step.
Since $\dfrac{35}{3\sqrt{145}} \approx 0.9688639$ is a positive number close to 1,
we should expect θ to be close to but greater than $0°$.

31. To show $\triangle ABC$ is right, we need only show one pair of its sides meet at a right angle.

So, we let $\mathbf{u} = \overrightarrow{AB}$, $\mathbf{v} = \overrightarrow{BC}$, and $\mathbf{w} = \overrightarrow{AC}$, then by the definition of *orthogonal* given prior to Example 1.16, we need only show $\mathbf{u} \cdot \mathbf{v}$, or $\mathbf{u} \cdot \mathbf{w}$, or $\mathbf{v} \cdot \mathbf{w} = 0$.

Following Example 1.1 of Section 1.1, we calculate the sides of $\triangle ABC$:
$$\mathbf{u} = \overrightarrow{AB} = [-3 - 1, 2 - 1, (-2) - (-1)] = [-4, 1, -1],$$
$$\mathbf{v} = \overrightarrow{BC} = [2 - (-3), 2 - 2, (-4 - (-2)] = [5, 0, -2],$$
$$\mathbf{w} = \overrightarrow{AC} = [2 - 1, 2 - 1, (-4) - (-1)] = [1, 1, -3].$$
Then $\mathbf{u} \cdot \mathbf{w} = (-4) \cdot 1 + 1 \cdot 1 + (-1) \cdot (-3) = -4 + 1 + 3 = 0 \Rightarrow$
The angle between $\mathbf{u} = \overrightarrow{AB}$ and $\mathbf{w} = \overrightarrow{AC}$ is $90° \Rightarrow \triangle ABC$ is a right triangle.

33. Following Example 1.15, we make a similar argument:

The dimensions do not matter, so we consider a cube with sides of length 1. Also, the cube is symmetric, so we need only consider one pair of diagonals.

Orient the cube relative to the coordinate axes in $\mathbb{R}^3$, as shown in Figure 1.31. Take the diagonals to be $[1, 1, 1]$ and $\mathbf{v} = [1, 1, -1]$ (from $(1, 1, 0)$ to $(0, 0, 1)$).

Then dot product between these two vectors satisfies:
$$\mathbf{u} \cdot \mathbf{v} = 1 \cdot 1 + 1 \cdot 1 + 1 \cdot (-1) = 1 + 1 - 1 = 1 \neq 0 \Rightarrow$$
The diagonals of a cube are not perpendicular. How might we generalize this result?

35. Following Example 1.17, we compute:

$$\mathbf{u}\cdot\mathbf{v} = \begin{bmatrix} 3/5 \\ -4/5 \end{bmatrix} \cdot \begin{bmatrix} 1 \\ 2 \end{bmatrix} = -1 \text{ and}$$

$$\mathbf{u}\cdot\mathbf{u} = \begin{bmatrix} 3/5 \\ -4/5 \end{bmatrix} \cdot \begin{bmatrix} 3/5 \\ -4/5 \end{bmatrix} = 1, \text{ so}$$

$$\operatorname{proj}_\mathbf{u}(\mathbf{v}) = \left(\frac{\mathbf{u}\cdot\mathbf{v}}{\mathbf{u}\cdot\mathbf{u}}\right)\mathbf{u} = \frac{-1}{1}\begin{bmatrix} 3/5 \\ -4/5 \end{bmatrix}$$

$$= \begin{bmatrix} -3/5 \\ 4/5 \end{bmatrix} = -\mathbf{u}.$$

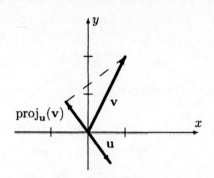

37. Following Example 1.17, we compute:

$$\mathbf{u}\cdot\mathbf{v} = \begin{bmatrix} 1 \\ -1 \\ 1 \\ -1 \end{bmatrix} \cdot \begin{bmatrix} 2 \\ -3 \\ -1 \\ -2 \end{bmatrix} = 6 \text{ and}$$

$$\mathbf{u}\cdot\mathbf{u} = \begin{bmatrix} 1 \\ -1 \\ 1 \\ -1 \end{bmatrix} \cdot \begin{bmatrix} 1 \\ -1 \\ 1 \\ -1 \end{bmatrix} = 4, \text{ so}$$

$$\operatorname{proj}_\mathbf{u}(\mathbf{v}) = \left(\frac{\mathbf{u}\cdot\mathbf{v}}{\mathbf{u}\cdot\mathbf{u}}\right)\mathbf{u} = \frac{6}{4}\begin{bmatrix} 1 \\ -1 \\ 1 \\ -1 \end{bmatrix}$$

$$= \begin{bmatrix} 3/2 \\ -3/2 \\ 3/2 \\ -3/2 \end{bmatrix} = \frac{3}{2}\mathbf{u}.$$

39. Following Example 1.17, we compute:

$$\mathbf{u}\cdot\mathbf{v} = \begin{bmatrix} 3.01 \\ -0.33 \\ 2.52 \end{bmatrix} \cdot \begin{bmatrix} 1.34 \\ 4.25 \\ -1.66 \end{bmatrix} = -1.5523 \text{ and}$$

$$\mathbf{u}\cdot\mathbf{u} = \begin{bmatrix} 3.01 \\ -0.33 \\ 2.52 \end{bmatrix} \cdot \begin{bmatrix} 3.01 \\ -0.33 \\ 2.52 \end{bmatrix} = 15.5194, \text{ so}$$

$$\operatorname{proj}_\mathbf{u}(\mathbf{v}) = \left(\frac{\mathbf{u}\cdot\mathbf{v}}{\mathbf{u}\cdot\mathbf{u}}\right)\mathbf{u} = \frac{-1.5523}{15.5194}\begin{bmatrix} 3.01 \\ -0.33 \\ 2.52 \end{bmatrix}$$

$$\approx \begin{bmatrix} -0.301 \\ 0.033 \\ -0.252 \end{bmatrix} \approx -\frac{1}{10}\mathbf{u}.$$

1.2 Length and Angle: The Dot Product

41. Let $\mathbf{u} = \overrightarrow{AB} = \begin{bmatrix} 4-3 \\ -2-(-1) \\ 6-4 \end{bmatrix} = \begin{bmatrix} 1 \\ -1 \\ 2 \end{bmatrix}$ and $\mathbf{v} = \overrightarrow{AC} = \begin{bmatrix} 5-3 \\ 0-(-1) \\ 2-4 \end{bmatrix} = \begin{bmatrix} 2 \\ 1 \\ -2 \end{bmatrix}$.

(a) We compute the necessary values ...

$\mathbf{u} \cdot \mathbf{v} = \begin{bmatrix} 1 \\ -1 \\ 2 \end{bmatrix} \cdot \begin{bmatrix} 2 \\ 1 \\ -2 \end{bmatrix} = -3,$

$\mathbf{u} \cdot \mathbf{u} = \begin{bmatrix} 1 \\ -1 \\ 2 \end{bmatrix} \cdot \begin{bmatrix} 1 \\ -1 \\ 2 \end{bmatrix} = 6 \ (\|\mathbf{u}\| = \sqrt{6}),$

$\text{proj}_\mathbf{u}(\mathbf{v}) = \left(\frac{\mathbf{u} \cdot \mathbf{v}}{\mathbf{u} \cdot \mathbf{u}}\right) \mathbf{u} = \begin{bmatrix} -1/2 \\ 1/2 \\ -1 \end{bmatrix} \Rightarrow$

$\mathbf{v} - \text{proj}_\mathbf{u}(\mathbf{v}) = \begin{bmatrix} 5/2 \\ 1/2 \\ -1 \end{bmatrix} \Rightarrow$

$\|\mathbf{v} - \text{proj}_\mathbf{u}(\mathbf{v})\| = \sqrt{\left(\frac{5}{2}\right)^2 + \left(\frac{1}{2}\right)^2 + (-1)^2}$

$= \frac{\sqrt{30}}{2}$

... then substitute into the formula for $\mathcal{A}$:

$\mathcal{A} = \frac{1}{2} \|\mathbf{u}\| \, \|\mathbf{v} - \text{proj}_\mathbf{u}(\mathbf{v})\|$

$= \frac{1}{2} \left(\sqrt{6}\right) \left(\frac{\sqrt{30}}{2}\right) = \frac{3\sqrt{5}}{2}.$

(b) We compute the necessary values ...

$\mathbf{u} \cdot \mathbf{v} = \begin{bmatrix} 1 \\ -1 \\ 2 \end{bmatrix} \cdot \begin{bmatrix} 2 \\ 1 \\ -2 \end{bmatrix} = -3,$

$\|\mathbf{u}\| = \sqrt{1^2 + (-1)^2 + 2^2} = \sqrt{6},$

$\|\mathbf{v}\| = \sqrt{2^2 + 1^2 + (-2)^2} = 3 \Rightarrow$

$\cos \theta = \frac{\mathbf{u} \cdot \mathbf{v}}{\|\mathbf{u}\| \, \|\mathbf{v}\|} = \frac{-3}{3\sqrt{6}} = -\frac{\sqrt{6}}{6} \Rightarrow$

$\sin \theta = \sqrt{1 - \cos^2 \theta} = \sqrt{1 - \left(\frac{-\sqrt{6}}{6}\right)^2} = \frac{\sqrt{30}}{6}$

... then substitute into the formula for $\mathcal{A}$:

$\mathcal{A} = \frac{1}{2} \|\mathbf{u}\| \, \|\mathbf{v}\| \sin \theta$

$= \frac{1}{2} \left(\sqrt{6}\right) (3) \left(\frac{\sqrt{30}}{6}\right) = \frac{3\sqrt{5}}{2}.$

43. Two vectors $\mathbf{u}$ and $\mathbf{v}$ are orthogonal **if and only if** [$\Leftrightarrow$] their dot product is zero. That is $\mathbf{u} \cdot \mathbf{v} = 0$. So, we set $\mathbf{u} \cdot \mathbf{v} = 0$ and solve for k:

$\mathbf{u} \cdot \mathbf{v} = \begin{bmatrix} 1 \\ -1 \\ 2 \end{bmatrix} \cdot \begin{bmatrix} k^2 \\ k \\ -3 \end{bmatrix} = 0 \Rightarrow k^2 - k - 6 = (k+2)(k-3) = 0 \Rightarrow k = -2, 3.$

Substituting k back into the expression for $\mathbf{v}$ we get:

When $k = -2$, $\mathbf{v}_1 = \begin{bmatrix} (-2)^2 \\ -2 \\ -3 \end{bmatrix} = \begin{bmatrix} 4 \\ -2 \\ -3 \end{bmatrix}$. When $k = 3$, $\mathbf{v}_2 = \begin{bmatrix} 3^2 \\ 3 \\ -3 \end{bmatrix} = \begin{bmatrix} 9 \\ 3 \\ -3 \end{bmatrix}$.

We check by computing $\mathbf{u} \cdot \mathbf{v}_1$ and $\mathbf{u} \cdot \mathbf{v}_2$ (they should both be zero):

$\mathbf{u} \cdot \mathbf{v}_1 = \begin{bmatrix} 1 \\ -1 \\ 2 \end{bmatrix} \cdot \begin{bmatrix} 4 \\ -2 \\ -3 \end{bmatrix} = 4 + 2 - 6 = 0$ and $\mathbf{u} \cdot \mathbf{v}_2 = \begin{bmatrix} 1 \\ -1 \\ 2 \end{bmatrix} \cdot \begin{bmatrix} 9 \\ 3 \\ -3 \end{bmatrix} = 9 - 3 - 6 = 0.$

45. As noted in the remarks just prior to Example 1.16:

The zero vector $\mathbf{0} = \begin{bmatrix} 0 \\ 0 \end{bmatrix} = \begin{bmatrix} a \\ b \end{bmatrix} \Rightarrow a = b = 0$ is orthogonal to all vectors in $\mathbb{R}^2$.

Having covered this case, we will now assume that at least one of a or $b \neq 0$.

Two vectors $\mathbf{u}$ and $\mathbf{v}$ are orthogonal ***if and only if*** [$\Leftrightarrow$] their dot product is zero. That is $\mathbf{u} \cdot \mathbf{v} = 0$. So, we set $\mathbf{u} \cdot \mathbf{v} = 0$ and solve for y in terms of x.

Case 1: $b \neq 0$: $\mathbf{u} \cdot \mathbf{v} = \begin{bmatrix} a \\ b \end{bmatrix} \cdot \begin{bmatrix} x \\ y \end{bmatrix} = 0 \Rightarrow ax + by = 0 \Rightarrow y = -\frac{a}{b}x$.

Substituting $y = -\frac{a}{b}x$ back into the expression for $\mathbf{v}$ we get: $\mathbf{v} = \begin{bmatrix} x \\ -\frac{a}{b}x \end{bmatrix} = x \begin{bmatrix} 1 \\ -\frac{a}{b} \end{bmatrix}$.

If we let $x = b$ we find $b \begin{bmatrix} 1 \\ -\frac{a}{b} \end{bmatrix} = \begin{bmatrix} b \\ -a \end{bmatrix}$, which clarifies the relationship between $\mathbf{u}$ and $\mathbf{v}$.

Case 2: $b = 0 \,(\Rightarrow a \neq 0)$: $\mathbf{u} \cdot \mathbf{v} = \begin{bmatrix} a \\ b \end{bmatrix} \cdot \begin{bmatrix} x \\ y \end{bmatrix} = 0 \Rightarrow ax + by = 0 \Rightarrow x = -\frac{b}{a}y$.

Substituting $x = -\frac{b}{a}y$ back into the expression for $\mathbf{v}$ we get: $\mathbf{v} = \begin{bmatrix} -\frac{b}{a}y \\ y \end{bmatrix} = y \begin{bmatrix} -\frac{b}{a} \\ 1 \end{bmatrix}$.

If we let $y = -a$ we find $-a \begin{bmatrix} -\frac{b}{a} \\ 1 \end{bmatrix} = \begin{bmatrix} b \\ -a \end{bmatrix}$ exactly as in *Case 1*.

Conclusion: Any vector orthogonal to $\begin{bmatrix} a \\ b \end{bmatrix} \left(\neq \begin{bmatrix} 0 \\ 0 \end{bmatrix} \right)$ must be a multiple of $\begin{bmatrix} b \\ -a \end{bmatrix}$.

Check: $\mathbf{u} \cdot \mathbf{v} = \begin{bmatrix} a \\ b \end{bmatrix} \cdot \begin{bmatrix} bx \\ -ax \end{bmatrix} = abx - bax = 0$ for all values of x.

1.2 Length and Angle: The Dot Product

47. We prove Theorem 1.2(b) by applying the definition of the dot product.
$$\begin{aligned}\mathbf{u}\cdot(\mathbf{v}+\mathbf{w}) &= u_1(v_1+w_1) + u_2(v_2+w_2) + \cdots + u_n(v_n+w_n) \\ &= u_1v_1 + u_1w_1 + u_2v_2 + u_2w_2 + \cdots + u_nv_n + u_nw_n \\ &= (u_1v_1 + u_2v_2 + \cdots + u_nw_n) + (u_1w_1 + u_2w_2 + \cdots + u_nw_n) \\ &= \mathbf{u}\cdot\mathbf{v} + \mathbf{u}\cdot\mathbf{w}.\end{aligned}$$

49. We need to show $d(\mathbf{u},\mathbf{v}) = \|\mathbf{u}-\mathbf{v}\| = \|\mathbf{v}-\mathbf{u}\| = d(\mathbf{v},\mathbf{u})$.
If we let $c = -1$ in Theorem 1.3(b), then $\|-\mathbf{w}\| = \|\mathbf{w}\|$. We use this **key fact** below.

PROOF:
$$\begin{aligned}d(\mathbf{u},\mathbf{v}) &= \|\mathbf{u}-\mathbf{v}\| & &\text{By definition} \\ &= \|-(\mathbf{v}-\mathbf{u})\| & &\text{By the fact that } (x-y) = -(y-x) \\ &= \|\mathbf{v}-\mathbf{u}\| & &\text{By } \|-\mathbf{w}\| = \|\mathbf{w}\| \text{ (key fact)} \\ &= d(\mathbf{v},\mathbf{u}). & &\text{By definition}\end{aligned}$$

51. We need to show $d(\mathbf{u},\mathbf{v}) = \|\mathbf{u}-\mathbf{v}\| = 0$ if and only if $\mathbf{u} = \mathbf{v}$.
This follows immediately from Theorem 1.3(a): $\|\mathbf{w}\| = 0$ if and only if $\mathbf{w} = \mathbf{0}$, with $\mathbf{w} = \mathbf{u}-\mathbf{v}$.

53. We need to show $\|\mathbf{u}-\mathbf{v}\| \geq \|\mathbf{u}\| - \|\mathbf{v}\|$. That is, $\|\mathbf{u}\| \leq \|\mathbf{u}-\mathbf{v}\| + \|\mathbf{v}\|$.
This follows immediately from Theorem 1.5, $\|\mathbf{x}+\mathbf{y}\| \leq \|\mathbf{x}\| + \|\mathbf{y}\|$, with $\mathbf{x} = \mathbf{u}-\mathbf{v}$ and $\mathbf{y} = \mathbf{v}$.

55. We need to show $(\mathbf{u}+\mathbf{v})\cdot(\mathbf{u}-\mathbf{v}) = \|\mathbf{u}\|^2 - \|\mathbf{v}\|^2$ for all vectors in $\mathbb{R}^n$.
Recall, by the definitions of the dot product and the norm, $\mathbf{w}\cdot\mathbf{w} = \|\mathbf{w}\|^2$.
We apply Theorem 1.2(b) and this **key fact** to complete our *PROOF*:

$$\begin{aligned}(\mathbf{u}+\mathbf{v})\cdot(\mathbf{u}-\mathbf{v}) &= \mathbf{u}\cdot\mathbf{u} - \mathbf{u}\cdot\mathbf{v} + \mathbf{v}\cdot\mathbf{u} - \mathbf{v}\cdot\mathbf{v} & &\text{By Theorem 1.2(b)} \\ &= \mathbf{u}\cdot\mathbf{u} - \mathbf{v}\cdot\mathbf{v} & &\text{By the fact that } -xy+yx=0 \\ &= \|\mathbf{u}\|^2 - \|\mathbf{v}\|^2. & &\text{By the fact that } \mathbf{w}\cdot\mathbf{w} = \|\mathbf{w}\|^2 \text{ (key fact)}\end{aligned}$$

57. Let $\mathbf{u}, \mathbf{v} \in \mathbb{R}^n$, and consider $\frac{1}{4}\|\mathbf{u}+\mathbf{v}\|^2 - \frac{1}{4}\|\mathbf{u}-\mathbf{v}\|^2$. By definition, we have:

$$\begin{aligned}\tfrac{1}{4}\|\mathbf{u}+\mathbf{v}\|^2 - \tfrac{1}{4}\|\mathbf{u}-\mathbf{v}\|^2 &= \tfrac{1}{4}[(\mathbf{u}+\mathbf{v})\cdot(\mathbf{u}+\mathbf{v}) + (\mathbf{u}-\mathbf{v})\cdot(\mathbf{u}-\mathbf{v})] \\ &= \tfrac{1}{4}[(\mathbf{u}\cdot\mathbf{u} + \mathbf{v}\cdot\mathbf{v} + 2\mathbf{u}\cdot\mathbf{v}) - (\mathbf{u}\cdot\mathbf{u} + \mathbf{v}\cdot\mathbf{v} - 2\mathbf{u}\cdot\mathbf{v})] \\ &= \tfrac{1}{4}[(\|\mathbf{u}\|^2 - \|\mathbf{u}\|^2) + (\|\mathbf{v}\|^2 - \|\mathbf{v}\|^2) + 4\mathbf{u}\cdot\mathbf{v}] = \mathbf{u}\cdot\mathbf{v}.\end{aligned}$$

59. (a) We need to show $(\mathbf{u} + \mathbf{v}) \cdot (\mathbf{u} - \mathbf{v}) = 0$ in $\mathbb{R}^n$ if and only if $\|\mathbf{u}\| = \|\mathbf{v}\|$.
By Exercise 55, $(\mathbf{u} + \mathbf{v}) \cdot (\mathbf{u} - \mathbf{v}) = \|\mathbf{u}\|^2 - \|\mathbf{v}\|^2$.
Therefore, $(\mathbf{u} + \mathbf{v}) \cdot (\mathbf{u} - \mathbf{v}) = \|\mathbf{u}\|^2 - \|\mathbf{v}\|^2 = 0$ if and only if $\|\mathbf{u}\|^2 = \|\mathbf{v}\|^2$.
It follows immediately that $\mathbf{u}+\mathbf{v}$ and $\mathbf{u}-\mathbf{v}$ are orthogonal in $\mathbb{R}^n$ if and only if $\|\mathbf{u}\| = \|\mathbf{v}\|$.

(b)

The proof in part (a) tells us that the diagonals of a parallelogram are perpendicular if and only if the lengths of its sides are equal.

61. We need to show $\|\mathbf{u}\| = 1$ and $\|\mathbf{v}\| = 2$ imply $\mathbf{u} \cdot \mathbf{v} \neq 3$.
From Theorem 1.4 (the Cauchy-Schwarz Inequality), we have $|\mathbf{x} \cdot \mathbf{y}| \leq \|\mathbf{x}\| \|\mathbf{y}\|$.
Substituting in the given values of $\|\mathbf{u}\| = 1$ and $\|\mathbf{v}\| = 2$ shows $|\mathbf{u} \cdot \mathbf{v}| \leq 2$.
Therefore, $-2 \leq \mathbf{u} \cdot \mathbf{v} \leq 2$. It follows immediately that $\mathbf{u} \cdot \mathbf{v} \neq 3$.

63. Two vectors ($\mathbf{u}$ and $\mathbf{v}$) are orthogonal if their dot product equals zero. So we evaluate:
$$\mathbf{u} \cdot (\mathbf{v} - \text{proj}_{\mathbf{u}}(\mathbf{v})) = \mathbf{u} \cdot \left(\mathbf{v} - \left(\frac{\mathbf{u} \cdot \mathbf{v}}{\mathbf{u} \cdot \mathbf{u}}\right)\mathbf{u}\right) = \mathbf{u} \cdot \mathbf{v} - \mathbf{u} \cdot \left(\frac{\mathbf{u} \cdot \mathbf{v}}{\mathbf{u} \cdot \mathbf{u}}\right)\mathbf{u}$$
$$= \mathbf{u} \cdot \mathbf{v} - \left(\frac{\mathbf{u} \cdot \mathbf{v}}{\mathbf{u} \cdot \mathbf{u}}\right)(\mathbf{u} \cdot \mathbf{u}) = \mathbf{u} \cdot \mathbf{v} - \mathbf{u} \cdot \mathbf{v} = 0.$$

65. (a) The Cauchy-Schwarz Inequality tells us $|\mathbf{u} \cdot \mathbf{v}| \le \|\mathbf{u}\|\|\mathbf{v}\|$.
Squaring both sides, we get $|\mathbf{u} \cdot \mathbf{v}|^2 \le \|\mathbf{u}\|^2\|\mathbf{v}\|^2$.
In $\mathbb{R}^2$ with $\mathbf{u} = \begin{bmatrix} u_1 \\ u_2 \end{bmatrix}$ and $\mathbf{v} = \begin{bmatrix} v_1 \\ v_2 \end{bmatrix}$, this becomes $(u_1v_1 + u_2v_2)^2 \le (u_1^2 + u_2^2)(v_1^2 + v_2^2) \Leftrightarrow$
$0 \le (u_1^2 + u_2^2)(v_1^2 + v_2^2) - (u_1v_1 + u_2v_2)^2 \Leftrightarrow 0 \le u_1^2v_2^2 + u_2^2v_1^2 - 2u_1u_2v_1v_2 \Leftrightarrow$
$0 \le \frac{1}{2}(u_1v_2 - u_2v_1)^2 + \frac{1}{2}(u_2v_1 - u_1v_2)^2$.
Since the final statement is true, all the statements are true.

(b) Let $\mathbf{u}$ and $\mathbf{v}$ be elements of $\mathbb{R}^3$. Then $|\mathbf{u} \cdot \mathbf{v}|^2 \le \|\mathbf{u}\|^2\|\mathbf{v}\|^2 \Leftrightarrow$
$(u_1v_1 + u_2v_2 + u_3v_3)^2 \le (u_1^2 + u_2^2 + u_3^2)(v_1^2 + v_2^2 + v_3^2) \Leftrightarrow$
$0 \le (u_1^2 + u_2^2 + u_3^2)(v_1^2 + v_2^2 + v_3^2) - (u_1v_1 + u_2v_2 + u_3v_3)^2 \Leftrightarrow$
$0 \le u_1^2v_2^2 + u_1^2v_3^2 + u_2^2v_1^2 + u_2^2v_3^2 + u_3^2v_1^2 + u_3^2v_2^2 - 2u_1v_1u_2v_2 - 2u_1v_1u_3v_3 - 2u_2v_2u_3v_3 \Leftrightarrow$
$0 \le \frac{1}{2}(u_1v_2 - u_2v_1)^2 + \frac{1}{2}(u_2v_1 - u_1v_2)^2 + \frac{1}{2}(u_1v_3 - u_3v_1)^2$
$\quad + \frac{1}{2}(u_3v_1 - u_1v_3)^2 + \frac{1}{2}(u_2v_3 - u_3v_2)^2 + \frac{1}{2}(u_3v_2 - u_2v_3)^2$.
Since the final statement is true, all the statements are true.

67. We have $\text{proj}_{\mathbf{u}}(\mathbf{v}) = c\mathbf{u}$. From the figure, we see that $\cos\theta = \dfrac{c\|\mathbf{u}\|}{\|\mathbf{v}\|}$, so $\mathbf{u} \cdot \mathbf{v} = \|\mathbf{u}\|\|\mathbf{v}\|\dfrac{c\|\mathbf{u}\|}{\|\mathbf{v}\|}$ which we solve for c to get $c = \dfrac{\mathbf{u} \cdot \mathbf{v}}{\|\mathbf{u}\|^2}$. Thus, $\text{proj}_{\mathbf{u}}(\mathbf{v}) = \left(\dfrac{\mathbf{u} \cdot \mathbf{v}}{\mathbf{u} \cdot \mathbf{u}}\right)\mathbf{u}$ since $\|\mathbf{u}\|^2 = \mathbf{u} \cdot \mathbf{u}$.

Exploration: Vectors and Geometry

Since Explorations are self-contained, only odd-numbered solutions will be provided.

1. Like Example 1.18: $\mathbf{m} - \mathbf{a} = \overrightarrow{AM} = \frac{1}{3}\overrightarrow{AB} = \frac{1}{3}(\mathbf{b} - \mathbf{a}) \Rightarrow \mathbf{m} = \mathbf{a} + \frac{1}{3}(\mathbf{b} - \mathbf{a}) = \frac{1}{3}(2\mathbf{a} + \mathbf{b})$.

   In general, $\mathbf{m} - \mathbf{a} = \overrightarrow{AM} = \frac{1}{n}\overrightarrow{AB} = \frac{1}{n}(\mathbf{b} - \mathbf{a}) \Rightarrow \mathbf{m} = \mathbf{a} + \frac{1}{n}(\mathbf{b} - \mathbf{a}) = \frac{1}{n}((n-1)\mathbf{a} + \mathbf{b})$.

   Note as $n \to \infty$, $\mathbf{m} \to \mathbf{a}$.

3. Draw in $\overrightarrow{AC}$. Then from Exercise 2, we have: $\overrightarrow{PQ} = \frac{1}{2}\overrightarrow{AB} = \overrightarrow{SR}$.

   Draw in $\overrightarrow{BD}$. Then from Exercise 2, we have: $\overrightarrow{PS} = \frac{1}{2}\overrightarrow{BD} = \overrightarrow{QR} \Rightarrow$

   $PQRS$ is a parallelogram (opposite sides are parallel and congruent).

5. We are given $\overrightarrow{AH}$ is orthogonal to $\overrightarrow{BC}$, that is $\overrightarrow{AH} \cdot \overrightarrow{BC} = 0$ and $\overrightarrow{BH}$ is orthogonal to $\overrightarrow{AC}$, that is $\overrightarrow{BH} \cdot \overrightarrow{AC} = 0$.

   We need to show $\overrightarrow{CH}$ is orthogonal to $\overrightarrow{AB}$, that is $\overrightarrow{CH} \cdot \overrightarrow{AB} = 0$.

   $\overrightarrow{AH} \cdot \overrightarrow{BC} = 0 \Rightarrow (\mathbf{h} - \mathbf{a}) \cdot (\mathbf{b} - \mathbf{c}) = 0 \quad \mathbf{h} \cdot \mathbf{b} - \mathbf{h} \cdot \mathbf{c} - \mathbf{a} \cdot \mathbf{b} + \mathbf{a} \cdot \mathbf{c} = 0$
   $\overrightarrow{BH} \cdot \overrightarrow{AC} = 0 \Rightarrow (\mathbf{h} - \mathbf{b}) \cdot (\mathbf{c} - \mathbf{a}) = 0 \Rightarrow \mathbf{h} \cdot \mathbf{c} - \mathbf{h} \cdot \mathbf{a} - \mathbf{b} \cdot \mathbf{c} + \mathbf{a} \cdot \mathbf{b} = 0 \Rightarrow$

   $0 = \mathbf{h} \cdot \mathbf{b} - \mathbf{h} \cdot \mathbf{a} - \mathbf{c} \cdot \mathbf{b} + \mathbf{a} \cdot \mathbf{c} = (\mathbf{h} - \mathbf{c}) \cdot (\mathbf{b} - \mathbf{a}) = \overrightarrow{CH} \cdot \overrightarrow{AB} = 0 \Rightarrow$

   $\overrightarrow{CH}$ is orthogonal to $\overrightarrow{AB}$, so all the altitudes intersect at the orthocenter, H.

7. Let O be the origin, then we have: $\mathbf{b} = -\mathbf{a}$ and $\|\mathbf{a}\|^2 = \|\mathbf{c}\|^2 = r^2$, r = radius of the circle.

   We need to show $\overrightarrow{AC}$ is orthogonal to $\overrightarrow{BC}$, that is $\overrightarrow{AC} \cdot \overrightarrow{BC} = 0 \Rightarrow$

   $(\mathbf{c} - \mathbf{a}) \cdot (\mathbf{c} - \mathbf{b}) = (\mathbf{c} - \mathbf{a}) \cdot (\mathbf{c} + \mathbf{a}) = \|\mathbf{c}\|^2 + \mathbf{c} \cdot \mathbf{a} - \|\mathbf{a}\|^2 - \mathbf{a} \cdot \mathbf{c} = (\mathbf{a} \cdot \mathbf{c} - \mathbf{a} \cdot \mathbf{c}) + (r^2 - r^2) = 0 \Rightarrow$

   $\overrightarrow{AC}$ is orthogonal to $\overrightarrow{BC}$, so $\angle ACB$ is a right angle.

1.3 Lines and Planes

Q: What is our big goal for Section 1.3?
A: To develop our intuition about lines and planes in $\mathbb{R}^2$ and $\mathbb{R}^3$.

Begin with careful application of the Examples to the Exercises.
Next pay close attention to the figures. Practice drawing them freehand.
Finally, substitute into the formulas and get ready to create some of our own.

Key Definitions and Concepts

distances the following two terms relate to distances:

point, line, $\mathbb{R}^2$	p.40	1.3	$d(B, \ell) = \dfrac{|ax_0 + by_0 - c|}{\sqrt{a^2 + b^2}}$
point, plane, $\mathbb{R}^3$	p.41	1.3	$d(B, \mathscr{P}) = \dfrac{|ax_0 + by_0 + cz_0 - d|}{\sqrt{a^2 + b^2 + c^2}}$

lines the following five terms relate to lines:

direction vector	p.32	1.3	$\mathbf{d}$, parallel to any vector on line ℓ
general form	p.33	1.3	$ax + by = c$ where $\mathbf{n} = \begin{bmatrix} a \\ b \end{bmatrix}$ is normal for ℓ
normal form	p.33	1.3	$\mathbf{n} \cdot (\mathbf{x} - \mathbf{p}) = 0$ or $\mathbf{n} \cdot \mathbf{x} = \mathbf{n} \cdot \mathbf{p}$
normal vector	p.31	1.3	$\mathbf{n}$, orthogonal to any vector on line ℓ
vector form	p.33	1.3	$\mathbf{x} = \mathbf{p} + t\mathbf{d}$ where $\mathbf{d}$ is a direction vector for ℓ
parametric equations	p.33	1.3	component equations from $\mathbf{x} = \mathbf{p} + t\mathbf{d}$

planes the following four terms relate to planes:

general form	p.35	1.3	$ax + by + cz = d$ where $\mathbf{n} = \begin{bmatrix} a \\ b \\ c \end{bmatrix}$ is normal for $\mathscr{P}$
normal form	p.35	1.3	$\mathbf{n} \cdot (\mathbf{x} - \mathbf{p}) = 0$ or $\mathbf{n} \cdot \mathbf{x} = \mathbf{n} \cdot \mathbf{p}$
normal vector	p.35	1.3	$\mathbf{n}$, orthogonal to any vector in plane $\mathscr{P}$
vector form	p.36	1.3	$\mathbf{x} = \mathbf{p} + s\mathbf{u} + t\mathbf{v}$ where $\mathbf{u}$ and $\mathbf{v}$ are parallel to $\mathscr{P}$

Theorems
There are no theorems in this section.

Additional Questions based on the Exercises

13. Q: How can we tell if two vectors are parallel?

A: They are scalar multiples of each other. Like $[1, 2, 3]$ and $[2, 4, 6] (= 2[1, 2, 3])$.

15. Q: Given any two non-parallel vectors, how do we find a plane they lie in?

A: Apply the key condition: both vectors have to be orthogonal to the normal vector.

Eg.: Let $\mathbf{u} = [1, 0, 0]$ and $\mathbf{v} = [0, 1, 0]$. We need a normal vector $\mathbf{n}$ such that $\mathbf{n} \cdot \mathbf{u} = \mathbf{n} \cdot \mathbf{v} = 0$.

The condition $\mathbf{n} \cdot \mathbf{u} = 0$ tells us $\mathbf{n}$ must be of the form $[0, b, c]$. Why?

Because $[a, b, c] \cdot [1, 0, 0] = a = 0$.

The condition $\mathbf{n} \cdot \mathbf{v} = 0$ tells us $\mathbf{n}$ must be of the form $[a, 0, c]$. Why?

Because $[a, b, c] \cdot [0, 1, 0] = b = 0$.

Putting those two conditions together, we see that $\mathbf{n}$ must be of the form $[0, 0, c]$.

So, one plane that contains both $\mathbf{u}$ and $\mathbf{v}$ is $0x + 0y + 1z = 0$, that is $z = 0$.

Q: Are $z = 0$ and $5z = 0$ the same plane or merely parallel?

A: These are the same plane because $z = 0$ if and only if $5z = 0$.

Q: Are $z = 1$ and $5z = 1$ the same plane or merely parallel?

A: These are merely parallel planes because $(0, 0, 1)$ lies in $z = 1$ but not $5z = 1$.

They are parallel because their normal vectors, $[0, 0, 1]$ and $[0, 0, 5]$, are parallel.

It is useful to note that $5z = 1$ and $z = \frac{1}{5}$ are the same plane. Prove that.

Applying these concepts to a Roller Coaster

Q: What key information do parametric equations, $\mathbf{x} = \mathbf{p} + s\mathbf{u} + t\mathbf{v}$, provide?

A: They tell us each component based on one value, t, usually thought of as time.

Like a roller coaster ...

The parametric equations would give us our location in space throughout the ride.
Find the parametric equations for a real roller coaster near you from the internet.
Map a rider's location in time using these equations.

Q: How quickly is a loop completed on this ride?

Q: How does the resultant vector $\mathbf{x}$ tell us a loop has been completed?

Q: If vectors along the ride are orthogonal to each other, what does that tell us about how we've moved along the ride?

1.3   Lines and Planes

17. Q: Given a line $ax + by = c$, is its direction vector $\mathbf{d} = [b, -a]$ or $[-b, a]$?
    A: Both of these are a direction vector. Why? They are parallel. How can we tell?
    They are scalar multiples of each other: $[b, -a] = -1[-b, a]$.

    Q: What direction vector has an obvious relationship to the slope of the line?
    A: For line $ax + by = c$, the slope is $-\frac{a}{b}$ and a direction vector is $[1, -\frac{a}{b}]$. Why?
    Since $\frac{1}{b}[b, -a] = [1, -\frac{a}{b}]$.

    Q: For a given line ℓ, what is the dot product of its normal vector and a direction vector?
    A: Not surprisingly, the dot product is zero.

    Taking the normal vector as $[a, b]$ and the direction vector $[b, -a]$ makes this obvious: $[a, b] \cdot [b, -a] = ab - ba = 0$.

    Q: A normal vector for line $ax + by = c$ is any vector parallel to $[a, b]$.
    A direction vector for line $ax + by = c$ is any vector parallel to $[b, -a]$.
    Show explicitly that the normal and direction vectors must be orthogonal.

19. Q: If two planes are parallel, must they have the same normal vector?
    A: Yes and no. The given normal vectors need only be parallel.
    However, any normal vector to one plane is also a normal vector to the parallel plane.

    Q: How do we tell if a given point lies in a given plane?
    A: We substitute the values of the coordinates into the given equation. How?
    For example, does $(1, 2, 3)$ lie in $x + y + z = 6$? Yes, because $1 + 2 + 3 = 6$.

25. When trying to understand something as complicated as planes in $\mathbb{R}^3$, it is useful to begin our investigation with planes that we can easily visualize. So, the discussion below focuses on planes parallel to $x = 0$, $y = 0$, and $z = 0$.

Q: Why is it obvious from the algebra alone that $x = 0$, $y = 0$, and $z = 0$ are planes?
A: The general form of the plane $ax + by + cz = d$ makes that obvious. Why?
That equation tells us any equation of the form $ax + by + cz = d$ is a plane $\mathbb{R}^3$ provided that a, b and c are not all zero at the same time.

Q: Why is it obvious from the algebra alone that $x = 0$ and $x = 1$ are parallel?
A: They have the same normal vector. Why is that obvious?
Comparing $x = 0$ and $x = 1$ to $ax + by + cz = d$ shows that they both have a normal vector of $\mathbf{n} = [a, b, c] = [1, 0, 0]$.

Q: Why is it obvious from the algebra alone that $x = 0$ and $y = 0$ are orthogonal?
A: $x = 0$ has normal vector $[1, 0, 0]$ and $y = 0$ has normal vector $[0, 1, 0]$.
$[1, 0, 0] \cdot [0, 1, 0] = 0$ tells us the normal vectors are orthogonal, so the planes are.

Q: We know $x = y$ is a line in $\mathbb{R}^2$. Is $x = y$ a plane in $\mathbb{R}^3$?
A: Yes. $x = y$ is a plane in $\mathbb{R}^3$. Why is this obvious?
Because we can put $x = y$ into the form $ax + by + cz = d$. Namely, $x - y = 0$.

Q: It seems like the plane $x - y = 0$ should meet the plane $x = 0$ at a 45°? Does it?
A: Yes. How can we prove that?
From Example 7 of Section 1.2, we note that $\cos \theta = \dfrac{|\mathbf{u} \cdot \mathbf{v}|}{\|\mathbf{u}\| \|\mathbf{v}\|}$.

So, given two planes $\mathscr{P}_1$ with $\mathbf{n}_1$ and $\mathscr{P}_2$ with $\mathbf{n}_2$, we have $\cos \theta = \dfrac{|\mathbf{n}_1 \cdot \mathbf{n}_2|}{\|\mathbf{n}_1\| \|\mathbf{n}_2\|}$.

Step 1. Since $\mathscr{P}_1$ has equation $x - y = 0$, $\mathbf{n}_1 = [1, -1, 0]$.
Since $\mathscr{P}_2$ has equation $x = 0$, $\mathbf{n}_2 = [1, 0, 0]$.

Step 2. Therefore, $\mathbf{n}_1 = [1, -1, 0] \cdot [1, 0, 0] = 1 \cdot 1 - 0 \cdot 1 + 0 \cdot 0 = 1$,
$\|\mathbf{n}_1\| = \sqrt{1^2 + (-1)^2 + 0^2} = \sqrt{2}$, and $\|\mathbf{n}_2\| = \sqrt{1^2 + 0^2 + 0^2} = 1$.

Step 3. So $\cos \theta = \dfrac{1}{\sqrt{2}}$ and $\theta = \cos^{-1} \left(\dfrac{1}{\sqrt{2}} \right) = 45°$.

1.3 Lines and Planes

The following questions are based on Exercises 37, 39, and 41.

When we derive a formula, like $d(B, \mathcal{P}) = \dfrac{|c_1 - c_2|}{\|\mathbf{n}\|}$, we should think about what it tells us.

Q: How could we adapt $d(B, \mathcal{P}) = \dfrac{|c_1 - c_2|}{\|\mathbf{n}\|}$ to compute the distance between planes?

A: Hint: consider our work in Exercise 37.

Comparing these two results, we see that the distance between two parallel planes is:
$$d(\mathcal{P}_1, \mathcal{P}_2) = \dfrac{|d_1 - d_2|}{\|\mathbf{n}\|} \text{ where } \ldots$$
the equation for $\mathcal{P}_1$ is $ax + by + cz = d_1$ and the equation for $\mathcal{P}_2$ is $ax + by + cz = d_2$. The normal vector for *both* planes is $\mathbf{n} = [a, b, c]$ because they are parallel.

Let's try to understand this result by considering the planes $z = 0$ and $z = 5$.

Q: Intuitively, what is the distance between $z = 0$ and $z = 5$?
A: Sketch these planes on the graph of $\mathbb{R}^3$. See how $z = 5$ is 5 units above $z = 0$?

Q: According to our formula, what is the distance between $z = 0$ and $z = 5$?
A: Substituting $d_1 = 0$, $d_2 = 5$, and $\|\mathbf{n}\| = \|[0, 0, 1]\| = 1$ into our formula, we get $\frac{5}{1} = 5$.

Q: What is the distance between $z = 1$ and $5z = 1$? $z = 1$ and $z = \frac{1}{5}$?

Q: If the norm of the normal to $\mathcal{P}_1$ and $\mathcal{P}_2$ is 1, what does our formula become?
A: Our formula reduces to $|d_1 - d_2|$.

Q: What does this simplified version of our formula tell us?

Q: Can we always find a normal to $\mathcal{P}_1$ and $\mathcal{P}_2$ that has a norm of 1?
A: Yes, because the normal is a vector so we can we normalize it.

Q: If we normalize a vector $\mathbf{v}$, is the vector $\mathbf{u}$ that results parallel to $\mathbf{v}$?
A: Consider the formula for the normalized vector, $\mathbf{u} = \frac{1}{\|\mathbf{v}\|}\mathbf{v}$.

    See how $\mathbf{v}$ has been multiplied by $\frac{1}{\|\mathbf{v}\|}$? But is $\frac{1}{\|\mathbf{v}\|}$ a scalar? Yes!

    So what do we conclude?

    If we normalize a vector $\mathbf{v}$, the vector $\mathbf{u}$ that results is parallel to $\mathbf{v}$.

Solutions to odd-numbered exercises from Section 1.3

1. Following Example 1.20, we will:
   (a) find the normal form by substituting into $\mathbf{n} \cdot \mathbf{x} = \mathbf{n} \cdot \mathbf{p}$ and
   (b) find the general form by computing those dot products.

   (a) $\mathbf{n} = \begin{bmatrix} 3 \\ 2 \end{bmatrix}$, $\mathbf{x} = \begin{bmatrix} x \\ y \end{bmatrix}$, and $\mathbf{p} = \begin{bmatrix} 0 \\ 0 \end{bmatrix} \Rightarrow$ The normal form is $\begin{bmatrix} 3 \\ 2 \end{bmatrix} \cdot \begin{bmatrix} x \\ y \end{bmatrix} = \begin{bmatrix} 3 \\ 2 \end{bmatrix} \cdot \begin{bmatrix} 0 \\ 0 \end{bmatrix} = 0$.

   (b) $\begin{bmatrix} 3 \\ 2 \end{bmatrix} \cdot \begin{bmatrix} x \\ y \end{bmatrix} = 3x + 2y$ and $\begin{bmatrix} 3 \\ 2 \end{bmatrix} \cdot \begin{bmatrix} 0 \\ 0 \end{bmatrix} = 0 \Rightarrow$ The general form is $3x + 2y = 0$.

3. Following Example 1.21, we will:
   (a) find the vector form by substituting into $\mathbf{x} = \mathbf{p} + t\mathbf{d}$ and
   (b) find the parametric form by equating components.

   (a) $\mathbf{x} = \begin{bmatrix} x \\ y \end{bmatrix}$, $\mathbf{p} = \begin{bmatrix} 1 \\ 0 \end{bmatrix}$, and $\mathbf{d} = \begin{bmatrix} -1 \\ 3 \end{bmatrix} \Rightarrow$ The vector form is $\begin{bmatrix} x \\ y \end{bmatrix} = \begin{bmatrix} 1 \\ 0 \end{bmatrix} + t \begin{bmatrix} -1 \\ 3 \end{bmatrix}$.

   (b) The vector form in (a) implies the parametric form is $\begin{matrix} x = 1 - t \\ y = 3t \end{matrix}$.

5. Following Example 1.21, we will:
   (a) find the vector form by substituting into $\mathbf{x} = \mathbf{p} + t\mathbf{d}$ and
   (b) find the parametric form by equating components.

   (a) $\mathbf{x} = \begin{bmatrix} x \\ y \\ z \end{bmatrix}$, $\mathbf{p} = \begin{bmatrix} 0 \\ 0 \\ 0 \end{bmatrix}$, and $\mathbf{d} = \begin{bmatrix} 1 \\ -1 \\ 4 \end{bmatrix} \Rightarrow$ The vector form is $\begin{bmatrix} x \\ y \\ z \end{bmatrix} = \begin{bmatrix} 0 \\ 0 \\ 0 \end{bmatrix} + t \begin{bmatrix} 1 \\ -1 \\ 4 \end{bmatrix}$.

   (b) The vector form in (a) implies the parametric form is $\begin{matrix} x = t \\ y = -t \\ z = 4t \end{matrix}$.

7. Following Example 1.23, we will:
   (a) find the normal form by substituting into $\mathbf{n} \cdot \mathbf{x} = \mathbf{n} \cdot \mathbf{p}$ and
   (b) find the general form by computing those dot products.

   (a) $\mathbf{n} = \begin{bmatrix} 3 \\ 2 \\ 1 \end{bmatrix}$, $\mathbf{x} = \begin{bmatrix} x \\ y \\ z \end{bmatrix}$, $\mathbf{p} = \begin{bmatrix} 0 \\ 1 \\ 0 \end{bmatrix} \Rightarrow$ The normal form is $\begin{bmatrix} 3 \\ 2 \\ 1 \end{bmatrix} \cdot \begin{bmatrix} x \\ y \\ z \end{bmatrix} = \begin{bmatrix} 3 \\ 2 \\ 1 \end{bmatrix} \cdot \begin{bmatrix} 0 \\ 1 \\ 0 \end{bmatrix} = 2$.

   (b) $\begin{bmatrix} 3 \\ 2 \\ 1 \end{bmatrix} \cdot \begin{bmatrix} x \\ y \\ z \end{bmatrix} = 3x + 2y + z$ and $\begin{bmatrix} 3 \\ 2 \\ 1 \end{bmatrix} \cdot \begin{bmatrix} 0 \\ 1 \\ 0 \end{bmatrix} = 2 \Rightarrow$ The general form is $3x + 2y + z = 2$.

1.3 Lines and Planes

9. Following Example 1.24, we will:
   (a) find the vector form by substituting into $\mathbf{x} = \mathbf{p} + s\mathbf{u} + t\mathbf{v}$ and
   (b) find the parametric form by equating components.

   (a) $\mathbf{x} = \begin{bmatrix} x \\ y \\ z \end{bmatrix}$, $\mathbf{p} = \begin{bmatrix} 0 \\ 0 \\ 0 \end{bmatrix}$, $\mathbf{u} = \begin{bmatrix} 2 \\ 1 \\ 2 \end{bmatrix}$, and $\mathbf{v} = \begin{bmatrix} -3 \\ 2 \\ 1 \end{bmatrix} \Rightarrow$

   The vector form is $\begin{bmatrix} x \\ y \\ z \end{bmatrix} = \begin{bmatrix} 0 \\ 0 \\ 0 \end{bmatrix} + s \begin{bmatrix} 2 \\ 1 \\ 2 \end{bmatrix} + t \begin{bmatrix} -3 \\ 2 \\ 1 \end{bmatrix}$.

   (b) The vector form in (a) implies the parametric form is $\begin{aligned} x &= 2s - 3t \\ y &= s + 2t \\ z &= 2s + t \end{aligned}$.

11. Following Example 1.24, we realize we may choose any point on ℓ, so we will use P (Q would also be fine).

    A convenient direction vector is $\mathbf{d} = \overrightarrow{PQ} = \begin{bmatrix} 2 \\ 2 \end{bmatrix}$ (or any scalar multiple of this).

    Thus we obtain: $\mathbf{x} = \mathbf{p} + t\mathbf{d}$
    $$= \begin{bmatrix} 1 \\ -2 \end{bmatrix} + t \begin{bmatrix} 2 \\ 2 \end{bmatrix}.$$

13. Following Example 1.24, we realize we need to find two direction vectors, $\mathbf{u}$ and $\mathbf{v}$. Since $P = (1, 1, 1)$, $Q = (4, 0, 2)$, and $R = (0, 1, -1)$ lie in plane $\mathscr{P}$, we compute:

    $$\mathbf{u} = \overrightarrow{PQ} = \mathbf{q} - \mathbf{p} = \begin{bmatrix} 3 \\ -1 \\ 1 \end{bmatrix} \text{ and } \mathbf{v} = \overrightarrow{PR} = \mathbf{r} - \mathbf{p} = \begin{bmatrix} -1 \\ 0 \\ -2 \end{bmatrix}.$$

    Since $\mathbf{u}$ and $\mathbf{v}$ are not scalar multiples of each other, they will serve as direction vectors. If $\mathbf{u}$ and $\mathbf{v}$ were scalar multiples of each other, we would not have a plane but simply a line. Therefore, we have the vector equation of $\mathscr{P}$:
    $$\begin{bmatrix} x \\ y \\ z \end{bmatrix} = \begin{bmatrix} 1 \\ 1 \\ 1 \end{bmatrix} + s \begin{bmatrix} 3 \\ -1 \\ 1 \end{bmatrix} + t \begin{bmatrix} -1 \\ 0 \\ -2 \end{bmatrix}.$$

15. The parametric equations and associated vector forms $\mathbf{x} = \mathbf{p} + t\mathbf{d}$ found below are *not* unique.

   (a) As in the remarks prior to Example 1.20, we begin by letting $x = t$.
   When we substitute $x = t$ into $y = 3x - 1$, we get $y = 3(t) - 1$. So, we have the following:

   Parametric equations $\begin{matrix} x = t \\ y = -1 + 3t \end{matrix}$ and vector form $\begin{bmatrix} x \\ y \end{bmatrix} = \begin{bmatrix} 0 \\ -1 \end{bmatrix} + t \begin{bmatrix} 1 \\ 3 \end{bmatrix}$.

   (b) In this case since the coefficient of y is 2, we begin by letting $x = 2t$.
   When we substitute $x = 2t$ into $3x + 2y = 5$, we get $3(2t) + 2y = 5$.
   Solving for y yields $y = -3t + 2.5$. So, we have the following:

   Parametric equations: $\begin{matrix} x = 2t \\ y = 2.5 - 3t \end{matrix}$ and vector form $\begin{bmatrix} x \\ y \end{bmatrix} = \begin{bmatrix} 0 \\ 2.5 \end{bmatrix} + t \begin{bmatrix} 2 \\ -3 \end{bmatrix}$.

   We discover the following pattern: if line ℓ has equation $ax + by = c$, then $\mathbf{d} = \begin{bmatrix} b \\ -a \end{bmatrix}$.

17. Need to show ℓ_1 with slope m_1 is perpendicular to ℓ_2 with slope m_2 if and only if $m_1 m_2 = -1$.
By definition, one possible form of the general equation for ℓ_1 with slope m_1 is $-m_1 x + y = b_1$.
So, the normal vector for ℓ_1 is $\mathbf{n}_1 = \begin{bmatrix} -m_1 \\ 1 \end{bmatrix}$ and the normal vector for ℓ_2 is $\mathbf{n}_2 = \begin{bmatrix} -m_2 \\ 1 \end{bmatrix}$.
Now we note ℓ_1 is perpendicular to line ℓ_2 if and only if $\mathbf{n}_1 \cdot \mathbf{n}_2 = 0$, so we have:

$$\mathbf{n}_1 \cdot \mathbf{n}_2 = \begin{bmatrix} -m_1 \\ 1 \end{bmatrix} \cdot \begin{bmatrix} -m_1 \\ 1 \end{bmatrix} = m_1 m_2 + 1 = 0 \text{ which implies } m_1 m_2 = -1 \text{ as we were to show.}$$

1.3 Lines and Planes

19. Given $\mathbf{n}_1$ is the normal vector of $\mathscr{P}_1$ and $\mathbf{n}$ is the normal vector of $\mathscr{P}$, we have:
If $\mathbf{n}_1$ and $\mathbf{n}$ are orthogonal which implies $\mathbf{n}_1 \cdot \mathbf{n} = 0$, then $\mathscr{P}_1$ is perpendicular to $\mathscr{P}$.
If $\mathbf{n}_1$ and $\mathbf{n}$ are parallel which implies $\mathbf{n}_1 = c\mathbf{n}$ (scalar multiples), then $\mathscr{P}_1$ is parallel to $\mathscr{P}$.

(a) Since the general form of $\mathscr{P}$ is $2x + 3y - z = 1$, its normal vector is $\mathbf{n} = \begin{bmatrix} 2 \\ 3 \\ -1 \end{bmatrix}$.

Since $\mathbf{n}_1 \cdot \mathbf{n} = \begin{bmatrix} 4 \\ -1 \\ 5 \end{bmatrix} \cdot \begin{bmatrix} 2 \\ 3 \\ -1 \end{bmatrix} = 4 \cdot 2 + (-1) \cdot 3 + 5 \cdot (-1) = 0$, $\mathscr{P}_1$ is perpendicular to $\mathscr{P}$.

(b) Since the general form of $\mathscr{P}$ is $4x - y + 5z = 0$, its normal vector is $\mathbf{n} = \begin{bmatrix} 4 \\ -1 \\ 5 \end{bmatrix}$.

Since $\mathbf{n}_1 = 1\mathbf{n}$, $\mathscr{P}_1$ is parallel to $\mathscr{P}$.

(c) Since the general form of $\mathscr{P}$ is $x - y - z = 3$, its normal vector is $\mathbf{n} = \begin{bmatrix} 1 \\ -1 \\ -1 \end{bmatrix}$.

Since $\mathbf{n}_1 \cdot \mathbf{n} = \begin{bmatrix} 4 \\ -1 \\ 5 \end{bmatrix} \cdot \begin{bmatrix} 1 \\ -1 \\ -1 \end{bmatrix} = 0$, $\mathscr{P}_1$ is perpendicular to $\mathscr{P}$.

(d) Since the general form of $\mathscr{P}$ is $4x + 6y - 2z = 0$, its normal vector is $\mathbf{n} = \begin{bmatrix} 4 \\ 6 \\ -2 \end{bmatrix}$.

Since $\mathbf{n}_1 \cdot \mathbf{n} = \begin{bmatrix} 4 \\ -1 \\ 5 \end{bmatrix} \cdot \begin{bmatrix} 4 \\ 6 \\ -2 \end{bmatrix} = 0$, $\mathscr{P}_1$ is perpendicular to $\mathscr{P}$.

21. Since the vector form is $\mathbf{x} = \mathbf{p} + t\mathbf{d}$, we use the given information to determine $\mathbf{p}$ and $\mathbf{d}$.

    The general equation of the given line is $2x - 3y = 1$, so its normal vector is $\mathbf{n} = \begin{bmatrix} 2 \\ -3 \end{bmatrix}$.

    Our line is parallel to the given line, so it has direction vector $\mathbf{d} = \begin{bmatrix} 3 \\ 2 \end{bmatrix}$.

    This comes from the solution of Exercise 45 in Section 1.2: $\mathbf{n} \cdot \mathbf{d} = \begin{bmatrix} a \\ b \end{bmatrix} \cdot \begin{bmatrix} b \\ -a \end{bmatrix} = ab - ab = 0$.

    Continuing, since our line passes through the point $P = (2, -1)$, we have $\mathbf{p} = \begin{bmatrix} 2 \\ -1 \end{bmatrix}$.

    So, the vector form of the line parallel to $2x - 3y = 1$ through the point $P = (2, -1)$ is
    $$\begin{bmatrix} x \\ y \end{bmatrix} = \begin{bmatrix} 2 \\ -1 \end{bmatrix} + t \begin{bmatrix} 3 \\ 2 \end{bmatrix}.$$

1.3 Lines and Planes

23. Since the vector form is $\mathbf{x} = \mathbf{p} + t\mathbf{d}$, we use the given information to determine $\mathbf{p}$ and $\mathbf{d}$.

A line with parametric equations $\begin{array}{l} x = a + et \\ y = b + ft \\ z = c + gt \end{array}$ has vector form $\begin{bmatrix} x \\ y \\ z \end{bmatrix} = \begin{bmatrix} a \\ b \\ c \end{bmatrix} + t \begin{bmatrix} e \\ f \\ g \end{bmatrix}$.

Therefore, its direction vector is $\mathbf{d} = \begin{bmatrix} e \\ f \\ g \end{bmatrix}$. We use this key observation below.

Since the given line has parametric equations $\begin{array}{l} x = 1 - t \\ y = 2 + 3t \\ z = -2 - t \end{array}$,

it has vector form $\begin{bmatrix} x \\ y \\ z \end{bmatrix} = \begin{bmatrix} 1 \\ 2 \\ -2 \end{bmatrix} + t \begin{bmatrix} -1 \\ 3 \\ -1 \end{bmatrix}$. So, its direction vector is $\begin{bmatrix} -1 \\ 3 \\ -1 \end{bmatrix}$.

Since our line is parallel to the given line, its direction vector is also $\mathbf{d} = \begin{bmatrix} -1 \\ 3 \\ -1 \end{bmatrix}$.

Furthermore, since our line passes through the point $P = (-1, 0, 3)$, we have $\mathbf{p} = \begin{bmatrix} -1 \\ 0 \\ 3 \end{bmatrix}$.

So, the vector form of the line parallel to the given line through $P = (-1, 0, 3)$ is

$$\begin{bmatrix} x \\ y \\ z \end{bmatrix} = \begin{bmatrix} -1 \\ 0 \\ 3 \end{bmatrix} + t \begin{bmatrix} -1 \\ 3 \\ -1 \end{bmatrix}.$$

25. Following Example 1.23, we will determine the general equations in two simple steps: First, we will use Figure 1.31 in Section 1.2 to find a normal vector **n** and a point vector **p**. Then we will substitute into $\mathbf{n}\cdot\mathbf{x} = \mathbf{n}\cdot\mathbf{p}$ and compute the dot products to find the equations.

(a) We start with $\mathscr{P}_1$ determined by the face of the cube in the yz-plane.

It is clear that a normal vector for $\mathscr{P}_1$ is $\mathbf{n} = \begin{bmatrix} 1 \\ 0 \\ 0 \end{bmatrix}$ or any vector parallel to the x-axis.

Also we see that $\mathscr{P}_1$ passes through the origin $P = (0,0,0)$, so we set $\mathbf{p} = \begin{bmatrix} 0 \\ 0 \\ 0 \end{bmatrix}$.

Substituting into $\mathbf{n}\cdot\mathbf{x} = \mathbf{n}\cdot\mathbf{p}$ yields $\begin{bmatrix} 1 \\ 0 \\ 0 \end{bmatrix} \cdot \begin{bmatrix} x \\ y \\ z \end{bmatrix} = \begin{bmatrix} 1 \\ 0 \\ 0 \end{bmatrix} \cdot \begin{bmatrix} 0 \\ 0 \\ 0 \end{bmatrix}$ or $1\cdot x + 0\cdot y + 0\cdot z = 0$.

So, the general equation for $\mathscr{P}_1$ determined by the face in the yz-plane is $x = 0$.

Likewise, the general equation for $\mathscr{P}_2$ determined by the face in the xz-plane is $y = 0$ and the general equation for $\mathscr{P}_3$ determined by the face in the xy-plane is $z = 0$.

We have found equations for the planes that pass through the origin.

We will use this information to find equations for the planes that pass through $(1,1,1)$.

We begin with $\mathscr{P}_4$ passing through the face parallel to the face in the yz-plane.

Since $\mathscr{P}_4$ is parallel to the face in the yz-plane, its normal vector is $\mathbf{n} = \begin{bmatrix} 1 \\ 0 \\ 0 \end{bmatrix}$.

As previously noted $\mathscr{P}_4$ passes through the point $P = (1,1,1)$, so we set $\mathbf{p} = \begin{bmatrix} 1 \\ 1 \\ 1 \end{bmatrix}$.

Substituting into $\mathbf{n}\cdot\mathbf{x} = \mathbf{n}\cdot\mathbf{p}$ yields $\begin{bmatrix} 1 \\ 0 \\ 0 \end{bmatrix} \cdot \begin{bmatrix} x \\ y \\ z \end{bmatrix} = \begin{bmatrix} 1 \\ 0 \\ 0 \end{bmatrix} \cdot \begin{bmatrix} 1 \\ 1 \\ 1 \end{bmatrix}$ or $1\cdot x + 0\cdot y + 0\cdot z = 1$.

So, the general equation for $\mathscr{P}_4$ is $x = 1$.

Likewise, the general equations for $\mathscr{P}_5$ and $\mathscr{P}_6$ are $y = 1$ and $z = 1$ respectively.

1.3 Lines and Planes

(b) We will use the given information to determine $\mathbf{n}$ and $\mathbf{p}$, then compute $\mathbf{n} \cdot \mathbf{x} = \mathbf{n} \cdot \mathbf{p}$.
We begin by observing the two key facts that will enable us to find $\mathbf{n}$ and $\mathbf{p}$:
Two planes $\mathscr{P}_1$, $\mathscr{P}$ are perpendicular if their normal vectors are orthogonal, so $\mathbf{n}_1 \cdot \mathbf{n} = 0$.
Every vector $\mathbf{u}$ in the plane $\mathscr{P}_1$ is orthogonal to its normal vector $\mathbf{n}_1$, so $\mathbf{n}_1 \cdot \mathbf{u} = 0$.
Condition 1: Our plane must be perpendicular to the xy-plane, so $\mathbf{n}_1 \cdot \mathbf{n} = 0$. From (a),

$$\mathbf{n} = \begin{bmatrix} 0 \\ 0 \\ 1 \end{bmatrix}, \text{ so } \mathbf{n}_1 \cdot \mathbf{n} = \begin{bmatrix} x \\ y \\ z \end{bmatrix} \cdot \begin{bmatrix} 0 \\ 0 \\ 1 \end{bmatrix} = 0 \Rightarrow z = 0. \text{ So, } \mathbf{n}_1 \text{ is of the form } \begin{bmatrix} x \\ y \\ 0 \end{bmatrix}.$$

Condition 2: $\mathbf{n}_1$ must be perpendicular to the vector $\mathbf{u}$ from the origin to $(1, 1, 1)$.

Since $\mathbf{u} = \begin{bmatrix} 1-0 \\ 1-0 \\ 1-0 \end{bmatrix} = \begin{bmatrix} 1 \\ 1 \\ 1 \end{bmatrix}$, we have $\mathbf{n}_1 \cdot \mathbf{n} = \begin{bmatrix} x \\ y \\ 0 \end{bmatrix} \cdot \begin{bmatrix} 1 \\ 1 \\ 1 \end{bmatrix} = 0 \Rightarrow x + y = 0 \Rightarrow y = -x.$

So, $\mathscr{P}_1$ must be of the form $\mathbf{n}_1 = \begin{bmatrix} x \\ -x \\ 0 \end{bmatrix} = x \begin{bmatrix} 1 \\ -1 \\ 0 \end{bmatrix}$. Letting $x = 1$ yields $\mathbf{n}_1 = \begin{bmatrix} 1 \\ -1 \\ 0 \end{bmatrix}$.

As previously noted $\mathscr{P}_1$ passes through the origin $P = (0, 0, 0)$, so we set $\mathbf{p} = \begin{bmatrix} 0 \\ 0 \\ 0 \end{bmatrix}$.

Now $\mathbf{n} \cdot \mathbf{x} = \mathbf{n} \cdot \mathbf{p}$ yields $\begin{bmatrix} 1 \\ -1 \\ 0 \end{bmatrix} \cdot \begin{bmatrix} x \\ y \\ z \end{bmatrix} = \begin{bmatrix} 1 \\ -1 \\ 0 \end{bmatrix} \cdot \begin{bmatrix} 0 \\ 0 \\ 0 \end{bmatrix}$ or $1 \cdot x + (-1) \cdot y + 0 \cdot z = 0.$

Therefore, the general equation for the plane perpendicular to the xy-plane and containing the diagonal from the origin to $(1, 1, 1)$ is $x - y = 0$.

(c) As above, use $\mathbf{u} = [0, 1, 1]$ and $\mathbf{v} = [1, 0, 1]$ from Example 1.15 of Section 1.2 to find $\mathbf{n}$.

From $\mathbf{n} \cdot \mathbf{u} = \begin{bmatrix} x \\ y \\ z \end{bmatrix} \cdot \begin{bmatrix} 0 \\ 1 \\ 1 \end{bmatrix} = 0 \Rightarrow y + z = 0 \Rightarrow y = -z.$

From $\mathbf{n} \cdot \mathbf{v} = \begin{bmatrix} x \\ -z \\ z \end{bmatrix} \cdot \begin{bmatrix} 1 \\ 0 \\ 1 \end{bmatrix} = 0 \Rightarrow x + z = 0 \Rightarrow x = -z.$

So, the normal vector $\mathbf{n} = \begin{bmatrix} -z \\ -z \\ z \end{bmatrix} = z \begin{bmatrix} -1 \\ -1 \\ 1 \end{bmatrix}$. When $z = -1$, we have $\mathbf{n} = \begin{bmatrix} 1 \\ 1 \\ -1 \end{bmatrix}$.

It is obvious the side diagonals pass through the origin $P = (0, 0, 0)$, so we set $\mathbf{p} = \begin{bmatrix} 0 \\ 0 \\ 0 \end{bmatrix}$.

Now $\mathbf{n} \cdot \mathbf{x} = \mathbf{n} \cdot \mathbf{p}$ yields $\begin{bmatrix} 1 \\ 1 \\ -1 \end{bmatrix} \cdot \begin{bmatrix} x \\ y \\ z \end{bmatrix} = \begin{bmatrix} 1 \\ 1 \\ -1 \end{bmatrix} \cdot \begin{bmatrix} 0 \\ 0 \\ 0 \end{bmatrix}$ or $1 \cdot x + 1 \cdot y + (-1) \cdot z = 0.$

The general equation for the plane containing the side diagonals is $x + y - z = 0$.

27. We will first follow Example 1.25, then use $d(Q, \ell) = \dfrac{|ax_0 + by_0 - c|}{\sqrt{a^2 + b^2}}$ and compare results.

Comparing $\begin{bmatrix} x \\ y \end{bmatrix} = \begin{bmatrix} -1 \\ 2 \end{bmatrix} + t \begin{bmatrix} 1 \\ -1 \end{bmatrix}$ to $\mathbf{x} = \mathbf{p} + t\mathbf{d}$, we see ℓ has $P = (-1, 2)$ and $\mathbf{d} = \begin{bmatrix} 1 \\ -1 \end{bmatrix}$.

As suggested by Figure 1.63, we need to calculate the length of $\overrightarrow{RQ}$, where R is the point on ℓ at the foot of the perpendicular from Q.

Now if we let $\mathbf{v} = \overrightarrow{PQ}$, then $\overrightarrow{PR} = \text{proj}_\mathbf{d}(\mathbf{v})$ and $\overrightarrow{RQ} = \mathbf{v} - \text{proj}_\mathbf{d}(\mathbf{v})$.

Step 1. $\mathbf{v} = \overrightarrow{PQ} = \mathbf{q} - \mathbf{p} = \begin{bmatrix} 2 \\ 2 \end{bmatrix} - \begin{bmatrix} -1 \\ 2 \end{bmatrix} = \begin{bmatrix} 3 \\ 0 \end{bmatrix}$.

Step 2. $\text{proj}_\mathbf{d}(\mathbf{v}) = \left(\dfrac{\mathbf{d} \cdot \mathbf{v}}{\mathbf{d} \cdot \mathbf{d}} \right) \mathbf{d} = \left(\dfrac{1 \cdot 3 + (-1) \cdot 0}{1 \cdot 1 + (-1) \cdot (-1)} \right) \begin{bmatrix} 1 \\ -1 \end{bmatrix} = \dfrac{3}{2} \begin{bmatrix} 1 \\ -1 \end{bmatrix} = \begin{bmatrix} 3/2 \\ -3/2 \end{bmatrix}$.

Step 3. The vector we want is $\mathbf{v} - \text{proj}_\mathbf{d}(\mathbf{v}) = \begin{bmatrix} 3 \\ 0 \end{bmatrix} - \begin{bmatrix} 3/2 \\ -3/2 \end{bmatrix} = \begin{bmatrix} 3/2 \\ 3/2 \end{bmatrix}$.

Step 4. The distance $d(Q, \ell)$ from Q to ℓ is $\|\mathbf{v} - \text{proj}_\mathbf{d}(\mathbf{v})\| = \left\| \begin{bmatrix} 3/2 \\ 3/2 \end{bmatrix} \right\|$.

So Theorem 1.3(b) implies $\|\mathbf{v} - \text{proj}_\mathbf{d}(\mathbf{v})\| = \dfrac{3}{2} \left\| \begin{bmatrix} 1 \\ 1 \end{bmatrix} \right\| = \dfrac{3}{2} \sqrt{1+1} = \dfrac{3\sqrt{2}}{2}$.

Now in order to calculate $d(Q, \ell) = \dfrac{|ax_0 + by_0 - c|}{\sqrt{a^2 + b^2}}$ we need to put ℓ into general form.

If $\mathbf{d} = \begin{bmatrix} a \\ b \end{bmatrix}$, then $\mathbf{n} = \begin{bmatrix} b \\ -a \end{bmatrix}$ because $\begin{bmatrix} a \\ b \end{bmatrix} \cdot \begin{bmatrix} b \\ -a \end{bmatrix} = 0$. For ℓ, $\mathbf{d} = \begin{bmatrix} 1 \\ -1 \end{bmatrix}$ so $\mathbf{n} = \begin{bmatrix} 1 \\ 1 \end{bmatrix}$.

From $\mathbf{n} \cdot \mathbf{x} = \mathbf{n} \cdot \mathbf{p}$ we have $\begin{bmatrix} 1 \\ 1 \end{bmatrix} \cdot \begin{bmatrix} x \\ y \end{bmatrix} = \begin{bmatrix} 1 \\ 1 \end{bmatrix} \cdot \begin{bmatrix} -1 \\ 2 \end{bmatrix}$ so $x + y = 1$ and $a = b = c = 1$.

Furthermore, since $Q = (2, 2) = (x_0, y_0)$ we have $x_0 = y_0 = 2$.

So $d(Q, \ell) = \dfrac{|2 + 2 - 1|}{\sqrt{1^2 + 1^2}} = \dfrac{3}{\sqrt{2}} = \dfrac{3\sqrt{2}}{2}$ exactly as we found by following Example 1.25.

1.3 Lines and Planes

29. We will follow Example 1.26, then use $d(Q, \mathscr{P}) = \dfrac{|ax_0 + by_0 + cz_0 - d|}{\sqrt{a^2 + b^2 + c^2}}$ and compare results.

By definition $ax + by + cz = d$ implies $\mathbf{n} = [a, b, c]$, so $x + y - z = 0$ implies $\mathbf{n} = [1, 1, -1]$.
As suggested by Figure 1.64, we need to calculate the length of $\overrightarrow{RQ} = \text{proj}_\mathbf{n}(\mathbf{v})$, where $\mathbf{v} = \overrightarrow{PQ}$.

Step 1. By trial and error, we find $P = (1, 0, 1)$ satisfies $x + y - z = 0$.

Step 2. $\mathbf{v} = \overrightarrow{PQ} = \mathbf{q} - \mathbf{p} = \begin{bmatrix} 2 \\ 2 \\ 2 \end{bmatrix} - \begin{bmatrix} 1 \\ 0 \\ 1 \end{bmatrix} = \begin{bmatrix} 1 \\ 2 \\ 1 \end{bmatrix}$.

Step 3. $\text{proj}_\mathbf{n}(\mathbf{v}) = \left(\dfrac{\mathbf{n} \cdot \mathbf{v}}{\mathbf{d} \cdot \mathbf{n}}\right) \mathbf{n} = \left(\dfrac{1 \cdot 1 + 1 \cdot 2 - 1 \cdot 1}{1^2 + 1^2 + (-1)^2}\right) \begin{bmatrix} 1 \\ 1 \\ -1 \end{bmatrix} = \dfrac{2}{3} \begin{bmatrix} 1 \\ 1 \\ -1 \end{bmatrix} = \begin{bmatrix} 2/3 \\ 2/3 \\ -2/3 \end{bmatrix}$.

Step 4. The distance from Q to $\mathscr{P}$ is $\|\text{proj}_\mathbf{n}(\mathbf{v})\| = \left\| \begin{bmatrix} 2/3 \\ 2/3 \\ -2/3 \end{bmatrix} \right\| = \dfrac{2}{3} \left\| \begin{bmatrix} 1 \\ 1 \\ -1 \end{bmatrix} \right\| = \dfrac{2\sqrt{3}}{3}$.

Now for $d(Q, \mathscr{P}) = \dfrac{|ax_0 + by_0 + cz_0 - d|}{\sqrt{a^2 + b^2 + c^2}}$ we need identify a, b, c, d, and x_0, y_0, z_0.

Since $x + y - z = 0$, $a = 1$, $b = 1$, $c = -1$, $d = 0$. From $Q = (2, 2, 2)$, $x_0 = y_0 = z_0 = 2$.

So $d(Q, \mathscr{P}) = \dfrac{|2 + 2 - 2 + 0|}{\sqrt{1^2 + 1^2 + (-1)^2}} = \dfrac{2}{\sqrt{3}} = \dfrac{2\sqrt{3}}{3}$ as we found by following Example 1.26.

31. Similar to Example 1.25, Figure 1.63 suggests we let $\mathbf{v} = \overrightarrow{PQ}$, then $\mathbf{w} = \overrightarrow{PR} = \text{proj}_\mathbf{d}(\mathbf{v})$.

Comparing $\begin{bmatrix} x \\ y \end{bmatrix} = \begin{bmatrix} -1 \\ 2 \end{bmatrix} + t \begin{bmatrix} 1 \\ -1 \end{bmatrix}$ to $\mathbf{x} = \mathbf{p} + t\mathbf{d}$, we see ℓ has $P = (-1, 2)$ and $\mathbf{d} = \begin{bmatrix} 1 \\ -1 \end{bmatrix}$.

Step 1. $\mathbf{v} = \overrightarrow{PQ} = \mathbf{q} - \mathbf{p} = \begin{bmatrix} 2 \\ 2 \end{bmatrix} - \begin{bmatrix} -1 \\ 2 \end{bmatrix} = \begin{bmatrix} 3 \\ 0 \end{bmatrix}$.

Step 2. $\mathbf{w} = \text{proj}_\mathbf{d}(\mathbf{v}) = \left(\dfrac{\mathbf{d} \cdot \mathbf{v}}{\mathbf{d} \cdot \mathbf{d}}\right) \mathbf{d} = \left(\dfrac{1 \cdot 3 + (-1) \cdot 0}{1 \cdot 1 + (-1) \cdot (-1)}\right) \begin{bmatrix} 1 \\ -1 \end{bmatrix} = \dfrac{3}{2} \begin{bmatrix} 1 \\ -1 \end{bmatrix} = \begin{bmatrix} 3/2 \\ -3/2 \end{bmatrix}$.

Step 3. So, $\mathbf{r} = \mathbf{p} + \overrightarrow{PR} = \mathbf{p} + \text{proj}_\mathbf{d}(\mathbf{v}) = \mathbf{p} + \mathbf{w} = \begin{bmatrix} -1 \\ 2 \end{bmatrix} + \begin{bmatrix} 3/2 \\ -3/2 \end{bmatrix} = \begin{bmatrix} 1/2 \\ 1/2 \end{bmatrix}$.

Therefore, the point R on ℓ that is closest to Q is $(\frac{1}{2}, \frac{1}{2})$.

33. Similar to Example 1.26, Figure 1.64 suggests we let $\mathbf{v} = \overrightarrow{PQ}$, then $\mathbf{w} = \overrightarrow{QR} = -\text{proj}_\mathbf{n}(\mathbf{v})$.

By definition $ax + by + cz = d$ implies $\mathbf{n} = [a, b, c]$, so $x + y - z = 0$ implies $\mathbf{n} = [1, 1, -1]$.

Step 1. By trial and error, we find $P = (1, 0, 1)$ satisfies $x + y - z = 0$.

Step 2. $\mathbf{v} = \overrightarrow{PQ} = \mathbf{q} - \mathbf{p} = \begin{bmatrix} 2 \\ 2 \\ 2 \end{bmatrix} - \begin{bmatrix} 1 \\ 0 \\ 1 \end{bmatrix} = \begin{bmatrix} 1 \\ 2 \\ 1 \end{bmatrix}$.

Step 3. $\mathbf{w} = \text{proj}_\mathbf{n}(\mathbf{v}) = \left(\dfrac{\mathbf{n} \cdot \mathbf{v}}{\mathbf{n} \cdot \mathbf{n}}\right)\mathbf{n} = \left(\dfrac{1 \cdot 1 + 1 \cdot 2 + (-1) \cdot 1}{1^2 + 1^2 + (-1)^2}\right)\begin{bmatrix} 1 \\ 1 \\ -1 \end{bmatrix} = \begin{bmatrix} 2/3 \\ 2/3 \\ -2/3 \end{bmatrix}$.

Step 4. So, $\mathbf{r} = \mathbf{p} + \overrightarrow{PQ} + \overrightarrow{QR} = \mathbf{p} + \mathbf{v} - \text{proj}_\mathbf{n}(\mathbf{v}) = \begin{bmatrix} 1 \\ 0 \\ 1 \end{bmatrix} + \begin{bmatrix} 1 \\ 2 \\ 1 \end{bmatrix} - \begin{bmatrix} 2/3 \\ 2/3 \\ -2/3 \end{bmatrix} = \begin{bmatrix} 4/3 \\ 4/3 \\ 8/3 \end{bmatrix}$.

Therefore, the point R in $\mathscr{P}$ that is closest to Q is $\left(\tfrac{4}{3}, \tfrac{4}{3}, \tfrac{8}{3}\right)$.

35. Since the given lines ℓ_1 and ℓ_2 are parallel, we can simply choose Q on ℓ_1, P on ℓ_2. Following Example 1.25, we have:

From ℓ_1, $Q = (1, 1)$. From ℓ_2, we have $P = (5, 4)$, $\mathbf{d} = [-2, 3]$, and $\mathbf{n} = [3, 2] = [a, b]$.

Step 1. $\mathbf{v} = \overrightarrow{PQ} = \mathbf{q} - \mathbf{p} = [1, 1] - [5, 4] = [-4, -3]$.

Step 2. $\text{proj}_\mathbf{d}(\mathbf{v}) = \left(\dfrac{\mathbf{d} \cdot \mathbf{v}}{\mathbf{d} \cdot \mathbf{d}}\right)\mathbf{d} = \left(\dfrac{(-2) \cdot (-4) + 3 \cdot (-3)}{(-2)^2 + 3^2}\right)\begin{bmatrix} -2 \\ 3 \end{bmatrix} = -\dfrac{1}{13}\begin{bmatrix} -2 \\ 3 \end{bmatrix} = \begin{bmatrix} 2/13 \\ -3/13 \end{bmatrix}$.

Step 3. The vector we want is $\mathbf{v} - \text{proj}_\mathbf{d}(\mathbf{v}) = \begin{bmatrix} -4 \\ -3 \end{bmatrix} - \begin{bmatrix} 2/13 \\ -3/13 \end{bmatrix} = \begin{bmatrix} -54/13 \\ -36/13 \end{bmatrix}$.

Step 4. The distance $d(Q, \ell_2)$ from ℓ_1 to ℓ_2 is $\|\mathbf{v} - \text{proj}_\mathbf{d}(\mathbf{v})\| = \left\| \begin{bmatrix} -54/13 \\ -36/13 \end{bmatrix} \right\|$.

So Theorem 1.3(b) implies $\|\mathbf{v} - \text{proj}_\mathbf{d}(\mathbf{v})\| = \dfrac{18}{13}\left\| \begin{bmatrix} 3 \\ 2 \end{bmatrix} \right\| = \dfrac{18}{13}\sqrt{4 + 9} = \dfrac{18\sqrt{13}}{13}$.

From $\mathbf{n} \cdot \mathbf{p} = \begin{bmatrix} 3 \\ 2 \end{bmatrix} \cdot \begin{bmatrix} 5 \\ 4 \end{bmatrix} = 23$, $c = 23$. Since $Q = (1, 1) = (x_0, y_0)$, we have $x_0 = y_0 = 1$.

Now compare: $d(\ell_1, \ell_2) = d(Q, \ell_2) = \dfrac{|ax_0 + by_0 - c|}{\sqrt{a^2 + b^2}} = \dfrac{|3 + 2 - 23|}{\sqrt{3^2 + 2^2}} = \dfrac{18}{\sqrt{13}} = \dfrac{18\sqrt{13}}{13}$.

1.3 Lines and Planes

37. Since the given planes $\mathscr{P}_1$ and $\mathscr{P}_2$ are parallel, we can simply choose Q in $\mathscr{P}_1$, P in $\mathscr{P}_2$. Following Example 1.26, we have:

Step 1. Since $2x + y - 2z = 0$, $Q = (0,0,0)$ is on $\mathscr{P}_1$.

Since $2x + y - 2z = 5$, $P = (0,5,0)$ is on $\mathscr{P}_2$ and $\mathbf{n} = [2,1,-2] = [a,b,c]$.

Step 2. $\mathbf{v} = \overrightarrow{PQ} = \mathbf{q} - \mathbf{p} = [0,0,0] - [0,5,0] = [0,-5,0]$

Step 3. $\text{proj}_\mathbf{n}(\mathbf{v}) = \left(\dfrac{\mathbf{n}\cdot\mathbf{v}}{\mathbf{n}\cdot\mathbf{n}}\right)\mathbf{n} = \left(\dfrac{2\cdot 0 - 1\cdot 5 - 2\cdot 0}{2^2 + 1^2 + (-2)^2}\right)\begin{bmatrix}2\\1\\-2\end{bmatrix} = -\dfrac{5}{9}\begin{bmatrix}2\\1\\-2\end{bmatrix} = \begin{bmatrix}-10/9\\-5/9\\10/9\end{bmatrix}$.

Step 4. The distance $d(Q,\mathscr{P}_2)$ from $\mathscr{P}_1$ to $\mathscr{P}_2$ is $\|\mathbf{v} - \text{proj}_\mathbf{n}(\mathbf{v})\| = \left\|\begin{bmatrix}-10/9\\-5/9\\10/9\end{bmatrix}\right\|$.

So Theorem 1.3(b) implies $\|\text{proj}_\mathbf{n}(\mathbf{v})\| = \dfrac{5}{9}\left\|\begin{bmatrix}2\\1\\-2\end{bmatrix}\right\| = \dfrac{5}{9}\sqrt{4+1+4} = \dfrac{5}{3}$.

From $\mathbf{n}\cdot\mathbf{p} = d = [2,1,-2]\cdot[0,5,0] = 5$. Since $Q = (0,0,0)$, $x_0 = y_0 = z_0 = 0$.

Now compare: $d(Q,\mathscr{P}_2) = \dfrac{|ax_0 + by_0 + cz_0 - d|}{\sqrt{a^2+b^2+c^2}} = \dfrac{|0+0+0-5|}{\sqrt{2^2+1^2+(-2)^2}} = \dfrac{5}{\sqrt{9}} = \dfrac{5}{3}$.

39. Will show $d(B,\ell) = \dfrac{|ax_0 + by_0 - c|}{\sqrt{a^2+b^2}}$, where $\mathbf{n} = \begin{bmatrix}a\\b\end{bmatrix}$, $\mathbf{n}\cdot\mathbf{a} = c$, and $B = (x_0,y_0)$.

Step 1. From Figure 1.61, we see $d(B,\ell) = \|\text{proj}_\mathbf{n}(\mathbf{v})\| = \left\|\left(\dfrac{\mathbf{n}\cdot\mathbf{v}}{\mathbf{n}\cdot\mathbf{n}}\right)\mathbf{n}\right\| = \dfrac{|\mathbf{n}\cdot\mathbf{v}|}{\|\mathbf{n}\|}$.

Step 2. Since $\mathbf{v} = \mathbf{b} - \mathbf{a}$, $\mathbf{n}\cdot\mathbf{v} = \mathbf{n}\cdot(\mathbf{b}-\mathbf{a}) = \mathbf{n}\cdot\mathbf{b} - \mathbf{n}\cdot\mathbf{a} = \begin{bmatrix}a\\b\end{bmatrix}\cdot\begin{bmatrix}x_0\\y_0\end{bmatrix} - c = ax_0 + by_0 - c$.

Step 3. So, $d(B,\ell) = \|\text{proj}_\mathbf{n}(\mathbf{v})\| = \left\|\left(\dfrac{\mathbf{n}\cdot\mathbf{v}}{\mathbf{n}\cdot\mathbf{n}}\right)\mathbf{n}\right\| = \dfrac{|\mathbf{n}\cdot\mathbf{v}|}{\|\mathbf{n}\|} = \dfrac{|ax_0 + by_0 - c|}{\sqrt{a^2+b^2}}$.

41. We will apply the formula from Exercise 39, $d(B,\ell) = \dfrac{|\mathbf{n}\cdot\mathbf{v}|}{\|\mathbf{n}\|}$.

Step 1. We select $B = (x_0,y_0)$ on ℓ_1 so that $\mathbf{n}\cdot\mathbf{b} = \begin{bmatrix}a\\b\end{bmatrix}\cdot\begin{bmatrix}x_0\\y_0\end{bmatrix} = ax_0 + by_0 = c_1$.

Step 2. We select A on ℓ_2 so that $\mathbf{n}\cdot\mathbf{a} = c_2$.

Step 3. Set $\mathbf{v} = \mathbf{b} - \mathbf{a}$, then $d(B,\mathscr{P}) = \dfrac{|\mathbf{n}\cdot\mathbf{v}|}{\|\mathbf{n}\|} = \dfrac{|\mathbf{n}\cdot(\mathbf{b}-\mathbf{a})|}{\|\mathbf{n}\|} = \dfrac{|\mathbf{n}\cdot\mathbf{b} - \mathbf{n}\cdot\mathbf{a}|}{\|\mathbf{n}\|} = \dfrac{|c_1 - c_2|}{\|\mathbf{n}\|}$.

43. As in Example 1.14 of Section 1.2, we note that $\cos\theta = \dfrac{|\mathbf{u}\cdot\mathbf{v}|}{\|\mathbf{u}\|\,\|\mathbf{v}\|}$.

So, given two planes $\mathscr{P}_1$ with $\mathbf{n}_1$ and $\mathscr{P}_2$ with $\mathbf{n}_2$, we have $\cos\theta = \dfrac{|\mathbf{n}_1\cdot\mathbf{n}_2|}{\|\mathbf{n}_1\|\,\|\mathbf{n}_2\|}$.

Step 1. Since $\mathscr{P}_1$ has equation $x+y+z=0$, $\mathbf{n}_1 = [1,1,1]$.
Since $\mathscr{P}_2$ has equation $2x+y-2z=0$, $\mathbf{n}_2 = [2,1,-2]$.

Step 2. Therefore, $\mathbf{n}_1 = [1,1,1]\cdot[2,1,-2] = 1\cdot 2 + 1\cdot 1 - 1\cdot 2 = 1$,
$\|\mathbf{n}_1\| = \sqrt{1^2+1^2+1^2} = \sqrt{3}$, and $\|\mathbf{n}_2\| = \sqrt{2^2+1^2+(-2)^2} = 3$.

Step 3. So $\cos\theta = \dfrac{1}{3\sqrt{3}}$ and $\theta = \cos^{-1}\left(\dfrac{1}{3\sqrt{3}}\right) \approx 78.9°$.

45. As in Example 1.14 of Section 1.2, we note that $\cos\theta = \dfrac{|\mathbf{u}\cdot\mathbf{v}|}{\|\mathbf{u}\|\,\|\mathbf{v}\|}$.

So, given $\mathscr{P}$ with $\mathbf{n}$ and ℓ with $\mathbf{d}$, we have $\cos\theta = \dfrac{|\mathbf{n}\cdot\mathbf{d}|}{\|\mathbf{n}\|\,\|\mathbf{d}\|}$.

Step 1. To show $\mathscr{P}$ and ℓ intersect, we note:
$x+y+2z = (2+t)+(1-2t)+2(3+t) = 9+t = 0$ implies $t = -9$.
So, $\mathscr{P}$ and ℓ intersect at the point $[2+(-9), 1-2(-9), 3+(-9)] = [-7, 19, -6]$.

Step 2. Since $\mathscr{P}$ has equation $x+y+2z=0$, $\mathbf{n} = [1,1,2]$.

Given $\begin{array}{c} x = 2+t \\ y = 1-2t \\ z = 3+t \end{array}$, ℓ satisfies $\begin{bmatrix} x \\ y \\ z \end{bmatrix} = \begin{bmatrix} 2 \\ 1 \\ 3 \end{bmatrix} + t\begin{bmatrix} 1 \\ -2 \\ 1 \end{bmatrix}$. So, $\mathbf{d} = \begin{bmatrix} 1 \\ -2 \\ 1 \end{bmatrix}$.

Step 3. Therefore, $\mathbf{n}\cdot\mathbf{d} = [1,1,2]\cdot[1,-2,1] = 1\cdot 1 - 1\cdot 2 + 2\cdot 1 = 1$,
$\|\mathbf{n}\| = \sqrt{1^2+1^2+2^2} = \sqrt{6}$, and $\|\mathbf{d}\| = \sqrt{1^2+(-2)^2+1^2} = \sqrt{6}$.

Step 4. So $\cos\theta = \dfrac{1}{\sqrt{6}\sqrt{6}} = \dfrac{1}{6}$ and $\theta = \cos^{-1}\left(\dfrac{1}{6}\right) \approx 80.4°$.

47. Will find an expression for $\mathbf{p}$ in terms of $\mathbf{v}$ and $\mathbf{n}$ given $\mathbf{n}$ is orthogonal to $\mathbf{p}$, that is $\mathbf{p}\cdot\mathbf{n} = 0$.

Step 1. We solve for c starting from the given equation $\mathbf{p} = \mathbf{v} - c\mathbf{n}$.

$\mathbf{p} = \mathbf{v} - c\mathbf{n} \quad\Rightarrow$	Given
$c\mathbf{n} = \mathbf{v} - \mathbf{p}$	By $x = y - z$ implies $z = y - x$
$(c\mathbf{n})\cdot\mathbf{n} = (\mathbf{v} - \mathbf{p})\cdot\mathbf{n} \quad\Rightarrow$	Taking the dot product of $\mathbf{n}$ with both sides.
$c(\mathbf{n}\cdot\mathbf{n}) = \mathbf{v}\cdot\mathbf{n} - \mathbf{p}\cdot\mathbf{n}$	By properties of the dot product
$c(\mathbf{n}\cdot\mathbf{n}) = \mathbf{n}\cdot\mathbf{v}$	By $\mathbf{v}\cdot\mathbf{n} = \mathbf{n}\cdot\mathbf{v}$ and $\mathbf{p}\cdot\mathbf{n} = 0$
$c = \dfrac{\mathbf{n}\cdot\mathbf{v}}{\mathbf{n}\cdot\mathbf{n}}$	Dividing both sides by $\mathbf{n}\cdot\mathbf{n}$ (a scalar)

Note: Figure 1.66 also shows $c\mathbf{n} = \text{proj}_\mathbf{n}(\mathbf{v}) = \left(\dfrac{\mathbf{n}\cdot\mathbf{v}}{\mathbf{n}\cdot\mathbf{n}}\right)\mathbf{n}$ which implies $c = \dfrac{\mathbf{n}\cdot\mathbf{v}}{\mathbf{n}\cdot\mathbf{n}}$.

Step 2. Letting $c = \dfrac{\mathbf{n}\cdot\mathbf{v}}{\mathbf{n}\cdot\mathbf{n}}$ in $\mathbf{p} = \mathbf{v} - c\mathbf{n}$, we have $\mathbf{p} = \mathbf{v} - \left(\dfrac{\mathbf{n}\cdot\mathbf{v}}{\mathbf{n}\cdot\mathbf{n}}\right)\mathbf{n}$.

Exploration: The Cross Product

Since Explorations are self-contained, only odd-numbered solutions will be provided.

1. (a) $\mathbf{u} \times \mathbf{v} = \begin{bmatrix} 1(2) - 1(-1) \\ 1(3) - 0(2) \\ 0(-1) - 1(3) \end{bmatrix} = \begin{bmatrix} 3 \\ 3 \\ -3 \end{bmatrix}$. (b) $\mathbf{u} \times \mathbf{v} = \begin{bmatrix} -3 \\ -3 \\ 3 \end{bmatrix}$.

   (c) $\mathbf{u} \times \mathbf{v} = \begin{bmatrix} 0 \\ 0 \\ 0 \end{bmatrix}$. (d) $\mathbf{u} \times \mathbf{v} = \begin{bmatrix} 1 \\ -2 \\ 1 \end{bmatrix}$.

3. Two vectors are orthogonal if their dot product equals zero. Check:

$$(\mathbf{u} \times \mathbf{v}) \cdot \mathbf{u} = \begin{bmatrix} u_2 v_3 - u_3 v_2 \\ u_3 v_1 - u_1 v_3 \\ u_1 v_2 - u_2 v_1 \end{bmatrix} \cdot \begin{bmatrix} u_1 \\ u_2 \\ u_3 \end{bmatrix}$$

$$= (u_2 v_3 - u_3 v_2) u_1 + (u_3 v_1 - u_1 v_3) u_2 + (u_1 v_2 - u_2 v_1) u_3$$
$$= (u_2 v_3 u_1 - u_1 v_3 u_2) + (u_3 v_1 u_2 - u_2 v_1 u_3) + (u_1 v_2 u_3 - u_3 v_2 u_1) = 0.$$

$$(\mathbf{u} \times \mathbf{v}) \cdot \mathbf{v} = \begin{bmatrix} u_2 v_3 - u_3 v_2 \\ u_3 v_1 - u_1 v_3 \\ u_1 v_2 - u_2 v_1 \end{bmatrix} \cdot \begin{bmatrix} v_1 \\ v_2 \\ v_3 \end{bmatrix}$$

$$= (u_2 v_3 - u_3 v_2) v_1 + (u_3 v_1 - u_1 v_3) v_2 + (u_1 v_2 - u_2 v_1) v_3$$
$$= (u_2 v_3 v_1 - u_2 v_1 v_3) + (u_3 v_1 v_2 - u_3 v_1 v_2) + (u_1 v_2 v_3 - u_1 v_3 v_2) = 0.$$

5. (a) $\mathbf{v} \times \mathbf{u} = \begin{bmatrix} v_2 u_3 - v_3 u_2 \\ v_3 u_1 - v_1 u_3 \\ v_1 u_2 - v_2 u_1 \end{bmatrix} = - \begin{bmatrix} u_2 v_3 - u_3 v_2 \\ u_3 v_1 - u_1 v_3 \\ u_1 v_2 - u_2 v_1 \end{bmatrix} = -(\mathbf{u} \times \mathbf{v}).$

(b) $\mathbf{u} \times \mathbf{0} = \begin{bmatrix} u_1 \\ u_2 \\ u_3 \end{bmatrix} \times \begin{bmatrix} 0 \\ 0 \\ 0 \end{bmatrix} \begin{bmatrix} u_2(0) - u_3(0) \\ u_3(0) - u_1(0) \\ u_1(0) - u_2(0) \end{bmatrix} = \begin{bmatrix} 0 \\ 0 \\ 0 \end{bmatrix} = \mathbf{0}.$

(c) $\mathbf{u} \times \mathbf{u} = \begin{bmatrix} u_2 u_3 - u_3 u_2 \\ u_3 u_1 - u_1 u_3 \\ u_1 u_2 - u_2 u_1 \end{bmatrix} = \begin{bmatrix} 0 \\ 0 \\ 0 \end{bmatrix} = \mathbf{0}.$

(d) $\mathbf{u} \times k\mathbf{v} = \begin{bmatrix} u_2 k v_3 - u_3 k v_2 \\ u_3 k v_1 - u_1 k v_3 \\ u_1 k v_2 - u_2 k v_1 \end{bmatrix} = k \begin{bmatrix} u_2 v_3 - u_3 v_2 \\ u_3 v_1 - u_1 v_3 \\ u_1 v_2 - u_2 v_1 \end{bmatrix} = k(\mathbf{u} \times \mathbf{v}).$

(e) $\mathbf{u} \times k\mathbf{u} = k(\mathbf{u} \times \mathbf{u}) = k(\mathbf{0}) = \mathbf{0}.$

(f) We compute the cross product as follows:

$$\mathbf{u} \times (\mathbf{v} + \mathbf{w}) = \begin{bmatrix} u_2(v_3 + w_3) - u_3(v_2 + w_2) \\ u_3(v_1 + w_1) - u_1(v_3 + w_3) \\ u_1(v_2 + w_2) - u_2(v_1 + w_1) \end{bmatrix}$$

$$= \begin{bmatrix} (u_2 v_3 - u_3 v_2) + (u_2 w_3 - u_3 w_2) \\ (u_3 v_1 - u_1 v_3) + (u_3 w_1 - u_1 w_3) \\ (u_1 v_2 - u_2 v_1) + (u_1 w_2 - u_2 w_1) \end{bmatrix}$$

$$= \begin{bmatrix} (u_2 v_3 - u_3 v_2) \\ (u_3 v_1 - u_1 v_3) \\ (u_1 v_2 - u_2 v_1) \end{bmatrix} + \begin{bmatrix} (u_2 w_3 - u_3 w_2) \\ (u_3 w_1 - u_1 w_3) \\ (u_1 w_2 - u_2 w_1) \end{bmatrix} = \mathbf{u} \times \mathbf{v} + \mathbf{u} \times \mathbf{w}.$$

7. *2*: $\mathbf{e}_1 \times (\mathbf{e}_2 \times \mathbf{e}_3) = (\mathbf{e}_1 \cdot \mathbf{e}_3)\mathbf{e}_2 - (\mathbf{e}_1 \cdot \mathbf{e}_2)\mathbf{e}_3$ [by 6(b)] $= 0$ since $\mathbf{e}_i \cdot \mathbf{e}_j = 0$ for $i \neq j$.
Thus, since the $\mathbf{e}_i$ have length 1, we must have $\mathbf{e}_1 = \mathbf{e}_2 \times \mathbf{e}_3$, by 6(c).
Show $\mathbf{e}_2 = \mathbf{e}_3 \times \mathbf{e}_1$ and $\mathbf{e}_3 = \mathbf{e}_1 \times \mathbf{e}_2$ by cyclically permuting the indices.

3: $\mathbf{u} \cdot (\mathbf{u} \times \mathbf{v}) = (\mathbf{u} \times \mathbf{u}) \cdot \mathbf{v}$ [by 10(a)] $= \mathbf{0} \cdot \mathbf{v}$ [by 9(c)] $= 0$, so $\mathbf{u}$ is orthogonal to $\mathbf{u} \times \mathbf{v}$.
Similarly, $\mathbf{v} \cdot (\mathbf{u} \times \mathbf{v}) = \mathbf{v} \cdot (-1)(\mathbf{v} \times \mathbf{u}) = (\mathbf{v} \times \mathbf{v}) \cdot \mathbf{u} = 0$. So $\mathbf{v}$ is orthogonal to $\mathbf{u} \times \mathbf{v}$.

1.4 Code Vectors and Modular Arithmetic

Q: What is our big goal for Section 1.4?

A: To explore a widespread and highly practical application of vectors and challenge our pre-conceived notions by the investigation of Modular Arithmetic.

Begin with binary arithmetic, then expand to general modular arithmetic.
Have fun encoding and decoding UPC and ISBN codes.

Key Definitions and Concepts

Note that the definition of *length* given below is applied only in the context of codes.

binary	p.47	1.4	see the text discussion of arithmetic in $\mathbb{Z}_2$
check digit	p.49	1.4	extra component added to vector to make its *parity* even
code	p.48	1.4	a set of vectors of the same *length* (m-ary)
decoding	p.48	1.4	converting code vectors into a message
encoding	p.48	1.4	converting a message into code vectors
length (m-ary)	p.51	1.4	the number of components in a vector
m-ary	p.51	1.4	see the text discussion of arithmetic in $\mathbb{Z}_m$
modular	p.50	1.4	see the text development of Modular Arithmetic
parity	p.49	1.4	the number of 1s in a code vector
ternary	p.50	1.4	see the text discussion of arithmetic in $\mathbb{Z}_3$
transposition	p.52	1.4	the interchange of two adjacent components

Theorems

There are no theorems in this section.

Additional Questions based on the Exercises

The following questions are intended to develop our intuition about Modular Arithmetic.

Q: Does $\mathbb{Z}_0$ exist? That is, does this notation make sense?
A: No. Notice the pattern: $\mathbb{Z}_2$ only contains the numbers 0 and 1.
What numbers would $\mathbb{Z}_0$ contain?

Q: Does $\mathbb{Z}_1$ exist? If so, what number or numbers does it contain?

Q: Can a number in $\mathbb{Z}_n$ have more than one multiplicative inverse?
A: No. Why not?
Pretend both c and d are multiplicative inverses for 5.
What would that tell us? Both $5c = 1$ and $d5 = 1$. That implies $c = d$. Why?
Multiplying $5c = 1$ by d, we get $(d5)c = 1d$. So $1c = c = d$.

Q: Why do many equations in modular arithmetic have more than one solution?
A: Because the modular equation is often several equations in all-in-one. Why?

Let's consider the following example: $4x = 0$ in $\mathbb{Z}_8$.
What are the four equations implied by this single equation?
$$4x = 0 + 0 \cdot 8 = 0 \Rightarrow x = 0,\ 4x = 0 + 1 \cdot 8 = 8 \Rightarrow x = 2,$$
$$4x = 0 + 2 \cdot 8 = 16 \Rightarrow x = 4,\text{ and } 4x = 0 + 3 \cdot 8 = 24 \Rightarrow x = 6.$$

Q: What is the multiplication table for $\mathbb{Z}_8$?
A: Hint: Here are the entries for 4: $0, 4, 0, 4, 0, 4, 0, 4$.

Q: What do those entries $0, 4, 0, 4, 0, 4, 0, 4$ tell us about $4x = 4$?
A: It has four solutions in $\mathbb{Z}_8$. Namely $x = 1, 3, 5$, and 7.
These entries also tell us that $4x = 1, 2, 3, 5, 6$, and 7 have no solution.

Q: What is the multiplication table for $\mathbb{Z}_{12}$?
A: Hint: Here are the entries for 8: $0, 8, 4, 0, 8, 4, 0, 8, 4, 0, 8, 4$.
Q: What do those entries tell us about solutions to $8x = ?$ in $\mathbb{Z}_{12}$?

37. Q: Given a binary vector $\mathbf{v}$ with even parity, what is $\mathbf{v} \cdot \mathbf{v}$?

    Q: What is the definition for length for an m-ary vector?
    A: The number of components in the vector! Not the *usual* definition.

1.4 Code Vectors and Modular Arithmetic

Solutions to odd-numbered exercises from Section 1.4

Exercises: Let's not let preconceived notions lead us to mistakes. How can we avoid that? By faithfully applying the rules.

1. $\mathbf{u} + \mathbf{v} = [0,1] + [1,1] = [1,0]$, $\mathbf{u} \cdot \mathbf{v} = 0 + 1 = 1$.

3. $\mathbf{u} + \mathbf{v} = [1,0,1,1] + [1,1,1,1] = [0,1,0,0]$, $\mathbf{u} \cdot \mathbf{v} = 1 + 0 + 1 + 1 = 1$.

5.

+	0	1	2	3
0	0	1	2	3
1	1	2	3	0
2	2	3	0	1
3	3	0	1	2

·	0	1	2	3
0	0	0	0	0
1	0	1	2	3
2	0	2	0	2
3	0	3	2	1

7. $2 + 2 + 2 = 6 = 0$ in $\mathbb{Z}_3$.

9. $2(2 + 1 + 2) = 2(2) = 3 \cdot 1 + 1 = 1$ in $\mathbb{Z}_3$.

11. $2 \cdot 3 \cdot 2 = 4 \cdot 3 + 0 = 0$ in $\mathbb{Z}_4$.

13. $2 + 1 + 2 + 2 + 1 = 2$ in $\mathbb{Z}_3$, $2 + 1 + 2 + 2 + 1 = 0$ in $\mathbb{Z}_4$, $2 + 1 + 2 + 2 + 1 = 3$ in $\mathbb{Z}_5$.

15. $8(6 + 4 + 3) = 8(4) = 5$ in $\mathbb{Z}_9$.

17. $[2,1,2] + [2,0,1] = [1,1,0]$ in $\mathbb{Z}_3^3$.

19. $[2,0,3,2] \cdot ([3,1,1,2] + [3,3,2,1]) = [2,0,3,2] \cdot [2,0,3,3] = 0 + 0 + 2 + 1 = 3$ in $\mathbb{Z}_4^4$.
$[2,0,3,2] \cdot ([3,1,1,2] + [3,3,2,1]) = [2,0,3,2] \cdot [1,4,3,3] = 2 + 0 + 4 + 1 = 2$ in $\mathbb{Z}_5^4$.

21. $x = 1 + (-5) = 1 + 1 = 2$ in $\mathbb{Z}_6$.

23. No solution. Why? Consider: $\frac{4}{2} = 2$.

25. $x = (3)^{-1} 4 = (2) 4 = 3$ in $\mathbb{Z}_5$.

27. No solution. Why? $\frac{6}{2} = 3$, $\frac{8}{2} = 4$.

29. $x = (2)^{-1}(2 + (-3)) = (3)(2 + 2) = 2$ in $\mathbb{Z}_5$.

31. Add 5 to both sides $\Rightarrow 6x = 6$, so $x = 1, 5$ (because $5 \cdot 8 = 40 = 8 \cdot 4 + 6$).

33. (a) All $a \neq 0$ in $\mathbb{Z}_5$ have a solution because 5 is a prime number.

(b) $a = 1, 5$ because they have no common factors with 6 other than 1.

(c) a and m can have no common factors other than 1, that is, the greatest common divisor (gcd) of a and m is 1.

35. We require that $[1,1,0,1,1,d] \cdot [1,1,1,1,1,1] = 0 \Rightarrow 1 + 1 + 1 + 1 + d = 0 \Rightarrow d = 0$. So the associated parity check code vector is $\mathbf{v} = [1,1,0,1,1,0]$.

37. We check if $\mathbf{v} \cdot \mathbf{1} = 0$: $[1,1,1,0,1,1] \cdot [1,1,1,1,1,1] = 1 + 1 + 1 + 1 + 1 = 1$. So a single error could have occurred.

39. We check if $\mathbf{v} \cdot \mathbf{1} = 0$: $[1,1,0,1,0,1,1,1] \cdot [1,1,1,1,1,1,1,1] = 1 + 1 + 1 + 1 + 1 + 1 = 0$. So a single error could not have occurred.

41. We require that $[3,4,2,3,d] \cdot [1,1,1,1,1] = 0$ in $\mathbb{Z}_5$. Thus $3+4+2+3+d = 0 \Rightarrow d = 3$.

43. We require that $[3,0,7,5,6,8,d] \cdot [1,1,1,1,1,1,1] = 0$ in $\mathbb{Z}_9$.
Thus $3+7+5+6+8+d = 0 \Rightarrow d = 7$.

45. $\mathbf{c} \cdot \mathbf{u} = 0 \Rightarrow 3(0+9+6+7+0+7) + (5+4+4+0+2) + d = 0 \Rightarrow 7+5+d = 0 \Rightarrow d = 8$.

47. Let $\mathbf{u} = [0,4,6,9,5,6,1,8,2,0,1,5]$ be a UPC vector.

   (a) We check if $\mathbf{c} \cdot \mathbf{u} = 0$:
   $$\begin{aligned}\mathbf{c} \cdot \mathbf{u} &= [3,1,3,1,3,1,3,1,3,1,3,1] \cdot [0,4,6,9,5,6,1,8,2,0,1,5] \\ &= 3(0+6+5+1+2+1) + (4+9+6+8+0) + 5 = 5+7+5 = 7 \neq 0\end{aligned}$$
   so, the UPC cannot be correct.

   (b) Now $\mathbf{u} = [0,4,u_3,9,5,6,1,8,2,0,1,5]$, and we require that $\mathbf{c} \cdot \mathbf{u} = 0$:
   $3(0+u_3+5+1+2+1) + (4+9+6+8+0) + 5 = 0 \Leftrightarrow$
   $3(u_3+9) + 2 = 0 \Leftrightarrow 3u_3 + 7 = 8 \Leftrightarrow 3u_3 = 1 \Leftrightarrow u_3 = 7$.
   So, the correct UPC is $[0,4,7,9,5,6,1,8,2,0,1,5]$.

49. (a) We check if $\mathbf{c} \cdot \mathbf{u}' = 0$:
   $$\begin{aligned}\mathbf{c} \cdot \mathbf{u}' &= [3,1,3,1,3,1,3,1,3,1,3,1] \cdot [0,4,7,9,2,7,0,2,0,9,4,6] \\ &= 3(0+7+2+0+0+4) + (4+9+7+2+9) + 6 = 9+1+6 = 6 \neq 0\end{aligned}$$
   so the error will be detected.

   (b) Transposing the third and fourth components of $\mathbf{u}$ gives $\mathbf{u}' = [0,7,9,4,2,7,0,2,0,9,4,6]$, while $\mathbf{c} \cdot \mathbf{u}' = 0$ in $\mathbb{Z}_{10}$, so the error will not be detected.

   (c) Assume there is a transposition error between the ith and $(i+1)^{th}$ components and that the new UPC $= \mathbf{u}' = \mathbf{u} + \mathbf{e}_i(u_{i+1} - u_i) + \mathbf{e}_{i+1}(u_i - u_{i+1})$ satisfies $\mathbf{c} \cdot \mathbf{u}' = 0$. So:
   $$\begin{aligned}0 &= \mathbf{c} \cdot (\mathbf{u} + \mathbf{e}_i(u_{i+1} - u_i) + \mathbf{e}_{i+1}(u_i - u_{i+1})) \\ &= \mathbf{c} \cdot \mathbf{u} + \mathbf{c} \cdot \mathbf{e}_i(u_{i+1} - u_i) + \mathbf{c} \cdot \mathbf{e}_{i+1}(u_i - u_{i+1}) \\ &= c_i(u_{i+1} - u_i) + c_{i+1}(u_i - u_{i+1}) = u_i(c_{i+1} - c_i) + u_{i+1}(c_i - c_{i+1})\end{aligned}$$

   Case 1: If i is even then $c_i = 1$, $c_{i+1} = 3$, and the constraint becomes $0 = 2u_i + 8u_{i+1}$.
   Case 2: If i is odd then $c_i = 3$, $c_{i+1} = 1$, and the constraint becomes $0 = 8u_i + 2u_{i+1}$.
   So regardless of i, if u_i and u_{i+1} are transposed, the odd component of u_i and u_{i+1} is, say, u_i, and the adjacent components satisfy $8u_i + 2u_{i+1} = 0$,
   the error will not be detected.
   Note, the trivial solution is $u_i = u_{i+1}$, that is the adjacent components are the same.

51. $\mathbf{c} \cdot \mathbf{u} = 0 \Rightarrow [10,9,8,7,6,5,4,3,2,1] \cdot [0,3,9,4,7,5,6,8,2,d] = 4 + d \Rightarrow d = 7$.

1.4 Code Vectors and Modular Arithmetic

53. (a) $\mathbf{u} = [0, 6, 7, 9, 7, 6, 2, 9, 0, 6]$ is a correct ISBN.
Consider a transposition between the fourth and fifth components.
Then $\mathbf{u}' = [0, 6, 7, 7, 9, 6, 2, 9, 0, 6]$.
Now check if $\mathbf{u}'$ is correct: $\mathbf{c} \cdot \mathbf{u}' = [10, 9, 8, 7, 6, 5, 4, 3, 2, 1] \cdot [0, 6, 7, 7, 9, 6, 2, 9, 0, 6] = 9 \neq 0$.
So, the error would be detected.

(b) In general, an ISBN vector $\mathbf{u}$ satisfies the check constraint $\sum_{i=1}^{10} (11 - i)(u_i) = 0$,

but in $\mathbb{Z}_{11}$ $(11 - i) = -i$, so the constraint becomes $\sum_{i=1}^{10} (-i)(u_i) = 0$.

Consider a transposition of any two adjacent elements (say the u_j and u_{j+1} elements).
This gives a new vector $\mathbf{u}' = \mathbf{u} + \mathbf{e}_j (u_{j+1} - u_j) + \mathbf{e}_{j+1} (u_j - u_{j+1})$.

The check constraint on $\mathbf{u}'$ becomes $\sum_{i=1}^{10} (-i)(u'_i) = 0$,

but $u'_i = u_i + (\mathbf{e}_j)_i (u_{j+1} - u_j) + (\mathbf{e}_{j+1})_i (u_j - u_{j+1})$, so this becomes

$$\sum_{i=1}^{10} (-i)(u'_i) = = \sum_{i=1}^{10} (-i)(u_i) + \sum_{i=1}^{10} (-i)(\mathbf{e}_j)_i (u_{j+1} - u_j) + \sum_{i=1}^{10} (-i)(\mathbf{e}_{j+1})_i (u_j - u_{j+1})$$

$$= \sum_{i=1}^{10} (-i)(\mathbf{e}_j)_i (u_{j+1} - u_j) + \sum_{i=1}^{10} (-i)(\mathbf{e}_{j+1})_i (u_j - u_{j+1})$$

$$= (-j)(u_{j+1} - u_j) + (-(j+1))(u_j - u_{j+1})$$

$$= (-j)(u_{j+1} - u_j) + (-j + 10)(u_j - u_{j+1}) = 10(u_j - u_{j+1})$$

So the constraint $\sum_{i=1}^{10} (-i)(u'_i) = 0$ can only be met if $u_j = u_{j+1}$,
in which case the transposed vector is the same as the original.
So the constraint $\sum_{i=1}^{10} (-i)(u_i) = 0$ cannot be broken by transposing adjacent components.
An error made in any two adjacent components of the ISBN will be detected.

(c) See the proof in part (b).

Chapter 1 Review

Key Definitions and Concepts

This list includes most but not all of the definitions listed at the end of each section.

angle	p.21	1.2	$\cos\theta = \frac{\mathbf{u}\cdot\mathbf{v}}{\|\mathbf{u}\|\|\mathbf{v}\|}$
binary	p.47	1.4	see the text discussion of arithmetic in $\mathbb{Z}_2$
check digit	p.49	1.4	extra component added to vector to make its *parity* even
code	p.48	1.4	a set of vectors of the same *length* (m-ary)
components	p.3	1.1	individual coordinates of a vector like 3, 2 of $[3,2]$
decoding	p.48	1.4	converting code vectors into a message
direction vector	p.32	1.3	$\mathbf{d}$, parallel to any vector on line ℓ
distance	p.20	1.2	$d(\mathbf{u},\mathbf{v}) = \|\mathbf{u}-\mathbf{v}\|$
dot product	p.15	1.2	$\mathbf{u}\cdot\mathbf{v} = u_1v_1 + u_2v_2 + \ldots + u_nv_n$
encoding	p.48	1.4	converting a message into code vectors
general form	p.33	1.3	$ax + by = c$ where $\mathbf{n} = \begin{bmatrix} a \\ b \end{bmatrix}$ is normal for ℓ
if and only if	p.17	1.2	worth repeating: signals a *double implication*
inner product	p.15	1.2	generalized notion of the dot product (Ch. 7)
length	p.17	1.2	$\|\mathbf{v}\| = \sqrt{\mathbf{v}\cdot\mathbf{v}}$ (Means the same thing as *norm*)
length (m-ary)	p.51	1.4	the number of components in in a vector
linear combination	p.12	1.1	$\mathbf{u} = c_1\mathbf{v}_1 + c_2\mathbf{v}_2 + \cdots + c_n\mathbf{v}_n$ where the c_i are scalars
m-ary	p.51	1.4	see the text discussion of arithmetic in $\mathbb{Z}_m$
modular	p.50	1.4	see the text development of Modular Arithmetic
norm	p.17	1.2	$\|\mathbf{v}\| = \sqrt{\mathbf{v}\cdot\mathbf{v}}$ (using the dot product)
normal form	p.33	1.3	$\mathbf{n}\cdot(\mathbf{x}-\mathbf{p}) = 0$ or $\mathbf{n}\cdot\mathbf{x} = \mathbf{n}\cdot\mathbf{p}$
normalizing	p.18	1.2	a unit vector in same direction ($\mathbf{u} = \frac{1}{\|\mathbf{v}\|}\mathbf{v}$)
normal vector	p.31	1.3	$\mathbf{n}$, orthogonal to any vector on line ℓ or plane $\mathscr{P}$

ordered	p.3	1.1	$[3,2]$ *vs.* $[2,3]$: these are *not* the same vector
orthogonal	p.23	1.2	$\mathbf{u}$ and $\mathbf{v}$ are orthogonal if $\mathbf{u} \cdot \mathbf{v} = 0$
parallel	p.8	1.1	if and only if scalar multiples of each other
parametric equations	p.33	1.3	component equations from $\mathbf{x} = \mathbf{p} + t\mathbf{d}$
parity	p.49	1.4	the number of 1s in a code vector
position vector	p.3	1.1	vector with tail at the origin O, i.e. $\overrightarrow{OA}$
projection	p.24	1.2	$\text{proj}_{\mathbf{u}}(\mathbf{v}) = \left(\frac{\mathbf{u} \cdot \mathbf{v}}{\mathbf{u} \cdot \mathbf{u}}\right)\mathbf{u}$
scalar	p.8	1.1	a real number c (that is, c is **not** a vector)
scalar product	p.15	1.2	another name for dot product (result is a scalar)
standard position	p.4	1.1	vector with its tail at the origin O, i.e. $\overrightarrow{OA}$
standard unit vectors	p.19	1.2	$\mathbf{e}_i$ with a 1 in the ith component and 0s elsewhere
ternary	p.50	1.4	see the text discussion of arithmetic in $\mathbb{Z}_3$
transposition	p.52	1.4	the interchange of two adjacent components
unit vector	p.18	1.2	vector of length 1
vector	p.3	1.1	*directed* line segment with *length* and *direction*
vector addition	p.5	1.1	$\mathbf{u} + \mathbf{v} = [u_1 + v_1, u_2 + v_2]$
vector form	p.33	1.3	$\mathbf{x} = \mathbf{p} + t\mathbf{d}$ where $\mathbf{d}$ is a direction vector for ℓ
zero vector	p.4	1.1	$\mathbf{0}$, *all* components are 0, so length is 0
zero vector	p.23	1.2	$\mathbf{0}$. Note: $\mathbf{0} \cdot \mathbf{v} = 0$ for *every* vector $\mathbf{v}$ in $\mathbf{R}^n$

Theorems

In the summary, we will list only the central result of the theorem.
For the complete statement of the theorem, refer to the text.

Thm 1.1	p.10	1.1	Algebraic Properties of Vectors in $\mathbb{R}^n$
Thm 1.2	p.16	1.2	Properties of the dot product ($\mathbf{u} \cdot \mathbf{v} = \mathbf{v} \cdot \mathbf{u}$ ...)
Thm 1.3	p.17	1.2	Properties of the norm ($\|\mathbf{v}\| = 0 \Leftrightarrow \mathbf{v} = \mathbf{0}$ ...)
Thm 1.4	p.19	1.2	Cauchy-Schwarz: $\|\mathbf{u} \cdot \mathbf{v}\| \leq \|\mathbf{u}\| \|\mathbf{v}\|$
Thm 1.5	p.19	1.2	Triangle Inequality: $\|\mathbf{u} + \mathbf{v}\| \leq \|\mathbf{u}\| + \|\mathbf{v}\|$
Thm 1.6	p.23	1.2	Pythagoras: $\|\mathbf{u} + \mathbf{v}\|^2 = \|\mathbf{u}\|^2 + \|\mathbf{v}\|^2 \Leftrightarrow \mathbf{u} \cdot \mathbf{v} = 0$
Cauchy-Schwarz	p.19	1.2	Thm 1.4: $\|\mathbf{u} \cdot \mathbf{v}\| \leq \|\mathbf{u}\| \|\mathbf{v}\|$
Pythagoras	p.23	1.2	Thm 1.6: $\|\mathbf{u} + \mathbf{v}\|^2 = \|\mathbf{u}\|^2 + \|\mathbf{v}\|^2 \Leftrightarrow \mathbf{u} \cdot \mathbf{v} = 0$
Triangle Inequality	p.19	1.2	Thm 1.5: $\|\mathbf{u} + \mathbf{v}\| \leq \|\mathbf{u}\| + \|\mathbf{v}\|$

Symbolism

Symbolism will not be included at the end of every chapter.

$\mathbf{v} = \mathbf{w}$	p.3	1.1	... if and only if *corresponding* components are equal
$\mathbf{u} + \mathbf{v}$	p.5	1.1	$\mathbf{u} + \mathbf{v} = [u_1 + v_1, u_2 + v_2]$ **vector addition**
$\mathbf{u} = \sum c_i \mathbf{v}_i$	p.12	1.1	the c_i are scalars (**linear combination**)
$\mathbf{u} - \mathbf{v}$	p.8	1.1	$\mathbf{u} - \mathbf{v} = \mathbf{u} + (-\mathbf{v})$ **vector subtraction**
$\mathbf{u} \| \mathbf{v}$	p.8	1.1	$\mathbf{u} \| \mathbf{v} \Leftrightarrow \mathbf{v} = c\mathbf{u}$ if and only if scalar multiples (**parallel**)
$\mathbf{v}$	p.3	1.1	a vector is denoted by a single, boldface, lowercase letter
$\|$	p.8	1.1	parallel (word is used not symbol)
$\Rightarrow$	p.4	1.1	implies (**if** vectors have the same direction, **then** they are parallel)
$\Leftrightarrow$	p.4	1.1	if and only if (this phrase is used on p. 4 of Section 1.1)
$\sum c_i \mathbf{v}_i$	p.12	1.1	shorthand for $c_1 \mathbf{v}_1 + c_2 \mathbf{v}_2 + \cdots + c_n \mathbf{v}_n$
$\overrightarrow{AB}$	p.3	1.1	an overhead arrow (vector $\overrightarrow{AB}$, differs from line segment $\overline{AB}$)

Solutions to odd-numbered exercises from Chapter 1 *Review*

1. We will explain and give counter examples to justify our answers below.

   (a) **True**. Follows from the properties of $\mathbb{R}^n$ listed in Theorem 1.1 in Section 1.1:

   $$\begin{aligned}
   \mathbf{u} &= \mathbf{u} + \mathbf{0} & &\text{Zero Property, Property (c)} \\
   &= \mathbf{u} + (\mathbf{w} + (-\mathbf{w})) & &\text{Additive Inverse Property, Property (d)} \\
   &= (\mathbf{u} + \mathbf{w}) + (-\mathbf{w}) & &\text{Distributive Property, Property (b)} \\
   &= (\mathbf{v} + \mathbf{w}) + (-\mathbf{w}) & &\text{By the given condition } \mathbf{u} + \mathbf{w} = \mathbf{v} + \mathbf{w} \\
   &= \mathbf{v} + (\mathbf{w} + (-\mathbf{w})) & &\text{Distributive Property, Property (b)} \\
   &= \mathbf{v} + \mathbf{0} & &\text{Additive Inverse Property, Property (d)} \\
   &= \mathbf{v} & &\text{Zero Property, Property (c)}
   \end{aligned}$$

   (b) **False**. See Example 1.16 and Exercise 54 in Section 1.2. Two key counter examples:
   Since $\mathbf{0} \cdot \mathbf{v} = 0$ for every vector $\mathbf{v}$ in $\mathbb{R}^3$, $\mathbf{0}$ is orthogonal to every vector.
   That is, if $\mathbf{u} = \mathbf{0}$, we know nothing about $\mathbf{v}$ and $\mathbf{w}$.
   Let $\mathbf{u}$ and $\mathbf{v}$ be orthogonal to $\mathbf{w}$ then $\mathbf{u} \cdot \mathbf{w} = \mathbf{v} \cdot \mathbf{w} = 0$.
   E.g., consider $\mathbf{u} = [a, 0, 0]$, $\mathbf{v} = [0, b, 0]$, and $\mathbf{w} = [0, 0, c]$.

   (c) **False**. Note this property is *not* listed in Theorem 1.2 in Section 1.2.
   Let $\mathbf{v} = \mathbf{0}$ then $\mathbf{u} \cdot \mathbf{0} = 0$ and $\mathbf{0} \cdot \mathbf{w} = 0$, but there is no restriction on $\mathbf{u}$ and $\mathbf{w}$.
   Let $\mathbf{u} = \mathbf{w}$ then $\mathbf{u} \cdot \mathbf{v} = 0$ and $\mathbf{v} \cdot \mathbf{u} = 0$, but $\mathbf{u} \cdot \mathbf{u} \neq 0$ unless $\mathbf{u} = \mathbf{0}$.
   E.g., consider $\mathbf{u} = [a, 0, 0]$, $\mathbf{v} = [0, b, 0]$, and $\mathbf{w} = [c, 0, 0]$, then $\mathbf{u} \cdot \mathbf{w} = ac$.

   (d) **False**. When a line is parallel to plane then $\mathbf{d} \cdot \mathbf{n} = 0$, that is $\mathbf{d}$ is *orthogonal* to $\mathbf{n}$.
   See Figure 1.57 in Section 1.3.

   (e) **True**. Every line in plane $\mathscr{P}$ and parallel to $\mathscr{P}$ is *orthogonal* to its normal vector $\mathbf{n}$.
   See Figure 1.62 in Section 1.3.

   (f) **True**. See the remarks following Example 1.24 in Section 1.3.

   (g) **False**. In $\mathbb{R}^3$ many non-parallel lines are *skew* (non-intersecting lines with $\mathbf{d}_1 \neq \mathbf{d}_2$).
   For example, ℓ_1 with $\mathbf{x} = t\mathbf{d}$ with $\mathbf{d} = [1, 0, 0]$ (the x-axis) and
   ℓ_2 with $\mathbf{x} = \mathbf{p} + t\mathbf{d}$ with $\mathbf{p} = [0, 0, 1]$ and $\mathbf{d} = [0, 1, 0]$
   (the line parallel to the y-axis through $[0, 0, 1]$).

   (h) **False**. See Examples 1.27 to 1.29 in Section 1.4.
   For example, $[1, 0, 1] \cdot [1, 0, 1] = 1 + 0 + 1 = 0$ in $\mathbb{Z}_2$.
   In general, the dot product of any binary vector with an even number of 1s is 0.

   (i) **True**. See Example 1.37 in Section 1.4.
   We have $\mathbf{c} \cdot \mathbf{u} = 3(0 + 1 + 7 + 5 + 7 + 8) + (4 + 7 + 1 + 2 + 0 + 2) = 100 = 0$ in $\mathbb{Z}_{10}$.

   (j) **False**. See Example 1.38 in Section 1.4.
   $\mathbf{c} \cdot \mathbf{u} = [10, 9, 8, 7, 6, 5, 4, 3, 2, 1] \cdot [0, 5, 3, 2, 3, 4, 1, 7, 4, 8] = 162 \neq 0$ in $\mathbb{Z}_{11}$.

Chapter 1 Review

3. See Example 1.5 in Section 1.1. **Note**: We should do the vector arithmetic first.

$$2\mathbf{x} + \mathbf{u} = 3(\mathbf{x} - \mathbf{v}) \text{ implies } \mathbf{x} = \mathbf{u} + 3\mathbf{v} = \begin{bmatrix} -1 \\ 5 \end{bmatrix} + 3 \begin{bmatrix} 3 \\ 2 \end{bmatrix} = \begin{bmatrix} 8 \\ 11 \end{bmatrix}.$$

5. We proceed as in Example 1.14 of Section 1.2.

    We have $\mathbf{u} \cdot \mathbf{v} = -1 \cdot 2 + 1 \cdot 1 - 2 \cdot 1 = -3$, $\|\mathbf{u}\| = \sqrt{(-1)^2 + 1^2 + 2^2} = \sqrt{6}$, and $\|\mathbf{v}\| = \sqrt{6}$.

    Therefore, $\cos \theta = -\dfrac{3}{\sqrt{6}\sqrt{6}} = -\dfrac{1}{2}$, so $\theta = \cos^{-1}\left(-\dfrac{1}{2}\right) = \dfrac{2\pi}{3}$ radians or $120°$.

7. We use the given conditions to find a unit vector in the xy-plane orthogonal to $\mathbf{v} = [1, 2, 3]$.

    Step 1. Figure 1.15 in Section 1.1 implies any vector in the xy-plane has a z-component of 0. So, the vector $\mathbf{u}$ we are looking for must be of the form $\mathbf{u} = [a, b, 0]$.

    Step 2. Like Exercise 42 in Section 1.2, since $\mathbf{u}$ is orthogonal to $\mathbf{v}$, we have

    $$\mathbf{u} \cdot \mathbf{v} = \begin{bmatrix} 1 \\ 2 \\ 3 \end{bmatrix} \cdot \begin{bmatrix} a \\ b \\ 0 \end{bmatrix} = a + 2b = 0. \text{ So } a = -2b \text{ and } \mathbf{u} = \begin{bmatrix} -2b \\ b \\ 0 \end{bmatrix} = b \begin{bmatrix} -2 \\ 1 \\ 0 \end{bmatrix}.$$

    Step 3. As in Example 1.12 of Section 1.2, we **normalize** $\mathbf{u}$ to create $\mathbf{w}$, the unit vector.

    Letting $b = 1$ above gives us $\mathbf{u} = \begin{bmatrix} -2 \\ 1 \\ 0 \end{bmatrix}$ and $\|\mathbf{u}\| = \sqrt{(-2)^2 + 1^2 + 0^2} = \sqrt{5}$.

    So *one* vector that works is $\mathbf{w} = \left(\dfrac{1}{\|\mathbf{u}\|}\right) \mathbf{u} = \dfrac{1}{\sqrt{5}} \begin{bmatrix} -2 \\ 1 \\ 0 \end{bmatrix} = \begin{bmatrix} -2/\sqrt{5} \\ 1/\sqrt{5} \\ 0 \end{bmatrix}.$

    Note: The fact that we got to choose a value for b implies there are infinitely many solutions.

9. Planes that are parallel have parallel normals, so since our plane is parallel to $2x + 3y - z = 0$, a normal to given plane and therefore our plane is $\mathbf{n} = \begin{bmatrix} 2 \\ 3 \\ -1 \end{bmatrix}$.

    As in Example 1.23 of Section 1.3, we find the plane through $P = (3, 2, 5)$ with $\mathbf{n} = \begin{bmatrix} 2 \\ 3 \\ -1 \end{bmatrix}$.

    With $\mathbf{p} = \begin{bmatrix} 3 \\ 2 \\ 5 \end{bmatrix}$ and $\mathbf{x} = \begin{bmatrix} x \\ y \\ z \end{bmatrix}$, we have $\mathbf{n} \cdot \mathbf{p} = 2 \cdot 3 + 3 \cdot 2 - 1 \cdot 5 = 7$.

    So the normal equation $\mathbf{n} \cdot \mathbf{x} = \mathbf{n} \cdot \mathbf{p}$ becomes the general equation $2x + 3y - z = 7$.

11. We proceed as in Exercise 41 of Section 1.2. See the notes prior to Exercise 41.

Let $\mathbf{u} = \overrightarrow{AB} = \begin{bmatrix} 1-1 \\ 0-1 \\ 1-0 \end{bmatrix} = \begin{bmatrix} 0 \\ -1 \\ 1 \end{bmatrix}$ and $\mathbf{v} = \overrightarrow{AC} = \begin{bmatrix} 0-1 \\ 1-1 \\ 2-0 \end{bmatrix} = \begin{bmatrix} -1 \\ 0 \\ 2 \end{bmatrix}$.

(a) We compute the necessary values ...

$\mathbf{u} \cdot \mathbf{v} = \begin{bmatrix} 0 \\ -1 \\ 1 \end{bmatrix} \cdot \begin{bmatrix} -1 \\ 0 \\ 2 \end{bmatrix} = 2,$

$\mathbf{u} \cdot \mathbf{u} = \begin{bmatrix} 0 \\ -1 \\ 1 \end{bmatrix} \cdot \begin{bmatrix} 0 \\ -1 \\ 1 \end{bmatrix} = 2 \, (\|\mathbf{u}\| = \sqrt{2}),$

$\operatorname{proj}_{\mathbf{u}}(\mathbf{v}) = \left(\frac{\mathbf{u} \cdot \mathbf{v}}{\mathbf{u} \cdot \mathbf{u}} \right) \mathbf{u} = \begin{bmatrix} 0 \\ -1 \\ 1 \end{bmatrix} \Rightarrow$

$\mathbf{v} - \operatorname{proj}_{\mathbf{u}}(\mathbf{v}) = \begin{bmatrix} -1 \\ 1 \\ 1 \end{bmatrix} \Rightarrow$

$\|\mathbf{v} - \operatorname{proj}_{\mathbf{u}}(\mathbf{v})\| = \sqrt{(-1)^2 + 1^2 + 1^2}$

$= \sqrt{3}$

... then substitute into the formula for $\mathcal{A}$:

$\mathcal{A} = \tfrac{1}{2} \|\mathbf{u}\| \, \|\mathbf{v} - \operatorname{proj}_{\mathbf{u}}(\mathbf{v})\|$

$= \tfrac{1}{2} \sqrt{2} \, \sqrt{3} = \tfrac{\sqrt{6}}{2}.$

(b) We compute the necessary values ...

$\mathbf{u} \cdot \mathbf{v} = \begin{bmatrix} 0 \\ -1 \\ 1 \end{bmatrix} \cdot \begin{bmatrix} -1 \\ 0 \\ 2 \end{bmatrix} = 2,$

$\|\mathbf{u}\| = \sqrt{0^2 + (-1)^2 + 1^2} = \sqrt{2},$

$\|\mathbf{v}\| = \sqrt{(-1)^2 + 0^2 + 2^2} = \sqrt{5} \Rightarrow$

$\cos \theta = \frac{\mathbf{u} \cdot \mathbf{v}}{\|\mathbf{u}\| \, \|\mathbf{v}\|} = \frac{2}{\sqrt{10}} = \frac{\sqrt{10}}{5} \Rightarrow$

$\sin \theta = \sqrt{1 - \cos^2 \theta} = \sqrt{1 - \left(\frac{\sqrt{10}}{5} \right)^2} = \frac{\sqrt{15}}{5}$

... then substitute into the formula for $\mathcal{A}$:

$\mathcal{A} = \tfrac{1}{2} \|\mathbf{u}\| \, \|\mathbf{v}\| \sin \theta$

$= \tfrac{1}{2} \sqrt{2} \, \sqrt{5} \, \tfrac{\sqrt{15}}{5} = \tfrac{\sqrt{6}}{2}.$

13. We proceed as in Exercise 61 from Section 1.2.
We need to show $\|\mathbf{u}\| = 2$ and $\|\mathbf{v}\| = 3$ imply $\mathbf{u} \cdot \mathbf{v} \neq -7$.
From Theorem 1.4 (the Cauchy-Schwarz Inequality), we have $|\mathbf{x} \cdot \mathbf{y}| \leq \|\mathbf{x}\| \, \|\mathbf{y}\|$.
Substituting in the given values of $\|\mathbf{u}\| = 2$ and $\|\mathbf{v}\| = 3$ shows $|\mathbf{u} \cdot \mathbf{v}| \leq 6$.
Therefore, $-6 \leq \mathbf{u} \cdot \mathbf{v} \leq 6$. It follows immediately that $\mathbf{u} \cdot \mathbf{v} \neq -7$.

Chapter 1 Review 59

15. We follow Example 1.27 in Section 1.3, then use $d(Q, \ell) = \dfrac{|ax_0 + by_0 + cz_0 - d|}{\sqrt{a^2 + b^2 + c^2}}$.

    Comparing $\begin{bmatrix} x \\ y \\ z \end{bmatrix} = \begin{bmatrix} 0 \\ 1 \\ 2 \end{bmatrix} + t \begin{bmatrix} 1 \\ 1 \\ 1 \end{bmatrix}$ to $\mathbf{x} = \mathbf{p} + t\mathbf{d}$, we see ℓ has $P = (0, 1, 2)$ and $\mathbf{d} = \begin{bmatrix} 1 \\ 1 \\ 1 \end{bmatrix}$.

    Now if we let $\mathbf{v} = \overrightarrow{PQ}$, then $\overrightarrow{PR} = \text{proj}_\mathbf{d}(\mathbf{v})$ and $\overrightarrow{RQ} = \mathbf{v} - \text{proj}_\mathbf{d}(\mathbf{v})$.

    Step 1. $\mathbf{v} = \overrightarrow{PQ} = \mathbf{q} - \mathbf{p} = [3, 2, 5] - [0, 1, 2] = [3, 1, 3]$.

    Step 2. $\text{proj}_\mathbf{d}(\mathbf{v}) = \left(\dfrac{\mathbf{d} \cdot \mathbf{v}}{\mathbf{d} \cdot \mathbf{d}}\right) \mathbf{d} = \left(\dfrac{1 \cdot 3 + 1 \cdot 1 + 1 \cdot 3}{1 \cdot 1 + 1 \cdot 1 + 1 \cdot 1}\right) \begin{bmatrix} 1 \\ 1 \\ 1 \end{bmatrix} = \dfrac{7}{3} \begin{bmatrix} 1 \\ 1 \\ 1 \end{bmatrix} = \begin{bmatrix} 7/3 \\ 7/3 \\ 7/3 \end{bmatrix}$.

    Step 3. The vector we want is $\mathbf{v} - \text{proj}_\mathbf{d}(\mathbf{v}) = \begin{bmatrix} 3 \\ 1 \\ 3 \end{bmatrix} - \begin{bmatrix} 7/3 \\ 7/3 \\ 7/3 \end{bmatrix} = \begin{bmatrix} 2/3 \\ -4/3 \\ 2/3 \end{bmatrix} = -\dfrac{2}{3} \begin{bmatrix} 1 \\ -2 \\ 1 \end{bmatrix}$.

    Step 4. The distance $d(Q, \ell)$ is $\|\mathbf{v} - \text{proj}_\mathbf{d}(\mathbf{v})\| = \left\| \begin{bmatrix} 2/3 \\ -4/3 \\ 2/3 \end{bmatrix} \right\| = \dfrac{2}{3} \left\| \begin{bmatrix} 1 \\ -2 \\ 1 \end{bmatrix} \right\| = \dfrac{2\sqrt{6}}{3}$.

    $\mathbf{n} \cdot \mathbf{x} = \mathbf{n} \cdot \mathbf{p}, \begin{bmatrix} 1 \\ -2 \\ 1 \end{bmatrix} \cdot \begin{bmatrix} x \\ y \\ z \end{bmatrix} = \begin{bmatrix} 1 \\ -2 \\ 1 \end{bmatrix} \cdot \begin{bmatrix} 0 \\ 1 \\ 2 \end{bmatrix}$ so $x - 2y + z = 0$ and $a = 1, b = -2, c = 1, d = 0$.

    So $d(Q, \ell) = \dfrac{|3 - 4 + 5 - 0|}{\sqrt{1^2 + (-2)^2 + 1^2}} = \dfrac{4}{\sqrt{6}} = \dfrac{2\sqrt{6}}{3}$ as in Example 1.27 of Section 1.3.

17. We begin by noting that $5 \cdot 3 = 15 = 1$ in $\mathbb{Z}_7$ so $3^{-1} = 5$ in $\mathbb{Z}_7$ since $5 \cdot 3 = 1 = 3 \cdot 3^{-1}$.

    $3(x + 2) = 5 \Rightarrow$ We multiply both sides by 5
    $5 \cdot 3(x + 2) = 5 \cdot 5 \Rightarrow$ because $3^{-1} = 5$ in $\mathbb{Z}_7$.
    $x + 2 = 25 = 4 \Rightarrow$ $25 = 21 + 4 = 7 \cdot 3 + 4 = 4$ in $\mathbb{Z}_7$.
    $x = 2$.   Simply subtract 2 from both sides.

    Note: We should check our answer: $3(2 + 2) = 12 = 7 \cdot 1 + 5 = 5$ in $\mathbb{Z}_7$.

19. To find the check digit d, we follow the remarks after Example 1.37 in Section 1.4. We find the value of d that makes $\mathbf{c} \cdot \mathbf{u} = 0$ performing all calculations in $\mathbb{Z}_{10}$.

    Recall the check vector $\mathbf{c}$ for UPC code is $\mathbf{c} = [3, 1, 3, 1, 3, 1, 3, 1, 3, 1, 3, 1]$ so we have:

    $\mathbf{c} \cdot \mathbf{u} = 3 \cdot (7 + 3 + 6 + 7 + 3 + 7) + 1 \cdot (3 + 9 + 1 + 0 + 1) + d = 3(3) + 1(4) + d = 3 + d = 0 \Rightarrow d = 7$.

    So, the check digit in UPC$[7, 3, 3, 9, 6, 1, 7, 0, 3, 1, 7, d]$ is 7.

    Check: $\mathbf{c} \cdot \mathbf{u} = 3 \cdot (7 + 3 + 6 + 7 + 3 + 7) + 1 \cdot (3 + 9 + 1 + 0 + 1 + 7) = 13 + 7 = 0$ in $\mathbb{Z}_{10}$.

Chapter 2

Systems of Linear Equations

2.1 Introduction to Systems of Linear Equations

Q: What is our goal for Section 2.1?

A: To understand linear systems and take a first look at how they relate to matrices.

We will compare systems with no solution, one solution, and many solutions.

We see how no solution corresponds no intersection and solutions relate to intersections.

Key Definitions and Concepts

augmented matrix	p.62	2.1	a coefficient matrix augmented by the constants
back substitution	p.62	2.1	the procedure used to solve Example 2.5 on p. 62
coefficient matrix	p.62	2.1	a matrix of coefficients taken from a linear system
coefficients	p.59	2.1	the a_i in $a_1x_1 + a_2x_2 + \cdots + a_nx_n = b$
consistent	p.61	2.1	a system of linear equations with at least one solution
constant term	p.58	2.1	b in $a_1x_1 + a_2x_2 + \cdots + a_nx_n = b$
equivalent	p.61	2.1	linear systems that have the same solution set
inconsistent	p.61	2.1	a system of linear equations with no solutions
linear equation	p.59	2.1	$a_1x_1 + a_2x_2 + \cdots + a_nx_n = b$
linear system	p.59	2.1	a set of linear equations with the same variables
matrices	p.62	2.1	the plural form of matrix
matrix	p.62	2.1	a rectangular array of numbers in rows and columns
solution	p.60	2.1	$[s_1, s_2, \ldots, s_n]$ such that $a_1s_1 + a_2s_2 + \cdots + a_ns_n = b$
solution set	p.60	2.1	the set of all solutions of the system

Theorems

There are no theorems in this section, but a key fact highlighted on p. 61 is listed below.

Box p.61 2.1 A system has either a unique, infinitely many, or no solution.

Discussion of Key Definitions and Concepts

Q: Why are equations like $a_1x_1 + a_2x_2 + \cdots + a_nx_n = b$ called *linear*?

A: Their associated graphs are often lines. For example $2x_1 - x_2 = 3$ in $\mathbb{R}^2$.

Q: What is the coefficient matrix associated with $2x - 5y = 7$, $-3x + 2y = -5$?

A: The associated coefficient matrix is $\begin{bmatrix} 2 & -5 \\ -3 & 2 \end{bmatrix}$.

Q: What does the first *column* of this matrix represent?

A: The coefficients of the *first* variable in each of the two equations *in order*.

Q: What does the first *row* of this matrix represent?

A: The coefficients of *both* variables *in order* from the first equation.

Q: What is the augmented matrix associated with $2x - 5y = 7$, $-3x + 2y = -5$?

A: The augmented matrix is $\left[\begin{array}{cc|c} 2 & -5 & 7 \\ -3 & 2 & -5 \end{array}\right]$.

Q: Why does a vertical line separate the last column in an augmented matrix?

A: To remind us that the last column is comprised of constants, *not* coefficients.

Discussion of Theorems (in this case, Box on p.61)

Q: Is $x = 3$, $x = 4$ a linear system? If so, is it consistent or inconsistent?

A: It is a inconsistent linear system since x cannot equal 3 and 4 at the same.
The vertical lines $x = 3$ and $x = 4$ are parallel vertical lines, so they do not intersect.
The graphs of inconsistent systems are often lines that do not intersect.

Q: Is $2x - 4y = 1$, $-x + 2y = 1$ consistent? If so, is the solution unique?

A: These lines are parallel, so the system is inconsistent. How can we see this algebraically?
Multiply the second equation by 2 to get $-2x + 4y = 2$, then add them together to get $0 = 3$.

Q: Is $2x - 4y = 0$, $-x + 2y = 0$ consistent or inconsistent?

A: This system is obviously consistent since letting $x = y = 0$ is clearly a solution.

Q: Is $ax + by = 0$, $cx + dy = 0$ consistent or inconsistent?

A: This system is obviously consistent since letting $x = y = 0$ is clearly a solution.
What does this tell us? If all the constants are zero, the system is always consistent.
Why? Because it has at least one solution: letting all the variables equal zero.

2.1 Introduction to Systems of Linear Equations

Solutions to odd-numbered exercises from Section 2.1

1. We follow Example 2.1 and justify our assertion by applying the definition of *linear*. $x - \pi y + (\sqrt[3]{5})z = 0$ *is* linear *because* power of z is 1 and π, $\sqrt[3]{5}$ are constants.

3. $x^{-1} + 7y + z = \sin\frac{\pi}{9}$ is *not* linear *because* x occurs to the power -1.

5. $3\cos x - 4y + z = \sqrt{3}$ is *not* linear *because* $\cos x$ is not linear.

7. As in Section 1.3, we put the equation of this line into general form $ax + by = c$.
$2x + y = 7 - 3y$ is equivalent to $2x + 4y = 7$ after adding $3y$ to both sides.
Note: When the equation is *linear* there is no restriction on x and y. Why?

9. We begin by determining the restrictions on the variables x and y.
Typical sources are 1) division, 2) square roots, and 3) domains (like $\log x \Rightarrow x > 0$).

   Step 1. Determine restriction *type*. With $\frac{1}{x} + \frac{1}{y} = \frac{4}{xy}$, it is division.

   Step 2. Set the denominators equal to zero to determine the restriction.
   We have $x = 0$, $y = 0$, and $xy = 0$. So, the *restriction* is $x, y \neq 0$.

   Step 3. Simplify the given equation using algebra.

   $$\frac{1}{x} + \frac{1}{y} = \frac{4}{xy} \xrightarrow{\text{common denominator}} \frac{y}{xy} + \frac{x}{xy} = \frac{4}{xy} \xrightarrow{\text{multiply both sides by } xy} x + y = 4.$$

   Note: This tells us the given function is equivalent to the line $x + y = 4$ provided $x, y \neq 0$.

11. As in Example 2.2(a), we set $x = t$ and solve for y.
Setting $x = t$ in $3x - 6y = 0$ gives us $3t - 6y = 0$. Solving for y yields $6y = 3t \Rightarrow y = \frac{1}{2}t$.
So, we see the complete set of solutions can be written in the parametric form $[t, \frac{1}{2}t]$.
Note: We could have set $y = t$ to get $3x - 6t = 0$ and solved for x so $x = 2t$ and $[2t, t]$.

13. As in Example 2.2(b), we set $y = s$, $z = t$ and solve for x. (Why is this a good choice?)
This substitution yields $x + 2s + 3t = 4$. Solving for x yields $x = 4 - 2s - 3t$.
So, a complete set of solutions written in parametric form is $[4 - 2s - 3t, s, t]$.

15. The lines intersect at $(3, -3)$, so the unique solution is $[3, -3]$.
To solve, subtract 2nd from 1st $\Rightarrow$
$-x = -3 \Leftrightarrow x = 3$,
so substitution $\Rightarrow y = -3$.

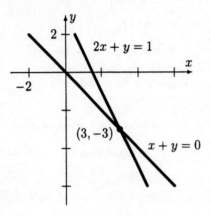

17. The lines are parallel $\Rightarrow$ no solution.
This system is inconsistent.
Add 3×2nd to 1st $\Rightarrow 0 = 6$.

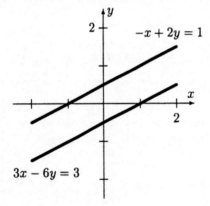

2.1 Introduction to Systems of Linear Equations

19. As in Example 2.5, we start from the last equation and work backward.
We find successively $y = 3$ and $x = 1 + 2(3) = 7$. So, the unique solution is $[x, y] = [7, 3]$.

21. We find the solution $[x, y, z] = \left[\frac{2}{3}, \frac{1}{3}, -\frac{1}{3}\right]$ using back substitution. Details below.

$3z = -1 \Rightarrow \qquad\qquad\qquad\qquad\qquad z = -\frac{1}{3}$

$2y - z = 1 \Rightarrow 2y = 1 + z \Rightarrow y = \frac{1}{2} + \frac{1}{2}z \Rightarrow y = \frac{1}{2} + \frac{1}{2}\left(-\frac{1}{3}\right) = \frac{1}{3}$

$x - y + z = 0 \Rightarrow x = y - z \Rightarrow \qquad\qquad x = \left(\frac{1}{3}\right) - \left(-\frac{1}{3}\right) = \frac{2}{3}$

23. We find the solution $[x_1, x_2, x_3, x_4] = [5, -2, 1, 1]$ using back substitution. Details below.
We find $x_3 = x_4 = 1$, $x_2 = -1 - 1 = -2$, and $x_1 = 1 - (-2) + 1 + 1 = 5$.

25. Working forward, we find $x = 2$, $y = -3 - 2(2) = -7$, and $z = -10 + 4(-7) + 3(2) = -32$.
So the unique solution to the system is $[x, y, z] = [2, -7, -32]$.

27. As in the solution to Example 2.6, we create the augmented matrix from the coefficients.

The system $\begin{array}{c} x - y = 0 \\ 2x + y = 3 \end{array}$ has $\left[\begin{array}{cc|c} 1 & -1 & 0 \\ 2 & 1 & 3 \end{array}\right]$ as its augmented matrix.

29. The system $\begin{array}{r} x + 5y = -1 \\ -x + y = -5 \\ 2x + 4y = 4 \end{array}$ has $\left[\begin{array}{cc|c} 1 & 5 & -1 \\ -1 & 1 & -5 \\ 2 & 4 & 4 \end{array}\right]$ as its augmented matrix.

31. The augmented matrix $\left[\begin{array}{ccc|c} 0 & 1 & 1 & 1 \\ 1 & -1 & 0 & 1 \\ 2 & -1 & 1 & 1 \end{array}\right]$ becomes $\begin{array}{r} y + z = 1 \\ x - y = 1 \\ 2x - y + z = 1 \end{array}$ as a system.

33. As in Example 2.4(a), we add $(x - y) + (2x + y) = 0 + 3$ to get $3x = 3 \Rightarrow x = 1$ and $y = 1$.
A quick check confirms that $[1, 1]$ is indeed the unique solution of the system.

35. As shown after Example 2.6, we row reduce the augmented matrix from Exercise 29.

$\left[\begin{array}{cc|c} 1 & 5 & -1 \\ -1 & 1 & -5 \\ 2 & 4 & 4 \end{array}\right] \xrightarrow[R_3 - 2R_1]{R_2 + R_1} \left[\begin{array}{cc|c} 1 & 5 & -1 \\ 0 & 6 & -6 \\ 0 & -6 & 6 \end{array}\right] \xrightarrow{R_3 + R_2} \left[\begin{array}{cc|c} 1 & 5 & -1 \\ 0 & 6 & -6 \\ 0 & 0 & 0 \end{array}\right] \Rightarrow$

$y = -1$ and $x = -1 - 5(-1) = 4$, so the solution is $[x, y] = [4, -1]$.

37. As shown after Example 2.6, we row reduce the augmented matrix from Exercise 31.

$\left[\begin{array}{ccc|c} 0 & 1 & 1 & 1 \\ 1 & -1 & 0 & 1 \\ 2 & -1 & 1 & 1 \end{array}\right] \xrightarrow[2R_2]{R_1 \leftrightarrow R_3} \left[\begin{array}{ccc|c} 2 & -1 & 1 & 1 \\ 2 & -2 & 0 & 2 \\ 0 & 1 & 1 & 1 \end{array}\right] \xrightarrow{R_2 - R_1} \left[\begin{array}{ccc|c} 2 & -1 & 1 & 1 \\ 0 & -1 & -1 & 1 \\ 0 & 1 & 1 & 1 \end{array}\right] \xrightarrow{R_3 + R_2} \left[\begin{array}{ccc|c} 2 & -1 & 1 & 1 \\ 0 & -1 & -1 & 1 \\ 0 & 0 & 0 & 2 \end{array}\right]$

$\Rightarrow 0 = 2 \Rightarrow$ No solution.

39. The key to this problem is simple substitution.

   (a) The fact that $x = t$ tells us that x is a free variable. What does that tell us?
   The linear equations we are looking for must be multiples of each other. Why?
   Substituting $t = x$ into $y = 3 - 2t$ yields $y = 3 - 2x \Rightarrow 2x + y = 3$.
   Any multiple of this equation will create the system we are looking for.
   For example, $2x + y = 3$ and $4x + 2y = 6$ (which is just $2\times$ the equation $2x + y = 3$).

   (b) Substituting $s = y$ into $y = 3 - 2x$ yields $s = 3 - 2x \Rightarrow x = \frac{3}{2} - \frac{1}{2}s$.
   The parametric solution then becomes $x = \frac{3}{2} - \frac{1}{2}s$ and $y = s$.

41. Let $u = \frac{1}{x}$, and $v = \frac{1}{y}$. Then the system of equations becomes $2u + 3v = 0$, $3u + 4v = 1$.
   Solving the second equation for v gives $v = \frac{1}{4} - \frac{3}{4}u$. So, substitution $\Rightarrow 2u + 3\left(\frac{1}{4} - \frac{3}{4}u\right) = 0$.
   Thus $u = 3$ and $v = \frac{1}{4} - \frac{3}{4}(3) = -2$. So, the solution is $[x, y] = \left[\frac{1}{3}, -\frac{1}{2}\right]$.

43. Let $u = \tan x$, $v = \sin y$, $w = \cos z \Rightarrow u - 2v = 2$, $u - v + w = 2$, $v - w = -1$.
   We form the augmented matrix and row reduce it to find the solution of the system:
   $$\begin{bmatrix} 1 & -2 & 0 & | & 2 \\ 1 & -1 & 1 & | & 2 \\ 0 & 1 & -1 & | & -1 \end{bmatrix} \xrightarrow{R_2 - R_1} \begin{bmatrix} 1 & -2 & 0 & | & 2 \\ 0 & 1 & 1 & | & 0 \\ 0 & 1 & -1 & | & -1 \end{bmatrix} \xrightarrow{R_3 - R_2} \begin{bmatrix} 1 & -2 & 0 & | & 2 \\ 0 & 1 & 1 & | & 0 \\ 0 & 0 & -2 & | & -1 \end{bmatrix}.$$
   Using back substitution $w = \frac{1}{2}$, $v = -\frac{1}{2}$, $u = 2 + 2\left(-\frac{1}{2}\right) = 1 \Rightarrow [u, v, w] = \left[1, -\frac{1}{2}, \frac{1}{2}\right]$.
   Since $x = \tan^{-1} u$, $y = \sin^{-1} v$, $z = \cos^{-1} w$, the solution is $[x, y, z] = \left[\frac{\pi}{4}, -\frac{\pi}{6}, \frac{\pi}{3}\right]$.

Exploration: Lies My Computer Told Me

Explorations are self-contained, so only odd-numbered solutions will be provided.

1. $\begin{aligned} x+y &= 0 \\ x+\tfrac{801}{800}y &= 1 \end{aligned} \Rightarrow \begin{aligned} -800x-800y &= 0 \\ 800x+801y &= 800 \end{aligned} \Rightarrow \begin{aligned} x &= -800 \\ y &= 800 \end{aligned}$

3. $\begin{aligned} x+y &= 0 \\ x+1.00y &= 1 \end{aligned} \Rightarrow \begin{aligned} -x-y &= 0 \\ x+1.00y &= 1 \end{aligned} \Rightarrow 0=1 \Rightarrow$ No solution.

2.2 Direct Methods for Solving Linear Systems

Q: What is our goal for Section 2.2?

A: To learn how to row reduce matrices using elementary row operations.
To understand what the rank of a matrix tells about the solution of a system.

Key Definitions and Concepts

$R_i \leftrightarrow R_j$	p.70	2.2	interchange two rows
kR_i	p.70	2.2	multiply a row by a nonzero constant
$R_i + kR_j$	p.70	2.2	add a multiple of a row to another row
elementary row operations	p.70	2.2	EROs: $R_i \leftrightarrow R_j$, kR_i, $R_i + kR_j$
free variable	p.75	2.2	a variable free to take on any value
Gauss-Jordan Elimination	p.76	2.2	see procedure described in box on p.77
Gaussian Elimination	p.72	2.2	see procedure described in box on p.72
homogeneous	p.79	2.2	system in which each constant term is 0
leading entry	p.68	2.2	the first nonzero entry in a row of a matrix
pivot	p.70	2.2	the entry chosen to become a leading entry
pivoting	p.70	2.2	see explanation in solution to Example 2.9 p.70
rank	p.75	2.2	number of nonzero rows in the REF of a matrix
reduced row echelon form	p.76	2.2	RREF: REF; leading entries, 1; all else, 0s
row echelon form	p.68	2.2	REF: zero rows, bottom; leading entries, left
row equivalent	p.72	2.2	matrix A can be converted into matrix B using EROs
row reduction	p.70	2.2	applying EROs to bring a matrix into REF

Theorems

The theorems focus on the relationship between number of equations and freedom of variables.

Rank Thm	p.74	2.2	Thm 2.2: number of free variables = $n - \text{rank}(A)$
Thm 2.1	p.72	2.2	A and B are row equivalent $\Leftrightarrow$ they reduce to same REF
Thm 2.2	p.74	2.2	Rank Thm: number of free variables = $n - \text{rank}(A)$
Thm 2.3	p.80	2.2	$[A|0]$: m equations $< n$ variables $\Rightarrow$ infinitely many solutions

Discussion of Key Definitions and Concepts

Q: What are the three basic elementary row operations (EROs) in your own words?

Q: What are three easy ways to avoid errors when using EROs?

1: Always add or subtract a multiple of row from an existing row (See Exercise 21).

For example, below we subtract $\frac{1}{2}$ of row 1 from the existing row 2:

$$\begin{bmatrix} 4 & 3 \\ 2 & 1 \end{bmatrix} \xrightarrow{R_2 - \frac{1}{2}R_1} \begin{bmatrix} 4 & 3 \\ 0 & -\frac{1}{2} \end{bmatrix}$$

2: Always multiply a row as a separate step and write down the new matrix.

For example, below we multiply row 1 by $\frac{1}{4}$ before subtracting $\frac{3}{4}$ of row 2 from it:

$$\begin{bmatrix} 4 & 3 \\ 0 & 1 \end{bmatrix} \xrightarrow{\frac{1}{4}R_1} \begin{bmatrix} 1 & \frac{3}{4} \\ 0 & 1 \end{bmatrix} \xrightarrow{R_1 - \frac{3}{4}R_2} \begin{bmatrix} 1 & 0 \\ 0 & 1 \end{bmatrix}$$

3: Whenever possible, multiply a row by a negative number and add rather than subtract.

Below we multiply row 2 by -2 before adding 3 times row 1 to it:

$$\begin{bmatrix} 2 & -4 & -2 & 6 \\ 3 & 1 & 6 & 6 \end{bmatrix} \xrightarrow{-2R_2} \begin{bmatrix} 2 & -4 & -2 & 6 \\ -6 & -2 & -12 & -12 \end{bmatrix} \xrightarrow{R_2 + 3R_1} \begin{bmatrix} 2 & -4 & -2 & 6 \\ 0 & -14 & -18 & 6 \end{bmatrix}$$

Q: What do we mean when we say a system is *consistent*?
A: We mean the system has a solution (See Section 2.1).

Q: What do we mean when we say a system is *homogenous*?
A: We mean that the constant term in each equation is 0.

Q: Why is every homogeneous system consistent?
A: Every homogenous system is consistent because it has at least one solution.

Q: What is the solution that every homogenous has in common?
A: The solution created by letting all of the variables equal 0. Why does this work?

Q: How can we understand this solution of letting all the variables equal 0 geometrically?
A: Lines with constant terms equal to 0 pass through the origin.

Therefore, all the lines pass through the origin. What does that tell us?

The origin is a common point of intersection for all the lines and therefore a solution.

What does the origin look like algebraically? All the coordinates (variables) are 0.

2.2 Direct Methods for Solving Linear Systems

Discussion of Key Definitions and Concepts, in particular, REF

Q: What is the importance of row echelon form (REF)? How does it help us solve systems?

A: REF allows us to identify the type of solution a system has without actually solving it. In Section 2.1, we learned that a system with more variables than equations always has infinitely many solutions. With REF, we can now build on that understanding. In particular, in systems where the number of variables is equal to the number of equations.

Q: In a matrix of coefficients, what does the number of columns correspond to?

A: The number of columns, usually designated by n, is equal to the number of variables.

Q: In a matrix of coefficients, what does the number of rows correspond to?

A: The number of rows, usually designated by m, is equal to the number of equations.

Q: In an augmented matrix, what does the number of columns correspond to?

A: The number of columns in this case is equal to the number of variables plus one. Why?

Q: In a system with an equal number of equations and variables, what does REF tell us?

A: In this case, REF tells exactly what type of system we have:

1: When REF has a zero row except for constant $\neq 0$, the system has no solution.

For example, consider the result when we row reduce the following augmented matrix:

$$\begin{bmatrix} 1 & 1 & 2 & 1 & | & 1 \\ 1 & -1 & -1 & 1 & | & 0 \\ 0 & 1 & 1 & 0 & | & -1 \\ 1 & 1 & 0 & 1 & | & 2 \end{bmatrix} \longrightarrow \begin{bmatrix} 1 & 0 & 0 & 1 & | & 0 \\ 0 & 1 & 0 & 0 & | & 0 \\ 0 & 0 & 1 & 0 & | & 0 \\ 0 & 0 & 0 & 0 & | & 1 \end{bmatrix} \Rightarrow$$

The fourth row is equivalent to the equation $0 = 1$ which clearly has no solution.

2: When REF has a zero row including constant $= 0$, the system has infinitely many solutions.

$$\begin{bmatrix} 1 & -3 & -2 & | & 0 \\ -1 & 2 & 1 & | & 0 \\ 2 & 4 & 6 & | & 0 \end{bmatrix} \xrightarrow{R_3 + 8R_1 + 10R_2} \begin{bmatrix} 1 & -3 & -2 & | & 0 \\ -1 & 2 & 1 & | & 0 \\ 0 & 0 & 0 & | & 0 \end{bmatrix}$$

The third row of $0 = 0$ tells us $x_3 = t$ is a free variable, so there are infinitely many solutions.

3: When REF has no zero rows, the system has a unique solution.

For example, consider the result when we row reduce the following augmented matrix:

$$\begin{bmatrix} \sqrt{2} & 1 & 2 & | & 1 \\ 0 & \sqrt{2} & -3 & | & -\sqrt{2} \\ 0 & -1 & \sqrt{2} & | & 1 \end{bmatrix} \longrightarrow \begin{bmatrix} 1 & 0 & 0 & | & \sqrt{2} \\ 0 & 1 & 0 & | & -1 \\ 0 & 0 & 1 & | & 0 \end{bmatrix} \Rightarrow \text{The solution is } \begin{bmatrix} x \\ y \\ z \end{bmatrix} = \begin{bmatrix} \sqrt{2} \\ -1 \\ 0 \end{bmatrix}.$$

Discussion of Key Definitions and Concepts, REF, RREF and Rank

Q: What is the definition of rank?

A: The number of nonzero rows in the row echelon form (REF) of a matrix.

Q: Would it be equivalent to define rank as the number of nonzero rows in RREF?

A: Yes, since the further reduction of REF to RREF does only two things:

1) It changes all the leading entries to 1s (instead of simply nonzero).

2) It uses those 1s to create 0s in all the rows above and below.

Neither one of these steps nor both together creates more rows of all zeros.

Q: Can the rank of a matrix ever be greater than the number of rows it has?

A: No. This is obvious from the definition since rank(A) = number of nonzero rows.

Q: Can the rank of a matrix ever be greater than the number of columns it has?

A: No. This is not as obvious, but it comes from the fact that the matrix must be in REF. In REF, the leading entry for each nonzero row must must occur in a unique column.

Work on developing this understanding through the rest of Chapters 2, 3, and 4.

Q: If we are given an $m \times n$ matrix A with $m \geq n$, what do we know about rank(A)?

A: We know that rank(A) must be less than equal to the number of columns in A, that is n.

Q: If we are given an $m \times n$ matrix A with $m \leq n$, what do we know about rank(A)?

A: We know that rank(A) must be less than equal to the number of rows in A, that is m.

Q: If we are given an $m \times n$ matrix A, what do we know about rank(A)?

A: We know that rank(A) must be less than equal to the lesser of m and n.

That is, we know that rank(A) must be less than or equal to the number of rows and the number of columns both. In particular, it is limited by the lesser number of the two.

2.2 Direct Methods for Solving Linear Systems

Discussion of Theorems

Q: Why are we interested in being able to identify matrices that are row equivalent?

A: Because that means the linear systems they represent have the same solution.

The following questions are under the assumption that the system is consistent.

Q: Why do we get free variables when rank(A) is less than the number of variables, n?

A: Rank(A) is the number of nonzero rows in REF. And what are those nonzero rows?

Those nonzero rows correspond to equations the variables have to satisfy.

If we have more variables than equations they have to satisfy, we get free variables.

Q: What happens when rank(A) is equal to the number of variables?

A: Since rank(A) $-n = n - n = 0$, there are no free variables.

What does that tell us? The solution is unique (it exists because the system is consistent).

For example, consider the result when we row reduce the augmented matrix from above:

$$\begin{bmatrix} \sqrt{2} & 1 & 2 & | & 1 \\ 0 & \sqrt{2} & -3 & | & -\sqrt{2} \\ 0 & -1 & \sqrt{2} & | & 1 \end{bmatrix} \longrightarrow \begin{bmatrix} 1 & 0 & 0 & | & \sqrt{2} \\ 0 & 1 & 0 & | & -1 \\ 0 & 0 & 1 & | & 0 \end{bmatrix} \Rightarrow \text{The solution is } \begin{bmatrix} x \\ y \\ z \end{bmatrix} = \begin{bmatrix} \sqrt{2} \\ -1 \\ 0 \end{bmatrix}.$$

Q: Can we give another explanation of why rank(A) $= n$ should imply a unique solution?

A: The number of variables = the number of equations they have to satisfy.

Q: Why does a matrix have to be in REF before we determine its rank?

A: Before we row reduce a matrix, we cannot tell how many rows will remain nonzero.

For example, consider the row reduction of the following matrix:

$$\begin{bmatrix} 1 & -3 & -2 \\ -1 & 2 & 1 \\ 2 & 4 & 6 \end{bmatrix} \xrightarrow{R_3 + 8R_1 + 10R_2} \begin{bmatrix} 1 & -3 & -2 \\ -1 & 2 & 1 \\ 0 & 0 & 0 \end{bmatrix}$$

Was it obvious before putting this matrix into REF that row 3 would become a row of zeroes?

Q: What does the number of nonzero rows tell us about the original system of linear equations?

A: It tells us the number of equations imposing unique conditions upon the variables.

Again let's consider the row reduction of the matrix we looked at above:

$$\begin{bmatrix} 1 & -3 & -2 \\ -1 & 2 & 1 \\ 2 & 4 & 6 \end{bmatrix} \xrightarrow{R_3 + 8R_1 + 10R_2} \begin{bmatrix} 1 & -3 & -2 \\ -1 & 2 & 1 \\ 0 & 0 & 0 \end{bmatrix}$$

So we see $R_3 = -8R_1 - 10R_2$. What does this tell us?

The original equation corresponding to row 3 did not impose a unique condition.

Instead, it was simply a combination of the conditions from equations 1 and 2.

Solutions to odd-numbered exercises from Section 2.2

1. No, this matrix is not in row echelon form. Why not? Give at least one reason.
   The leading entry in row 3 appears to the left of the leading entry in row 2.

3. This matrix is in row echelon form, and also reduced row echelon form. Why is the 3 okay?
   The 3 occurs in a column that does not contain a leading 1.

5. No, this matrix is not in row echelon form. Why not? Give a reason.
   The row of all zeroes is not at the bottom.

7. No, this matrix is not in row echelon form. Why not? Give a reason.
   The leading entry in row 2 appears underneath the leading entry in row 1.

9. (a) $\begin{bmatrix} 0 & 0 & 1 \\ 0 & 1 & 1 \\ 1 & 1 & 1 \end{bmatrix} \xrightarrow{R_1 \leftrightarrow R_3} \begin{bmatrix} 1 & 1 & 1 \\ 0 & 1 & 1 \\ 0 & 0 & 1 \end{bmatrix}$.

   (b) $\ldots \begin{bmatrix} 1 & 1 & 1 \\ 0 & 1 & 1 \\ 0 & 0 & 1 \end{bmatrix} \xrightarrow[R_2-R_3]{R_1-R_2} \begin{bmatrix} 1 & 0 & 0 \\ 0 & 1 & 0 \\ 0 & 0 & 1 \end{bmatrix}$.

11. (a) $\begin{bmatrix} 3 & 5 \\ 5 & -2 \\ 2 & 4 \end{bmatrix} \xrightarrow[R_3-\frac{2}{3}R_1]{R_2-\frac{5}{3}R_1} \begin{bmatrix} 3 & 5 \\ 0 & -\frac{31}{3} \\ 0 & \frac{2}{3} \end{bmatrix} \xrightarrow[\frac{3}{2}R_3]{-\frac{3}{31}R_2} \begin{bmatrix} 3 & 5 \\ 0 & 1 \\ 0 & 1 \end{bmatrix} \xrightarrow{R_3-R_2} \begin{bmatrix} 3 & 5 \\ 0 & 1 \\ 0 & 0 \end{bmatrix}$.

   (b) Continuing from (a): $\begin{bmatrix} 3 & 5 \\ 0 & 1 \\ 0 & 0 \end{bmatrix} \xrightarrow{R_1-5R_2} \begin{bmatrix} 3 & 0 \\ 0 & 1 \\ 0 & 0 \end{bmatrix} \xrightarrow{\frac{1}{3}R_1} \begin{bmatrix} 1 & 0 \\ 0 & 1 \\ 0 & 0 \end{bmatrix}$.

13. (a) $\begin{bmatrix} 3 & -2 & -1 \\ 2 & -1 & -1 \\ 4 & -3 & -1 \end{bmatrix} \xrightarrow[-3R_3]{-3R_2} \begin{bmatrix} 3 & -2 & -1 \\ -6 & 3 & 3 \\ -12 & 9 & 3 \end{bmatrix} \xrightarrow[R_3+4R_1]{R_2+2R_1} \begin{bmatrix} 3 & -2 & -1 \\ 0 & -1 & 1 \\ 0 & 1 & -1 \end{bmatrix} \xrightarrow{R_3+R_2} \begin{bmatrix} 3 & -2 & -1 \\ 0 & 1 & -1 \\ 0 & 0 & 0 \end{bmatrix}$.

   (b) Continuing from (a): $\begin{bmatrix} 3 & -2 & -1 \\ 0 & 1 & -1 \\ 0 & 0 & 0 \end{bmatrix} \xrightarrow{R_1+2R_2} \begin{bmatrix} 3 & 0 & -3 \\ 0 & 1 & -1 \\ 0 & 0 & 0 \end{bmatrix} \xrightarrow{\frac{1}{3}R_1} \begin{bmatrix} 1 & 0 & -1 \\ 0 & 1 & -1 \\ 0 & 0 & 0 \end{bmatrix}$.

2.2 Direct Methods for Solving Linear Systems

15. $\begin{bmatrix} 1 & 2 & -4 & -4 & 5 \\ 0 & -1 & 10 & 9 & -5 \\ 0 & 0 & 1 & 1 & -1 \\ 0 & 0 & 0 & 0 & 24 \end{bmatrix} \xrightarrow{R_4+29R_3} \begin{bmatrix} 1 & 2 & -4 & -4 & 5 \\ 0 & -1 & 10 & 9 & -5 \\ 0 & 0 & 1 & 1 & -1 \\ 0 & 0 & 29 & 29 & -5 \end{bmatrix} \xrightarrow{8R_3} \begin{bmatrix} 1 & 2 & -4 & -4 & 5 \\ 0 & -1 & 10 & 9 & -5 \\ 0 & 0 & 8 & 8 & -8 \\ 0 & 0 & 29 & 29 & -5 \end{bmatrix}$

$\xrightarrow{R_4-3R_2} \begin{bmatrix} 1 & 2 & -4 & -4 & 5 \\ 0 & -1 & 10 & 9 & -5 \\ 0 & 0 & 8 & 8 & -8 \\ 0 & 3 & -1 & 2 & 10 \end{bmatrix} \xrightarrow{R_3 \leftrightarrow R_2} \begin{bmatrix} 1 & 2 & -4 & -4 & 5 \\ 0 & 0 & 8 & 8 & -8 \\ 0 & -1 & 10 & 9 & -5 \\ 0 & 3 & -1 & 2 & 10 \end{bmatrix} \xrightarrow[\substack{R_2+2R_1 \\ R_3+2R_1 \\ R_4-R_1}]{} \begin{bmatrix} 1 & 2 & -4 & -4 & 5 \\ 2 & 4 & 0 & 0 & 2 \\ 2 & 3 & 2 & 1 & 5 \\ -1 & 1 & 3 & 6 & 5 \end{bmatrix}$.

17. $A = \begin{bmatrix} 1 & 2 \\ 3 & 4 \end{bmatrix} \xrightarrow{R_2-2R_1} \begin{bmatrix} 1 & 2 \\ 1 & 0 \end{bmatrix} \xrightarrow{-\frac{1}{2}R_1} \begin{bmatrix} -\frac{1}{2} & -1 \\ 1 & 0 \end{bmatrix} \xrightarrow{R_1+\frac{7}{2}R_2} \begin{bmatrix} 3 & -1 \\ 1 & 0 \end{bmatrix} = B.$

So A and B are row equivalent. Convert A into B by $R_2 - 2R_1$, $-\frac{1}{2}R_1$, $R_1 + \frac{7}{2}R_2$.

19. Performing $R_2 + R_1$ and $R_1 + R_2$ does *not* leave rows 1 and 2 identical.
    After performing $R_2 + R_1$ the second row is now $R'_2 = R_2 + R_1$.
    So $R_1 + R_2$ is actually $R_1 + R'_2 = R_1 + (R_2 + R_1) = 2R_1 + R_2$.
    Performing $R_2 + R_1$ and $R_1 + R_2$ simultaneously annuls their linearity.

21. Our first task is to show that $\begin{bmatrix} 3 & 1 \\ 2 & 4 \end{bmatrix} \xrightarrow{3R_2-2R_1} \begin{bmatrix} 3 & 1 \\ 0 & 10 \end{bmatrix}$ is *not* an elementary row operation.

    Compare $3R_2 - 2R_1$ to the elementary row operations $R_i \leftrightarrow R_j$, kR_i, $R_i + kR_j$.
    Clearly, $3R_2 - 2R_1$ is a combination of kR_i and $R_i + kR_j$ done at the same time.
    Performing row operations simultaneously annuls their linearity.

    One way to achieve the result is: $\begin{bmatrix} 3 & 1 \\ 2 & 4 \end{bmatrix} \xrightarrow{R_2-\frac{2}{3}R_1} \begin{bmatrix} 3 & 1 \\ 0 & \frac{10}{3} \end{bmatrix} \xrightarrow{3R_2} \begin{bmatrix} 3 & 1 \\ 0 & 10 \end{bmatrix}$.

23. Since **rank** = the number of nonzero rows in the row echelon form of a matrix, before we answer we should put each of the matrices into row echelon form.

(1) Since this matrix A is not in its row echelon form B, we must row reduce A first.
$$\begin{bmatrix} 1 & 0 & 1 \\ 0 & 0 & 3 \\ 0 & 1 & 0 \end{bmatrix} \xrightarrow{R_2 \leftrightarrow R_3} \begin{bmatrix} 1 & 0 & 1 \\ 0 & 1 & 0 \\ 0 & 0 & 3 \end{bmatrix}.$$ So, rank A = the number of nonzero rows in $B = 3$.

(2) A is in row echelon form, so we need only count the number of its nonzero rows.
Since $\begin{bmatrix} 7 & 0 & 1 & 0 \\ 0 & 1 & -1 & 4 \\ 0 & 0 & 0 & 0 \end{bmatrix}$ has two nonzero rows, rank $A = 2$.

(3) A is in row echelon form, so we need only count the number of its nonzero rows.
Since $\begin{bmatrix} 0 & 1 & 3 & 0 \\ 0 & 0 & 0 & 1 \end{bmatrix}$ has two nonzero rows, rank $A = 2$.

(4) A is in row echelon form, so we need only count the number of its nonzero rows.
Since $\begin{bmatrix} 0 & 0 & 0 \\ 0 & 0 & 0 \\ 0 & 0 & 0 \end{bmatrix}$ has no nonzero rows, rank $A = 0$.

(5) Since this matrix A is not in its row echelon form B, we must row reduce A first.
$$\begin{bmatrix} 1 & 0 & 3 & -4 & 0 \\ 0 & 0 & 0 & 0 & 0 \\ 0 & 1 & 5 & 0 & 1 \end{bmatrix} \xrightarrow{R_2 \leftrightarrow R_3} \begin{bmatrix} 1 & 0 & 3 & -4 & 0 \\ 0 & 1 & 5 & 0 & 1 \\ 0 & 0 & 0 & 0 & 0 \end{bmatrix}.$$ So, rank $A = 2$.

(6) Since this matrix A is not in its row echelon form B, we must row reduce A first.
$$\begin{bmatrix} 0 & 0 & 1 \\ 0 & 1 & 0 \\ 1 & 0 & 0 \end{bmatrix} \xrightarrow{R_1 \leftrightarrow R_3} \begin{bmatrix} 1 & 0 & 0 \\ 0 & 1 & 0 \\ 0 & 0 & 1 \end{bmatrix}.$$ So, rank $A = 3$.

(7) Since this matrix A is not in its row echelon form B, we must row reduce A first.
$$\begin{bmatrix} 1 & 2 & 3 \\ 1 & 0 & 0 \\ 0 & 1 & 1 \\ 0 & 0 & 1 \end{bmatrix} \xrightarrow{\substack{R_2 - R_1 \\ R_3 - \frac{1}{2}R_1 + \frac{1}{2}R_2 \\ R_4 - R_1 + R_2 + 2R_3}} \begin{bmatrix} 1 & 2 & 3 \\ 0 & -2 & -3 \\ 0 & 0 & -\frac{1}{2} \\ 0 & 0 & 0 \end{bmatrix}.$$ So, rank $A = 3$.

(8) A is in row echelon form, so we need only count the number of its nonzero rows.
Since $\begin{bmatrix} 2 & 1 & 3 & 5 \\ 0 & 0 & 1 & -1 \\ 0 & 0 & 0 & 3 \\ 0 & 0 & 0 & 0 \end{bmatrix}$ has 3 nonzero rows, rank $A = 3$.

2.2   Direct Methods for Solving Linear Systems

25. We have the following system of equations: $\begin{bmatrix} 1 & 2 & -3 \\ 2 & -1 & 1 \\ 4 & -1 & 1 \end{bmatrix} \begin{bmatrix} x_1 \\ x_2 \\ x_3 \end{bmatrix} = \begin{bmatrix} 9 \\ 0 \\ 4 \end{bmatrix}$.

We form the augmented matrix and row reduce it as follows:

$\begin{bmatrix} 1 & 2 & -3 & | & 9 \\ 2 & -1 & 1 & | & 0 \\ 4 & -1 & 1 & | & 4 \end{bmatrix} \xrightarrow[R_3 - 2R_2]{R_1 + 3R_3} \begin{bmatrix} 13 & -1 & 0 & | & 21 \\ 2 & -1 & 1 & | & 0 \\ 0 & 1 & -1 & | & 4 \end{bmatrix} \xrightarrow[-R_3]{R_1 \leftrightarrow R_2} \begin{bmatrix} 2 & -1 & 1 & | & 0 \\ 13 & -1 & 0 & | & 21 \\ 0 & -1 & 1 & | & -4 \end{bmatrix}$

$\xrightarrow[-R_2]{R_1 - R_3} \begin{bmatrix} 2 & 0 & 0 & | & 4 \\ -13 & 1 & 0 & | & -21 \\ 0 & -1 & 1 & | & 4 \end{bmatrix} \xrightarrow{\frac{1}{2}R_1} \begin{bmatrix} 1 & 0 & 0 & | & 2 \\ -13 & 1 & 0 & | & -21 \\ 0 & -1 & 1 & | & -4 \end{bmatrix} \xrightarrow{R_2 + 13R_1} \begin{bmatrix} 1 & 0 & 0 & | & 2 \\ 0 & 1 & 0 & | & 5 \\ 0 & -1 & 1 & | & -4 \end{bmatrix}$

$\xrightarrow{R_3 + R_2} \begin{bmatrix} 1 & 0 & 0 & | & 2 \\ 0 & 1 & 0 & | & 5 \\ 0 & 0 & 1 & | & 1 \end{bmatrix}$. So, the solution is $\begin{bmatrix} x_1 \\ x_2 \\ x_3 \end{bmatrix} = \begin{bmatrix} 2 \\ 5 \\ 1 \end{bmatrix}$.

27. We form the augmented matrix and row reduce it as follows:

$\begin{bmatrix} 1 & -3 & -2 & | & 0 \\ -1 & 2 & 1 & | & 0 \\ 2 & 4 & 6 & | & 0 \end{bmatrix} \xrightarrow{R_3 + 8R_1 + 10R_2} \begin{bmatrix} 1 & -3 & -2 & | & 0 \\ -1 & 2 & 1 & | & 0 \\ 0 & 0 & 0 & | & 0 \end{bmatrix}$

$R_3 = -8R_1 - 10R_2$ (excluding constants) does not cause a problem here? Why?
Since the system is homogeneous (all constants $= 0$), the system has at least one solution.

$\xrightarrow{R_2 + R_1} \begin{bmatrix} 1 & -3 & -2 & | & 0 \\ 0 & -1 & -1 & | & 0 \\ 0 & 0 & 0 & | & 0 \end{bmatrix} \xrightarrow{-R_2} \begin{bmatrix} 1 & -3 & -2 & | & 0 \\ 0 & 1 & 1 & | & 0 \\ 0 & 0 & 0 & | & 0 \end{bmatrix} \xrightarrow{R_1 + 3R_2} \begin{bmatrix} 1 & 0 & 1 & | & 0 \\ 0 & 1 & 1 & | & 0 \\ 0 & 0 & 0 & | & 0 \end{bmatrix}$.

The third row of $0 = 0$ tells us that $x_3 = t$ is a free variable.
Back substituting, we have $x_2 + t = 0 \Rightarrow x_2 = -t$ and $x_1 + t = 0 \Rightarrow x_1 = -t$.

So, the solution is $\begin{bmatrix} x_1 \\ x_2 \\ x_3 \end{bmatrix} = t \begin{bmatrix} -1 \\ -1 \\ 1 \end{bmatrix}$ or equivalently $\begin{bmatrix} x_1 \\ x_2 \\ x_3 \end{bmatrix} = t \begin{bmatrix} 1 \\ 1 \\ -1 \end{bmatrix}$.

29. Note that there are 3 equations but only 2 variables to satisfy them.
It is helpful, therefore, to begin by noting $R_3 = 9R_1 - 4R_2$.

$\begin{bmatrix} 2 & 1 & | & 3 \\ 4 & 1 & | & 7 \\ 2 & 5 & | & -1 \end{bmatrix} \xrightarrow{R_3 - 9R_1 + 4R_2} \begin{bmatrix} 2 & 1 & | & 3 \\ 4 & 1 & | & 7 \\ 0 & 0 & | & 0 \end{bmatrix} \xrightarrow{R_2 - 2R_1} \begin{bmatrix} 2 & 1 & | & 3 \\ 0 & -1 & | & 1 \\ 0 & 0 & | & 0 \end{bmatrix} \to \cdots \to \begin{bmatrix} 1 & 0 & | & 2 \\ 0 & 1 & | & -1 \\ 0 & 0 & | & 0 \end{bmatrix}$

So, the solution is $\begin{bmatrix} r \\ s \end{bmatrix} = \begin{bmatrix} 2 \\ -1 \end{bmatrix}$.

31. From the beginning, we know this system has infinitely many solutions. Why? Because this system has 5 variables and only 3 equations they have to satisfy.

We form the augmented matrix and row reduce it as follows:
$$\begin{bmatrix} \frac{1}{2} & 1 & -1 & -6 & 0 & | & 2 \\ \frac{1}{6} & \frac{1}{2} & 0 & -3 & 1 & | & -1 \\ \frac{1}{3} & 0 & -2 & 0 & -4 & | & 8 \end{bmatrix} \longrightarrow \begin{bmatrix} 1 & 0 & -6 & 0 & -12 & | & 24 \\ 0 & 1 & 2 & -6 & 6 & | & -10 \\ 0 & 0 & 0 & 0 & 0 & | & 0 \end{bmatrix}$$

Since rank $A = 2$ and $5 - 2 = 3$, we get 3 free variables: $x_3 = r$, $x_4 = s$, and $x_5 = t$.
Back substituting, we get $x_2 = -10 - 2r + 6s - 6t$, $x_1 = 24 + 6r + 12t$.

So, the solution is $\begin{bmatrix} x_1 \\ x_2 \\ x_3 \\ x_4 \\ x_5 \end{bmatrix} = \begin{bmatrix} 24 \\ -10 \\ 0 \\ 0 \\ 0 \end{bmatrix} + r \begin{bmatrix} 6 \\ -2 \\ 1 \\ 0 \\ 0 \end{bmatrix} + s \begin{bmatrix} 0 \\ 6 \\ 0 \\ 1 \\ 0 \end{bmatrix} + t \begin{bmatrix} 12 \\ -6 \\ 0 \\ 0 \\ 1 \end{bmatrix}$.

33. We form the augmented matrix and row reduce it as follows:
$$\begin{bmatrix} 1 & 1 & 2 & 1 & | & 1 \\ 1 & -1 & -1 & 1 & | & 0 \\ 0 & 1 & 1 & 0 & | & -1 \\ 1 & 1 & 0 & 1 & | & 2 \end{bmatrix} \longrightarrow \begin{bmatrix} 1 & 0 & 0 & 1 & | & 0 \\ 0 & 1 & 0 & 0 & | & 0 \\ 0 & 0 & 1 & 0 & | & 0 \\ 0 & 0 & 0 & 0 & | & 1 \end{bmatrix} \Rightarrow$$

The fourth row is equivalent to the equation $0 = 1$ which clearly has no solution. Therefore, the system is inconsistent.

Q: Rank $A = 3$. How does that relate to the fact that there is no solution?
A: The system has no solution when A has a zero row with corresponding constant $\neq 0$.

35. Begin by thinking of this system as $[A|\mathbf{x}]$, then determine rank A by inspection.
Mentally performing $R_1 \leftrightarrow R_3$ to put matrix A into row echelon form,
makes it obvious that rank $A = 3$ (because A has 3 nonzero rows).

Since rank $A = 3$, this is a system of 3 equations and 3 variables.
Therefore, the system has a unique solution because there are $3 - 3 = 0$ free variables.

37. Since this system has 4 variables and at most 3 equations,
it has infinitely many solutions. Why? There is at least one free variable.

2.2 Direct Methods for Solving Linear Systems

39. We need only show that the condition $ad - bc \neq 0$ implies that rank $A = 2$. Why?
If rank $A = 2$, there are $2 - 2 = 0$ free variables so the system has a unique solution.

Case 1: $a = 0$, which implies both $b \neq 0$ and $c \neq 0$. Why? Because $0d - bc = -bc \neq 0$.

Row reduce A: $\begin{bmatrix} 0 & b \\ c & d \end{bmatrix} \xrightarrow{R_1 \leftrightarrow R_2} \begin{bmatrix} c & d \\ 0 & b \end{bmatrix}$.

A is now in row echelon form with 2 nonzero rows. Therefore, rank $A = 2$.

Case 2: $c = 0$, which implies both $a \neq 0$ and $d \neq 0$. Why? Because $ad - b0 = ad \neq 0$.

Row reduce A: $\begin{bmatrix} a & b \\ 0 & d \end{bmatrix}$.

A is now in row echelon form with 2 nonzero rows. Therefore, rank $A = 2$.

Case 3: $a \neq 0$ and $c \neq 0$.

Row reduce A: $\begin{bmatrix} a & b \\ c & d \end{bmatrix} \xrightarrow[aR_2]{cR_1} \begin{bmatrix} ac & bc \\ ac & ad \end{bmatrix} \xrightarrow{R_2 - R_1} \begin{bmatrix} ac & bc \\ 0 & ad - bc \end{bmatrix}$.

A is now in row echelon form with 2 nonzero rows. Therefore, rank $A = 2$.

41. First row reduce the system $[A|\mathbf{x}]$ and then answers parts (a), (b), and (c).

$\begin{bmatrix} 1 & k & | & 1 \\ k & 1 & | & 1 \end{bmatrix} \xrightarrow{R_2 - kR_1} \begin{bmatrix} 1 & k & | & 1 \\ 0 & 1-k^2 & | & 1-k \end{bmatrix}$

(a) When $k = -1$, this system has no solution. Why?
The system has no solution when A has a zero row with corresponding constant $\neq 0$.
$1 - k^2 = 0 \Rightarrow k = \pm 1$ create a zero row in A.
And $k = -1 \Rightarrow$ the constant $1 - k = 1 + 1 \neq 0$.

(b) When $k \neq \pm 1$, this system has a unique solution. Why?
From (a), we see when $k \neq \pm 1$ then rank $A = 2$. So, there are $2 - 2 = 0$ free variables.

(c) When $k = 1$, this system has infinitely many solutions.
The system has infinitely many solutions when A has a zero row with constant $= 0$.
From (a), we see this is exactly the case when $k = 1$.

43. First row reduce the system $[A|\mathbf{x}]$ and then answers parts (a), (b), and (c).
$$\begin{bmatrix} 1 & 1 & k & | & 1 \\ 1 & k & 1 & | & 1 \\ k & 1 & 1 & | & -2 \end{bmatrix} \longrightarrow \begin{bmatrix} 1 & 1 & k & | & 1 \\ 0 & k-1 & 1-k & | & 0 \\ 0 & 0 & k^2+k-2 & | & k+2 \end{bmatrix}$$

(a) When $k = 1$, this system has no solution. Why?

The system has no solution when A has a zero row with corresponding constant $\neq 0$.
$k^2 + k - 2 = 0 \Rightarrow k = 1$ makes the bottom row 0, but the constant $k + 2 = 1 + 2 = 3 \neq 0$.

(b) When $k \neq 1, -2$, this system has a unique solution. Why?

When $k \neq 1, -2$, rank $A = 3$. So, there are $3 - 3 = 0$ free variables.

(c) When $k = -2$, this system has infinitely many solutions.

The system has infinitely many solutions when A has a zero row with constant $= 0$.
$k^2 + k - 2 = 0 \Rightarrow k = -2$ makes the bottom row 0 and the constant $k + 2 = -2 + 2 = 0$.

45. As in Example 2.14, find the line of intersection of $3x + 2y + z = -1$, $2x - y + 4z = 5$.

First, observe that there will be a line of intersection. Why?
The normal vectors of the two planes, $[3, 2, 1]$ and $[2, -1, 4]$ are not parallel.

The points that lie in the intersection of the two planes correspond to the points in the solution the system: $\begin{array}{rcl} 3x + 2y + z &=& -1 \\ 2x - y + 4z &=& 5 \end{array}$

Gauss-Jordan elimination yields: $\begin{bmatrix} 3 & 2 & 1 & | & -1 \\ 2 & -1 & 4 & | & 5 \end{bmatrix} \longrightarrow \begin{bmatrix} 1 & 0 & \frac{9}{7} & | & \frac{9}{7} \\ 0 & 1 & -\frac{10}{7} & | & -\frac{17}{7} \end{bmatrix}$

Replacing variables, we have: $\begin{array}{rcl} x + \frac{9}{7}z &=& \frac{9}{7} \\ y - \frac{10}{7}z &=& -\frac{17}{7} \end{array} \Rightarrow \begin{array}{rcl} z &=& 1 - \frac{7}{9}x \\ y &=& -\frac{17}{7} + \frac{10}{7}z \end{array}$

To eliminate fractions we set $x = 9t$, so $z = 1 - \frac{7}{9}(9t) = 1 - 7t$.

Substituting $z = 1 - 7t$ into $y = -\frac{17}{7} + \frac{10}{7}z$ yields: $y = -\frac{17}{7} + \frac{10}{7}(1 - 7t) = -1 - 10t$.

Summarizing, we now have $x = 9t$, $y = -1 - 10t$, and $z = 1 - 7t$.

Therefore, the line is $\begin{bmatrix} x \\ y \\ z \end{bmatrix} = \begin{bmatrix} 0 \\ -1 \\ 1 \end{bmatrix} + t \begin{bmatrix} 1 \\ -10 \\ -9 \end{bmatrix}$.

2.2 Direct Methods for Solving Linear Systems

47. When looking for examples, begin with familiar planes like $x = 0$, $y = 0$, and $z = 0$.

(a) Let's start with $x = 0$ and $y = 0$. These planes obviously intersect in the z-axis. Why?

As in Exercise 45: $\begin{array}{l} x + 0y + 0z = 0 \\ 0x + y + 0z = 0 \end{array} \Rightarrow \left[\begin{array}{ccc|c} 1 & 0 & 0 & 0 \\ 0 & 1 & 0 & 0 \end{array}\right] \Rightarrow x = 0, y = 0, z = t \Rightarrow$

The line of intersection of $x = 0$ and $y = 0$ is $\begin{bmatrix} x \\ y \\ z \end{bmatrix} = \begin{bmatrix} 0 \\ 0 \\ 0 \end{bmatrix} + t \begin{bmatrix} 0 \\ 0 \\ 1 \end{bmatrix}$, the z-axis.

All we need is one other plane that passes through the z-axis to complete our example. It may help to sketch $\mathbb{R}^2$ and look for a line that passes through the origin. One such line is $x = y$ which corresponds to the plane $x - y = 0$ in $\mathbb{R}^3$.

Sketch these three planes in $\mathbb{R}^3$ to confirm they intersect in the z-axis.

Q: How do we confirm these three planes intersect in the z-axis algebraically?
A: Check the intersection between $x = 0$ and $x - y = 0$. Why is that enough?

As in Exercise 45: $\begin{array}{l} x + 0y + 0z = 0 \\ x - y + 0z = 0 \end{array} \Rightarrow \left[\begin{array}{ccc|c} 1 & 0 & 0 & 0 \\ 1 & 1 & 0 & 1 \end{array}\right] \Rightarrow x = y = 0, \text{ and } z = t \Rightarrow$

The line of intersection of $x = 0$ and $x - y = 0$ is $\begin{bmatrix} x \\ y \\ z \end{bmatrix} = \begin{bmatrix} 0 \\ 0 \\ 0 \end{bmatrix} + t \begin{bmatrix} 0 \\ 0 \\ 1 \end{bmatrix}$.

Q: Start with $y = 0$ and $z = 0$ and then $x = 1$ and $y = 1$. Is there a pattern?

(b) Begin with $x = 0$ and $y = 0$. We need one plane that crosses across both of these. It may help to visualize $\mathbb{R}^2$ and look for a line that cuts across the first quadrant. It is obvious that the plane $x + y = 1$ will complete the example?

Sketch these three planes in $\mathbb{R}^3$ to confirm they intersect in pairs.
For example, $x = 0$ and $x + y = 1$ intersect in the line $[x, y, z] = [0, 1, 0] + t[0, 0, 1]$.

As in Exercise 45: $\begin{array}{l} x + 0y + 0z = 0 \\ x + y + 0z = 1 \end{array} \Rightarrow \left[\begin{array}{ccc|c} 1 & 0 & 0 & 0 \\ 1 & 1 & 0 & 1 \end{array}\right] \Rightarrow x = 0, y = 1, \text{ and } z = t \Rightarrow$

The line of intersection of $x = 0$ and $x + y = 1$ is $\begin{bmatrix} x \\ y \\ z \end{bmatrix} = \begin{bmatrix} 0 \\ 1 \\ 0 \end{bmatrix} + t \begin{bmatrix} 0 \\ 0 \\ 1 \end{bmatrix}$.

(c) An obvious example is $x = 0$, $x = 1$, and $y = 0$. Why?
The normal vector for $x = 0$ and $x = 1$ is $[1, 0, 0]$
while the normal vector for $y = 0$ is $[0, 1, 0]$.

(d) The most obvious example is $x = 0$, $y = 0$, and $z = 0$.
Note that any example of $x = a$, $y = a$, and $z = a$ will work.
Are there any other obvious pattern examples that will work?

49. As in Example 2.15, if these lines intersect we need to determine the point of intersection. As pointed out in that example, we need to change the parameter for the first line to s. We want to find an $\mathbf{x} = [x, y, z]$ that satisfies both equations simultaneously. That is, we want $\mathbf{x} = \mathbf{p} + s\mathbf{u} = \mathbf{q} + t\mathbf{v}$ or $s\mathbf{u} - t\mathbf{v} = \mathbf{q} - \mathbf{p}$.

Substituting the given $\mathbf{p}$, $\mathbf{q}$, $\mathbf{u}$, and $\mathbf{v}$ into $s\mathbf{u} - t\mathbf{v} = \mathbf{q} - \mathbf{p}$, we obtain the equations:

$$s\begin{bmatrix} 1 \\ 0 \\ 1 \end{bmatrix} - t\begin{bmatrix} 2 \\ 3 \\ 1 \end{bmatrix} = \begin{bmatrix} -1 \\ 1 \\ -1 \end{bmatrix} - \begin{bmatrix} 3 \\ 1 \\ 0 \end{bmatrix} \Rightarrow \begin{array}{rcr} s - 2t &=& -4 \\ -3t &=& 0 \\ s - t &=& -1 \end{array}$$

From this, there is clearly no solution since $t = 0$ implies $s = -4$ and -1 at the same time. Therefore, we conclude that these lines do not intersect.

51. Recall the definition of the dot product: $\mathbf{u} \times \mathbf{v} = u_1 v_1 + u_2 v_2 + u_3 v_3$.
So, vectors that satisfy $\mathbf{u} \cdot \mathbf{x} = 0$ and $\mathbf{v} \cdot \mathbf{x} = 0$ correspond to the system:

$\begin{array}{l} u_1 x_1 + u_2 x_2 + u_3 x_3 = 0 \\ v_1 x_1 + v_2 x_2 + v_3 x_3 = 0 \end{array}$ which leads to the augmented matrix $\begin{bmatrix} u_1 & u_2 & u_3 & | & 0 \\ v_1 & v_2 & v_3 & | & 0 \end{bmatrix}$.

Gauss-Jordan elimination yields:

$$\begin{bmatrix} u_1 & u_2 & u_3 & | & 0 \\ v_1 & v_2 & v_3 & | & 0 \end{bmatrix} \xrightarrow{u_1 R_2} \begin{bmatrix} u_1 & u_2 & u_3 & | & 0 \\ u_1 v_1 & u_1 v_2 & u_1 v_3 & | & 0 \end{bmatrix} \xrightarrow{R_2 - v_1 R_1} \begin{bmatrix} u_1 & u_2 & u_3 & | & 0 \\ 0 & u_1 v_2 - u_2 v_1 & u_1 v_3 - u_3 v_1 & | & 0 \end{bmatrix}$$

The second row implies: $(u_1 v_2 - u_2 v_1) x_2 = (u_3 v_1 - u_1 v_3) x_3 \Rightarrow x_2 = \dfrac{u_3 v_1 - u_1 v_3}{u_1 v_2 - u_2 v_1} x_3$.

To clear the fraction, we let $x_3 = (u_1 v_2 - u_2 v_1) t \Rightarrow x_2 = (u_3 v_1 - u_1 v_3) t$

The first row implies: $u_1 x_1 = -u_2 x_2 - u_3 x_3$.

So, $u_1 x_1 = -u_2 (u_3 v_1 - u_1 v_3) t - u_3 (u_1 v_2 - u_2 v_1) t = (u_1 u_2 v_3 - u_1 u_3 v_2) t$.

Dividing both sides by u_1 yields: $x_1 = (u_2 v_3 - u_3 v_2) t$.

Therefore, as was to be shown: $\begin{bmatrix} x_1 \\ x_2 \\ x_3 \end{bmatrix} = \begin{bmatrix} u_2 v_3 - u_3 v_2 \\ u_3 v_1 - u_1 v_3 \\ u_1 v_2 - u_2 v_1 \end{bmatrix} t.$

53. Following Example 2.16, we note that we need only use addition and multiplication. We form augmented matrix and perform Gauss-Jordan elimination in $\mathbb{Z}_3$:

$$\begin{bmatrix} 1 & 2 & | & 1 \\ 1 & 1 & | & 2 \end{bmatrix} \xrightarrow{R_2 + 2R_1} \begin{bmatrix} 1 & 2 & | & 1 \\ 0 & 2 & | & 1 \end{bmatrix} \xrightarrow{2R_2} \begin{bmatrix} 1 & 2 & | & 1 \\ 0 & 1 & | & 2 \end{bmatrix} \xrightarrow{R_1 + R_2} \begin{bmatrix} 1 & 0 & | & 0 \\ 0 & 1 & | & 2 \end{bmatrix}$$

So the solution is $\begin{bmatrix} x \\ y \end{bmatrix} = \begin{bmatrix} 0 \\ 2 \end{bmatrix}$.

2.2 Direct Methods for Solving Linear Systems

55. Following Example 2.17, we note that we need only use addition and multiplication. We form the augmented matrix and perform Gauss-Jordan elimination in $\mathbb{Z}_3$:

$$\begin{bmatrix} 1 & 1 & 0 & | & 1 \\ 0 & 1 & 1 & | & 0 \\ 1 & 0 & 1 & | & 1 \end{bmatrix} \xrightarrow{R_3+2R_1} \begin{bmatrix} 1 & 1 & 0 & | & 1 \\ 0 & 1 & 1 & | & 0 \\ 0 & 2 & 1 & | & 0 \end{bmatrix} \xrightarrow{R_3+R_2} \begin{bmatrix} 1 & 1 & 0 & | & 1 \\ 0 & 1 & 1 & | & 0 \\ 0 & 0 & 2 & | & 0 \end{bmatrix} \xrightarrow{2R_3} \begin{bmatrix} 1 & 1 & 0 & | & 1 \\ 0 & 1 & 1 & | & 0 \\ 0 & 0 & 1 & | & 0 \end{bmatrix}$$

$$\xrightarrow{R_2+2R_3} \begin{bmatrix} 1 & 1 & 0 & | & 1 \\ 0 & 1 & 0 & | & 0 \\ 0 & 0 & 1 & | & 0 \end{bmatrix} \xrightarrow{R_1+2R_2} \begin{bmatrix} 1 & 0 & 0 & | & 1 \\ 0 & 1 & 0 & | & 0 \\ 0 & 0 & 1 & | & 0 \end{bmatrix}$$

So the solution is: $\begin{bmatrix} x \\ y \\ z \end{bmatrix} = \begin{bmatrix} 1 \\ 0 \\ 0 \end{bmatrix}$.

57. Following Example 2.17, we note that we need only use addition and multiplication. We form the augmented matrix and perform Gauss-Jordan elimination in $\mathbb{Z}_7$:

$$\begin{bmatrix} 3 & 2 & | & 1 \\ 1 & 4 & | & 1 \end{bmatrix} \xrightarrow{R_2+2R_1} \begin{bmatrix} 3 & 2 & | & 1 \\ 0 & 1 & | & 3 \end{bmatrix} \xrightarrow{R_1+5R_2} \begin{bmatrix} 3 & 0 & | & 2 \\ 0 & 1 & | & 3 \end{bmatrix} \xrightarrow{5R_1} \begin{bmatrix} 1 & 0 & | & 3 \\ 0 & 1 & | & 3 \end{bmatrix}$$

So the solution is $\begin{bmatrix} x \\ y \end{bmatrix} = \begin{bmatrix} 3 \\ 3 \end{bmatrix}$.

59. Recall the Rank Theorem (which applies to systems over $\mathbb{Z}_p$ not just $\mathbb{R}^n$) says: If the system is consistent, then: number of free variables $= n-$ rank (A).

In $\mathbb{Z}_p$, however, each free variable can only take on p different values.
If there is 1 free variable, there are $p^1 = p$ solutions as in Exercise 58.
If there are 2 free variables, there are $p \times p = p^2$ solutions.
In general, the total number of solutions is $p^{\text{number of free variables}} = p^{n-\text{rank}(A)}$.

Exploration: Partial Pivoting

Explorations are self-contained, so only odd-numbered solutions will be provided.

1. (a) Solving $0.00021x = 1$ to 5 significant digits, we have: $x = \dfrac{1}{0.00021} \approx 4761.9$.

   (b) Solving $0.0002x = 1$ to 4 significant digits, we have: $x = \dfrac{1}{0.0002} \approx 5000$.
   So, the effect of an 0.00001 error is $5000 - 4761.9 = 238.1$.

3. (a) Without partial pivoting:

Pivoting on 0.001 to 3 significant digits,
we first divide the first row by 0.001 ($\Rightarrow$ 1   995   1000),
then multiply it by -10.2 and subtract the result from row 2 $\Rightarrow$

$$\begin{bmatrix} 0.001 & 0.995 & | & 1.00 \\ -10.2 & 1.00 & | & -50.0 \end{bmatrix} \longrightarrow \begin{bmatrix} 0.001 & 0.995 & | & 1.00 \\ 0 & -10,100 & | & -10,200 \end{bmatrix} \Rightarrow y = 1.01.$$

Back substituting, we get $x = \dfrac{1.00 - 1.00}{0.001} = 0.$

This error was introduced because 1.00 and -50.0 were ignored
when reducing to 3 significant digits.
They were overwhelmed by $-10,100$ and $-10,200$ respectively.

With partial pivoting:

Pivoting on -10.2 to 3 significant digits,
we first divide the first row by -10.2 ($\Rightarrow$ 1   0.098   4.90),
then multiply it by 0.001 and subtract the result from row 2 $\Rightarrow$

$$\begin{bmatrix} -10.2 & 1.00 & | & -50.0 \\ 0.001 & 0.995 & | & 1.00 \end{bmatrix} \longrightarrow \begin{bmatrix} -10.2 & 1.00 & | & -50.0 \\ 0 & 0.995 & | & 1.00 \end{bmatrix} \Rightarrow y = 1.00.$$

Back substituting, we get $x = \dfrac{-50.0 - 1.00}{-10.2} = \dfrac{51}{10.2} = 5.$

Pivoting on the largest absolute value reduces the error in our solution.

(b) Without partial pivoting:

Pivoting on 10 to 3 significant digits,
we first divide the first row by 10 ($\Rightarrow$ 1 -0.7 0   0.7),
then multiply it by 3 and add the result to row 2 and
also multiply it by 5 and subtract the result from row 3.

$$\begin{bmatrix} 10.0 & -7.00 & 0.00 & | & 7.00 \\ -3.0 & 2.09 & 6.00 & | & 3.91 \\ 5.0 & 1.00 & 5.00 & | & 6.00 \end{bmatrix} \longrightarrow \begin{bmatrix} 10.0 & -7.00 & 0.00 & | & 7.00 \\ 0.0 & -0.01 & 6.00 & | & 6.01 \\ 0.0 & 2.50 & 5.00 & | & 2.50 \end{bmatrix}.$$

Pivoting on -0.01 to 3 significant digits,
we first divide the second row by -0.01 ($\Rightarrow$ 0   1 -600 -601),
then multiply it by 2.5 and and subtract the result from row 3.

$$\begin{bmatrix} 10.0 & -7.00 & 0.00 & | & 7.00 \\ 0.0 & -0.01 & 6.00 & | & 6.01 \\ 0.0 & 2.50 & 5.00 & | & 2.50 \end{bmatrix} \longrightarrow \begin{bmatrix} 10.0 & -7.00 & 0.00 & | & 7.00 \\ 0.0 & -0.01 & 6.00 & | & 6.01 \\ 0.0 & 0 & -1500 & | & -1500 \end{bmatrix} \Rightarrow$$

Back substituting, we get $z = 1.00$, $y = 1.00$, and $x = 0.00.$
Repeat by interchanging rows 2 and 3 to make the second pivot 2.5.

Exploration: An Introduction to the Analysis of Algorithms

Explorations are self-contained, so only odd-numbered solutions will be provided.

1. We will count the number of operations, one step at a time.
$$\begin{bmatrix} 2 & 4 & 6 & | & 8 \\ 3 & 9 & 6 & | & 12 \\ -1 & 1 & -1 & | & 1 \end{bmatrix} \longrightarrow \begin{bmatrix} 1 & 2 & 3 & | & 4 \\ 3 & 9 & 6 & | & 12 \\ -1 & 1 & -1 & | & 1 \end{bmatrix}.$$

In this step, we performed 3 operations, $\frac{1}{2} \cdot 4 = 2$, $\frac{1}{2} \cdot 6 = 3$, and $\frac{1}{2} \cdot 8 = 4$.

Note, we don't count $\frac{1}{2} \cdot 2 = 1$ because that's automatic once we decide to multiply by $\frac{1}{2}$.

$$\begin{bmatrix} 1 & 2 & 3 & | & 4 \\ 3 & 9 & 6 & | & 12 \\ -1 & 1 & -1 & | & 1 \end{bmatrix} \longrightarrow \begin{bmatrix} 1 & 2 & 3 & | & 4 \\ 0 & 3 & -3 & | & 0 \\ -1 & 1 & -1 & | & 1 \end{bmatrix}.$$

In this step, we performed 3 operations, $3 \cdot 2 = 6$, $3 \cdot 3 = 9$, and $3 \cdot 4 = 12$, for a total of 6.

Note, we don't count the subtractions (or additions) and again we don't count $3 \cdot 1 = 3$ that created the zero because this is automatic once we decide to multiply by 3.

$$\begin{bmatrix} 1 & 2 & 3 & | & 4 \\ 0 & 3 & -3 & | & 0 \\ -1 & 1 & -1 & | & 1 \end{bmatrix} \longrightarrow \begin{bmatrix} 1 & 2 & 3 & | & 4 \\ 0 & 3 & -3 & | & 0 \\ 0 & 3 & 2 & | & 5 \end{bmatrix}.$$

In this step, we performed 3 operations, $1 \cdot 2 = 2$, $1 \cdot 3 = 3$, and $1 \cdot 4 = 4$, for a total of 9.

$$\begin{bmatrix} 1 & 2 & 3 & | & 4 \\ 0 & 3 & -3 & | & 0 \\ 0 & 3 & 2 & | & 5 \end{bmatrix} \longrightarrow \begin{bmatrix} 1 & 2 & 3 & | & 4 \\ 0 & 1 & -1 & | & 0 \\ 0 & 3 & 2 & | & 5 \end{bmatrix}.$$

In this step, we performed 2 operations, $(\frac{1}{3})(-3) = -1$ and $\frac{1}{3} \cdot 0 = 0$, for a total of 11.

$$\begin{bmatrix} 1 & 2 & 3 & | & 4 \\ 0 & 1 & -1 & | & 0 \\ 0 & 3 & 2 & | & 5 \end{bmatrix} \longrightarrow \begin{bmatrix} 1 & 2 & 3 & | & 4 \\ 0 & 1 & -1 & | & 0 \\ 0 & 0 & 5 & | & 5 \end{bmatrix}.$$

In this step, we performed 2 operations, $(-3)(-1) = 3$ and $(-3)(0) = 0$, for a total of 13.

$$\begin{bmatrix} 1 & 2 & 3 & | & 4 \\ 0 & 1 & -1 & | & 0 \\ 0 & 0 & 5 & | & 5 \end{bmatrix} \longrightarrow \begin{bmatrix} 1 & 2 & 3 & | & 4 \\ 0 & 1 & -1 & | & 0 \\ 0 & 0 & 1 & | & 1 \end{bmatrix}.$$

In this step, we performed 1 operation, $\frac{1}{5} \cdot 5 = 5$, for a total of 14.

Finally, to complete the back substitution, we need only 3 more operations. See that? Namely, $-x_3$ to find x_2, and $2x_2$, $3x_3$ to find x_1, for a total of 17.

3. (a) There are n operations required to create the first leading 1 because we have to divide every entry in the first row (except a_{11}) by a_{11}.

   We don't have to divide a_{11} by a_{11} because the resulting 1 results from the choice itself.

   There are n operations required to create the first zero in column 1 because we have to multiply every entry in the first row (except the leading 1) by a_{21}.

   We don't have to multiply 1 by a_{21} because the resulting 0 results from the choice itself.

   There are n operations required to create each zero in column 1 because we have to multiply every entry in the first row (except the leading 1) by a_{k1}.

   There are $n-1$ rows (excluding the first row), so $n + (n-1)n$ operations are required. Recall, the first n operations created the leading 1 in row 1.

   (b) As above, we see it takes $n-1$ operations to create the leading 1 in the second row and then $n-1$ operations to create the zeros in column 2.

   Now there are only $n-2$ rows left to create zeroes in since rows 1 and 2 are excluded.

   So, $(n-1) + (n-2)(n-1)$ operations are required to create the leading 1 in row 2 and all the zeros beneath it.

   Continuing this process, we see:
   $[n + (n-1)n] + [(n-1) + (n-2)(n-1)] + \cdots + [2 + 1 \cdot 2] + 1$.
   This simplifies to: $n^2 + (n-1)^2 + \cdots + 2^2 + 1^2$.
   This simplification follows quickly from the following observation:
   $n + (n-1)n = n[1 + (n-1)] = n \cdot n = n^2$.

   (c) There is 1 operation required to find x_{n-1} (involving x_n).
   There are 2 operations required to find x_{n-2} (involving x_n and x_{n-1})
   Continuing this reasoning, we get the total: $1 + 2 + \cdots + (n-1)$.

   (d) Exercises 45 and 46 in Section 2.4 and Appendix B give us the following equations:
   $n^2 + (n-1)^2 + \cdots + 2^2 + 1^2 = \frac{n(n+1)(2n+1)}{6} = \frac{1}{3}n^3 + \frac{1}{2}n^2 + \frac{1}{6}n$ and
   $1 + 2 + \cdots + (n-1) = \frac{n(n-1)}{2} = \frac{1}{2}n^2 - \frac{1}{2}n$.
   Adding these together we get the total number of operations is:
   $\left(\frac{1}{3}n^3 + \frac{1}{2}n^2 + \frac{1}{6}n\right) + \left(\frac{1}{2}n^2 - \frac{1}{2}n\right) = \frac{1}{3}n^3 + n^2 - \frac{1}{3}n$.
   So, for large values of n the total number of operations required is $T(n) \approx \frac{1}{3}n^3$.

2.3 Spanning Sets and Linear Independence

Q: What is our goal for Section 2.3?

A: To learn how to row reduce matrices using elementary row operations.
To understand how linear independence relates to the solutions of corresponding systems.

Key Definitions and Concepts

A_c		2.3	Created by taking vectors v_i as its columns (see below)
A_r		2.3	Created by taking vectors v_i as its rows (see below)
$b = c_1 v_1 + \cdots + c_n v_n$	p.638	A.5	b is a linear combination of v_i (see Appendix A)
$b = \sum c_i v_i$	p.638	A.5	b is a linear combination of v_i (see Appendix A)
a_i	p.90	2.3	rows or columns of A written as vectors
$[A\|b]$	p.91	2.3	augmented matrix of a linear system
$[A\|0]$	p.97	2.3	augmented matrix of a homogenous linear system
linear combination	p.95	2.3	$v = c_1 v_1 + \cdots + c_n v_n$ $v = \sum c_i v_i$
linearly dependent set of vectors	p.94	2.3	vector can be written as linear combination of others $c_1 v_1 + \cdots + c_n v_n = 0$ with at least one $c_i \neq 0$ $\sum c_i v_i = 0$ with at least one $c_i \neq 0$
linearly independent set of vectors	p.94	2.3	no vector can be written as linear combination of others $c_1 v_1 + \cdots + c_n v_n = 0 \Leftrightarrow$ all the $c_i = 0$ $\sum c_i v_i = 0 \Leftrightarrow$ all the $c_i = 0$
span	p.92	2.3	the set of all linear combinations of a set of vectors
span(S)	p.92	2.3	all linear combinations of $S = \{s_1, s_2, \ldots, s_n\}$
spanning set for $\mathbb{R}^n$	p.92	2.3	set S such that span(S) $= \mathbb{R}^n$
spanning set for $\mathbb{R}^n$	p.92	2.3	set S such that v in $\mathbb{R}^n \Rightarrow v = \sum c_i s_i$

Theorems

The theorems relate linear independence to the matrices to A_c and A_r (see definitions above).

Thm 2.4	p.90	2.3	$[A\|b]$ is consistent $\Leftrightarrow b = \sum c_i a_i$
Thm 2.5	p.95	2.3	linearly dependent set $\Leftrightarrow$ linear combination of the others
Thm 2.6	p.97	2.3	linearly dependent set $\Leftrightarrow [A_c\|0]$ has nontrivial solution
Thm 2.7	p.98	2.3	linearly dependent set $\Leftrightarrow$ rank(A_r) $< m$ where A_r is $m \times n$
Thm 2.8	p.99	2.3	m vectors in $\mathbb{R}^n$ are linearly dependent if $m > n$

Discussion of Key Definitions and Concepts

Before we begin our discussion, we should study parts of Appendix A.
Examples 1 through 3 present a thorough overview of set notation.
Read the remarks following Example 9 regarding the **contrapositive** of a statement.

Q: What does it mean for one vector to be a linear combination of some other vectors?

A: In somewhat general terms, if $\mathbf{v} = c_1\mathbf{v}_1 + c_2\mathbf{v}_2$, then $\mathbf{v}$ is a linear combination of $\mathbf{v}_1$ and $\mathbf{v}_2$.

Example: Since $\begin{bmatrix} 2 \\ 3 \end{bmatrix} = 2\begin{bmatrix} 1 \\ 0 \end{bmatrix} + 3\begin{bmatrix} 0 \\ 1 \end{bmatrix}$, $\begin{bmatrix} 2 \\ 3 \end{bmatrix}$ is a linear combination of $\begin{bmatrix} 1 \\ 0 \end{bmatrix}$ and $\begin{bmatrix} 0 \\ 1 \end{bmatrix}$.

Q: For a given vector $\mathbf{v}$, are the vectors it can be written as a linear combination of unique?

A: No. In general, there are infinitely many such vectors. Let's revisit our example above.

Since $\begin{bmatrix} 2 \\ 3 \end{bmatrix} = \begin{bmatrix} 2 \\ 0 \end{bmatrix} + \begin{bmatrix} 0 \\ 3 \end{bmatrix}$, $\begin{bmatrix} 2 \\ 3 \end{bmatrix}$ is a linear combination of $\begin{bmatrix} 2 \\ 0 \end{bmatrix}$ and $\begin{bmatrix} 0 \\ 3 \end{bmatrix}$.

In fact, $\begin{bmatrix} 2 \\ 3 \end{bmatrix}$ is a linear combination of $\left\{\begin{bmatrix} a \\ 0 \end{bmatrix}, \begin{bmatrix} 0 \\ b \end{bmatrix}\right\}$ $(a \neq 0, b \neq 0)$.

Q: What is the only vector that is a linear combination of every vector?

A: The zero vector, $\mathbf{0}$. Why? Because $0\mathbf{v} = \mathbf{0}$ for all vectors $\mathbf{v}$.

Q: Why is any set of vectors that contains the zero vector linearly dependent?

A: To help us understand why this must be true, let's consider an example:

Since $0\begin{bmatrix} 1 \\ 0 \end{bmatrix} + 0\begin{bmatrix} 0 \\ 1 \end{bmatrix} + \begin{bmatrix} 0 \\ 0 \end{bmatrix} = \begin{bmatrix} 0 \\ 0 \end{bmatrix} = \mathbf{0}$ (the zero vector),

the set $S = \left\{\begin{bmatrix} 1 \\ 0 \end{bmatrix}, \begin{bmatrix} 0 \\ 1 \end{bmatrix}, \begin{bmatrix} 0 \\ 0 \end{bmatrix}\right\}$ is linearly dependent.

In general, if $0\mathbf{v}_1 + 0\mathbf{v}_2 + \mathbf{0} = \mathbf{0}$, then the set $S = \{\mathbf{v}_1, \mathbf{v}_2, \mathbf{0}\}$ is linearly dependent.

Why is this enough?

Because we have written the zero vector as a linear combination of the vectors in S with at least one of the coefficients $\neq 0$.

Q: Theorem 2.5 says any set of vectors is linearly dependent if and only if one vector among the set can be written as a linear combination of the other vectors in the set. Why does this imply that any set containing the zero vector must be linearly dependent?

A: As remarked above, the zero vector is a linear combination of every vector. Therefore, any set containing the zero vector contains a vector that can be written as a linear combination of the other vectors in the set. Namely, the zero vector.

2.3 Spanning Sets and Linear Independence

Discussion of Key Definitions and Concepts, continued

Q: If set S is linearly independent, is there any subset $R \subseteq S$ that is linearly dependent?

A: No. Why not? We are asked to prove this in Exercise 46. We discuss it here.

For example: Let $S = \{\mathbf{v}_1, \mathbf{v}_2, \mathbf{v}_3\}$ be a linearly independent set of vectors. Then by definition, $c_1\mathbf{v}_1 + c_2\mathbf{v}_2 + c_3\mathbf{v}_3 = \mathbf{0} \Leftrightarrow c_1 = c_2 = c_3 = 0$.

Now let d_1, d_2 be any scalars such that $d_1\mathbf{v}_1 + d_2\mathbf{v}_2 = \mathbf{0}$.
We will show $d_1 = d_2 = 0$ and conclude that $R = \{\mathbf{v}_1, \mathbf{v}_2\} \subseteq S$ is linearly independent.

$$d_1\mathbf{v}_1 + d_2\mathbf{v}_2 = \mathbf{0} \Rightarrow d_1\mathbf{v}_1 + d_2\mathbf{v}_2 + 0\mathbf{v}_3 = \mathbf{0} \Rightarrow d_1 = d_2 = 0.$$

Therefore, $R = \{\mathbf{v}_1, \mathbf{v}_2\} \subseteq S$ is a linearly independent set of vectors.

Q: If set S is linearly dependent, are all sets T such that $S \subseteq T$ linearly dependent?

A: Yes. Why?

Since S is linearly dependent, there is a vector $\mathbf{v}$ in S that can be written as a linear combination of the other vectors in S. Since $S \subseteq T$, both $\mathbf{v}$ and those other vectors are in T as well.

So, by Theorem 2.5, T is linearly dependent.

Q: What does span(S) mean?

A: All the vectors that can be written as a linear combination of the vectors in S.

Q: What is span($\mathbf{0}$)? That is, what is the span of the zero vector?

A: Obviously, span($\mathbf{0}$) = $\mathbf{0}$. Why? Because $c\mathbf{0} = \mathbf{0}$ for all scalars c.

Q: What is span($\begin{bmatrix} 1 \\ 0 \end{bmatrix}$)?

A: Obviously, span($\begin{bmatrix} 1 \\ 0 \end{bmatrix}$) = $\left\{\begin{bmatrix} c \\ 0 \end{bmatrix}\right\}$ Why? Because $c\begin{bmatrix} 1 \\ 0 \end{bmatrix} = \begin{bmatrix} c \\ 0 \end{bmatrix}$ for all scalars c.

Q: What is the only vector in the span of every set?

A: The zero vector. Why? Because $0\mathbf{v} = \mathbf{0}$ for every vector $\mathbf{v}$.

Discussion of Theorems

Q: What does the statement of Theorem 2.4 mean: $[A|b]$ is consistent $\Leftrightarrow b = \sum c_i a_i$?

A: Consider both sides of the implication in the following example:

To say $[A|b] = \begin{bmatrix} 1 & 0 & | & 2 \\ 0 & 1 & | & 3 \end{bmatrix}$ is consistent means that this system has a solution.

That means there must exist c_1 and c_2 such that $c_1 \begin{bmatrix} 1 \\ 0 \end{bmatrix} + c_2 \begin{bmatrix} 0 \\ 1 \end{bmatrix} = \begin{bmatrix} 2 \\ 3 \end{bmatrix}$.

On the other hand, to say $b = \begin{bmatrix} 2 \\ 3 \end{bmatrix}$ is a linear combination of the columns of A,

$\begin{bmatrix} 1 \\ 0 \end{bmatrix}$ and $\begin{bmatrix} 0 \\ 1 \end{bmatrix}$, means there must exist c_1 and c_2 such that $c_1 \begin{bmatrix} 1 \\ 0 \end{bmatrix} + c_2 \begin{bmatrix} 0 \\ 1 \end{bmatrix} = \begin{bmatrix} 2 \\ 3 \end{bmatrix}$.

This example makes it clear why the two conditions are equivalent.

Q: Let a_i be the columns of A written as vectors.
Then the contrapositive of Theorem 2.4 is: $[A|b]$ is inconsistent $\Leftrightarrow b$ is not in span(a_i).
Does this restatement make it obvious that any homogeneous system ($b = 0$) is consistent?

A: Yes. Why? Because the zero vector ($b = 0$) is in the span of every set of vectors.

Q: How might we restate Theorem 2.4 using the definition of a linearly dependent set?

A: Let $\{v_i\}$ be the set of linearly independent columns of A and and $T = \{v_i, b\}$. Then

$[A|b]$ is consistent $\Leftrightarrow T$ is linearly dependent.

Why is this equivalent?

T is linearly dependent means b must be a linear combination of the other vectors in T. But those other vectors are the linearly independent columns of A.

Q: Does this restatement make it obvious that any homogeneous system ($b = 0$) is consistent?

A: Yes. Why? Because any set T containing the zero vector ($b = 0$) is linearly dependent.

Q: What happens when $b = a_i = 0$?

A: In this case, there are no linearly independent columns of A,
but the inclusion of $b = 0$ still makes T is a linearly dependent set.

Q: How might we restate Theorem 2.4 using the definition of a linearly independent set?

A: $[A|b]$ is inconsistent $\Leftrightarrow S = \{a_i, b\}$ are linearly independent. Why is this equivalent?
S is linearly independent means b cannot be written as a linear combination of the a_i.

2.3 Spanning Sets and Linear Independence

Discussion of Theorems, continued

Q: What is the point of Theorem 2.5?

A: To see that a linearly dependent set must have at least one vector that can be written as a linear combination of the other vectors in the set.

Consider $S = \{\mathbf{v}_1, \mathbf{v}_2, \mathbf{v}_3\}$. To say S is linearly dependent means:

There must exist $c_1\mathbf{v}_1 + c_2\mathbf{v}_2 + c_3\mathbf{v}_3 = \mathbf{0}$ with at least one $c_i \neq 0$.

We will assume $c_1 \neq 0$ and show $\mathbf{v}_1$ can be written as a linear combination of $\mathbf{v}_2$ and $\mathbf{v}_3$.

$$c_1\mathbf{v}_1 + c_2\mathbf{v}_2 + c_3\mathbf{v}_3 = \mathbf{0} \text{ and } c_1 \neq 0 \Rightarrow \mathbf{v}_1 = -\tfrac{c_2}{c_1}\mathbf{v}_2 - \tfrac{c_3}{c_1}\mathbf{v}_3.$$

Though we assumed $c_1 \neq 0$, the argument works equally well for $c_2 \neq 0$ or $c_3 \neq 0$.

Q: From above, we know $S = \left\{ \begin{bmatrix} 2 \\ 3 \end{bmatrix}, \begin{bmatrix} 1 \\ 0 \end{bmatrix}, \begin{bmatrix} 0 \\ 1 \end{bmatrix} \right\}$ is linearly dependent. Does that imply one of its vectors is a linear combination of the others? If so, write out an example.

A: Yes. E.g.: Since $\begin{bmatrix} 2 \\ 3 \end{bmatrix} = 2\begin{bmatrix} 1 \\ 0 \end{bmatrix} + 3\begin{bmatrix} 0 \\ 1 \end{bmatrix}$, $\begin{bmatrix} 2 \\ 3 \end{bmatrix}$ is a linear combination of $\begin{bmatrix} 1 \\ 0 \end{bmatrix}$ and $\begin{bmatrix} 0 \\ 1 \end{bmatrix}$.

Q: Can we write $\begin{bmatrix} 1 \\ 0 \end{bmatrix}$ as a linear combination of $\begin{bmatrix} 2 \\ 3 \end{bmatrix}$ and $\begin{bmatrix} 0 \\ 1 \end{bmatrix}$?

Q: Does this linear combination alone imply $S = \left\{ \begin{bmatrix} 2 \\ 3 \end{bmatrix}, \begin{bmatrix} 1 \\ 0 \end{bmatrix}, \begin{bmatrix} 0 \\ 1 \end{bmatrix} \right\}$ is linearly dependent?

A: Yes. Use Theorem 2.5 to prove that.

Q: Can we repeat this argument with $\begin{bmatrix} 0 \\ 1 \end{bmatrix}$ as a linear combination of $\begin{bmatrix} 2 \\ 3 \end{bmatrix}$ and $\begin{bmatrix} 1 \\ 0 \end{bmatrix}$?

A: Yes.

Discussion of Theorems, continued

Q: Is Theorem 2.6 equivalent to the statement:
   v_i are linearly independent $\Leftrightarrow$ $[A_c|0]$ has only the trivial solution?

A: Yes. In fact, this statement is the *contrapositive* of Theorem 2.6.
   See the remarks following Example 9 in Appendix A for the definition of contrapositive.
   The contrapositive of Theorem 2.6 is useful in Exercise 42.

   Let's consider what both sides of the implication mean in the following example:

   To say $S = \left\{ \begin{bmatrix} 1 \\ 0 \end{bmatrix}, \begin{bmatrix} 0 \\ 1 \end{bmatrix} \right\}$ is linearly independent means:

   $$c_1 \begin{bmatrix} 1 \\ 0 \end{bmatrix} + c_2 \begin{bmatrix} 0 \\ 1 \end{bmatrix} = \begin{bmatrix} 0 \\ 0 \end{bmatrix} = 0 \Leftrightarrow c_1 = c_2 = 0.$$

   On the other hand to say, $[A_c|0] = \begin{bmatrix} 1 & 0 & | & 0 \\ 0 & 1 & | & 0 \end{bmatrix}$ has only the trivial solution means:

   The only linear combination of the columns of A,

   $$\begin{bmatrix} 1 \\ 0 \end{bmatrix} \text{ and } \begin{bmatrix} 0 \\ 1 \end{bmatrix}, \text{ that equals } 0 \text{ is } c_1 = c_2 = 0.$$

   That is, $c_1 \begin{bmatrix} 1 \\ 0 \end{bmatrix} + c_2 \begin{bmatrix} 0 \\ 1 \end{bmatrix} = \begin{bmatrix} 0 \\ 0 \end{bmatrix} = 0 \Leftrightarrow c_1 = c_2 = 0.$

   This example makes it clear that the two conditions are equivalent.

Q: Solve $[A|0] = \begin{bmatrix} 2 & 0 & | & 0 \\ 3 & 1 & | & 0 \end{bmatrix}$. What does this tell us?

Q: Since the only solution is $\mathbf{b} = 0$, we conclude $\begin{bmatrix} 2 \\ 3 \end{bmatrix}$ and $\begin{bmatrix} 0 \\ 1 \end{bmatrix}$ are linearly independent.

Q: Is Theorem 2.7 equivalent to the statement:
   v_i are linearly independent $\Leftrightarrow$ rank$(A_r) = m$ where A_r is an $m \times n$ matrix?

A: Yes. In fact, this statement is the *contrapositive* of Theorem 2.7.
   See the remarks following Example 9 in Appendix A for the definition of contrapositive.

   Why?

   Because rank$(A_r) = m \Leftrightarrow$ no one row is a linear combination of the other rows
   $\Leftrightarrow$ the rows of $A_r = v_i$ are linearly independent.

   The contrapositive of Theorem 2.7 is useful in Exercises 32 thorough 41.

2.3 Spanning Sets and Linear Independence

Discussion of Theorems, continued

Q: Given $A = \begin{bmatrix} 1 & 0 \\ 0 & 1 \\ 2 & 3 \end{bmatrix}$, what is rank($A$)? What do we conclude?

A: Since $A_r = \begin{bmatrix} 1 & 0 \\ 0 & 1 \\ 2 & 3 \end{bmatrix} \longrightarrow \begin{bmatrix} 1 & 0 \\ 0 & 1 \\ 0 & 0 \end{bmatrix}$, we have rank($A_r$) = 2 < 3 = m.

So, we conclude the rows of $A = [1\ 0], [0\ 1], [2\ 3]$ are linearly dependent.

Note: rank(A_r) = 2 < 3 = m precisely because when we row reduce A we get a row of zeroes. We get that row of zeroes because $R_3 - 2R_1 - 3R_3 = 0$, but that implies $R_3 = 2R_1 + 3R_2$. This equation makes it obvious that R_3 is a linear combination of R_1 and R_2.

Q: Does Theorem 2.8 imply a set of $m \leq n$ vectors in $\mathbb{R}^n$ is always linearly independent?

A: No. There is no minimum number of vectors that will guarantee linear independence. Why not? Because the zero vector even all by itself is a linearly dependent set.

Q: Can a set with only one vector be linearly dependent?

A: Yes. But only if that vector is the zero vector.

Q: What is an obvious spanning set for $\mathbb{R}^2$? That is, a set S such that span(S) = $\mathbb{R}^2$?

A: Perhaps, the most obvious spanning set for $\mathbb{R}^2$ is the following:

$S = \left\{ \begin{bmatrix} 1 \\ 0 \end{bmatrix}, \begin{bmatrix} 0 \\ 1 \end{bmatrix} \right\}$ since any vector in $\mathbb{R}^2$ can be written as $\begin{bmatrix} c_1 \\ c_2 \end{bmatrix} = c_1 \begin{bmatrix} 1 \\ 0 \end{bmatrix} + c_2 \begin{bmatrix} 0 \\ 1 \end{bmatrix}$.

Q: How can we see Theorem 2.8 is true in the special case when there are exactly n linearly independent vectors in a set S of $m > n$ vectors in $\mathbb{R}^n$?

A: Consider the following example: In $\mathbb{R}^2$, let the vectors $\begin{bmatrix} 1 \\ 0 \end{bmatrix}$ and $\begin{bmatrix} 0 \\ 1 \end{bmatrix}$ be in S.

Since we are assuming $n = 2 < m$, S contains a vector $\mathbf{v}$ other than $\begin{bmatrix} 1 \\ 0 \end{bmatrix}$ and $\begin{bmatrix} 0 \\ 1 \end{bmatrix}$.

But, it is obvious that vector $\mathbf{v}$ can be written as a linear combination of $\begin{bmatrix} 1 \\ 0 \end{bmatrix}$ and $\begin{bmatrix} 0 \\ 1 \end{bmatrix}$.

Therefore, Theorem 2.4, this tells us that S is linearly dependent.

Q: Why is Theorem 2.8 true when there are less than n linearly independent vectors in S?

A: Because that implies there is a set of vectors $R \subseteq S$ that is linearly dependent. As argued above, since S contains a linearly dependent set, S must be linearly dependent.

Discussion of Theorems, continued

Q: We saw above that $S = \left\{ \begin{bmatrix} 1 \\ 0 \end{bmatrix}, \begin{bmatrix} 0 \\ 1 \end{bmatrix} \right\}$ is a linearly independent set of two vectors. Is any set of two linearly independent vectors a spanning set for $\mathbb{R}^2$?

A: Yes. Let's show that explicitly using Theorem 2.8. We need to show that any vector $\mathbf{v}$ in $\mathbb{R}^2$ can be written as a linear combination of $\mathbf{v}_1$ and $\mathbf{v}_2$.

Since $\mathbf{v}_1$ and $\mathbf{v}_2$ are linearly independent,

Theorem 2.8 implies $S = \left\{ \mathbf{v}_1, \mathbf{v}_2, \begin{bmatrix} 1 \\ 0 \end{bmatrix} \right\}$ is linearly dependent.

Why? Because S is a set of 3 vectors in $\mathbb{R}^2$ and $3 > 2$. What does that tell us?

That tells us $\begin{bmatrix} 1 \\ 0 \end{bmatrix}$ is a linear combination of $\mathbf{v}_1$ and $\mathbf{v}_2$: $\begin{bmatrix} 1 \\ 0 \end{bmatrix} = a_1 \mathbf{v}_1 + a_2 \mathbf{v}_2$.

Likewise, $\begin{bmatrix} 0 \\ 1 \end{bmatrix}$ is a linear combination of $\mathbf{v}_1$ and $\mathbf{v}_2$: $\begin{bmatrix} 0 \\ 1 \end{bmatrix} = b_1 \mathbf{v}_1 + b_2 \mathbf{v}_2$.

So, given any vector $\mathbf{v}$ in $\mathbb{R}^2$, we have: $\mathbf{v} = \begin{bmatrix} x \\ y \end{bmatrix} = x \begin{bmatrix} 1 \\ 0 \end{bmatrix} + y \begin{bmatrix} 0 \\ 1 \end{bmatrix} =$

$x(a_1 \mathbf{v}_1 + a_2 \mathbf{v}_2) + y(b_1 \mathbf{v}_1 + b_2 \mathbf{v}_2) = (xa_1 + yb_1)\mathbf{v}_1 + (xa_2 + yb_2)\mathbf{v}_2 = c_1 \mathbf{v}_1 + c_2 \mathbf{v}_2$.

Therefore, $\mathbf{v}$ can be written as a linear combination of $\mathbf{v}_1$ and $\mathbf{v}_2$.

2.3 Spanning Sets and Linear Independence

Solutions to odd-numbered exercises from Section 2.3

1. As in Example 2.18, we want to find scalars x and y such that:

$$x \begin{bmatrix} 1 \\ -1 \end{bmatrix} + y \begin{bmatrix} 2 \\ -1 \end{bmatrix} = \begin{bmatrix} 1 \\ 2 \end{bmatrix}$$ Expanding, we obtain the system: $\begin{aligned} x + 2y &= 1 \\ -x - y &= 2 \end{aligned}$

We then row reduce the associated augmented matrix: $\begin{bmatrix} 1 & 2 & | & 1 \\ -1 & -1 & | & 2 \end{bmatrix} \longrightarrow \begin{bmatrix} 1 & 0 & | & -5 \\ 0 & 1 & | & 3 \end{bmatrix}$

So the solution is $x = -5$, $y = 2$, and the linear combination is $-5 \begin{bmatrix} 1 \\ -1 \end{bmatrix} + 3 \begin{bmatrix} 2 \\ -1 \end{bmatrix} = \begin{bmatrix} 1 \\ 2 \end{bmatrix}$.

3. As in Example 2.18, we want to find scalars x and y such that:

$$x \begin{bmatrix} 1 \\ 1 \\ 0 \end{bmatrix} + y \begin{bmatrix} 0 \\ 1 \\ 1 \end{bmatrix} = \begin{bmatrix} 1 \\ 2 \\ 3 \end{bmatrix}$$ Expanding, we obtain the system: $\begin{aligned} x &= 1 \\ x + y &= 2 \\ y &= 3 \end{aligned}$

Since $x = 1$ and $y = 3$ implies $x + y \ne 2$, this system clearly has no solution.
Therefore, **v** is not a linear combination of $\mathbf{u}_1$ and $\mathbf{u}_2$.

5. Similar to Example 2.18, we want to find scalars x, y, and z such that:

$$x \begin{bmatrix} 1 \\ 1 \\ 0 \end{bmatrix} + y \begin{bmatrix} 0 \\ 1 \\ 1 \end{bmatrix} + z \begin{bmatrix} 1 \\ 0 \\ 1 \end{bmatrix} = \begin{bmatrix} 1 \\ 2 \\ 3 \end{bmatrix}$$ Expanding, we obtain the system: $\begin{aligned} x \phantom{{}+y} + z &= 1 \\ x + y \phantom{{}+z} &= 2 \\ y + z &= 3 \end{aligned}$

Since $z = 1$ and $y = 2$ implies $y + z = 3$, the solution is $x = 0$, $y = 2$, $z = 1$.

Row reduce the augmented matrix to confirm that: $\begin{bmatrix} 1 & 0 & 1 & | & 1 \\ 1 & 1 & 0 & | & 2 \\ 0 & 1 & 1 & | & 3 \end{bmatrix} \longrightarrow \begin{bmatrix} 1 & 0 & 0 & | & 0 \\ 0 & 1 & 0 & | & 2 \\ 0 & 0 & 1 & | & 1 \end{bmatrix}$

So the solution is $x = 0$, $y = 2$, $z = 1$, and the linear combination is $2 \begin{bmatrix} 0 \\ 1 \\ 1 \end{bmatrix} + 1 \begin{bmatrix} 1 \\ 0 \\ 1 \end{bmatrix} = \begin{bmatrix} 1 \\ 2 \\ 3 \end{bmatrix}$.

7. Applying Theorem 2.4, we check to see if $[A|\mathbf{b}]$ is consistent. Why?
Theorem 2.4 says $[A|\mathbf{b}]$ is consistent $\Leftrightarrow$ **b** is a linear combination of the columns of A.
That is exactly what is required for **b** to be in the span of the columns of A.

So we row reduce $[A|\mathbf{b}] = \begin{bmatrix} 1 & 2 & | & 5 \\ 3 & 4 & | & 6 \end{bmatrix} \longrightarrow \begin{bmatrix} 1 & 0 & | & -4 \\ 0 & 1 & | & \frac{9}{2} \end{bmatrix}$ to see that it is consistent.

What do we conclude? The vector **b** is in the span of the columns of A.

In particular, the solution tells us the linear combination is $-4 \begin{bmatrix} 1 \\ 3 \end{bmatrix} + \frac{9}{2} \begin{bmatrix} 2 \\ 4 \end{bmatrix} = \begin{bmatrix} 5 \\ 6 \end{bmatrix}$.

9. As in Example 2.19, we must show $x \begin{bmatrix} 1 \\ 1 \end{bmatrix} + y \begin{bmatrix} 1 \\ -1 \end{bmatrix} = \begin{bmatrix} a \\ b \end{bmatrix}$ can always be solved.

The augmented matrix is $\begin{bmatrix} 1 & 1 & | & a \\ 1 & -1 & | & b \end{bmatrix}$, and row reduction produces:

$$\begin{bmatrix} 1 & 1 & | & a \\ 1 & -1 & | & b \end{bmatrix} \xrightarrow{R_1+R_2} \begin{bmatrix} 2 & 0 & | & a+b \\ 1 & -1 & | & b \end{bmatrix} \xrightarrow[-R_2]{1/2 R_1} \begin{bmatrix} 1 & 0 & | & (a+b)/2 \\ 0 & 1 & | & -b \end{bmatrix} \xrightarrow{R_2+R_1} \begin{bmatrix} 1 & 0 & | & (a+b)/2 \\ -1 & 1 & | & (a-b)/2 \end{bmatrix}$$

We see that $x = (a+b)/2$ and $y = (a-b)/2$, so for any choice of a and b we have

$$\left(\frac{a+b}{2}\right)\begin{bmatrix} 1 \\ 1 \end{bmatrix} + \left(\frac{a-b}{2}\right)\begin{bmatrix} 1 \\ -1 \end{bmatrix} = \begin{bmatrix} a \\ b \end{bmatrix} \quad \text{Check this!}$$

11. Similar to Example 2.19, we must show $x \begin{bmatrix} 1 \\ 0 \\ 1 \end{bmatrix} + y \begin{bmatrix} 1 \\ 1 \\ 0 \end{bmatrix} + z \begin{bmatrix} 0 \\ 1 \\ 1 \end{bmatrix} = \begin{bmatrix} a \\ b \\ c \end{bmatrix}$ can always be solved.

The augmented matrix is $\begin{bmatrix} 1 & 1 & 0 & | & a \\ 0 & 1 & 1 & | & b \\ 1 & 0 & 1 & | & c \end{bmatrix}$, and row reduction produces:

$$\begin{bmatrix} 1 & 1 & 0 & | & a \\ 0 & 1 & 1 & | & b \\ 1 & 0 & 1 & | & c \end{bmatrix} \xrightarrow{R_3-R_1+R_2} \begin{bmatrix} 1 & 1 & 0 & | & a \\ 0 & 1 & 1 & | & b \\ 0 & 0 & 2 & | & -a+b+c \end{bmatrix} \xrightarrow{\frac{1}{2}R_3} \begin{bmatrix} 1 & 1 & 0 & | & a \\ 0 & 1 & 1 & | & b \\ 0 & 0 & 1 & | & (-a+b+c)/2 \end{bmatrix}$$

$$\xrightarrow{R_2-R_3} \begin{bmatrix} 1 & 1 & 0 & | & a \\ 0 & 1 & 0 & | & (a+b-c)/2 \\ 0 & 0 & 1 & | & (-a+b+c)/2 \end{bmatrix} \xrightarrow{R_1-R_2} \begin{bmatrix} 1 & 0 & 0 & | & (a-b+c)/2 \\ 0 & 1 & 0 & | & (a+b-c)/2 \\ 0 & 0 & 1 & | & (-a+b+c)/2 \end{bmatrix}$$

We see that $x = (a-b+c)/2$, $y = (a+b-c)/2$, and $z = (-a+b+c)/2$. So for any choice of a, b, and c we have:

$$\left(\frac{a-b+c}{2}\right)\begin{bmatrix} 1 \\ 0 \\ 1 \end{bmatrix} + \left(\frac{a+b-c}{2}\right)\begin{bmatrix} 1 \\ 1 \\ 0 \end{bmatrix} + \left(\frac{-a+b+c}{2}\right)\begin{bmatrix} 0 \\ 1 \\ 1 \end{bmatrix} = \begin{bmatrix} a \\ b \\ c \end{bmatrix} \quad \text{Check this!}$$

2.3 Spanning Sets and Linear Independence

13. We should describe the span of the given vectors (a) geometrically and (b) algebraically.

(a) Geometrically, we can see that the set of all linear combinations of $\begin{bmatrix} 2 \\ -4 \end{bmatrix}$ and $\begin{bmatrix} -1 \\ 2 \end{bmatrix}$ is just the line through the origin with $\begin{bmatrix} -1 \\ 2 \end{bmatrix}$ as direction vector.

Why do we not have to consider $\begin{bmatrix} 2 \\ -4 \end{bmatrix}$? Because $\begin{bmatrix} 2 \\ -4 \end{bmatrix} = -2 \begin{bmatrix} -1 \\ 2 \end{bmatrix}$.

(b) Algebraically, the vector equation of this line is $\begin{bmatrix} x \\ y \end{bmatrix} = \begin{bmatrix} -1 \\ 2 \end{bmatrix} t$.

That is just another way of saying that $\begin{bmatrix} x \\ y \end{bmatrix}$ is in the span of $\begin{bmatrix} -1 \\ 2 \end{bmatrix}$.

Suppose we want to obtain the general equation of this line.
One method is to use the system of equations arising from the vector equation:

$\begin{bmatrix} x \\ y \end{bmatrix} = \begin{bmatrix} -1 \\ 2 \end{bmatrix} t \Rightarrow \begin{matrix} x = -t \\ y = 2t \end{matrix}$ So $y = 2(-x) = -2x \Rightarrow 2x + y = 0$.

15. We should describe the span of the given vectors (a) geometrically and (b) algebraically.

(a) Geometrically, we can see that the set of all linear combinations of $\begin{bmatrix} 1 \\ 2 \\ 0 \end{bmatrix}$ and $\begin{bmatrix} 3 \\ 2 \\ -1 \end{bmatrix}$ is just the plane through the origin with $\begin{bmatrix} 1 \\ 2 \\ 0 \end{bmatrix}$ and $\begin{bmatrix} 3 \\ 2 \\ -1 \end{bmatrix}$ as direction vectors.

(b) Algebraically, the vector equation of this plane is $\begin{bmatrix} x \\ y \\ z \end{bmatrix} = s \begin{bmatrix} 1 \\ 2 \\ 0 \end{bmatrix} + t \begin{bmatrix} 3 \\ 2 \\ -1 \end{bmatrix}$.

That is just another way of saying that $\begin{bmatrix} x \\ y \\ z \end{bmatrix}$ is in the span of $\begin{bmatrix} 1 \\ 2 \\ 0 \end{bmatrix}$ and $\begin{bmatrix} 3 \\ 2 \\ -1 \end{bmatrix}$.

Suppose we want to obtain the general equation of this plane.
One method is to use the system of equations arising from the vector equation:

$\begin{matrix} s + 3t = x \\ 2s + 2t = y \\ -t = z \end{matrix} \Rightarrow \left[\begin{array}{cc|c} 1 & 3 & x \\ 2 & 2 & y \\ 0 & -1 & z \end{array} \right] \longrightarrow \left[\begin{array}{cc|c} 1 & 3 & x \\ 0 & -4 & -2x+y \\ 0 & 0 & (2x-y+4z)/4 \end{array} \right]$

We know this system is consistent, since $\begin{bmatrix} x \\ y \\ z \end{bmatrix}$ is in the span of $\begin{bmatrix} 1 \\ 2 \\ 0 \end{bmatrix}$ and $\begin{bmatrix} 3 \\ 2 \\ -1 \end{bmatrix}$.

So, we *must* have $2x - y + 4z = 0$, giving us the general equation we seek.

Note: Both $\begin{bmatrix} 1 \\ 2 \\ 0 \end{bmatrix}$ and $\begin{bmatrix} 3 \\ 2 \\ -1 \end{bmatrix}$ are orthogonal to $\begin{bmatrix} 2 \\ -1 \\ 4 \end{bmatrix}$. Should they be?

17. Since the three points $(1,0,3)$, $(-1,1,-3)$, and $(0,0,0)$ must lie in the plane, $(1,0,3)$ and $(-1,1,-3)$ must satisfy the equation of a plane through the origin $ax+by+cz=0$.

We substitute the two nonzero points into $ax+by+cz=0$ to create a homogenous system:
$$\begin{array}{r} a \quad\quad + 3c = 0 \\ -a + b - 3c = 0 \end{array} \Rightarrow \begin{bmatrix} 1 & 0 & 3 & | & 0 \\ -1 & 1 & -3 & | & 0 \end{bmatrix} \xrightarrow{R_2+R_1} \begin{bmatrix} 1 & 0 & 3 & | & 0 \\ 0 & 1 & 0 & | & 0 \end{bmatrix}$$

So, we have $b=0$ and $a=-3c$.

Letting $c=-1 \Rightarrow a=3$ and $b=0$ yields $3x+0y-z=3x-z=0$ is the general equation.

Q: Does the free variable c imply infinitely many planes contain these three points?
A: Hint: We can divide the general solution $-3cx+cz=0$ by $-c \neq 0 \Rightarrow 3x-z=0$.

19. To show $\mathbf{u}$, $\mathbf{v}$, and $\mathbf{w}$ are in span($\mathbf{u}$, $\mathbf{u}+\mathbf{v}$, $\mathbf{u}+\mathbf{v}+\mathbf{w}$), we must show that $\mathbf{u}$, $\mathbf{v}$, and $\mathbf{w}$ can be written as a linear combination of $\mathbf{u}$, $\mathbf{u}+\mathbf{v}$, $\mathbf{u}+\mathbf{v}+\mathbf{w}$.

Q: Can we simply let $\mathbf{u}=\mathbf{u}$, $\mathbf{v}=\mathbf{v}$, and $\mathbf{w}=\mathbf{w}$?
A: No. Why not? These vectors, except for $\mathbf{u}$, are not explicitly listed in the spanning set.

Instead, we need linear combinations of $\mathbf{u}$, $\mathbf{u}+\mathbf{v}$, $\mathbf{u}+\mathbf{v}+\mathbf{w}$ that yield $\mathbf{u}$, $\mathbf{v}$, and $\mathbf{w}$. So:
$\mathbf{u}=\mathbf{u}$, $\mathbf{v}=-\mathbf{u}+(\mathbf{u}+\mathbf{v})$, and $\mathbf{w}=-(\mathbf{u}+\mathbf{v})+(\mathbf{u}+\mathbf{v}+\mathbf{w})$.

Note: We have now shown that we can use $\mathbf{u}$, $\mathbf{v}$, and $\mathbf{w}$. How?

2.3 Spanning Sets and Linear Independence

21. When proving something for n, first let $n = 1$ or 2 to look for the underlying pattern.
    Assume that there are only two vectors $\mathbf{u}_1$ and $\mathbf{u}_2$ and two vectors $\mathbf{v}_1$ and $\mathbf{v}_2$.
    We are told $\mathbf{w}$ is a linear combination of $\mathbf{u}_1$ and $\mathbf{u}_2$. So: $\mathbf{w} = w_1\mathbf{u}_1 + w_2\mathbf{u}_2$.
    We are also told that both $\mathbf{u}_1$ and $\mathbf{u}_2$ are linear combinations of $\mathbf{v}_1$ and $\mathbf{v}_2$.
    So, we have both: $\mathbf{u}_1 = v_{11}\mathbf{v}_1 + v_{12}\mathbf{v}_2$ and $\mathbf{u}_2 = v_{21}\mathbf{v}_1 + v_{22}\mathbf{v}_2$.
    We need to show these assumptions imply $\mathbf{w}$ is a linear combination of $\mathbf{v}_1$ and $\mathbf{v}_2$. How?
    Let $\mathbf{u}_1 = u_{1_1}\mathbf{v}_1 + u_{1_2}\mathbf{v}_2$ and $\mathbf{u}_2 = u_{2_1}\mathbf{v}_1 + u_{2_2}\mathbf{v}_2$ in $\mathbf{w} = w_1\mathbf{u}_1 + w_2\mathbf{u}_2$.
    This substitution yields: $\mathbf{w} = w_1(u_{1_1}\mathbf{v}_1 + u_{1_2}\mathbf{v}_2) + w_2(u_{2_1}\mathbf{v}_1 + u_{2_2}\mathbf{v}_2)$.
    It is now obvious that $\mathbf{w}$ is a linear combination of $\mathbf{v}_1$ and $\mathbf{v}_2$. Why?
    Observe that this reasoning holds for any n and proceed to the proof.

    (a) Let $\mathbf{w} = w_1\mathbf{u}_1 + w_1\mathbf{u}_2 + \cdots + w_1\mathbf{u}_k$,
    and assume that each $\mathbf{u}_i$ is a linear combination of vectors $\mathbf{v}_1, \mathbf{v}_2, \ldots, \mathbf{v}_m$.
    Then each $\mathbf{u}_i = u_{i_1}\mathbf{v}_1 + u_{i_2}\mathbf{v}_2 + \cdots + u_{i_m}\mathbf{v}_m$, and

    $$\begin{aligned}
    \mathbf{w} &= w_1\mathbf{u}_1 + w_2\mathbf{u}_2 + \cdots + w_k\mathbf{u}_k \\
    &= w_1(u_{1_1}\mathbf{v}_1 + u_{1_2}\mathbf{v}_2 + \cdots + u_{1_m}\mathbf{v}_m) + w_2(u_{2_1}\mathbf{v}_1 + u_{2_2}\mathbf{v}_2 + \cdots + u_{2_m}\mathbf{v}_m) + \cdots \\
    &\quad \cdots + w_k(u_{k_1}\mathbf{v}_1 + u_{k_2}\mathbf{v}_2 + \cdots + u_{k_m}\mathbf{v}_m) \\
    &= (w_1u_{1_1} + w_2u_{2_1} + \cdots + w_ku_{k_1})\mathbf{v}_1 + (w_1u_{1_2} + w_2u_{2_2} + \cdots + w_ku_{k_2})\mathbf{v}_2 + \cdots \\
    &\quad \cdots + (w_1u_{1_m} + w_2u_{2_m} + \cdots + w_ku_{k_m})\mathbf{v}_m \\
    &= w'_1\mathbf{v}_1 + w'_2\mathbf{v}_2 + \cdots + w'_m\mathbf{v}_m.
    \end{aligned}$$

    So, any vector $\mathbf{w} \in \mathrm{span}\,(\mathbf{u}_1, \mathbf{u}_2, \ldots, \mathbf{u}_k)$ is also in $\mathrm{span}\,(\mathbf{v}_1, \mathbf{v}_2, \ldots, \mathbf{v}_m)$,
    and $\mathrm{span}\,(\mathbf{u}_1, \mathbf{u}_2, \ldots, \mathbf{u}_k) \subseteq \mathrm{span}\,(\mathbf{v}_1, \mathbf{v}_2, \ldots, \mathbf{v}_m)$.

    (b) Suppose that in addition to (a), each $\mathbf{v}_j$ is a linear combination of $\mathbf{u}_1, \mathbf{u}_2, \ldots, \mathbf{u}_k$.
    Let $\mathbf{w}$ be an arbitrary vector in $\mathrm{span}\,(\mathbf{v}_1, \mathbf{v}_2, \ldots, \mathbf{v}_m)$.
    Then $\mathbf{w} = w'_1\mathbf{v}_1 + w'_2\mathbf{v}_2 + \cdots + w'_m\mathbf{v}_m$, but each $\mathbf{v}_j = v_{j_1}\mathbf{u}_1 + v_{j_2}\mathbf{u}_2 + \cdots + v_{j_k}\mathbf{u}_k$, so

    $$\begin{aligned}
    \mathbf{w} &= w'_1(v_{1_1}\mathbf{u}_1 + v_{1_2}\mathbf{u}_2 + \cdots + v_{1_k}\mathbf{u}_k) + w'_2(v_{2_1}\mathbf{u}_1 + v_{2_2}\mathbf{u}_2 + \cdots + v_{2_k}\mathbf{u}_k) + \cdots \\
    &\quad \cdots + w'_m(v_{m_1}\mathbf{u}_1 + v_{m_2}\mathbf{u}_2 + \cdots + v_{m_k}\mathbf{u}_k) \\
    &= (w'_1v_{1_1} + w'_2v_{2_1} + \cdots + w'_mv_{m_1})\mathbf{u}_1 + (w'_1v_{1_2} + w'_2v_{2_2} + \cdots + w'_mv_{m_2})\mathbf{u}_2 + \cdots \\
    &\quad \cdots + (w'_1v_{1_k} + w'_2v_{2_k} + \cdots + w'_mv_{m_k})\mathbf{u}_k \\
    &= w_1\mathbf{u}_1 + w_1\mathbf{u}_2 + \cdots + w_k\mathbf{u}_k.
    \end{aligned}$$

    So, any vector $\mathbf{w} \in \mathrm{span}\,(\mathbf{v}_1, \mathbf{v}_2, \ldots, \mathbf{v}_m)$ is also in $\mathrm{span}\,(\mathbf{u}_1, \mathbf{u}_2, \ldots, \mathbf{u}_k)$,
    and $\mathrm{span}\,(\mathbf{v}_1, \mathbf{v}_2, \ldots, \mathbf{v}_m) \subseteq \mathrm{span}\,(\mathbf{u}_1, \mathbf{u}_2, \ldots, \mathbf{u}_k)$.
    But we already had $\mathrm{span}\,(\mathbf{u}_1, \mathbf{u}_2, \ldots, \mathbf{u}_k) \subseteq \mathrm{span}\,(\mathbf{v}_1, \mathbf{v}_2, \ldots, \mathbf{v}_m)$,
    so $\mathrm{span}\,(\mathbf{u}_1, \mathbf{u}_2, \ldots, \mathbf{u}_k) = \mathrm{span}\,(\mathbf{v}_1, \mathbf{v}_2, \ldots, \mathbf{v}_m)$.

    (c) Need only show $\mathbf{e}_1, \mathbf{e}_2, \mathbf{e}_3$ are linear combinations of $\mathbf{v}_1 = \begin{bmatrix} 1 \\ 0 \\ 0 \end{bmatrix}$, $\mathbf{v}_2 = \begin{bmatrix} 1 \\ 1 \\ 0 \end{bmatrix}$, $\mathbf{v}_3 = \begin{bmatrix} 1 \\ 1 \\ 1 \end{bmatrix}$.

    That's obvious since $\mathbf{e}_1 = \mathbf{v}_1$, $\mathbf{e}_2 = \mathbf{v}_2 - \mathbf{v}_1$, and $\mathbf{e}_3 = \mathbf{v}_3 - \mathbf{v}_2$. Why is that enough?
    Because then we have $\mathbb{R}^3 = \mathrm{span}(\mathbf{e}_1, \mathbf{e}_2, \mathbf{e}_3) = \mathrm{span}(\mathbf{v}_1, \mathbf{v}_2, \mathbf{v}_3)$.

23. Since there is no obvious dependence relation here, we follow Example 2.23.

Find scalars c_1, c_2, and c_3 such that: $c_1 \begin{bmatrix} 1 \\ 1 \\ 1 \end{bmatrix} + c_2 \begin{bmatrix} 1 \\ 2 \\ 3 \end{bmatrix} + c_3 \begin{bmatrix} 1 \\ -1 \\ 2 \end{bmatrix} = \begin{bmatrix} 0 \\ 0 \\ 0 \end{bmatrix}$.

Form the linear system, its associated augmented matrix, and row reduce to solve:

$$\begin{matrix} c_1 + c_2 + c_3 = 0 \\ c_1 + 2c_2 - c_3 = 0 \\ c_1 + 3c_2 + 2c_3 = 0 \end{matrix} \Rightarrow \begin{bmatrix} 1 & 1 & 1 & 0 \\ 1 & 2 & -1 & 0 \\ 1 & 3 & 2 & 0 \end{bmatrix} \longrightarrow \begin{bmatrix} 1 & 0 & 0 & 0 \\ 0 & 1 & 0 & 0 \\ 0 & 0 & 1 & 0 \end{bmatrix}$$

Since $c_1 = c_2 = c_3 = 0$ is the unique solution, the vectors are linearly independent.

25. The vectors $\mathbf{v}_1 = \begin{bmatrix} 0 \\ 1 \\ 2 \end{bmatrix}$, $\mathbf{v}_2 = \begin{bmatrix} 2 \\ 1 \\ 3 \end{bmatrix}$ and $\mathbf{v}_3 = \begin{bmatrix} 2 \\ 0 \\ 1 \end{bmatrix}$ are linearly dependent.

This can be determined by inspection because $\mathbf{v}_1 - \mathbf{v}_2 + \mathbf{v}_3 = \mathbf{0}$.

27. The vectors $\mathbf{v}_1 = \begin{bmatrix} 3 \\ 4 \\ 5 \end{bmatrix}$, $\mathbf{v}_2 = \begin{bmatrix} 6 \\ 7 \\ 8 \end{bmatrix}$, $\mathbf{v}_3 = \begin{bmatrix} 0 \\ 0 \\ 0 \end{bmatrix}$ are linearly dependent.

This can be determined by inspection because $\mathbf{v}_3$ is the zero vector. Why is that enough? Because $0\mathbf{v}_1 + 0\mathbf{v}_2 + \mathbf{v}_3 = \mathbf{0}$.
Any set of vectors containing the zero vector is linearly dependent. Why?

29. Since there is no obvious dependence relation here, we follow Example 2.23.

Find scalars c_1, c_2, c_3, c_4 such that: $c_1 \begin{bmatrix} 1 \\ -1 \\ 1 \\ 0 \end{bmatrix} + c_2 \begin{bmatrix} -1 \\ 1 \\ 0 \\ 1 \end{bmatrix} + c_3 \begin{bmatrix} 1 \\ 0 \\ 1 \\ -1 \end{bmatrix} + c_4 \begin{bmatrix} 0 \\ 1 \\ -1 \\ 1 \end{bmatrix} = \begin{bmatrix} 0 \\ 0 \\ 0 \\ 0 \end{bmatrix}$.

Form the linear system, its associated augmented matrix, and row reduce to solve:

$$\begin{matrix} c_1 - c_2 + c_3 = 0 \\ -c_1 + c_2 + c_4 = 0 \\ c_1 + c_3 - c_4 = 0 \\ c_2 - c_3 + c_4 = 0 \end{matrix} \Rightarrow \begin{bmatrix} 1 & -1 & 1 & 0 & 0 \\ -1 & 1 & 0 & 1 & 0 \\ 1 & 0 & 1 & -1 & 0 \\ 0 & 1 & -1 & 1 & 0 \end{bmatrix} \longrightarrow \begin{bmatrix} 1 & 0 & 0 & 0 & 0 \\ 0 & 1 & 0 & 0 & 0 \\ 0 & 0 & 1 & 0 & 0 \\ 0 & 0 & 0 & 1 & 0 \end{bmatrix}$$

Since $c_1 = c_2 = c_3 = c_4 = 0$ is the unique solution, the vectors are linearly independent.

31. The vectors $\mathbf{v}_1 = \begin{bmatrix} 3 \\ -1 \\ 1 \\ -1 \end{bmatrix}$, $\mathbf{v}_2 = \begin{bmatrix} -1 \\ 3 \\ 1 \\ -1 \end{bmatrix}$, $\mathbf{v}_3 = \begin{bmatrix} 1 \\ 1 \\ 3 \\ 1 \end{bmatrix}$, and $\mathbf{v}_4 = \begin{bmatrix} -1 \\ -1 \\ 1 \\ 3 \end{bmatrix}$ are linearly dependent.

This can be determined by inspection because $\mathbf{v}_1 + \mathbf{v}_2 - \mathbf{v}_3 + \mathbf{v}_4 = \mathbf{0}$.

2.3 Spanning Sets and Linear Independence

33. Exercises 32 through 41 provide a check on our solutions to Exercises 22 through 31. How? In these exercises the directions tell us to follow Example 2.25 and apply Theorem 2.7:

We construct a matrix with these vectors as its rows and proceed to reduce it to echelon form. Each time a row changes, we denote the new row by adding a prime symbol:

$$A = \begin{bmatrix} \mathbf{v}_1 \\ \mathbf{v}_2 \\ \mathbf{v}_3 \end{bmatrix} = \begin{bmatrix} 1 & 1 & 1 \\ 1 & 2 & 3 \\ 1 & -1 & 2 \end{bmatrix} \xrightarrow{\substack{R'_2 = R_2 - R_1 \\ R'_3 = R_3 - R_1}} \begin{bmatrix} 1 & 1 & 1 \\ 0 & 1 & 2 \\ 0 & -2 & 1 \end{bmatrix} \xrightarrow{R''_3 = R'_3 + 2R'_2} \begin{bmatrix} 1 & 1 & 1 \\ 0 & 1 & 2 \\ 0 & 0 & 5 \end{bmatrix}$$

We can stop. Why? We have put A into row echelon form. How can we tell?
Since the rank of a matrix is the number of nonzero rows in its row echelon form, rank$(A) = 3$.
What do we conclude? We conclude $\mathbf{v}_1$, $\mathbf{v}_2$, and $\mathbf{v}_3$ are linearly independent. How?

Theorem 2.7 states that $\mathbf{v}_1, \mathbf{v}_2, \ldots, \mathbf{v}_m$ are linearly dependent if and only if rank$(A) < m$. But that implies the following: If rank$(A) \geq m$, $\mathbf{v}_1, \mathbf{v}_2, \ldots, \mathbf{v}_m$ are linearly independent.

In this case, therefore, we argue as follows:
Since rank$(A) \geq 3$, Theorem 2.7 implies $\mathbf{v}_1$, $\mathbf{v}_2$, and $\mathbf{v}_3$ are linearly independent.

Does the agree with the solution we found in Exercise 23? It should.
Which method was easier for this Exercise? Why?

35. Exercises 32 through 41 provide a check on our solutions to Exercises 22 through 31. How? In these exercises the directions tell us to follow Example 2.25 and apply Theorem 2.7:

We construct a matrix with these vectors as its rows and proceed to reduce it to echelon form. Each time a row changes, we denote the new row by adding a prime symbol:

$$A = \begin{bmatrix} \mathbf{v}_1 \\ \mathbf{v}_2 \\ \mathbf{v}_3 \end{bmatrix} = \begin{bmatrix} 0 & 1 & 2 \\ 2 & 1 & 3 \\ 2 & 0 & 1 \end{bmatrix} \xrightarrow{\substack{R'_1 = R_2 \\ R'_2 = R_1}} \begin{bmatrix} 2 & 1 & 3 \\ 0 & 1 & 2 \\ 2 & 0 & 1 \end{bmatrix} \xrightarrow{R'_3 = R_3 - R'_1} \begin{bmatrix} 2 & 1 & 3 \\ 0 & 1 & 2 \\ 0 & -1 & -2 \end{bmatrix} \xrightarrow{R''_3 = R'_3 + R'_2} \begin{bmatrix} 2 & 1 & 3 \\ 0 & 1 & 2 \\ 0 & 0 & 0 \end{bmatrix}$$

We can stop. Why? We have created a zero row. What does that tell us?
Since the rank of a matrix is the number of nonzero rows in its row echelon form, rank$(A) = 2$.
What do we conclude? We conclude $\mathbf{v}_1$, $\mathbf{v}_2$, and $\mathbf{v}_3$ are linearly dependent. How?

Theorem 2.7 states that $\mathbf{v}_1, \mathbf{v}_2, \ldots, \mathbf{v}_m$ are linearly dependent if and only if rank$(A) < m$.
So, since rank$(A) = 2 < 3$, Theorem 2.7 implies $\mathbf{v}_1$, $\mathbf{v}_2$, and $\mathbf{v}_3$ are linearly dependent.

From the row reduction, we see: $\mathbf{0} = R'''_3 = R'_3 + R'_2 = (R'_3 - R'_1) + R'_2 = R_3 - R_2 + R_1$.

This equation yields a dependence relation among the original vectors:

$$\begin{bmatrix} 0 \\ 1 \\ 2 \end{bmatrix} - \begin{bmatrix} 2 \\ 1 \\ 3 \end{bmatrix} + \begin{bmatrix} 2 \\ 0 \\ 1 \end{bmatrix} = \begin{bmatrix} 0 \\ 0 \\ 0 \end{bmatrix}.$$ Compare this result to Exercise 25. Does it agree?

37. Exercises 32 through 41 provide a check on our solutions to Exercises 22 through 31. How? In these exercises the directions tell us to follow Example 2.25 and apply Theorem 2.7:

We construct a matrix with these vectors as its rows and proceed to reduce it to echelon form. Each time a row changes, we denote the new row by adding a prime symbol:

$$A = \begin{bmatrix} \mathbf{v}_1 \\ \mathbf{v}_2 \\ \mathbf{v}_3 \end{bmatrix} = \begin{bmatrix} 3 & 4 & 5 \\ 6 & 7 & 8 \\ 0 & 0 & 0 \end{bmatrix}$$

We can stop. Why? The matrix A has a zero row. What does that tell us?
Since the rank of a matrix is the number of nonzero rows in its row echelon form, $\text{rank}(A) \leq 2$. What do we conclude? We conclude $\mathbf{v}_1$, $\mathbf{v}_2$, and $\mathbf{v}_3$ are linearly dependent. How?

Theorem 2.7 states that $\mathbf{v}_1, \mathbf{v}_2, \ldots, \mathbf{v}_m$ are linearly dependent if and only if $\text{rank}(A) < m$. So, since $\text{rank}(A) \leq 2 < 3$, Theorem 2.7 implies $\mathbf{v}_1$, $\mathbf{v}_2$, and $\mathbf{v}_3$ are linearly dependent.

Furthermore, we have the obvious dependence relation among the original vectors:

$$0 \begin{bmatrix} 3 \\ 4 \\ 5 \end{bmatrix} + 0 \begin{bmatrix} 6 \\ 7 \\ 8 \end{bmatrix} + \begin{bmatrix} 0 \\ 0 \\ 0 \end{bmatrix} = \begin{bmatrix} 0 \\ 0 \\ 0 \end{bmatrix}.$$ Compare this result to Exercise 37. Does it agree?

This can be determined by inspection because $\mathbf{v}_3$ is the zero vector. Why is that enough? Any set of vectors containing the zero vector is linearly dependent. Why?

39. Exercises 32 through 41 provide a check on our solutions to Exercises 22 through 31. How? In these exercises the directions tell us to follow Example 2.25 and apply Theorem 2.7:

We construct a matrix with these vectors as its rows and proceed to reduce it to echelon form. Each time a row changes, we denote the new row by adding a prime symbol:

$$A = \begin{bmatrix} \mathbf{v}_1 \\ \mathbf{v}_2 \\ \mathbf{v}_3 \\ \mathbf{v}_4 \end{bmatrix} = \begin{bmatrix} 1 & -1 & 1 & 0 \\ -1 & 1 & 0 & 1 \\ 1 & 0 & 1 & -1 \\ 0 & 1 & -1 & 1 \end{bmatrix} \xrightarrow{\substack{R_2'=R_2+R_1 \\ R_3'=R_3-R_1}} \begin{bmatrix} 1 & -1 & 1 & 0 \\ 0 & 0 & 1 & 1 \\ 0 & 1 & 0 & -1 \\ 0 & 1 & -1 & 1 \end{bmatrix} \xrightarrow{R_4'=R_4-R_3'} \begin{bmatrix} 1 & -1 & 1 & 0 \\ 0 & 0 & 1 & 1 \\ 0 & 1 & 0 & -1 \\ 0 & 0 & -1 & 2 \end{bmatrix}$$

$$\xrightarrow{\substack{R_2''=R_3' \\ R_3''=R_2'}} \begin{bmatrix} 1 & -1 & 1 & 0 \\ 0 & 1 & 0 & -1 \\ 0 & 0 & 1 & 1 \\ 0 & 0 & -1 & 2 \end{bmatrix} \xrightarrow{R_4'''=R_4'+R_3''} \begin{bmatrix} 1 & -1 & 1 & 0 \\ 0 & 1 & 0 & -1 \\ 0 & 0 & 1 & 1 \\ 0 & 0 & 0 & 3 \end{bmatrix}$$

We can stop. Why? We have put A into row echelon form. How can we tell?
Since the rank of a matrix is the number of nonzero rows in its row echelon form, $\text{rank}(A) = 4$. What do we conclude? We conclude $\mathbf{v}_1$, $\mathbf{v}_2$, $\mathbf{v}_3$, and $\mathbf{v}_4$ are linearly independent.

Theorem 2.7 states that $\mathbf{v}_1, \mathbf{v}_2, \ldots, \mathbf{v}_m$ are linearly dependent if and only if $\text{rank}(A) < m$. But that implies the following: If $\text{rank}(A) \geq m$, $\mathbf{v}_1, \mathbf{v}_2, \ldots, \mathbf{v}_m$ are linearly independent.

In this case, therefore, we argue as follows:
Since $\text{rank}(A) \geq 4$, Theorem 2.7 implies $\mathbf{v}_1$, $\mathbf{v}_2$, $\mathbf{v}_3$, and $\mathbf{v}_4$ are linearly independent.

Does the agree with the solution we found in Exercise 29? It should.
Which method was easier for this Exercise? Why?

2.3 Spanning Sets and Linear Independence

41. Exercises 32 through 41 provide a check on our solutions to Exercises 22 through 31. How? In these exercises the directions tell us to follow Example 2.25 and apply Theorem 2.7:

We construct a matrix with these vectors as its rows and proceed to reduce it to echelon form. Each time a row changes, we denote the new row by adding a prime symbol:

$$A = \begin{bmatrix} \mathbf{v}_1 \\ \mathbf{v}_2 \\ \mathbf{v}_3 \\ \mathbf{v}_4 \end{bmatrix} = \begin{bmatrix} 3 & -1 & 1 & -1 \\ -1 & 3 & 1 & -1 \\ 1 & 1 & 3 & 1 \\ -1 & -1 & 1 & 3 \end{bmatrix} \xrightarrow[R'_3=R_1]{R'_1=R_3} \begin{bmatrix} 1 & 1 & 3 & 1 \\ -1 & 3 & 1 & -1 \\ 3 & -1 & 1 & -1 \\ -1 & -1 & 1 & 3 \end{bmatrix} \xrightarrow[R'''_3=R'_3-3R'_1]{R'_2=R_2+R'_1} \begin{bmatrix} 1 & 1 & 3 & 1 \\ 0 & 4 & 4 & 0 \\ 0 & -4 & -8 & -4 \\ 0 & 0 & 4 & 4 \end{bmatrix}$$

$$\xrightarrow{R'''_3=R''_3+R'_2} \begin{bmatrix} 1 & 1 & 3 & 1 \\ 0 & 4 & 4 & 0 \\ 0 & 0 & -4 & -4 \\ 0 & 0 & 4 & 4 \end{bmatrix} \xrightarrow{R''''_4=R'_4+R'''_3} \begin{bmatrix} 1 & 1 & 3 & 1 \\ 0 & 4 & 4 & 0 \\ 0 & 0 & -4 & -4 \\ 0 & 0 & 0 & 0 \end{bmatrix}$$

We can stop. Why? We have created a zero row. What does that tell us?
Since the rank of a matrix is the number of nonzero rows in its row echelon form, rank$(A) = 3$.
What do we conclude? We conclude $\mathbf{v}_1$, $\mathbf{v}_2$, $\mathbf{v}_3$, and $\mathbf{v}_4$ are linearly dependent.

Theorem 2.7 states that $\mathbf{v}_1, \mathbf{v}_2, \ldots, \mathbf{v}_m$ are linearly dependent if and only if rank$(A) < m$. So, since rank$(A) = 3 < 4$, Theorem 2.7 implies $\mathbf{v}_1$, $\mathbf{v}_2$, $\mathbf{v}_3$, and $\mathbf{v}_3$ are linearly dependent.

From the row reduction above, we see: $\mathbf{0} = R''''_4 = R'_4 + R'''_3$. Substituting, we have:

$$\mathbf{0} = (R_4 + R'_1) + (R''_3 + R'_2) = (R_4 + R'_1) + (R'_3 - 3R'_1) + (R_2 + R'_1)$$
$$= R_4 + R'_3 + R_2 - R'_1 = R_4 + R_1 + R_2 - R_3.$$

This equation yields a dependence relation among the original vectors:

$$\begin{bmatrix} 3 \\ -1 \\ 1 \\ -1 \end{bmatrix} + \begin{bmatrix} -1 \\ 3 \\ 1 \\ -1 \end{bmatrix} - \begin{bmatrix} 1 \\ 1 \\ 3 \\ 1 \end{bmatrix} + \begin{bmatrix} -1 \\ -1 \\ 1 \\ 3 \end{bmatrix} = \begin{bmatrix} 0 \\ 0 \\ 0 \\ 0 \end{bmatrix}.$$ Does this agree with Exercise 31?

43. We apply the definition of linear independence and Examples 2.23 and 2.25 to prove our claims.

   (a) We will show that $\mathbf{u}+\mathbf{v}$, $\mathbf{v}+\mathbf{w}$, and $\mathbf{u}+\mathbf{w}$ are linearly independent.
   Given $c_1(\mathbf{u}+\mathbf{v})+c_2(\mathbf{v}+\mathbf{w})+c_3(\mathbf{u}+\mathbf{w})=0$, we will show $c_1=c_2=c_3=0$.
   Multiplying and gathering like terms yields: $(c_1+c_3)\mathbf{u}+(c_1+c_2)\mathbf{v}+(c_2+c_3)\mathbf{w}=0$.
   Since $\mathbf{u}$, $\mathbf{v}$, and $\mathbf{w}$ are linearly independent, $c_1+c_3=c_1+c_2=c_2+c_3=0$.
   We create the matrix of coefficients A and row reduce to determine its rank:
   $$\begin{array}{rl} c_1 \phantom{{}+c_2} + c_3 & = 0 \\ c_1 + c_2 \phantom{{}+c_3} & = 0 \\ \phantom{c_1+{}} c_2 + c_3 & = 0 \end{array} \Rightarrow \begin{bmatrix} 1 & 0 & 1 \\ 1 & 1 & 0 \\ 0 & 1 & 1 \end{bmatrix} \longrightarrow \begin{bmatrix} 1 & 0 & 1 \\ 0 & 1 & -1 \\ 0 & 0 & 2 \end{bmatrix}$$
   Since $\text{rank}(A)=3$ the only solution is the trivial one, so $c_1=c_2=c_3=0$.

   (b) We will show that $\mathbf{u}-\mathbf{v}$, $\mathbf{v}-\mathbf{w}$, and $\mathbf{u}-\mathbf{w}$ are linearly dependent.
   Given $c_1(\mathbf{u}-\mathbf{v})+c_2(\mathbf{v}-\mathbf{w})+c_3(\mathbf{u}-\mathbf{w})=0$, we will show $c_1=c_2=-c_3$.
   Multiplying and gathering like terms yields: $(c_1+c_3)\mathbf{u}+(-c_1+c_2)\mathbf{v}+(-c_2-c_3)\mathbf{w}=0$.
   Since $\mathbf{u}$, $\mathbf{v}$, and $\mathbf{w}$ are linearly independent, $c_1+c_3=-c_1+c_2=-c_2-c_3=0$.
   We form the augmented matrix and row reduce to solve:
   $$\begin{array}{rl} c_1 \phantom{{}+c_2} + c_3 & = 0 \\ -c_1 + c_2 \phantom{{}+c_3} & = 0 \\ \phantom{-c_1+{}} -c_2 - c_3 & = 0 \end{array} \Rightarrow \left[\begin{array}{ccc|c} 1 & 0 & 1 & 0 \\ -1 & 1 & 0 & 0 \\ 0 & -1 & -1 & 0 \end{array}\right] \longrightarrow \left[\begin{array}{ccc|c} 1 & 0 & 1 & 0 \\ 0 & 1 & 1 & 0 \\ 0 & 0 & 0 & 0 \end{array}\right]$$
   This clearly has the solution $c_1=c_2=-c_3$ as we were to show.

45. We will follow the proof of Theorem 2.8 and use the assumption that $n<m$.

   PROOF: Let A be the $m \times n$ matrix with vectors $\mathbf{v}_1, \mathbf{v}_2, \ldots, \mathbf{v}_m$ in $\mathbb{R}^n$ as its rows.

   By Theorem 2.7, $\mathbf{v}_1, \mathbf{v}_2, \ldots, \mathbf{v}_m$ are linearly dependent if and only if $\text{rank}(A)<m$.
   By definition, $\text{rank}(A)$ = number of nonzero rows its its row echelon form. But, the definition of row echelon form implies that $\text{rank}(A)$ is always $\leq$ number of columns in $A=n$. Why?

   So, we have $\text{rank}(A) \leq n < m$, as required to show $\mathbf{v}_1, \mathbf{v}_2, \ldots, \mathbf{v}_m$ are linearly dependent.

47. As suggested in the hint, we will use the result of Exercise 21(b).

   PROOF:

   Since $S'=\{\mathbf{v}_1,\ldots,\mathbf{v}_k\} \subseteq S=\{\mathbf{v}_1,\ldots,\mathbf{v}_k,\mathbf{v}\}$, Exercise 21(b) implies $\text{span}(S') \subseteq \text{span}(S)$.
   So, we need only show $\text{span}(S) \subseteq \text{span}(S')$ to prove $\text{span}(S)=\text{span}(S')$.

   Since $\mathbf{v}$ is a linear combination of $\mathbf{v}_1,\ldots,\mathbf{v}_k$, Exercise 21(b) implies $\text{span}(S) \subseteq \text{span}(S')$.
   Therefore, $\text{span}(S)=\text{span}(S')$.

2.4 Applications

Q: What is our goal for Section 2.4?

A: To see how matrices apply to economics, chemistry, networking and games.
Since these applications are fully explained in the text, we simply present solutions here.

Key Definitions and Concepts

branch	p.104	2.4	a directed edge of a network
conservation of flow	p.104	2.4	at each node, the flow in equals the flow out
Current Law (nodes)	p.106	2.4	sum of currents flowing into a node equals sum out
Kirchoff's Laws	p.106	2.4	Current Law (nodes) and Voltage Law (circuits)
network	p.104	2.4	nodes connected by a series of branches
node	p.104	2.4	point where branches of a network meet
Ohm's Law	p.106	2.4	force = resistance × current, $E = RI$
Voltage Law	p.106	2.4	voltage drops around a circuit equals total voltage

Solutions to odd-numbered exercises from Section 2.4

1. Let x_1, x_2, and x_3 be the number of bacteria of strains I, II, and III, respectively. Then from the consumption of A, B, and C, we get the following system:

$$\begin{array}{rl} x_1 + 2x_2 & = 400 \\ 2x_1 + x_2 + x_3 & = 600 \\ x_1 + x_2 + 2x_2 & = 600 \end{array} \Rightarrow \begin{bmatrix} 1 & 2 & 0 & | & 400 \\ 2 & 1 & 1 & | & 600 \\ 1 & 1 & 2 & | & 600 \end{bmatrix} \longrightarrow \begin{bmatrix} 1 & 0 & 0 & | & 160 \\ 0 & 1 & 0 & | & 120 \\ 0 & 0 & 1 & | & 160 \end{bmatrix}.$$

So, 160, 120, and 160 bacteria of strains I, II, and III respectively can coexist.

3. Let x_1, x_2, and x_3 be the number of small, medium, and large arrangements. Then from the consumption of flowers in each arrangement we get:

$$\begin{array}{rl} x_1 + 2x_2 + 4x_3 & = 24 \\ 3x_1 + 4x_2 + 8x_3 & = 50 \\ 3x_1 + 6x_2 + 6x_2 & = 48 \end{array} \Rightarrow \begin{bmatrix} 1 & 2 & 4 & | & 24 \\ 3 & 4 & 8 & | & 50 \\ 3 & 6 & 6 & | & 48 \end{bmatrix} \longrightarrow \begin{bmatrix} 1 & 0 & 0 & | & 2 \\ 0 & 1 & 0 & | & 3 \\ 0 & 0 & 1 & | & 4 \end{bmatrix}.$$

So, 2 small, 3 medium, and 4 large arrangements were sold that day.

5. Let x_1, x_2, and x_3 be the number of house, special, and gourmet blends. Then from the consumption of beans in each blend we get

$$\begin{array}{rl} 300x_1 + 200x_2 + 100x_3 & = 30,000 \\ 200x_2 + 200x_3 & = 15,000 \\ 200x_1 + 100x_2 + 200x_3 & = 25,000 \end{array}$$

Form augmented matrix and reduce it: $\begin{bmatrix} 300 & 200 & 100 & | & 30,000 \\ 0 & 200 & 200 & | & 15,000 \\ 200 & 100 & 200 & | & 25,000 \end{bmatrix} \longrightarrow \begin{bmatrix} 1 & 0 & 0 & | & 65 \\ 0 & 1 & 0 & | & 30 \\ 0 & 0 & 1 & | & 45 \end{bmatrix}.$

The merchant should make 65 house blend, 30 special blend, and 45 gourmet blend.

7. Let x, y, z, and w be the number of FeS_2, O_2, Fe_2O_3, and SO_2 molecules respectively. Then, compare the number of iron, sulfur, and oxygen atoms in reactants and products:

$$\begin{array}{rl} \text{Iron}: & x = 2z \\ \text{Sulfur}: & 2x = w \\ \text{Oxygen}: & 2y = 3z + 2w \end{array} \Rightarrow \begin{bmatrix} 1 & 0 & -2 & 0 & | & 0 \\ 2 & 0 & 0 & -1 & | & 0 \\ 0 & 2 & -3 & -2 & | & 0 \end{bmatrix} \longrightarrow \begin{bmatrix} 1 & 0 & 0 & -\frac{1}{2} & | & 0 \\ 0 & 1 & 0 & -\frac{11}{8} & | & 0 \\ 0 & 0 & 1 & -\frac{1}{4} & | & 0 \end{bmatrix}.$$

Thus $z = \frac{1}{4}w$, $y = \frac{11}{8}w$, and $x = \frac{1}{2}w$.
The smallest positive value of w that will produce integer values for all four variables is the least common denominator of $\frac{1}{2}$, $\frac{11}{8}$, $\frac{1}{2}$ $\Rightarrow w = 8$, $x = 4$, $y = 11$, and $z = 2$.
Therefore, the balanced chemical equation is $4FeS_2 + 11O_2 \longrightarrow 2Fe_2O_3 + 8SO_2$.

2.4 Applications

9. $2C_4H_{10} + 13O_2 \longrightarrow 8CO_2 + 10H_2O$

11. $2C_5H_{11}OH + 15O_2 \longrightarrow 12H_2O + 10CO_2$

13. $Na_2CO_3 + 4C + N_2 \longrightarrow 2NaCN + 3CO$

15. (a) By applying the conservation of flow rule to each node we obtain the system of equations

$$f_1 + f_2 = 20 \qquad f_2 - f_3 = -10 \qquad f_1 + f_3 = 30$$

We form the augmented matrix and perform Gauss-Jordan elimination to get

$$\begin{bmatrix} 1 & 1 & 0 & | & 20 \\ 0 & 1 & -1 & | & -10 \\ 1 & 0 & 1 & | & 30 \end{bmatrix} \longrightarrow \begin{bmatrix} 1 & 0 & 1 & | & 30 \\ 0 & 1 & -1 & | & -10 \\ 0 & 0 & 0 & | & 0 \end{bmatrix}$$

Letting $f_3 = t$, the possible flows are $f_1 = 30 - t$, $f_2 = t - 10$, $f_3 = t$.

(b) In this case $f_2 = 5$, but $f_2 = t - 10$ so $t = 15$. Then the other flows are $f_1 = f_3 = 15$.

(c) Each flow must be nonnegative $\Rightarrow t \le 30$, $t \ge 10$, $t \ge 0$
Thus $10 \le t \le 30$, so $0 \le f_1 \le 20$, $0 \le f_2 \le 20$, $10 \le f_3 \le 30$.

(d) A negative flow, if allowed, would indicate a transport in the opposite direction. For example, a negative flow into a node is the same as a positive flow out of a node. So, if $f_2 < 0$, then the arrow on f_2 could be changed and the flow taken as positive.

17. (a) By applying the conservation of flow rule to each node we obtain the system of equations

$$f_1 + f_2 = 100 \qquad f_2 + f_3 = f_4 + 150 \qquad f_4 + f_5 = 150 \qquad f_1 + 200 = f_3 + f_5$$

We rearrange the equations, and perform Gauss-Jordan elimination

$$\begin{bmatrix} 1 & 1 & 0 & 0 & 0 & | & 100 \\ 0 & 1 & 1 & -1 & 0 & | & 150 \\ 0 & 0 & 0 & 1 & 1 & | & 150 \\ 1 & 0 & -1 & 0 & -1 & | & -200 \end{bmatrix} \longrightarrow \begin{bmatrix} 1 & 0 & -1 & 0 & -1 & | & -200 \\ 0 & 1 & 1 & 0 & 1 & | & 300 \\ 0 & 0 & 0 & 1 & 1 & | & 150 \\ 0 & 0 & 0 & 0 & 0 & | & 0 \end{bmatrix}$$

We parameterize the solution by letting $f_5 = t$, and $f_3 = s$, so the possible flows are $f_1 = -200 + s + t$, $f_2 = 300 - s - t$, $f_3 = s$, $f_4 = 150 - t$, $f_5 = t$.

(b) If DC is closed then $f_5 = t = 0$, so the flow through DB must be $200 \le f_3 \le 300 \; \frac{liters}{day}$.

(c) If DB were closed, then f_5 must carry away at least $200 \; \frac{liters}{day}$.
But node C has a maximum outflow of $150 \; \frac{liters}{day}$, so, it will not be able to handle the inflow from f_5.
Thus DB cannot be closed.

From the solution in (a), with DB closed $f_3 = s = 0$, and the solution becomes $f_1 = -200 + t$, $f_2 = 300 - t$, $f_3 = 0$, $f_4 = 150 - t$, $f_5 = t$.
But each flow must be positive, so we get the following constraints on t:
$t \ge 200$, $t \le 300$, $t \le 150$, $t \ge 0$.
But this gives rise to a contradiction since it demands $t \le 150$ and $t \ge 200$, which is impossible, and again we see that DB cannot be closed.

(d) Each flow must be nonnegative, so the solutions give the following constraints:
$s \ge 200 - t$, $s \le 300 - t$, $s \ge 0$, $t \le 150$, $t \ge 0$.
From these we see that $0 \le t \le 150$ and $50 \le s \le 300$.
Combining the constraints on s and t with the five solutions,
we see that $0 \le f_1 \le 100$, $0 \le f_2 \le 100$, $50 \le f_3 \le 300$, $0 \le f_4 \le 150$, $0 \le f_5 \le 150$.

19. Applying the current law to node A gives the equation $I_1 + I_3 = I_2$ or $I_1 - I_2 + I_3 = 0$.
Applying the voltage law to the top circuit (circuit $ABCA$) gives $-I_2 - I_1 + 8 = 0$.
Similarly, for the circuit $ABDA$ we obtain $-I_2 + 13 - 4I_3 = 0$.
We get the following system:

$$\begin{array}{c} I_1 - I_2 + I_3 = 0 \\ I_1 + I_2 = 8 \\ I_2 + 4I_3 = 13 \end{array} \Rightarrow \begin{bmatrix} 1 & -1 & 1 & | & 0 \\ 1 & 1 & 0 & | & 8 \\ 0 & 1 & 4 & | & 13 \end{bmatrix} \longrightarrow \begin{bmatrix} 1 & 0 & 0 & | & 3 \\ 0 & 1 & 0 & | & 5 \\ 0 & 0 & 1 & | & 2 \end{bmatrix} \Rightarrow$$

$I_1 = 3$ amps, $I_2 = 5$ amps, and $I_3 = 2$ amps.

2.4 Applications

21. (a) Applying the current and voltage laws to the circuit gives the system:

Node B : $I = I_1 + I_4$
Node C : $I_1 = I_1 + I_4$
Node D : $I_2 + I_5 = I$
Node E : $I_3 + I_4 = I_5$

$\Rightarrow$

Circuit ABEDA : $-2I_4 - I_5 + 14 = 0$
Circuit BCEB : $-I_1 - I_3 + 2I_4 = 0$
Circuit CDEC : $-2I_2 + I_5 + I_3 = 0$

$\Rightarrow$

Gauss-Jordan elimination gives

$$\begin{bmatrix} 1 & -1 & 0 & 0 & -1 & 0 & | & 0 \\ 0 & 1 & -1 & -1 & 0 & 0 & | & 0 \\ 1 & 0 & -1 & 0 & 0 & -1 & | & 0 \\ 0 & 0 & 0 & 1 & 1 & -1 & | & 0 \\ 0 & 0 & 0 & 0 & 2 & 1 & | & 14 \\ 0 & 1 & 0 & 1 & -2 & 0 & | & 0 \\ 0 & 0 & 2 & -1 & 0 & -1 & | & 0 \end{bmatrix} \longrightarrow \begin{bmatrix} 1 & 0 & 0 & 0 & 0 & 0 & | & 10 \\ 0 & 1 & 0 & 0 & 0 & 0 & | & 6 \\ 0 & 0 & 1 & 0 & 0 & 0 & | & 4 \\ 0 & 0 & 0 & 1 & 0 & 0 & | & 2 \\ 0 & 0 & 0 & 0 & 1 & 0 & | & 4 \\ 0 & 0 & 0 & 0 & 0 & 1 & | & 6 \\ 0 & 0 & 0 & 0 & 0 & 0 & | & 0 \end{bmatrix}$$

So, the currents are $I = 10$, $I_1 = 6$, $I_2 = 4$, $I_3 = 2$, $I_4 = 4$, and $I_5 = 6$ amps.

(b) From Ohm's Law, the effective resistance is found to be $R_{eff} = \frac{V}{I} = \frac{14}{10} = \frac{7}{5}$ ohms.

(c) In this case we force $I_3 = 0$, and assign r to the resistance in branch BC.
We then get the following system of equations:

Node B : $I = I_1 + I_4$
Node C : $I_1 = I_2$
Node D : $I_2 + I_5 = I$
Node E : $I_4 = I_5$

$\Rightarrow$

Circuit ABEDA : $-2I_4 - I_5 + 14 = 0$
Circuit BCEB : $-rI_1 + 2I_4 = 0$
Circuit CDEC : $-2I_2 + I_5 = 0$

$\Rightarrow$

Partial row reduction gives

$$\begin{bmatrix} 1 & -1 & 0 & 0 & -1 & 0 & | & 0 \\ 0 & 1 & -1 & 0 & 0 & 0 & | & 0 \\ 1 & 0 & -1 & 0 & 0 & -1 & | & 0 \\ 0 & 0 & 0 & 0 & 1 & -1 & | & 0 \\ 0 & 0 & 0 & 0 & 2 & 1 & | & 14 \\ 0 & r & 0 & 0 & -2 & 0 & | & 0 \\ 0 & 0 & 2 & 0 & 0 & -1 & | & 0 \end{bmatrix} \longrightarrow \begin{bmatrix} 1 & -1 & 0 & 0 & -1 & 0 & | & 0 \\ 0 & 1 & -1 & 0 & 0 & 0 & | & 0 \\ 0 & 0 & 1 & 0 & 0 & -1/2 & | & 0 \\ 0 & 0 & 0 & 0 & 1 & -1 & | & 0 \\ 0 & 0 & 0 & 0 & 0 & 1 & | & 14/3 \\ 0 & r & 0 & 0 & -2 & 0 & | & 0 \\ 0 & 0 & 0 & 0 & 0 & 0 & | & 0 \end{bmatrix}$$

So, substitution $\Rightarrow I_4 = I_5 = \frac{14}{3}$ and $I_1 = I_2 = \frac{7}{3} \Rightarrow \frac{7}{3}r - 2\frac{14}{3} = 0 \Rightarrow r = 4$.
So, if we let the resistance in $BC = r = 4$, then the current in $CE = 0$.

23. (a) Over $\mathbb{Z}_2$, we need to solve $x_1\mathbf{a} + x_2\mathbf{b} + \cdots + x_5\mathbf{e} = \mathbf{t} + \mathbf{s}$:

$$\mathbf{s} = \begin{bmatrix} 0 \\ 0 \\ 0 \\ 0 \\ 0 \end{bmatrix}, \mathbf{t} = \begin{bmatrix} 0 \\ 1 \\ 0 \\ 1 \\ 0 \end{bmatrix} \Rightarrow \left[\begin{array}{ccccc|c} 1 & 1 & 0 & 0 & 0 & 0 \\ 1 & 1 & 1 & 0 & 0 & 1 \\ 0 & 1 & 1 & 1 & 0 & 0 \\ 0 & 0 & 1 & 1 & 1 & 1 \\ 0 & 0 & 0 & 1 & 1 & 0 \end{array}\right] \longrightarrow \left[\begin{array}{ccccc|c} 1 & 0 & 0 & 0 & 1 & 1 \\ 0 & 1 & 0 & 0 & 1 & 1 \\ 0 & 0 & 1 & 0 & 0 & 1 \\ 0 & 0 & 0 & 1 & 1 & 0 \\ 0 & 0 & 0 & 0 & 0 & 0 \end{array}\right]$$

Thus, x_5 is a free variable; hence there are exactly two solutions.

Solving for the other variables, we get: $x_1 = 1 + x_5$, $x_2 = 1 + x_5$, $x_3 = 1$, $x_4 = x_5 \Rightarrow$

$x_5 = 0$ and $x_5 = 1 \Rightarrow$ the solutions $\begin{bmatrix} x_1 \\ x_2 \\ x_3 \\ x_4 \\ x_5 \end{bmatrix} = \begin{bmatrix} 1 \\ 1 \\ 1 \\ 0 \\ 0 \end{bmatrix}$ and $\begin{bmatrix} x_1 \\ x_2 \\ x_3 \\ x_4 \\ x_5 \end{bmatrix} = \begin{bmatrix} 0 \\ 0 \\ 1 \\ 1 \\ 1 \end{bmatrix}$.

So, push switches 1, 2, and 3 or switches 3, 4, and 5.

(b) In this case, $\mathbf{t} = \mathbf{e}_2$ over $\mathbb{Z}_2 \Rightarrow$:

$$\left[\begin{array}{ccccc|c} 1 & 1 & 0 & 0 & 0 & 0 \\ 1 & 1 & 1 & 0 & 0 & 1 \\ 0 & 1 & 1 & 1 & 0 & 0 \\ 0 & 0 & 1 & 1 & 1 & 0 \\ 0 & 0 & 0 & 1 & 1 & 0 \end{array}\right] \longrightarrow \left[\begin{array}{ccccc|c} 1 & 0 & 0 & 0 & 1 & 0 \\ 0 & 1 & 0 & 0 & 1 & 0 \\ 0 & 0 & 1 & 0 & 0 & 0 \\ 0 & 0 & 0 & 1 & 1 & 0 \\ 0 & 0 & 0 & 0 & 0 & 1 \end{array}\right]$$

This shows that there is no solution in this case.

That is, it is impossible to start with all lights off and turn only the second light on.

25. The possible configurations are

$$\begin{bmatrix} x_1 \\ x_2 \\ x_3 \\ x_4 \\ x_5 \end{bmatrix} \in \left\{ \begin{bmatrix} 0\\0\\0\\0\\0\\1\\0\\1\\1\\0 \end{bmatrix}, \begin{bmatrix} 1\\1\\0\\0\\0\\1\\0\\0\\0\\1 \end{bmatrix}, \begin{bmatrix} 1\\1\\1\\0\\0\\1\\0\\1\\0\\1 \end{bmatrix}, \begin{bmatrix} 0\\1\\1\\1\\0\\1\\0\\0\\1\\0 \end{bmatrix}, \begin{bmatrix} 0\\0\\1\\1\\1\\1\\1\\0\\1\\0 \end{bmatrix}, \begin{bmatrix} 0\\0\\0\\1\\1\\1\\1\\1\\0\\1 \end{bmatrix}, \begin{bmatrix} 0\\0\\1\\0\\0\\1\\1\\0\\0\\1 \end{bmatrix}, \begin{bmatrix} 0\\1\\0\\1\\0\\1\\1\\1\\1\\0 \end{bmatrix}, \begin{bmatrix} 0\\1\\1\\0\\1\\1\\1\\0\\1\\1 \end{bmatrix}, \begin{bmatrix} 0\\1\\0\\0\\1\\1\\1\\1\\1\\1 \end{bmatrix} \right\}$$

2.4 Applications

27. In this situation the switches correspond to the vectors

$$\mathbf{a} = \begin{bmatrix} 1 \\ 1 \\ 0 \\ 0 \\ 0 \end{bmatrix}, \mathbf{b} = \begin{bmatrix} 1 \\ 1 \\ 1 \\ 0 \\ 0 \end{bmatrix}, \mathbf{c} = \begin{bmatrix} 0 \\ 1 \\ 1 \\ 1 \\ 0 \end{bmatrix}, \mathbf{d} = \begin{bmatrix} 0 \\ 0 \\ 1 \\ 1 \\ 1 \end{bmatrix}, \mathbf{e} = \begin{bmatrix} 0 \\ 0 \\ 0 \\ 1 \\ 1 \end{bmatrix}$$

In $\mathbb{Z}_3^5$, with $\mathbf{s} = \mathbf{0}$, we need to solve $x_1\mathbf{a} + x_2\mathbf{b} + \cdots + x_5\mathbf{e} = \mathbf{t}$.

$$\mathbf{t} = \begin{bmatrix} 2 \\ 1 \\ 2 \\ 1 \\ 2 \end{bmatrix} \Rightarrow \left[\begin{array}{ccccc|c} 1 & 1 & 0 & 0 & 0 & 2 \\ 1 & 1 & 1 & 0 & 0 & 1 \\ 0 & 1 & 1 & 1 & 0 & 2 \\ 0 & 0 & 1 & 1 & 1 & 1 \\ 0 & 0 & 0 & 1 & 1 & 2 \end{array}\right] \longrightarrow \left[\begin{array}{ccccc|c} 1 & 0 & 0 & 0 & 1 & 1 \\ 0 & 1 & 0 & 0 & 2 & 1 \\ 0 & 0 & 1 & 0 & 0 & 2 \\ 0 & 0 & 0 & 1 & 1 & 2 \\ 0 & 0 & 0 & 0 & 0 & 0 \end{array}\right]$$

So, x_5 is free and there are exactly three solutions ($x_5 = 0, 1, 2$). Solving for the other variables in terms of x_5 (over $\mathbb{Z}_3$), we get:

$$\begin{bmatrix} x_1 \\ x_2 \\ x_3 \\ x_4 \\ x_5 \end{bmatrix} = \begin{bmatrix} 1 \\ 1 \\ 2 \\ 2 \\ 0 \end{bmatrix}, \begin{bmatrix} x_1 \\ x_2 \\ x_3 \\ x_4 \\ x_5 \end{bmatrix} = \begin{bmatrix} 0 \\ 2 \\ 2 \\ 1 \\ 1 \end{bmatrix}, \begin{bmatrix} x_1 \\ x_2 \\ x_3 \\ x_4 \\ x_5 \end{bmatrix} = \begin{bmatrix} 2 \\ 0 \\ 2 \\ 0 \\ 2 \end{bmatrix}.$$

29. (a) The matrix representing actions of touching the squares is $S = \begin{bmatrix} \mathbf{a} & \mathbf{b} & \mathbf{c} & \mathbf{d} & \mathbf{e} & \mathbf{f} & \mathbf{g} & \mathbf{h} & \mathbf{i} \end{bmatrix}$. Over $\mathbb{Z}_2$, we need to solve $\mathbf{s} + x_1\mathbf{a} + x_2\mathbf{b} + \cdots + x_5\mathbf{e} = \mathbf{t}$ or $x_1\mathbf{a} + x_2\mathbf{b} + \cdots + x_9\mathbf{i} = \mathbf{t} - \mathbf{s} = \mathbf{s}$.

$$\mathbf{s} = \begin{bmatrix} 0 \\ 1 \\ 1 \\ 1 \\ 0 \\ 1 \\ 1 \\ 1 \\ 0 \end{bmatrix}, \mathbf{t} = \begin{bmatrix} 0 \\ 0 \\ 0 \\ 0 \\ 0 \\ 0 \\ 0 \\ 0 \\ 0 \end{bmatrix} \Rightarrow \left[\begin{array}{ccccccccc|c} 1 & 1 & 0 & 1 & 0 & 0 & 0 & 0 & 0 & 0 \\ 1 & 1 & 1 & 0 & 1 & 0 & 0 & 0 & 0 & 1 \\ 0 & 1 & 1 & 0 & 0 & 1 & 0 & 0 & 0 & 1 \\ 1 & 0 & 0 & 1 & 1 & 0 & 1 & 0 & 0 & 1 \\ 1 & 0 & 1 & 0 & 1 & 0 & 1 & 0 & 1 & 0 \\ 0 & 0 & 1 & 0 & 1 & 1 & 0 & 0 & 1 & 1 \\ 0 & 0 & 0 & 1 & 0 & 0 & 1 & 1 & 0 & 1 \\ 0 & 0 & 0 & 0 & 1 & 0 & 1 & 1 & 1 & 1 \\ 0 & 0 & 0 & 0 & 0 & 1 & 0 & 1 & 1 & 0 \end{array}\right] \longrightarrow \left[\begin{array}{ccccccccc|c} 1 & 0 & 0 & 0 & 0 & 0 & 0 & 0 & 0 & 0 \\ 0 & 1 & 0 & 0 & 0 & 0 & 0 & 0 & 0 & 0 \\ 0 & 0 & 1 & 0 & 0 & 0 & 0 & 0 & 0 & 1 \\ 0 & 0 & 0 & 1 & 0 & 0 & 0 & 0 & 0 & 0 \\ 0 & 0 & 0 & 0 & 1 & 0 & 0 & 0 & 0 & 0 \\ 0 & 0 & 0 & 0 & 0 & 1 & 0 & 0 & 0 & 0 \\ 0 & 0 & 0 & 0 & 0 & 0 & 1 & 0 & 0 & 1 \\ 0 & 0 & 0 & 0 & 0 & 0 & 0 & 1 & 0 & 0 \\ 0 & 0 & 0 & 0 & 0 & 0 & 0 & 0 & 1 & 0 \end{array}\right]$$

showing that touching the third and seventh squares will turn all nine squares black.

(b) Since $S \longrightarrow I_9$, we can always find a solution to the system of equations $x_1\mathbf{a} + x_2\mathbf{b} + \cdots + x_9\mathbf{i} = \mathbf{s}$ in part (a).

31. Let Grace's and Hans' ages be g and h. Then we have $g = 3h$ and $g + 5 = 2(h+5) \Leftrightarrow g = 2(h+5) - 5$, so $3h = 2h + 5 \Leftrightarrow h = 5$ and $g = 15$. So, Hans is 5 and Grace is 15.

33. Let the areas be a and b. We have the following equations:

$$a + b = 1800$$
$$\frac{2}{3}a + \frac{1}{2}b = 1100$$

We solve these to find that $a = 1200$ square yards and $b = 600$ square yards.

35. (a) From the addition table we get: $a + c = 2$, $a + d = 4$, $b + c = 3$, $b + d = 5$.

$$\begin{bmatrix} 1 & 0 & 1 & 0 & | & 2 \\ 1 & 0 & 0 & 1 & | & 4 \\ 0 & 1 & 1 & 0 & | & 3 \\ 0 & 1 & 0 & 1 & | & 5 \end{bmatrix} \longrightarrow \begin{bmatrix} 1 & 0 & 0 & 1 & | & 4 \\ 0 & 1 & 0 & 1 & | & 5 \\ 0 & 0 & 1 & -1 & | & -2 \\ 0 & 0 & 0 & 0 & | & 0 \end{bmatrix}$$

So we see that d is a free variable. Solving for the other variables in terms of d we obtain $a = 4 - d$, $b = 5 - d$, and $c = -2 + d$. Hence there are an infinite number of solutions.

(b) We have: $a + c = 3$, $a + d = 4$, $b + c = 6$, and $b + d = 5$

$$\begin{bmatrix} 1 & 0 & 1 & 0 & | & 3 \\ 1 & 0 & 0 & 1 & | & 4 \\ 0 & 1 & 1 & 0 & | & 6 \\ 0 & 1 & 0 & 1 & | & 5 \end{bmatrix} \longrightarrow \begin{bmatrix} 1 & 0 & 0 & 1 & | & 0 \\ 0 & 1 & 0 & 1 & | & 0 \\ 0 & 0 & 1 & -1 & | & 0 \\ 0 & 0 & 0 & 0 & | & 1 \end{bmatrix}$$

So we see that this is an inconsistent system.

37. (a) From the addition table we get the following system of equations:

$$a + d = 3 \qquad b + d = 2 \qquad c + d = 1$$
$$a + e = 5 \qquad b + e = 4 \qquad c + e = 3$$
$$a + f = 4 \qquad b + f = 3 \qquad c + f = 1$$

We form the augmented matrix and perform Gauss-Jordan elimination to get

$$\begin{bmatrix} 1 & 0 & 0 & 1 & 0 & 0 & | & 3 \\ 1 & 0 & 0 & 0 & 1 & 0 & | & 5 \\ 1 & 0 & 0 & 0 & 0 & 1 & | & 4 \\ 0 & 1 & 0 & 1 & 0 & 0 & | & 2 \\ 0 & 1 & 0 & 0 & 1 & 0 & | & 4 \\ 0 & 1 & 0 & 0 & 0 & 1 & | & 3 \\ 0 & 0 & 1 & 1 & 0 & 0 & | & 1 \\ 0 & 0 & 1 & 0 & 1 & 0 & | & 3 \\ 0 & 0 & 1 & 0 & 0 & 1 & | & 1 \end{bmatrix} \longrightarrow \begin{bmatrix} 1 & 0 & 0 & 0 & 0 & 1 & | & 0 \\ 0 & 1 & 0 & 0 & 0 & 1 & | & 0 \\ 0 & 0 & 1 & 0 & 0 & 1 & | & 0 \\ 0 & 0 & 0 & 1 & 0 & -1 & | & 0 \\ 0 & 0 & 0 & 0 & 1 & -1 & | & 0 \\ 0 & 0 & 0 & 0 & 0 & 0 & | & 1 \\ 0 & 0 & 0 & 0 & 0 & 0 & | & 0 \\ 0 & 0 & 0 & 0 & 0 & 0 & | & 0 \\ 0 & 0 & 0 & 0 & 0 & 0 & | & 0 \end{bmatrix}$$

So, we see that this is an inconsistent system.

(b) From the addition table we get the following system of equations:

$$a + d = 1 \qquad b + d = 2 \qquad c + d = 3$$
$$a + e = 3 \qquad b + e = 4 \qquad c + e = 5$$
$$a + f = 4 \qquad b + f = 5 \qquad c + f = 6$$

We form the augmented matrix and perform Gauss-Jordan elimination to get

$$\begin{bmatrix} 1 & 0 & 0 & 1 & 0 & 0 & | & 1 \\ 1 & 0 & 0 & 0 & 1 & 0 & | & 3 \\ 1 & 0 & 0 & 0 & 0 & 1 & | & 4 \\ 0 & 1 & 0 & 1 & 0 & 0 & | & 2 \\ 0 & 1 & 0 & 0 & 1 & 0 & | & 4 \\ 0 & 1 & 0 & 0 & 0 & 1 & | & 5 \\ 0 & 0 & 1 & 1 & 0 & 0 & | & 3 \\ 0 & 0 & 1 & 0 & 1 & 0 & | & 5 \\ 0 & 0 & 1 & 0 & 0 & 1 & | & 6 \end{bmatrix} \longrightarrow \begin{bmatrix} 1 & 0 & 0 & 0 & 0 & 1 & | & 4 \\ 0 & 1 & 0 & 0 & 0 & 1 & | & 5 \\ 0 & 0 & 1 & 0 & 0 & 1 & | & 6 \\ 0 & 0 & 0 & 1 & 0 & -1 & | & -3 \\ 0 & 0 & 0 & 0 & 1 & -1 & | & -1 \\ 0 & 0 & 0 & 0 & 0 & 0 & | & 0 \\ 0 & 0 & 0 & 0 & 0 & 0 & | & 0 \\ 0 & 0 & 0 & 0 & 0 & 0 & | & 0 \\ 0 & 0 & 0 & 0 & 0 & 0 & | & 0 \end{bmatrix}$$

We see that f is a free variable, and there are an infinite number of solutions of the form $a = 4 - f$, $b = 5 - f$, $c = 6 - f$, $d = -3 + f$, $e = -1 + f$.

39. (a) We know that the three points $(0,1)$, $(-1,4)$, and $(2,1)$ must satisfy the equation $y = ax^2 + bx + c$. Substitution $\Rightarrow c = 1$, $a - b + c = 4$, and $4a + 2b + c = 1$. So:

$$\begin{bmatrix} 0 & 0 & 1 & | & 1 \\ 1 & -1 & 1 & | & 4 \\ 4 & 2 & 1 & | & 1 \end{bmatrix} \longrightarrow \begin{bmatrix} 1 & 0 & 0 & | & 1 \\ 0 & 1 & 0 & | & -2 \\ 0 & 0 & 1 & | & 1 \end{bmatrix}$$

Thus $a = 1$, $b = -2$, and $c = 1$, and the equation of the parabola is $y = x^2 - 2x + 1$.

(b) We know that the three points $(-3,1)$, $(-2,2)$, and $(-1,5)$ must satisfy the equation $y = ax^2 + bx + c$. Substitution $\Rightarrow 9a - 3b + c = 1$, $4a - 2b + c = 2$, and $a - b + c = 5$. So:

$$\begin{bmatrix} 9 & -3 & 1 & | & 1 \\ 4 & -2 & 1 & | & 2 \\ 1 & 1 & 1 & | & 5 \end{bmatrix} \longrightarrow \begin{bmatrix} 1 & 0 & 0 & | & 1 \\ 0 & 1 & 0 & | & 6 \\ 0 & 0 & 1 & | & 10 \end{bmatrix}$$

Thus $a = 1$, $b = 6$, and $c = 10$, and the equation of the parabola is $y = x^2 + 6x + 10$.

41. We have: $\frac{3x+1}{x^2+2x-3} = \frac{A}{x-1} + \frac{B}{x+3} \Leftrightarrow (x+3)A + (x-1)B = 3x+1 \Leftrightarrow$
$x(A+B) + (3A - B) = 3x + 1$.
Equating the coefficients of x and constants we get:

$A + B = 3$, $3A - B = 1 \Rightarrow \begin{bmatrix} 1 & 1 & | & 3 \\ 3 & -1 & | & 1 \end{bmatrix} \longrightarrow \begin{bmatrix} 1 & 0 & | & 1 \\ 0 & 1 & | & 2 \end{bmatrix} \Rightarrow A = 1, B = 2 \Rightarrow$

The partial fraction decomposition is $\frac{3x+1}{x^2+2x-3} = \frac{1}{x-1} + \frac{2}{x+3}$.

2.4 Applications

43. We have: $\frac{x-1}{(x+1)(x^2+1)(x^2+4)} = \frac{A}{x+1} + \frac{Bx+C}{x^2+1} + \frac{Dx+E}{x^2+4} \Leftrightarrow$

$(x^4 + 5x^2 + 4) A + (x^3 + 4x + x^2 + 4)(Bx + C) + (x^3 + x + x^2 + 1)(Dx + E) = x - 1 \Leftrightarrow$

$x^4 (A + B + D) + x^3 (B + C + D + E) + x^2 (5A + 4B + C + D + E)$
$\qquad + x(4C + 4B + E + D) + (4A + 4C + E) = x - 1$

Equating coefficients, we get

$$A + B + D = 0$$
$$B + C + D + E = 0$$
$$5A + 4B + C + D + E = 0$$
$$4B + 4C + D + E = 1$$
$$4A + 4C + E = -1$$

From these, we form the augmented matrix and perform Gauss-Jordan elimination

$$\begin{bmatrix} 1 & 1 & 0 & 1 & 0 & 0 \\ 0 & 1 & 1 & 1 & 1 & 0 \\ 5 & 4 & 1 & 1 & 1 & 0 \\ 0 & 4 & 4 & 1 & 1 & 1 \\ 4 & 0 & 4 & 0 & 1 & -1 \end{bmatrix} \longrightarrow \begin{bmatrix} 1 & 0 & 0 & 0 & 0 & -\frac{1}{5} \\ 0 & 1 & 0 & 0 & 0 & \frac{1}{3} \\ 0 & 0 & 1 & 0 & 0 & 0 \\ 0 & 0 & 0 & 1 & 0 & -\frac{2}{15} \\ 0 & 0 & 0 & 0 & 1 & -\frac{1}{5} \end{bmatrix}$$

So, $A = -\frac{1}{5}$, $B = \frac{1}{3}$, $C = 0$, $D = -\frac{2}{15}$, and $E = -\frac{1}{5}$, and we have

$$\frac{x-1}{(x+1)(x^2+1)(x^2+4)} = -\frac{\frac{1}{5}}{x+1} + \frac{\frac{1}{3}x}{x^2+1} - \frac{\frac{1}{15}(2x+3)}{x^2+4}.$$

45. Assume $1 + 2 + \cdots + n = an^2 + bn + c$, and let $n = 0, 1, 2$ to get

$$c = 0$$
$$a + b + c = 1$$
$$4a + 2b + c = 3$$

From these we form the augmented matrix and perform Gauss-Jordan elimination to get

$$\begin{bmatrix} 0 & 0 & 1 & | & 0 \\ 1 & 1 & 1 & | & 1 \\ 4 & 2 & 1 & | & 3 \end{bmatrix} \longrightarrow \begin{bmatrix} 1 & 0 & 0 & | & \frac{1}{2} \\ 0 & 1 & 0 & | & \frac{1}{2} \\ 0 & 0 & 1 & | & 0 \end{bmatrix}$$

Thus $a = \frac{1}{2}$, $b = \frac{1}{2}$, $c = 0$, and we find that $1 + 2 + \cdots + n = \frac{1}{2}n^2 + \frac{1}{2}n = \frac{1}{2}n(n+1)$.

47. Assume $1^3 + 2^3 + \cdots + n^3 = an^4 + bn^3 + cn^2 + dn + e$, and let $n = 0, 1, 2, 3, 4$ to get

$$e = 0$$
$$a + b + c + d + e = 1$$
$$16a + 8b + 4c + 2d + e = 9$$
$$81a + 27b + 9c + 3d + e = 36$$
$$256a + 64b + 16c + 4d + e = 100$$

From these we form the augmented matrix and perform Gauss-Jordan elimination to get

$$\begin{bmatrix} 0 & 0 & 0 & 0 & 1 & | & 0 \\ 1 & 1 & 1 & 1 & 1 & | & 1 \\ 16 & 8 & 4 & 2 & 1 & | & 9 \\ 81 & 27 & 9 & 3 & 1 & | & 36 \\ 256 & 64 & 16 & 4 & 1 & | & 100 \end{bmatrix} \longrightarrow \begin{bmatrix} 1 & 0 & 0 & 0 & 0 & | & \frac{1}{4} \\ 0 & 1 & 0 & 0 & 0 & | & \frac{1}{2} \\ 0 & 0 & 1 & 0 & 0 & | & \frac{1}{4} \\ 0 & 0 & 0 & 1 & 0 & | & 0 \\ 0 & 0 & 0 & 0 & 1 & | & 0 \end{bmatrix}$$

Thus $a = \frac{1}{4}$, $b = \frac{1}{2}$, $c = \frac{1}{4}$, $d = 0$, $e = 0$, and we find that

$$1^3 + 2^3 + \cdots + n^3 = \frac{1}{4}n^4 + \frac{1}{2}n^3 + \frac{1}{4}n^2 = \frac{1}{4}n^2(n+1)^2 = \left(\frac{n(n+1)}{2}\right)^2.$$

2.5 Iterative Methods for Solving Linear Systems

Q: What is our goal for Section 2.5?

A: To learn how to use matrices to solve iterative systems.
Since this process is fully explained in the text, we simply present solutions here.

Key Definitions and Concepts

Gauss-Seidel method	p.122	2.5	This process is applied in Example 2.35
Jacobi's method	p.122	2.5	This process is applied in Example 2.35
converges	p.123	2.5	when iterates approach a solution
divergence	p.125	2.5	when iterates do not approach a solution
iterates	p.123	2.5	vectors found through the iterative process
strictly diagonally dominant	p.126	2.5	$\|a_{11}\| > \|a_{12}\| + \|a_{13}\| + \cdots + \|a_{1n}\|$ $\|a_{22}\| > \|a_{21}\| + \|a_{23}\| + \cdots + \|a_{2n}\|$ and $\|a_{nn}\| > \|a_{n1}\| + \|a_{n2}\| + \cdots + \|a_{n,n-1}\|$

Theorems

Thm 2.9	p.124	2.5	*A strictly diagonally dominant* $\Rightarrow$ *iterates converge*
Thm 2.10	p.124	2.5	*methods converge* $\Rightarrow$ *they converge to the solution*

Solutions to odd-numbered exercises from Section 2.5

1. Begin by solving the first equation for x_1 and the second equation for x_2 to obtain

$$x_1 = \frac{6}{7} + \frac{1}{7}x_2$$
$$x_2 = \frac{4}{5} + \frac{1}{5}x_1$$

Using the initial vector $[x_1, x_2] = [0, 0]$, we get a sequence of approximations:

n	0	1	2	3	4	5
x_1	0	0.857	0.971	0.996	0.999	1.000
x_2	0	0.800	0.971	0.994	0.999	0.999

The exact solution to this system is $[x_1, x_2] = [1, 1]$.

3.

n	0	1	2	3	4	5
x_1	0	0.2222	0.2540	0.2610	0.2620	0.2623
x_2	0	0.2857	0.3492	0.3583	0.3603	0.3606

The exact solution is $x = \frac{16}{61}, y = \frac{22}{61}$.

5.

n	0	1	2	3	4	5	6	7	8
x_1	0	0.3333	0.2500	0.3055	0.2916	0.3009	0.2986	0.3001	0.2997
x_2	0	0.2500	0.08337	0.1250	0.09722	0.1042	0.09957	0.1008	0.09997
x_3	0	0.3333	0.2500	0.3055	0.2916	0.3009	0.2986	0.3001	0.2997

The exact solution is $[x_1, x_2, x_3] = [0.3, 0.1, 0.3]$

7.

n	0	1	2	3	4
x_1	0	0.8571	0.9959	0.9999	1.000
x_2	0	0.9714	0.9992	0.9999	1.000

The Gauss-Seidel method takes 4 steps while the Jacobi method takes 5.

9.

n	0	1	2	3	4
x_1	0	0.2222	0.2610	0.2623	0.2623
x_2	0	0.3492	0.3603	0.3607	0.3607

The Gauss-Seidel method takes 4 steps while the Jacobi method takes 5.

11.

n	0	1	2	3	4	5
x_1	0	0.3333	0.2778	0.2963	0.2994	0.2999
x_2	0	0.1667	0.1111	0.1019	0.1003	0.1001
x_3	0	0.2778	0.2963	0.2994	0.2999	0.2999

The Gauss-Seidel method takes 5 steps while the Jacobi method takes 8.

13.

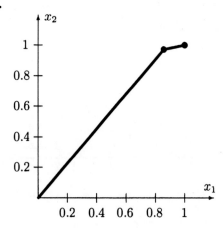

15. Applying the Gauss-Seidel method to $x_1 - 2x_2 = 3$, $3x_1 + 2x_2 = 1$ gives:

n	0	1	2	3	4
x_1	0	3	-5	19	-53
x_2	0	-4	8	-28	80

which evidently diverges. If, however, we swap the two equations to get $3x_1 + 2x_2 = 1$, $x_1 - 2x_2 = 3$ and use the Gauss-Seidel method on this system we get the table:

n	0	1	2	3	4	5	6	7	8	9
x_1	0.	0.333	1.222	0.926	1.025	0.992	1.003	0.999	1.000	1.000
x_2	0	-1.334	-0.889	-1.037	-0.988	-1.004	-0.999	-1.000	-1.000	-1.000

Thus the solution to the system of equations is approximately $[x_1, x_2] = [1.000, -1.000]$. The exact solution is $[1, -1]$.

17.

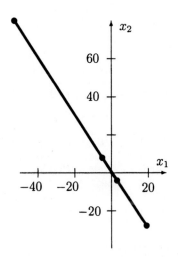

19. Applying the Gauss-Seidel method to the system of equations

$$5x_1 - 2x_2 + 3x_3 = -8$$
$$x_1 + 4x_2 - 4x_3 = 102$$
$$-2x_1 - 2x_2 + 4x_3 = -90$$

gives the following table of iterations:

n	0	1	2	3	4	5	6
x_1	0	-1.60	14.97	8.55	10.74	9.84	10.12
x_2	0	25.9	11.41	14.05	11.62	11.72	11.25
x_3	0	-10.35	-9.31	-11.20	-11.32	-11.72	-11.82

n	7	8	9	10	11	12	13
x_1	9.99	10.02	10.00	10.01	10.00	10.00	10.00
x_2	11.18	11.08	11.05	11.02	11.01	11.00	11.00
x_3	-11.91	-11.95	-11.98	-11.99	-12.00	-12.00	-12.00

Thus the solution to the system of equations is approximately $[x_1, x_2, x_3] = [10.00, 11.00, -12.00]$. The exact solution is $[10, 11, -12]$.

21. Continuing iterations of Exercise 19 to achieve a solution accurate to within 0.001:

n	0	1	$\cdots$	13	14	15	16	17
x_1	0	-1.60	$\cdots$	10.001	10.000	10.000	10.000	10.000
x_2	0	25.9	$\cdots$	11.004	11.002	11.001	11.000	11.000
x_3	0	-10.35	$\cdots$	-11.998	-11.999	-12.000	-12.000	-12.000

23. From the temperature-averaging property we get the system of four equations

$$t_1 = \frac{1}{4}(t_2 + t_3)$$
$$t_2 = \frac{1}{4}(t_1 + t_4)$$
$$t_3 = \frac{1}{4}(t_1 + t_4 + 200)$$
$$t_4 = \frac{1}{4}(t_2 + t_3 + 200)$$

upon which we apply the Gauss-Seidel method. With $t_1 = t_2 = t_3 = t_4 = 0$ we get:

n	0	1	2	3	4	5	6	7	8	9	10
t_1	0	0	12.5	21.876	24.220	24.806	24.952	24.988	24.998	25.000	25.000
t_2	0	0	18.75	23.438	24.610	24.904	24.976	24.994	25.000	25.000	25.000
t_3	0	50	68.75	73.438	74.610	74.904	74.976	74.994	75.000	75.000	75.000
t_4	0	62.5	71.876	74.220	74.806	74.952	74.988	74.998	75.000	75.000	75.000

Thus the equilibrium temperatures at the interior points are $t_1 = 25.000$, $t_2 = 25.000$, $t_4 = 75.000$ (to an accuracy of 0.001).

2.5 Iterative Methods for Solving Linear Systems

25. Here the equations are

$$t_1 = \frac{1}{4}(t_2 + 80)$$

$$t_2 = \frac{1}{4}(t_1 + t_3 + t_4)$$

$$t_3 = \frac{1}{4}(t_2 + t_5 + 80)$$

$$t_4 = \frac{1}{4}(t_2 + t_5 + 5)$$

$$t_5 = \frac{1}{4}(t_3 + t_4 + t_6 + 5)$$

$$t_6 = \frac{1}{4}(t_5 + 85)$$

Following the same procedure as in the previous exercises, we get the following table:

n	0	1	2	3	4	5	6
t_1	0	20	21.250	22.813	23.330	23.660	23.773
t_2	0	5	11.250	13.321	14.639	15.093	15.237
t_3	0	21.25	24.609	26.988	27.731	27.963	28.035
t_4	0	2.5	5.859	8.238	8.981	9.213	9.285
t_5	0	7.188	14.629	16.283	16.758	16.904	16.949
t_6	0	23.047	24.907	25.321	25.440	25.476	25.487

n	7	8	9	10	11	12
t_1	23.809	23.821	23.824	23.825	23.826	23.826
t_2	15.282	15.297	15.301	15.302	15.304	15.304
t_3	28.058	28.065	28.067	28.068	28.069	28.069
t_4	9.308	9.315	9.317	9.318	9.319	9.319
t_5	16.963	16.968	16.969	16.970	16.970	16.970
t_6	25.491	25.492	25.492	25.493	25.493	25.493

So the equilibrium temperatures at the interior points are found to be about $t_1 = 23.826$, $t_2 = 15.304$, $t_3 = 28.069$, $t_4 = 9.319$, $t_5 = 16.970$, and $t_6 = 25.493$.

27. (a) Let x_1 correspond to the left end of the paper and x_2 to the right end, and let n be the number of folds. Then the first six values of $[x_1, x_2]$ are

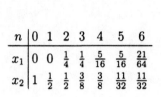

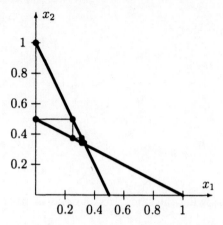

(b) Two linear equations that determine the new values of the endpoints at each iteration are $x_2 = -\frac{1}{2}x_1 + \frac{1}{2}$ and $x_1 = -\frac{1}{2}x_2 + \frac{1}{2}$. These two lines are plotted in part (a).

(c) Switching to decimal representation, we continue applying the Gauss-Seidel method to approximate the point of convergence, giving rise to the sequence of endpoints:

n	0	1	2	3	4	5	6	7	8	9	10	11	12
x_1	0	0	$\frac{1}{4}$	$\frac{1}{4}$	$\frac{5}{16}$	$\frac{5}{16}$	$\frac{21}{64}$	0.328	0.332	0.332	0.333	0.333	0.333
x_2	1	$\frac{1}{2}$	$\frac{1}{2}$	$\frac{3}{8}$	$\frac{3}{8}$	$\frac{11}{32}$	$\frac{11}{32}$	0.336	0.336	0.334	0.334	0.333	0.333

(d) We have the system of equations $x_2 = -\frac{1}{2}x_1 + \frac{1}{2}$, $x_1 = -\frac{1}{2}x_2 + \frac{1}{2}$ $\Rightarrow$

$$\begin{bmatrix} \frac{1}{2} & 1 & | & \frac{1}{2} \\ 1 & \frac{1}{2} & | & \frac{1}{2} \end{bmatrix} \longrightarrow \begin{bmatrix} 1 & 0 & | & \frac{1}{3} \\ 0 & 1 & | & \frac{1}{3} \end{bmatrix}$$

Hence the exact solution to the system of equations is $[x_1, x_2] = \left[\frac{1}{3}, \frac{1}{3}\right]$.

The ends of the paper converge at $\frac{1}{3}$.

Chapter 2 Review

Key Definitions and Concepts

This list includes most but not all of the definitions listed at the end of each section.

augmented matrix	p.62	2.1	a coefficient matrix augmented by the constants
back substitution	p.62	2.1	the procedure used to solve Example 2.5 on p. 62
branch	p.104	2.4	a directed edge of a network
coefficient matrix	p.62	2.1	a matrix of coefficients taken from a linear system
coefficients	p.59	2.1	the a_i in $a_1 x_1 + a_2 x_2 + \cdots + a_n x_n = b$
conservation of flow	p.104	2.4	at each node, the flow in equals the flow out
consistent	p.61	2.1	a system of linear equations with at least one solution
constant term	p.59	2.1	b in $a_1 x_1 + a_2 x_2 + \cdots + a_n x_n = b$
converges	p.123	2.5	when iterates approach a solution
Current Law (nodes)	p.106	2.4	sum of currents flowing into a node equals sum out
divergence	p.125	2.5	when iterates do not approach a solution
elementary row operations	p.70	2.2	EROs: $R_i \leftrightarrow R_j$, kR_i, $R_i + kR_j$
equivalent	p.61	2.1	linear systems that have the same solution set
free variable	p.75	2.2	a variable free to take on any value
Gauss-Jordan Elimination	p.76	2.2	see procedure described in box on p.77
Gauss-Seidel method	p.122	2.5	This process is applied in Example 2.35
Gaussian Elimination	p.72	2.2	see procedure described in box on p.72
homogeneous	p.79	2.2	system in which each constant term is 0
inconsistent	p.61	2.1	a system of linear equations with no solutions
iterates	p.123	2.5	vectors found through the iterative process
Jacobi's method	p.122	2.5	This process is applied in Example 2.35
Kirchoff's Laws	p.106	2.4	Current Law (nodes) and Voltage Law (circuits)

Key Definitions and Concepts, continued

leading entry	p.68	2.2	the first nonzero entry in a row of a matrix
linear combination	p.95	2.3	$\mathbf{v} = c_1\mathbf{v}_1 + \cdots + c_n\mathbf{v}_n$ $\mathbf{v} = \sum c_i \mathbf{v}_i$
linearly dependent set of vectors	p.94	2.3	vector can be written as linear combination of others $c_1\mathbf{v}_1 + \cdots + c_n\mathbf{v}_n = 0$ with at least one $c_i \neq 0$ $\sum c_i \mathbf{v}_i = 0$ with at least one $c_i \neq 0$
linear equation	p.59	2.1	$a_1 x_1 + a_2 x_2 + \cdots + a_n x_n = b$
linearly independent set of vectors	p.94	2.3	no vector can be written as linear combination of others $c_1\mathbf{v}_1 + \cdots + c_n\mathbf{v}_n = 0 \Leftrightarrow$ all the $c_i = 0$ $\sum c_i \mathbf{v}_i = 0 \Leftrightarrow$ all the $c_i = 0$
linear system	p.59	2.1	a set of linear equations with the same variables
matrix	p.62	2.1	a rectangular array of numbers in rows and columns
network	p.104	2.4	nodes connected by a series of branches
Ohm's Law	p.106	2.4	force = resistance × current, $E = RI$
pivoting	p.70	2.2	see explanation in solution to Example 2.9 p.70
rank	p.75	2.2	number of nonzero rows in the REF of a matrix
reduced row echelon form	p.76	2.2	RREF: REF; leading entries, 1; all else, 0s
row echelon form	p.68	2.2	REF: zero rows, bottom; leading entries, left
row equivalent	p.72	2.2	matrix A can be converted into matrix B using EROs
row reduction	p.70	2.2	applying EROs to bring a matrix into REF
solution	p.60	2.1	$[s_1, s_2, \ldots, s_n]$ such that $a_1 s_1 + a_2 s_2 + \cdots + a_n s_n = b$
span(S)	p.92	2.3	all linear combinations of $S = \{\mathbf{s}_1, \mathbf{s}_2, \ldots, \mathbf{s}_n\}$
spanning set for $\mathbb{R}^n$	p.92	2.3	set S such that span$(S) = \mathbb{R}^n$, that is set S such that $\mathbf{v}$ in $\mathbb{R}^n \Rightarrow \mathbf{v} = \sum c_i \mathbf{s}_i$
strictly diagonally dominant	p.126	2.5	$\|a_{11}\| > \|a_{12}\| + \|a_{13}\| + \cdots + \|a_{1n}\|$ $\|a_{22}\| > \|a_{21}\| + \|a_{23}\| + \cdots + \|a_{2n}\|$ and $\|a_{nn}\| > \|a_{n1}\| + \|a_{n2}\| + \cdots + \|a_{n,n-1}\|$
Voltage Law	p.106	2.4	voltage drops around a circuit equals total voltage

Chapter 2 Review

Theorems

Thm 2.1	p.72	2.2	A and B are row equivalent $\Leftrightarrow$ they reduce to same REF
Thm 2.2	p.74	2.2	**Rank Thm**: number of free variables $= n - \text{rank}(A)$
Thm 2.3	p.80	2.2	$[A|0]$: m equations $< n$ variables $\Rightarrow$ infinitely many solutions
Thm 2.4	p.91	2.3	$[A|\mathbf{b}]$ is consistent $\Leftrightarrow \mathbf{b} = \sum c_i \mathbf{a}_i$
Thm 2.5	p.95	2.3	linearly dependent set $\Leftrightarrow$ linear combination of the others
Thm 2.6	p.97	2.3	linearly dependent set $\Leftrightarrow [A_c|0]$ has nontrivial solution
Thm 2.7	p.98	2.3	linearly dependent set $\Leftrightarrow \text{rank}(A_r) < m$ where A_r is $m \times n$
Thm 2.8	p.99	2.3	m vectors in $\mathbb{R}^n$ are linearly dependent if $m > n$
Thm 2.9	p.124	2.5	A strictly diagonally dominant $\Rightarrow$ iterates converge
Thm 2.10	p.124	2.5	methods converge $\Rightarrow$ they converge to the solution
Rank Thm	p.75	2.2	number of free variables $= n - \text{rank}(A)$

Symbolism

$R_i \leftrightarrow R_j$	p.70	2.2	interchange two rows
kR_i	p.70	2.2	multiply a row by a nonzero constant
$R_i + kR_j$	p.70	2.2	add a multiple of a row to another row
A_c		2.3	Created by taking vectors $\mathbf{v}_i$ as its columns
A_r		2.3	Created by taking vectors $\mathbf{v}_i$ as its rows
$\mathbf{b} = c_1\mathbf{v}_1 + \cdots + c_n\mathbf{v}_n$	p.638	A.5	$\mathbf{b}$ is a linear combination of $\mathbf{v}_i$ (see Appendix A)
$\mathbf{b} = \sum c_i \mathbf{v}_i$	p.638	A.5	$\mathbf{b}$ is a linear combination of $\mathbf{v}_i$ (see Appendix A)
$\mathbf{a}_i$	p.90	2.3	rows or columns of A written as vectors
$[A|\mathbf{b}]$	p.91	2.3	augmented matrix of a linear system
$[A|0]$	p.97	2.3	augmented matrix of a homogenous linear system

Solutions to selected exercises from Chapter 2 *Review*

1. We will explain and give counter examples to justify our answers below.

    (a) **False**. In Section 2.1, see the definition of an *inconsistent* system.
    A useful counter example is parallel lines in $\mathbb{R}^2$, like $x + y = 0$ and $x + y = 1$.

    (b) **True**. In Section 2.2, see remarks prior to the definition of an *homogenous* system.
    Q: Why does a homogenous system guarantee the associated lines intersect?
    A: Because all the associated lines pass through the origin.

    (c) **False**. In Section 2.2, see Theorem 2.6 (system must be homogenous).
    When there are fewer conditions than variables,
    we can solve a homogenous system.

    (d) **False**. In Section 2.2, see remarks under *Homogenous Systems* on p78.
    When a system has more equations than variables,
    it can either have a unique solution, infinitely many solutions, or no solution.

    (e) **True**. In Section 2.3, see Theorem 2.4 ($[A|\mathbf{b}]$ is consistent $\Leftrightarrow \mathbf{b} = \sum c_i \mathbf{a}_i$).
    Q: How might we state an informal proof of this theorem?
    A: The fact that $\mathbf{b} = \sum c_i \mathbf{a}_i$ says $\mathbf{x} = [c_i]$ is a solution of $[A|\mathbf{b}]$.

    (f) **False**. We need an additional condition to make this true. Which one?
    Q: If $\mathbf{u} \neq \mathbf{0}$ and $\mathbf{v}$ are linearly dependent, what is span$(\mathbf{u}, \mathbf{v})$?
    A: Then span$(\mathbf{u}, \mathbf{v}) = $ span$(\mathbf{u}) = c\mathbf{u}$, a line through the origin.
    Q: If $\mathbf{u}$ and $\mathbf{v}$ are linearly independent, what is span$(\mathbf{u}, \mathbf{v})$?
    A: Then span$(\mathbf{u}, \mathbf{v}) = c\mathbf{u} + d\mathbf{v}$, a plane a line through the origin.

    (g) **True**. Show this by proving the *contrapositive* (See Example 9 of Appendix A).
    Q: What is the *contrapositive* of this statement?
    A: If $\mathbf{u}$ and $\mathbf{v}$ are parallel, then they are linearly dependent.
    Q: How might we prove this statement is true?
    A: If $\mathbf{u}$ and $\mathbf{v}$ are parallel, then $\mathbf{v} = c\mathbf{u} \Rightarrow -c\mathbf{u} + \mathbf{v} = \mathbf{0}$.
    Q: Why does the fact that $\mathbf{u}$ and $\mathbf{v}$ are parallel imply $\mathbf{v} = c\mathbf{u}$?
    A: Vectors are defined to be *parallel* if they are scalar multiples of each other.
    Q: Why does $-c\mathbf{u} + \mathbf{v} = \mathbf{0}$ imply $\mathbf{u}$ and $\mathbf{v}$ are linearly dependent?
    A: Two vectors are linearly dependent if $c_1 \mathbf{v}_1 + c_2 \mathbf{v}_2 = \mathbf{0}$ (one $c_i \neq 0$).

    (h) **True**. Why? A *closed* path means there has been no displacement.
    Q: Since there is no displacement, what do know about the sum of the vectors?
    A: Since there is no displacement, we have $\sum \mathbf{v}_i = \mathbf{0}$.
    Q: Why does $\sum \mathbf{v}_i = \mathbf{0}$ imply $\mathbf{v}_i$ are linearly dependent?
    A: Vectors are linearly dependent if $\sum c_i \mathbf{v}_i = \mathbf{0}$ (at least one $c_i \neq 0$).

    (i) **False**. This pairwise condition is much *weaker* than linear independence.
    Consider this counter example: $\mathbf{u} = [1, 0, 0]$, $\mathbf{v} = [0, 1, 0]$, and $\mathbf{w} = [1, 1, 0]$.
    Geometrically, consider 3 lines in the same plane none of which are parallel.

    (j) **True**. In Section 2.3, see Thm 2.8 (m vectors in $\mathbb{R}^n$ are linearly dependent if $m > n$).
    Q: What is one way of stating Theorem 2.8 informally in our own words?
    A: When there are more vectors than entries, we can solve $\sum c_i \mathbf{v}_i = \mathbf{0}$.

Chapter 2 Review

3. As in Example 2.12 of Section 2.2, we form the augmented matrix and row reduce to solve.

$$\begin{bmatrix} 1 & 1 & -2 & | & 4 \\ 1 & 3 & -1 & | & 7 \\ 2 & 1 & -5 & | & 7 \end{bmatrix} \xrightarrow[R_3-2R_1]{R_2-R_1} \begin{bmatrix} 1 & 1 & -2 & | & 4 \\ 0 & 2 & 1 & | & 3 \\ 0 & -1 & -1 & | & -1 \end{bmatrix} \xrightarrow{-2R_3} \begin{bmatrix} 1 & 1 & -2 & | & 4 \\ 0 & 2 & 1 & | & 3 \\ 0 & 2 & 2 & | & 2 \end{bmatrix} \xrightarrow{R_3-R_2} \begin{bmatrix} 1 & 1 & -2 & | & 4 \\ 0 & 2 & 1 & | & 3 \\ 0 & 0 & 1 & | & -1 \end{bmatrix}$$

$$\xrightarrow[R_2+R_3]{R_1-2R_3} \begin{bmatrix} 1 & 1 & 0 & | & 2 \\ 0 & 2 & 0 & | & 4 \\ 0 & 0 & 1 & | & -1 \end{bmatrix} \xrightarrow{\frac{1}{2}R_2} \begin{bmatrix} 1 & 1 & 0 & | & 2 \\ 0 & 1 & 0 & | & 2 \\ 0 & 0 & 1 & | & -1 \end{bmatrix} \xrightarrow{R_1-R_2} \begin{bmatrix} 1 & 0 & 0 & | & 0 \\ 0 & 1 & 0 & | & 2 \\ 0 & 0 & 1 & | & -1 \end{bmatrix} \Rightarrow \begin{bmatrix} x_1 \\ x_2 \\ x_3 \end{bmatrix} = \begin{bmatrix} 0 \\ 2 \\ -1 \end{bmatrix}.$$

5. As in Example 2.16 of Section 2.2, we form the augmented matrix and row reduce over $\mathbb{Z}_7$. Since we are using modular arithmetic, we need only add and multiply. Why?

$$\begin{bmatrix} 2 & 3 & | & 4 \\ 1 & 2 & | & 3 \end{bmatrix} \xrightarrow{R_2+3R_1} \begin{bmatrix} 2 & 3 & | & 4 \\ 0 & 4 & | & 1 \end{bmatrix} \xrightarrow{2R_2} \begin{bmatrix} 2 & 3 & | & 4 \\ 0 & 1 & | & 2 \end{bmatrix} \xrightarrow{R_1+4R_2} \begin{bmatrix} 2 & 0 & | & 5 \\ 0 & 1 & | & 2 \end{bmatrix} \xrightarrow{4R_1} \begin{bmatrix} 1 & 0 & | & 6 \\ 0 & 1 & | & 2 \end{bmatrix}$$

So, the solution is $\begin{bmatrix} x \\ y \end{bmatrix} = \begin{bmatrix} 6 \\ 2 \end{bmatrix}$. We can check this: $\begin{matrix} 2(6) + 3(2) = 18 = 4 \\ 6 + 2(2) = 10 = 3 \end{matrix}$ in $\mathbb{Z}_7$.

7. As in Exercise 40 of Section 2.2, we row reduce to find the restrictions on k.
Note: The system has no solution when A has a zero row with corresponding constant $\neq 0$.

$$\begin{bmatrix} k & 2 & | & 1 \\ 1 & 2k & | & 1 \end{bmatrix} \xrightarrow{R_1 \leftrightarrow R_1} \begin{bmatrix} 1 & 2k & | & 1 \\ k & 2 & | & 1 \end{bmatrix} \xrightarrow{R_2-kR_1} \begin{bmatrix} 1 & 2k & | & 1 \\ 0 & 2-2k^2 & | & 1-k \end{bmatrix} \xrightarrow{-R_2} \begin{bmatrix} 1 & 2k & | & 1 \\ 0 & 2(k-1)(k+1) & | & k-1 \end{bmatrix}$$

So, the only value of k that creates a zero row with corresponding constant $\neq 0$ is $k = -1$.
That is, the only value of k for which this system is inconsistent is $k = -1$.

9. As in Example 2.15 of Section 2.2, we need to determine the point of intersection. We want to find an $\mathbf{x} = [x, y, z]$ that satisfies both equations simultaneously. That is, we want $\mathbf{x} = \mathbf{p} + s\mathbf{u} = \mathbf{q} + t\mathbf{v}$ or $s\mathbf{u} - t\mathbf{v} = \mathbf{q} - \mathbf{p}$.

Substituting the given $\mathbf{p}$, $\mathbf{q}$, $\mathbf{u}$, and $\mathbf{v}$ into $s\mathbf{u} - t\mathbf{v} = \mathbf{q} - \mathbf{p}$, we obtain the equations:

$$s\begin{bmatrix} 1 \\ -1 \\ 2 \end{bmatrix} - t\begin{bmatrix} -1 \\ 1 \\ 1 \end{bmatrix} = \begin{bmatrix} 5 \\ -2 \\ -4 \end{bmatrix} - \begin{bmatrix} 1 \\ 2 \\ 3 \end{bmatrix} \Rightarrow \begin{array}{r} s + t = 4 \\ -s - t = -4 \\ 2s - t = -7 \end{array}$$

We form the augmented matrix and row reduce to find values for s and t.

$$\begin{bmatrix} 1 & 2 & | & 4 \\ -1 & -1 & | & -4 \\ 2 & -1 & | & -7 \end{bmatrix} \longrightarrow \begin{bmatrix} 1 & 0 & | & -1 \\ 0 & 0 & | & 0 \\ 0 & 1 & | & 5 \end{bmatrix} \Rightarrow \text{So, } s = -1 \text{ and } t = 5.$$

Therefore, the point of intersection is: $\begin{bmatrix} x \\ y \\ z \end{bmatrix} = \begin{bmatrix} 1 \\ 2 \\ 3 \end{bmatrix} - 1 \begin{bmatrix} 1 \\ -1 \\ 2 \end{bmatrix} = \begin{bmatrix} 0 \\ 3 \\ 1 \end{bmatrix}.$

Check that substituting $t = 5$ into the other equation gives the same point.

11. As in Example 2.21 of Section 2.3, the equation of the plane we are looking for is:

$$\begin{bmatrix} x \\ y \\ z \end{bmatrix} = s\begin{bmatrix} 1 \\ 1 \\ 1 \end{bmatrix} + t\begin{bmatrix} 3 \\ 2 \\ 1 \end{bmatrix} \Rightarrow \begin{array}{r} s + 3t = x \\ s + 2t = y \\ s + t = z \end{array}$$

We form the augmented matrix and row reduce to find conditions for x, y, and z.

$$\begin{bmatrix} 1 & 3 & | & x \\ 1 & 2 & | & y \\ 1 & 1 & | & z \end{bmatrix} \xrightarrow{\substack{R_2 - R_1 \\ R_3 - R_1}} \begin{bmatrix} 1 & 3 & | & x \\ 0 & -1 & | & -x+y \\ 0 & -2 & | & -x+z \end{bmatrix} \xrightarrow{-R_2} \begin{bmatrix} 1 & 3 & | & x \\ 0 & 1 & | & x-y \\ 0 & -2 & | & -x+z \end{bmatrix} \xrightarrow{R_3 + 2R_2} \begin{bmatrix} 1 & 3 & | & x \\ 0 & 1 & | & x-y \\ 0 & 0 & | & x-2y+z \end{bmatrix}$$

By assumption the system is consistent so $x - 2y + z = 0$, the equation of the plane we sought.

Q: How can we verify that both these vectors lie in the plane?
A: By computing $1 - 2(1) + 1 = 0$ and $3 - 2(2) + 1 = 0$.

Q: What is the cross product of the given vectors?
A: $[-1, 2, -1]$. Should this agree with the normal of our plane? Does it?

Chapter 2 Review

13. We use Exercise 21 of Section 2.3 to determine whether or not $\mathbb{R}^3 = \text{span}(\mathbf{u}, \mathbf{v}, \mathbf{w})$.

    (a) We need only show $\mathbf{e}_1, \mathbf{e}_2, \mathbf{e}_3$ are linear combinations of $\mathbf{u}, \mathbf{v}$, and $\mathbf{w}$.
    Why is that enough? Because then we have $\mathbb{R}^3 = \text{span}(\mathbf{e}_1, \mathbf{e}_2, \mathbf{e}_3) = \text{span}(\mathbf{u}, \mathbf{v}, \mathbf{w})$.
    We begin with $\mathbf{e}_1$. We want to find scalars x, y, and z such that:

    $$x \begin{bmatrix} 1 \\ 1 \\ 0 \end{bmatrix} + y \begin{bmatrix} 1 \\ 0 \\ 1 \end{bmatrix} + z \begin{bmatrix} 0 \\ 1 \\ 1 \end{bmatrix} = \mathbf{e}_1 = \begin{bmatrix} 1 \\ 0 \\ 0 \end{bmatrix}$$

    We form the augmented matrix and row reduce to find values for x, y, and z.

    $$\begin{bmatrix} 1 & 1 & 0 & | & 1 \\ 1 & 0 & 1 & | & 0 \\ 0 & 1 & 1 & | & 0 \end{bmatrix} \longrightarrow \begin{bmatrix} 1 & 0 & 0 & | & \frac{1}{2} \\ 0 & 1 & 0 & | & \frac{1}{2} \\ 0 & 0 & 1 & | & -\frac{1}{2} \end{bmatrix} \Rightarrow \mathbf{e}_1 = \frac{1}{2}\mathbf{u} + \frac{1}{2}\mathbf{v} - \frac{1}{2}\mathbf{w}.$$

    Likewise for $\mathbf{e}_2$, we form the augmented matrix and row reduce.

    $$\begin{bmatrix} 1 & 1 & 0 & | & 0 \\ 1 & 0 & 1 & | & 1 \\ 0 & 1 & 1 & | & 0 \end{bmatrix} \longrightarrow \begin{bmatrix} 1 & 0 & 0 & | & \frac{1}{2} \\ 0 & 1 & 0 & | & -\frac{1}{2} \\ 0 & 0 & 1 & | & \frac{1}{2} \end{bmatrix} \Rightarrow \mathbf{e}_1 = \frac{1}{2}\mathbf{u} - \frac{1}{2}\mathbf{v} + \frac{1}{2}\mathbf{w}.$$

    Finally for $\mathbf{e}_3$, we form the augmented matrix and row reduce.

    $$\begin{bmatrix} 1 & 1 & 0 & | & 0 \\ 1 & 0 & 1 & | & 0 \\ 0 & 1 & 1 & | & 1 \end{bmatrix} \longrightarrow \begin{bmatrix} 1 & 0 & 0 & | & -\frac{1}{2} \\ 0 & 1 & 0 & | & \frac{1}{2} \\ 0 & 0 & 1 & | & \frac{1}{2} \end{bmatrix} \Rightarrow \mathbf{e}_1 = -\frac{1}{2}\mathbf{u} + \frac{1}{2}\mathbf{v} + \frac{1}{2}\mathbf{w}.$$

    What do we conclude? $\mathbb{R}^3 = \text{span}(\mathbf{e}_1, \mathbf{e}_2, \mathbf{e}_3) = \text{span}(\mathbf{u}, \mathbf{v}, \mathbf{w})$.
    We could also have used our intuition to solve this problem. How?

    $$\begin{bmatrix} 1 \\ 1 \\ 0 \end{bmatrix} + \begin{bmatrix} 1 \\ 0 \\ 1 \end{bmatrix} + \begin{bmatrix} 0 \\ 1 \\ 1 \end{bmatrix} = \begin{bmatrix} 2 \\ 2 \\ 2 \end{bmatrix}. \text{ So } \frac{1}{2}\begin{bmatrix} 2 \\ 2 \\ 2 \end{bmatrix} = \begin{bmatrix} 1 \\ 1 \\ 1 \end{bmatrix} \Rightarrow \begin{bmatrix} 1 \\ 1 \\ 1 \end{bmatrix} - \begin{bmatrix} 1 \\ 1 \\ 0 \end{bmatrix} = \begin{bmatrix} 0 \\ 0 \\ 1 \end{bmatrix} = \mathbf{e}_3 \ldots$$

    (b) These are clearly linearly dependent because $\mathbf{w} = \mathbf{u} + \mathbf{v}$.
    Therefore, $\mathbb{R}^3$ is not equal to $\text{span}(\mathbf{u}, \mathbf{v}, \mathbf{w})$.
    In order for a set to span $\mathbb{R}^3$ is must have 3 linearly independent vectors.

15. Since $\mathbf{a}_1, \mathbf{a}_2$, and $\mathbf{a}_3$ are linearly dependent, $\text{rank}(A) \leq 2 < 3$.
    Since $\mathbf{a}_1, \mathbf{a}_2$, and $\mathbf{a}_3$ are not zero, $0 < \text{rank}(A)$.
    Combining these two conditions, we see that $\text{rank}(A)$ must be 1 or 2.

17. As in Exercise 43 of Section 2.3, we apply the definition of linear independence.

    We will show that $\mathbf{u} + \mathbf{v}$ and $\mathbf{u} - \mathbf{v}$ are linearly independent.
    Given $c_1(\mathbf{u} + \mathbf{v}) + c_2(\mathbf{u} - \mathbf{v}) = 0$, we will show $c_1 = c_2 = 0$.

    Multiplying and gathering like terms yields: $(c_1 + c_2)\mathbf{u} + (c_1 - c_2)\mathbf{v} = 0$.
    Since $\mathbf{u}$ and $\mathbf{v}$ are linearly independent, $c_1 + c_2 = c_1 - c_2 = 0 \Rightarrow c_1 = c_2 = 0$.

    Also we could create the matrix of coefficients A and row reduce to determine its rank:

    $$\begin{matrix} c_1 + c_2 = 0 \\ c_1 - c_2 = 0 \end{matrix} \Rightarrow \begin{bmatrix} 1 & 1 \\ 1 & -1 \end{bmatrix} \longrightarrow \begin{bmatrix} 1 & 0 \\ 0 & 1 \end{bmatrix}$$

    Since $\text{rank}(A) = 2$ the only solution is the trivial one, so $c_1 = c_2 = 0$.

19. In order for $[A \mid \mathbf{b}]$ to be consistent, rank(A) must equal rank$([A \mid \mathbf{b}])$. Why? A system has no solution when A has a zero row with corresponding constant $\neq 0$.

Note that rank$([A \mid \mathbf{b}])$ cannot be less than rank(A). Why not?

Chapter 3

Matrices

3.1 Matrix Operations

Q: What is our goal for Section 3.1?

A: To develop our intuition about matrix operations, especially matrix multiplication. We need to get comfortable with the pattern of how matrix multiplication is performed. Note that it is *not* commutative. That is, AB is *not* always equal to BA.

Key Definitions and Concepts

A^T	p.149	3.1	A-transpose, created by switching rows and columns
$\mathbf{e}_i$	p.142	3.1	standard $1 \times m$ unit vector
$\mathbf{e}_j$	p.143	3.1	standard $n \times 1$ unit vector
induction	p.147	3.1	See Appendix B, *Mathematical Induction*
matrix product	p.139	3.1	$C = AB \Rightarrow c_{ij} = a_{i1}b_{1j} + a_{i2}b_{2j} + \cdots + a_{in}b_{nj}$
outer product	p.145	3.1	see description of this process after Example 3.10
symmetric	p.149	3.1	$A^T = A$
transpose	p.148	3.1	A^T, create by switching rows and columns

Theorems

Thm 3.1   p.142   3.1 $\mathbf{e}_i A$ is row of A, $A\mathbf{e}_j$ is column of A

Discussion of Induction (Other concepts and theorems are discussed with the Exercises.)

Given $A = \begin{bmatrix} 1 & 1 \\ 0 & 1 \end{bmatrix}$, we will prove $A^n = \begin{bmatrix} 1 & n \\ 0 & 1 \end{bmatrix}$ for $n \geq 1$ using *induction*.
See Appendix B for discussion and examples of *Mathematical Induction*.

1: $A^1 = \begin{bmatrix} 1 & 1 \\ 0 & 1 \end{bmatrix}$. This is obvious, so there is nothing to show.

n: $A^n = \begin{bmatrix} 1 & n \\ 0 & 1 \end{bmatrix}$. This is the induction hypothesis.

$n+1$: $A^{n+1} = \begin{bmatrix} 1 & n+1 \\ 0 & 1 \end{bmatrix}$. This is the statement we must prove using the induction hypothesis.

$$A^{n+1} = A^1 A^n \stackrel{\text{by induction}}{=} \begin{bmatrix} 1 & 1 \\ 0 & 1 \end{bmatrix} \begin{bmatrix} 1 & n \\ 0 & 1 \end{bmatrix} \stackrel{\text{by matrix multiplication}}{=} \begin{bmatrix} 1 & n+1 \\ 0 & 1 \end{bmatrix}$$

We have proven (by induction) that $A^n = \begin{bmatrix} 1 & n \\ 0 & 1 \end{bmatrix}$ for $n \geq 1$.

Solutions to odd-numbered exercises from Section 3.1

1. Following Examples 3.1 through 3.5, we have:
$$A + 2D = \begin{bmatrix} 3 & 0 \\ -1 & 5 \end{bmatrix} + 2 \begin{bmatrix} 0 & -3 \\ -2 & 1 \end{bmatrix} = \begin{bmatrix} 3 + 2(0) & 0 + 2(-3) \\ -1 + 2(-2) & 5 + 2(1) \end{bmatrix} = \begin{bmatrix} 3 & -6 \\ -5 & 7 \end{bmatrix}.$$

3. $B - C$ is not possible. Why not? B is a 2×3 matrix and C is a 3×2 matrix. We can only add and subtract matrices of the same size.

5. By the definition of the matrix product, $C = AB$, and Example 3.6, we have: $AB =$
$$\begin{bmatrix} 3 & 0 \\ -1 & 5 \end{bmatrix} \begin{bmatrix} 4 & -2 & 1 \\ 0 & 2 & 3 \end{bmatrix} = \begin{bmatrix} 3(4) + 0(0) & 3(-2) + 0(2) & 3(1) + 0(3) \\ (-1)(4) + 5(0) & 3(-2) + (-1)(2) & 3(1) + (-1)(3) \end{bmatrix} = \begin{bmatrix} 12 & -6 & 3 \\ -4 & 12 & 14 \end{bmatrix}.$$

7. We begin by applying the definition of matrix multiplication to see if BC is possible:
Since B is $[2 \times 3]$ and C is $[3 \times 2]$, BC, $[2 \times 3][3 \times 2]$, is possible. Why?
Because the *inner* numbers match. What does that tell us?
The number of columns in $B = 3 =$ the number of rows in C.

Furthermore, since BC is $[2 \times 3][3 \times 2]$, BC will be a 2×2 matrix.
Since D is also a 2×2 matrix, we can add them together. That is, $D + BC$ is possible.

$$\text{First, } BC = \begin{bmatrix} 4 & -2 & 1 \\ 0 & 2 & 3 \end{bmatrix} \begin{bmatrix} 1 & 2 \\ 3 & 4 \\ 5 & 6 \end{bmatrix} = \begin{bmatrix} 4(1) - 2(3) + 1(5) & 4(2) - 2(4) + 1(6) \\ 0(1) + 2(3) + 3(5) & 0(2) + 2(4) + 3(6) \end{bmatrix} = \begin{bmatrix} 3 & 6 \\ 21 & 26 \end{bmatrix}.$$

$$\text{So, } D + BC = \begin{bmatrix} 0 & -3 \\ -2 & 1 \end{bmatrix} + \begin{bmatrix} 3 & 6 \\ 21 & 26 \end{bmatrix} = \begin{bmatrix} 3 & 3 \\ 19 & 27 \end{bmatrix}.$$

9. Before we begin, we should determine if AF and $E(AF)$ are possible.

Since A is $[2 \times 2]$ and F is $[2 \times 1]$, AF, $[2 \times 2][2 \times 1]$, is possible. Why?
Because the *inner* numbers match. What does that tell us?
The number of columns in $A = 2 =$ the number of rows in F.
Furthermore, since AF is $[2 \times 2][2 \times 1]$, AF will be a 2×1 matrix.

Since E is $[1 \times 2]$ and AF is $[2 \times 1]$, $E(AF)$, $[1 \times 2][2 \times 1]$, is possible. Why?
Because the *inner* numbers match. What does that tell us?
The number of columns in $E = 1 =$ the number of rows in AF.
Furthermore, since $E(AF)$ is $[1 \times 2][2 \times 1]$, $E(AF)$ will be a 1×1 matrix.

$$\text{First, } AF = \begin{bmatrix} 3 & 0 \\ -1 & 5 \end{bmatrix} \begin{bmatrix} -1 \\ 2 \end{bmatrix} = \begin{bmatrix} 3(-1) + 0(2) \\ (-1)(-1) + 5(2) \end{bmatrix} = \begin{bmatrix} -3 \\ 11 \end{bmatrix}.$$

$$\text{So } E(AF) = [4 \ 2] \begin{bmatrix} -3 \\ 11 \end{bmatrix} = [4(-3) + 2(11)] = [10].$$

11. Since FE, $[2 \times 1][1 \times 2]$, is possible and yields a 2×2 matrix, we have:
$$FE = \begin{bmatrix} -1 \\ 2 \end{bmatrix} [4 \ 2] = \begin{bmatrix} -1(4) & (-1)(2) \\ 2(4) & 2(2) \end{bmatrix} = \begin{bmatrix} -4 & -2 \\ 8 & 4 \end{bmatrix}.$$

Q: Does FE equal EF?
A: No. In fact, note that EF, $[1 \times 2][2 \times 1]$, is possible and yields a 1×1 matrix.
This is a good example of the general fact that matrix multiplication does not commute.

3.1 Matrix Operations

Since Exercises 13 through 15 use the skills above, we simply present the answers below.

13. $B^T C^T - (CB)^T = \begin{bmatrix} 0 & 0 & 0 \\ 0 & 0 & 0 \\ 0 & 0 & 0 \end{bmatrix}.$

15. $A^3 = \begin{bmatrix} 27 & 0 \\ -49 & 125 \end{bmatrix}.$

17. We should use this exercise to increase our understanding of matrix multiplication. To help us find an example, we should first take a look at the general pattern.

$A^2 = AA = \begin{bmatrix} a & b \\ c & d \end{bmatrix} \begin{bmatrix} a & b \\ c & d \end{bmatrix} = \begin{bmatrix} a^2 + bc & ab + bd \\ ac + cd & bc + d^2 \end{bmatrix} = \begin{bmatrix} 0 & 0 \\ 0 & 0 \end{bmatrix}.$ What do we have?

We have $ac + cd = 0 \Rightarrow ac = -cd \Rightarrow a = -d$, provided $c \neq 0$. Let $a = 1$. Then $d = -1 \Rightarrow 1 + bc = 0 \Rightarrow bc = -1$, so $b = 1$ and $c = -1$ should work.

Check: $A^2 = AA = \begin{bmatrix} 1 & 1 \\ -1 & -1 \end{bmatrix} \begin{bmatrix} 1 & 1 \\ -1 & -1 \end{bmatrix} = \begin{bmatrix} 1-1 & 1-1 \\ -1+1 & -1+1 \end{bmatrix} = \begin{bmatrix} 0 & 0 \\ 0 & 0 \end{bmatrix}.$

So, $A = \begin{bmatrix} 1 & 1 \\ -1 & -1 \end{bmatrix}$ works.

Q: What other choices for a, b, c, and d work? Is there a pattern we can see?
A: Any matrix all of whose entries have the same absolute value where $\mathbf{A}_1 = -\mathbf{A}_2$ will work.

19. The number of units of each product shipped to each warehouse is given by $A = \begin{bmatrix} 200 & 75 \\ 150 & 100 \\ 100 & 125 \end{bmatrix}.$

The cost of shipping one unit of each product is given by $B = \begin{bmatrix} 1.50 & 1.00 & 2.00 \\ 1.75 & 1.50 & 1.00 \end{bmatrix}$

(where b_{ij} is the cost of shipping a unit of product j by $i = 1$ truck, $i = 2$ train).
Compare the cost of shipping the products to each of the warehouses:

$BA = \begin{bmatrix} 1.50 & 1.00 & 2.00 \\ 1.75 & 1.50 & 1.00 \end{bmatrix} \begin{bmatrix} 200 & 75 \\ 150 & 100 \\ 100 & 125 \end{bmatrix} = \begin{bmatrix} 650.00 & 462.50 \\ 675.00 & 406.25 \end{bmatrix} \Rightarrow$

It is cheaper to ship the products to warehouse 1 by truck, but to warehouse 2 by train.

21. $\begin{bmatrix} 1 & -2 & 3 \\ 2 & 1 & -5 \end{bmatrix} \begin{bmatrix} x_1 \\ x_2 \\ x_3 \end{bmatrix} = \begin{bmatrix} 0 \\ 4 \end{bmatrix}.$

23. $A\mathbf{b}_1 = 2 \begin{bmatrix} 1 \\ -3 \\ 2 \end{bmatrix} + \begin{bmatrix} 0 \\ 1 \\ 0 \end{bmatrix} - \begin{bmatrix} -2 \\ 1 \\ -1 \end{bmatrix} = \begin{bmatrix} 4 \\ -6 \\ 5 \end{bmatrix},$ $A\mathbf{b}_2 = 3 \begin{bmatrix} 1 \\ -3 \\ 2 \end{bmatrix} - \begin{bmatrix} 0 \\ 1 \\ 0 \end{bmatrix} + 6 \begin{bmatrix} -2 \\ 1 \\ -1 \end{bmatrix} = \begin{bmatrix} -9 \\ -4 \\ 0 \end{bmatrix},$

and $A\mathbf{b}_3 = \begin{bmatrix} 0 \\ 1 \\ 0 \end{bmatrix} + 4 \begin{bmatrix} -2 \\ 1 \\ -1 \end{bmatrix} = \begin{bmatrix} -8 \\ 5 \\ -4 \end{bmatrix}.$ Therefore, $AB = \begin{bmatrix} 4 & -9 & -8 \\ -6 & -4 & 5 \\ 5 & 0 & -4 \end{bmatrix}.$

25. The outer product expansion of AB is

$$\mathbf{a}_1\mathbf{B}_1 + \mathbf{a}_2\mathbf{B}_2 + \mathbf{a}_3\mathbf{B}_3 = \begin{bmatrix} 2 & 3 & 0 \\ -6 & -9 & 0 \\ 4 & 6 & 0 \end{bmatrix} + \begin{bmatrix} 0 & 0 & 0 \\ 1 & -1 & 1 \\ 0 & 0 & 0 \end{bmatrix} + \begin{bmatrix} 2 & -12 & -8 \\ -1 & 6 & 4 \\ 1 & -6 & -4 \end{bmatrix}$$

$$= \begin{bmatrix} 4 & -9 & -8 \\ -6 & -4 & 5 \\ 5 & 0 & -4 \end{bmatrix}.$$

27. $\mathbf{B}_1 A = 2\begin{bmatrix} 1 & 0 & -2 \end{bmatrix} + 3\begin{bmatrix} -3 & 1 & 1 \end{bmatrix} = \begin{bmatrix} -7 & 3 & -1 \end{bmatrix}$,

$\mathbf{B}_2 A = 1\begin{bmatrix} 1 & 0 & -2 \end{bmatrix} - \begin{bmatrix} -3 & 1 & 1 \end{bmatrix} + \begin{bmatrix} 2 & 0 & -1 \end{bmatrix} = \begin{bmatrix} 6 & -1 & -4 \end{bmatrix}$, and

$\mathbf{B}_3 A = -\begin{bmatrix} 1 & 0 & -2 \end{bmatrix} + 6\begin{bmatrix} -3 & 1 & 1 \end{bmatrix} + 4\begin{bmatrix} 2 & 0 & -1 \end{bmatrix} = \begin{bmatrix} -11 & 6 & 4 \end{bmatrix}$.

Therefore, $BA = \begin{bmatrix} -7 & 3 & -1 \\ 6 & -1 & -4 \\ -11 & 6 & 4 \end{bmatrix}$.

29. Assume that the columns of $B = \begin{bmatrix} \mathbf{b}_1 & \mathbf{b}_2 & \cdots & \mathbf{b}_n \end{bmatrix}$ are linearly dependent $\Rightarrow$
There exists a solution to $x_1\mathbf{b}_1 + x_2\mathbf{b}_2 + \cdots + x_n\mathbf{b}_n = \mathbf{0}$ (at least one $x_i \neq 0$).
Now consider the partition of AB in terms of column vectors:
$AB = A\begin{bmatrix} \mathbf{b}_1 & \mathbf{b}_2 & \cdots & \mathbf{b}_n \end{bmatrix} = \begin{bmatrix} A\mathbf{b}_1 & A\mathbf{b}_2 & \cdots & A\mathbf{b}_n \end{bmatrix}$.
But then $A(x_1\mathbf{b}_1 + x_2\mathbf{b}_2 + \cdots + x_n\mathbf{b}_n) = x_1(A\mathbf{b}_1) + x_2(A\mathbf{b}_2) + \cdots + x_n(A\mathbf{b}_n) = \mathbf{0}$,
showing that the columns of AB are linearly dependent.

31. For matrices A, B we have the block structure $A = \begin{bmatrix} A_{11} & A_{12} \\ A_{21} & A_{22} \end{bmatrix}$ and $B = \begin{bmatrix} B_{11} & B_{12} \\ B_{21} & B_{22} \end{bmatrix} \Rightarrow$

$$AB = \begin{bmatrix} A_{11} & A_{12} \\ A_{21} & A_{22} \end{bmatrix}\begin{bmatrix} B_{11} & B_{12} \\ B_{21} & B_{22} \end{bmatrix} = \begin{bmatrix} A_{11}B_{11} + A_{12}B_{21} & A_{11}B_{12} + A_{12}B_{22} \\ A_{21}B_{11} + A_{22}B_{21} & A_{21}B_{12} + A_{22}B_{22} \end{bmatrix}$$

$$= \begin{bmatrix} \begin{bmatrix} 1 & -1 \\ 0 & 1 \end{bmatrix}\begin{bmatrix} 2 & 3 \\ -1 & 1 \end{bmatrix} + \begin{bmatrix} 0 & 0 \\ 0 & 0 \end{bmatrix}\begin{bmatrix} 0 & 0 \\ 0 & 0 \end{bmatrix} & \begin{bmatrix} 1 & -1 \\ 0 & 1 \end{bmatrix}\begin{bmatrix} 0 \\ 0 \end{bmatrix} + \begin{bmatrix} 0 & 0 \\ 0 & 0 \end{bmatrix}\begin{bmatrix} 1 \\ 1 \end{bmatrix} \\ \begin{bmatrix} 0 & 0 \end{bmatrix}\begin{bmatrix} 2 & 3 \\ -1 & 1 \end{bmatrix} + \begin{bmatrix} 2 & 3 \end{bmatrix}\begin{bmatrix} 0 & 0 \\ 0 & 0 \end{bmatrix} & \begin{bmatrix} 0 & 0 \end{bmatrix}\begin{bmatrix} 0 \\ 0 \end{bmatrix} + \begin{bmatrix} 2 & 3 \end{bmatrix}\begin{bmatrix} 1 \\ 1 \end{bmatrix} \end{bmatrix}$$

$$= \begin{bmatrix} 3 & 2 & 0 \\ -1 & 1 & 0 \\ 0 & 0 & 5 \end{bmatrix}.$$

33. $AB = \begin{bmatrix} 1 & 2 & 2 & 0 \\ 3 & 4 & 5 & 3 \\ 1 & 0 & 1 & 2 \\ 0 & 1 & 0 & -1 \end{bmatrix}.$

3.1 Matrix Operations

35. (a) Computing the powers of matrix A as required, we have:
$$A^2 = \begin{bmatrix} -1 & 1 \\ -1 & 0 \end{bmatrix}, \ A^3 = \begin{bmatrix} -1 & 0 \\ 0 & -1 \end{bmatrix}, \ A^4 = \begin{bmatrix} 0 & -1 \\ 1 & -1 \end{bmatrix},$$
$$A^5 = \begin{bmatrix} 1 & -1 \\ 1 & 0 \end{bmatrix}, \ A^6 = \begin{bmatrix} 1 & 0 \\ 0 & 1 \end{bmatrix}, \ A^7 = \begin{bmatrix} 0 & 1 \\ -1 & 1 \end{bmatrix} = A!$$

(b) From our work in (a), we see that $A^1 = A^7 = A^{1 \cdot 6+1}$.
So the powers of A that actually create *distinct* matrices act like $\mathbb{Z}_6$.

See Section 1.4, Examples 1.32 through 1.35.

So, to determine A^{2001}, we should first determine the value of 2001 in $\mathbb{Z}_6$.
How? Divide 2001 by 6 and look at the remainder: $2001 = 333 \cdot 6 + 3 = 3$ in $\mathbb{Z}_6$.

Therefore $A^{2001} = A^{333 \cdot 6+3} = A^3 = \begin{bmatrix} -1 & 0 \\ 0 & -1 \end{bmatrix}$.

37. Given $A = \begin{bmatrix} 1 & 1 \\ 0 & 1 \end{bmatrix}$, we will prove $A^n = \begin{bmatrix} 1 & n \\ 0 & 1 \end{bmatrix}$ for $n \geq 1$ using *induction*.

See Appendix B for discussion and examples of *Mathematical Induction*.

1: $A^1 = \begin{bmatrix} 1 & 1 \\ 0 & 1 \end{bmatrix}$. This is obvious, so there is nothing to show.

n: $A^n = \begin{bmatrix} 1 & n \\ 0 & 1 \end{bmatrix}$. This is the induction hypothesis.

$n+1$: $A^{n+1} = \begin{bmatrix} 1 & n+1 \\ 0 & 1 \end{bmatrix}$. This is the statement we must prove using the induction hypothesis.

$$A^{n+1} = A^1 A^n \stackrel{\text{by induction}}{=} \begin{bmatrix} 1 & 1 \\ 0 & 1 \end{bmatrix} \begin{bmatrix} 1 & n \\ 0 & 1 \end{bmatrix} \stackrel{\text{by matrix multiplication}}{=} \begin{bmatrix} 1 & n+1 \\ 0 & 1 \end{bmatrix}$$

We have proven (by induction) that $A^n = \begin{bmatrix} 1 & n \\ 0 & 1 \end{bmatrix}$ for $n \geq 1$.

39. (a) $A = \begin{bmatrix} 1 & -1 & 1 & -1 \\ -1 & 1 & -1 & 1 \\ 1 & -1 & 1 & -1 \\ -1 & 1 & -1 & 1 \end{bmatrix}$.

(b) $A = \begin{bmatrix} 0 & 1 & 2 & 3 \\ -1 & 0 & 1 & 2 \\ -2 & -1 & 0 & 1 \\ -3 & -2 & -1 & 0 \end{bmatrix}$.

(c) $A = \begin{bmatrix} 0 & 0 & 0 & 0 \\ 1 & 1 & 1 & 1 \\ 2 & 4 & 8 & 16 \\ 3 & 9 & 27 & 81 \end{bmatrix}$.

(d) $A = \begin{bmatrix} \frac{1}{2}\sqrt{2} & 1 & \frac{1}{2}\sqrt{2} & 0 \\ 1 & \frac{1}{2}\sqrt{2} & 0 & -\frac{1}{2}\sqrt{2} \\ \frac{1}{2}\sqrt{2} & 0 & -\frac{1}{2}\sqrt{2} & -1 \\ 0 & -\frac{1}{2}\sqrt{2} & -1 & -\frac{1}{2}\sqrt{2} \end{bmatrix}$.

41. Let A be an $m \times n$ matrix, and $\mathbf{e}_i$ a $1 \times m$ standard unit vector.
If $a_1, a_2, \ldots, a_m$ are the rows of A then the product $\mathbf{e}_i A$ can be written
$\mathbf{e}_i A = 0 \cdot a_1 + 0 \cdot a_2 + \cdots + 1 \cdot a_i + \cdots + 0 \cdot a_m = a_i$ which is the ith row of A.
We could also prove this by direct calculation:

$$\mathbf{e}_i A = \begin{bmatrix} 0 & \cdots & 1 & \cdots & 0 \end{bmatrix} \begin{bmatrix} a_{11} & a_{12} & \cdots & a_{1n} \\ \vdots & \vdots & \vdots & \vdots \\ a_{i1} & a_{i2} & \cdots & a_{in} \\ \vdots & \vdots & \vdots & \vdots \\ a_{m1} & a_{m2} & \cdots & a_{mn} \end{bmatrix} = \begin{bmatrix} a_{i1} & a_{i2} & \cdots & a_{in} \end{bmatrix}$$

since the 1 in $\mathbf{e}_i$ is the ith entry.

3.2 Matrix Algebra

Q: What are our main goals for Section 3.2?

A: To see that algebra with matrices works as expected with a couple of important exceptions:
1) For addition, subtraction, and scalar multiplication, the matrices must be the same size.
2) For multiplication, only when A is $[m \times n]$ and B is $[n \times p]$ is AB possible. Why?
3) Since AB is *not* always equal to BA, we have left and right distribution.

Since addition, subtraction and scalar multiplication operate as they did for vectors, we can define linear combination, dependence, and independence precisely as we did for vectors in Section 2.3. This is a key understanding to develop in this section.

Key Definitions and Concepts

A^T	p.149	3.1	A-transpose, created by switching rows and columns
O	p.152	3.2	O is commonly used to stand for the zero matrix
induction	p.147	3.1	See Appendix B, *Mathematical Induction*
linear combination	p.152	2.3	$\mathbf{B} = c_1 \mathbf{A}_1 + \cdots + c_n \mathbf{A}_n$ $\mathbf{B} = \sum c_i \mathbf{A}_i$
linearly dependent	p.155	3.2	matrix can be written as linear combination of others $c_1 \mathbf{A}_1 + \cdots + c_n \mathbf{A}_n = O$ with at least one $c_i \neq 0$
linearly independent	p.155	3.2	no matrix can be written as linear combination of others $c_1 \mathbf{A}_1 + \cdots + c_n \mathbf{A}_n = O \Leftrightarrow$ all the $c_i = 0$
matrix product	p.139	3.1	$C = AB \Rightarrow c_{ij} = a_{i1}b_{1j} + a_{i2}b_{2j} + \cdots + a_{in}b_{nj}$
skew-symmetric	p.160	3.2	$A^T = -A$
symmetric	p.149	3.1	$A^T = A$
trace	p.160	3.2	$\text{tr}(A) = u_{11} + u_{22} + \cdots + u_{nn}$ That is, the *trace* is the sum of the diagonal entries
transpose	p.148	3.1	A^T, create by switching rows and columns
upper triangular	p.160	3.2	matrix whose entries below the main diagonal are all zero

Theorems

Thm 3.2	p.152	3.2	Properties of Matrix Addition and Scalar Multiplication
Thm 3.2	p.156	3.2	Properties of Matrix Multiplication
Thm 3.2	p.157	3.2	Properties of the Transpose, A^T
Thm 3.2.4	p.159	3.2	a. $A + A^T$ is symmetric if A is square AA^T and $A^T A$ b. AA^T and $A^T A$ are always symmetric

Discussion of Key Definitions and Concepts

Q: Do addition, subtraction, and scalar multiplication work for matrices in the same way they work for the variables we are used to and like they do for vectors?

A: Yes, provided those matrices are the same size. For example:

Given $A = \begin{bmatrix} 1 & 2 \\ 3 & 4 \end{bmatrix}$ and $B = \begin{bmatrix} -1 & 0 \\ 1 & 1 \end{bmatrix}$, we can solve $X - 2A + 3B = 0$ for X.

We find $X = 2A - 3B$ and then substitute to find $X = 2\begin{bmatrix} 1 & 2 \\ 3 & 4 \end{bmatrix} - 3\begin{bmatrix} -1 & 0 \\ 1 & 1 \end{bmatrix} = \begin{bmatrix} 5 & 4 \\ 3 & 5 \end{bmatrix}$.

Q: What does it mean to say matrix B is a linear combination of matrices A_1 and A_2?

A: This means $B = c_1 A_1 + c_2 A_2$ exactly as it does for vectors (see Section 2.3).

Q: If B commutes with A_1 and A_2, does B commute with $M = c_1 A_1 + c_2 A_2$?

A: $BM = B(c_1 A_1 + c_2 A_2) = c_1 B A_1 + c_2 B A_2 = c_1 A_1 B + c_2 A_2 B = (c_1 A_1 + c_2 A_2)B = MB$.
What does this show? Yes, B does commute with M. This is useful in Exercises 26 and 27.

Q: What does it mean to say matrices A_1, A_2, and A_3 are linearly dependent?

A: This means $c_1 A_1 + c_2 A_2 + c_3 A_3 = O$ with at least one $c_i \neq 0$.
Again, this is exactly the same thing it means for vectors (see Section 2.3).
Note that we use O to stand for the zero matrix. This is a common usage.

Q: What does it mean to say matrices A_1, A_2, and A_3 are linearly independent?

A: This means $c_1 A_1 + c_2 A_2 + c_3 A_3 = O$ if and only if $c_1 = c_2 = c_3 = 0$.
Once more, this is exactly the same thing it means for vectors (see Section 2.3).
Note again that we use O to stand for the zero matrix. This is a common usage.

Another Example of Induction

We need to show $(A^r)^T = (A^T)^r$. We will prove this using *induction*.
See Appendix B for discussion and examples of *Mathematical Induction*.

1: $(A^1)^T = (A^T)^1$. This is obvious, so there is nothing to show.

r: $(A^r)^T = (A^T)^r$. This is the induction hypothesis.

$r+1$: $(A^{r+1})^T = (A^T)^{r+1}$. This is the statement we must prove using the induction hypothesis.

$$(A^{r+1})^T = (A \cdot A^r)^T \overset{\substack{\text{by Thm 3d} \\ (AB)^T = B^T A^T}}{=} (A^r)^T A^T \overset{\substack{\text{by induction} \\ (A^r)^T = (A^T)^r}}{=} (A^T)^r A^T = (A^T)^{r+1}.$$

Induction can seem a little bit like *magic* at first glance.
Pay close attention to how the induction hypothesis is used to in the proof.

3.2 Matrix Algebra

Discussion of Key Definitions and Concepts, continued

Q: How can we test if matrix B is a linear combination of matrices A_1 and A_2?

A: Rewrite A_1, A_2 and B as column vectors, then test by row reduction as in Section 2.3. Let's walk through the process in Example 3.16 to see why this is a good and true method.

Given $A_1 = \begin{bmatrix} 1 & 2 \\ -1 & 1 \end{bmatrix}$, $A_2 = \begin{bmatrix} 0 & 1 \\ 2 & 1 \end{bmatrix}$, and $B = \begin{bmatrix} 2 & 5 \\ 0 & 3 \end{bmatrix}$.

We need scalars such that $c_1 A_1 + c_2 A_2 = B$. So, $c_1 \begin{bmatrix} 1 & 2 \\ -1 & 1 \end{bmatrix} + c_2 \begin{bmatrix} 0 & 1 \\ 2 & 1 \end{bmatrix} = \begin{bmatrix} 2 & 5 \\ 0 & 3 \end{bmatrix}$.

The left-hand side of this equation can be rewritten as $\begin{bmatrix} c_1 & 2c_1 + c_2 \\ -c_1 + 2c_2 & c_1 + c_2 \end{bmatrix}$

Comparing entries and the definition of matrix equality yields
$$\begin{aligned} c_1 &= 2 \\ 2c_1 + c_2 &= 5 \\ -c_1 + 2c_2 &= 0 \\ c_1 + c_2 &= 3 \end{aligned}$$

Gauss-Jordan elimination easily gives $\begin{bmatrix} 1 & 0 & | & 2 \\ 2 & 1 & | & 5 \\ -1 & 2 & | & 0 \\ 1 & 1 & | & 3 \end{bmatrix} \longrightarrow \begin{bmatrix} 1 & 0 & | & 2 \\ 0 & 1 & | & 1 \\ 0 & 0 & | & 0 \\ 0 & 0 & | & 0 \end{bmatrix}$

So, $c_1 = 2$ and $c_2 = 1$. Thus, $2A_1 + A_2 = B$, which can be easily checked.

As claimed, we write the matrices as column vectors and row reduce as in Section 2.3. It is critical that we write the entries from each matrix in the same order each time. Why? So the entries that corresponded in the matrices also correspond in the columns.

Q: How can we test if matrix B is in span(A_1, A_2)?

A: This is the same thing as asking if B is a linear combination of A_1 and A_2.

Q: How can we test if matrices A_1, A_2, and A_3 are linearly independent?

A: Rewrite A_1, A_2 and A_3 as column vectors, then test by row reduction as in Section 2.3.

For example: Given $A_1 = \begin{bmatrix} 1 & 0 \\ 0 & 1 \end{bmatrix}$, $A_2 = \begin{bmatrix} 0 & -1 \\ 1 & 0 \end{bmatrix}$, and $A_3 = \begin{bmatrix} 1 & 1 \\ 0 & 1 \end{bmatrix}$.

We want to show $c_1 A_1 + c_2 A_2 + c_3 A_3 = O$ if and only if $c_1 = c_2 = c_3 = 0$.

$\begin{bmatrix} 1 & 0 & 1 & | & 0 \\ 0 & -1 & 1 & | & 0 \\ 0 & 1 & 0 & | & 0 \\ 1 & 0 & 1 & | & 0 \end{bmatrix} \longrightarrow \begin{bmatrix} 1 & 0 & 0 & | & 0 \\ 0 & 1 & 0 & | & 0 \\ 0 & 0 & 1 & | & 0 \\ 0 & 0 & 0 & | & 0 \end{bmatrix}$

Clearly, the unique solution to this system is $c_1 = c_2 = c_3 = 0$.

So, we conclude that matrices A_1, A_2, and A_3 are linearly independent.

It is critical that we write the entries from each matrix in the same order each time. Why?

Q: How can we test if matrices A_1, A_2, and A_3 are linearly dependent?

A: Rewrite A_1, A_2 and A_3 as column vectors and test as we did in Section 2.3. For an example of this see Exercise 14 solved in detail below.

Solutions to odd-numbered exercises from Section 3.2

1. Following remarks prior to Example 3.16, the key assumption is matrices are the same size. Then add, subtract, and multiply (by scalars only) as in *normal* algebra.
$$X - 2A + 3B = 0 \Rightarrow X = 2A - 3B = \begin{bmatrix} 5 & 4 \\ 3 & 5 \end{bmatrix}.$$

3. $X = \frac{2}{3}(A + 2B) = \begin{bmatrix} -\frac{2}{3} & \frac{4}{3} \\ \frac{10}{3} & 4 \end{bmatrix}.$

5. As in Example 3.16, we want to find scalars c_1 and c_2 such that $c_1 A_1 + c_2 A_2 = B$.
$$c_1 \begin{bmatrix} 1 & 2 \\ -1 & 1 \end{bmatrix} + c_2 \begin{bmatrix} 0 & 1 \\ 2 & 1 \end{bmatrix} = \begin{bmatrix} 2 & 5 \\ 0 & 3 \end{bmatrix}$$

The left-hand side of this equation can be rewritten as $\begin{bmatrix} c_1 & 2c_1 + c_2 \\ -c_1 + 2c_2 & c_1 + c_2 \end{bmatrix}$

Comparing entries and the definition of matrix equality yields
$$\begin{array}{rcl} c_1 & = & 2 \\ 2c_1 + c_2 & = & 5 \\ -c_1 + 2c_2 & = & 0 \\ c_1 + c_2 & = & 3 \end{array}$$

Gauss-Jordan elimination easily gives $\begin{bmatrix} 1 & 0 & | & 2 \\ 2 & 1 & | & 5 \\ -1 & 2 & | & 0 \\ 1 & 1 & | & 3 \end{bmatrix} \longrightarrow \begin{bmatrix} 1 & 0 & | & 2 \\ 0 & 1 & | & 1 \\ 0 & 0 & | & 0 \\ 0 & 0 & | & 0 \end{bmatrix}$

So, $c_1 = 2$ and $c_2 = 1$. Thus, $2A_1 + A_2 = B$, which can be easily checked.

Having walked through the process, we note this pattern in our augmented matrix: the first column is the entries of A_1, the second column is the entries of A_2, and the third column, the augmented column, is the entries of B. Make use of this pattern!

7. As in Example 3.16, we form the augmented matrix and row reduce to solve. As in Exercise 5, the first column is the entries of A_1, the second column is the entries of A_2, the third column is the entries of A_3, and the augmented column is the entries of B.
$$\begin{bmatrix} 1 & -1 & 1 & | & 3 \\ 0 & 2 & 1 & | & 1 \\ -1 & 0 & 1 & | & 1 \\ 0 & 0 & 0 & | & 0 \\ 1 & 1 & 0 & | & 1 \\ 0 & 0 & 0 & | & 0 \end{bmatrix} \longrightarrow \begin{bmatrix} 1 & 0 & 0 & | & 0 \\ 0 & 1 & 0 & | & 0 \\ 0 & 0 & 1 & | & 0 \\ 0 & 0 & 0 & | & 1 \\ 0 & 0 & 0 & | & 0 \\ 0 & 0 & 0 & | & 0 \end{bmatrix}$$

Since there is no solution, B is *not* a linear combination of A_1, A_2, and A_3.

Q: Why is it obvious that this system has no solution?
A: After row reduction, we have a row of zeroes with a corresponding constant term $\neq 0$.

3.2   Matrix Algebra

9. As in Example 3.17, we write out a general linear combination of A_1 and A_2.
As in Exercise 5, the first column is the entries of A_1, the second column is the entries of A_2, but the augmented column is now the variables of w, x, y, and z.

$$\begin{bmatrix} 1 & 0 & | & w \\ 2 & 1 & | & x \\ -1 & 2 & | & y \\ 1 & 1 & | & z \end{bmatrix} \longrightarrow \begin{bmatrix} 1 & 0 & | & w \\ 0 & 1 & | & x - 2w \\ 0 & 0 & | & y + 5w - 2x \\ 0 & 0 & | & z + w - x \end{bmatrix}$$

Two restrictions come from the last two rows: $y + 5w - 2x = 0$ and $z + w - x = 0$.

So, the span of A_1 and A_2 consists of $\begin{bmatrix} w & x \\ y & z \end{bmatrix}$ for which $y = 2x - 5w$ and $z = x - w$.

That is, $\text{span}(A_1, A_2) = \left\{ \begin{bmatrix} w & x \\ 2x - 5w & x - w \end{bmatrix} \right\}$.

Q: Why must $y + 5w - 2x = 0$ and $z + w - x = 0$?
A: After row reduction, a row of zeroes must have a corresponding constant term of zero.

Q: Is matrix B from Exercise 5 in $\text{span}(A_1, A_2)$? Should it be?
A: Yes, as well it should be since B is a linear combination of A_1 and A_2.

Q: How can we tell that B is in the $\text{span}(A_1, A_2)$?
A: By noting that $2(5) - 5(2) = 0$ and $5 - 2 = 3$.

11. Following Example 3.17, we create an augmented matrix and row reduce to find restrictions.
As in Exercise 7, the first column is the entries of A_1, the second column is the entries of A_2, the third column is the entries of A_3, but the augmented column is a, b, c, d, e, and f.

$$\begin{bmatrix} 1 & -1 & 1 & | & a \\ 0 & 2 & 1 & | & b \\ -1 & 0 & 1 & | & c \\ 0 & 0 & 0 & | & d \\ 1 & 1 & 0 & | & e \\ 0 & 0 & 0 & | & f \end{bmatrix} \longrightarrow \begin{bmatrix} 1 & 1 & 0 & | & e \\ 0 & 1 & 1 & | & c + e \\ 0 & 0 & -1 & | & b - 2c - 2e \\ 0 & 0 & 0 & | & a + 3b - 4c - 5e \\ 0 & 0 & 0 & | & d \\ 0 & 0 & 0 & | & f \end{bmatrix}$$

The restrictions are $a + 3b - 4c - 5e = 0$ and $d = f = 0$.

So, $\text{span}(A_1, A_2, A_3) = \left\{ \begin{bmatrix} -3b + 4c + 5e & b & c \\ 0 & e & 0 \end{bmatrix} \right\}$.

Q: Why did we not have to continue to reduced row echelon form?
A: The first 3 rows are obviously linearly independent, but restrictions come from zero rows.

Q: Is matrix B from Exercise 7 in $\text{span}(A_1, A_2, A_3)$? Should it be?
A: No. It shouldn't be since B is *not* a linear combination of A_1, A_2, and A_3.

Q: How can we tell that B is *not* in the $\text{span}(A_1, A_2)$?
A: By noting that $3 + 3(1) - 4(c) - 5(1) \neq 0$.

13. Following Example 3.18, we create an augmented matrix and row reduce to solve.
    As in Exercise 8, the first column is the entries of A_1, the second column is the entries of A_2, but now the augmented column is all zeroes.
    $$\begin{bmatrix} 1 & 4 & | & 0 \\ 2 & 3 & | & 0 \\ 3 & 2 & | & 0 \\ 4 & 1 & | & 0 \end{bmatrix} \longrightarrow \begin{bmatrix} 1 & 0 & | & 0 \\ 0 & 1 & | & 0 \\ 0 & 0 & | & 0 \\ 0 & 0 & | & 0 \end{bmatrix}$$
    Clearly, the only solution is $c_1 = c_2 = 0$. What do we conclude?
    We conclude that A_1 and A_2 are linearly independent.

15. Following Example 3.18, we create an augmented matrix and row reduce to solve.
    As in Exercise 8, the first column is the entries of A_1, the second column is the entries of A_2, the third column is the entries of A_3, the fourth column is the entries of A_4, but now the augmented column is all zeroes.
    $$\begin{bmatrix} 0 & 1 & -2 & -1 & | & 0 \\ 1 & 0 & -1 & -3 & | & 0 \\ 5 & 2 & 0 & 1 & | & 0 \\ 2 & 3 & 1 & 9 & | & 0 \\ -1 & 1 & 0 & 4 & | & 0 \\ 0 & 1 & 2 & 5 & | & 0 \end{bmatrix} \longrightarrow \begin{bmatrix} 1 & 0 & 0 & 0 & | & 0 \\ 0 & 1 & 0 & 0 & | & 0 \\ 0 & 0 & 1 & 0 & | & 0 \\ 0 & 0 & 0 & 1 & | & 0 \\ 0 & 0 & 0 & 0 & | & 0 \\ 0 & 0 & 0 & 0 & | & 0 \end{bmatrix}$$
    Clearly, the only solution is $c_1 = c_2 = c_3 = c_4 = 0$. What do we conclude?
    We conclude that A_1, A_2, A_3, and A_4 are linearly independent.

3.2 Matrix Algebra

17. Let A, B, and C be matrices of the same size ($m \times n$) and let c and d be scalars.

(a) $A + B = \begin{bmatrix} a_{11} & \cdots & a_{1n} \\ \vdots & \ddots & \vdots \\ a_{m1} & \cdots & a_{mn} \end{bmatrix} + \begin{bmatrix} b_{11} & \cdots & b_{1n} \\ \vdots & \ddots & \vdots \\ b_{m1} & \cdots & b_{mn} \end{bmatrix} = \begin{bmatrix} a_{11}+b_{11} & \cdots & a_{1n}+b_{1n} \\ \vdots & \ddots & \vdots \\ a_{m1}+b_{m1} & \cdots & a_{mn}+b_{mn} \end{bmatrix}$.

But, a_{ij} and b_{ij} are scalars, which commute $\Rightarrow$

$= \begin{bmatrix} b_{11}+a_{11} & \cdots & b_{1n}+a_{1n} \\ \vdots & \ddots & \vdots \\ b_{m1}+a_{m1} & \cdots & b_{mn}+a_{mn} \end{bmatrix} = B + A.$

(b) $(A+B)+C = \left(\begin{bmatrix} a_{11} & \cdots & a_{1n} \\ \vdots & \ddots & \vdots \\ a_{m1} & \cdots & a_{mn} \end{bmatrix} + \begin{bmatrix} b_{11} & \cdots & b_{1n} \\ \vdots & \ddots & \vdots \\ b_{m1} & \cdots & b_{mn} \end{bmatrix} \right) + \begin{bmatrix} c_{11} & \cdots & c_{1n} \\ \vdots & \ddots & \vdots \\ c_{m1} & \cdots & c_{mn} \end{bmatrix}$

$= \begin{bmatrix} b_{11}+a_{11} & \cdots & b_{1n}+a_{1n} \\ \vdots & \ddots & \vdots \\ b_{m1}+a_{m1} & \cdots & b_{mn}+a_{mn} \end{bmatrix} + \begin{bmatrix} c_{11} & \cdots & c_{1n} \\ \vdots & \ddots & \vdots \\ c_{m1} & \cdots & c_{mn} \end{bmatrix}$

$= \begin{bmatrix} a_{11}+b_{11}+c_{11} & \cdots & a_{1n}+b_{1n}+c_{1n} \\ \vdots & \ddots & \vdots \\ a_{m1}+b_{m1}+c_{m1} & \cdots & a_{mn}+b_{mn}+c_{mn} \end{bmatrix}$

$= \begin{bmatrix} a_{11} & \cdots & a_{1n} \\ \vdots & \ddots & \vdots \\ a_{m1} & \cdots & a_{mn} \end{bmatrix} + \left(\begin{bmatrix} b_{11} & \cdots & b_{1n} \\ \vdots & \ddots & \vdots \\ b_{m1} & \cdots & b_{mn} \end{bmatrix} + \begin{bmatrix} c_{11} & \cdots & c_{1n} \\ \vdots & \ddots & \vdots \\ c_{m1} & \cdots & c_{mn} \end{bmatrix} \right)$

$= A + (B + C).$

(c) $A + 0 = \begin{bmatrix} a_{11} & \cdots & a_{1n} \\ \vdots & \ddots & \vdots \\ a_{m1} & \cdots & a_{mn} \end{bmatrix} + \begin{bmatrix} 0 & \cdots & 0 \\ \vdots & \ddots & \vdots \\ 0 & \cdots & 0 \end{bmatrix} = \begin{bmatrix} a_{11}+0 & \cdots & a_{1n}+0 \\ \vdots & \ddots & \vdots \\ a_{m1}+0 & \cdots & a_{mn}+0 \end{bmatrix} = A.$

(d) $A + (-A) = \begin{bmatrix} a_{11} & \cdots & a_{1n} \\ \vdots & \ddots & \vdots \\ a_{m1} & \cdots & a_{mn} \end{bmatrix} + \left(- \begin{bmatrix} a_{11} & \cdots & a_{1n} \\ \vdots & \ddots & \vdots \\ a_{m1} & \cdots & a_{mn} \end{bmatrix} \right)$

$= \begin{bmatrix} a_{11} & \cdots & a_{1n} \\ \vdots & \ddots & \vdots \\ a_{m1} & \cdots & a_{mn} \end{bmatrix} + \left(\begin{bmatrix} -a_{11} & \cdots & -a_{1n} \\ \vdots & \ddots & \vdots \\ -a_{m1} & \cdots & -a_{mn} \end{bmatrix} \right)$

$= \begin{bmatrix} a_{11}-a_{11} & \cdots & a_{1n}-a_{1n} \\ \vdots & \ddots & \vdots \\ a_{m1}-a_{m1} & \cdots & a_{mn}-a_{mn} \end{bmatrix} = \begin{bmatrix} 0 & \cdots & 0 \\ \vdots & \ddots & \vdots \\ 0 & \cdots & 0 \end{bmatrix} = O.$

19. Let A, B, and C be matrices of appropriate dimensions. Then

$$(A+B)C = \left(\begin{bmatrix} a_{11} & \cdots & a_{1n} \\ \vdots & \ddots & \vdots \\ a_{m1} & \cdots & a_{mn} \end{bmatrix} + \begin{bmatrix} b_{11} & \cdots & b_{1n} \\ \vdots & \ddots & \vdots \\ b_{m1} & \cdots & b_{mn} \end{bmatrix}\right) \begin{bmatrix} c_{11} & \cdots & c_{1n} \\ \vdots & \ddots & \vdots \\ c_{m1} & \cdots & c_{mn} \end{bmatrix}$$

$$= \begin{bmatrix} a_{11}+b_{11} & \cdots & a_{1n}+b_{1n} \\ \vdots & \ddots & \vdots \\ a_{m1}+b_{m1} & \cdots & a_{mn}+b_{mn} \end{bmatrix} \begin{bmatrix} c_{11} & \cdots & c_{1n} \\ \vdots & \ddots & \vdots \\ c_{m1} & \cdots & c_{mn} \end{bmatrix}$$

$$= \begin{bmatrix} (a_{11}+b_{11})c_{11}+\cdots+(a_{m1}+b_{m1})c_{1n} & \cdots & (a_{1n}+b_{1n})c_{11}+\cdots+(a_{mn}+b_{mn})c_{1n} \\ \vdots & \ddots & \vdots \\ (a_{11}+b_{11})c_{m1}+\cdots+(a_{m1}+b_{m1})c_{mn} & \cdots & (a_{1n}+b_{1n})c_{m1}+\cdots+(a_{mn}+b_{mn})c_{mn} \end{bmatrix}$$

$$= \begin{bmatrix} a_{11}c_{11}+b_{11}c_{11}+\cdots+a_{m1}c_{1n}+b_{m1}c_{1n} & \cdots & a_{1n}c_{11}+b_{1n}c_{11}+\cdots+a_{mn}c_{1n}+b_{mn}c_{1n} \\ \vdots & \ddots & \vdots \\ a_{11}c_{m1}+b_{11}c_{m1}+\cdots+a_{m1}c_{mn}+b_{m1}c_{mn} & \cdots & a_{1n}c_{m1}+b_{1n}c_{m1}+\cdots+a_{mn}c_{mn}+b_{mn}c_{mn} \end{bmatrix}$$

$$= \begin{bmatrix} (a_{11}c_{11}+\cdots+a_{m1}c_{1n})+ & & (a_{1n}c_{11}+\cdots+a_{mn}c_{1n})+ \\ +(b_{11}c_{11}+\cdots+b_{m1}c_{1n}) & \cdots & +(b_{1n}c_{11}+\cdots+b_{mn}c_{1n}) \\ \vdots & \ddots & \vdots \\ (a_{11}c_{m1}+\cdots+a_{m1}c_{mn})+ & & \\ +(b_{11}c_{m1}+\cdots+b_{m1}c_{mn}) & \cdots & \\ (a_{1n}c_{m1}+\ldots++a_{mn}c_{mn})+ & & \\ +(b_{1n}c_{m1}+\cdots+b_{mn}c_{mn}) & & \end{bmatrix}$$

$$= \begin{bmatrix} (a_{11}c_{11}+\cdots+a_{m1}c_{1n}) & \cdots & (a_{1n}c_{11}+\cdots+a_{mn}c_{1n}) \\ \vdots & \ddots & \vdots \\ (a_{11}c_{m1}+\cdots+a_{m1}c_{mn}) & \cdots & (a_{1n}c_{m1}+\cdots+a_{mn}c_{mn}) \end{bmatrix}$$
$$+ \begin{bmatrix} (b_{11}c_{11}+\cdots+b_{m1}c_{1n}) & \cdots & (b_{1n}c_{11}+\cdots b_{mn}c_{1n}) \\ \vdots & \ddots & \vdots \\ (b_{11}c_{m1}+\cdots+b_{m1}c_{mn}) & \cdots & (b_{1n}c_{m1}+\cdots b_{mn}c_{mn}) \end{bmatrix}$$

$$= AC + BC.$$

3.2 Matrix Algebra

21. To prove $I_m A = A$, note $I_m = \begin{bmatrix} e_1 \\ e_2 \\ \vdots \\ e_m \end{bmatrix} \Rightarrow I_m A = \begin{bmatrix} e_1 A_1 \\ e_2 A_2 \\ \vdots \\ e_m A_m \end{bmatrix} = \begin{bmatrix} A_1 \\ A_2 \\ \vdots \\ A_m \end{bmatrix} = A.$

23. We will compute AB and BA, then equate entries to find the conditions on a, b, c, and d.

$$AB = \begin{bmatrix} 1 & 1 \\ 0 & 1 \end{bmatrix} \begin{bmatrix} a & b \\ c & d \end{bmatrix} = \begin{bmatrix} a+c & b+d \\ c & d \end{bmatrix} = \begin{bmatrix} a & a+b \\ c & c+d \end{bmatrix} = \begin{bmatrix} a & b \\ c & d \end{bmatrix} \begin{bmatrix} 1 & 1 \\ 0 & 1 \end{bmatrix} = BA$$

Equating entries gives us the following four equations (conditions on a, b, c, and d):
$a + c = a$, $b + d = a + b$, $c = c$, and $d = c + d \Rightarrow$ The conditions are $a = d$ and $c = 0$.

25. We will compute AB and BA, then equate entries to find the conditions on a, b, c, and d.

$$AB = \begin{bmatrix} 1 & 2 \\ 3 & 4 \end{bmatrix} \begin{bmatrix} a & b \\ c & d \end{bmatrix} = \begin{bmatrix} a+2c & b+2d \\ 3a+4c & 3b+4d \end{bmatrix} = \begin{bmatrix} a+3b & 2a+4b \\ c+3d & 2c+4d \end{bmatrix} = \begin{bmatrix} a & b \\ c & d \end{bmatrix} \begin{bmatrix} 1 & 2 \\ 3 & 4 \end{bmatrix} = BA$$

Equating entries gives us the following four equations (conditions on a, b, c, and d):
$a + 2c = a + 3b$, $b + 2d = 2a + 4b$, $3a + 4c = c + 3d$, and $3b + 4d = 2c + 4d$.
So, the conditions on a, b, c, and d are $3b = 2c$ and $a = d - c$.

27. We should use linear combinations and our work in Exercise 26 to answer this question. How?

Consider $A_1 = \begin{bmatrix} 1 & 0 \\ 0 & 0 \end{bmatrix}$, $A_2 = \begin{bmatrix} 0 & 1 \\ 0 & 0 \end{bmatrix}$, $A_3 = \begin{bmatrix} 0 & 0 \\ 1 & 0 \end{bmatrix}$, and $A_4 = \begin{bmatrix} 0 & 0 \\ 0 & 1 \end{bmatrix}$. Then we have:

$$M = \begin{bmatrix} a & b \\ c & d \end{bmatrix} = aA_1 + bA_1 + cA_3 + dA_4 = a\begin{bmatrix} 1 & 0 \\ 0 & 0 \end{bmatrix} + b\begin{bmatrix} 0 & 1 \\ 0 & 0 \end{bmatrix} + c\begin{bmatrix} 0 & 0 \\ 1 & 0 \end{bmatrix} + d\begin{bmatrix} 0 & 0 \\ 0 & 1 \end{bmatrix}.$$

What does this tell us? Our matrix B need only commute with A_1, A_2, A_3, and A_4.
We will compute $A_2 B$ and BA_2, then equate entries to find conditions on a, b, c, and d.
Repeat the process for $A_3 B$ and BA_3, then combine with our answer from Exercise 26.

$$A_2 B = \begin{bmatrix} 0 & 1 \\ 0 & 0 \end{bmatrix} \begin{bmatrix} a & b \\ c & d \end{bmatrix} = \begin{bmatrix} c & d \\ 0 & 0 \end{bmatrix} = \begin{bmatrix} 0 & a \\ 0 & c \end{bmatrix} = \begin{bmatrix} a & b \\ c & d \end{bmatrix} \begin{bmatrix} 0 & 1 \\ 0 & 0 \end{bmatrix} = BA_2$$

Equating entries gives us the following four equations (conditions on a, b, c, and d):
$c = 0$, $d = a$, $0 = 0$, and $c = 0 \Rightarrow$ The A_2 conditions are $a = d$ and $c = 0$.
Repeating the process for $A_3 B$ and BA_3 yields:

$$A_3 B = \begin{bmatrix} 0 & 0 \\ 1 & 0 \end{bmatrix} \begin{bmatrix} a & b \\ c & d \end{bmatrix} = \begin{bmatrix} 0 & 0 \\ a & b \end{bmatrix} = \begin{bmatrix} b & 0 \\ d & 0 \end{bmatrix} = \begin{bmatrix} a & b \\ c & d \end{bmatrix} \begin{bmatrix} 0 & 0 \\ 1 & 0 \end{bmatrix} = BA_3$$

Equating entries gives us the following four equations (conditions on a, b, c, and d):
$0 = b$, $0 = 0$, $a = d$, and $b = 0 \Rightarrow$ The A_3 conditions are $a = d$ and $b = 0$.

Combining the conditions for A_1, A_2, A_3, and A_4 gives us:
The required conditions so that B will commute with any 2×2 matrix are $a = d$ and $b = c = 0$.

29. Let A and B be two upper triangular $n \times n$ matrices.

Then, $A = \begin{bmatrix} a_{11} & a_{12} & a_{13} & \cdots & a_{1n} \\ 0 & a_{22} & a_{23} & \cdots & a_{2n} \\ 0 & 0 & a_{33} & \cdots & a_{3n} \\ \vdots & \vdots & \vdots & \ddots & \vdots \\ 0 & 0 & \cdots & 0 & a_{nn} \end{bmatrix}$ and $B = \begin{bmatrix} b_{11} & b_{12} & b_{13} & \cdots & b_{1n} \\ 0 & b_{22} & b_{23} & \cdots & b_{2n} \\ 0 & 0 & b_{33} & \cdots & b_{3n} \\ \vdots & \vdots & \vdots & \ddots & \vdots \\ 0 & 0 & \cdots & 0 & b_{nn} \end{bmatrix} \Rightarrow$

$$AB = \begin{bmatrix} a_{11} & a_{12} & a_{13} & \cdots & a_{1n} \\ 0 & a_{22} & a_{23} & \cdots & a_{2n} \\ 0 & 0 & a_{33} & \cdots & a_{3n} \\ \vdots & \vdots & \vdots & \ddots & \vdots \\ 0 & 0 & \cdots & 0 & a_{nn} \end{bmatrix} \begin{bmatrix} b_{11} & b_{12} & b_{13} & \cdots & b_{1n} \\ 0 & b_{22} & b_{23} & \cdots & b_{2n} \\ 0 & 0 & b_{33} & \cdots & b_{3n} \\ \vdots & \vdots & \vdots & \ddots & \vdots \\ 0 & 0 & \cdots & 0 & b_{nn} \end{bmatrix}$$

$$= \begin{bmatrix} a_{11}b_{11} & a_{11}b_{12} + a_{12}b_{22} & a_{11}b_{13} + a_{12}b_{23} + a_{13}b_{33} & \cdots & a_{11}b_{1n} + a_{12}b_{2n} + \cdots + a_{nn}b_{nn} \\ 0 & a_{22}b_{22} & a_{22}b_{23} + a_{23}b_{33} & \cdots & a_{22}b_{2n} + a_{23}b_{3n} + \cdots + a_{2n}b_{nn} \\ 0 & 0 & a_{33}b_{33} & \cdots & a_{33}b_{3n} + \cdots + a_{3n}b_{nn} \\ \vdots & \vdots & \vdots & \ddots & \vdots \\ 0 & 0 & 0 & 0 & a_{nn}b_{nn} \end{bmatrix}$$

So we see that AB is also an upper triangular matrix.

31. We need to show $(A^r)^T = (A^T)^r$. We will prove this using *induction*.
See Appendix B for discussion and examples of *Mathematical Induction*.

1: $(A^1)^T = (A^T)^1$. This is obvious, so there is nothing to show.

r: $(A^r)^T = (A^T)^r$. This is the induction hypothesis.

$r+1$: $(A^{r+1})^T = (A^T)^{r+1}$. This is the statement we must prove using the induction hypothesis.

$$(A^{r+1})^T = (A \cdot A^r)^T \stackrel{\substack{\text{by Thm 3.4d} \\ (AB)^T = B^T A^T}}{=} (A^r)^T A^T \stackrel{\substack{\text{by induction} \\ (A^r)^T = (A^T)^r}}{=} (A^T)^r A^T = (A^T)^{r+1}.$$

Induction can seem a little bit like *magic* at first glance.
Pay close attention to how the induction hypothesis is used in the proof.

3.2 Matrix Algebra

33. We need to show $(A_1 A_2 \cdots A_n)^T = A_n^T \cdots A_2^T A_1^T$ for $n \geq 1$.
We will prove this using *induction*.
See Appendix B for discussion and examples of *Mathematical Induction*.

1: $(A_1)^T = A_1^T$. This is obvious, so there is nothing to show.

k: $(A_1 A_2 \cdots A_k)^T = A_k^T \cdots A_2^T A_1^T$.
This is the induction hypothesis, so there is nothing to show.

$k+1$: $(A_1 + A_2 \cdots A_k A_{k+1})^T = A_{k+1}^T A_k^T \cdots A_2^T A_1^T$
This is the statement we must prove using the induction hypothesis.

$$(A_1 A_2 \cdots A_k A_{k+1})^T = ((A_1 A_2 \cdots A_k) A_{k+1})^T = A_{k+1}^T (A_1 A_2 \cdots A_k)^T \quad [by\ Thm\ 3.4d]$$
$$= A_{k+1}^T (A_k^T \cdots A_2^T A_1^T) \quad [by\ induction]$$

We have shown the pattern holds for $k+1$. What does that mean?

We have proven (by induction) that $(A_1 A_2 \cdots A_n)^T = A_n^T \cdots A_2^T A_1^T$.

35. Let A and B be symmetric $n \times n$ matrices and let k be a scalar.

(a) To prove $A + B$ is symmetric, we need to show $(A+B)^T = A+B$.

$$(A+B)^T \underset{\substack{by \\ Thm\ 3.4b}}{=} A^T + B^T \underset{\substack{A\ and\ B\ are \\ symmetric}}{=} A+B.$$

(b) To prove kA is symmetric, we need to show $(kA)^T = kA$.

$$(kA)^T \underset{\substack{by \\ Thm\ 3.4c}}{=} kA^T \underset{\substack{A\ is \\ symmetric}}{=} kA.$$

37. For each matrix, we will simply check to see if $A^T = -A$ is satisfied.

(a) Since $A^T = \begin{bmatrix} 1 & -2 \\ 2 & 3 \end{bmatrix} \neq -\begin{bmatrix} 1 & 2 \\ -2 & 3 \end{bmatrix} = -A$, A is *not* skew-symmetric.

(b) Since $A^T = \begin{bmatrix} 0 & 1 \\ -1 & 0 \end{bmatrix} = -\begin{bmatrix} 0 & -1 \\ 1 & 0 \end{bmatrix} = -A$, A is skew-symmetric.

(c) Since $A^T = \begin{bmatrix} 0 & -3 & 1 \\ 3 & 0 & -2 \\ -1 & 2 & 0 \end{bmatrix} = -\begin{bmatrix} 0 & 3 & -1 \\ -3 & 0 & 2 \\ 1 & -2 & 0 \end{bmatrix} = -A$, A is skew-symmetric.

(d) Since $A^T = \begin{bmatrix} 0 & -1 & 2 \\ 1 & 0 & 5 \\ 2 & 5 & 0 \end{bmatrix} \neq -\begin{bmatrix} 0 & 1 & 2 \\ -1 & 0 & 5 \\ 2 & 5 & 0 \end{bmatrix} = -A$, A is *not* skew-symmetric.

39. If A is skew-symmetric ($A^T = -A$), then the diagonal entries must be zero ($a_{ii} = 0$).
$A^T = -A \;\Rightarrow\; [A^T]_{ij} = [-A]_{ij} \;\Rightarrow\; [A]_{ji} = -[A]_{ij}$
So $a_{ji} = -a_{ij} \;\Rightarrow\; a_{ii} = -a_{ii} \;\Rightarrow\; 2a_{ii} = 0 \;\Rightarrow\; a_{ii} = 0$

41. Let A and B be skew-symmetric 2×2 matrices, so $A = \begin{bmatrix} 0 & a \\ -a & 0 \end{bmatrix}$ and $B = \begin{bmatrix} 0 & b \\ -b & 0 \end{bmatrix}$.

Demanding that AB be skew-symmetric gives us the equation:

$$(AB)^T = -AB \Leftrightarrow \left(\begin{bmatrix} 0 & a \\ -a & 0 \end{bmatrix} \begin{bmatrix} 0 & b \\ -b & 0 \end{bmatrix} \right)^T = -\begin{bmatrix} 0 & a \\ -a & 0 \end{bmatrix} \begin{bmatrix} 0 & b \\ -b & 0 \end{bmatrix} \Leftrightarrow$$

$$\begin{bmatrix} -ab & 0 \\ 0 & -ab \end{bmatrix}^T = \begin{bmatrix} ab & 0 \\ 0 & ab \end{bmatrix} \Leftrightarrow \begin{bmatrix} -ab & 0 \\ 0 & -ab \end{bmatrix} = \begin{bmatrix} ab & 0 \\ 0 & ab \end{bmatrix}.$$

So, $-ab = ab \Leftrightarrow ab = 0$. Letting $O =$ the zero matrix, we get:
AB will be skew-symmetric provided either $A = O$ or $B = O$ (or both).

43. We will prove this claim in (a) and demonstrate it with an example in (b).

(a) If A is $n \times n$, then $A = B + C$, where B is symmetric and C is skew-symmetric.

$$S \underset{\text{Thm 3.5a}}{\overset{\text{symmetric by}}{=}} A + A^T \text{ and } S' \underset{\text{Exercise 42}}{\overset{\text{skew-symmetric by}}{=}} A - A^T$$

Now simply note $A = \tfrac{1}{2}(A + A^T) + \tfrac{1}{2}(A - A^T) = \tfrac{1}{2}S + \tfrac{1}{2}S'$

Q: When S is symmetric and S' is skew-symmetric, are cS and cS' also?
A: Yes, since $(cS)^T = cS^T = cS$ and $(cS')^T = c(S')^T = -cS'$.

(b) $A = \begin{bmatrix} 1 & 2 & 3 \\ 4 & 5 & 6 \\ 7 & 8 & 9 \end{bmatrix} = \tfrac{1}{2}\left(\begin{bmatrix} 1 & 2 & 3 \\ 4 & 5 & 6 \\ 7 & 8 & 9 \end{bmatrix} + \begin{bmatrix} 1 & 4 & 7 \\ 2 & 5 & 8 \\ 3 & 6 & 9 \end{bmatrix} \right) + \tfrac{1}{2}\left(\begin{bmatrix} 1 & 2 & 3 \\ 4 & 5 & 6 \\ 7 & 8 & 9 \end{bmatrix} - \begin{bmatrix} 1 & 4 & 7 \\ 2 & 5 & 8 \\ 3 & 6 & 9 \end{bmatrix} \right)$

$= \begin{bmatrix} 1 & 3 & 5 \\ 3 & 5 & 7 \\ 5 & 7 & 9 \end{bmatrix} + \begin{bmatrix} 0 & -1 & -2 \\ 1 & 0 & -1 \\ 2 & 1 & 0 \end{bmatrix}.$

45. Let A and B be $n \times n$ matrices. Then

$$\begin{aligned}
\operatorname{tr}(AB) &= (a_{11}b_{11} + a_{12}b_{21} + \cdots + a_{1n}b_{n1}) + (a_{21}b_{12} + a_{22}b_{22} + \cdots + a_{2n}b_{n2}) + \\
&\qquad \cdots + (a_{n1}b_{1n} + a_{n2}b_{2n} + \cdots + a_{nn}b_{nn}) \\
&= (b_{11}a_{11} + b_{12}a_{21} + \cdots + b_{1n}a_{n1}) + (b_{21}a_{12} + b_{22}a_{22} + \cdots + b_{2n}a_{n2}) + \\
&\qquad \cdots + (b_{n1}a_{1n} + b_{n2}a_{2n} + \cdots + b_{nn}a_{nn}) \\
&= \operatorname{tr}(BA)
\end{aligned}$$

47. If A, B are 2×2, we will show $\operatorname{tr}(AB - BA) \neq \operatorname{tr}(I_2)$ which will imply $AB - BA \neq I_2$.

$$\operatorname{tr}(AB - BA) \underset{\text{44 and 45}}{\overset{\text{by Exercises}}{=}} \operatorname{tr}(AB) - \operatorname{tr}(BA) = 0 \neq 2 = \operatorname{tr}(I_2) \Rightarrow AB - BA \neq I_2$$

Q: We have shown $AB - BA$ must equal a matrix of trace zero. Is this true for $n \times n$ matrices?
A: Yes, since $\operatorname{tr}(AB - BA) = 0$ for any square matrix.

3.3 The Inverse of a Matrix

Q: What are our main goals for Section 3.3?
A: To learn how to compute inverses, especially with E_i, and use inverses to solve systems.

Key Definitions and Concepts

A^{-1}	p.162	3.3	A-inverse: satisfies $AA^{-1} = I$ and is unique
A^{-n}	p.164	3.3	This is defined to be $(A^{-1})^n = (A^n)^{-1}$ and is unique
E_i	p.169	3.3	Matrix created by an elementary row operation on I
O	p.152	3.2	O is commonly used to stand for the zero matrix
det A	p.163	3.3	The determinant of A for 2×2 matrices det $A = ad - bc$
determinant	p.163	3.3	The determinant of A for 2×2 matrices det $A = ad - bc$
elementary matrix	p.169	3.3	Any matrix that can be obtained by performing an elementary row operation on an identity matrix
induction	p.147	3.1	See Appendix B, *Mathematical Induction*
inverse	p.161	3.3	A-inverse: satisfies $AA^{-1} = I$ and is unique
invertible	p.161	3.3	If A^{-1} exists, A is called *invertible*.

Theorems

Thm 3.6	p.162	3.3	If A is invertible, then its inverse is unique. If $AA^{-1} = I$, then A^{-1} is unique.
Thm 3.7	p.162	3.3	$A\mathbf{x} = \mathbf{b}$ has unique solution $\mathbf{x} = A^{-1}\mathbf{b}$ The inverse A^{-1} exists if and only if det $A = ad - bc \neq 0$
Thm 3.8	p.163	3.3	Formula for A^{-1} for 2×2 matrices using det A
Thm 3.9	p.166	3.3	Key Properties of A^{-1}, for example $(A^{-1})^{-1} = I$
Thm 3.10	p.170	3.3	Perform row operation E on A is equivalent to EA
Thm 3.11	p.171	3.3	E^{-1} is created by undoing the operation that created E
Thm 3.12	p.171	3.3	Five conditions that are equivalent to A being invertible
Thm 3.13	p.173	3.3	If $AB = I$ or $BA = I$, then $B = A^{-1}$
Thm 3.14	p.173	3.3	If $E_n E_{n-1} \cdots E_1 A = I$, then $A^{-1} = E_n E_{n-1} \cdots E_1$

Discussion of Key Definitions and Concepts

Q: Can we create a matrix that is not invertible by multiplying invertible matrices together?
A: No. If A, B are invertible, then so is AB and $(AB)^{-1} = B^{-1}A^{-1}$. See Theorem 3.9.

Q: Can we create a matrix that is invertible by multiplying matrices that are not invertible?
A: No. If AB is invertible, both A and B are invertible. See the solution to Exercise 47.

The rest of the discussion is coupled with solutions to particular exercises.

Solutions to odd-numbered exercises from Section 3.3

1. As in Example 3.24, we begin by computing the determinant of A, det A. Why? Since if det $A = 0$, then A is not invertible.

   Since $A = \begin{bmatrix} a & b \\ c & d \end{bmatrix} = \begin{bmatrix} 4 & 7 \\ 1 & 2 \end{bmatrix}$, det $A = ad - bc = 4(2) - 1(7) = 1$, A is invertible.

   So, by Theorem 3.8, $A^{-1} = \frac{1}{ad-bc} \begin{bmatrix} d & -b \\ -c & a \end{bmatrix} = \frac{1}{\det A} \begin{bmatrix} d & -b \\ -c & a \end{bmatrix} = \frac{1}{1} \begin{bmatrix} 2 & -7 \\ -1 & 4 \end{bmatrix} = \begin{bmatrix} 2 & -7 \\ -1 & 4 \end{bmatrix}$.

   Q: How can we check our answer for A^{-1}?
   A: By applying the definition. That is, by checking that $AA^{-1} = I$. Let's do that:

   $AA^{-1} = \begin{bmatrix} 4 & 7 \\ 1 & 2 \end{bmatrix} \begin{bmatrix} 2 & -7 \\ -1 & 4 \end{bmatrix} = \begin{bmatrix} 4(2)+7(-1) & 1(2)+2(-1) \\ 4(-7)+7(4) & 1(-7)+2(4) \end{bmatrix} = \begin{bmatrix} 1 & 0 \\ 0 & 1 \end{bmatrix} = I.$

   Q: Should $A^{-1}A = I$, too?
   A: Yes. We should check that as well. On the calculator, these are quick and simple to do.

3. As is Example 3.24, we begin by computing the determinant of A, det A. Why?

   Since $A = \begin{bmatrix} 3 & 4 \\ 6 & 8 \end{bmatrix}$, det $A = 3(8) - 4(6) = 0$, A is not invertible.

5. Since det $A = \frac{3}{4}(\frac{2}{3}) - \frac{3}{5}(\frac{5}{6}) = 0$, A is not invertible.

7. Since det $A = (-1.5)(2.4) - (-4.2)(0.5) = -1.5$, A is invertible. So, by Theorem 3.8:

   $A^{-1} = -\frac{1}{1.5} \begin{bmatrix} 2.4 & 4.2 \\ -0.5 & -1.5 \end{bmatrix} = \begin{bmatrix} -1.6 & -2.8 \\ 0.\overline{3} & 1 \end{bmatrix}$

   Check: $AA^{-1} = \begin{bmatrix} -1.5 & -4.2 \\ 0.5 & 2.4 \end{bmatrix} \begin{bmatrix} -1.6 & -2.8 \\ 0.\overline{3} & 1 \end{bmatrix} = \begin{bmatrix} 1 & 0 \\ 0 & 1 \end{bmatrix} = I.$

9. Since det $A = (a)(a) - (-b)(b) = a^2 + b^2$, provided a and b are not both zero, A is invertible.

   $A^{-1} = \frac{1}{a^2+b^2} \begin{bmatrix} a & b \\ -b & a \end{bmatrix} = \begin{bmatrix} \frac{a}{a^2+b^2} & \frac{b}{a^2+b^2} \\ -\frac{b}{a^2+b^2} & \frac{a}{a^2+b^2} \end{bmatrix}$ (a and b not both zero)

   Check: $AA^{-1} = \begin{bmatrix} a & -b \\ b & a \end{bmatrix} \begin{bmatrix} \frac{a}{a^2+b^2} & \frac{b}{a^2+b^2} \\ -\frac{b}{a^2+b^2} & \frac{a}{a^2+b^2} \end{bmatrix} = \begin{bmatrix} \frac{a^2+b^2}{a^2+b^2} & \frac{ab-ba}{a^2+b^2} \\ \frac{ba-ab}{a^2+b^2} & \frac{b^2+a^2}{a^2+b^2} \end{bmatrix} = \begin{bmatrix} 1 & 0 \\ 0 & 1 \end{bmatrix} = I.$

11. As in Example 3.25, we use the inverse of the coefficient matrix A to solve the system. That is, the reasoning we will employ here is: $A\mathbf{x} = \mathbf{b} \Rightarrow \mathbf{x} = A^{-1}\mathbf{b}$.

    Since $A = \begin{bmatrix} 2 & 1 \\ 5 & 3 \end{bmatrix}$, we have $A^{-1} = \frac{1}{1} \begin{bmatrix} 3 & -1 \\ -5 & 2 \end{bmatrix} = \begin{bmatrix} 3 & -1 \\ -5 & 2 \end{bmatrix}$.

    Therefore, since $\mathbf{b} = \begin{bmatrix} -1 \\ 2 \end{bmatrix}$, we have $\mathbf{x} = A^{-1}\mathbf{b} = \begin{bmatrix} 3 & -1 \\ -5 & 2 \end{bmatrix} \begin{bmatrix} -1 \\ 2 \end{bmatrix} = \begin{bmatrix} -5 \\ 9 \end{bmatrix}$.

    Q: How can we check our answer for $\mathbf{x}$?
    A: By applying the condition. That is, by checking that $A\mathbf{x} = \mathbf{b}$. Let's do that:

    Check: $A\mathbf{x} = \begin{bmatrix} 2 & 1 \\ 5 & 3 \end{bmatrix} \begin{bmatrix} -5 \\ 9 \end{bmatrix} = \begin{bmatrix} -1 \\ 2 \end{bmatrix} = \mathbf{b}.$

3.3 The Inverse of a Matrix

13. As in Example 3.25, we use the inverse of the coefficient matrix A to solve the systems.

Since $A = \begin{bmatrix} 1 & 2 \\ 2 & 6 \end{bmatrix}$, we have $A^{-1} = \frac{1}{2} \begin{bmatrix} 6 & -2 \\ -2 & 1 \end{bmatrix} = \begin{bmatrix} 3 & -1 \\ -1 & \frac{1}{2} \end{bmatrix}$.

Also, note that we are given $\mathbf{b}_1 = \begin{bmatrix} 3 \\ 5 \end{bmatrix}$, $\mathbf{b}_2 = \begin{bmatrix} -1 \\ 2 \end{bmatrix}$, and $\mathbf{b}_3 = \begin{bmatrix} 2 \\ 0 \end{bmatrix}$.

(a) Since $A\mathbf{x}_i = \mathbf{b}_i$, the solution in each case is $\mathbf{x}_i = A^{-1}\mathbf{b}_i$. So we have:

$$\mathbf{x}_1 = A^{-1}\mathbf{b}_1 = \begin{bmatrix} 3 & -1 \\ -1 & \frac{1}{2} \end{bmatrix} \begin{bmatrix} 3 \\ 5 \end{bmatrix} = \begin{bmatrix} 4 \\ -\frac{1}{2} \end{bmatrix}$$

$$\mathbf{x}_2 = A^{-1}\mathbf{b}_2 = \begin{bmatrix} 3 & -1 \\ -1 & \frac{1}{2} \end{bmatrix} \begin{bmatrix} -1 \\ 2 \end{bmatrix} = \begin{bmatrix} -5 \\ 2 \end{bmatrix}$$

$$\mathbf{x}_3 = A^{-1}\mathbf{b}_3 = \begin{bmatrix} 3 & -1 \\ -1 & \frac{1}{2} \end{bmatrix} \begin{bmatrix} 2 \\ 0 \end{bmatrix} = \begin{bmatrix} 6 \\ -2 \end{bmatrix}$$

(b) We form the augmented matrix $[\,A \mid \mathbf{b}_1 \ \mathbf{b}_2 \ \mathbf{b}_3\,]$ and row reduce to solve.

$$\begin{bmatrix} 1 & 2 & | & 3 & -1 & 2 \\ 2 & 6 & | & 5 & 2 & 0 \end{bmatrix} \longrightarrow \begin{bmatrix} 1 & 0 & | & 4 & -5 & 6 \\ 0 & 1 & | & -\frac{1}{2} & 2 & -2 \end{bmatrix}.$$

(c) The A^{-1} method requires 7 multiplications, while row reduction requires only 6.

15. To prove X is the inverse of A, all we have to show is $AX = I$.
Theorem 3.9d. asserts $(A^T)^{-1} = (A^{-1})^T$, so all we need to show is $(A^T)(A^{-1})^T = I$.

$$(A^T)(A^{-1})^T \overset{\substack{\text{by Thm 3.4d} \\ B^T A^T = (AB)^T}}{=} (A^{-1}A)^T = I^T \overset{\substack{\text{This is obvious.} \\ \text{Why?}}}{=} I$$

17. In (a) we will give an example, and in (b) we will prove $(AB)^{-1} = A^{-1}B^{-1} \Leftrightarrow AB = BA$.

(a) We choose A, B, then compute $(AB)^{-1}$, $A^{-1}B^{-1}$ to show $(AB)^{-1} \neq A^{-1}B^{-1}$.
Let $A = \begin{bmatrix} 1 & 0 \\ 2 & 1 \end{bmatrix}$ and $B = \begin{bmatrix} 1 & 0 \\ -2 & 1 \end{bmatrix}$. Then

$$(AB)^{-1} = \left(\begin{bmatrix} 1 & 0 \\ 2 & 1 \end{bmatrix}\begin{bmatrix} 1 & 0 \\ 1 & -1 \end{bmatrix}\right)^{-1} = \left(\begin{bmatrix} 1 & 0 \\ 3 & -1 \end{bmatrix}\right)^{-1} = \begin{bmatrix} 1 & 0 \\ 3 & -1 \end{bmatrix},$$

$$A^{-1}B^{-1} = \begin{bmatrix} 1 & 0 \\ 2 & 1 \end{bmatrix}^{-1}\begin{bmatrix} 1 & 0 \\ 1 & -1 \end{bmatrix}^{-1} = \begin{bmatrix} 1 & 0 \\ -2 & 1 \end{bmatrix}\begin{bmatrix} 1 & 0 \\ 1 & -1 \end{bmatrix} = \begin{bmatrix} 1 & 0 \\ -1 & -1 \end{bmatrix}.$$

So, clearly $(AB)^{-1} \neq A^{-1}B^{-1}$.

(b) Since this is an *if and only if* statement, we have two claims to prove.

if: If $(AB)^{-1} = A^{-1}B^{-1}$, then $AB = BA$. (Note $(AB)^{-1} = A^{-1}B^{-1}$ is *given*).

$$AB \stackrel{\text{by Thm 3.9a}}{=} ((AB)^{-1})^{-1} \stackrel{\text{by given}}{=} (A^{-1}B^{-1})^{-1} \stackrel{\text{by Thm 3.9c}}{=} (B^{-1})^{-1}(A^{-1})^{-1} \stackrel{\text{by Thm 3.9a}}{=} BA$$

only if: If $AB = BA$, then $(AB)^{-1} = A^{-1}B^{-1}$.

$$(AB)^{-1} \stackrel{\text{by given } AB=BA}{=} (BA)^{-1} \stackrel{\text{by Thm 3.9c}}{=} A^{-1}B^{-1}$$

19. Let $A = \begin{bmatrix} 1 & 0 \\ 0 & 1 \end{bmatrix}$ and $B = \begin{bmatrix} 1 & 2 \\ 0 & 1 \end{bmatrix}$. Then

$$(A+B)^{-1} = \left(\begin{bmatrix} 1 & 0 \\ 0 & 1 \end{bmatrix} + \begin{bmatrix} 1 & 2 \\ 0 & 1 \end{bmatrix}\right)^{-1} = \begin{bmatrix} 2 & 2 \\ 0 & 2 \end{bmatrix}^{-1} = \begin{bmatrix} \frac{1}{2} & -\frac{1}{2} \\ 0 & \frac{1}{2} \end{bmatrix}$$

$$A^{-1} + B^{-1} = \begin{bmatrix} 1 & 0 \\ 0 & 1 \end{bmatrix}^{-1} + \begin{bmatrix} 1 & 2 \\ 0 & 1 \end{bmatrix}^{-1} = \begin{bmatrix} 1 & 0 \\ 0 & 1 \end{bmatrix} + \begin{bmatrix} 1 & -2 \\ 0 & 1 \end{bmatrix} = \begin{bmatrix} 2 & -2 \\ 0 & 2 \end{bmatrix}$$

So $(A+B)^{-1} \neq A^{-1} + B^{-1}$.

21. $AXB = (BA)^2 \Rightarrow A^{-1}(AXB)B^{-1} = A^{-1}(BA)^2B^{-1} \Rightarrow (A^{-1}A)X(BB^{-1}) = A^{-1}(BA)^2B^{-1}$
So $X = A^{-1}(BA)^2B^{-1}$.

23. $ABXA^{-1}B^{-1} = I + A \Rightarrow X = B^{-1}A^{-1}(I+A)BA = B^{-1}A^{-1}BA + B^{-1}A^{-1}ABA$
So $X = (AB)^{-1}BA + A$.

25. $E = \begin{bmatrix} 0 & 0 & 1 \\ 0 & 1 & 0 \\ 1 & 0 & 0 \end{bmatrix}$.

27. $E = \begin{bmatrix} 1 & 0 & 0 \\ 0 & 1 & 0 \\ -1 & 0 & 1 \end{bmatrix}$.

29. $E = \begin{bmatrix} 1 & 0 & 0 \\ 0 & 1 & 2 \\ 0 & 0 & 1 \end{bmatrix}$.

31. $\begin{bmatrix} 3 & 0 \\ 0 & 1 \end{bmatrix}^{-1} = \begin{bmatrix} \frac{1}{3} & 0 \\ 0 & 1 \end{bmatrix}$.

33. $\begin{bmatrix} 0 & 1 \\ 1 & 0 \end{bmatrix}^{-1} = \begin{bmatrix} 0 & 1 \\ 1 & 0 \end{bmatrix}$.

3.3 The Inverse of a Matrix

35. $\begin{bmatrix} 1 & 0 & 0 \\ 0 & 1 & -2 \\ 0 & 0 & 1 \end{bmatrix}^{-1} = \begin{bmatrix} 1 & 0 & 0 \\ 0 & 1 & 2 \\ 0 & 0 & 1 \end{bmatrix}.$

37. $\begin{bmatrix} 1 & 0 & 0 \\ 0 & c & 0 \\ 0 & 0 & 1 \end{bmatrix}^{-1} = \begin{bmatrix} 1 & 0 & 0 \\ 0 & \frac{1}{c} & 0 \\ 0 & 0 & 1 \end{bmatrix}.$

39. As in Example 3.29, we attempt to express A as a product of elementary matrices. Why? To compute A^{-1} and to *factor* A, that is to see how A is created by elementary row operations. As we row reduce A, we use the elementary operations involved to create E_i.

$$A = \begin{bmatrix} 1 & 0 \\ -1 & -2 \end{bmatrix} \xrightarrow{R_2+R_1} \begin{bmatrix} 1 & 0 \\ 0 & -2 \end{bmatrix} \xrightarrow{-\frac{1}{2}R_2} \begin{bmatrix} 1 & 0 \\ 0 & 1 \end{bmatrix}$$

We recreate those steps on the identity matrix I to create E_1 and E_2.
This calculation can be done mentally, but we write it here to demonstrate the process.

Since $I = \begin{bmatrix} 1 & 0 \\ 0 & 1 \end{bmatrix} \xrightarrow{R_2+R_1} \begin{bmatrix} 1 & 0 \\ 1 & 1 \end{bmatrix}$, we get $E_1 = \begin{bmatrix} 1 & 0 \\ 1 & 1 \end{bmatrix}$.

Since $I = \begin{bmatrix} 1 & 0 \\ 0 & 1 \end{bmatrix} \xrightarrow{-\frac{1}{2}R_2} \begin{bmatrix} 1 & 0 \\ 0 & -\frac{1}{2} \end{bmatrix}$, we get $E_2 = \begin{bmatrix} 1 & 0 \\ 0 & -\frac{1}{2} \end{bmatrix}$.

Since $(E_2 E_1) A = I$, by definition $A^{-1} = E_2 E_1$.
So, Theorem 3.9a implies $A = (A^{-1})^{-1} = (E_2 E_1)^{-1} = E_1^{-1} E_2^{-1}$.
But E_1^{-1} and E_2^{-1} can be written down without doing any calculation. How?
Since E_1 was created by $R_2 + R_1$, we create E_1^{-1} by $R_2 - R_1$.
Likewise, since E_2 was created by $-\frac{1}{2} R_2$, we create E_2^{-1} by $-2R_2$.

Therefore, $E_1^{-1} = \begin{bmatrix} 1 & 0 \\ -1 & 1 \end{bmatrix}$ and $E_2^{-1} = \begin{bmatrix} 1 & 0 \\ 0 & -2 \end{bmatrix}.$

We could verify this by performing the indication operation on the identity matrix I. Or we could simply verify the claim that $A = E_1^{-1} E_2^{-1}$ directly.

Check: $A = \begin{bmatrix} 1 & 0 \\ -1 & -2 \end{bmatrix} = \begin{bmatrix} 1 & 0 \\ -1 & 1 \end{bmatrix} \begin{bmatrix} 1 & 0 \\ 0 & -2 \end{bmatrix} = E_1^{-1} E_2^{-1}$

This answer is not unique. For example, we could have row reduced A as follows.

$$A = \begin{bmatrix} 1 & 0 \\ -1 & -2 \end{bmatrix} \xrightarrow{-\frac{1}{2}R_2} \begin{bmatrix} 1 & 0 \\ \frac{1}{2} & 1 \end{bmatrix} \xrightarrow{R_2-\frac{1}{2}R_1} \begin{bmatrix} 1 & 0 \\ 0 & 1 \end{bmatrix}$$

So $E_1 = \begin{bmatrix} 1 & 0 \\ 0 & -\frac{1}{2} \end{bmatrix}$, $E_2 = \begin{bmatrix} 1 & 0 \\ -\frac{1}{2} & 1 \end{bmatrix}$ and $E_1^{-1} = \begin{bmatrix} 1 & 0 \\ 0 & -2 \end{bmatrix}$, $E_2^{-1} = \begin{bmatrix} 1 & 0 \\ \frac{1}{2} & 1 \end{bmatrix}.$

Verify that $A = E_1^{-1} E_2^{-1}$ in this case as well.

41. Suppose $AB = I$. Then, consider: $B\mathbf{x} = \mathbf{0}$. Left-multiplying by $A \Rightarrow AB\mathbf{x} = A\mathbf{0}$. This implies that $I\mathbf{x} = \mathbf{x} = \mathbf{0}$. Thus $B\mathbf{x} = \mathbf{0}$ has the unique solution $\mathbf{x} = \mathbf{0} \Rightarrow$ The equivalence of (c) and (a) in the Fundamental Theorem (3.12) $\Rightarrow B$ is invertible. If we now right-multiply both sides of $AB = I$ by B^{-1}, we obtain
$$ABB^{-1} = IB^{-1} \Leftrightarrow AI = B^{-1} \Leftrightarrow A = B^{-1} \Leftrightarrow A^{-1} = B.$$
Thus the inverse of A exists and $A^{-1} = B$.

43. (a) Suppose that A is invertible (thus A^{-1} exists) and $BA = CA$.
    If we now right-multiply both sides of $BA = CA$ by A^{-1}, we obtain
    $BAA^{-1} = CAA^{-1} \Leftrightarrow BI = CI \Leftrightarrow B = C$.

    (b) Let $A = \begin{bmatrix} 1 & 2 \\ 2 & 4 \end{bmatrix}$, $B = \begin{bmatrix} 1 & 1 \\ 1 & 1 \end{bmatrix}$, and $C = \begin{bmatrix} 3 & 0 \\ 1 & 1 \end{bmatrix}$. Then

    $BA = \begin{bmatrix} 1 & 1 \\ 1 & 1 \end{bmatrix} \begin{bmatrix} 1 & 2 \\ 2 & 4 \end{bmatrix} = \begin{bmatrix} 3 & 6 \\ 3 & 6 \end{bmatrix} = \begin{bmatrix} 3 & 0 \\ 1 & 1 \end{bmatrix} \begin{bmatrix} 1 & 2 \\ 2 & 4 \end{bmatrix} = CA$. So $BA = CA$, but $B \neq C$.

45. To prove X is the inverse of A, all we have to show is $AX = I$.
    We claim $A^{-1} = 2I - A$, so all we need to show is $A(2I - A) = I$.

    $$A^2 - 2A + I = O \overset{\text{given}}{\Rightarrow} 2A - A^2 = I \overset{\text{factor}}{\Rightarrow} A(2I - A) = I$$

47. Let A and B be square matrices and assume that AB is invertible.
    Then $AB(AB)^{-1} = A\left(B(AB)^{-1}\right) = I$, showing that A is invertible with $A^{-1} = B(AB)^{-1}$.
    Therefore $B(AB)^{-1}A = I$, showing that B is invertible with $B^{-1} = (AB)^{-1}A$.

49. As in Example 3.30, we adjoin the identity matrix to A then row reduce $[\,A\,|\,I\,]$ to $[\,I\,|\,A^{-1}\,]$.

    $[\,A\,|\,I\,] = \begin{bmatrix} -2 & 4 & | & 1 & 0 \\ 3 & -1 & | & 0 & 1 \end{bmatrix} \longrightarrow \begin{bmatrix} 1 & 0 & | & \frac{1}{10} & \frac{2}{5} \\ 0 & 1 & | & \frac{3}{10} & \frac{1}{5} \end{bmatrix} = [\,I\,|\,A^{-1}\,] \Rightarrow A^{-1} = \begin{bmatrix} \frac{1}{10} & \frac{2}{5} \\ \frac{3}{10} & \frac{1}{5} \end{bmatrix}$

51. As in Example 3.30, we adjoin the identity matrix to A then row reduce $[\,A\,|\,I\,]$ to $[\,I\,|\,A^{-1}\,]$.

    $[\,A\,|\,I\,] = \begin{bmatrix} 1 & a & | & 1 & 0 \\ -a & 1 & | & 0 & 1 \end{bmatrix} \longrightarrow \begin{bmatrix} 1 & 0 & | & \frac{1}{a^2+1} & -\frac{a}{a^2+1} \\ 0 & 1 & | & \frac{a}{a^2+1} & \frac{1}{a^2+1} \end{bmatrix} = [\,I\,|\,A^{-1}\,] \Rightarrow A^{-1} = \begin{bmatrix} \frac{1}{a^2+1} & -\frac{a}{a^2+1} \\ \frac{a}{a^2+1} & \frac{1}{a^2+1} \end{bmatrix}$

    Q: What is the restriction on a?
    A: Since $a^2 + 1 = 0 \Rightarrow a^2 = -1$, there is no restriction on a. Why not?

    Q: Does this agree with the formula for A^{-1} given in Theorem 3.8, that is $A^{-1} = \frac{1}{ad-bc}\begin{bmatrix} d & -b \\ -c & a \end{bmatrix}$?

53. As in Example 3.30, we adjoin the identity matrix to A then row reduce $[\,A\,|\,I\,]$ to $[\,I\,|\,A^{-1}\,]$.

    $[\,A\,|\,I\,] = \begin{bmatrix} 1 & -1 & 2 & | & 1 & 0 & 0 \\ 3 & 1 & 2 & | & 0 & 1 & 0 \\ 2 & 3 & -1 & | & 0 & 0 & 1 \end{bmatrix} \longrightarrow \begin{bmatrix} 1 & -1 & 2 & | & 1 & 0 & 0 \\ 0 & 4 & -4 & | & -3 & 1 & 0 \\ 0 & 5 & -5 & | & -2 & 0 & 1 \end{bmatrix} \neq [\,I\,|\,A^{-1}\,] \Rightarrow$

    Since the left matrix cannot be reduced to I (why?), we conclude A^{-1} does not exist.

3.3 The Inverse of a Matrix

55. As in Example 3.30, we adjoin the identity matrix to A then row reduce $[\,A\,|\,I\,]$ to $[\,I\,|\,A^{-1}\,]$.

$$[\,A\,|\,I\,] = \begin{bmatrix} a & 0 & 0 & 1 & 0 & 0 \\ 1 & a & 0 & 0 & 1 & 0 \\ 0 & 1 & a & 0 & 0 & 1 \end{bmatrix} \longrightarrow \begin{bmatrix} 1 & 0 & 0 & \frac{1}{a} & 0 & 0 \\ 0 & 1 & 0 & -\frac{1}{a^2} & \frac{1}{a} & 0 \\ 0 & 0 & 1 & \frac{1}{a^3} & -\frac{1}{a^2} & \frac{1}{a} \end{bmatrix} = [\,I\,|\,A^{-1}\,]$$

Q: The entries of A^{-1} imply $a \neq 0$. Why is that obvious in the original matrix A?

A: If $a = 0$ in A, we have $A = \begin{bmatrix} 0 & 0 & 0 \\ 1 & 0 & 0 \\ 0 & 1 & 0 \end{bmatrix}$, which is obviously not invertible. Why?

57. $A^{-1} = \begin{bmatrix} -11 & -2 & 5 & -4 \\ 4 & 1 & -2 & 2 \\ 5 & 1 & -2 & 2 \\ 9 & 2 & -4 & 3 \end{bmatrix}$.

59. As in Example 3.30, we adjoin the identity matrix to A then row reduce $[\,A\,|\,I\,]$ to $[\,I\,|\,A^{-1}\,]$.

$$[\,A\,|\,I\,] = \begin{bmatrix} 1 & 0 & 0 & 0 & 1 & 0 & 0 & 0 \\ 0 & 1 & 0 & 0 & 0 & 1 & 0 & 0 \\ 0 & 0 & 1 & 0 & 0 & 0 & 1 & 0 \\ a & b & c & d & 0 & 0 & 0 & 1 \end{bmatrix} \longrightarrow \begin{bmatrix} 1 & 0 & 0 & 0 & 1 & 0 & 0 & 0 \\ 0 & 1 & 0 & 0 & 0 & 1 & 0 & 0 \\ 0 & 0 & 1 & 0 & 0 & 0 & 1 & 0 \\ 0 & 0 & 0 & 1 & -\frac{a}{d} & -\frac{b}{d} & -\frac{c}{d} & \frac{1}{d} \end{bmatrix} = [\,I\,|\,A^{-1}\,]$$

Q: The entries of A^{-1} imply $d \neq 0$. Why is that obvious in the original matrix A?

A: If $d = 0$ in A, we have $A = \begin{bmatrix} 1 & 0 & 0 & 0 \\ 0 & 1 & 0 & 0 \\ 0 & 0 & 1 & 0 \\ a & b & c & 0 \end{bmatrix}$, which is obviously not invertible. Why?

61. As in Example 3.32, we adjoin the identity matrix to A then row reduce $[\,A\,|\,I\,]$ to $[\,I\,|\,A^{-1}\,]$. Recall that we need only use addition and multiplication in $\mathbb{Z}_5$.

$$[\,A\,|\,I\,] = \begin{bmatrix} 4 & 2 & 1 & 0 \\ 3 & 4 & 0 & 1 \end{bmatrix} \longrightarrow \begin{bmatrix} 1 & 0 & 1 & 0 \\ 0 & 0 & 3 & 1 \end{bmatrix} \neq [\,I\,|\,A^{-1}\,] \Rightarrow$$

Since the left matrix cannot be reduced to I (why?), we conclude A^{-1} does not exist.

Q: How else could we have determined A is not invertible?
A: By noting that $\det A = 4 \cdot 4 - 2 \cdot 3 = 10 = 0$ in $\mathbb{Z}_5$. Why?

63. As in Example 3.32, we adjoin the identity matrix to A then row reduce $[\,A\,|\,I\,]$ to $[\,I\,|\,A^{-1}\,]$. Recall that we need only use addition and multiplication in $\mathbb{Z}_7$.

$$[\,A\,|\,I\,] = \begin{bmatrix} 1 & 5 & 0 & 1 & 0 & 0 \\ 1 & 2 & 4 & 0 & 1 & 0 \\ 3 & 6 & 1 & 0 & 0 & 1 \end{bmatrix} \longrightarrow \begin{bmatrix} 1 & 0 & 0 & 4 & 6 & 4 \\ 0 & 1 & 0 & 5 & 3 & 2 \\ 0 & 0 & 1 & 0 & 6 & 5 \end{bmatrix} = [\,I\,|\,A^{-1}\,]$$

Q: How many 3×3 matrices are there in $\mathbb{Z}_7$?
A: There are $40{,}453{,}607$ 3×3 matrices in $\mathbb{Z}_7$. Why? Because $7^9 = 40{,}453{,}607$.

65. $\begin{bmatrix} O & B \\ C & I \end{bmatrix} \begin{bmatrix} O & B \\ C & I \end{bmatrix}^{-1} = \begin{bmatrix} O & B \\ C & I \end{bmatrix} \begin{bmatrix} -(BC)^{-1} & (BC)^{-1}B \\ C(BC)^{-1} & I - C(BC)^{-1}B \end{bmatrix}$

$= \begin{bmatrix} BC(BC)^{-1} & B - BC(BC)^{-1}B \\ -C(BC)^{-1} + C(BC)^{-1} & C(BC)^{-1}B + I - C(BC)^{-1}B \end{bmatrix}$

$= \begin{bmatrix} I & B - B \\ O & I \end{bmatrix} = \begin{bmatrix} I & O \\ O & I \end{bmatrix}.$

67. $\begin{bmatrix} O & B \\ C & D \end{bmatrix} \begin{bmatrix} O & B \\ C & D \end{bmatrix}^{-1} = \begin{bmatrix} O & B \\ C & D \end{bmatrix} \begin{bmatrix} -(BD^{-1}C)^{-1} & (BD^{-1}C)^{-1}BD^{-1} \\ D^{-1}C(BD^{-1}C)^{-1} & D^{-1} - D^{-1}C(BD^{-1}C)^{-1}BD^{-1} \end{bmatrix}$

$= \begin{bmatrix} BD^{-1}C(BD^{-1}C)^{-1} & BD^{-1} - BD^{-1}C(BD^{-1}C)^{-1}BD^{-1} \\ -C(BD^{-1}C)^{-1} + DD^{-1}C(BD^{-1}C)^{-1} & C(BD^{-1}C)^{-1}BD^{-1} + DD^{-1} + \\ & \left(-DD^{-1}C(BD^{-1}C)^{-1}BD^{-1}\right) \end{bmatrix}$

$= \begin{bmatrix} I & O \\ O & I \end{bmatrix}.$

69. We partition the matrix $\begin{bmatrix} 1 & 0 & 0 & 0 \\ 0 & 1 & 0 & 0 \\ 2 & 3 & 1 & 0 \\ 1 & 2 & 0 & 1 \end{bmatrix}$ into the form $\begin{bmatrix} I_2 & B \\ C & I_2 \end{bmatrix}$ as in Exercise 66, where $B = O$.

Then $(I - BC)^{-1} = I$, $(I - BC)^{-1}B = OB = O$, $-C(I - BC)^{-1} = -CI = -C$, and

$I + C(I - BC)^{-1}B = I$, so $\begin{bmatrix} 1 & 0 & 0 & 0 \\ 0 & 1 & 0 & 0 \\ 2 & 3 & 1 & 0 \\ 1 & 2 & 0 & 1 \end{bmatrix}^{-1} = \begin{bmatrix} 1 & 0 & 0 & 0 \\ 0 & 1 & 0 & 0 \\ -2 & -3 & 1 & 0 \\ -1 & -2 & 0 & 1 \end{bmatrix}.$

71. We partition the matrix $\begin{bmatrix} 0 & 0 & 1 & 1 \\ 0 & 0 & 1 & 0 \\ 0 & -1 & 1 & 0 \\ 1 & 1 & 0 & 1 \end{bmatrix}$ into the form $\begin{bmatrix} O & B \\ C & I \end{bmatrix}$ as in Exercise 65. Then

$-(BC)^{-1} = -\left(\begin{bmatrix} 1 & 1 \\ 1 & 0 \end{bmatrix} \begin{bmatrix} 0 & -1 \\ 1 & 1 \end{bmatrix}\right)^{-1} = -\left(\begin{bmatrix} 1 & 0 \\ 0 & -1 \end{bmatrix}\right)^{-1} = \begin{bmatrix} -1 & 0 \\ 0 & 1 \end{bmatrix},$

$(BC)^{-1}B = \begin{bmatrix} 1 & 0 \\ 0 & -1 \end{bmatrix} \begin{bmatrix} 1 & 1 \\ 1 & 0 \end{bmatrix} = \begin{bmatrix} 1 & 1 \\ -1 & 0 \end{bmatrix},$

$C(BC)^{-1} = \begin{bmatrix} 0 & -1 \\ 1 & 1 \end{bmatrix} \begin{bmatrix} 1 & 0 \\ 0 & -1 \end{bmatrix} = \begin{bmatrix} 0 & 1 \\ 1 & -1 \end{bmatrix},$ and

$I - C(BC)^{-1}B = I - \begin{bmatrix} 0 & 1 \\ 1 & -1 \end{bmatrix} \begin{bmatrix} 1 & 1 \\ 1 & 0 \end{bmatrix} = \begin{bmatrix} 0 & 0 \\ 0 & 0 \end{bmatrix}.$

Thus, $\begin{bmatrix} 0 & 0 & 1 & 1 \\ 0 & 0 & 1 & 0 \\ 0 & -1 & 1 & 0 \\ 1 & 1 & 0 & 1 \end{bmatrix}^{-1} = \begin{bmatrix} -1 & 0 & 1 & 1 \\ 0 & 1 & -1 & 0 \\ 0 & 1 & 0 & 0 \\ 1 & -1 & 0 & 0 \end{bmatrix}.$

3.4 The LU Factorization

Q: What are our main goals for Section 3.4?
A: To learn how to factor A into LU and use that factorization to solve systems.
   Also, we are introduced to proofs of existence and uniqueness. See Exercises 33 and 34.

Key Definitions and Concepts

Term	Page	Sec	Description
LU factorization	p.179	3.4	$A = LU$ with L unit lower triangular and and U upper triangular
LDU factorization	p.189	3.4	$A = LDU$ same as LU except D is diagonal and U unit upper triangular
L	p.179	3.4	Often used to stand for a unit lower triangular matrix
P	p.185	3.4	Often used to stand for a permutation matrix
U	p.179	3.4	Often used to stand for an upper triangular matrix
$P^T LU$ factorization	p.184	3.4	Adjustment to the LU factorization when we have to permute the rows of A
back substitution	p.180	3.4	See Example 5 in Section 2.1. Used when solving $U\mathbf{x} = \mathbf{y}$
forward substitution	p.180	3.4	See Exercises 25 and 26 in Section 2.1. Used when solving $L\mathbf{y} = \mathbf{b}$
multiplier	p.181	3.4	the scalar k in $R_i - kR_j$
permutation matrix	p.185	3.4	$P = P_k \cdots P_2 P_1$ where multiplication by P_i performs a row interchange like $R_i \leftrightarrow R_j$
upper diagonal	p.179	3.4	U where all the entries below the diagonal are 0. That is, $i < j \Rightarrow a_{ij} = 0$.
unit lower diagonal	p.179	3.4	L with all the entries above the diagonal are 0 and all the entries on the diagonal are 1. That is, $i > j \Rightarrow a_{ij} = 0$ and $a_{ii} = 1$.

Theorems

	Page	Sec	
Thm 3.15	p.180	3.4	If $A \longrightarrow$ REF without $R_i \leftrightarrow Rj$, then $A = LU$ exists.
Thm 3.16	p.184	3.4	If A is invertible, then we have the following: If $A = L_1 U_1$ and $A = L_2 U_2$ then $L_1 = L_2$ and $U_1 = U_2$. That is, if an LU factorization exists, it is unique.
Thm 3.17	p.185	3.4	If P is a permutation matrix, then $P^{-1} = P^T$.
Thm 3.18	p.186	3.4	If A is square, then $A = P^T LU$ exists.

Discussion of Key Definitions, Concepts, and Theorems

Q: How is the $P^T LU$ factorization of A related to the E_i factorization of A?
A: $A = E_1 E_2 \cdots E_n$ only if A is invertible, but $A = P^T LU$ for any square matrix. So the $P^T LU$ factorization is an extension of the E_i factorization.
Furthermore, if A is invertible and $A = LU$ exists, that LU factorization is unique. This is not true for the E_i factorization. See Exercises 39 and 40 in Section 3.3.

Q: How does the LU factorization help us to solve systems?
A: We want to solve $A\mathbf{x} = L(U\mathbf{x}) = \mathbf{b}$, so we let $U\mathbf{x} = \mathbf{y}$. Why?
A: Because $L\mathbf{y} = \mathbf{b}$ is easy to solve using *forward substitution* and $U\mathbf{x} = \mathbf{y}$ is easy to solve using *back substitution*.

Q: Where do the entries in L of $A = LU$ come from?
A: They are the opposites of the multipliers we use when row reducing A to U. Like:
$$A = \begin{bmatrix} 1 & 2 \\ -3 & -1 \end{bmatrix} \xrightarrow{R_2' = R_2 + 3R_1} \begin{bmatrix} 1 & 2 \\ 0 & 5 \end{bmatrix} = U \Rightarrow l_{21} = -3 \Rightarrow L = \begin{bmatrix} 1 & 0 \\ -3 & 1 \end{bmatrix}$$

Q: We added 3 times row 1 to row 2 above, so why is $l_{21} = -3$ instead of 3 in L?
A: Since multiplying U by L gives us A ($A = LU$), L must do the opposite of row reduction.

Q: Why is L in $A = LU$ factorization lower triangular?
A: It is *lower* triangular because we only add rows above to rows below.

Q: Why is L in $A = LU$ not just lower triangular, but *unit* lower triangular?
A: It is *unit* lower triangular because although we add (and subtract) multiples of rows we never replace any row by a multiple of itself.

Q: Why is row 1 of L always equal to row 1 of I in the LU factorization?
A: Row 1 is the top row, so we do not add or subtract any rows from it. It is unchanged. This pattern is demonstrated nicely, for instance, in Exercises 1 through 6.

Q: Why is row 1 of U always equal to row 1 of A in the LU factorization?
A: Row 1 is the top row, so we do not add or subtract any rows from it. It is unchanged. This pattern is demonstrated nicely, for instance, in Exercises 1 through 6.

Here is an example that demonstrates the four patterns we just discussed:
$$A = \begin{bmatrix} 1 & 2 & 3 \\ 4 & 5 & 6 \\ 8 & 7 & 9 \end{bmatrix} \xrightarrow[R_3' = R_3 - 8R_1]{R_2' = R_2 - 4R_1} \begin{bmatrix} 1 & 2 & 3 \\ 0 & -3 & -6 \\ 0 & -9 & -15 \end{bmatrix} \xrightarrow{R_3'' = R_3' - 3R_2'} \begin{bmatrix} 1 & 2 & 3 \\ 0 & -3 & -6 \\ 0 & 0 & 3 \end{bmatrix} = U \Rightarrow$$

$$\begin{matrix} l_{21} = 4 \\ l_{31} = 8 \quad l_{32} = 3 \end{matrix} \Rightarrow L = \begin{bmatrix} 1 & 0 & 0 \\ 4 & 1 & 0 \\ 8 & 3 & 1 \end{bmatrix}$$

See how we never added rows below to rows above?
See how we never replaced a row with a multiple of itself?
See how row 1 of L, $\begin{bmatrix} 1 & 0 & 0 \end{bmatrix}$, is identical to row 1 of I_3?
See how row 1 of U, $\begin{bmatrix} 1 & 2 & 3 \end{bmatrix}$, is identical to row 1 of A?

3.4 The LU Factorization

Discussion of Key Definitions, Concepts, and Theorems, continued

Q: How can we determine if our LU factorization of A is correct?
A: By checking that $LU = A$, just as we checked A^{-1} by verifying that $AA^{-1} = I$. Like:

$$\text{Check: } LU = \begin{bmatrix} 1 & 0 & 0 \\ 4 & 1 & 0 \\ 8 & 3 & 1 \end{bmatrix} \begin{bmatrix} 1 & 2 & 3 \\ 0 & -3 & -6 \\ 0 & 0 & 3 \end{bmatrix} = \begin{bmatrix} 1 & 2 & 3 \\ 4 & 5 & 6 \\ 8 & 7 & 9 \end{bmatrix} = A$$

Q: What prevents a square matrix A from having an LU factorization?
A: Its leading entries do not occur from right to left as we proceed up A from bottom to top.

Q: How can we correct this problem?
A: Simply by permuting the rows of A. Therefore, this problem can always be fixed.
That is why any square matrix has an $P^T LU$ factorization, as Theorem 3.18 asserts.

Q: Do we have to know how to permute the rows of A at the beginning of the problem?
A: No, though it is often obvious we do not have to know at the beginning.
We can row reduce until it becomes obvious, then permute the rows as necessary.

Q: If we know how to permute the rows of A, how do we find compute P^T?
A: First we compute P by permuting the rows of I
in exactly the same way as permuted the rows of A. That gives us P.
Then to create P^T, we simply switch the rows and columns of P. For example:

$$\text{If we permute } A = \begin{bmatrix} 0 & 1 & 2 \\ -1 & 3 & 5 \\ 1 & 0 & -5 \end{bmatrix} \text{ to } \begin{bmatrix} 1 & 0 & -5 \\ 0 & 1 & 2 \\ -1 & 3 & 5 \end{bmatrix} \text{ then}$$

$$P = \begin{bmatrix} 0 & 0 & 1 \\ 1 & 0 & 0 \\ 0 & 1 & 0 \end{bmatrix}. \text{ So } P^T = \begin{bmatrix} 0 & 1 & 0 \\ 0 & 0 & 1 \\ 1 & 0 & 0 \end{bmatrix}.$$

Q: Since $PA = LU \Rightarrow A = P^{-1}LU$, why is the factorization called $A = P^T LU$?
A: Because for any permutation matrix P, we have $P^{-1} = P^T$.
This is precisely the assertion of Theorem 3.17.

Q: Why can we solve $A\mathbf{x} = P^T LU\mathbf{x} = \mathbf{b}$ by solving $LU\mathbf{x} = P\mathbf{b}$?
A: First of all, $P^T LU\mathbf{x} = \mathbf{b} \Rightarrow LU\mathbf{x} = P\mathbf{b}$ because $P^{-1} = P^T$.
Secondly, all multiplication by P does is permute the entries of $\mathbf{b}$.
That is equivalent to writing the original equations in a different order.
But, obviously, that does not affect the solution. See Exercise 27.

Discussion of Key Definitions, Concepts, and Theorems, continued

Q: In Exercise 31, why does the introduction of D to create $A = LDU$ make it possible for U to be a *unit* upper triangular matrix?

A: Multiplication by D multiplies row i by diagonal entry d_{ii}.
In effect, we let the diagonal entries of U migrate to D.

Let's look at an example:

We begin by factoring $U = DU_1$, where D is diagonal and U_1 is *unit* upper triangular. To factor $U = DU_1$, we adapt the *multiplier* method of Example 3.35.

In this case, however, our goal is to create 1s on the diagonals. So:

$$\begin{bmatrix} -2 & 1 \\ 0 & 6 \end{bmatrix} \xrightarrow{R'_1 = -\frac{1}{2}R_1} \begin{bmatrix} 1 & -\frac{1}{2} \\ 0 & 6 \end{bmatrix} \xrightarrow{R'_2 = \frac{1}{6}R_2} \begin{bmatrix} 1 & -\frac{1}{2} \\ 0 & 1 \end{bmatrix} = U_1 \text{ unit upper triangular}$$

$$\begin{matrix} l_{11} = -2 \\ l_{22} = 6 \end{matrix} \Rightarrow D = \begin{bmatrix} -2 & 0 \\ 0 & 6 \end{bmatrix}$$

See how the diagonal entries of U, namely -2 and 6, are now the diagonal entries of D? Let's check it and at the same time see what happens when we multiply U_1 by D:

$$\text{Check: } DU_1 = \begin{bmatrix} -2 & 0 \\ 0 & 6 \end{bmatrix} \begin{bmatrix} 1 & -\frac{1}{2} \\ 0 & 1 \end{bmatrix} = \begin{bmatrix} -2 & 1 \\ 0 & 6 \end{bmatrix} = U$$

See how row 1 of U_1 was multiplied by -2 and row 2 of U_1 was multiplied by 6?

Q: Theorem 3.18 makes the following assertion:
If A is invertible and an LU factorization of A exists, then it is unique.
Does that imply that if A in not invertible, the LU factorization is not unique?

A: No. Not exactly. We should be careful. This is not the contrapositive statement.

Q: What is the contrapositive statement of Theorem 3.18?

A: If the LU factorization is not unique for A, then A is not invertible.
That being said, it is often the case that a matrix that is not invertible does indeed have more than one LU factorization.

Q: The matrix in Exercise 14 is not invertible.
Can we find another LU factorization of it?

A: Hint: In the first step use row 2 to create a zero in the first column of row 3.
Compare the resulting L and U to those found in the solution included below.

3.4 The LU Factorization

Discussion of Proofs of Uniqueness

We discuss this topic in the context of the solution to Exercises 33:

33. We need to show if $A = LDU$ is symmetric and invertible, then $U = L^T$.

    The key to a successful proof is strong application of the given conditions. For example:

    Q: Since $A = LDU$ is symmetric, what do we know about A?
    A: We know $LDU = A = A^T = (LDU)^T = U^T D^T L^T = U^T D L^T$. Does that help?

    Q: Since $A = LDU$ is invertible, what do we know about A?
    A: Theorem 3.16 $\Rightarrow$ If $A = L_1 U_1$ and $A = L_2 U_2$, then $L_1 = L_2$ and $U_1 = U_2$.

    Q: How can we put these two facts together to create a proof?
    A: We use the fact that A is symmetric to create $A = L_1 U_1$ and $A = L_2 U_2$.
    Then we use the fact that A is invertible to equate factors and see how $U = L^T$ results.

    PROOF: $A = L(DU)$ and $A = U^T(DL^T)$ (because A is symmetric)
    $\Rightarrow DU = DL^T$ (because A is invertible so $U_1 = U_2$)
    $\Rightarrow U = L^T$ (because D is invertible so $D^{-1}DM = M$)

    There are several underlying facts in this proof which we examine in detail below:

    Q: If U is unit upper triangular, is U^T unit lower triangular?
    A: Yes, since in U if $i > j \Rightarrow a_{ij} = 0 = a_{ji}$ if $j < i$ in U^T.
    Also, it is obvious that $a_{ii} = 1$ in both U and U^T.

    Q: Where did we use this fact in our proof?
    A: When claiming that $A = U^T(DL^T)$ is an LU factorization of A.
    In order for that to be true, U^T must be unit lower triangular. Why?

    Q: If L is lower triangular, is L^T upper triangular?
    A: Yes, since in L if $i < j \Rightarrow a_{ij} = 0 = a_{ji}$ if $j > i$ in L^T.

    Q: If D is diagonal and U is upper triangular is DU upper triangular?
    A: Yes. We proved this for lower triangular matrices in Exercise 29.

    Q: Where did we use these two facts in our proof?
    A: When claiming that $A = U^T(DL^T)$ is an LU factorization of A.
    In order for that to be true, DL^T must be upper triangular. Why?

    Q: When $A = LDU$ is invertible, is D invertible?
    A: Yes. Since $A^{-1}A = (A^{-1}LD)U = I \Rightarrow U$ is invertible and
    $AA^{-1} = L(DUA^{-1}) = I \Rightarrow L$ is invertible. So we have:
    $D = L^{-1}AU^{-1} \Rightarrow D(UA^{-1}L) = (L^{-1}AU^{-1})(UA^{-1}L) = I \Rightarrow D$ is invertible.
    In fact, if $A_1 A_2 \cdots A_n$ is invertible, then each A_i is invertible.

    Q: Where did we use this fact in our proof?
    A: When claiming that $DU = DL^T \Rightarrow U = L^T$ since $U = D^{-1}DU = D^{-1}DL^T = L^T$.

Solutions to odd-numbered exercises from Section 3.4

1. Since $A = LU$, where L is unit lower triangular and U is upper triangular, we have:

$$A\mathbf{x} = \mathbf{b} \overset{A=LU}{\Rightarrow} (LU)\mathbf{x} = \mathbf{b} \overset{(AB)C=A(BC)}{\Rightarrow} L(U\mathbf{x}) = \mathbf{b} \overset{U\mathbf{x}=\mathbf{y}}{\Rightarrow} L\mathbf{y} = \mathbf{b}$$

So we solve the system by the two-step method outlined after Theorem 3.15:

1) Solve $L\mathbf{y} = \mathbf{b}$ by *forward substitution* (see Exercises 25 and 26 in Section 2.1) and
2) Solve $U\mathbf{x} = \mathbf{y}$ by *back substitution* (see Example 2.5 in Section 2.1)

Since $A = \begin{bmatrix} -2 & 1 \\ 2 & 5 \end{bmatrix} = \begin{bmatrix} 1 & 0 \\ -1 & 1 \end{bmatrix} \begin{bmatrix} -2 & 1 \\ 0 & 6 \end{bmatrix} = LU$ and $\mathbf{b} = \begin{bmatrix} 5 \\ 1 \end{bmatrix}$, $L\mathbf{y} = \mathbf{b} \Rightarrow$

$\begin{bmatrix} 1 & 0 \\ -1 & 1 \end{bmatrix} \begin{bmatrix} y_1 \\ y_2 \end{bmatrix} = \begin{bmatrix} 5 \\ 1 \end{bmatrix} \Rightarrow \begin{matrix} y_1 = 5 \\ -y_1 + y_2 = 1 \end{matrix} \Rightarrow \begin{matrix} y_1 = 5 \\ y_2 = y_1 + 1 \end{matrix}$ So, $\mathbf{y} = \begin{bmatrix} 5 \\ 6 \end{bmatrix}$.

Likewise, since $U = \begin{bmatrix} -2 & 1 \\ 0 & 6 \end{bmatrix}$ and $\mathbf{y} = \begin{bmatrix} 5 \\ 6 \end{bmatrix}$, $U\mathbf{x} = \mathbf{y} \Rightarrow$

$\begin{bmatrix} -2 & 1 \\ 0 & 6 \end{bmatrix} \begin{bmatrix} x_1 \\ x_2 \end{bmatrix} = \begin{bmatrix} 5 \\ 6 \end{bmatrix} \Rightarrow \begin{matrix} -2x_1 + x_2 = 5 \\ 6x_2 = 6 \end{matrix} \Rightarrow \begin{matrix} x_1 = \frac{1}{2}x_2 - \frac{5}{2} \\ x_2 = 1 \end{matrix}$ So, $\mathbf{x} = \begin{bmatrix} -2 \\ 1 \end{bmatrix}$.

Check: $A\mathbf{x} = \begin{bmatrix} -2 & 1 \\ 2 & 5 \end{bmatrix} \begin{bmatrix} -2 \\ 1 \end{bmatrix} = \begin{bmatrix} 5 \\ 1 \end{bmatrix} = \mathbf{b}$.

3.4 The LU Factorization

3. Since $A = LU$, where L is unit lower triangular and U is upper triangular, we have:

$$A\mathbf{x} = \mathbf{b} \overset{A=LU}{\Rightarrow} (LU)\mathbf{x} = \mathbf{b} \overset{(AB)C=A(BC)}{\Rightarrow} L(U\mathbf{x}) = \mathbf{b} \overset{U\mathbf{x}=\mathbf{y}}{\Rightarrow} L\mathbf{y} = \mathbf{b}$$

So we solve the system by the two-step method outlined after Theorem 3.15:

1) Solve $L\mathbf{y} = \mathbf{b}$ by *forward substitution* (see Exercises 25 and 26 in Section 2.1) and
2) Solve $U\mathbf{x} = \mathbf{y}$ by *back substitution* (see Example 2.5 in Section 2.1)

$$A = \begin{bmatrix} 2 & 1 & -2 \\ 2 & 3 & -4 \\ 4 & -3 & 0 \end{bmatrix} = \begin{bmatrix} 1 & 0 & 0 \\ -1 & 1 & 0 \\ 2 & -\tfrac{5}{4} & 1 \end{bmatrix} \begin{bmatrix} 2 & 1 & -2 \\ 0 & 4 & -6 \\ 0 & 0 & -\tfrac{7}{2} \end{bmatrix} = LU \text{ and } \mathbf{b} = \begin{bmatrix} -3 \\ 1 \\ 0 \end{bmatrix}, \text{ so } L\mathbf{y} = \mathbf{b} \Rightarrow$$

$$\begin{bmatrix} 1 & 0 & 0 \\ -1 & 1 & 0 \\ 2 & -\tfrac{5}{4} & 1 \end{bmatrix} \begin{bmatrix} y_1 \\ y_2 \\ y_3 \end{bmatrix} = \begin{bmatrix} -3 \\ 1 \\ 0 \end{bmatrix} \Rightarrow \begin{matrix} y_1 & & & = -3 \\ -y_1 + & y_2 & & = 1 \\ 2y_1 - & \tfrac{5}{4}y_2 & + y_3 & = 0 \end{matrix} \Rightarrow \begin{matrix} y_1 = & -3 \\ y_2 = & y_1 + 1 \\ y_3 = -2y_1 + \tfrac{5}{4}y_2 \end{matrix} \Rightarrow$$

$$\begin{matrix} y_1 = -3 \\ y_2 = -3 \; + \; 1 = -2 \\ y_3 = -2(-3) + \tfrac{5}{4}(-2) = \tfrac{7}{2} \end{matrix} \qquad \text{So, } \mathbf{y} = \begin{bmatrix} -3 \\ -2 \\ \tfrac{7}{2} \end{bmatrix}$$

Likewise, $U = \begin{bmatrix} 2 & 1 & -2 \\ 0 & 4 & -6 \\ 0 & 0 & -\tfrac{7}{2} \end{bmatrix}$ and $\mathbf{y} = \begin{bmatrix} -3 \\ -2 \\ \tfrac{7}{2} \end{bmatrix}$, so $U\mathbf{x} = \mathbf{y} \Rightarrow$

$$\begin{bmatrix} 2 & 1 & -2 \\ 0 & 4 & -6 \\ 0 & 0 & -\tfrac{7}{2} \end{bmatrix} \begin{bmatrix} x_1 \\ x_2 \\ x_3 \end{bmatrix} = \begin{bmatrix} -3 \\ -2 \\ \tfrac{7}{2} \end{bmatrix} \Rightarrow \begin{matrix} 2x_1 + x_2 - 2x_3 = -3 \\ 4x_2 - 6x_3 = -2 \\ -\tfrac{7}{2}x_3 = \tfrac{7}{2} \end{matrix} \Rightarrow \begin{matrix} x_1 = -\tfrac{1}{2}x_2 + x_3 - \tfrac{3}{2} \\ x_2 = \tfrac{3}{2}x_3 - \tfrac{1}{2} \\ x_3 = -1 \end{matrix} \Rightarrow$$

$$\begin{matrix} x_1 = -\tfrac{1}{2}(-2) - 1 - \tfrac{3}{2} = -\tfrac{3}{2} \\ x_2 = \tfrac{3}{2}(-1) - \tfrac{1}{2} = -2 \\ x_3 = -1 \end{matrix} \qquad \Rightarrow \qquad \text{So, } \mathbf{x} = \begin{bmatrix} -\tfrac{3}{2} \\ -2 \\ -1 \end{bmatrix}$$

Check: $A\mathbf{x} = \begin{bmatrix} 2 & 1 & -2 \\ 2 & 3 & -4 \\ 4 & -3 & 0 \end{bmatrix} \begin{bmatrix} -\tfrac{3}{2} \\ -2 \\ -1 \end{bmatrix} = \begin{bmatrix} -3 \\ 1 \\ 0 \end{bmatrix} = \mathbf{b}.$

5. We solve the system by the two-step method outlined after Theorem 3.15:

$$A\mathbf{x} = \mathbf{b} \xRightarrow{A=LU} (LU)\mathbf{x} = \mathbf{b} \xRightarrow{U\mathbf{x}=\mathbf{y}} L\mathbf{y} = \mathbf{b}$$

$$L\mathbf{y} = \mathbf{b} \Rightarrow \begin{bmatrix} 1 & 0 & 0 & 0 \\ 3 & 1 & 0 & 0 \\ 4 & 0 & 1 & 0 \\ 2 & -1 & 5 & 1 \end{bmatrix} \begin{bmatrix} y_1 \\ y_2 \\ y_3 \\ y_4 \end{bmatrix} = \begin{bmatrix} 1 \\ 2 \\ 2 \\ 1 \end{bmatrix} \Rightarrow \mathbf{y} = \begin{bmatrix} y_1 \\ y_2 \\ y_3 \\ y_4 \end{bmatrix} = \begin{bmatrix} 1 \\ -1 \\ -2 \\ 8 \end{bmatrix} \Rightarrow$$

$$U\mathbf{x} = \mathbf{y} \Rightarrow \begin{bmatrix} 2 & -1 & 0 & 0 \\ 0 & -1 & 5 & -3 \\ 0 & 0 & 1 & 0 \\ 0 & 0 & 0 & 4 \end{bmatrix} \begin{bmatrix} x_1 \\ x_2 \\ x_3 \\ x_4 \end{bmatrix} = \begin{bmatrix} 1 \\ -1 \\ -2 \\ 8 \end{bmatrix} \Rightarrow \mathbf{x} = \begin{bmatrix} x_1 \\ x_2 \\ x_3 \\ x_4 \end{bmatrix} = \begin{bmatrix} -7 \\ -15 \\ -2 \\ 2 \end{bmatrix}$$

$$\text{Check: } A\mathbf{x} = \begin{bmatrix} 2 & -1 & 0 & 0 \\ 6 & -4 & 5 & -3 \\ 8 & -4 & 1 & 0 \\ 4 & -1 & 0 & 7 \end{bmatrix} \begin{bmatrix} -7 \\ -15 \\ -2 \\ 2 \end{bmatrix} = \begin{bmatrix} 1 \\ 2 \\ 2 \\ 1 \end{bmatrix} = \mathbf{b}$$

7. Following the *multiplier* method of Example 3.35, we find the LU factorization of A:

$$A = \begin{bmatrix} 1 & 2 \\ -3 & -1 \end{bmatrix} \xrightarrow{R_2' = R_2 + 3R_1} \begin{bmatrix} 1 & 2 \\ 0 & 5 \end{bmatrix} = U \Rightarrow l_{21} = -3 \Rightarrow L = \begin{bmatrix} 1 & 0 \\ -3 & 1 \end{bmatrix}$$

$$\text{Check: } LU = \begin{bmatrix} 1 & 0 \\ -3 & 1 \end{bmatrix} \begin{bmatrix} 1 & 2 \\ 0 & 5 \end{bmatrix} = \begin{bmatrix} 1 & 2 \\ -3 & -1 \end{bmatrix} = A$$

9. Following the *multiplier* method of Example 3.35, we find the LU factorization of A:

$$A = \begin{bmatrix} 1 & 2 & 3 \\ 4 & 5 & 6 \\ 8 & 7 & 9 \end{bmatrix} \xrightarrow[R_3' = R_2 - 8R_1]{R_2' = R_2 - 4R_1} \begin{bmatrix} 1 & 2 & 3 \\ 0 & -3 & -6 \\ 0 & -9 & -15 \end{bmatrix} \xrightarrow{R_3'' = R_3' - 3R_2'} \begin{bmatrix} 1 & 2 & 3 \\ 0 & -3 & -6 \\ 0 & 0 & 3 \end{bmatrix} = U \Rightarrow$$

$$\begin{matrix} l_{21} = 4 \\ l_{31} = 8 \quad l_{32} = 3 \end{matrix} \Rightarrow L = \begin{bmatrix} 1 & 0 & 0 \\ 4 & 1 & 0 \\ 8 & 3 & 1 \end{bmatrix}$$

$$\text{Check: } LU = \begin{bmatrix} 1 & 0 & 0 \\ 4 & 1 & 0 \\ 8 & 3 & 1 \end{bmatrix} \begin{bmatrix} 1 & 2 & 3 \\ 0 & -3 & -6 \\ 0 & 0 & 3 \end{bmatrix} = \begin{bmatrix} 1 & 2 & 3 \\ 4 & 5 & 6 \\ 8 & 7 & 9 \end{bmatrix} = A$$

3.4 The LU Factorization

11. Following the *multiplier* method of Example 3.35, we find the LU factorization of A:

$$A = \begin{bmatrix} 1 & 2 & 3 & -1 \\ 2 & 6 & 3 & 0 \\ 0 & 6 & -6 & 7 \\ -1 & -2 & -9 & 0 \end{bmatrix} \xrightarrow[R'_4 = R_4 + R_1]{R'_2 = R_2 - 2R_1} \begin{bmatrix} 1 & 2 & 3 & -1 \\ 0 & 4 & -3 & 2 \\ 0 & 6 & -6 & 7 \\ 0 & 0 & -6 & -1 \end{bmatrix} \xrightarrow{R'_3 = R_3 - \frac{3}{2}R'_2} \begin{bmatrix} 1 & 2 & 3 & -1 \\ 0 & 4 & -3 & 2 \\ 0 & 0 & -\frac{3}{2} & 4 \\ 0 & 0 & -6 & -1 \end{bmatrix}$$

$$\xrightarrow{R''_4 = R'_4 - 4R'_3} \begin{bmatrix} 1 & 2 & 3 & -1 \\ 0 & 4 & -3 & 2 \\ 0 & 0 & -\frac{3}{2} & 4 \\ 0 & 0 & 0 & -17 \end{bmatrix} = U \Rightarrow \begin{matrix} l_{21} = 2 \\ l_{31} = 0 \\ l_{41} = -1 \end{matrix} \begin{matrix} l_{32} = \frac{3}{2} \\ l_{42} = 0 \end{matrix} \begin{matrix} l_{43} = 4 \end{matrix} \Rightarrow L = \begin{bmatrix} 1 & 0 & 0 & 0 \\ 2 & 1 & 0 & 0 \\ 0 & \frac{3}{2} & 1 & 0 \\ -1 & 0 & 4 & 1 \end{bmatrix}$$

$$\text{Check: } LU = \begin{bmatrix} 1 & 0 & 0 & 0 \\ 2 & 1 & 0 & 0 \\ 0 & \frac{3}{2} & 1 & 0 \\ -1 & 0 & 4 & 1 \end{bmatrix} \begin{bmatrix} 1 & 2 & 3 & -1 \\ 0 & 4 & -3 & 2 \\ 0 & 0 & -\frac{3}{2} & 4 \\ 0 & 0 & 0 & -17 \end{bmatrix} = \begin{bmatrix} 1 & 2 & 3 & -1 \\ 2 & 6 & 3 & 0 \\ 0 & 6 & -6 & 7 \\ -1 & -2 & -9 & 0 \end{bmatrix}$$

13. By adapting the *multiplier* method of Example 3.35, we find the LU factorization of A. However, in this case, we simply note that A is already upper triangular so $L = I_3$.

$$\text{Check: } LU = \begin{bmatrix} 1 & 0 & 0 \\ 0 & 1 & 0 \\ 0 & 0 & 1 \end{bmatrix} \begin{bmatrix} 1 & 0 & 1 & -2 \\ 0 & 3 & 3 & 1 \\ 0 & 0 & 0 & 5 \end{bmatrix} = \begin{bmatrix} 1 & 0 & 1 & -2 \\ 0 & 3 & 3 & 1 \\ 0 & 0 & 0 & 5 \end{bmatrix}$$

Q: Why is $L = I_3$, the 3 × 3 identity matrix instead of I_4, the 4 × 4 identity matrix?
A: Because A has 3 rows. Considering only size, we have $A = LU = [3 \times 3][3 \times 4] = [3 \times 4]$.

15. Since $A = LU \Rightarrow A^{-1} = U^{-1}L^{-1}$, we use the LU factorization of A to find A^{-1}:

$$\text{Since } A = \begin{bmatrix} -2 & 1 \\ 2 & 5 \end{bmatrix} = \begin{bmatrix} 1 & 0 \\ -1 & 1 \end{bmatrix} \begin{bmatrix} -2 & 1 \\ 0 & 6 \end{bmatrix} = LU, \text{ we have:}$$

$$L^{-1} = \frac{1}{1} \begin{bmatrix} 1 & 0 \\ 1 & 1 \end{bmatrix} = \begin{bmatrix} 1 & 0 \\ 1 & 1 \end{bmatrix} \text{ and } U^{-1} = -\frac{1}{12} \begin{bmatrix} 6 & -1 \\ 0 & -2 \end{bmatrix} = \begin{bmatrix} -\frac{1}{2} & \frac{1}{12} \\ 0 & \frac{1}{6} \end{bmatrix}.$$

$$\text{Therefore, } A^{-1} = U^{-1}L^{-1} = \begin{bmatrix} -\frac{1}{2} & \frac{1}{12} \\ 0 & \frac{1}{6} \end{bmatrix} \begin{bmatrix} 1 & 0 \\ 1 & 1 \end{bmatrix} = \begin{bmatrix} -\frac{5}{12} & \frac{1}{12} \\ \frac{1}{6} & \frac{1}{6} \end{bmatrix}$$

$$\text{Check: } AA^{-1} = \begin{bmatrix} -2 & 1 \\ 2 & 5 \end{bmatrix} \begin{bmatrix} -\frac{5}{12} & \frac{1}{12} \\ \frac{1}{6} & \frac{1}{6} \end{bmatrix} = \begin{bmatrix} 1 & 0 \\ 0 & 1 \end{bmatrix} = I.$$

Note: To compute L^{-1} and U^{-1}, we used the formula: $A^{-1} = \frac{1}{ad-bc} \begin{bmatrix} d & -b \\ -c & a \end{bmatrix}$.

17. We compute A^{-1} one column at a time using the method outlined below:

$$(LU)\mathbf{x}_1 = \mathbf{e}_1 \overset{U\mathbf{x}_1=\mathbf{y}_1}{\Rightarrow} L\mathbf{y}_1 = \mathbf{e}_1 \quad \text{and} \quad (LU)\mathbf{x}_2 = \mathbf{e}_2 \overset{U\mathbf{x}_2=\mathbf{y}_2}{\Rightarrow} L\mathbf{y}_2 = \mathbf{e}_2$$

We begin by computing column 1 of A^{-1}:

$$L\mathbf{y}_1 = \mathbf{e}_1 \Rightarrow \begin{bmatrix} 1 & 0 \\ -1 & 1 \end{bmatrix} \begin{bmatrix} y_{11} \\ y_{21} \end{bmatrix} = \begin{bmatrix} 1 \\ 0 \end{bmatrix} \Rightarrow \mathbf{y}_1 = \begin{bmatrix} 1 \\ 1 \end{bmatrix} \Rightarrow$$

$$U\mathbf{x}_1 = \mathbf{y}_1 \Rightarrow \begin{bmatrix} -2 & 1 \\ 0 & 6 \end{bmatrix} \begin{bmatrix} x_{11} \\ x_{21} \end{bmatrix} = \begin{bmatrix} 1 \\ 1 \end{bmatrix} \Rightarrow \mathbf{x}_1 = \begin{bmatrix} -\frac{5}{12} \\ \frac{1}{6} \end{bmatrix}$$

We repeat this process to compute column 2 of A^{-1}:

$$L\mathbf{y}_2 = \mathbf{e}_2 \Rightarrow \begin{bmatrix} 1 & 0 \\ -1 & 1 \end{bmatrix} \begin{bmatrix} y_{12} \\ y_{22} \end{bmatrix} = \begin{bmatrix} 0 \\ 1 \end{bmatrix} \Rightarrow \mathbf{y}_2 = \begin{bmatrix} 0 \\ 1 \end{bmatrix} \Rightarrow$$

$$U\mathbf{x}_2 = \mathbf{y}_2 \Rightarrow \begin{bmatrix} -2 & 1 \\ 0 & 6 \end{bmatrix} \begin{bmatrix} x_{12} \\ x_{22} \end{bmatrix} = \begin{bmatrix} 0 \\ 1 \end{bmatrix} \Rightarrow \mathbf{x}_2 = \begin{bmatrix} \frac{1}{12} \\ \frac{1}{6} \end{bmatrix}$$

Therefore, $A^{-1} = \begin{bmatrix} \mathbf{x}_1 & \mathbf{x}_2 \end{bmatrix} = \begin{bmatrix} -\frac{5}{12} & \frac{1}{12} \\ \frac{1}{6} & \frac{1}{6} \end{bmatrix}$ exactly as we found in Exercise 15.

19. Since we need $R_1 \to R_2$, $R_2 \to R_3$, and $R_3 \to R_1$ one possibility is:

$$P = \begin{bmatrix} 0 & 0 & 1 \\ 1 & 0 & 0 \\ 0 & 1 & 0 \end{bmatrix} = E_{13}E_{12} = \begin{bmatrix} 0 & 0 & 1 \\ 0 & 1 & 0 \\ 1 & 0 & 0 \end{bmatrix} \begin{bmatrix} 0 & 1 & 0 \\ 1 & 0 & 0 \\ 0 & 0 & 1 \end{bmatrix}.$$

Q: Is this factorization of P or could we have found another one?
A: No, it is not unique. For instance, we could have chosen $P = E_{12}E_{23}$.

21. We find one possible factorization by tracing the row interchanges: $P = E_{13}E_{23}E_{34}$.

3.4 The LU Factorization

23. To find the $P^T LU$ factorization of A, we begin by permuting the rows of A:

Since $A = \begin{bmatrix} 0 & 1 & 4 \\ -1 & 2 & 1 \\ 1 & 3 & 3 \end{bmatrix}$, we have $PA = \begin{bmatrix} 0 & 1 & 0 \\ 1 & 0 & 0 \\ 0 & 0 & 1 \end{bmatrix} \begin{bmatrix} 0 & 1 & 4 \\ -1 & 2 & 1 \\ 1 & 3 & 3 \end{bmatrix} = \begin{bmatrix} -1 & 2 & 1 \\ 0 & 1 & 4 \\ 1 & 3 & 3 \end{bmatrix}$.

Now we follow the *multiplier* method of Example 3.35 to find the LU factorization of PA:

$$PA = \begin{bmatrix} -1 & 2 & 1 \\ 0 & 1 & 4 \\ 1 & 3 & 3 \end{bmatrix} \xrightarrow{R_3' = R_3 + R_1} \begin{bmatrix} -1 & 2 & 1 \\ 0 & 1 & 4 \\ 0 & 5 & 4 \end{bmatrix} \xrightarrow{R_3'' = R_3' - 5R_2} \begin{bmatrix} -1 & 2 & 1 \\ 0 & 1 & 4 \\ 0 & 0 & -16 \end{bmatrix} = U \Rightarrow$$

$$\begin{array}{l} l_{21} = 0 \\ l_{31} = -1 \quad l_{32} = 5 \end{array} \Rightarrow L = \begin{bmatrix} 1 & 0 & 0 \\ 0 & 1 & 0 \\ -1 & 5 & 1 \end{bmatrix}$$

Check: $LU = \begin{bmatrix} 1 & 0 & 0 \\ 0 & 1 & 0 \\ -1 & 5 & 1 \end{bmatrix} \begin{bmatrix} -1 & 2 & 1 \\ 0 & 1 & 4 \\ 0 & 0 & -16 \end{bmatrix} = \begin{bmatrix} -1 & 2 & 1 \\ 0 & 1 & 4 \\ 1 & 3 & 3 \end{bmatrix} = PA$

Now $PA = LU \Rightarrow A = P^{-1}LU$, but $P^{-1} = P^T$, so we have $A = P^T LU$.

Check: $P^T LU = \begin{bmatrix} 0 & 1 & 0 \\ 1 & 0 & 0 \\ 0 & 0 & 1 \end{bmatrix} \begin{bmatrix} 1 & 0 & 0 \\ 0 & 1 & 0 \\ -1 & 5 & 1 \end{bmatrix} \begin{bmatrix} -1 & 2 & 1 \\ 0 & 1 & 4 \\ 0 & 0 & -16 \end{bmatrix} = \begin{bmatrix} 0 & 1 & 4 \\ -1 & 2 & 1 \\ 1 & 3 & 3 \end{bmatrix} = A$

25. To find the $P^T LU$ factorization of A, we begin by permuting the rows of A:

Since $A = \begin{bmatrix} 0 & -1 & 1 & 3 \\ -1 & 1 & 1 & 2 \\ 0 & 1 & -1 & 1 \\ 0 & 0 & 1 & 1 \end{bmatrix}$, we have $PA = \begin{bmatrix} 0 & 1 & 0 & 0 \\ 1 & 0 & 0 & 0 \\ 0 & 0 & 0 & 1 \\ 0 & 0 & 1 & 0 \end{bmatrix} \begin{bmatrix} 0 & -1 & 1 & 3 \\ -1 & 1 & 1 & 2 \\ 0 & 1 & -1 & 1 \\ 0 & 0 & 1 & 1 \end{bmatrix} = \begin{bmatrix} -1 & 1 & 1 & 2 \\ 0 & -1 & 1 & 3 \\ 0 & 0 & 1 & 1 \\ 0 & 1 & -1 & 1 \end{bmatrix}$.

Now we follow the *multiplier* method of Example 3.35 to find the LU factorization of PA:

$$PA = \begin{bmatrix} -1 & 1 & 1 & 2 \\ 0 & -1 & 1 & 3 \\ 0 & 0 & 1 & 1 \\ 0 & 1 & -1 & 1 \end{bmatrix} \xrightarrow{R_4' = R_4 + R_2} \begin{bmatrix} -1 & 1 & 1 & 2 \\ 0 & -1 & 1 & 3 \\ 0 & 0 & 1 & 1 \\ 0 & 0 & 0 & 4 \end{bmatrix} = U \Rightarrow$$

$$\begin{array}{l} l_{21} = 0 \\ l_{31} = 0 \quad l_{32} = 0 \\ l_{41} = 0 \quad l_{42} = -1 \quad l_{43} = 0 \end{array} \Rightarrow L = \begin{bmatrix} 1 & 0 & 0 & 0 \\ 0 & 1 & 0 & 0 \\ 0 & 0 & 1 & 0 \\ 0 & -1 & 0 & 1 \end{bmatrix}$$

Check: $LU = \begin{bmatrix} 1 & 0 & 0 & 0 \\ 0 & 1 & 0 & 0 \\ 0 & 0 & 1 & 0 \\ 0 & -1 & 0 & 1 \end{bmatrix} \begin{bmatrix} -1 & 1 & 1 & 2 \\ 0 & -1 & 1 & 3 \\ 0 & 0 & 1 & 1 \\ 0 & 0 & 0 & 4 \end{bmatrix} = \begin{bmatrix} -1 & 1 & 1 & 2 \\ 0 & -1 & 1 & 3 \\ 0 & 0 & 1 & 1 \\ 0 & 1 & -1 & 1 \end{bmatrix} = PA$

Now $PA = LU \Rightarrow A = P^{-1}LU$, but $P^{-1} = P^T$, so we have $A = P^T LU$.

Check: $P^T LU = \begin{bmatrix} 0 & 1 & 0 & 0 \\ 1 & 0 & 0 & 0 \\ 0 & 0 & 0 & 1 \\ 0 & 0 & 1 & 0 \end{bmatrix} \begin{bmatrix} 1 & 0 & 0 & 0 \\ 0 & 1 & 0 & 0 \\ 0 & 0 & 1 & 0 \\ 0 & -1 & 0 & 1 \end{bmatrix} \begin{bmatrix} -1 & 1 & 1 & 2 \\ 0 & -1 & 1 & 3 \\ 0 & 0 & 1 & 1 \\ 0 & 0 & 0 & 4 \end{bmatrix} = \begin{bmatrix} 0 & -1 & 1 & 3 \\ -1 & 1 & 1 & 2 \\ 0 & 1 & -1 & 1 \\ 0 & 0 & 1 & 1 \end{bmatrix} = A$

27. We solve the system $LU\mathbf{x} = P\mathbf{b}$ by the two-step method outlined after Theorem 3.15:
$$PA\mathbf{x} = P\mathbf{b} \stackrel{PA=LU}{\Rightarrow} (LU)\mathbf{x} = P\mathbf{b} \stackrel{U\mathbf{x}=\mathbf{y}}{\Rightarrow} L\mathbf{y} = \mathbf{b}$$

Since $P = \begin{bmatrix} 0 & 1 & 0 \\ 1 & 0 & 0 \\ 0 & 0 & 1 \end{bmatrix}$ and $\mathbf{b} = \begin{bmatrix} 1 \\ 1 \\ 5 \end{bmatrix}$, we have: $P\mathbf{b} = \begin{bmatrix} 0 & 1 & 0 \\ 1 & 0 & 0 \\ 0 & 0 & 1 \end{bmatrix} \begin{bmatrix} 1 \\ 1 \\ 5 \end{bmatrix} = \begin{bmatrix} 1 \\ 1 \\ 5 \end{bmatrix}$.

$$L\mathbf{y} = P\mathbf{b} \Rightarrow \begin{bmatrix} 1 & 0 & 0 \\ 0 & 1 & 0 \\ \frac{1}{2} & -\frac{1}{2} & 1 \end{bmatrix} \begin{bmatrix} y_1 \\ y_2 \\ y_3 \end{bmatrix} = \begin{bmatrix} 1 \\ 1 \\ 5 \end{bmatrix} \Rightarrow \mathbf{y} = \begin{bmatrix} y_1 \\ y_2 \\ y_3 \end{bmatrix} = \begin{bmatrix} 1 \\ 1 \\ 5 \end{bmatrix} \Rightarrow$$

$$U\mathbf{x} = \mathbf{y} \Rightarrow \begin{bmatrix} 2 & 3 & 2 \\ 0 & 1 & -1 \\ 0 & 0 & -\frac{5}{2} \end{bmatrix} \begin{bmatrix} x_1 \\ x_2 \\ x_3 \end{bmatrix} = \begin{bmatrix} 1 \\ 1 \\ 5 \end{bmatrix} \Rightarrow \mathbf{x} = \begin{bmatrix} x_1 \\ x_2 \\ x_3 \end{bmatrix} = \begin{bmatrix} 4 \\ -1 \\ -2 \end{bmatrix} \Rightarrow$$

Check: $A\mathbf{x} = \begin{bmatrix} 0 & 1 & -1 \\ 2 & 3 & 2 \\ 1 & 1 & -1 \end{bmatrix} \begin{bmatrix} 4 \\ -1 \\ -2 \end{bmatrix} = \begin{bmatrix} 1 \\ 1 \\ 5 \end{bmatrix} = \mathbf{b}$

29. Let L and L' be unit lower diagonal matrices and $A = LL'$. Then:

$$a_{ij} = \sum_{k=1}^{n} l_{ik}l'_{ki} = \sum_{k=1}^{j-1} l_{ik}l'_{ki} + l_{ij}l'_{ji} + \sum_{k=j+1}^{n} l_{ik}l'_{ki} = \begin{cases} 0+1+0=1 & i=j \\ *+0+0=* & i>j \\ 0+0+0=0 & i<j \end{cases} \Rightarrow$$

A is unit lower diagonal.

31. We begin by factoring $U = DU_1$, where D is diagonal and U_1 is *unit* upper triangular. To factor $U = DU_1$, we adapt the *multiplier* method of Example 3.35.

In this case, however, our goal is to create 1s on the diagonals. So:

$$\begin{bmatrix} -2 & 1 \\ 0 & 6 \end{bmatrix} \stackrel{R'_1 = -\frac{1}{2}R_1}{\longrightarrow} \begin{bmatrix} 1 & -\frac{1}{2} \\ 0 & 6 \end{bmatrix} \stackrel{R'_2 = \frac{1}{6}R_2}{\longrightarrow} \begin{bmatrix} 1 & -\frac{1}{2} \\ 0 & 1 \end{bmatrix} = U_1 \text{ unit upper triangular} \Rightarrow$$

$\begin{matrix} l_{11} = -2 \\ l_{22} = 6 \end{matrix} \Rightarrow D = \begin{bmatrix} -2 & 0 \\ 0 & 6 \end{bmatrix}$

Check: $DU_1 = \begin{bmatrix} -2 & 0 \\ 0 & 6 \end{bmatrix} \begin{bmatrix} 1 & -\frac{1}{2} \\ 0 & 1 \end{bmatrix} = \begin{bmatrix} -2 & 1 \\ 0 & 6 \end{bmatrix} = U$

Since $A = LU = \begin{bmatrix} 1 & 0 \\ -1 & 1 \end{bmatrix} \begin{bmatrix} -2 & 1 \\ 0 & 6 \end{bmatrix}$, $A = LDU_1 = \begin{bmatrix} 1 & 0 \\ -1 & 1 \end{bmatrix} \begin{bmatrix} -2 & 0 \\ 0 & 6 \end{bmatrix} \begin{bmatrix} 1 & -\frac{1}{2} \\ 0 & 1 \end{bmatrix}$.

Q: Why is $l_{11} = -2$ in D? $l_{22} = 6$?

A: -2 is the multiplicative inverse of $-\frac{1}{2}$, so $(-2)(-\frac{1}{2}) = 1$ ...

3.4 The LU Factorization

33. We need to show if $A = LDU$ is symmetric and invertible, then $U = L^T$.

    The key to a successful proof is strong application of the given conditions. For example:

    Q: Since $A = LDU$ is symmetric, what do we know about A?
    A: We know $LDU = A = A^T = (LDU)^T = U^T D^T L^T = U^T DL^T$. Does that help?

    Q: Since $A = LDU$ is invertible, what do we know about A?
    A: Theorem 3.16 $\Rightarrow$ If $A = L_1 U_1$ and $A = L_2 U_2$, then $L_1 = L_2$ and $U_1 = U_2$.

    Q: How can we put these two facts together to create a proof?
    A: We use the fact that A is symmetric to create $A = L_1 U_1$ and $A = L_2 U_2$.
    Then we use the fact that A is invertible to equate factors and see how $U = L^T$ results.

    PROOF: $A = L(DU)$ and $A = U^T(DL^T)$ (because A is symmetric)
    $\Rightarrow DU = DL^T$ (because A is invertible so $U_1 = U_2$)
    $\Rightarrow U = L^T$ (because D is invertible so $D^{-1}DM = M$)

    There are several underlying facts in this proof which we examine in detail below:

    Q: If U is unit upper triangular, is U^T unit lower triangular?
    A: Yes, since in U if $i > j \Rightarrow a_{ij} = 0 = a_{ji}$ if $j < i$ in U^T.
    Also, it is obvious that $a_{ii} = 1$ in both U and U^T.

    Q: Where did we use this fact in our proof?
    A: When claiming that $A = U^T(DL^T)$ is an LU factorization of A.
    In order for that to be true, U^T must be unit lower triangular. Why?

    Q: If L is lower triangular, is L^T upper triangular?
    A: Yes, since in L if $i < j \Rightarrow a_{ij} = 0 = a_{ji}$ if $j > i$ in L^T.

    Q: If D is diagonal and U is upper triangular is DU upper triangular?
    A: Yes. We proved this for lower triangular matrices in Exercise 29.

    Q: Where did we use these two facts in our proof?
    A: When claiming that $A = U^T(DL^T)$ is an LU factorization of A.
    In order for that to be true, DL^T must be upper triangular. Why?

    Q: When $A = LDU$ is invertible, is D invertible?
    A: Yes. Since $A^{-1}A = (A^{-1}LD)U = I \Rightarrow U$ is invertible and
    $AA^{-1} = L(DUA^{-1}) = I \Rightarrow L$ is invertible. So we have:
    $D = L^{-1}AU^{-1} \Rightarrow D(UA^{-1}L) = (L^{-1}AU^{-1})(UA^{-1}L) = I \Rightarrow D$ is invertible.
    In fact, if $A_1 A_2 \cdots A_n$ is invertible, then each A_i is invertible.

    Q: Where did we use this fact in our proof?
    A: When claiming that $DU = DL^T \Rightarrow U = L^T$ since $U = D^{-1}DU = D^{-1}DL^T = L^T$.

3.5 Subspaces, Basis, Dimension, and Rank

Q: What are our main goals for Section 3.5?

A: To take our first look at subspaces, especially the big three related to matrices: the row space, the column space and null space. These ideas are profound. We will continue to work with them through the rest of the book. The concepts of basis, linear independence, and subspaces take time to truly understand.

Key Definitions and Concepts

$[\mathbf{v}]_B$	p.206	3.5	coordinate vector with respect to B
col(A)	p.193	3.5	column space of A spanned by the columns of A
dim(S)	p.201	3.5	the number of vectors in a basis called its *dimension*
null(A)	p.195	3.5	null space of A, $\mathbf{x}$ such that $A\mathbf{x} = \mathbf{0}$
nullity(A)	p.203	3.5	the dimension of null(A)
row(A)	p.193	3.5	row space of A spanned by the rows of A
basis	p.196	3.5	set of vectors that spans S and is linearly independent
column space	p.193	3.5	col(A) spanned by the columns of A
dimension	p.201	3.5	dim(S), the number of vectors in a basis
null space	p.195	3.5	null(A), $\mathbf{x}$ such that $A\mathbf{x} = \mathbf{0}$
nullity	p.203	3.5	the dimension of null(A)
rank	p.202	3.5	the dimension of row(A) or col(A) (they are the same)
row space	p.193	3.5	row(A) spanned by the rows of A
subspace	p.190	3.5	$\mathbf{0}$, $\mathbf{u} + \mathbf{v}$, and $c\mathbf{v}$ are in S

Theorems

Basis Thm	p.200	3.5	Any two bases have the same number of vectors (*dimension*)
Fund Thm	p.204	3.5	**Fundamental Theorem: Version 2**
Rank Thm	p.203	3.5	**Rank Theorem:** rank(A) + nullity(A) = n
Thm 3.19	p.190	3.5	span($\mathbf{v}_1, \mathbf{v}_2, \ldots, \mathbf{v}_n$) is a subspace
Thm 3.20	p.194	3.5	If $A \longrightarrow B$, then row(A) = row(B)
Thm 3.21	p.194	3.5	null(A), $\mathbf{x}$ such that $A\mathbf{x} = \mathbf{0}$, is a subspace
Thm 3.22	p.195	3.5	$A\mathbf{x} = \mathbf{b}$ has exactly none, one, or infinitely many solutions
Thm 3.23	p.200	3.5	Any two bases have the same number of vectors (*dimension*)
Thm 3.24	p.202	3.5	row(A) and col(A) have the same dimension
Thm 3.25	p.202	3.5	rank(A) = rank(A^T)
Thm 3.26	p.202	3.5	rank(A) + nullity(A) = n
Thm 3.27	p.204	3.5	Fundamental Theorem: Version 2
Thm 3.28	p.205	3.5	rank($A^T A$) = rank(A) and $A^T A$ invertible $\Leftrightarrow$ rank(A) = n
Thm 3.29	p.206	3.5	If $\{\mathbf{v}_1, \ldots, \mathbf{v}_k\}$ is a basis, then $\mathbf{v} = \sum c_i \mathbf{v}_i$ is unique

Discussion of Key Definitions, Concepts, and Theorems

The next several questions are based on Exercises 1 through 10.
There are additional questions and discussion included with the Exercises themselves.
Because of the richness of the section, there is much discussion included with the Exercises.

How to prove or disprove S is a subspace

Q: How do we prove that a given set S is a subspace?
A: We have two methods: 1) apply the definition or 2) show that $S = \text{span}(\mathbf{v}_i)$.
   Method 1) amounts to showing $\mathbf{0}$, $\mathbf{u} + \mathbf{v}$, and $c\mathbf{v}$ are in S.
   Method 2) reveals more about S, but finding $\{\mathbf{v}_i\}$ can be difficult sometimes.
   For examples, see Exercises 1 through 10.

Q: How do we prove that a given set S is *not* a subspace?
A: This is much simpler: all we have to show is S fails to have one of the required properties.
   Specifically, we have to show $\mathbf{0}$ or $\mathbf{u} + \mathbf{v}$ or $c\mathbf{v}$ are *not* in S. One failure is enough.
   For examples, see Exercises 1 through 10.

Lines and Planes as subspaces in $\mathbb{R}^n$

Q: Is any line through the origin a subspace in $\mathbb{R}^n$?
A: Yes, since any line through the origin is equal to the span of its direction vector.
   For a full discussion and example, see Exercise 9 and Table 1 in Section 1.3.

Q: Is any line *not* through the origin a subspace in $\mathbb{R}^n$?
A: No, because the zero vector is *not* on the line.

Q: Is any plane through the origin a subspace in $\mathbb{R}^n$?
A: Yes, since any plane through the origin is equal to the span of its two direction vectors.
   For a full discussion and example, see Exercise 5 and Table 2 in Section 1.3.

Q: Above we claim that any plane through the origin is a subspace because
   any plane through the origin is equal to the span of two vectors.
   If $ax + by + cz = 0$ and $a, b, c \neq 0$, find two vectors that span the subspace S it defines.

A: Since $x = -\frac{b}{a}y - \frac{c}{a}z$, $\begin{bmatrix} x \\ y \\ z \end{bmatrix} = y \begin{bmatrix} -b \\ a \\ 0 \end{bmatrix} + z \begin{bmatrix} -c \\ 0 \\ a \end{bmatrix} = y\mathbf{u} + z\mathbf{v}$, $S = \text{span}(\mathbf{u}, \mathbf{v})$.

Q: If $x + y + z = 0$, can we find two vectors that span the subspace S it defines?

A: From the previous question, we have: $S = \text{span}\left(\begin{bmatrix} -1 \\ 1 \\ 0 \end{bmatrix}, \begin{bmatrix} -1 \\ 0 \\ 1 \end{bmatrix} \right)$.

Q: Does this answer imply $\mathbf{v} = \begin{bmatrix} -1 \\ 2 \\ -1 \end{bmatrix}$ is a linear combination of $\left\{ \begin{bmatrix} -1 \\ 1 \\ 0 \end{bmatrix}, \begin{bmatrix} -1 \\ 0 \\ 1 \end{bmatrix} \right\}$?

A: Yes, since $(-1) + (2) + (-1) = 0$, $\mathbf{v}$ lies in the plane. Find the linear combination.
   For a full discussion and example, see Exercise 5 and Table 2 in Section 1.3.

3.5 Subspaces, Basis, Dimension, and Rank

Discussion of Key Definitions, Concepts, and Theorems, continued

How to prove or disprove b is in col(A)

Q: What is the key insight to answering the question: Is $\mathbf{b}$ in $\text{col}(A)$?
A: To say $\mathbf{b}$ is in $\text{col}(A)$ means $\mathbf{b}$ is a linear combination of the columns of A.
We did this in Example 2.18 of Section 2.3. Form the augmented matrix and row reduce. This is shown in detail in the solution to Exercise 11.

How to prove or disprove w is in row(A)

Q: What is the key insight to answering the question: Is $\mathbf{w}$ in $\text{row}(A)$?
A: To say $\mathbf{w}$ is in $\text{row}(A)$ means $\mathbf{w}$ is a linear combination of the rows of A.
This is shown in detail in the solution to Exercise 11 below.

11. As in Example 3.41, we will determine whether $\mathbf{b}$ is in $\text{col}(A)$ and $\mathbf{w}$ is in $\text{row}(A)$.

$\text{col}(A)$: To say $\mathbf{b}$ is in $\text{col}(A)$ means $\mathbf{b}$ is a linear combination of the columns of A. So:

$\mathbf{b} = \begin{bmatrix} 3 \\ 2 \end{bmatrix}$ is in the column space of $A = \begin{bmatrix} 1 & 0 & -1 \\ 1 & 1 & 1 \end{bmatrix}$ if the system $A\mathbf{x} = \mathbf{b}$ is consistent.

We row reduce the augmented matrix: $\begin{bmatrix} 1 & 0 & -1 & | & 3 \\ 1 & 1 & 1 & | & 2 \end{bmatrix} \longrightarrow \begin{bmatrix} 1 & 0 & -1 & | & 3 \\ 0 & 1 & 2 & | & -1 \end{bmatrix} \Rightarrow$

The system is consistent. So, $\mathbf{b} \in \text{col}(A)$. In particular, $\mathbf{b} = 3\mathbf{a}_1 - \mathbf{a}_2$.

$\text{row}(A)$: To say $\mathbf{w}$ is in $\text{row}(A)$ means $\mathbf{w}$ is a linear combination of the rows of A. So:

$\mathbf{w} = \begin{bmatrix} -1 & 1 & 1 \end{bmatrix} \in \text{row}(A)$ if $\begin{bmatrix} A \\ \hline \mathbf{w} \end{bmatrix} \longrightarrow \begin{bmatrix} A \\ \hline \mathbf{0} \end{bmatrix}$

by elementary row operations *excluding* row interchanges involving the *last* row.

So, we have $\begin{bmatrix} A \\ \hline \mathbf{w} \end{bmatrix} = \begin{bmatrix} 1 & 0 & -1 \\ 1 & 1 & 1 \\ -1 & 1 & 1 \end{bmatrix} \xrightarrow[R_3+R_1]{R_2-R_1} \begin{bmatrix} 1 & 0 & -1 \\ 0 & 1 & 2 \\ 0 & 1 & 0 \end{bmatrix} \xrightarrow{R_3-R_2} \begin{bmatrix} 1 & 0 & -1 \\ 0 & 1 & 2 \\ 0 & 0 & -2 \end{bmatrix} \Rightarrow$

We cannot make the last row all zeroes $\Rightarrow \mathbf{w} \notin \text{row}(A)$.

How to prove or disprove x is in null(A)

Q: What is the key insight to answering the question: Is $\mathbf{x}$ in $\text{null}(A)$?
A: If $\mathbf{x}$ is in $\text{null}(A)$ then $A\mathbf{x} = \mathbf{0}$, so all we have to do is multiply.
This is shown in detail in the solution to Exercise 15 below.

15. Since $A\mathbf{v} = \mathbf{0}$ implies $\mathbf{v}$ is in $\text{null}(A)$, we simply multiply $A\mathbf{v}$ to check:

$A\mathbf{v} = \begin{bmatrix} 1 & 0 & -1 \\ 1 & 1 & 1 \end{bmatrix} \begin{bmatrix} -1 \\ 3 \\ -1 \end{bmatrix} = \begin{bmatrix} 0 \\ 1 \end{bmatrix} \neq \mathbf{0} \Rightarrow \mathbf{v} \notin \text{null}(A)$.

Discussion of Key Definitions, Concepts, and Theorems, continued

How to find bases for $\text{col}(A)$, $\text{row}(A)$, or $\text{null}(A)$

Q: How do we find bases for $\text{col}(A)$, $\text{row}(A)$, or $\text{null}(A)$?
A: This is outlined beautifully in the Box on p.200 of Section 3.5.
For examples of the application of this process, see Exercises 17 and 19.

Discussion of $\text{col}(A)$ and how to find a basis from among its columns

Q: Let $\{\mathbf{a}_i\}$ be the columns of A. Does $\text{col}(A) = \text{span}(\mathbf{a}_i)$?
A: Yes. If $\mathbf{x}$ is in $\text{col}(A)$, then $\mathbf{x} = \sum c_i \mathbf{a}_i$.

Q: Can we always form a basis for $\text{col}(A)$ from a subset S of the columns of A?
A: Yes, but we must have $\text{span}(S) = \text{col}(A)$ and S must be linearly independent.

Q: What does that tell us about the number of vectors in a basis for $\text{col}(A)$?
A: Since $\{\mathbf{a}_i\}$ contains a basis for $\text{col}(A)$,
the number of vectors in a basis for $\text{col}(A) \leq$ the number of columns of A.

Q: Is that subset S unique?
A: No. To see that consider the following matrix A:

$$A = \begin{bmatrix} \mathbf{a}_1 & \mathbf{a}_2 & \mathbf{a}_3 \end{bmatrix} = \begin{bmatrix} 1 & 1 & 1 \\ 1 & 0 & 1 \\ 0 & 1 & 1 \end{bmatrix}$$

Then $S_{12} = \{\mathbf{a}_1, \mathbf{a}_2\}$, $S_{23} = \{\mathbf{a}_2, \mathbf{a}_3\}$, and $S_{23} = \{\mathbf{a}_1, \mathbf{a}_3\}$ are all bases for $\text{col}(A)$.

Q: Do we have to form a basis for $\text{col}(A)$ from the columns of A?
A: No. $T_{13} = \{\mathbf{a}_1, \mathbf{e}_3\}$, $T_{22} = \{\mathbf{a}_2, \mathbf{e}_2\}$, and $T_{32} = \{\mathbf{a}_3, \mathbf{e}_2\}$ are all bases for $\text{col}(A)$.

Q: Summarizing the above, what is one way to find a basis for $\text{col}(A)$?
A: By finding a linearly independent subset of the columns of A that spans the columns.

Q: Is the number of vectors in a linearly independent subset of the columns of $A \leq \text{rank}(A)$?
A: Yes. In fact, if that subset spans the columns, then it is a basis for $\text{col}(A)$.
If it does not span, then it must contain fewer vectors than a basis for A.
So, number of vectors in that subset $\leq$ number of vectors in a basis for $\text{col}(A) = \text{rank}(A)$.

Q: How many bases for $\text{col}(A)$ are there?
A: Infinitely many. Why? If $\{\mathbf{v}_i\}$ is a basis so is $\{c_i \mathbf{v}_i\}$ provided $c_i \neq 0$. Prove this.

Q: In the light of this discussion, why is it obvious that $\text{rank}(A) \leq$ the number of columns?
A: Recall, that the number of vectors in a basis for $\text{col}(A) = \dim(\text{col}(A))$.
Now note the columns contain a basis for $\text{col}(A)$, so $\dim(\text{col}(A)) \leq$ number of columns.
So, $\text{rank}(A) = \dim(\text{col}(A)) \leq$ the number of columns.

3.5 Subspaces, Basis, Dimension, and Rank

Discussion of Key Definitions, Concepts, and Theorems, continued

Discussion of row(A) *and how to find a basis from among its rows*

Q: Let $\{\mathbf{A}_i\}$ be the rows of A. Does row(A) = span($\mathbf{A}_i$)?
A: Yes. If $\mathbf{x}$ is in row(A), then $\mathbf{x} = \sum c_i \mathbf{A}_i$.

Q: Can we always form a basis for row(A) from a subset S of the rows of A?
A: Yes, but we must have span(S) = row(A) and S must be linearly independent.

Q: What does that tell us about the number of vectors in a basis for row(A)?
A: Since $\{\mathbf{A}_i\}$ contains a basis for row(A),
the number of vectors in a basis for row(A) $\leq$ the number of rows of A.

Q: Is that subset S unique?
A: No. To see that consider the following matrix A:

$$A = \begin{bmatrix} \mathbf{A}_1 \\ \mathbf{A}_2 \\ \mathbf{A}_3 \end{bmatrix} = \begin{bmatrix} 1 & 1 & 1 \\ 1 & 0 & 1 \\ 0 & 1 & 0 \end{bmatrix}$$

Then $S_{12} = \{\mathbf{A}_1, \mathbf{A}_2\}$, $S_{23} = \{\mathbf{A}_2, \mathbf{A}_3\}$, and $S_{23} = \{\mathbf{A}_1, \mathbf{A}_3\}$ are all bases for row(A).

Q: Do we have to form a basis for row(A) from the rows of A?
A: No. Consider the discussion below $A \longrightarrow R$.

Q: How many bases for row(A) are there?
A: Infinitely many. Why? If $\{\mathbf{V}_i\}$ is a basis so is $\{c_i \mathbf{V}_i\}$ provided $c_i \neq 0$. Prove this.

Q: Summarizing the above, what is one way to find a basis for row(A)?
A: By finding a linearly independent subset of the rows of A that spans the rows.

Q: When $A \longrightarrow R$, it is obvious that the nonzero rows of R are linearly independent.
But, do the nonzero rows of R span the rows of A?
A: Yes, since the only rows that become zero are linear combinations of the other rows.

Q: Provided $A \longrightarrow R$ uses no row interchanges, then the rows of A corresponding to the nonzero rows of R are also linearly independent. Why?
A: Because the rows of R are linear combinations of the rows of A. See Exercise 25.

Q: Why is it sufficient to reduce A only to row echelon form U?
A: As the remark following Example 3.46 explains and then demonstrates,
the nonzero rows of U are linearly independent. That is all that is required. Why?
Again, provided no row exchanges the corresponding rows in A are linearly independent.

Q: If $A \longrightarrow R$ uses row interchanges, can we identify the linearly independent rows of A?
A: Yes, simply by tracking those row interchanges. Is this also true when $A \longrightarrow U$? Yes.

Q: Is the number of vectors in a linearly independent subset of the rows of $A \leq$ rank(A)?
A: Yes. In fact, if that subset spans the rows, then it is a basis for row(A).
If it does not span, then it must contain fewer vectors than a basis for A.
So, number of vectors in that subset $\leq$ number of vectors in a basis for row(A) = rank(A).

Q: In the light of the above discussion, why is it obvious that rank(A) $\leq$ the number of rows?
A: Recall, that the number of vectors in a basis for row(A) = dim(row(A)).
Now note the rows contain a basis for row(A), so dim(row(A)) $\leq$ number of rows.
So, rank(A) = dim(row(A)) $\leq$ the number of rows.

Discussion of Key Definitions, Concepts, and Theorems, continued

Discussion of rank(A)

Q: In Exercises 17 through 24, we note the following pattern:
The number of vectors in the basis for row(A) is always
equal to the number of vectors in the basis for col(A).
Is this a coincidence?

A: No, those are examples of the following extremely important principle:

Rank must satisfy the following two conditions simultaneously:
  1) Rank = the number of vectors in a basis for row(A)
  2) Rank = the number of vectors in a basis for col(A)
Therefore, rank must be less than or equal to the smaller of these two numbers.

Q: Can rank ever be greater than the number of rows?
A: No. Since the rows of A, $\{\mathbf{A}_i\}$, contain a basis for row(A),
the number of vectors in a basis for row(A) $\leq$ the number of rows of A.

Q: Can rank ever be greater than the number of columns?
A: No. Since the columns of A, $\{\mathbf{a}_i\}$, contain a basis for col(A),
the number of vectors in a basis for col(A) $\leq$ the number of columns of A.

Putting these together we have the following extremely important principle:

Rank must satisfy the following two conditions simultaneously:
  1) Rank $\leq$ the number of rows
  2) Rank $\leq$ the number of columns
Therefore, rank must be less than or equal to the smaller of these two numbers.

Q: In Exercise 17, A is 2×3 and $\dim(\text{row}(A)) + \dim(\text{null}(A)) = 2 + 1 = 3$.
Is this a coincidence?
A: No. This an example of the Rank Theorem: if A is $m \times n$, then $\text{rank}(A) + \text{nullity}(A) = n$.
How? Because $\dim(\text{row}(A)) = \text{rank}(A)$ and $\text{nullity}(A) = \dim(\text{null}(A))$, both by definition.
Identify this pattern in Exercise 19.

Contrapositive Implications of the Fundamental Theorem

The Fundamental Theorem is rich with implications. Examples:

Q: Why does Theorem 3.27 imply if nullity(A) > 0, the columns of A are linearly dependent?
A: h. implies i. means: If the columns of A are linearly independent, then nullity(A) = 0.
The contrapositive is: If nullity(A) > 0, then the columns of A are linearly dependent.

Q: Why does Theorem 3.27 imply if rank(A) < n, the columns of A are linearly dependent?
A: h. implies f. means: If the columns of A are linearly independent, then rank(A) = n.
The contrapositive is: If rank(A) < n, then the columns of A are linearly dependent.

Q: Why does Theorem 3.27 imply if nullity(A) > 0, A is *not* invertible?
A: a. implies g. means: If A is invertible, then nullity(A) = 0.
The contrapositive is: If nullity(A) > 0, then A is *not* invertible.

Q: Why does Theorem 3.27 imply if rank(A) < n, A is *not* invertible?
A: a. implies f. means: If A is invertible, then rank(A) = n.
The contrapositive is: If rank(A) < n, then A is *not* invertible.

3.5 Subspaces, Basis, Dimension, and Rank

Solutions to odd-numbered exercises from Section 3.5

1. As in Example 3.38, substituting the condition $x = 0$ into $\begin{bmatrix} x \\ y \end{bmatrix}$ yields $\begin{bmatrix} 0 \\ y \end{bmatrix} = y \begin{bmatrix} 0 \\ 1 \end{bmatrix}$.

   Since y is arbitrary, $S = \text{span}\left(\begin{bmatrix} 0 \\ 1 \end{bmatrix}\right)$. So, S is a subspace of $\mathbb{R}^2$ by Theorem 3.19.

   Q: Geometrically speaking, what is $x = 0$?
   A: The equation $x = 0$ is a line through the origin. This should come as no surprise. Why? Because a line through the origin is a subspace in $\mathbb{R}^2, \mathbb{R}^3, \ldots, \mathbb{R}^n$.

3. As in Example 3.38, substituting the condition $y = 2x$ into $\begin{bmatrix} x \\ y \end{bmatrix}$ yields $\begin{bmatrix} x \\ 2x \end{bmatrix} = x \begin{bmatrix} 1 \\ 2 \end{bmatrix}$.

   Since x is arbitrary, $S = \text{span}\left(\begin{bmatrix} 1 \\ 2 \end{bmatrix}\right)$. So, S is a subspace of $\mathbb{R}^2$ by Theorem 3.19.

   Q: Geometrically speaking, what is $y = 2x$?
   A: Since $y = 2x \Rightarrow 2x - y = 0$, this is obviously a line through the origin.

5. As in Example 3.38, substituting the condition $x = y = z$ into $\begin{bmatrix} x \\ y \\ z \end{bmatrix}$ yields $\begin{bmatrix} x \\ x \\ x \end{bmatrix} = x \begin{bmatrix} 1 \\ 1 \\ 1 \end{bmatrix}$.

   Since x is arbitrary, $S = \text{span}\left(\begin{bmatrix} 1 \\ 1 \\ 1 \end{bmatrix}\right)$. So, S is a subspace of $\mathbb{R}^3$ by Theorem 3.19.

   Q: Geometrically speaking, what is $x = y = z$?
   A: Since $x = y = z \Rightarrow x - y = 0$ and $x - z = 0$, this is obviously a line through the origin. Why? Because $x - y = 0$ and $x - z = 0$ are planes through the origin that intersect in that line.
   Note: a plane through the origin is a subspace in $\mathbb{R}^3, \mathbb{R}^4, \cdots, \mathbb{R}^n$.

7. We should recognize right away that set S defined by $x - y + z = 1$ is *not* a subspace. Why? Because $x - y + z = 1$ is a plane that does *not* pass through the origin. So we conclude:

   Property (1) (**0** is in S) fails so S is not a subspace of $\mathbb{R}^3$.

   Q: How can we tell that $x - y + z = 1$? is a plane that does *not* pass through the origin?
   A: The equation of a plane through the origin must be of the form $ax + by + cz = 0$. Why? Because the zero vector must satisfy the equation. So, we must have: $a0 + b0 + c0 = 0$.

9. Every line ℓ through the origin in $\mathbb{R}^3$ is a subspace of $\mathbb{R}^3$ because if ℓ has equation $\mathbf{x} = \mathbf{0} + t\mathbf{d}$ then $\ell = \text{span}(\mathbf{d})$.
Therefore, once again, Theorem 3.19 implies that ℓ is a subspace of $\mathbb{R}^3$.

We might also have stated our proof this way:

ℓ is a line $\Rightarrow$	ℓ has equation $\mathbf{x} = \mathbf{p} + t\mathbf{d}$
ℓ passes through the origin $\Rightarrow$	$\mathbf{x} = \mathbf{0} + t\mathbf{d} = t\mathbf{d}$
The definition of span $\Rightarrow$	$\ell = \text{span}(\mathbf{d})$
So, Theorem 3.19 $\Rightarrow$	ℓ is a subspace of $\mathbb{R}^3$

Q: Did we use the fact that ℓ passes through the origin?
A: Yes, in Step 2. Otherwise $\mathbf{0}$ will not be in ℓ as required.

Q: Did we use the fact that ℓ is a line in $\mathbb{R}^3$?
A: No. The parametric equation is the same in $\mathbb{R}^n$, so our conclusion is true in $\mathbb{R}^n$.

11. As in Example 3.41, we will determine whether $\mathbf{b}$ is in $\text{col}(A)$ and $\mathbf{w}$ is in $\text{row}(A)$.

$\text{col}(A)$: To say $\mathbf{b}$ is in $\text{col}(A)$ means $\mathbf{b}$ is a linear combination of the columns of A. So:

$\mathbf{b} = \begin{bmatrix} 3 \\ 2 \end{bmatrix}$ is in the column space of $A = \begin{bmatrix} 1 & 0 & -1 \\ 1 & 1 & 1 \end{bmatrix}$ if the system $A\mathbf{x} = \mathbf{b}$ is consistent.

We row reduce the augmented matrix: $\begin{bmatrix} 1 & 0 & -1 & | & 3 \\ 1 & 1 & 1 & | & 2 \end{bmatrix} \longrightarrow \begin{bmatrix} 1 & 0 & -1 & | & 3 \\ 0 & 1 & 2 & | & -1 \end{bmatrix} \Rightarrow$

The system is consistent. So, $\mathbf{b} \in \text{col}(A)$. In particular, $\mathbf{b} = 3\mathbf{a}_1 - \mathbf{a}_2$.

$\text{row}(A)$: To say $\mathbf{w}$ is in $\text{row}(A)$ means $\mathbf{w}$ is a linear combination of the rows of A. So:

$\mathbf{w} = \begin{bmatrix} -1 & 1 & 1 \end{bmatrix} \in \text{row}(A)$ if $\begin{bmatrix} A \\ \mathbf{w} \end{bmatrix} \longrightarrow \begin{bmatrix} A \\ \mathbf{0} \end{bmatrix}$

by elementary row operations *excluding* row interchanges involving the *last* row.

So, we have $\begin{bmatrix} A \\ \mathbf{w} \end{bmatrix} = \begin{bmatrix} 1 & 0 & -1 \\ 1 & 1 & 1 \\ -1 & 1 & 1 \end{bmatrix} \xrightarrow[R_3+R_1]{R_2-R_1} \begin{bmatrix} 1 & 0 & -1 \\ 0 & 1 & 2 \\ 0 & 1 & 0 \end{bmatrix} \xrightarrow{R_3-R_2} \begin{bmatrix} 1 & 0 & -1 \\ 0 & 1 & 2 \\ 0 & 0 & -2 \end{bmatrix} \Rightarrow$

We cannot make the last row all zeroes $\Rightarrow \mathbf{w} \notin \text{row}(A)$.

13. As in the remarks following Example 3.41, we determine whether $\mathbf{w}$ is in $\text{row}(A)$ using $\text{col}(A^T)$.

To say $\mathbf{w}$ is in $\text{row}(A)$ means $\mathbf{w}^T$ is a linear combination of the *columns* of A^T. So:

We row reduce to check: $\begin{bmatrix} A^T & | & \mathbf{w}^T \end{bmatrix} = \begin{bmatrix} 1 & 1 & | & -1 \\ 0 & 1 & | & 1 \\ -1 & 1 & | & 1 \end{bmatrix} \longrightarrow \begin{bmatrix} 1 & 0 & | & 0 \\ 0 & 1 & | & 0 \\ 0 & 0 & | & 1 \end{bmatrix} \Rightarrow$

Since the last row is $\begin{bmatrix} 0 & 0 & | & 1 \end{bmatrix}$, this system is inconsistent. So, $\mathbf{w} \notin \text{row}(A)$.

15. Since $A\mathbf{v} = \mathbf{0}$ implies $\mathbf{v}$ is in $\text{null}(A)$, we simply multiply $A\mathbf{v}$ to check:

$A\mathbf{v} = \begin{bmatrix} 1 & 0 & -1 \\ 1 & 1 & 1 \end{bmatrix} \begin{bmatrix} -1 \\ 3 \\ -1 \end{bmatrix} = \begin{bmatrix} 0 \\ 1 \end{bmatrix} \neq \mathbf{0} \Rightarrow \mathbf{v} \notin \text{null}(A).$

3.5 Subspaces, Basis, Dimension, and Rank

17. We find bases for row(A), col(A), and null(A) as in Examples 3.45, 3.47, and 3.48 respectively.

row(A): A basis for row(A) must span the rows of A and be linearly independent.
Given $A \longrightarrow R$, Theorem 3.20 asserts that the rows of R span the rows of A. Why? Because the rows of A are linear combinations of the rows of R (and vice-versa). Finally, we simply observe that the nonzero rows of R are linearly independent.

Since $A = \begin{bmatrix} 1 & 0 & -1 \\ 1 & 1 & 1 \end{bmatrix} \longrightarrow \begin{bmatrix} 1 & 0 & -1 \\ 0 & 1 & 2 \end{bmatrix} = R,$

we conclude that $\{\begin{bmatrix} 1 & 0 & -1 \end{bmatrix}, \begin{bmatrix} 0 & 1 & 2 \end{bmatrix}\}$ is a basis for row(A).
We should also note that provided $A \longrightarrow R$ uses no row interchanges, the corresponding rows in A are also linearly independent. Whence, it is obvious that those rows form a basis for row(A).

col(A): A basis for col(A) must span the columns of A and be linearly independent.
When $A \longrightarrow R$, the columns with leading 1s in R are linearly independent. As shown in Example 3.47, the corresponding columns in A are also linearly independent. Whence, it is obvious that those columns form a basis for col(A).

Since $A = \begin{bmatrix} 1 & 0 & -1 \\ 1 & 1 & 1 \end{bmatrix} \longrightarrow \begin{bmatrix} 1 & 0 & -1 \\ 0 & 1 & 2 \end{bmatrix} = R,$

we conclude that $\left\{\begin{bmatrix} 1 \\ 1 \end{bmatrix}, \begin{bmatrix} 0 \\ 1 \end{bmatrix}\right\}$ is a basis for col(A).

null(A): Since $A\mathbf{v} = \mathbf{0}$ implies $\mathbf{v}$ is in null(A), we solve $[\,A\,|\,\mathbf{0}\,] \longrightarrow [\,R\,|\,\mathbf{0}\,]$ to find the conditions:

$[\,R\,|\,\mathbf{0}\,] = \begin{bmatrix} 1 & 0 & -1 & | & 0 \\ 0 & 1 & 2 & | & 0 \end{bmatrix} \Rightarrow \begin{array}{l} x_1 - x_3 = 0 \\ x_2 + 2x_3 = 0 \\ x_3 \text{ free} \end{array} \Rightarrow \begin{array}{l} x_1 = 1s \\ x_2 = -2s \\ x_3 = 1s \end{array}$

Since t is arbitrary, null(A) = span $\left(\begin{bmatrix} 1 \\ -2 \\ 1 \end{bmatrix}\right)$. So, $\left\{\begin{bmatrix} 1 \\ -2 \\ 1 \end{bmatrix}\right\}$ is a basis for null(A).

19. We find bases for row(A), col(A), and null(A) as in Examples 3.45, 3.47, and 3.48 respectively.

row(A): A basis for row(A) must span the rows of A and be linearly independent. Given $A \longrightarrow R$, Theorem 3.20 asserts that the rows of R span the rows of A. Why? Because the rows of A are linear combinations of the rows of R (and vice-versa). Finally, we simply observe that the nonzero rows of R are linearly independent.

$$\text{Since } A = \begin{bmatrix} 1 & 1 & 0 & 1 \\ 0 & 1 & -1 & 1 \\ 0 & 1 & -1 & -1 \end{bmatrix} \longrightarrow \begin{bmatrix} 1 & 0 & 1 & 0 \\ 0 & 1 & -1 & 0 \\ 0 & 0 & 0 & 1 \end{bmatrix} = R,$$

we conclude that $\{[1\ 0\ 1\ 0], [0\ 1\ -1\ 0], [0\ 0\ 0\ 1]\}$ is a basis for row(A).

We should also note that provided $A \longrightarrow R$ uses no row interchanges, the corresponding rows in A are also linearly independent. Whence, it is obvious that those rows form a basis for row(A).

col(A): A basis for col(A) must span the columns of A and be linearly independent. When $A \longrightarrow R$, the columns with leading 1s in R are linearly independent. As shown in Example 3.47, the corresponding columns in A are also linearly independent. Whence, it is obvious that those columns form a basis for col(A).

$$\text{Since } A = \begin{bmatrix} 1 & 1 & 0 & 1 \\ 0 & 1 & -1 & 1 \\ 0 & 1 & -1 & -1 \end{bmatrix} \longrightarrow \begin{bmatrix} 1 & 0 & 1 & 0 \\ 0 & 1 & -1 & 0 \\ 0 & 0 & 0 & 1 \end{bmatrix} = R,$$

we conclude that $\left\{ \begin{bmatrix} 1 \\ 0 \\ 0 \end{bmatrix}, \begin{bmatrix} 1 \\ 1 \\ 1 \end{bmatrix}, \begin{bmatrix} 1 \\ 1 \\ -1 \end{bmatrix} \right\}$ is a basis for col(A).

null(A): Since $A\mathbf{v} = \mathbf{0}$ implies $\mathbf{v}$ is in null(A), we solve $[\,A\,|\,\mathbf{0}\,] \longrightarrow [\,R\,|\,\mathbf{0}\,]$ to find the conditions.

$$[\,R\,|\,\mathbf{0}\,] = \begin{bmatrix} 1 & 0 & 1 & 0 & | & 0 \\ 0 & 1 & -1 & 0 & | & 0 \\ 0 & 0 & 0 & 1 & | & 0 \end{bmatrix} \Rightarrow \begin{matrix} x_1 + x_3 = 0 \\ x_2 - x_3 = 0 \\ x_3 \text{ free} \\ x_4 = 0 \end{matrix} \Rightarrow \begin{matrix} x_1 = -s \\ x_2 = s \\ x_3 = s \\ x_4 = 0 \end{matrix}$$

Since s is arbitrary, null(A) = span $\left(\begin{bmatrix} -1 \\ 1 \\ 1 \\ 0 \end{bmatrix} \right)$. So, $\left\{ \begin{bmatrix} -1 \\ 1 \\ 1 \\ 0 \end{bmatrix} \right\}$ is a basis for null(A).

3.5 Subspaces, Basis, Dimension, and Rank

21. We find bases for row(A) and col(A) following Examples 3.45 and 3.47 respectively.

row(A): A basis for col(A) must span the columns of A and be linearly independent.
Clearly, the linearly independent *columns* of A^T do just that.
When $A^T \longrightarrow R$, the columns with leading 1s in R are linearly independent.
As in Example 3.47, the corresponding columns in A^T are also linearly independent.
Whence, it is obvious that the *transposes* of those columns form a basis for row(A).

$$\text{Since } A^T = \begin{bmatrix} 1 & 1 \\ 0 & 1 \\ -1 & 1 \end{bmatrix} \longrightarrow \begin{bmatrix} 1 & 0 \\ 0 & 1 \\ 0 & 0 \end{bmatrix} = R$$

we conclude that $\{ [\,1\ 0\ -1\,], [\,1\ 1\ 1\,] \}$ is a basis for row(A).

col(A): A basis for col(A) must span the columns of A and be linearly independent.
When $A^T \longrightarrow R$, the linearly independent *rows* (the nonzero rows) of R do just that.
Whence, it is obvious that the *transposes* of those rows form a basis for col(A).

$$\text{Since } A^T = \begin{bmatrix} 1 & 1 \\ 0 & 1 \\ -1 & 1 \end{bmatrix} \longrightarrow \begin{bmatrix} 1 & 0 \\ 0 & 1 \\ 0 & 0 \end{bmatrix} = R$$

we conclude that $\left\{ \begin{bmatrix} 1 \\ 0 \end{bmatrix}, \begin{bmatrix} 0 \\ 1 \end{bmatrix} \right\}$ is a basis for col(A).

We should also note that provided $A^T \longrightarrow R$ uses no row interchanges,
the corresponding rows in A^T are also linearly independent.
Whence, it is obvious that the *transposes* of those rows form a basis for col(A).

23. We find bases for row(A) and col(A) following Examples 3.45 and 3.47 respectively.

row(A): A basis for col(A) must span the columns of A and be linearly independent.
Clearly, the linearly independent *columns* of A^T do just that.
When $A^T \longrightarrow R$, the columns with leading 1s in R are linearly independent.
As in Example 3.47, the corresponding columns in A^T are also linearly independent.
Whence, it is obvious that the *transposes* of those columns form a basis for row(A).

Since $A^T = \begin{bmatrix} 1 & 0 & 0 \\ 1 & 1 & 1 \\ 0 & -1 & -1 \\ 1 & 1 & -1 \end{bmatrix} \longrightarrow \begin{bmatrix} 1 & 0 & 0 \\ 0 & 1 & 0 \\ 0 & 0 & 1 \\ 0 & 0 & 0 \end{bmatrix} = R,$

we conclude that $\{[\,1\ 1\ 0\ 1\,], [\,0\ 1\ -1\ 1\,], [\,0\ 1\ -1\ -1\,]\}$ is a basis for row(A).

col(A): A basis for col(A) must span the columns of A and be linearly independent.
When $A^T \longrightarrow R$, the linearly independent *rows* (the nonzero rows) of R do just that.
Whence, it is obvious that the *transposes* of those rows form a basis for col(A).

Since $A^T = \begin{bmatrix} 1 & 0 & 0 \\ 1 & 1 & 1 \\ 0 & -1 & -1 \\ 1 & 1 & -1 \end{bmatrix} \longrightarrow \begin{bmatrix} 1 & 0 & 0 \\ 0 & 1 & 0 \\ 0 & 0 & 1 \\ 0 & 0 & 0 \end{bmatrix} = R,$

we conclude that $\left\{ \begin{bmatrix} 1 \\ 0 \\ 0 \end{bmatrix}, \begin{bmatrix} 0 \\ 1 \\ 0 \end{bmatrix}, \begin{bmatrix} 0 \\ 0 \\ 1 \end{bmatrix} \right\}$ is a basis for col(A).

We should also note that provided $A^T \longrightarrow R$ uses no row interchanges, the corresponding rows in A^T are also linearly independent.
Whence, it is obvious that the *transposes* of those rows form a basis for col(A).

3.5 Subspaces, Basis, Dimension, and Rank

25. Our answers to Exercises 17 and 21 appear different because we used different methods. Let's compare the methods and answers for row(A) and col(A) from each of the exercises.

row(A): In Exercise 17, we found the basis for row(A) as follows:
Given $A \longrightarrow R$, the linearly independent (nonzero) rows of R span the rows of A. Thus:

Since $A = \begin{bmatrix} 1 & 0 & -1 \\ 1 & 1 & 1 \end{bmatrix} \longrightarrow \begin{bmatrix} 1 & 0 & -1 \\ 0 & 1 & 2 \end{bmatrix} = R$,

we conclude that $\{\begin{bmatrix} 1 & 0 & -1 \end{bmatrix}, \begin{bmatrix} 0 & 1 & 2 \end{bmatrix}\}$ is a basis for row(A).

In Exercise 21, on the other hand, we found the basis for row(A) as follows:
When $A^T \longrightarrow R$, the *transposes* of the *columns* in A^T corresponding to the columns with leading 1s in R form a basis for row(A). Thus:

Since $A^T = \begin{bmatrix} 1 & 1 \\ 0 & 1 \\ -1 & 1 \end{bmatrix} \longrightarrow \begin{bmatrix} 1 & 0 \\ 0 & 1 \\ 0 & 0 \end{bmatrix} = R$

we conclude that $\{\begin{bmatrix} 1 & 0 & -1 \end{bmatrix}, \begin{bmatrix} 1 & 1 & 1 \end{bmatrix}\}$ is a basis for row(A).

So, first we used rows of R as our basis, then we used rows of A.
Do the rows of A corresponding to the nonzero rows of R form a basis?
Yes. Since $A \longrightarrow R$ uses no row interchanges, those rows are linearly independent.
We prove this explicitly by showing the spans of these two sets are equal.
By Exercise 21 of Section 2.3, we need only observe:

$$-\begin{bmatrix} 1 & 0 & -1 \end{bmatrix} + \begin{bmatrix} 1 & 1 & 1 \end{bmatrix} = \begin{bmatrix} 0 & 1 & 2 \end{bmatrix}.$$

Why is this enough? The basis vectors in each set are linear combinations of each other.

col(A): In Exercise 17, we found the basis for col(A) as follows:
When $A \longrightarrow R$, the columns in A corresponding to the columns with leading 1s in R form a basis for col(A). Thus:

Since $A = \begin{bmatrix} 1 & 0 & -1 \\ 1 & 1 & 1 \end{bmatrix} \longrightarrow \begin{bmatrix} 1 & 0 & -1 \\ 0 & 1 & 2 \end{bmatrix} = R$,

we conclude that $\left\{ \begin{bmatrix} 1 \\ 1 \end{bmatrix}, \begin{bmatrix} 0 \\ 1 \end{bmatrix} \right\}$ is a basis for col(A).

In Exercise 21, on the other hand, we found the basis for col(A) as follows:
Given $A^T \longrightarrow R$, the *transposes* of the linearly independent (nonzero) *rows* of R span the columns of A. Thus:

Since $A^T = \begin{bmatrix} 1 & 1 \\ 0 & 1 \\ -1 & 1 \end{bmatrix} \longrightarrow \begin{bmatrix} 1 & 0 \\ 0 & 1 \\ 0 & 0 \end{bmatrix} = R$

we conclude that $\left\{ \begin{bmatrix} 1 \\ 0 \end{bmatrix}, \begin{bmatrix} 0 \\ 1 \end{bmatrix} \right\}$ is a basis for col(A).

So, first we used columns of A as our basis, then we used *transposes* of the rows of R. Notice, however, those rows correspond to the columns of A found in Exercise 17.

For example, $\begin{bmatrix} 1 & 0 \end{bmatrix}$ of R corresponds to $\begin{bmatrix} 1 \\ 1 \end{bmatrix}$ of A.

Explicitly, it is obvious that the span of both these sets is $\mathbb{R}^2$. Why?
Since both sets contain two vectors and the dimension of $\mathbb{R}^2$ is obviously 2.

27. As in Example 3.46, given $S = \{\mathbf{u}, \mathbf{v}, \mathbf{w}\}$ we form matrix B and row reduce:

$$\text{Since } B = \begin{bmatrix} \mathbf{u}^T \\ \mathbf{v}^T \\ \mathbf{w}^T \end{bmatrix} = \begin{bmatrix} 1 & -1 & 0 \\ -1 & 0 & 1 \\ 0 & 1 & -1 \end{bmatrix} \xrightarrow{R_3 + R_1 + R_2} \begin{bmatrix} 1 & -1 & 0 \\ -1 & 0 & 1 \\ 0 & 0 & 0 \end{bmatrix},$$

we conclude that $\{\mathbf{u}, \mathbf{v}\} = \left\{ \begin{bmatrix} 1 \\ -1 \\ 0 \end{bmatrix}, \begin{bmatrix} -1 \\ 0 \\ 1 \end{bmatrix} \right\}$ is a basis for span(S).

Note: We rearrange and transpose the vectors of S to simplify the row reduction.
As noted in the remark following Example 3.46, we need only reduce to row echelon form. Then, we find the basis by identifying linearly independent vectors in the original set.

Q: Does $\{\mathbf{u}, \mathbf{w}\}$ also form a basis for S? What about $\{\mathbf{v}, \mathbf{w}\}$?
A: Yes, since no 2 vectors are multiples of each other. Why is that enough?

29. As in Example 3.46, given $S = \{\mathbf{u}, \mathbf{v}, \mathbf{w}\}$ we form matrix B and row reduce:

$$\text{Since } B = \begin{bmatrix} \mathbf{v} \\ \mathbf{u} \\ \mathbf{w} \end{bmatrix} = \begin{bmatrix} 1 & -1 & 0 \\ 2 & -3 & 1 \\ 4 & -4 & 1 \end{bmatrix} \xrightarrow[R_3 - 4R_1]{R_2 - 2R_1} \begin{bmatrix} 1 & -1 & 0 \\ 0 & -1 & 1 \\ 0 & 0 & 1 \end{bmatrix} = U,$$

we conclude that $\left\{ \begin{bmatrix} 1 \\ -1 \\ 0 \end{bmatrix}, \begin{bmatrix} 0 \\ -1 \\ 1 \end{bmatrix}, \begin{bmatrix} 0 \\ 0 \\ 1 \end{bmatrix} \right\}$ is a basis for span(S).

Furthermore, since we have 3 linearly independent vectors in $\mathbb{R}^3$,

we can also conclude that $\left\{ \begin{bmatrix} 1 \\ 0 \\ 0 \end{bmatrix}, \begin{bmatrix} 0 \\ 1 \\ 0 \end{bmatrix}, \begin{bmatrix} 0 \\ 0 \\ 1 \end{bmatrix} \right\}$ is a basis for span(S).

Q: How could we have reached this conclusion using row reduction?
A: By continuing to reduce to B to row reduced echelon form. So, $B \longrightarrow I$.

31. As in Example 3.46, given $S = \{\mathbf{u}, \mathbf{v}, \mathbf{w}\}$ we form matrix B and row reduce:

$$\text{Since } B = \begin{bmatrix} \mathbf{v} \\ \mathbf{u} \\ \mathbf{w} \end{bmatrix} = \begin{bmatrix} 1 & -1 & 0 \\ 2 & -3 & 1 \\ 4 & -4 & 1 \end{bmatrix} \xrightarrow[R_3 - 4R_1]{R_2 - 2R_1} \begin{bmatrix} 1 & -1 & 0 \\ 0 & -1 & 1 \\ 0 & 0 & 1 \end{bmatrix} = U,$$

we conclude that $\{\mathbf{u}, \mathbf{v}, \mathbf{w}\} = \left\{ \begin{bmatrix} 2 \\ -3 \\ 1 \end{bmatrix}, \begin{bmatrix} 1 \\ -1 \\ 0 \end{bmatrix}, \begin{bmatrix} 4 \\ -4 \\ 1 \end{bmatrix} \right\}$ is a basis for span(S).

Q: Why can we conclude $\{\mathbf{u}, \mathbf{v}, \mathbf{w}\}$ is a basis for S after only one step?
A: Because the rows of B corresponding to the linearly independent (nonzero) rows of U are linearly independent. This is only true because we performed no row interchanges.

33. Let R be a matrix in echelon form. Then row(R) = span(the rows of R) by definition. Nonzero rows of R are linearly independent (first entries are in different columns) $\Rightarrow$ Nonzero rows of R form a basis for row(R), by definition.

3.5 Subspaces, Basis, Dimension, and Rank

35. We use our work from Exercise 17 to determine rank(A) and nullity(A) below.

rank(A) = number of nonzero rows in R = number of vectors in a basis for row(A) or col(A)

Since $A = \begin{bmatrix} 1 & 0 & -1 \\ 1 & 1 & 1 \end{bmatrix} \longrightarrow \begin{bmatrix} 1 & 0 & -1 \\ 0 & 1 & 2 \end{bmatrix} = R$,

and $\{[\,1\ 0\ -1\,], [\,0\ 1\ 2\,]\}$ is a basis for row(A), we have rank(A) = 2.

nullity(A) = n − rank(A) = number of nonzero vectors in a basis for null(A)

From Exercise 17, $\left\{ \begin{bmatrix} 1 \\ -2 \\ 1 \end{bmatrix} \right\}$ is a basis for null(A), so nullity(A) = 3 − 2 = 1.

37. We use Exercise 19 to determine rank(A) and The Rank Theorem to determine nullity(A).

rank(A) = number of vectors in a basis for col(A) or row(A)

Since $\left\{ \begin{bmatrix} 1 \\ 0 \\ 0 \end{bmatrix}, \begin{bmatrix} 1 \\ 1 \\ 1 \end{bmatrix}, \begin{bmatrix} 1 \\ 1 \\ -1 \end{bmatrix} \right\}$ is a basis for col(A), we have rank(A) = 3.

nullity(A) = n − rank(A) (by Theorem 3.26, The Rank Theorem)
Since $n = 4$ (because A is 3 × 4) and rank(A) = 3, nullity(A) = 4 − 3 = 1.
The basis for null(A) we found in Exercise 20 should have 3 vectors in it. Why?

39. If nullity(A) > 0, then the columns of A are linearly dependent.
Though we could prove this using theorems, it is instructive to prove it directly.

If nullity(A) > 0, then there exists a vector $\mathbf{x} \neq \mathbf{0}$ such that $A\mathbf{x} = \mathbf{0}$.

Let $A = [\,\mathbf{a}_1\ \mathbf{a}_2\ \cdots\ \mathbf{a}_n\,]$ and $\mathbf{x}^T = [\,c_1\ c_2\ \cdots\ c_n\,]$.

Since $\mathbf{x} \neq \mathbf{0}$ at least one $c_i \neq 0$. Then $A\mathbf{x} = \sum c_i \mathbf{a}_i = \mathbf{0}$ where at least one $c_i \neq 0$.
Therefore, the columns of A are linearly dependent.

So all we have to show is: If A is a 3 × 5 matrix, then nullity(A) > 0.

$$\underset{\text{Rank Thm}}{\text{nullity}(A) =} n - \text{rank}(A) \underset{A \text{ is } 3\times 5}{=} 5 - \text{rank}(A) \underset{\text{rank}(A)\leq 3}{\geq} 5 - 3 = 2 > 0$$

Q: If A is 3 × 5, why is it obvious that rank(A) ≤ 3?
A: Recall, that the number of vectors in a basis for row(A) = dim(row(A)).
Now note the rows contain a basis for row(A), so dim(row(A)) ≤ number of rows.
So, rank(A) = dim(row(A)) ≤ the number of rows = 3.

41. Rank must satisfy the following two conditions simultaneously:
1) rank(A) = dim(row(A)) ≤ the number of rows
2) rank(A) = dim(col(A)) ≤ the number of columns

Therefore, rank must be less than or equal to the smaller of these two numbers.

Since A is 3 × 5, rank(A) can equal 0, 1, 2, or 3.
Therefore, since $n = 5$ and nullity(A) = n − rank(A), we have:

nullity(A) = 5 − 3 = 2, 5 − 2 = 3, 5 − 1 = 4, or 5 − 0 = 5

43. $A = \begin{bmatrix} 1 & 2 & a \\ -2 & 4a & 2 \\ a & -2 & 1 \end{bmatrix} \xrightarrow[R_3-aR_1]{R_2+2R_1} \begin{bmatrix} 1 & 2 & a \\ 0 & a+1 & \frac{a+1}{2} \\ 0 & 0 & \frac{(a+1)(a-2)}{2} \end{bmatrix}$.

If $a = -1$, then $A \longrightarrow \begin{bmatrix} 1 & 2 & -1 \\ 0 & 0 & 0 \\ 0 & 0 & 0 \end{bmatrix} \Rightarrow \text{rank}(A) = 1$.

If $a = 2$, then $A \longrightarrow \begin{bmatrix} 1 & 2 & 2 \\ 0 & 3 & \frac{3}{2} \\ 0 & 0 & 0 \end{bmatrix} \Rightarrow \text{rank}(A) = 2$. Otherwise, $\text{rank}(A) = 3$.

45. As in Example 3.52, $\{\mathbf{u}, \mathbf{v}, \mathbf{w}\}$ form a basis for $\mathbb{R}^3 \Leftrightarrow$ When $A = \begin{bmatrix} \mathbf{u}^T \\ \mathbf{v}^T \\ \mathbf{w}^T \end{bmatrix}$, $\text{rank}(A) = 3$.

$A = \begin{bmatrix} \mathbf{u}^T \\ \mathbf{v}^T \\ \mathbf{w}^T \end{bmatrix} = \begin{bmatrix} 1 & 1 & 0 \\ 1 & 0 & 1 \\ 0 & 1 & 1 \end{bmatrix} \xrightarrow{R_2-R_1} \begin{bmatrix} 1 & 1 & 0 \\ 0 & -1 & 1 \\ 0 & 1 & 1 \end{bmatrix} \xrightarrow{R_3+R_2} \begin{bmatrix} 1 & 1 & 0 \\ 0 & -1 & 1 \\ 0 & 0 & 2 \end{bmatrix}$

Since $\text{rank}(A) = 3$, $\{\mathbf{u}, \mathbf{v}, \mathbf{w}\}$ form a basis for $\mathbb{R}^3$.

47. As in Example 3.52, $\{\mathbf{x}, \mathbf{u}, \mathbf{v}, \mathbf{w}\}$ form a basis for $\mathbb{R}^4 \Leftrightarrow$ When $A = \begin{bmatrix} \mathbf{x}^T \\ \mathbf{u}^T \\ \mathbf{v}^T \\ \mathbf{w}^T \end{bmatrix}$, $\text{rank}(A) = 4$.

$A = \begin{bmatrix} \mathbf{x}^T \\ \mathbf{u}^T \\ \mathbf{v}^T \\ \mathbf{w}^T \end{bmatrix} = \begin{bmatrix} 1 & 1 & 1 & 0 \\ 1 & 1 & 0 & 1 \\ 1 & 0 & 1 & 1 \\ 0 & 1 & 1 & 1 \end{bmatrix} \xrightarrow[R_3-R_1]{R_2-R_1} \begin{bmatrix} 1 & 1 & 1 & 0 \\ 0 & 0 & -1 & 1 \\ 0 & -1 & 0 & 1 \\ 0 & 1 & 1 & 1 \end{bmatrix} \xrightarrow{R_3+R_2+R_4} \begin{bmatrix} 1 & 1 & 1 & 0 \\ 0 & 0 & -1 & 1 \\ 0 & 0 & 0 & 3 \\ 0 & 1 & 1 & 1 \end{bmatrix}$

Since $\text{rank}(A) = 4$, $\{\mathbf{x}, \mathbf{u}, \mathbf{v}, \mathbf{w}\}$ form a basis for $\mathbb{R}^4$.

49. We find the coordinate vector $[\mathbf{w}]_B$ by finding c_1 and c_1 such that $\mathbf{w} = c_1\mathbf{b}_1 + c_1\mathbf{b}_2$. As in Example 2.18 of Section 2.3, we form the matrix $A = [\,\mathbf{b}_1 \ \mathbf{b}_2\,|\,\mathbf{w}\,]$ and row reduce.

$A = [\,\mathbf{b}_1 \ \mathbf{b}_2\,|\,\mathbf{w}\,] = \begin{bmatrix} 1 & 1 & | & 1 \\ 2 & 0 & | & 6 \\ 0 & -1 & | & 2 \end{bmatrix} \longrightarrow \begin{bmatrix} 1 & 0 & | & 3 \\ 0 & 1 & | & -2 \\ 0 & 0 & | & 0 \end{bmatrix}$

Since $\mathbf{w} = 3\mathbf{b}_1 - 2\mathbf{b}_2$, we have the coordinate vector $[\mathbf{w}]_B = \begin{bmatrix} 3 \\ -2 \end{bmatrix}$.

51. We row reduce over $\mathbb{Z}_2$ to find $\text{rank}(A)$, then $\text{nullity}(A) = n - \text{rank}(A)$.

$A = \begin{bmatrix} 1 & 1 & 0 \\ 0 & 1 & 1 \\ 1 & 0 & 1 \end{bmatrix} \xrightarrow{R_3+R_1} \begin{bmatrix} 1 & 1 & 0 \\ 0 & 1 & 1 \\ 0 & 1 & 1 \end{bmatrix}$

Since $\text{rank}(A) = 2$, we have $\text{nullity}(A) = 3 - 2 = 1$.

3.5 Subspaces, Basis, Dimension, and Rank

53. We row reduce over $\mathbb{Z}_5$ to find rank(A), then nullity(A) = n − rank(A).

$$A = \begin{bmatrix} 1 & 3 & 1 & 4 \\ 2 & 3 & 0 & 1 \\ 1 & 0 & 4 & 0 \end{bmatrix} \xrightarrow[R_3+4R_1]{R_2+3R_1} \begin{bmatrix} 1 & 3 & 1 & 4 \\ 0 & 2 & 4 & 2 \\ 0 & 2 & 3 & 1 \end{bmatrix}$$

Since rank(A) = 3, we have nullity(A) = 4 − 3 = 1.

55. We need to show if $\mathbf{v}$ is in row(A) and $\mathbf{x}$ is in null(A), then $\mathbf{v} \cdot \mathbf{x} = 0$.

    First, we will show if $\mathbf{x}$ is in null(A) and $\mathbf{A}_i$ is the ith row of A, then $\mathbf{A}_i \cdot \mathbf{x} = 0$.

    Let $A = \begin{bmatrix} \mathbf{A}_1 \\ \mathbf{A}_2 \\ \vdots \\ \mathbf{A}_m \end{bmatrix}$ and $\mathbf{x} = \begin{bmatrix} x_1 \\ x_2 \\ \vdots \\ x_n \end{bmatrix}$. Then we have: If $\mathbf{x}$ is in null(A), then $A\mathbf{x} = \mathbf{0}$.

    So, $A\mathbf{x} = \begin{bmatrix} a_{11}x_1 + a_{12}x_2 + \cdots + a_{1n}x_n \\ a_{21}x_1 + a_{22}x_2 + \cdots + a_{2n}x_n \\ \vdots \\ a_{m1}x_1 + a_{m2}x_2 + \cdots + a_{mn}x_n \end{bmatrix} = \begin{bmatrix} \mathbf{A}_1 \cdot \mathbf{x} \\ \mathbf{A}_2 \cdot \mathbf{x} \\ \vdots \\ \mathbf{A}_m \cdot \mathbf{x} \end{bmatrix} = \begin{bmatrix} 0 \\ 0 \\ \vdots \\ 0 \end{bmatrix}$.

    Therefore, if $\mathbf{x}$ is in null(A), then $\mathbf{A}_i \cdot \mathbf{x} = 0$.

    If $\mathbf{v}$ is in row(A), then $\mathbf{v}$ is a linear combination of the rows of A so $\mathbf{v} = \sum c_i \mathbf{A}_i$.
    So, we have: $\mathbf{v} \cdot \mathbf{x} = (\sum c_i \mathbf{A}_i) \cdot \mathbf{x} = \sum c_i (\mathbf{A}_i \cdot \mathbf{x}) = \sum c_i 0 = 0$.

    Q: What is the idea behind the proof of this exercise?
    A: If a vector is orthogonal to a set of vectors,
        it is orthogonal to all linear combinations of those vectors.

57. We will prove part (a) using the idea we suggested by Exercise 29 of Section 3.1.

    (a) Let $\{\mathbf{A}\mathbf{b}_k\}$ be a basis for col(AB) formed from the columns of AB.
        Then rank(AB) = number of vectors in $\{\mathbf{A}\mathbf{b}_k\}$ = number of vectors in $\{\mathbf{b}_k\}$.
        First we show: If $\{\mathbf{A}\mathbf{b}_k\}$ is linearly independent, then $\{\mathbf{b}_k\}$ is linearly independent.
            That is, if $\sum c_k \mathbf{b}_k = \mathbf{0}$, we need to show all the $c_k = 0$.
            $\sum c_k \mathbf{b}_k = \mathbf{0} \Rightarrow A(\sum c_k \mathbf{b}_k) = \mathbf{0} \Rightarrow \sum c_k (\mathbf{A}\mathbf{b}_k) = \mathbf{0}$
        Since $\{\mathbf{A}\mathbf{b}_k\}$ is linearly independent by assumption, all the $c_k = 0$ as required.
        Now, since $\{\mathbf{b}_k\}$ is a linearly independent subspace of the columns of B,
        the number of vectors in $\{\mathbf{b}_k\} \leq$ rank(B).
        Therefore, rank(AB) = number of vectors in $\{\mathbf{b}_k\} \leq$ rank(B).

    (b) Let $A = \begin{bmatrix} 1 & 1 \\ 0 & 0 \end{bmatrix}$ and $B = \begin{bmatrix} 1 & 0 \\ 1 & 1 \end{bmatrix}$. Then B has rank 2, but $AB = \begin{bmatrix} 2 & 1 \\ 0 & 0 \end{bmatrix}$ has rank 1.

59. We prove this Exercise using the results of Exercise 57 and Exercise 58.

   (a) By Exercise 53, it always true that $\text{rank}(UA) \leq \text{rank}(A)$.
   So, it suffices to show $\text{rank}(A) \leq \text{rank}(UA)$ to conclude $\text{rank}(UA) = \text{rank}(A)$.
   But $\text{rank}(A) = \text{rank}(U^{-1}(UA)) \leq \text{rank}(UA)$ as required.

   (b) By Exercise 58, it always true that $\text{rank}(AV) \leq \text{rank}(A)$.
   So, it suffices to show $\text{rank}(A) \leq \text{rank}(AV)$ to conclude $\text{rank}(AV) = \text{rank}(A)$.
   But $\text{rank}(A) = \text{rank}((AV)V^{-1}) \leq \text{rank}(AV)$ as required.

61. We will show $A = \sum \mathbf{u}_i \mathbf{v}_i^T$, where $\text{rank}(\mathbf{u}_i \mathbf{v}_i^T) = 1$ by Exercise 60.

   Since $\text{rank}(A) = r$, a basis for $\text{col}(A)$ has r vectors, $\{\mathbf{u}_i\}$, in $\mathbb{R}^m$. So:

   $$A = \begin{bmatrix} c_{11}\mathbf{u}_1 & & c_{n1}\mathbf{u}_1 \\ + & & + \\ \vdots & \cdots & \vdots \\ + & & + \\ c_{1r}\mathbf{u}_r & & c_{nr}\mathbf{u}_r \end{bmatrix} = \sum \begin{bmatrix} c_{1i}\mathbf{u}_i & \cdots & c_{ni}\mathbf{u}_i \end{bmatrix}$$

   Now if we let $\mathbf{v}_i^T = \begin{bmatrix} c_{1i} & \cdots & c_{ni} \end{bmatrix}$, then $A = \sum \mathbf{u}_i \mathbf{v}_i^T$ as required.

63. If $A^2 = O$, then $\text{col}(A) \subseteq \text{null}(A)$. That is, if $\mathbf{x} = \sum c_i \mathbf{a}_i$, then $A\mathbf{x} = \mathbf{0}$.

   Recall that $\mathbf{a}_i = A\mathbf{e}_i$.

   So, $A\mathbf{x} = A(\sum c_i \mathbf{a}_i) = \sum c_i (A\mathbf{a}_i) = \sum c_i (A(A\mathbf{e}_i)) = \sum c_i (A^2 \mathbf{e}_i) = \sum c_i (O\mathbf{e}_i) = \mathbf{0}$.

   Therefore, since $\text{col}(A) \subseteq \text{null}(A)$, $\text{rank}(A) = \dim(\text{col}(A)) \leq \dim(\text{null}(A)) = \text{nullity}(A)$.

   So, $\text{rank}(A) + \text{rank}(A) \leq \text{rank}(A) + \text{nullity}(A) = n \Rightarrow 2\,\text{rank}(A) \leq n \Rightarrow \text{rank}(A) \leq \frac{n}{2}$.

3.6 Introduction to Linear Transformations

Q: What are our main goals for Section 3.6?

A: To take our first look at linear transformations and their relationship to matrices.
T is linear if $T(c\mathbf{v}) = cT(\mathbf{v})$ and $T(\mathbf{u}+\mathbf{v}) = T(\mathbf{u}) + T(\mathbf{v})$.
If T is linear, then there is a matrix A such that $T(\mathbf{v}) = A\mathbf{v}$.
At the end of the section we see the range of T is equal to the column space of A.

Key Definitions and Concepts

linear transformation	p.211	3.6	$T(c\mathbf{v}) = cT(\mathbf{v})$ and $T(\mathbf{u}+\mathbf{v}) = T(\mathbf{u}+\mathbf{v})$ If T is linear, then there is an A such that $T(\mathbf{v}) = A\mathbf{v}$.
inverse transformation	p.219	3.6	$S \circ T = I_n$ and $T \circ S = I_n$ $S(T(\mathbf{v})) = [S][T]\mathbf{v} = \mathbf{v} \Rightarrow [T] = [S]^{-1}$

Theorems

Thm 3.30	p.212	3.6	Given A, then $T_A(\mathbf{v}) = A\mathbf{v}$ is linear
Thm 3.31	p.214	3.6	$[T] = \begin{bmatrix} T(\mathbf{e}_1) & T(\mathbf{e}_2) & \ldots & T(\mathbf{e}_n) \end{bmatrix}$
Thm 3.32	p.218	3.6	Given $S \circ T$, $S(T(\mathbf{v})) = [S][T]\mathbf{v}$
Thm 3.33	p.220	3.6	$S \circ T = I_n$, $S(T(\mathbf{v})) = [S][T]\mathbf{v} = \mathbf{v} \Rightarrow [T] = [S]^{-1}$

Discussion of Key Definitions, Concepts, and Theorems

How to prove or disprove T is a linear transformation

Q: How do we prove that a given transformation T is linear?
A: We prove T is a linear transformation by showing that $T(c_1\mathbf{v}_1+c_2\mathbf{v}_2) = c_1T(\mathbf{v}_1)+c_2T(\mathbf{v}_2)$.
Or by showing $T(c\mathbf{v}) = cT(\mathbf{v})$ and $T(\mathbf{u}+\mathbf{v}) = T(\mathbf{u})+T(\mathbf{v})$.
See Exercises 3 through 6.

Q: Is there any other method for proving that a given transformation T is linear?
Yes. By showing $T(\mathbf{v}) = A\mathbf{v}$ for some matrix A.
See Exercises 11 through 18.

Q: Is there any advantage to finding the matrix $A = [T]$ for a linear transformation T?
Yes. We then can then compute $T(\mathbf{v})$ for any vector $\mathbf{v}$ by multiplying $A\mathbf{v}$.

Q: How do we prove that a given transformation T is *not* linear?
A: We prove T is *not* a linear transformation by showing that one of the properties fails.
That is, by showing $T(c\mathbf{v}) \neq cT(\mathbf{v})$ or $T(\mathbf{u}+\mathbf{v}) \neq T(\mathbf{u}+\mathbf{v})$.
It is important to note that this property only has to fail for one vector $\mathbf{v}$.
See Exercises 7 through 9.

Why linear transformations map the zero vector to the zero vector

Q: If T is linear, does $T(\mathbf{0}) = \mathbf{0}$. If so, why?
A: Yes. Because $T(\mathbf{0}) = T(0(\mathbf{0})) = 0(T(\mathbf{0})) = \mathbf{0}$.

Q: Does this give us another way of proving a transformation is *not* linear?
A: Yes. If $T(\mathbf{0}) \neq \mathbf{0}$, then T is not linear.

Q: What does $T(\mathbf{0}) = \mathbf{0}$ mean geometrically?
A: Geometrically speaking this says: a linear transformation does not move the origin.

Q: What does $T(\mathbf{0}) \neq \mathbf{0}$ mean geometrically?
A: Geometrically speaking this says: a transformation that moves the origin is not linear.

Q: With this in mind, explain why we might at least hope that rotations are linear?
A: Rotations leave the origin fixed.

Q: What about reflection in a line that passes through the origin?
A: Reflection in a line that pass through the origin leaves the origin fixed.
So, reflection in a line that passes through the origin *might* be linear.

Q: What about reflection in a line that does not pass through the origin?
A: Reflection in a line that does not pass through the origin moves the origin.
So, reflection in a line that does not pass through the origin is definitely not linear.

Q: If the only vector that T maps to the zero vector is the zero vector, what can we conclude?
A: If $T(\mathbf{v}) = \mathbf{0} \Rightarrow \mathbf{v} = \mathbf{0}$, then T is invertible. Why?
Because that is equivalent to saying $[T]\mathbf{v} = \mathbf{0}$ has only the trivial solution.

3.6 Introduction to Linear Transformations

Discussion of Key Definitions, Concepts, and Theorems, continued

Transformations that fail to be linear

Q: Why should we suspect that $S\begin{bmatrix} x \\ y \end{bmatrix} = \begin{bmatrix} x^2 \\ y \end{bmatrix}$ is not linear?

A: We should suspect that S is *not* linear because x^2 is *not* linear.

Q: What generalization does this observation lead us to?

A: Both $T\begin{bmatrix} x \\ y \end{bmatrix} = \begin{bmatrix} x^n \\ y \end{bmatrix}$ and $S\begin{bmatrix} x \\ y \end{bmatrix} = \begin{bmatrix} x \\ y^n \end{bmatrix}$ are not linear. Why?

Hint: Consider what happens to scalars c. That is, does $T(c\mathbf{v}) = cT(\mathbf{v})$?

Q: How can we tell immediately that $S\begin{bmatrix} x \\ y \end{bmatrix} = \begin{bmatrix} x+1 \\ y+1 \end{bmatrix}$ is not linear?

A: Because S moves the origin from $\begin{bmatrix} 0 \\ 0 \end{bmatrix}$ to $\begin{bmatrix} 1 \\ 1 \end{bmatrix}$.

Q: What generalization does this observation lead us to?

A: Any transformation of the form $S\begin{bmatrix} x \\ y \end{bmatrix} = \begin{bmatrix} x+c \\ y+d \end{bmatrix}$ is not linear (unless $c = d = 0$). Why?

Q: What about $S\begin{bmatrix} x \\ y \end{bmatrix} = \begin{bmatrix} x+cx \\ y+dy \end{bmatrix}$ or $T\begin{bmatrix} x \\ y \end{bmatrix} = \begin{bmatrix} x+cy \\ y+dx \end{bmatrix}$?

A: There is some hope at least because these transformations do not move the origin.

Q: What about $T\begin{bmatrix} x \\ y \end{bmatrix} = \begin{bmatrix} ax \\ by \end{bmatrix}$?

A: By now, we should strongly suspect that all such transformations are linear. Find $[T]$.

Since $T(\mathbf{e}_1) = T\begin{bmatrix} 1 \\ 0 \end{bmatrix} = \begin{bmatrix} a \\ 0 \end{bmatrix}$ and $T(\mathbf{e}_2) = T\begin{bmatrix} 0 \\ 1 \end{bmatrix} = \begin{bmatrix} 0 \\ b \end{bmatrix}$, we have: $[T] = \begin{bmatrix} a & 0 \\ 0 & b \end{bmatrix}$.

Q: What about $T\begin{bmatrix} x \\ y \end{bmatrix} = \begin{bmatrix} ax+dy \\ by+cx \end{bmatrix}$?

A: By now, we should strongly suspect that all such transformations are linear. Find $[T]$.

Discussion of Theorems: Transformations that are always linear: $T(\mathbf{v}) = A\mathbf{v}$

Q: Given any matrix A, is $T(\mathbf{v}) = A\mathbf{v}$ always linear?
A: Yes. Every matrix corresponds to a linear transformation.

Q: Given a matrix A and $T(\mathbf{v}) = A\mathbf{v}$, how do we find the formula for T?
A: Simply by multiplying $A\mathbf{v}$.

Consider: $A = \begin{bmatrix} 2 & 3 \\ 4 & 5 \end{bmatrix}$, then $A\mathbf{v} = \begin{bmatrix} 2 & 3 \\ 4 & 5 \end{bmatrix} \begin{bmatrix} x \\ y \end{bmatrix} = \begin{bmatrix} 2x+3y \\ 4x+5y \end{bmatrix} = T(\mathbf{v})$.

Discussion of Theorems: The matrix of a linear transformation

Q: What does $[T] = [\,T(\mathbf{e}_1)\ T(\mathbf{e}_2)\ \ldots\ T(\mathbf{e}_n)\,]$ tell us?
A: This seemingly simple equation tells us two very important things:
1) If T is a linear transformation, then there is a matrix $[T]$ such that $T(\mathbf{v}) = [T]\mathbf{v}$.
2) Not only that, but it explicitly tells us how to find that matrix $[T]$.
Find the vectors $T(\mathbf{e}_i)$ and let those vectors be the columns of the matrix $[T]$.

Consider the following example: Let $T\begin{bmatrix} x \\ y \end{bmatrix} = \begin{bmatrix} 2x \\ y \end{bmatrix}$.

Since $T(\mathbf{e}_1) = T\begin{bmatrix} 1 \\ 0 \end{bmatrix} = \begin{bmatrix} 2 \\ 0 \end{bmatrix}$ and $T(\mathbf{e}_2) = T\begin{bmatrix} 0 \\ 1 \end{bmatrix} = \begin{bmatrix} 0 \\ 1 \end{bmatrix}$, we have: $[T] = \begin{bmatrix} 2 & 0 \\ 0 & 1 \end{bmatrix}$.

Q: Why does this matrix $[T]$ work for any other vector $\mathbf{v}$?
That is, why does $T(\mathbf{v}) = [T]\mathbf{v}$?
A: Because $T(\mathbf{v}) = T(\sum c_i \mathbf{e}_i) = \sum c_i T(\mathbf{e}_i) = [\,T(\mathbf{e}_1)\ T(\mathbf{e}_2)\ \ldots\ T(\mathbf{e}_n)\,]\mathbf{v} = [T]\mathbf{v}$.

Consider the following example. Let $\mathbf{v} = \begin{bmatrix} 3 \\ 5 \end{bmatrix}$.

$$T(\mathbf{v}) = T(3\mathbf{e}_1 + 5\mathbf{e}_2) = 3T(\mathbf{e}_1) + 5T(\mathbf{e}_2) = [\,T(\mathbf{e}_1)\ T(\mathbf{e}_2)\,]\begin{bmatrix} 3 \\ 5 \end{bmatrix} = \begin{bmatrix} 2 & 0 \\ 0 & 1 \end{bmatrix}\begin{bmatrix} 3 \\ 5 \end{bmatrix} = [T]\mathbf{v}.$$

Q: How might we put the equation $T(\sum c_i \mathbf{e}_i) = \sum c_i T(\mathbf{e}_i)$ in our own words?
A: The components of $\mathbf{v}$, the c_i, become the coefficients of $T(\mathbf{e}_i)$.

Discussion of Theorems: The inverse of a linear transformation

Q: What does $S \circ T = I_n$ if and only if $[T]^{-1} = [S]$ tell us?
A: These statements combine to give us two very important insights:
1) T has an inverse if and only if $[T]$ has an inverse matrix $[T]^{-1}$.
2) Moreover, if $S \circ T = I_n$, then that inverse matrix is $[S]$.

If $T\begin{bmatrix} x \\ y \end{bmatrix} = \begin{bmatrix} 2x \\ y \end{bmatrix}$, then $[T] = \begin{bmatrix} 2 & 0 \\ 0 & 1 \end{bmatrix}$ and $[T]^{-1} = \frac{1}{2}\begin{bmatrix} 1 & 0 \\ 0 & 2 \end{bmatrix} = \begin{bmatrix} \frac{1}{2} & 0 \\ 0 & 1 \end{bmatrix}$.

So $S\begin{bmatrix} x \\ y \end{bmatrix} = \begin{bmatrix} \frac{1}{2} & 0 \\ 0 & 1 \end{bmatrix}\begin{bmatrix} x \\ y \end{bmatrix} = \begin{bmatrix} \frac{1}{2}x \\ y \end{bmatrix}$ should be the inverse function of T.

Check: $S \circ T \begin{bmatrix} x \\ y \end{bmatrix} = S\left(T\begin{bmatrix} x \\ y \end{bmatrix}\right) = S\begin{bmatrix} 2x \\ y \end{bmatrix} = \begin{bmatrix} x \\ y \end{bmatrix}$, so $S \circ T = I_2$ as required.

Q: Are all rotations R_θ invertible?
A: Yes, since $[R_\theta] = \begin{bmatrix} \cos\theta & -\sin\theta \\ \sin\theta & \cos\theta \end{bmatrix}$ implies $\det[R_\theta] = \cos^2\theta + \sin^2\theta = 1$. So $[R_\theta]^{-1}$ exists.

Discussion of Theorems: Composition of linear transformations

Q: What does $S(T(\mathbf{v})) = [S][T]\mathbf{v}$ tell us?
A: To compose linear transformations all we have to do is multiply their matrices together.

As above: $S(T(\mathbf{v})) = [S][T]\mathbf{v} = \begin{bmatrix} \frac{1}{2} & 0 \\ 0 & 1 \end{bmatrix}\begin{bmatrix} 2 & 0 \\ 0 & 1 \end{bmatrix}\begin{bmatrix} x \\ y \end{bmatrix} = \begin{bmatrix} 1 & 0 \\ 0 & 1 \end{bmatrix}\begin{bmatrix} x \\ y \end{bmatrix} = I\mathbf{v}$.

3.6 Introduction to Linear Transformations

Solutions to odd-numbered exercises from Section 3.6

1. Since T is the linear transformation corresponding to matrix A, $T(\mathbf{x}) = A\mathbf{x}$. So:
$$T(\mathbf{u}) = \begin{bmatrix} 2 & -1 \\ 3 & 4 \end{bmatrix} \begin{bmatrix} 1 \\ 2 \end{bmatrix} = \begin{bmatrix} 0 \\ 11 \end{bmatrix}, T(\mathbf{v}) = \begin{bmatrix} 2 & -1 \\ 3 & 4 \end{bmatrix} \begin{bmatrix} 3 \\ -2 \end{bmatrix} = \begin{bmatrix} 8 \\ 1 \end{bmatrix}.$$

3. We prove T is a linear transformation by showing that $T(c_1\mathbf{v}_1 + c_2\mathbf{v}_2) = c_1 T(\mathbf{v}_1) + c_2 T(\mathbf{v}_2)$.

Let $T\begin{bmatrix} x \\ y \end{bmatrix} = \begin{bmatrix} x+y \\ x-y \end{bmatrix}$, $\mathbf{v}_1 = \begin{bmatrix} x_1 \\ y_1 \end{bmatrix}$ and $\mathbf{v}_2 = \begin{bmatrix} x_2 \\ y_2 \end{bmatrix}$. Then:

$$\begin{aligned}
T(c_1\mathbf{v}_1 + c_2\mathbf{v}_2) &= T\left(c_1 \begin{bmatrix} x_1 \\ y_1 \end{bmatrix} + c_2 \begin{bmatrix} x_2 \\ y_2 \end{bmatrix}\right) = T\left(\begin{bmatrix} c_1 x_1 + c_2 x_2 \\ c_1 y_1 + c_2 y_2 \end{bmatrix}\right) \\
&= \begin{bmatrix} c_1 x_1 + c_2 x_2 + c_1 y_1 + c_2 y_2 \\ c_1 x_1 + c_2 x_2 - c_1 y_1 - c_2 y_2 \end{bmatrix} = \begin{bmatrix} c_1 x_1 + c_1 y_1 \\ c_1 x_1 - c_1 y_1 \end{bmatrix} + \begin{bmatrix} c_2 x_2 + c_2 y_2 \\ c_2 x_2 - c_2 y_2 \end{bmatrix} \\
&= c_1 \begin{bmatrix} x_1 + y_1 \\ x_1 - y_1 \end{bmatrix} + c_2 \begin{bmatrix} x_2 + y_2 \\ x_2 - y_2 \end{bmatrix} = c_1 T\begin{bmatrix} x_1 \\ y_1 \end{bmatrix} + c_2 T \begin{bmatrix} x_2 \\ y_2 \end{bmatrix} \\
&= c_1 T(\mathbf{v}_1) + c_2 T(\mathbf{v}_2).
\end{aligned}$$

Therefore, we conclude that T is a linear transformation.

5. We prove T is a linear transformation by showing that $T(c_1\mathbf{v}_1 + c_2\mathbf{v}_2) = c_1 T(\mathbf{v}_1) + c_2 T(\mathbf{v}_2)$.

Let $T\begin{bmatrix} x \\ y \\ z \end{bmatrix} = \begin{bmatrix} x - y + z \\ 2x + y - 3z \end{bmatrix}$, $\mathbf{v}_1 = \begin{bmatrix} x_1 \\ y_1 \\ z_1 \end{bmatrix}$ and $\mathbf{v}_2 = \begin{bmatrix} x_2 \\ y_2 \\ z_2 \end{bmatrix}$. Then

$$\begin{aligned}
T(c_1\mathbf{v}_1 + c_2\mathbf{v}_2) &= T\left(c_1 \begin{bmatrix} x_1 \\ y_1 \\ z_1 \end{bmatrix} + c_2 \begin{bmatrix} x_2 \\ y_2 \\ z_2 \end{bmatrix}\right) = T\left(\begin{bmatrix} c_1 x_1 + c_2 x_2 \\ c_1 y_1 + c_2 y_2 \\ c_1 z_1 + c_2 z_2 \end{bmatrix}\right) \\
&= \begin{bmatrix} c_1 x_1 + c_2 x_2 - c_1 y_1 - c_2 y_2 + c_1 z_1 + c_2 z_2 \\ 2 c_1 x_1 + 2 c_2 x_2 + c_1 y_1 + c_2 y_2 - 3 c_1 z_1 - 3 c_2 z_2 \end{bmatrix} \\
&= \begin{bmatrix} c_1 x_1 - c_1 y_1 + c_1 z_1 \\ 2 c_1 x_1 + c_1 y_1 - 3 c_1 z_1 \end{bmatrix} + \begin{bmatrix} c_2 x_2 - c_2 y_2 + c_2 z_2 \\ 2 c_2 x_2 + c_2 y_2 - 3 c_2 z_2 \end{bmatrix} \\
&= c_1 \begin{bmatrix} x_1 - y_1 + z_1 \\ 2x_1 + y_1 - 3z_1 \end{bmatrix} + c_2 \begin{bmatrix} x_2 - y_2 + z_2 \\ 2x_2 + y_2 - 3z_2 \end{bmatrix} = c_1 T \begin{bmatrix} x_1 \\ y_1 \\ z_1 \end{bmatrix} + c_2 T \begin{bmatrix} x_1 \\ y_1 \\ z_1 \end{bmatrix} \\
&= c_1 T(\mathbf{v}_1) + c_2 T(\mathbf{v}_2).
\end{aligned}$$

7. We prove T is *not* a linear transformation by showing that $T(c\mathbf{v}) \neq cT(\mathbf{v})$ (property (2) fails).

   Let $T \begin{bmatrix} x \\ y \end{bmatrix} = \begin{bmatrix} y \\ x^2 \end{bmatrix}$ and $\mathbf{v} = \begin{bmatrix} x \\ y \end{bmatrix}$. Then

   $$T(c\mathbf{v}) = T\left(c \begin{bmatrix} x \\ y \end{bmatrix}\right) = T \begin{bmatrix} cx \\ cy \end{bmatrix} = \begin{bmatrix} cy \\ c^2 x^2 \end{bmatrix} \neq \begin{bmatrix} cy \\ cx^2 \end{bmatrix} = c \begin{bmatrix} y \\ cx^2 \end{bmatrix} = cT \begin{bmatrix} x \\ y \end{bmatrix} = cT(\mathbf{v})$$

   Since $T(c\mathbf{v}) \neq cT(\mathbf{v})$ (property (2) fails), T is *not* a linear transformation.

   Q: Is $S \begin{bmatrix} x \\ y \end{bmatrix} = \begin{bmatrix} x^2 \\ y \end{bmatrix}$ linear?

   A: No, by a very similar argument to the one given above.
   We should suspect both T and S are *not* linear because x^2 is *not* linear.

9. We prove T is *not* a linear transformation by showing that $T(c\mathbf{v}) \neq cT(\mathbf{v})$ (property (2) fails).

   Let $T \begin{bmatrix} x \\ y \end{bmatrix} = \begin{bmatrix} xy \\ x+y \end{bmatrix}$ and $\mathbf{v} = \begin{bmatrix} x \\ y \end{bmatrix}$. Then

   $$T(c\mathbf{v}) = T \begin{bmatrix} cx \\ cy \end{bmatrix} = \begin{bmatrix} cxcy \\ cx+cy \end{bmatrix} = c \begin{bmatrix} cxy \\ x+y \end{bmatrix} \neq c \begin{bmatrix} xy \\ x+y \end{bmatrix} = cT \begin{bmatrix} x \\ y \end{bmatrix} = cT(\mathbf{v})$$

   Since $T(c\mathbf{v}) \neq cT(\mathbf{v})$ (property (2) fails), T is *not* a linear transformation.

   Q: Is there any reason to suspect that T is not linear before completing the proof?
   A: Yes, by a very similar argument to the one given above.
   We should suspect T is *not* linear because xy is *not* linear.

11. As on p.212, we confirm T is a linear transformation by finding A such that $T(\mathbf{v}) = A\mathbf{v}$.

    We have $T \begin{bmatrix} x \\ y \end{bmatrix} = \begin{bmatrix} x+y \\ x-y \end{bmatrix} = \begin{bmatrix} 1 \\ 1 \end{bmatrix} x + \begin{bmatrix} 1 \\ -1 \end{bmatrix} y = \begin{bmatrix} 1 & 1 \\ 1 & -1 \end{bmatrix} \begin{bmatrix} x \\ y \end{bmatrix}$.

    So $T = T_A$ where $A = \begin{bmatrix} 1 & 1 \\ 1 & -1 \end{bmatrix}$.

13. As on p.212, we confirm T is a linear transformation by finding A such that $T(\mathbf{v}) = A\mathbf{v}$.

    We have $T \begin{bmatrix} x \\ y \\ z \end{bmatrix} = \begin{bmatrix} x-y+z \\ 2x+y-3z \end{bmatrix} = \begin{bmatrix} 1 \\ 2 \end{bmatrix} x + \begin{bmatrix} -1 \\ 1 \end{bmatrix} y + \begin{bmatrix} 1 \\ -3 \end{bmatrix} z = \begin{bmatrix} 1 & -1 & 1 \\ 2 & 1 & -3 \end{bmatrix} \begin{bmatrix} x \\ y \\ z \end{bmatrix}$.

    So $T = T_A$ where $A = \begin{bmatrix} 1 & -1 & 1 \\ 2 & 1 & -3 \end{bmatrix}$.

15. As in Example 3.56, we confirm T is a linear transformation by finding A such that $T(\mathbf{v}) = A\mathbf{v}$.

    We have $F \begin{bmatrix} x \\ y \end{bmatrix} = \begin{bmatrix} -x \\ y \end{bmatrix} = \begin{bmatrix} -1 \\ 0 \end{bmatrix} x + \begin{bmatrix} 0 \\ 1 \end{bmatrix} y = \begin{bmatrix} -1 & 0 \\ 0 & 1 \end{bmatrix} \begin{bmatrix} x \\ y \end{bmatrix}$.

    So we identify $F = \begin{bmatrix} -1 & 0 \\ 0 & 1 \end{bmatrix}$ as the matrix performing the desired transformation.

3.6 Introduction to Linear Transformations

17. As in Example 3.57, we confirm T is a linear transformation by finding A such that $T(\mathbf{v}) = A\mathbf{v}$.

We have $D \begin{bmatrix} x \\ y \end{bmatrix} = \begin{bmatrix} 2x \\ 3y \end{bmatrix} = \begin{bmatrix} 2 \\ 0 \end{bmatrix} x + \begin{bmatrix} 0 \\ 3 \end{bmatrix} y = \begin{bmatrix} 2 & 0 \\ 0 & 3 \end{bmatrix} \begin{bmatrix} x \\ y \end{bmatrix}$.

So, $D = \begin{bmatrix} 2 & 0 \\ 0 & 3 \end{bmatrix}$ is the matrix performing the desired transformation.

19. Let $A_1 = \begin{bmatrix} k & 0 \\ 0 & 1 \end{bmatrix}$, $A_2 = \begin{bmatrix} 1 & 0 \\ 0 & k \end{bmatrix}$, $B = \begin{bmatrix} 0 & 1 \\ 1 & 0 \end{bmatrix}$, $C_1 = \begin{bmatrix} 1 & k \\ 0 & 1 \end{bmatrix}$, and $C_2 = \begin{bmatrix} 1 & 0 \\ k & 1 \end{bmatrix}$.

$A_1 \begin{bmatrix} x \\ y \end{bmatrix} = \begin{bmatrix} k & 0 \\ 0 & 1 \end{bmatrix} \begin{bmatrix} x \\ y \end{bmatrix} = \begin{bmatrix} kx \\ y \end{bmatrix}$: A_1 stretches vectors horizontally by a factor of k.

$A_2 \begin{bmatrix} x \\ y \end{bmatrix} = \begin{bmatrix} 1 & 0 \\ 0 & k \end{bmatrix} \begin{bmatrix} x \\ y \end{bmatrix} = \begin{bmatrix} x \\ ky \end{bmatrix}$: A_2 stretches vectors vertically by a factor of k.

$B \begin{bmatrix} x \\ y \end{bmatrix} = \begin{bmatrix} 0 & 1 \\ 1 & 0 \end{bmatrix} \begin{bmatrix} y \\ x \end{bmatrix} = \begin{bmatrix} x \\ ky \end{bmatrix}$: B reflects vectors in the line $y = x$.

$C_1 \begin{bmatrix} x \\ y \end{bmatrix} = \begin{bmatrix} 1 & k \\ 0 & 1 \end{bmatrix} \begin{bmatrix} x \\ y \end{bmatrix} = \begin{bmatrix} x + ky \\ y \end{bmatrix}$: C_1 extends vectors horizontally by ky.

$C_2 \begin{bmatrix} x \\ y \end{bmatrix} = \begin{bmatrix} 1 & 0 \\ k & 1 \end{bmatrix} \begin{bmatrix} x \\ y \end{bmatrix} = \begin{bmatrix} x \\ y + kx \end{bmatrix}$: C_1 extends vectors vertically by kx.

The effects of these transformations are illustrated below (with $\mathbf{a} = \begin{bmatrix} 3 \\ 1 \end{bmatrix}$ and $k = 2$).

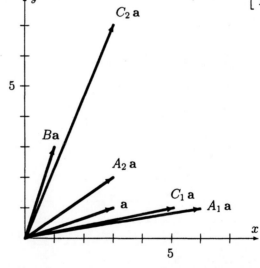

21. Using the formula from Example 3.58, we compute the matrix for a rotation of $-30° = 330°$.

$R_{330°}(\mathbf{e}_1) = \begin{bmatrix} \cos 330° \\ \sin 330° \end{bmatrix} = \begin{bmatrix} \frac{\sqrt{3}}{2} \\ -\frac{1}{2} \end{bmatrix}$ and $R_{330°}(\mathbf{e}_2) = \begin{bmatrix} -\sin 330° \\ \cos 330° \end{bmatrix} = \begin{bmatrix} \frac{1}{2} \\ \frac{\sqrt{3}}{2} \end{bmatrix}$,

So by Theorem 3.31, we have: $A = \begin{bmatrix} R_{330°}(\mathbf{e}_1) & R_{330°}(\mathbf{e}_2) \end{bmatrix} = \begin{bmatrix} \frac{\sqrt{3}}{2} & \frac{1}{2} \\ -\frac{1}{2} & \frac{\sqrt{3}}{2} \end{bmatrix}$.

23. We let $\mathbf{d} = \begin{bmatrix} 1 \\ -1 \end{bmatrix}$. Then $P_\ell(\mathbf{e}_1) = \left(\frac{\mathbf{d}\cdot\mathbf{e}_1}{\mathbf{d}\cdot\mathbf{d}}\right)\mathbf{d} = \frac{1}{1+1}\begin{bmatrix} 1 \\ -1 \end{bmatrix} = \begin{bmatrix} \frac{1}{2} \\ -\frac{1}{2} \end{bmatrix}$ and

$P_\ell(\mathbf{e}_2) = \left(\frac{\mathbf{d}\cdot\mathbf{e}_2}{\mathbf{d}\cdot\mathbf{d}}\right)\mathbf{d} = \frac{-1}{1+1}\begin{bmatrix} 1 \\ -1 \end{bmatrix} = \begin{bmatrix} -\frac{1}{2} \\ \frac{1}{2} \end{bmatrix}$.

Thus the standard matrix of this projection is $A = \begin{bmatrix} P_\ell(\mathbf{e}_1) & P_\ell(\mathbf{e}_2) \end{bmatrix} = \begin{bmatrix} \frac{1}{2} & -\frac{1}{2} \\ -\frac{1}{2} & \frac{1}{2} \end{bmatrix}$.

25. Reflection in the line $y = -x$ is given by $R\begin{bmatrix} x \\ y \end{bmatrix} = \begin{bmatrix} -y \\ -x \end{bmatrix}$.

$R(\mathbf{e}_1) = \begin{bmatrix} 0 \\ -1 \end{bmatrix}$ and $R(\mathbf{e}_2) = \begin{bmatrix} -1 \\ 0 \end{bmatrix}$.

Thus the standard matrix of this projection is $A = \begin{bmatrix} R(\mathbf{e}_1) & R(\mathbf{e}_2) \end{bmatrix} = \begin{bmatrix} 0 & -1 \\ -1 & 0 \end{bmatrix}$.

27. We want to reflect about the line $y = 2x$ which lies in the direction of vector $\mathbf{d} = \begin{bmatrix} 1 \\ 2 \end{bmatrix}$

so that $F_\ell(\mathbf{x}) = \frac{1}{1+4}\begin{bmatrix} 1-4 & 2(2) \\ 2(2) & -1+4 \end{bmatrix} = \begin{bmatrix} -\frac{3}{5} & \frac{4}{5} \\ \frac{4}{5} & \frac{3}{5} \end{bmatrix}$.

Then $F_\ell(\mathbf{e}_1) = \begin{bmatrix} -\frac{3}{5} & \frac{4}{5} \\ \frac{4}{5} & \frac{3}{5} \end{bmatrix}\begin{bmatrix} 1 \\ 0 \end{bmatrix} = \begin{bmatrix} -\frac{3}{5} \\ \frac{4}{5} \end{bmatrix}$ and $F_\ell(\mathbf{e}_2) = \begin{bmatrix} -\frac{3}{5} & \frac{4}{5} \\ \frac{4}{5} & \frac{3}{5} \end{bmatrix}\begin{bmatrix} 0 \\ 1 \end{bmatrix} = \begin{bmatrix} \frac{4}{5} \\ \frac{3}{5} \end{bmatrix}$

and the standard matrix of this transformation is $A = \begin{bmatrix} F_\ell(\mathbf{e}_1) & F_\ell(\mathbf{e}_2) \end{bmatrix} = \begin{bmatrix} -\frac{3}{5} & \frac{4}{5} \\ \frac{4}{5} & \frac{3}{5} \end{bmatrix}$.

3.6 Introduction to Linear Transformations

29. We have the two linear transformations defined by $T\begin{bmatrix} x_1 \\ x_2 \end{bmatrix} = \begin{bmatrix} x_1 \\ 2x_1 - x_2 \\ 3x_1 + 4x_2 \end{bmatrix}$ and

$S\begin{bmatrix} y_1 \\ y_2 \\ y_3 \end{bmatrix} = \begin{bmatrix} 2y_1 + y_3 \\ 3y_2 - y_3 \\ y_1 - y_2 \\ y_1 + y_2 + y_3 \end{bmatrix}$. To find $(S \circ T)\begin{bmatrix} x_1 \\ x_2 \end{bmatrix}$ directly, we calculate

$$(S \circ T)\begin{bmatrix} x_1 \\ x_2 \end{bmatrix} = S\begin{bmatrix} x_1 \\ 2x_1 - x_2 \\ 3x_1 + 4x_2 \end{bmatrix} = \begin{bmatrix} 2(x_1) + (3x_1 + 4x_2) \\ 3(2x_1 - x_2) - (3x_1 + 4x_2) \\ x_1 - (2x_1 - x_2) \\ x_1 + (2x_1 - x_2) + (3x_1 + 4x_2) \end{bmatrix}$$

$$= \begin{bmatrix} 5x_1 + 4x_2 \\ 3x_1 - 7x_2 \\ -x_1 + x_2 \\ 6x_1 + 3x_2 \end{bmatrix} = \begin{bmatrix} 5 & 4 \\ 3 & -7 \\ -1 & 1 \\ 6 & 3 \end{bmatrix} \begin{bmatrix} x_1 \\ x_2 \end{bmatrix}.$$

Finally, we identify the standard matrix of $(S \circ T)$ as $A = \begin{bmatrix} 5 & 4 \\ 3 & -7 \\ -1 & 1 \\ 6 & 3 \end{bmatrix}$.

31. (a) $(S \circ T)\begin{bmatrix} x_1 \\ x_2 \end{bmatrix} = S\begin{bmatrix} x_1 + 2x_2 \\ -3x_1 + x_2 \end{bmatrix} = \begin{bmatrix} (x_1 + 2x_2) + 3(-3x_1 + x_2) \\ (x_1 + 2x_2) - (-3x_1 + x_2) \end{bmatrix}$

$= \begin{bmatrix} -8x_1 + 5x_2 \\ 4x_1 + x_2 \end{bmatrix} = \begin{bmatrix} -8 & 5 \\ 4 & 1 \end{bmatrix} \begin{bmatrix} x_1 \\ x_2 \end{bmatrix}$

So by direct substitution we identify the matrix of $S \circ T$ as $\begin{bmatrix} -8 & 5 \\ 4 & 1 \end{bmatrix}$.

(b) We see that the standard matrices are $[S] = \begin{bmatrix} 1 & 3 \\ 1 & -1 \end{bmatrix}$ and $[T] = \begin{bmatrix} 1 & 2 \\ -3 & 1 \end{bmatrix}$.

So Theorem 3.32 gives $[S \circ T] = [S][T] = \begin{bmatrix} 1 & 3 \\ 1 & -1 \end{bmatrix} \begin{bmatrix} 1 & 2 \\ -3 & 1 \end{bmatrix} = \begin{bmatrix} -8 & 5 \\ 4 & 1 \end{bmatrix}$

which is the same result as obtained by direct substitution.

33. (a) $(S \circ T) \begin{bmatrix} x_1 \\ x_2 \\ x_3 \end{bmatrix} = S \begin{bmatrix} x_1 + x_2 - x_3 \\ 2x_1 - x_2 + x_3 \end{bmatrix} = \begin{bmatrix} 4(x_1 + x_2 - x_3) - 2(2x_1 - x_2 + x_3) \\ -(x_1 + x_2 - x_3) + (2x_1 - x_2 + x_3) \end{bmatrix}$

$= \begin{bmatrix} 6x_2 - 6x_3 \\ x_1 - 2x_2 + 2x_3 \end{bmatrix} = \begin{bmatrix} 0 & 6 & -6 \\ 1 & -2 & 2 \end{bmatrix} \begin{bmatrix} x_1 \\ x_2 \\ x_3 \end{bmatrix}.$

So by direct substitution we identify the matrix of $S \circ T$ as $\begin{bmatrix} 0 & 6 & -6 \\ 1 & -2 & 2 \end{bmatrix}.$

(b) We see that the standard matrices are $[S] = \begin{bmatrix} 4 & -2 \\ -1 & 1 \end{bmatrix}$ and $[T] = \begin{bmatrix} 1 & 1 & -1 \\ 2 & -1 & 1 \end{bmatrix}$. So Theorem 3.32 gives

$[S \circ T] = [S][T] = \begin{bmatrix} 4 & -2 \\ -1 & 1 \end{bmatrix} \begin{bmatrix} 1 & 1 & -1 \\ 2 & -1 & 1 \end{bmatrix} = \begin{bmatrix} 0 & 6 & -6 \\ 1 & -2 & 2 \end{bmatrix}$

which is the same result as obtained by direct substitution.

35. (a) $(S \circ T) \begin{bmatrix} x_1 \\ x_2 \\ x_3 \end{bmatrix} = S \begin{bmatrix} x_1 + x_2 \\ x_2 + x_3 \\ x_1 + x_3 \end{bmatrix} = \begin{bmatrix} x_1 + x_2 - (x_2 + x_3) \\ x_2 + x_3 - (x_1 + x_3) \\ -(x_1 + x_2) + x_1 + x_3 \end{bmatrix}$

$= \begin{bmatrix} x_1 - x_3 \\ -x_1 + x_2 \\ -x_2 + x_3 \end{bmatrix} = \begin{bmatrix} 1 & 0 & -1 \\ -1 & 1 & 0 \\ 0 & -1 & 1 \end{bmatrix} \begin{bmatrix} x_1 \\ x_2 \\ x_3 \end{bmatrix}.$

So by direct substitution we identify the matrix of $S \circ T$ as $\begin{bmatrix} 1 & 0 & -1 \\ -1 & 1 & 0 \\ 0 & -1 & 1 \end{bmatrix}.$

(b) We see that the standard matrices are $[S] = \begin{bmatrix} 1 & -1 & 0 \\ 0 & 1 & -1 \\ -1 & 0 & 1 \end{bmatrix}$ and $[T] = \begin{bmatrix} 1 & 1 & 0 \\ 0 & 1 & 1 \\ 1 & 0 & 1 \end{bmatrix}.$

So Theorem 3.32 gives $[S \circ T] = [S][T] = \begin{bmatrix} 1 & -1 & 0 \\ 0 & 1 & -1 \\ -1 & 0 & 1 \end{bmatrix} \begin{bmatrix} 1 & 1 & 0 \\ 0 & 1 & 1 \\ 1 & 0 & 1 \end{bmatrix} = \begin{bmatrix} 1 & 0 & -1 \\ -1 & 1 & 0 \\ 0 & -1 & 1 \end{bmatrix}$

which is the same result as we obtained by direct substitution.

37. A reflection in the y-axis is given by $T = \begin{bmatrix} -1 & 0 \\ 0 & 1 \end{bmatrix}$,

while a clockwise rotation through $30°$ is given by $S = \begin{bmatrix} \cos(-30°) & -\sin(-30°) \\ \sin(-30°) & \cos(-30°) \end{bmatrix}.$

Then, by Theorem 3.32, the composite transformation is given by

$[S \circ T] = [S][T] = \begin{bmatrix} \cos 30° & \sin 30° \\ -\sin 30° & \cos 30° \end{bmatrix} \begin{bmatrix} -1 & 0 \\ 0 & 1 \end{bmatrix} = \begin{bmatrix} -\cos 30° & \sin 30° \\ \sin 30° & \cos 30° \end{bmatrix} = \begin{bmatrix} -\frac{\sqrt{3}}{2} & \frac{1}{2} \\ \frac{1}{2} & \frac{\sqrt{3}}{2} \end{bmatrix}.$

3.6 Introduction to Linear Transformations

39. A reflection in the line $y = x$ with $\mathbf{d} = \begin{bmatrix} 1 \\ 1 \end{bmatrix}$ is given by $T = \frac{1}{2}\begin{bmatrix} 0 & 2 \\ 2 & 0 \end{bmatrix} = \begin{bmatrix} 0 & 1 \\ 1 & 0 \end{bmatrix}$.

A counterclockwise rotation through $30°$ is given by $S = \begin{bmatrix} \cos 30° & -\sin 30° \\ \sin 30° & \cos 30° \end{bmatrix} = \begin{bmatrix} \frac{\sqrt{3}}{2} & -\frac{1}{2} \\ \frac{1}{2} & \frac{\sqrt{3}}{2} \end{bmatrix}$.

A reflection in the line $y = -x$ with $\mathbf{d} = \begin{bmatrix} 1 \\ -1 \end{bmatrix}$ is given by $U = \frac{1}{2}\begin{bmatrix} 0 & -2 \\ -2 & 0 \end{bmatrix} = \begin{bmatrix} 0 & -1 \\ -1 & 0 \end{bmatrix}$.

Then, by Theorem 3.32, the composite transformation is given by

$$[U \circ S \circ T] = [U][S][T] = \begin{bmatrix} 0 & -1 \\ -1 & 0 \end{bmatrix}\begin{bmatrix} \frac{\sqrt{3}}{2} & -\frac{1}{2} \\ \frac{1}{2} & \frac{\sqrt{3}}{2} \end{bmatrix}\begin{bmatrix} 0 & 1 \\ 1 & 0 \end{bmatrix} = \begin{bmatrix} -\frac{\sqrt{3}}{2} & -\frac{1}{2} \\ \frac{1}{2} & -\frac{\sqrt{3}}{2} \end{bmatrix}.$$

41. Let θ be the angle between:

Line ℓ with direction vector $\mathbf{l} = \begin{bmatrix} \cos \alpha \\ \sin \alpha \end{bmatrix}$ and line m with the direction vector $\mathbf{m} = \begin{bmatrix} \cos \beta \\ \sin \beta \end{bmatrix}$

Now, we need to consider two cases: 1) $\alpha > \beta$ and 2) $\alpha < \beta$.

$\alpha > \beta$: If $\alpha > \beta$ then $\theta = \alpha - \beta$ and

$$\begin{aligned} F_m \circ F_\ell &= \begin{bmatrix} \cos 2\beta & \sin 2\beta \\ \sin 2\beta & -\cos 2\beta \end{bmatrix}\begin{bmatrix} \cos 2\alpha & \sin 2\alpha \\ \sin 2\alpha & -\cos 2\alpha \end{bmatrix} \\ &= \begin{bmatrix} \cos 2\beta \cos 2\alpha + \sin 2\beta \sin 2\alpha & \cos 2\beta \sin 2\alpha - \sin 2\beta \cos 2\alpha \\ \sin 2\beta \cos 2\alpha - \cos 2\beta \sin 2\alpha & \cos 2\beta \cos 2\alpha + \sin 2\beta \sin 2\alpha \end{bmatrix} \\ &= \begin{bmatrix} \cos(-2\beta + 2\alpha) & \sin(-2\beta + 2\alpha) \\ -\sin(-2\beta + 2\alpha) & \cos(-2\beta + 2\alpha) \end{bmatrix} = \begin{bmatrix} \cos 2\theta & \sin 2\theta \\ -\sin 2\theta & \cos 2\theta \end{bmatrix} = R_{-2\theta}. \end{aligned}$$

$\alpha < \beta$: If $\alpha < \beta$ then $\theta = \beta - \alpha$.
Making a similar calculation to the above, we get $F_m \circ F_\ell = R_{+2\theta}$.

43. Let ℓ, m, and n are three lines through the origin with angles from the x-axis given by θ, α, and β respectively. Then

$$\begin{aligned} F_n \circ F_m \circ F_\ell &= \begin{bmatrix} \cos 2\theta & \sin 2\theta \\ \sin 2\theta & -\cos 2\theta \end{bmatrix}\begin{bmatrix} \cos 2\alpha & \sin 2\alpha \\ \sin 2\alpha & -\cos 2\alpha \end{bmatrix}\begin{bmatrix} \cos 2\beta & \sin 2\beta \\ \sin 2\beta & -\cos 2\beta \end{bmatrix} \\ &= \begin{bmatrix} \cos(2\theta - 2\alpha + 2\beta) & \sin(2\theta - 2\alpha + 2\beta) \\ \sin(2\theta - 2\alpha + 2\beta) & -\cos(2\theta - 2\alpha + 2\beta) \end{bmatrix} \\ &= \begin{bmatrix} \cos(2(\theta - \alpha + \beta)) & \sin(2(\theta - \alpha + \beta)) \\ \sin(2(\theta - \alpha + \beta)) & -\cos(2(\theta - \alpha + \beta)) \end{bmatrix}. \end{aligned}$$

So the reflection about the three lines ℓ, m, and n is the same as one reflection about the line which makes an angle with the x-axis of $\theta - \alpha + \beta$.

45. By Exercise 44, we know that T maps ℓ, $\mathbf{p} + t\mathbf{d}$, and m, $\mathbf{q} + s\mathbf{d}$, onto ℓ', $T(\mathbf{p}+t\mathbf{d}) = T(\mathbf{p}) + tT(\mathbf{d})$ and m', $T(\mathbf{q}+s\mathbf{d}) = T(\mathbf{q}) + sT(\mathbf{d})$.

If $T(\mathbf{d}) \neq \mathbf{0}$ and $T(\mathbf{p}) \neq T(\mathbf{q})$, then ℓ' and m' are distinct parallel lines.
If $T(\mathbf{d}) \neq \mathbf{0}$ and $T(\mathbf{p}) = T(\mathbf{q})$, then ℓ' and m' are the same line.
If $T(\mathbf{d}) = \mathbf{0}$ and $T(\mathbf{p}) \neq T(\mathbf{q})$, then ℓ' and m' are distinct points $T(\mathbf{p})$ and $T(\mathbf{q})$.
If $T(\mathbf{d}) = \mathbf{0}$ and $T(\mathbf{p}) = T(\mathbf{q})$, then ℓ' and m' are a single point $T(\mathbf{p}) = T(\mathbf{q})$.

47. $D\begin{bmatrix}-1\\1\end{bmatrix} = \begin{bmatrix}2 & 0\\0 & 3\end{bmatrix}\begin{bmatrix}-1\\1\end{bmatrix} = \begin{bmatrix}-2\\3\end{bmatrix}$,

$D\begin{bmatrix}1\\1\end{bmatrix} = \begin{bmatrix}2\\3\end{bmatrix}$,

$D\begin{bmatrix}1\\-1\end{bmatrix} = \begin{bmatrix}2\\-3\end{bmatrix}$,

$D\begin{bmatrix}-1\\-1\end{bmatrix} = \begin{bmatrix}-2\\-3\end{bmatrix}$.

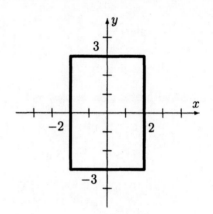

49. $P\begin{bmatrix}-1\\1\end{bmatrix} = \begin{bmatrix}\frac{1}{5} & \frac{2}{5}\\\frac{2}{5} & \frac{4}{5}\end{bmatrix}\begin{bmatrix}-1\\1\end{bmatrix} = \begin{bmatrix}\frac{1}{5}\\\frac{2}{5}\end{bmatrix}$,

$P\begin{bmatrix}1\\1\end{bmatrix} = \begin{bmatrix}\frac{3}{5}\\\frac{6}{5}\end{bmatrix}$,

$P\begin{bmatrix}1\\-1\end{bmatrix} = \begin{bmatrix}-\frac{1}{5}\\-\frac{2}{5}\end{bmatrix}$, $P\begin{bmatrix}-1\\-1\end{bmatrix} = \begin{bmatrix}-\frac{3}{5}\\-\frac{6}{5}\end{bmatrix}$.

51. $T\begin{bmatrix}-1\\1\end{bmatrix} = \begin{bmatrix}\frac{1}{2}\sqrt{3}+\frac{1}{2}\\-\frac{1}{2}+\frac{1}{2}\sqrt{3}\end{bmatrix}$,

$T\begin{bmatrix}1\\1\end{bmatrix} = \begin{bmatrix}\frac{1}{2}-\frac{1}{2}\sqrt{3}\\\frac{1}{2}\sqrt{3}+\frac{1}{2}\end{bmatrix}$,

$T\begin{bmatrix}1\\-1\end{bmatrix} = \begin{bmatrix}-\frac{1}{2}-\frac{1}{2}\sqrt{3}\\\frac{1}{2}-\frac{1}{2}\sqrt{3}\end{bmatrix}$,

$T\begin{bmatrix}-1\\-1\end{bmatrix} = \begin{bmatrix}-\frac{1}{2}+\frac{1}{2}\sqrt{3}\\-\frac{1}{2}-\frac{1}{2}\sqrt{3}\end{bmatrix}$.

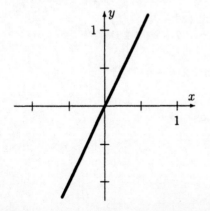

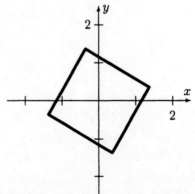

3.6 Introduction to Linear Transformations

53. Since this is an if and only if statement, there are two statements to prove.

if: If T is linear, then $T(c_1\mathbf{v}_1 + c_2\mathbf{v}_2) = c_1T(\mathbf{v}_1) + c_2T(\mathbf{v}_2)$.

$$T(c_1\mathbf{v}_1 + c_2\mathbf{v}_2) \stackrel{\substack{T(\mathbf{u}+\mathbf{v})\\=T(\mathbf{u})+T(\mathbf{v})}}{=} T(c_1\mathbf{v}_1) + T(c_2\mathbf{v}_2) \stackrel{\substack{T(c\mathbf{v})\\=cT(\mathbf{v})}}{=} c_1T(\mathbf{v}_1) + c_2T(\mathbf{v}_2)$$

only if: If $T(c_1\mathbf{v}_1 + c_2\mathbf{v}_2) = c_1T(\mathbf{v}_1) + c_2T(\mathbf{v}_2)$, then T is linear.

$$c_1 = c_2 = 1 \Rightarrow T(\mathbf{v}_1 + \mathbf{v}_2) = T(\mathbf{v}_1) + T(\mathbf{v}_2)$$
$$c_2 = 0 \Rightarrow T(c_1\mathbf{v}_1) = c_1T(\mathbf{v}_1)$$

Since $T(\mathbf{u} + \mathbf{v}) = T(\mathbf{u}) + T(\mathbf{v})$ and $T(c\mathbf{v}) = cT(\mathbf{v})$, T is linear.

55. The Fundamental Theorem implies every invertible matrix is a product of elementary matrices. The matrices in Exercise 19 represent the 3 types of elementary 2×2 matrices.
So, T_A must be a composition of the three types of transformation represented in Exercise 19.

3.7 Applications

Q: What are our main goals for Section 3.7?

A: To see how matrices apply to Markov chains, population growth, graphs, and codes. Since these applications are fully explained in the text, we simply present solutions here.

Key Definitions and Concepts

adjacency matrix	p.236	3.7	$A(G)$ where $a_{ij} = 1$ if edge, 0 otherwise
circuit	p.237	3.7	a closed *path* (ends and begins at the same *vertex*)
code	p.240	3.7	See discussion prior to Example 3.696 and Section 1.4
generator matrix	p.240	3.7	G: generates a *code*. See definition before Theorem 3.34
graph	p.236	3.7	finite set of *points* and *edges*
Hamming code	p.242	3.7	See Example 3.70 for construction
length	p.237	3.7	number of edges a *path* contains
Leslie matrix	p.234	3.7	Population matrix L described after Example 3.66
Markov chain	p.228	3.7	See description given prior to Example 3.64
parity matrix	p.240	3.7	P: checks a *code*. See definition before Theorem 3.34
path	p.237	3.7	set of *edges* that connects one *vertex* to another
probability vector	p.229	3.7	vector with nonnegative components that add to 1
simple path	p.237	3.7	a *path* that does not contain the same *edge* twice
state vector	p.229	3.7	$\mathbf{x}_k$: $\mathbf{x}_k = P^k \mathbf{x}_0$ in a Markov Chain
steady state vector	p.231	3.7	$\mathbf{x}$: $\mathbf{x} = P\mathbf{x}$ in a Markov Chain
transition matrix	p.229	3.7	P: $\mathbf{x}_k = P^k \mathbf{x}_0$ in a Markov Chain

Theorems

Thm 3.34	p.241	3.7	Gives conditions on G and P for a given *code*

Solutions to odd-numbered exercises from Section 3.7

1. $\mathbf{x}_1 = P\mathbf{x}_0 = \begin{bmatrix} 0.5 & 0.3 \\ 0.5 & 0.7 \end{bmatrix} \begin{bmatrix} 0.5 \\ 0.5 \end{bmatrix} = \begin{bmatrix} 0.4 \\ 0.6 \end{bmatrix}$, $\mathbf{x}_2 = P\mathbf{x}_1 = \begin{bmatrix} 0.5 & 0.3 \\ 0.5 & 0.7 \end{bmatrix} \begin{bmatrix} 0.4 \\ 0.6 \end{bmatrix} = \begin{bmatrix} 0.38 \\ 0.62 \end{bmatrix}$.

3. Since there are 2 steps involved, we calculate: $P^2 = \begin{bmatrix} 0.5 & 0.3 \\ 0.5 & 0.7 \end{bmatrix}^2 = \begin{bmatrix} 0.40 & 0.36 \\ 0.60 & 0.64 \end{bmatrix}$.

    So, the probability of the state 2 population being in state 2 is $[P^2]_{22} = 0.64$.

5. $\mathbf{x}_1 = P\mathbf{x}_0 = \begin{bmatrix} \frac{1}{2} & \frac{1}{3} & \frac{1}{3} \\ 0 & \frac{1}{3} & \frac{2}{3} \\ \frac{1}{2} & \frac{1}{3} & 0 \end{bmatrix} \begin{bmatrix} 120 \\ 180 \\ 90 \end{bmatrix} = \begin{bmatrix} 150 \\ 120 \\ 120 \end{bmatrix}$, $\mathbf{x}_2 = P\mathbf{x}_1 = \begin{bmatrix} \frac{1}{2} & \frac{1}{3} & \frac{1}{3} \\ 0 & \frac{1}{3} & \frac{2}{3} \\ \frac{1}{2} & \frac{1}{3} & 0 \end{bmatrix} \begin{bmatrix} 150 \\ 120 \\ 120 \end{bmatrix} = \begin{bmatrix} 155 \\ 120 \\ 115 \end{bmatrix}$.

7. Since there are 2 steps involved, we calculate: $P^2 = \begin{bmatrix} \frac{1}{2} & \frac{1}{3} & \frac{1}{3} \\ 0 & \frac{1}{3} & \frac{2}{3} \\ \frac{1}{2} & \frac{1}{3} & 0 \end{bmatrix}^2 = \begin{bmatrix} \frac{5}{12} & \frac{7}{18} & \frac{7}{18} \\ \frac{1}{3} & \frac{1}{3} & \frac{2}{9} \\ \frac{1}{4} & \frac{5}{18} & \frac{7}{18} \end{bmatrix}$.

    So, the probability of the state 3 population being in state 2 is $[P^2]_{32} = \frac{5}{18}$.

9. Let A be a dry day and B a wet one. We are given the following probabilities:
    Tomorrow will be wet is 0.662 if today is wet and 0.250 if today is dry.
    Tomorrow will be dry is 0.750 if today is dry and 0.338 if today is wet.

    (a) The transition matrix for this Markov chain is $P = \begin{bmatrix} 0.750 & 0.338 \\ 0.250 & 0.662 \end{bmatrix}$ with $\mathbf{x}_i = \begin{bmatrix} x_1 \\ x_2 \end{bmatrix}$, where x_1, x_2 denote the probability that tomorrow will be dry or wet respectively.

    (b) We are told that Monday is a dry day; we can describe this state as $\mathbf{x}_0 = \begin{bmatrix} 1 \\ 0 \end{bmatrix}$.

    The probability that Wednesday will be wet is then given by
    $$\mathbf{x}_2 = P(P\mathbf{x}_0) = \begin{bmatrix} 0.750 & 0.338 \\ 0.250 & 0.662 \end{bmatrix} \begin{bmatrix} 0.750 & 0.338 \\ 0.250 & 0.662 \end{bmatrix} \begin{bmatrix} 1 \\ 0 \end{bmatrix} = \begin{bmatrix} 0.647 \\ 0.353 \end{bmatrix}.$$
    So, given that Monday is dry, the probability that Wednesday will be wet is 0.353.

    (c) To find the distribution of wet and dry days in the long run, we need $P\mathbf{x} = \mathbf{x}$.
    So, we find $\begin{bmatrix} 1-0.750 & -0.338 & | & 0 \\ -0.250 & 1-0.662 & | & 0 \end{bmatrix} \longrightarrow \begin{bmatrix} 1 & -1.352 & | & 0 \\ 0 & 0 & | & 0 \end{bmatrix}$.
    We parametrize the solution by letting $x_2 = t$, so $x_1 = 1.352t$.
    However, a day will either be wet or dry so we impose the constraint:
    $1 = x_1 + x_2 = 1.352t + t \Rightarrow t = \frac{1}{2.352} \approx 0.425$.
    Thus, in the long run roughly 57.5% of days will be dry and 42.5% will be wet.

3.7 Applications

11. Let states A, B, and C denote a good, fair, or poor pine nut crop.
If one year's crop is good, the following crop will be good, fair, or poor with probabilities 0.08, 0.07, and 0.85, respectively.
If one year's crop is fair then the following year's crop will be good, fair, or poor with probabilities 0.09, 0.11, and 0.80, respectively.
If a crop is good then the following year's crop will be good, fair, or poor with probabilities 0.11, 0.05, and 0.84, respectively.

(a) The transition matrix for this Markov chain is $P = \begin{bmatrix} 0.08 & 0.09 & 0.11 \\ 0.07 & 0.11 & 0.05 \\ 0.85 & 0.80 & 0.84 \end{bmatrix}$.

(b) We are told that the 1940 crop is good, so $\mathbf{x}_0 = \begin{bmatrix} 1 \\ 0 \\ 0 \end{bmatrix}$.

The probabilities of a good crop in the following five years are as follows:

$$
\begin{aligned}
1941: & \quad P_{11} = 0.0800 \\
1942: & \quad (P)^2_{11} = 0.1062 \\
1943: & \quad (P)^3_{11} = 0.1057 \\
1944: & \quad (P)^4_{11} = 0.1057 \\
1945: & \quad (P)^5_{11} = 0.1057
\end{aligned}
$$

(c) To find the proportion of the crops in the long run, we need $P\mathbf{x} = \mathbf{x}$.
So, we have $[\, I - P \mid \mathbf{0}\,]$:

$$
\begin{bmatrix} 1-0.08 & -0.09 & -0.11 & | & 0 \\ -0.07 & 1-0.11 & -0.05 & | & 0 \\ -0.85 & -0.80 & 1-0.84 & | & 0 \end{bmatrix} \longrightarrow \begin{bmatrix} 1 & 0 & -\frac{1024}{8125} & | & 0 \\ 0 & 1 & -\frac{537}{8125} & | & 0 \\ 0 & 0 & 0 & | & 0 \end{bmatrix}
$$

So, x_3 is free and the parametric solution is $x_1 = 0.1260t$, $x_2 = 0.0661t$, $x_3 = t$.
But x_i denotes a probability, so the state components must be normalized by $1 = x_1 + x_2 + x_3 = 1.1921t \Rightarrow t = 0.839$.
Thus, after a long time 10.6% of the crops are good, 5.5% are fair, and 83.9% are poor.

13. Let $P = [\, \mathbf{p}_1 \; \cdots \; \mathbf{p}_n \,]$ be a stochastic matrix (so the elements of each $\mathbf{p}_i$ sum to 1), and let $\mathbf{j}$ be a row vector consisting entirely of 1s.
Then $\mathbf{j}P = [\, \mathbf{j}\cdot\mathbf{p}_1 \; \cdots \; \mathbf{j}\cdot\mathbf{p}_n\,] = [\, \sum(\mathbf{p}_1)_i \; \cdots \; \sum(\mathbf{p}_n)_i\,] = [\, 1 \; \cdots \; 1\,] = \mathbf{j}$.
Conversely, suppose that $\mathbf{j}P = \mathbf{j}$.
Then we have $\mathbf{j}\cdot\mathbf{p}_i = 1$ for each i, so $\sum_j (\mathbf{p}_i)_j = 1$ for all i, showing P is stochastic.

15. From Exercise 9 we have the following:
If Monday is a dry day, the expected number of days until a wet day
can be determined by first deleting the second row and column of the transition matrix
$P = \begin{bmatrix} 0.750 & 0.338 \\ 0.250 & 0.662 \end{bmatrix}$ giving $Q = [0.750]$.

So, the expected number of days until the next wet day is given by:
$(I - Q)^{-1} = (1 - 0.750)^{-1} = 4.0.$

Thus, we expect four days until the next wet day.

17. From Exercise 11 we have the following:
If the pine nut crop is fair one year the expected number of years until
a good crop occurs is given by deleting the first row and column of the transition matrix
$P = \begin{bmatrix} 0.08 & 0.09 & 0.11 \\ 0.07 & 0.11 & 0.05 \\ 0.85 & 0.80 & 0.84 \end{bmatrix}$ giving $Q = \begin{bmatrix} 0.11 & 0.05 \\ 0.80 & 0.84 \end{bmatrix}.$

Summing on the first (fair) column of $(I - Q)^{-1} = \begin{bmatrix} 1.562 & 0.488 \\ 7.812 & 8.691 \end{bmatrix} = 9.374.$

So, it will be roughly 9 or 10 years before a good crop occurs.

19. A population with three age classes has a Leslie matrix $L = \begin{bmatrix} 1 & 1 & 3 \\ 0.7 & 0 & 0 \\ 0 & 0.5 & 0 \end{bmatrix}.$

The initial population vector is $\mathbf{x}_0 = \begin{bmatrix} 100 \\ 100 \\ 100 \end{bmatrix}$, so

$\mathbf{x}_1 = L\mathbf{x}_0 = \begin{bmatrix} 1 & 1 & 3 \\ 0.7 & 0 & 0 \\ 0 & 0.5 & 0 \end{bmatrix} \begin{bmatrix} 100 \\ 100 \\ 100 \end{bmatrix} = \begin{bmatrix} 500 \\ 70.0 \\ 50.0 \end{bmatrix}$ and

$\mathbf{x}_2 = L\mathbf{x}_1 = \begin{bmatrix} 1 & 1 & 3 \\ 0.7 & 0 & 0 \\ 0 & 0.5 & 0 \end{bmatrix} \begin{bmatrix} 500 \\ 70.0 \\ 50.0 \end{bmatrix} = \begin{bmatrix} 720.0 \\ 350.0 \\ 35.0 \end{bmatrix}.$

Alternatively,

$\mathbf{x}_2 = L\mathbf{x}_1 = LL\mathbf{x}_1 = \begin{bmatrix} 1 & 1 & 3 \\ 0.7 & 0 & 0 \\ 0 & 0.5 & 0 \end{bmatrix} \begin{bmatrix} 1 & 1 & 3 \\ 0.7 & 0 & 0 \\ 0 & 0.5 & 0 \end{bmatrix} \begin{bmatrix} 100 \\ 100 \\ 100 \end{bmatrix} = \begin{bmatrix} 720.0 \\ 350.0 \\ 35.0 \end{bmatrix}$ and

$\mathbf{x}_3 = L\mathbf{x}_2 = \begin{bmatrix} 1 & 1 & 3 \\ 0.7 & 0 & 0 \\ 0 & 0.5 & 0 \end{bmatrix} \begin{bmatrix} 720.0 \\ 350.0 \\ 35.0 \end{bmatrix} = \begin{bmatrix} 1175.0 \\ 504.0 \\ 175.0 \end{bmatrix}.$

3.7 Applications

21. A certain species with two year-long age classes is described by two possible Leslie matrices

$$L_1 = \begin{bmatrix} 0 & 5 \\ 0.8 & 0 \end{bmatrix} \text{ and } L_2 = \begin{bmatrix} 4 & 1 \\ 0.8 & 0 \end{bmatrix}.$$

(a) Starting with $\mathbf{x}_0 = \begin{bmatrix} 10 \\ 10 \end{bmatrix}$ the following table gives $\mathbf{x}_1, \ldots, \mathbf{x}_{10}$ for both cases.

	$\mathbf{x}_0$	$\mathbf{x}_1$	$\mathbf{x}_2$	$\mathbf{x}_3$	$\mathbf{x}_4$	$\mathbf{x}_5$
L_1	$\begin{bmatrix} 10 \\ 10 \end{bmatrix}$	$\begin{bmatrix} 50 \\ 8.0 \end{bmatrix}$	$\begin{bmatrix} 40.0 \\ 40.0 \end{bmatrix}$	$\begin{bmatrix} 200.0 \\ 32.0 \end{bmatrix}$	$\begin{bmatrix} 160.0 \\ 160.0 \end{bmatrix}$	$\begin{bmatrix} 800.0 \\ 128.0 \end{bmatrix}$
L_2	$\begin{bmatrix} 10 \\ 10 \end{bmatrix}$	$\begin{bmatrix} 50 \\ 8.0 \end{bmatrix}$	$\begin{bmatrix} 208.0 \\ 40.0 \end{bmatrix}$	$\begin{bmatrix} 872.0 \\ 166.4 \end{bmatrix}$	$\begin{bmatrix} 3654.4 \\ 697.6 \end{bmatrix}$	$\begin{bmatrix} 15,315.2 \\ 2923.52 \end{bmatrix}$

	$\mathbf{x}_6$	$\mathbf{x}_7$	$\mathbf{x}_8$	$\mathbf{x}_9$	$\mathbf{x}_{10}$
L_1	$\begin{bmatrix} 640.0 \\ 640.0 \end{bmatrix}$	$\begin{bmatrix} 3200.0 \\ 512.0 \end{bmatrix}$	$\begin{bmatrix} 2560.0 \\ 2560.0 \end{bmatrix}$	$\begin{bmatrix} 12,800.0 \\ 2048.0 \end{bmatrix}$	$\begin{bmatrix} 10,240.0 \\ 10,240.0 \end{bmatrix}$
L_2	$\begin{bmatrix} 64,184.32 \\ 12,252.16 \end{bmatrix}$	$\begin{bmatrix} 268,989.44 \\ 51,347.456 \end{bmatrix}$	$\begin{bmatrix} 1.127 \times 10^6 \\ 215,191.552 \end{bmatrix}$	$\begin{bmatrix} 4.724 \times 10^6 \\ 901,844.17 \end{bmatrix}$	$\begin{bmatrix} 1.979 \times 10^7 \\ 3.779 \times 10^6 \end{bmatrix}$

(b)

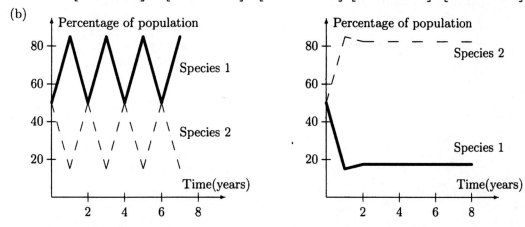

The graphs suggest that the first Leslie matrix does not have a steady state, while the second Leslie matrix quickly approaches its steady-state value.

23. Let the Leslie matrix for the VW Beetle be $L = \begin{bmatrix} 0 & 0 & 20 \\ s & 0 & 0 \\ 0 & 0.5 & 0 \end{bmatrix}$.

To understand the effect of varying the survival probability s of the young beetles, we consider the evolution of the initial state vector over several years.

If the initial state is $\mathbf{x}_0 = \begin{bmatrix} 10 \\ 10 \\ 10 \end{bmatrix}$, then

$$\mathbf{x}_1 = L\mathbf{x}_0 = \begin{bmatrix} 0 & 0 & 20 \\ s & 0 & 0 \\ 0 & 0.5 & 0 \end{bmatrix} \begin{bmatrix} 10 \\ 10 \\ 10 \end{bmatrix} = \begin{bmatrix} 200 \\ 10s \\ 5.0 \end{bmatrix}, \mathbf{x}_2 = \begin{bmatrix} 0 & 0 & 20 \\ s & 0 & 0 \\ 0 & 0.5 & 0 \end{bmatrix} \begin{bmatrix} 200 \\ 10s \\ 5.0 \end{bmatrix} = \begin{bmatrix} 100.0 \\ 200s \\ 5.0s \end{bmatrix},$$

$$\mathbf{x}_3 = \begin{bmatrix} 0 & 0 & 20 \\ s & 0 & 0 \\ 0 & 0.5 & 0 \end{bmatrix} \begin{bmatrix} 100.0 \\ 200s \\ 5.0s \end{bmatrix} = \begin{bmatrix} 100.0s \\ 100.0s \\ 100.0s \end{bmatrix}.$$

At this point it is apparent that after three years of evolution, the relative abundance of each class is equal to the initial distribution.

As well, we see the following patterns:

If $s < 0.1$, then the overall population declines after three years.
If $s = 0.1$, the overall population stays the same after three years.
If $s > 0.1$, the overall population increases after three years.

25. $A = \begin{bmatrix} 0 & 1 & 0 & 1 \\ 1 & 0 & 1 & 0 \\ 0 & 1 & 0 & 1 \\ 1 & 0 & 1 & 0 \end{bmatrix}$.

27. $A = \begin{bmatrix} 0 & 1 & 1 & 1 & 1 \\ 1 & 0 & 1 & 0 & 0 \\ 1 & 1 & 0 & 1 & 0 \\ 1 & 0 & 1 & 0 & 1 \\ 1 & 0 & 0 & 1 & 0 \end{bmatrix}$.

29.
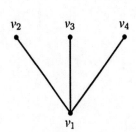

31.
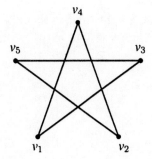

3.7   Applications

33. $A = \begin{bmatrix} 0 & 1 & 1 & 0 \\ 0 & 0 & 0 & 0 \\ 0 & 1 & 0 & 1 \\ 1 & 0 & 0 & 0 \end{bmatrix}.$

35. $A = \begin{bmatrix} 0 & 1 & 0 & 1 & 0 \\ 1 & 0 & 0 & 1 & 0 \\ 1 & 1 & 0 & 0 & 0 \\ 1 & 0 & 0 & 0 & 1 \\ 1 & 0 & 0 & 0 & 0 \end{bmatrix}.$

37.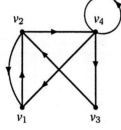

39.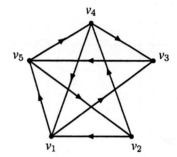

41. In Exercise 30, $A = \begin{bmatrix} 0 & 1 & 0 & 1 \\ 1 & 1 & 1 & 1 \\ 0 & 1 & 0 & 1 \\ 1 & 1 & 1 & 0 \end{bmatrix} \Rightarrow A^2 = \begin{bmatrix} 2 & 2 & 2 & 1 \\ 2 & 4 & 2 & 3 \\ 2 & 2 & 2 & 1 \\ 1 & 3 & 1 & 3 \end{bmatrix}.$

The number of paths of length 2 between v_1 and v_2 is given by the element $[A^2]_{12}$. So, we see that there are two paths of length 2 between vertices v_1 and v_2.

43. In Exercise 30, $A = \begin{bmatrix} 0 & 1 & 0 & 1 \\ 1 & 1 & 1 & 1 \\ 0 & 1 & 0 & 1 \\ 1 & 1 & 1 & 0 \end{bmatrix} \Rightarrow A^3 = \begin{bmatrix} 3 & 7 & 3 & 6 \\ 7 & 11 & 7 & 8 \\ 3 & 7 & 3 & 6 \\ 6 & 8 & 6 & 5 \end{bmatrix}.$

The number of paths of length 3 between v_1 and v_3 is given by the element $[A^3]_{13}$. So, we see that there are three paths of length 3 between vertices v_1 and v_3.

45. In Exercise 37, $A = \begin{bmatrix} 0 & 1 & 0 & 0 \\ 1 & 0 & 0 & 1 \\ 0 & 1 & 0 & 0 \\ 1 & 0 & 1 & 1 \end{bmatrix} \Rightarrow A^2 = \begin{bmatrix} 1 & 0 & 0 & 1 \\ 1 & 1 & 1 & 1 \\ 1 & 0 & 0 & 1 \\ 1 & 2 & 1 & 1 \end{bmatrix}.$

The number of paths of length 2 between v_1 and v_3 is given by the element $[A^2]_{13}$. So, we see that there are no paths of length 2 between vertex v_1 and v_3.

47. In Exercise 40, $A = \begin{bmatrix} 0 & 1 & 0 & 0 & 1 \\ 0 & 0 & 0 & 1 & 0 \\ 1 & 0 & 0 & 1 & 1 \\ 1 & 0 & 1 & 0 & 0 \\ 1 & 1 & 0 & 0 & 0 \end{bmatrix} \Rightarrow A^3 = \begin{bmatrix} 1 & 1 & 1 & 1 & 1 \\ 1 & 1 & 0 & 1 & 2 \\ 2 & 3 & 0 & 3 & 3 \\ 3 & 3 & 1 & 1 & 1 \\ 2 & 1 & 1 & 1 & 0 \end{bmatrix}.$

The number of paths of length 3 from vertex v_4 to vertex v_1 is given by the element $[A^3]_{41}$. So, we see that there are three paths of length 3 from vertex v_4 to vertex v_1.

49. Let A be the adjacency matrix of a graph G.

  (a) Assume that row i of A contains only zeros.
  This implies that there are no edges joining vertex i to any of the other vertices.
  We can conclude that this is a disconnected graph.

  (b) Assume that column j of A contains only zeros.
  This implies that there are no edges joining vertex j to any of the other vertices.
  We can conclude that this is a disconnected graph.

51. Since $A = \begin{bmatrix} 0 & 0 & 0 & 0 & 1 & 0 \\ 1 & 0 & 1 & 0 & 1 & 1 \\ 1 & 0 & 0 & 1 & 1 & 0 \\ 1 & 1 & 0 & 0 & 1 & 0 \\ 0 & 0 & 0 & 0 & 0 & 1 \\ 1 & 0 & 1 & 1 & 0 & 0 \end{bmatrix}$ and $\mathbf{j} = \begin{bmatrix} 1 \\ 1 \\ 1 \\ 1 \\ 1 \\ 1 \end{bmatrix}$, we have: $A\mathbf{j} = \begin{bmatrix} 0 & 0 & 0 & 0 & 1 & 0 \\ 1 & 0 & 1 & 0 & 1 & 1 \\ 1 & 0 & 0 & 1 & 1 & 0 \\ 1 & 1 & 0 & 0 & 1 & 0 \\ 0 & 0 & 0 & 0 & 0 & 1 \\ 1 & 0 & 1 & 1 & 0 & 0 \end{bmatrix} \begin{bmatrix} 1 \\ 1 \\ 1 \\ 1 \\ 1 \\ 1 \end{bmatrix} = \begin{bmatrix} 1 \\ 4 \\ 3 \\ 3 \\ 1 \\ 3 \end{bmatrix}$.

The number of wins that each player had is given by $A\mathbf{j} = \begin{bmatrix} 1 & 4 & 3 & 3 & 1 & 3 \end{bmatrix}^T$, so the ranking is: first, P_2; second: P_3, P_4, P_6 (tie); third: P_1, P_5 (tie).

If we use the notion of combined wins and indirect wins
the ranking will be determined by $(A + A^2)\mathbf{j} = \begin{bmatrix} 2 & 12 & 8 & 9 & 4 & 10 \end{bmatrix}^T$.
So, the ranking is: first: P_2; second: P_6; third: P_4.

3.7 Applications

53. The email list for this system defines a digraph with 5 vertices.
We let vertices 1–5 correspond to Annie, Bert, Carla, Daria, and Ehaz respectively.

(a) The adjacency matrix for this digraph is $A = \begin{bmatrix} 0 & 0 & 1 & 0 & 1 \\ 0 & 0 & 1 & 1 & 0 \\ 0 & 0 & 0 & 0 & 1 \\ 1 & 0 & 1 & 0 & 0 \\ 0 & 1 & 0 & 0 & 0 \end{bmatrix}$.

(b) The component of the adjacency matrix $[A^k]_{ij}$ has a value of 1 if there is a path of length k starting at i and ending at j.
Thus, if Bert hears a rumor we need to calculate $A'_m = \sum_{k=1}^m A^k$ (until each column in row 2 has a nonzero value).
So, we add all paths of increasing length until each person has heard the message.
Calculating the powers of A' gives:

$A'_1 = A^1 = A = \begin{bmatrix} 0 & 0 & 1 & 0 & 1 \\ 0 & 0 & 1 & 1 & 0 \\ 0 & 0 & 0 & 0 & 1 \\ 1 & 0 & 1 & 0 & 0 \\ 0 & 1 & 0 & 0 & 0 \end{bmatrix}$ and $A'_2 = A^1 + A^2 = \begin{bmatrix} 0 & 1 & 1 & 0 & 2 \\ 1 & 0 & 2 & 1 & 1 \\ 0 & 1 & 0 & 0 & 1 \\ 1 & 0 & 2 & 0 & 2 \\ 0 & 1 & 1 & 1 & 0 \end{bmatrix}$.

So, it takes two steps until everyone (other than Bert) has heard the rumor.
From A'_2 we see that by this time Carla has heard the message twice,
while everyone else has heard it once.

(c) If Annie hears a rumor we need to calculate A'_m:
$A'_m = \sum_{k=1}^m A^k$ (until each column in row 1 has a nonzero value).
So, we add all paths of increasing length until each person has heard the message.

Continuing the previous calculation of the powers of A' gives $A'_3 = \begin{bmatrix} 0 & 2 & 2 & 1 & 2 \\ 1 & 1 & 3 & 1 & 3 \\ 0 & 1 & 1 & 1 & 1 \\ 1 & 2 & 2 & 0 & 3 \\ 1 & 1 & 2 & 1 & 1 \end{bmatrix}$.

So, it takes three steps until everyone (other than Annie) has heard the rumor.

(d) If A is the adjacency matrix of a digraph with n vertices,
we can tell if vertex i is connected to vertex j by a path (of some length) by calculating
$A'_n = \sum_{k=1}^n A^k$ (recall n is the number of vertices).
If there is an off-diagonal component with a value of zero,
then we know that these two vertices are not connected.

55. The key thing to notice is the following:
$a_{ij}a_{jk} \neq 0 \Leftrightarrow$ there is an edge from v_i to v_k and v_j to v_k at the same time.
So, $(AA^T)_{ij} = \sum a_{ij}a_{jk}$ = number of vertices simultaneously adjacent to v_i and v_j.

57. $U = \{v_1, v_2, v_3\}$ and $V = \{v_4, v_5\}$.   **59.** $U = \{v_1, v_2, v_4\}$ and $V = \{v_3, v_5, v_6\}$.

61. We have the following set of vectors in $\mathbb{Z}_2^2$ and their encoded vectors as:
$[0,0] \longrightarrow [0,0,0,0]$, $[0,1] \longrightarrow [0,1,0,1]$, $[1,0] \longrightarrow [1,0,1,0]$, and $[1,1] \longrightarrow [1,1,1,1]$.
This is not an error-correcting code.

For example, if we transmit $[0,0,0,0]$ and an error occurs in the last component, then $[0,0,0,1]$ is received and an error is detected since this is not a legal code vector. However, the receiver cannot correct the error since $[0,0,0,1]$ would also be the result of an error in the second component if $[0,1,0,1]$ had been transmitted.

63. $\mathbf{c} = \begin{bmatrix} 1 \\ 1 \\ 0 \\ 0 \\ 0 \\ 1 \\ 1 \end{bmatrix}$ **65.** $\mathbf{c} = \begin{bmatrix} 1 \\ 1 \\ 1 \\ 1 \\ 1 \\ 1 \\ 1 \end{bmatrix}$

67. We compute $P\mathbf{c}' = \begin{bmatrix} 1 & 1 & 0 & 1 & 1 & 0 & 0 \\ 1 & 0 & 1 & 1 & 0 & 1 & 0 \\ 0 & 1 & 1 & 1 & 0 & 0 & 1 \end{bmatrix} \begin{bmatrix} 1 \\ 1 \\ 0 \\ 0 \\ 0 \\ 1 \\ 0 \end{bmatrix} = \begin{bmatrix} 1 \\ 0 \\ 1 \end{bmatrix}$ which we recognize as column 2 of P.

The error is in the third component of $\mathbf{c}'$. Changing it, we recover the correct code vector $\mathbf{c}$. Thus, the original message was $\mathbf{x} = [1,0,0,0]^T$.

69. The parity check code in Example 1.31 is a code $\mathbb{Z}_2^6 \longrightarrow \mathbb{Z}_2^7$.

(a) For some arbitrary vector $\mathbf{a} = [a_1, a_2, a_3, a_4, a_5, a_6]^T \in \mathbb{Z}_2^6$
the encoded vector has a 7th component a_7 such that $a_1 + a_2 + a_3 + a_4 + a_5 + a_6 + a_7 = 0$.
So, if we take $P = [1,1,1,1,1,1,1]$, then $P[\mathbf{a}, a_7] = 0$,
where $[\mathbf{a}, a_7] = [a_1, a_2, a_3, a_4, a_5, a_6, a_7]^T$.
Thus P is $\begin{bmatrix} 1 & 1 & 1 & 1 & 1 & 1 & 1 \end{bmatrix}$.

(b) A standard generator matrix for this code is $G = \begin{bmatrix} 1 & 0 & 0 & 0 & 0 & 0 \\ 0 & 1 & 0 & 0 & 0 & 0 \\ 0 & 0 & 1 & 0 & 0 & 0 \\ 0 & 0 & 0 & 1 & 0 & 0 \\ 0 & 0 & 0 & 0 & 1 & 0 \\ 0 & 0 & 0 & 0 & 0 & 1 \\ 1 & 1 & 1 & 1 & 1 & 1 \end{bmatrix}$.

This is not an error-correcting code because the columns of P are not distinct.

3.7 Applications

71. Define a code $\mathbb{Z}_2^3 \longrightarrow \mathbb{Z}_2^6$ using the standard generator matrix $G = \begin{bmatrix} 1 & 0 & 0 \\ 0 & 1 & 0 \\ 0 & 0 & 1 \\ 1 & 0 & 0 \\ 1 & 1 & 0 \\ 1 & 1 & 1 \end{bmatrix}$.

(a) The eight code words are $\begin{bmatrix} 1 \\ 1 \\ 1 \\ 1 \\ 0 \\ 1 \end{bmatrix}, \begin{bmatrix} 1 \\ 1 \\ 0 \\ 1 \\ 0 \\ 0 \end{bmatrix}, \begin{bmatrix} 1 \\ 0 \\ 1 \\ 1 \\ 1 \\ 0 \end{bmatrix}, \begin{bmatrix} 0 \\ 1 \\ 1 \\ 0 \\ 1 \\ 0 \end{bmatrix}, \begin{bmatrix} 1 \\ 0 \\ 0 \\ 1 \\ 1 \\ 1 \end{bmatrix}, \begin{bmatrix} 0 \\ 1 \\ 0 \\ 0 \\ 1 \\ 1 \end{bmatrix}, \begin{bmatrix} 0 \\ 0 \\ 1 \\ 0 \\ 0 \\ 1 \end{bmatrix}$, and $\begin{bmatrix} 0 \\ 0 \\ 0 \\ 0 \\ 0 \\ 0 \end{bmatrix}$.

(b) The standard parity check matrix for this code is $P = \begin{bmatrix} 1 & 0 & 0 & 1 & 0 & 0 \\ 1 & 1 & 0 & 0 & 1 & 0 \\ 1 & 1 & 1 & 0 & 0 & 1 \end{bmatrix}$.

This is not an error-correcting code since columns three and six are the same.

73. We wish to construct the standard parity check and generator matrices for the $(15, 11)$ code.
Thus, we have $n = 15$, $k = 11$, so that there are $n - k = 4$ parity check equations.
The parity check equations give rise to the last four columns of P and
by Theorem 3.34, the 15 columns of P need to be nonzero and distinct.
Thus, the columns of P consist of $2^4 - 1 = 15$ unique vectors of $\mathbb{Z}_2^{n-k} = \mathbb{Z}_2^4$.

One such candidate for P is $P = \begin{bmatrix} 1 & 1 & 1 & 1 & 0 & 1 & 1 & 1 & 0 & 0 & 0 & 1 & 0 & 0 & 0 \\ 1 & 1 & 1 & 0 & 1 & 1 & 0 & 0 & 1 & 1 & 0 & 0 & 1 & 0 & 0 \\ 1 & 1 & 0 & 1 & 1 & 0 & 1 & 0 & 1 & 0 & 1 & 0 & 0 & 1 & 0 \\ 1 & 0 & 1 & 1 & 1 & 0 & 0 & 1 & 0 & 1 & 1 & 0 & 0 & 0 & 1 \end{bmatrix}$.

Then, by Theorem 3.34, we identify $A = \begin{bmatrix} 1 & 1 & 1 & 1 & 0 & 1 & 1 & 1 & 0 & 0 & 0 \\ 1 & 1 & 1 & 0 & 1 & 1 & 0 & 0 & 1 & 1 & 0 \\ 1 & 1 & 0 & 1 & 1 & 0 & 1 & 0 & 1 & 0 & 1 \\ 1 & 0 & 1 & 1 & 1 & 0 & 0 & 1 & 0 & 1 & 1 \end{bmatrix}$.

So, the generator matrix for the Hamming $(15, 11)$ code is $G = \begin{bmatrix} 1 & 0 & 0 & 0 & 0 & 0 & 0 & 0 & 0 & 0 & 0 \\ 0 & 1 & 0 & 0 & 0 & 0 & 0 & 0 & 0 & 0 & 0 \\ 0 & 0 & 1 & 0 & 0 & 0 & 0 & 0 & 0 & 0 & 0 \\ 0 & 0 & 0 & 1 & 0 & 0 & 0 & 0 & 0 & 0 & 0 \\ 0 & 0 & 0 & 0 & 1 & 0 & 0 & 0 & 0 & 0 & 0 \\ 0 & 0 & 0 & 0 & 0 & 1 & 0 & 0 & 0 & 0 & 0 \\ 0 & 0 & 0 & 0 & 0 & 0 & 1 & 0 & 0 & 0 & 0 \\ 0 & 0 & 0 & 0 & 0 & 0 & 0 & 1 & 0 & 0 & 0 \\ 0 & 0 & 0 & 0 & 0 & 0 & 0 & 0 & 1 & 0 & 0 \\ 0 & 0 & 0 & 0 & 0 & 0 & 0 & 0 & 0 & 1 & 0 \\ 0 & 0 & 0 & 0 & 0 & 0 & 0 & 0 & 0 & 0 & 1 \\ 1 & 1 & 1 & 1 & 0 & 1 & 1 & 1 & 0 & 0 & 0 \\ 1 & 1 & 1 & 0 & 1 & 1 & 0 & 0 & 1 & 1 & 0 \\ 1 & 1 & 0 & 1 & 1 & 0 & 1 & 0 & 1 & 0 & 1 \\ 1 & 0 & 1 & 1 & 1 & 0 & 0 & 1 & 0 & 1 & 1 \end{bmatrix}$.

75. Let $G = \begin{bmatrix} I_k \\ A \end{bmatrix}$ be a standard generator matrix and
$P = \begin{bmatrix} A & I_{n-k} \end{bmatrix}$ be a standard parity check matrix for the same binary code.
Let $\mathbf{x}$ be a message vector in $\mathbb{Z}_2^k$ and let the corresponding code vector be $c = G\mathbf{x}$.
Then $P\mathbf{c} = \mathbf{0}$.

Suppose there has been an error in the ith component, resulting in the vector $\mathbf{c}'$.
It follows that $\mathbf{c}' = \mathbf{c} + \mathbf{e}_i$.

Also, assume that two columns of P are the same.
Then $\mathbf{p}_i = \mathbf{p}_j$. We now compute $P\mathbf{c}' = P(\mathbf{c} + \mathbf{e}_i) = P\mathbf{c} + P\mathbf{e}_i = \mathbf{0} + \mathbf{p}_i = \mathbf{p}_i$.

Normally, this would pinpoint the error to be in the ith component.
However, since $\mathbf{p}_i = \mathbf{p}_j$ we also get $P\mathbf{c}' = \mathbf{p}_i = \mathbf{p}_j$,
so that the error could have occurred in the jth column as well.

Chapter 3 Review

Key Definitions and Concepts

This list includes most but not all of the definitions listed at the end of each section. We begin the list with key symbols and symbol-based definitions.

Symbol	Page	Section	Description
A^T	p.149	3.1	A-transpose, created by switching rows and columns
A^{-1}	p.162	3.3	A-inverse: satisfies $AA^{-1} = I$ and is unique
A^{-n}	p.167	3.3	This is defined to be $(A^{-1})^n = (A^n)^{-1}$ and is unique
$\mathbf{e}_i$	p.142	3.1	standard $1 \times m$ unit vector
$\mathbf{e}_j$	p.142	3.1	standard $n \times 1$ unit vector
E_i	p.169	3.3	Matrix created by an elementary row operation on I
L	p.179	3.4	Often used to stand for a unit lower triangular matrix
O	p.152	3.2	O is commonly used to stand for the zero matrix
P	p.185	3.4	Often used to stand for a permutation matrix
U	p.179	3.4	Often used to stand for an upper triangular matrix
LU factorization	p.179	3.4	$A = LU$ with L unit lower triangular and and U upper triangular
LDU factorization	p.189	3.4	$A = LDU$ same as LU except D is diagonal and U unit upper triangular
$P^T LU$ factorization	p.184	3.4	Adjustment to the LU factorization when we have to permute the rows of A
$[\mathbf{v}]_{\mathcal{B}}$	p.198	3.5	coordinate vector with respect to $\mathcal{B}$
col(A)	p.193	3.5	column space of A spanned by the columns of A
dim(S)	p.201	3.5	the number of vectors in a basis called its *dimension*
null(A)	p.195	3.5	null space of A, $\mathbf{x}$ such that $A\dot{\mathbf{x}} = \mathbf{0}$
nullity(A)	p.203	3.5	the dimension of null(A)
row(A)	p.193	3.5	row space of A spanned by the rows of A
det A	p.163	3.3	The determinant of A for 2×2 matrices det $A = ad - bc$

Key Definitions and Concepts, continued

adjacency matrix	p.236	3.7	$A(G)$ where $a_{ij} = 1$ if edge, 0 otherwise
basis	p.196	3.5	set of vectors that spans S and is linearly independent
circuit	p.237	3.7	a closed *path* (ends and begins at the same *vertex*
code	p.240	3.7	See discussion prior to Example 3.69 and Section 1.4
column space	p.193	3.5	col(A) spanned by the columns of A
determinant	p.163	3.3	The determinant of A for 2×2 matrices $\det A = ad - bc$
dimension	p.201	3.5	$\dim(S)$, the number of vectors in a basis
elementary matrix	p.169	3.3	Any matrix that can be obtained by performing an elementary row operation on an identity matrix
generator matrix	p.240	3.7	G: generates a *code*. See definition before Theorem 3.34
graph	p.236	3.7	finite set of *points* and *edges*
Hamming code	p.242	3.7	See Example 3.70 for construction
induction	p.147	3.1	See Appendix B, *Mathematical Induction*
inverse	p.161	3.3	A-inverse: satisfies $AA^{-1} = I$ and is unique
inverse transformation	p.219	3.6	$S \circ T = I_n$ and $T \circ S = I_n$ $S(T(\mathbf{v})) = [S][T]\mathbf{v} = \mathbf{v} \Rightarrow [T] = [S]^{-1}$
invertible	p.161	3.3	If A^{-1} exists, A is called *invertible*.
length	p.237	3.7	number of edges a *path* contains
Leslie matrix	p.234	3.7	Population matrix L described after Example 3.66
linear combination	p.152	2.3	$\mathbf{B} = c_1 \mathbf{A}_1 + \cdots + c_n \mathbf{A}_n$ $\mathbf{B} = \sum c_i \mathbf{A}_i$
linear transformation	p.211	3.6	$T(c\mathbf{v}) = cT(\mathbf{v})$ and $T(\mathbf{u} + \mathbf{v}) = T(\mathbf{u} + \mathbf{v})$ If T is linear, then there is an A such that $T(\mathbf{v}) = A\mathbf{v}$.
linearly dependent	p.155	3.2	matrix can be written as linear combination of others $c_1 \mathbf{A}_1 + \cdots + c_n \mathbf{A}_n = O$ with at least one $c_i \neq 0$
linearly independent	p.155	3.2	no matrix can be written as linear combination of others $c_1 \mathbf{A}_1 + \cdots + c_n \mathbf{A}_n = O \Leftrightarrow$ all the $c_i = 0$

Chapter 3 Review

Key Definitions and Concepts, continued

Markov chain	p.228	3.7	See description given prior to Example 3.64
matrix product	p.139	3.1	$C = AB \Rightarrow c_{ij} = a_{i1}b_{1j} + a_{i2}b_{2j} + \cdots + a_{in}b_{nj}$
multiplier	p.181	3.4	the scalar k in $R_i - kR_j$
null space	p.195	3.5	null(A), $\mathbf{x}$ such that $A\mathbf{x} = \mathbf{0}$
nullity	p.203	3.5	the dimension of null(A)
outer product	p.145	3.1	see description of this process after Example 3.10
parity matrix	p.240	3.7	P: checks a *code*. See definition before Theorem 3.34
path	p.237	3.7	set of *edges* that connects one *vertex* to another
permutation matrix	p.185	3.4	$P = P_k \cdots P_2 P_1$ where multiplication by P_i performs a row interchange like $R_i \leftrightarrow R_j$
probability vector	p.229	3.7	vector with nonnegative components that add to 1
rank	p.202	3.5	the dimension of row(A) or col(A) (they are the same)
row space	p.193	3.5	row(A) spanned by the rows of A
simple path	p.237	3.7	a *path* that does not contain the same *edge* twice
skew-symmetric	p.160	3.2	$A^T = -A$
state vector	p.229	3.7	$\mathbf{x}_k$: $\mathbf{x}_k = P^k \mathbf{x}_0$ in a Markov Chain
steady state vector	p.231	3.7	$\mathbf{x}$: $\mathbf{x} = P\mathbf{x}$ in a Markov Chain
subspace	p.190	3.5	$\mathbf{0}$, $\mathbf{u} + \mathbf{v}$, and $c\mathbf{v}$ are in S
symmetric	p.149	3.1	$A^T = A$
trace	p.160	3.2	tr$(A) = a_{11} + a_{22} + \cdots + a_{nn}$ That is, the *trace* is the sum of the diagonal entries
transition matrix	p.229	3.7	P: $\mathbf{x}_k = P^k \mathbf{x}_0$ in a Markov Chain
transpose	p.148	3.1	A^T, create by switching rows and columns
unit lower diagonal	p.179	3.4	L with all the entries above the diagonal are 0 and all the entries on the diagonal are 1. That is, $i > j \Rightarrow a_{ij} = 0$ and $a_{ii} = 1$.
upper diagonal	p.179	3.4	U where all the entries below the diagonal are 0. That is, $i < j \Rightarrow a_{ij} = 0$.
upper triangular	p.160	3.2	matrix whose entries below the main diagonal are all zero

Theorems

Thm 3.1	p.142	3.1	$e_i A$ is row of A, Ae_j is column of A
Thm 3.2	p.152	3.2	Properties of Matrix Addition and Scalar Multiplication
Thm 3.3	p.156	3.2	Properties of Matrix Multiplication
Thm 3.4	p.157	3.2	Properties of the Transpose, A^T
Thm 3.5	p.159	3.2	a. $A + A^T$ is symmetric if A is square AA^T and $A^T A$ b. AA^T and $A^T A$ are always symmetric
Thm 3.6	p.162	3.3	If A is invertible, then its inverse is unique. If $AA^{-1} = I$, then A^{-1} is unique.
Thm 3.7	p.162	3.3	$A\mathbf{x} = \mathbf{b}$ has unique solution $\mathbf{x} = A^{-1}\mathbf{b}$ The inverse A^{-1} exists if and only if $\det A = ad - bc \neq 0$
Thm 3.8	p.163	3.3	Formula for A^{-1} for 2×2 matrices using $\det A$
Thm 3.9	p.166	3.3	Key Properties of A^{-1}, for example $(A^{-1})^{-1} = I$
Thm 3.10	p.170	3.3	Perform row operation E on A is equivalent to EA
Thm 3.11	p.171	3.3	E^{-1} is created by undoing the operation that created E
Thm 3.12	p.171	3.3	Five conditions that are equivalent to A being invertible
Thm 3.13	p.173	3.3	If $AB = I$ or $BA = I$, then $B = A^{-1}$
Thm 3.14	p.173	3.3	If $E_n E_{n-1} \cdots E_1 A = I$, then $A^{-1} = E_n E_{n-1} \cdots E_1$
Thm 3.15	p.180	3.4	If $A \longrightarrow$ REF without $R_i \leftrightarrow Rj$, then $A = LU$ exists.
Thm 3.16	p.184	3.4	If A is invertible, then we have the following: If $A = L_1 U_1$ and $A = L_2 U_2$ then $L_1 = L_2$ and $U_1 = U_2$. That is, if an LU factorization exists, it is unique.
Thm 3.17	p.185	3.4	If P is a permutation matrix, then $P^{-1} = P^T$.
Thm 3.18	p.186	3.4	If A is square, then $A = P^T LU$ exists.
Thm 3.19	p.190	3.5	$\text{span}(\mathbf{v}_1, \mathbf{v}_2, \ldots, \mathbf{v}_n)$ is a subspace
Thm 3.20	p.194	3.5	If $A \longrightarrow B$, then $\text{row}(A) = \text{row}(B)$
Thm 3.21	p.194	3.5	$\text{null}(A)$, $\mathbf{x}$ such that $A\mathbf{x} = \mathbf{0}$, is a subspace
Thm 3.22	p.195	3.5	$A\mathbf{x} = \mathbf{b}$ has exactly none, one, or infinitely many solutions
Thm 3.23	p.200	3.5	Any two bases have the same number of vectors (*dimension*)
Thm 3.24	p.202	3.5	$\text{row}(A)$ and $\text{col}(A)$ have the same dimension
Thm 3.25	p.202	3.5	$\text{rank}(A) = \text{rank}(A^T)$
Thm 3.26	p.203	3.5	**Rank Theorem**: $\text{rank}(A) + \text{nullity}(A) = n$
Thm 3.27	p.204	3.5	**Fundamental Theorem**: Version 2
Thm 3.28	p.205	3.5	$\text{rank}(A^T A) = \text{rank}(A)$ and $A^T A$ invertible $\Leftrightarrow \text{rank}(A) = n$
Thm 3.29	p.206	3.5	If $\{\mathbf{v}_1, \ldots, \mathbf{v}_k\}$ is a basis, then $\mathbf{v} = \sum c_i \mathbf{v}_i$ is unique
Thm 3.30	p.212	3.6	Given A, then $T_A(\mathbf{v}) = A\mathbf{v}$ is linear
Thm 3.31	p.214	3.6	$[T] = \begin{bmatrix} T(\mathbf{e}_1) & T(\mathbf{e}_2) & \ldots & T(\mathbf{e}_n) \end{bmatrix}$
Thm 3.32	p.218	3.6	Given $S \circ T$, $S(T(\mathbf{v})) = [S][T]\mathbf{v}$
Thm 3.33	p.220	3.6	$S \circ T = I_n$, $S(T(\mathbf{v})) = [S][T]\mathbf{v} = \mathbf{v} \Rightarrow [T] = [S]^{-1}$
Thm 3.34	p.241	3.7	Gives conditions on G and P for a given *code*

Chapter 3 Review

Solutions to odd-number exercises from Chapter 3 Review

1. We will explain and give counter examples to justify our answers below.

    (a) **True.** See Exercise 28 in Section 3.2.
    Let A be an $m \times n$ matrix, then A^T is an $n \times m$ matrix.
    Since AA^T is $[m \times n][n \times m]$, AA^T is an $m \times m$ matrix.
    Since $A^T A$ is $[n \times m][m \times n]$, $A^T A$ is an $n \times n$ matrix.
    So not only are AA^T and $A^T A$ defined, they are square matrices.

    (b) **False.** See Theorem 3.27 of Section 3.5.
    Theorem 3.27 of Section 3.5 implies $AB = O \Rightarrow B = O$ if and only if A is invertible.
    Since vectors are matrices, when $AB = O \Rightarrow B = O$, what do we know about null(A)?
    Since $B = O$ means null(A) = $\mathbf{0}$, Theorem 3.27 of Section 3.5 implies A is invertible.
    When A^{-1} exists, we have: $AB = O \Rightarrow (A^{-1}A)B = A^{-1}O = O \Rightarrow B = O$.
    As a counterexample, consider: $\begin{bmatrix} 1 & 0 \\ 0 & 0 \end{bmatrix} \begin{bmatrix} 0 & 0 \\ 0 & 1 \end{bmatrix} = \begin{bmatrix} 0 & 0 \\ 0 & 0 \end{bmatrix}$.

    (c) **False.** See Exercises 20 through 23 in Section 3.3.
    What is the source of the problem? Matrix multiplication does *not* commute.
    Instead, we should solve for X using A^{-1}.
    If $XA = B$ then $X(AA^{-1}) = BA^{-1}$, so $X = BA^{-1}$ (not always equal to $A^{-1}B$).
    When $X = \begin{bmatrix} 1 & 0 \\ -2 & 1 \end{bmatrix}$, $A = \begin{bmatrix} 1 & 0 \\ 1 & -1 \end{bmatrix}$, and $B = \begin{bmatrix} 1 & 0 \\ -1 & -1 \end{bmatrix}$,
    then $A^{-1}B = \begin{bmatrix} 1 & 0 \\ 2 & 1 \end{bmatrix} \neq \begin{bmatrix} 1 & 0 \\ -2 & 1 \end{bmatrix} = BA^{-1}$.
    So, $XA = B \Rightarrow X = A^{-1}B$ if and only if $A^{-1}B = BA^{-1}$.
    Does $XA = AX$ if and only if $A^{-1}B = BA^{-1}$? Why or why not?

    (d) **True.** See Theorem 3.11 in Section 3.3.
    What is the idea? E performs exactly 1 operation and E^{-1} undoes it.
    For example, if E performs kR_i, then E^{-1} performs $\frac{1}{k}R_i$.
    If E performs $R_i + R_j$, then E^{-1} performs $R_i - R_j$.
    Let $E = \begin{bmatrix} 1 & 1 \\ 0 & 1 \end{bmatrix}$, then $E^{-1} = \begin{bmatrix} 1 & -1 \\ 0 & 1 \end{bmatrix}$. Confirm this by showing $EE^{-1} = I$.

    (e) **True.** See Section 3.3. Prove this by considering each type of E separately.
    Since E is obtained from I, we need only consider the entries that differ from I.

    If E performs kR_i, then $[E]_{ii} = k = [E^T]_{ii}$, so $E^T = E$ performs kR_i.
    If E performs $R_i + kR_j$, then $[E]_{ij} = k = [E^T]_{ji}$.
    So, E^T is an elementary matrix that performs $R_j + kR_i$.
    If E performs $R_i \leftrightarrow R_j$, then $[E]_{ii} = [E]_{jj} = 0$ and $[E]_{ij} = [E]_{ji} = 1$.
    So E^T performs $R_i \leftrightarrow R_j$, too since $[E^T]_{ii} = [E]_{ii}$ and $[E^T]_{ji} = [E]_{ij}$.

    So, when E performs kR_i or $R_i \leftrightarrow R_j$, then $E^T = E$.
    Furthermore, when E performs $R_i \leftrightarrow R_j$, $E^T = E^{-1} = E$.

1. We explain and give counter examples to justify our answers below (continued).

    (f) **False**. See Section 3.3.

    An elementary matrix can only perform *one* elementary row operation on I.
    But $E_2 E_1$ performs *two* elementary row operations on I.
    When $E_2 = \begin{bmatrix} 0 & 1 \\ 1 & 0 \end{bmatrix}$ and $E_1 = \begin{bmatrix} 2 & 0 \\ 0 & 1 \end{bmatrix}$, then $E_2 E_1 = \begin{bmatrix} 0 & 1 \\ 2 & 0 \end{bmatrix}$.

    (g) **True**. See Theorem 3.21 in Section 3.5.

    We should be able to easily recreate the proof in the text.
    Is the following enough? If $\mathbf{u}$, $\mathbf{v}$ are in null(A), then
    $A(c\mathbf{u} + d\mathbf{v}) = c(A\mathbf{u}) + d(A\mathbf{v}) = \mathbf{0} + \mathbf{0} = \mathbf{0}$ is in null(A).

    (h) **False**. See the discussions of Exercises 1 through 6 in Section 3.5.

    What is the problem? Every subspace must contain the zero vector.
    What condition do we need to add make the statement true?
    Every plane *that passes through the origin* is a subspace.
    Is the dimension of the plane actually two?
    Yes. A basis for the subspace is the two direction vectors in the parametric form.
    For that form, $\mathbf{x} = \mathbf{p} + s\mathbf{u} + t\mathbf{v}$, and further discussion see Section 1.3.

    (i) **True**. See the definition of a linear transformation in Section 3.6.

    We verify that $T(\mathbf{x}) = -\mathbf{x}$ satisfies the two necessary conditions.
    $$T(\mathbf{u} + \mathbf{v}) = -(\mathbf{u} + \mathbf{v}) = (-\mathbf{u}) + (-\mathbf{v}) = T(\mathbf{u}) + T(\mathbf{v})$$
    $$T(c\mathbf{u}) = -(c\mathbf{u}) = c(-\mathbf{u}) = cT(\mathbf{u})$$
    Theorems 3.30 and 3.31 of Section 3.6 imply an A with $A\mathbf{x} = -\mathbf{x}$ is also enough.
    Note that $A = \begin{bmatrix} T(\mathbf{e}_1) & T(\mathbf{e}_2) \end{bmatrix} = \begin{bmatrix} -1 & 0 \\ 0 & -1 \end{bmatrix}$ does just that.
    Finally, note that $A = -I$. Why does that make sense?

    (j) **False**. See Theorems 3.30 and 3.31 of Section 3.6.

    The matrix A must be 5×4 not 4×5.

Chapter 3 Review

3. Since B^2 is $[2 \times 3][2 \times 3]$, B^2 is not possible.

   Q: If A is not a square matrix, is A^2 ever defined?
   A: No, because $[m \times n][m \times n]$ is not possible unless $m = n$.

5. Since BB^T is $[2 \times 3][3 \times 2]$, BB^T is $[2 \times 2]$.

   $BB^T = \begin{bmatrix} 2 & 0 & -1 \\ 3 & -3 & 4 \end{bmatrix} \begin{bmatrix} 2 & 3 \\ 0 & -3 \\ -1 & 4 \end{bmatrix} = \begin{bmatrix} 5 & 2 \\ 2 & 34 \end{bmatrix}$. So, $(BB^T)^{-1} = \frac{1}{166}\begin{bmatrix} 34 & -2 \\ -2 & 5 \end{bmatrix} = \begin{bmatrix} \frac{17}{83} & -\frac{1}{83} \\ -\frac{1}{83} & \frac{5}{166} \end{bmatrix}$.

7. As in Example 3.11 of Section 3.1, we compute the outer product expansion of AA^T.

   $\mathbf{a}_1 A_1^T = \begin{bmatrix} 1 \\ 3 \end{bmatrix}[\,1\ 3\,] = \begin{bmatrix} 1 & 3 \\ 3 & 9 \end{bmatrix}$. and $\mathbf{a}_2 A_2^T = \begin{bmatrix} 2 \\ 5 \end{bmatrix}[\,2\ 5\,] = \begin{bmatrix} 4 & 10 \\ 10 & 25 \end{bmatrix}$.

   So $AA^T = \begin{bmatrix} 1 & 3 \\ 3 & 9 \end{bmatrix} + \begin{bmatrix} 4 & 10 \\ 10 & 25 \end{bmatrix} = \begin{bmatrix} 5 & 13 \\ 13 & 34 \end{bmatrix}$.

   Note: The outer product expansion of AA^T is always defined.

9. As in Exercises 22 and 52 in Section 3.3, since $AX = B$ we have $X = A^{-1}B$.

   So, by Theorem 3.8 of Section 3.3, $A = (A^{-1})^{-1} = \frac{2}{1}\begin{bmatrix} 4 & 1 \\ \frac{3}{2} & \frac{1}{2} \end{bmatrix} = \begin{bmatrix} 8 & 2 \\ 3 & 1 \end{bmatrix}$.

   As in Example 3.31 of Section 3.3, we try to row reduce $[\,A\,|\,I\,]$ into $[\,I\,|\,A^{-1}\,]$.

   $[\,A\,|\,I\,] = \begin{bmatrix} 1 & 0 & -1 & | & 1 & 0 & 0 \\ 2 & 3 & -1 & | & 0 & 1 & 0 \\ 0 & 1 & 1 & | & 0 & 0 & 1 \end{bmatrix} \longrightarrow \begin{bmatrix} 1 & 0 & 0 & | & 2 & -\frac{1}{2} & \frac{3}{2} \\ 0 & 1 & 0 & | & -1 & \frac{1}{2} & -\frac{1}{2} \\ 0 & 0 & 1 & | & 1 & -\frac{1}{2} & \frac{3}{2} \end{bmatrix} = [\,I\,|\,A^{-1}\,]$

   So $X = A^{-1}B = \begin{bmatrix} 2 & -\frac{1}{2} & \frac{3}{2} \\ -1 & \frac{1}{2} & -\frac{1}{2} \\ 1 & -\frac{1}{2} & \frac{3}{2} \end{bmatrix}\begin{bmatrix} -1 & -3 \\ 5 & 0 \\ 3 & -2 \end{bmatrix} = \begin{bmatrix} 0 & -9 \\ 2 & 4 \\ 1 & -6 \end{bmatrix}$.

   Check: $AX = \begin{bmatrix} 1 & 0 & -1 \\ 2 & 3 & -1 \\ 0 & 1 & 1 \end{bmatrix}\begin{bmatrix} 0 & -9 \\ 2 & 4 \\ 1 & -6 \end{bmatrix} = \begin{bmatrix} -1 & -3 \\ 5 & 0 \\ 3 & -2 \end{bmatrix}$.

   We could also have solved the problem directly as follows:

   $[\,A\,|\,AX\,] = \begin{bmatrix} 1 & 0 & -1 & | & -1 & -3 \\ 2 & 3 & -1 & | & 5 & 0 \\ 0 & 1 & 1 & | & 3 & -2 \end{bmatrix} \longrightarrow \begin{bmatrix} 1 & 0 & 0 & | & 0 & -9 \\ 0 & 1 & 0 & | & 2 & 4 \\ 0 & 0 & 1 & | & 1 & -6 \end{bmatrix} = [\,I\,|\,X\,]$

11. As in Exercise 45 of Section 3.3, to prove X is the inverse of B we show $BX = I$.
    We claim $(I - A)^{-1} = I + A + A^2$, so we show $(I - A)(I + A + A^2) = I$.

    $(I - A)(I + A + A^2) = I + A + A^2 - A - A^2 - A^3 = I - A^3 \stackrel{\substack{A^3 = O \\ \text{given}}}{=} I$

13. Find bases for row(A), col(A), and null(A) by Examples 3.45, 3.47, and 3.48 in Section 3.5.

row(A): A basis for row(A) must span the rows of A and be linearly independent.
The linearly independent rows (which are simply the nonzero rows) of U do just that.

Since $A = \begin{bmatrix} 2 & -4 & 5 & 8 & 5 \\ 1 & -2 & 2 & 3 & 1 \\ 4 & -8 & 3 & 2 & 6 \end{bmatrix} \longrightarrow \begin{bmatrix} 2 & -4 & 5 & 8 & 5 \\ 0 & 0 & 1 & 2 & 3 \\ 0 & 0 & 0 & 0 & 1 \end{bmatrix} = U$

we conclude $\{[\,2\ -4\ 5\ 8\ 5\,], [\,0\ 0\ 1\ 2\ 3\,], [\,0\ 0\ 0\ 0\ 1\,]\}$ is a basis for row(A).

Q: In $A \longrightarrow U$, why is sufficient to reduce A only to row echelon form U?
A: As the remark after Example 3.46 in Section 3.5 explains and demonstrates,
the nonzero rows of U are linearly independent. That is all that is required. Why?

We should also note that provided $A \longrightarrow U$ uses no row interchanges,
the corresponding rows in A are also linearly independent.
Whence, it is obvious that those rows form a basis for row(A).

col(A): A basis for col(A) must span the columns of A and be linearly independent.
When $A \longrightarrow U$, the columns with leading entries in U are linearly independent.
As in Example 3.47, the corresponding columns in A are also linearly independent.
Whence, it is obvious that those columns form a basis for col(A).

Since $A = \begin{bmatrix} 2 & -4 & 5 & 8 & 5 \\ 1 & -2 & 2 & 3 & 1 \\ 4 & -8 & 3 & 2 & 6 \end{bmatrix} \longrightarrow \begin{bmatrix} 2 & -4 & 5 & 8 & 5 \\ 0 & 0 & 1 & 2 & 3 \\ 0 & 0 & 0 & 0 & 1 \end{bmatrix} = U$

we conclude that $\left\{ \begin{bmatrix} 2 \\ 1 \\ 4 \end{bmatrix}, \begin{bmatrix} 5 \\ 2 \\ 3 \end{bmatrix}, \begin{bmatrix} 5 \\ 1 \\ 6 \end{bmatrix} \right\}$ is a basis for col(A).

null(A): Since $A\mathbf{v} = \mathbf{0}$ implies $\mathbf{v}$ is in null(A), we solve $[\,A\,|\,\mathbf{0}\,] \longrightarrow [\,R\,|\,\mathbf{0}\,]$ to find the conditions.

Note $A = \begin{bmatrix} 2 & -4 & 5 & 8 & 5 \\ 1 & -2 & 2 & 3 & 1 \\ 4 & -8 & 3 & 2 & 6 \end{bmatrix} \longrightarrow \begin{bmatrix} 2 & -4 & 5 & 8 & 5 \\ 0 & 0 & 1 & 2 & 3 \\ 0 & 0 & 0 & 0 & 1 \end{bmatrix} \longrightarrow \begin{bmatrix} 1 & -2 & 0 & -1 & 0 \\ 0 & 0 & 1 & 2 & 0 \\ 0 & 0 & 0 & 0 & 1 \end{bmatrix} = R$

$[\,R\,|\,\mathbf{0}\,] = \begin{bmatrix} 1 & -2 & 0 & -1 & 0 & | & 0 \\ 0 & 0 & 1 & 2 & 0 & | & 0 \\ 0 & 0 & 0 & 0 & 1 & | & 0 \end{bmatrix} \Rightarrow \begin{array}{l} x_1 = 2s + 1t \\ x_2 = s \\ x_3 = -2t \\ x_4 = t \\ x_5 = 0 \end{array}$

Therefore, $\left\{ \begin{bmatrix} 2 \\ 1 \\ 0 \\ 0 \\ 0 \end{bmatrix}, \begin{bmatrix} 1 \\ 0 \\ -2 \\ 1 \\ 0 \end{bmatrix} \right\}$ is a basis for null(A).

Chapter 3 Review

15. We consider when A is invertible and when A is *not* invertible separately below.

    A^{-1}: By Theorem 3.9 in Section 3.3, if A is invertible then so is A^T.
    Therefore, by Theorem 3.27 of Section 3.5, $\text{null}(A) = \text{null}(A^T) = \mathbf{0}$.
    This conclusion is not stated directly but following immediately from c. or g. Why?
    From c.: If $A\mathbf{x} = \mathbf{0}$ has only the trivial solution, then $\mathbf{x} = \mathbf{0} = \text{null}(A)$.
    From g.: If $\text{nullity}(A) = 0$ then $\mathbf{0}$ is a basis for $\text{null}(A)$ which implies $\text{null}(A) = \mathbf{0}$.

    no A^{-1}: Since the column space is affected by row reduction, we suspect $\text{null}(A) \neq \text{null}(A^T)$.
    We prove $\text{null}(A) \neq \text{null}(A^T)$ with the following counterexample.

    Let $A = \begin{bmatrix} 1 & 1 & 1 \\ 0 & 0 & 0 \\ 0 & 0 & 0 \end{bmatrix}$ then $A^T = \begin{bmatrix} 1 & 0 & 0 \\ 1 & 0 & 0 \\ 1 & 0 & 0 \end{bmatrix}$. Note $A^T \longrightarrow \begin{bmatrix} 1 & 0 & 0 \\ 0 & 0 & 0 \\ 0 & 0 & 0 \end{bmatrix}$.

    Then, $\text{null}(A) = \text{span}\left(\begin{bmatrix} -1 \\ 1 \\ 0 \end{bmatrix}, \begin{bmatrix} -1 \\ 0 \\ 1 \end{bmatrix}\right)$ and $\text{null}(A^T) = \text{span}\left(\begin{bmatrix} 0 \\ 1 \\ 0 \end{bmatrix}, \begin{bmatrix} 0 \\ 0 \\ 1 \end{bmatrix}\right)$.

    Then $\begin{bmatrix} -1 \\ 1 \\ 0 \end{bmatrix}$ is in $\text{null}(A)$, but $\begin{bmatrix} -1 \\ 1 \\ 0 \end{bmatrix}$ is *not* in $\text{null}(A^T)$, so $\text{null}(A) \neq \text{null}(A^T)$.

17. We consider $A^T A$ and AA^T separately below.

    $A^T A$: Since A has n linearly independent columns, $\text{rank}(A) = n$.
    By Theorem 3.28 of Section 3.5, $\text{rank}(A^T A) = \text{rank}(A) = n$, so $A^T A$ is invertible.

    AA^T: Since m may be greater than n, we should suspect that AA^T is not necessarily invertible.
    We prove AA^T is not necessarily invertible with the following counterexample.

    Let $A = \begin{bmatrix} 1 & 0 \\ 0 & 1 \\ 1 & 0 \end{bmatrix}$ then $A^T A = \begin{bmatrix} 1 & 0 & 1 \\ 0 & 1 & 0 \end{bmatrix} \begin{bmatrix} 1 & 0 \\ 0 & 1 \\ 1 & 0 \end{bmatrix} = \begin{bmatrix} 2 & 0 \\ 0 & 1 \end{bmatrix}$ is obviously invertible.

    On the other hand, $AA^T = \begin{bmatrix} 1 & 0 \\ 0 & 1 \\ 1 & 0 \end{bmatrix} \begin{bmatrix} 1 & 0 & 1 \\ 0 & 1 & 0 \end{bmatrix} = \begin{bmatrix} 1 & 0 & 1 \\ 0 & 1 & 0 \\ 1 & 0 & 1 \end{bmatrix}$ reduces to $\begin{bmatrix} 1 & 0 & 1 \\ 0 & 1 & 0 \\ 0 & 0 & 0 \end{bmatrix}$.

    Since $\text{rank}(AA^T) = 2 < 3$, AA^T is *not* invertible.

    Q: In this case, we have $\text{rank}(AA^T) \leq \text{rank}(A)$. Is that always true?
    A: Yes, since we saw in Exercise 57 of Section 3.5 that $\text{rank}(AB) \leq \text{rank}(A)$.

    Q: In this case, we saw $\text{rank}(AA^T) = \text{rank}(A^T A)$. Is that always true?
    A: Hint: Consider the proof of Theorem 3.28 in Section 3.5.

19. We find rotation $[R]$, projection $[P]$. Then by Theorem 3.32 of Section 3.6, $[P \circ R] = [P][R]$.

$[R]$: From Example 3.58 in Section 3.6, the rotation $[R]$ has matrix: $[R] = \begin{bmatrix} \cos\theta & -\sin\theta \\ \sin\theta & \cos\theta \end{bmatrix}$.

So, a rotation of $45°$ is given by $[R] = \begin{bmatrix} \cos 45° & -\sin 45° \\ \sin 45° & \cos 45° \end{bmatrix} = \begin{bmatrix} \frac{\sqrt{2}}{2} & -\frac{\sqrt{2}}{2} \\ \frac{\sqrt{2}}{2} & \frac{\sqrt{2}}{2} \end{bmatrix}$.

$[P]$: From Example 3.59 in Section 3.6, $[P]$ through ℓ with $\mathbf{d}$ is: $[P] = \frac{1}{d_1^2 + d_2^2} \begin{bmatrix} d_1^2 & d_1 d_2 \\ d_1 d_2 & d_2^2 \end{bmatrix}$.

So, for $y = -2x$ with $\mathbf{d} = \begin{bmatrix} 1 \\ -2 \end{bmatrix}$ we have: $[P] = \frac{1}{5}\begin{bmatrix} 1 & -2 \\ -2 & 4 \end{bmatrix} = \begin{bmatrix} \frac{1}{5} & -\frac{2}{5} \\ -\frac{2}{5} & \frac{4}{5} \end{bmatrix}$.

So $[P][R] = \begin{bmatrix} \frac{1}{5} & -\frac{2}{5} \\ -\frac{2}{5} & \frac{4}{5} \end{bmatrix} \begin{bmatrix} \frac{\sqrt{2}}{2} & -\frac{\sqrt{2}}{2} \\ \frac{\sqrt{2}}{2} & \frac{\sqrt{2}}{2} \end{bmatrix} = \begin{bmatrix} \frac{\sqrt{2}-2\sqrt{2}}{10} & \frac{-\sqrt{2}-2\sqrt{2}}{10} \\ \frac{-2\sqrt{2}+4\sqrt{2}}{10} & \frac{2\sqrt{2}+4\sqrt{2}}{10} \end{bmatrix} = \begin{bmatrix} \frac{-\sqrt{2}}{10} & \frac{-3\sqrt{2}}{10} \\ \frac{\sqrt{2}}{5} & \frac{3\sqrt{2}}{5} \end{bmatrix}$.

Chapter 4

Eigenvalues and Eigenvectors

4.1 Introduction to Eigenvalues and Eigenvectors

Q: What are our main goals for Section 4.1?

A: To take our first look at eigenvalues and eigenvectors, primarily of 2×2 matrices. The key equation an eigenvector satisfies is $A\mathbf{x} = \lambda \mathbf{x}$ with eigenvalue λ. Geometrically, the key idea is that eigenvectors are mapped to parallel vectors.

Key Definitions and Concepts

λ	p.253	4.1	λ often used to denote an eigenvalue of A
null(A)	p.195	3.5	the set of vectors $\mathbf{x}$ such that $A\mathbf{x} = \mathbf{0}$
E_λ	p.255	4.1	$E_\lambda = \text{null}(A - \lambda I) = \{\text{eigenvectors of } \lambda\} \cup \{\text{the zero vector, } \mathbf{0}\}$
eigenvalue	p.253	4.1	if $A\mathbf{x} = \lambda\mathbf{x}$, then A has eigenvalue λ
eigenvector	p.253	4.1	if $A\mathbf{x} = \lambda\mathbf{x}$, then $\mathbf{x}$ is an eigenvector of A
eigenspace	p.255	4.1	$E_\lambda = \text{null}(A - \lambda I) = \{\text{eigenvectors of } \lambda\} \cup \{\text{the zero vector, } \mathbf{0}\}$
			That is all vectors such that $(A - \lambda I)\mathbf{x} = \mathbf{0}$ or $A\mathbf{x} = \lambda\mathbf{x}$
null space	p.195	3.5	the set of vectors $\mathbf{x}$ such that $A\mathbf{x} = \mathbf{0}$

Theorems
There are no theorems in this section.

Discussion of Key Definitions and Concepts

This section is brief but important, so there is much discussion included with the Exercises.

How to tell if x is an eigenvector of A

Q: How can we tell if a given vector $\mathbf{x}$ is an eigenvector of a matrix A?

A: We simply multiply $A\mathbf{x} = \mathbf{y}$. What are we looking for?
We are looking to see if the resultant vector $\mathbf{y}$ is a multiple of $\mathbf{x}$. Why?
Because if it is, then we conclude the vector $\mathbf{x}$ is a eigenvector of A.
That is, if $A\mathbf{x} = \lambda\mathbf{x}$, then $\mathbf{x}$ is an eigenvector of A with eigenvalue λ.
See Exercises 1 through 6.

Q: What is the only vector that can *never* be an eigenvector of *any* matrix?
A: The zero vector, $\mathbf{0}$. Why? Because eigenvectors are *nonzero* by definition.

Q: If $\mathbf{x}$ is an eigenvector of A, is $c\mathbf{x}$ ($c \neq 0$) also an eigenvector?
A: Yes. Why? Because if $A\mathbf{x} = \lambda\mathbf{x}$, then $A(c\mathbf{x}) = c(A\mathbf{x}) = c\lambda\mathbf{x}$.

Q: If $\mathbf{x}$ is an eigenvector of A with eigenvalue λ, what is the eigenvalue of $c\mathbf{x}$ ($c \neq 0$)?
A: $c\lambda$. Why? Because if $A\mathbf{x} = \lambda\mathbf{x}$, then $A(c\mathbf{x}) = c(A\mathbf{x}) = c\lambda\mathbf{x}$.

Discussion of Key Definitions and Concepts, continued

How to tell if a given scalar λ is an eigenvalue of A

Q: How can we tell if a given scalar λ is an eigenvalue of a matrix A?
A: We have to show that $\text{null}(A - \lambda I) \neq \mathbf{0}$. See Exercises 7 through 12.

Q: How can we show that $\text{null}(A - \lambda I) \neq \mathbf{0}$?
A: Similar to Example 3.51 of Section 3.5, we row reduce the matrix $A - \lambda I$ until it is obvious that its rank is less than number of rows or columns. So, all we need is an obvious dependence relation or a zero row. Why is that enough?

Q: Why does $\text{rank}(A - \lambda I) <$ number of rows or columns imply $\text{null}(A - \lambda I) \neq \mathbf{0}$?
A: Because $\text{rank}(A - \lambda I) <$ number of rows or columns implies $\text{nullity}(A - \lambda I) > 0$. Furthermore, $\text{nullity}(A - \lambda I) > 0$ means that $\text{null}(A - \lambda I) \neq \mathbf{0}$. See Theorems 3.26 (Rank Theorem) and 3.27 (Fundamental Theorem) in Section 3.5. Note this conclusion relies partly on the fact that the matrix A is square. Why?

How do we find the eigenvectors associated with a given eigenvalue λ

Q: How do we find the eigenvectors associated with a given eigenvalue λ?
A: We row reduce and solve $[A - \lambda I | \mathbf{0}]$. See Exercises 7 through 12.

Q: This is exactly the same process used in Example 3.48 in Section 3.5 to find a basis for the null space of a matrix. Why should these be the same process?
A: Because the eigenvectors of λ form a basis for $\text{null}(A - \lambda I)$.

Discussion of eigenspace, E_λ

Q: What is an eigenspace, E_λ?
A: An eigenspace is the set of all vectors such that $(A - \lambda I)\mathbf{x} = \mathbf{0}$ or $A\mathbf{x} = \lambda\mathbf{x}$. That is, $E_\lambda = \text{null}(A - \lambda I) = \{\text{eigenvectors of } \lambda\} \cup \{\text{the zero vector, } \mathbf{0}\}$.

Q: Is an eigenspace E_λ a subspace such as we learned about in Section 3.5?
A: Yes, in fact, it is the null space of a particular matrix. Namely, $A - \lambda I$.

Q: Why do we have to add in the zero vector? That is, $E_\lambda = \{\text{eigenvectors of } \lambda\} \cup \{\text{the zero vector, } \mathbf{0}\}$?
A: We have to add the zero vector because eigenvectors are nonzero by definition and in order for E_λ to be a subspace it must contain the zero vector.

Q: What vector is in E_0 for any matrix A with eigenvalue 0?
A: The zero vector. Because the zero vector is in every subspace.

Q: What is the only vector in every eigenspace E_λ?
A: The zero vector. Because the zero vector is in every subspace. So the zero vector can never be an eigenvector, but it must be in every eigenspace.

4.1 Introduction to Eigenvalues and Eigenvectors

Discussion of Key Definitions and Concepts, continued

Discussion of the geometry of eigenvectors

Q: How can we tell if **x** is an eigenvector of A from a geometric point-of-view?
A: From the remarks prior to Example 4.4, we have the following key insight:
x is an eigenvector of A if and only if A transforms **x** to a parallel vector.
So, lines that do *not* bend at the unit circle represent eigenvectors.
Here, however, all lines bend at the unit circle, so we conclude there are no eigenvectors.
See Exercises 13 through 22.

Q: Consider the projection onto a line ℓ with direction vector **d**. How does thinking about it geometrically help us predict what the eigenvectors should be?
A: Vectors that are parallel to **d** are unchanged by projection onto ℓ. Why?
So, those vectors are mapped to parallel vectors (namely themselves).
Vectors that are perpendicular to **d** are mapped to the zero vector. Why?
What happens to vectors that are neither parallel nor perpendicular to **d**?
See Exercises 19 and 20.

Q: What about a rotation? How does thinking about it geometrically help us predict what the eigenvectors should be?
A: Every nonzero vector is rotated, so none of them are mapped to a parallel vector.

Q: Based upon the above, what types of transformations have no eigenvectors?
A: Rotations. See Exercise 18. Is the graph in Exercise 22 suggestive of a rotation?

How do we find all eigenvalues associated with a given matrix A

Q: How do we find all eigenvalues associated with a given matrix A?
A: We solve $\det(A - \lambda I) = 0$ for all values of λ. Those are the eigenvalues of A.
See Exercises 23 through 38.

Q: If $\text{rank}(A) <$ the number of rows, what eigenvalue does A have to have?
A: Zero. Why? Because $(A - 0I)\mathbf{x} = A\mathbf{x} = \mathbf{0}$ has nontrivial solutions. Why?
See Theorems 3.26 (Rank Theorem) and 3.27 (Fundamental Theorem) in Section 3.5.
Note this conclusion relies partly on the fact that the matrix A is square. Why?

Q: If A is not invertible, what eigenvalue does A have to have?
A: Zero. Why? Because $(A - 0I)\mathbf{x} = A\mathbf{x} = \mathbf{0}$ has nontrivial solutions. Why?
See Exercises 7 through 12.

Q: If A is invertible, what eigenvalue does A definitely not have?
A: Zero. Why? Because $(A - 0I)\mathbf{x} = A\mathbf{x} = \mathbf{0}$ has only the trivial solution. Why?
Why does the zero vector not count as an eigenvector? Eigenvectors must be *nonzero*.

Q: What do these conclusion tell us about projections?
A: A projection matrix always has rank 1. Why?
Geometrically, vectors perpendicular to the line of projection are mapped to **0**.
What does this tell us? All those vectors have eigenvalue 0.

Q: What do these conclusion tell us about rotations?
A: Every rotation matrix is invertible. Why?
Geometrically, all nonzero vectors are rotated and so do not map to parallel vectors.
What does this tell us? Rotations have no eigenvectors with eigenvalue 0.

Solutions to odd-numbered exercises from Section 4.1

1. If $A\mathbf{x} = \lambda\mathbf{x}$, then $\mathbf{x}$ is an eigenvector of A corresponding to λ.
   So, as in Example 4.1, since $A\mathbf{v} = \begin{bmatrix} 0 & 3 \\ 3 & 0 \end{bmatrix}\begin{bmatrix} 1 \\ 1 \end{bmatrix} = \begin{bmatrix} 3 \\ 3 \end{bmatrix} = 3\begin{bmatrix} 1 \\ 1 \end{bmatrix} = 3\mathbf{v}$,
   we see $\mathbf{v}$ is an eigenvector of A corresponding to (the eigenvalue) 3.

3. We compute $A\mathbf{v} = \begin{bmatrix} -1 & 1 \\ 6 & 0 \end{bmatrix}\begin{bmatrix} 1 \\ -2 \end{bmatrix} = \begin{bmatrix} -3 \\ 6 \end{bmatrix} = -3\begin{bmatrix} 1 \\ -2 \end{bmatrix} = -3\mathbf{v}$,
   we see $\mathbf{v}$ is an eigenvector of A corresponding to (the eigenvalue) -3.

5. We compute $A\mathbf{v} = \begin{bmatrix} 3 & 0 & 0 \\ 0 & 1 & -2 \\ 1 & 0 & 1 \end{bmatrix}\begin{bmatrix} 2 \\ -1 \\ 1 \end{bmatrix} = \begin{bmatrix} 6 \\ -3 \\ 3 \end{bmatrix} = 3\begin{bmatrix} 2 \\ -1 \\ 1 \end{bmatrix} = 3\mathbf{v}$,
   so $\mathbf{v}$ is an eigenvector of A corresponding to the eigenvalue 3.

7. As in Example 4.2, we show $\text{null}(A - 3I) \neq \mathbf{0}$ then compute $\text{null}(A - 3I)$ to find $\mathbf{x}$.

   Since $A\mathbf{x} = 3\mathbf{x}$ implies $(A - 3I)\mathbf{x} = \mathbf{0}$, we have:
   $$A - 3I = \begin{bmatrix} 2 & 2 \\ 2 & -1 \end{bmatrix} - \begin{bmatrix} 3 & 0 \\ 0 & 3 \end{bmatrix} = \begin{bmatrix} -1 & 2 \\ 2 & -4 \end{bmatrix}$$
   Since the columns of $A - 3I$ are clearly linearly dependent (because $\mathbf{a}_2 = -2\mathbf{a}_1$), the Fundamental Theorem of Invertible Matrices implies that $\text{null}(A - 3I) \neq \mathbf{0}$. That is $A\mathbf{x} = 3\mathbf{x}$ has a nontrivial solution, so 3 is an eigenvalue of A.

   Since $A\mathbf{x} = 3\mathbf{x}$ implies $(A - 3I)\mathbf{x} = \mathbf{0}$, we now compute $\text{null}(A - 3I)$.
   $$[A - 3I \mid \mathbf{0}] = \begin{bmatrix} -1 & 2 & | & 0 \\ 2 & -4 & | & 0 \end{bmatrix} \longrightarrow \begin{bmatrix} 1 & -2 & | & 0 \\ 0 & 0 & | & 0 \end{bmatrix}$$
   So, if $\mathbf{x} = \begin{bmatrix} x_1 \\ x_2 \end{bmatrix}$ is an eigenvector corresponding to the eigenvalue 3, then $x_1 = 2x_2$.

   These eigenvectors are of the form $\mathbf{x} = \begin{bmatrix} 2x_2 \\ x_2 \end{bmatrix}$. That is nonzero multiples of $\mathbf{x} = \begin{bmatrix} 2 \\ 1 \end{bmatrix}$.

   Q: What does this tell us about $\text{null}(A - 3I)$? What about E_3?

   A: The above shows $\text{null}(A - 3I) = \text{span}\left(\begin{bmatrix} 2 \\ 1 \end{bmatrix}\right) = E_3$, the *eigenspace* of 3.

4.1 Introduction to Eigenvalues and Eigenvectors

9. As in Example 4.2, we show null$(A - I) \neq 0$ then compute null$(A - I)$ to find **x**.

Since $A\mathbf{x} = \mathbf{x}$ implies $(A - I)\mathbf{x} = \mathbf{0}$, we have:
$$A - I = \begin{bmatrix} 0 & 4 \\ -1 & 5 \end{bmatrix} - \begin{bmatrix} 1 & 0 \\ 0 & 1 \end{bmatrix} = \begin{bmatrix} -1 & 4 \\ -1 & 4 \end{bmatrix}$$

Since the columns of $A - I$ are clearly linearly dependent (because $\mathbf{a}_2 = -4\mathbf{a}_1$), the Fundamental Theorem of Invertible Matrices implies that null$(A - I) \neq 0$. That is $A\mathbf{x} = \mathbf{x}$ has a nontrivial solution, so 1 is an eigenvalue of A.

Since $A\mathbf{x} = \mathbf{x}$ implies $(A - I)\mathbf{x} = \mathbf{0}$, we now compute null$(A - I)$.
$$[A - I \mid \mathbf{0}] = \begin{bmatrix} -1 & 4 & | & 0 \\ -1 & 4 & | & 0 \end{bmatrix} \longrightarrow \begin{bmatrix} -1 & 4 & | & 0 \\ 0 & 0 & | & 0 \end{bmatrix}$$

So, if $\mathbf{x} = \begin{bmatrix} x_1 \\ x_2 \end{bmatrix}$ is an eigenvector corresponding to the eigenvalue 1, then $x_1 = 4x_2$.

These eigenvectors are of the form $\mathbf{x} = \begin{bmatrix} 4x_2 \\ x_2 \end{bmatrix}$. That is nonzero multiples of $\mathbf{x} = \begin{bmatrix} 4 \\ 1 \end{bmatrix}$.

Q: What does this tell us about null$(A - I)$? What about E_1?

A: The above shows null$(A - I) = \text{span}\left(\begin{bmatrix} 4 \\ 1 \end{bmatrix}\right) = E_1$, the *eigenspace* of 1.

11. As in Example 4.2, we show null$(A + I) \neq 0$ then compute null$(A + I)$ to find **x**.

Since $A\mathbf{x} = -1\mathbf{x}$ implies $(A + I)\mathbf{x} = \mathbf{0}$, we have:
$$A + I = \begin{bmatrix} 1 & 0 & 2 \\ -1 & 1 & 1 \\ 2 & 0 & 1 \end{bmatrix} + \begin{bmatrix} 1 & 0 & 0 \\ 0 & 1 & 0 \\ 0 & 0 & 1 \end{bmatrix} = \begin{bmatrix} 2 & 0 & 2 \\ -1 & 2 & 1 \\ 2 & 0 & 2 \end{bmatrix}$$

Since the columns of $A + I$ are clearly linearly dependent (because $\mathbf{a}_3 = \mathbf{a}_1 + \mathbf{a}_2$), the Fundamental Theorem of Invertible Matrices implies that null$(A + I) \neq 0$. That is $A\mathbf{x} = -1\mathbf{x}$ has a nontrivial solution, so -1 is an eigenvalue of A.

Since $A\mathbf{x} = -1\mathbf{x}$ implies $(A + I)\mathbf{x} = \mathbf{0}$, we now compute null$(A + I)$.
$$[A + I \mid \mathbf{0}] = \begin{bmatrix} 2 & 0 & 2 & | & 0 \\ -1 & 2 & 1 & | & 0 \\ 2 & 0 & 2 & | & 0 \end{bmatrix} \longrightarrow \begin{bmatrix} 1 & 0 & 1 \\ 0 & 1 & 1 \\ 0 & 0 & 0 \end{bmatrix}$$

If $\mathbf{x} = \begin{bmatrix} x_1 \\ x_2 \\ x_3 \end{bmatrix}$ is an eigenvector corresponding to the eigenvalue -1, then $x_1 = x_2 = -x_3$.

These eigenvectors are of the form $\mathbf{x} = \begin{bmatrix} -x_3 \\ -x_3 \\ x_3 \end{bmatrix}$, nonzero multiples of $\mathbf{x} = \begin{bmatrix} -1 \\ -1 \\ 1 \end{bmatrix}$.

Q: What does this tell us about null$(A + I)$? What about E_{-1}?

A: The above shows null$(A + I) = \text{span}\left(\begin{bmatrix} -1 \\ -1 \\ 1 \end{bmatrix}\right) = E_{-1}$, the *eigenspace* of -1.

13. Since A reflects F in the y-axis, the only vectors that F maps parallel to themselves are: vectors parallel to the x-axis (i.e. multiples of $\begin{bmatrix} 1 \\ 0 \end{bmatrix}$), which are reversed (eigenvalue -1), and vectors parallel to the y-axis (multiples of $\begin{bmatrix} 0 \\ 1 \end{bmatrix}$), which are sent to themselves (eigenvalue 1).

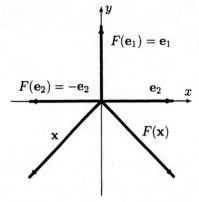

So, $\lambda = \pm 1$ are the eigenvalues of A with eigenspaces

$$E_{-1} = \text{span}\left(\begin{bmatrix} 1 \\ 0 \end{bmatrix}\right), \text{ and}$$

$$E_1 = \text{span}\left(\begin{bmatrix} 0 \\ 1 \end{bmatrix}\right).$$

15. From the remarks prior to Example 4.4, we have the following key insight:
$\mathbf{x}$ is an eigenvector of A if and only if A transforms $\mathbf{x}$ to a parallel vector.
Why? Because then $A\mathbf{x}$ and $\mathbf{x}$ are multiples of each other. That is, $A\mathbf{x} = \lambda\mathbf{x}$.
Recall that $E_\lambda = \text{null}(A - \lambda I) = \{\text{eigenvectors of } \lambda\} \cup \{\text{the zero vector, } \mathbf{0}\}$.
We have to add the zero vector because eigenvectors are nonzero by definition.

Since $A\mathbf{x} = \begin{bmatrix} 1 & 0 \\ 0 & 0 \end{bmatrix} \begin{bmatrix} x \\ y \end{bmatrix} = \begin{bmatrix} x \\ 0 \end{bmatrix}$, A is the matrix of projection P onto the x-axis.

Consider vectors $\mathbf{v}$ parallel to the x-axis, parallel to the y-axis, and not parallel to either axis.

x-axis: If $\mathbf{v}$ is parallel to the x-axis, P transforms $\mathbf{v}$ to itself. That is, $P(\mathbf{v}) = \mathbf{v}$.
So, all nonzero vectors parallel to the x-axis are eigenvectors of A corresponding to 1.

y-axis: If $\mathbf{v}$ is parallel to the y-axis, P transforms $\mathbf{v}$ to $\mathbf{0}$. That is, $P(\mathbf{v}) = \mathbf{0}$.
So, all nonzero vectors parallel to the y-axis are eigenvectors of A corresponding to 0.

neither: If $\mathbf{v}$ is not parallel to either axis, P transforms $\mathbf{v}$ to a nonparallel vector.
So, all nonzero vectors not parallel to either axis are not eigenvectors of A.

So $E_1 = \text{span}(x\text{-axis}) = \text{span}\left(\begin{bmatrix} 1 \\ 0 \end{bmatrix}\right)$ and $E_0 = \text{span}(y\text{-axis}) = \text{span}\left(\begin{bmatrix} 0 \\ 1 \end{bmatrix}\right)$.

Q: Given that the x-axis is a line, how might we generalize this result?
A: Hint: Consider vectors parallel, perpendicular and neither to the given line.

4.1 Introduction to Eigenvalues and Eigenvectors

17. From the remarks prior to Example 4.4, we have the following key insight:
x is an eigenvector of A if and only if A transforms **x** to a parallel vector.
Why? Because then $A\mathbf{x}$ and **x** are multiples of each other. That is, $A\mathbf{x} = \lambda\mathbf{x}$.
Recall that $E_\lambda = \text{null}(A - \lambda I) = \{\text{eigenvectors of } \lambda\} \cup \{\text{the zero vector}, \mathbf{0}\}$.
We have to add the zero vector because eigenvectors are nonzero by definition.

Since $A\mathbf{x} = \begin{bmatrix} 2 & 0 \\ 0 & 3 \end{bmatrix} \begin{bmatrix} x \\ y \end{bmatrix} = \begin{bmatrix} 2x \\ 3y \end{bmatrix}$, A is the matrix of stretching S.

Consider vectors **v** parallel to the x-axis, parallel to the y-axis, and not parallel to either axis.

- x-axis: If **v** is parallel to the x-axis, S transforms **v** to twice itself. That is, $S(\mathbf{v}) = 2\mathbf{v}$.
  So, all nonzero vectors parallel to the x-axis are eigenvectors of A corresponding to 2.
- y-axis: If **v** is parallel to the y-axis, S transforms **v** to thrice itself. That is, $S(\mathbf{v}) = 3\mathbf{v}$.
  So, all nonzero vectors parallel to the y-axis are eigenvectors of A corresponding to 3.
- *neither*: If **v** is not parallel to either axis, S transforms **v** to a nonparallel vector.
  So, all vectors not parallel to either axis are not eigenvectors of A.

So $E_2 = \text{span}(x\text{-axis}) = \text{span}\left(\begin{bmatrix} 1 \\ 0 \end{bmatrix}\right)$ and $E_3 = \text{span}(y\text{-axis}) = \text{span}\left(\begin{bmatrix} 0 \\ 1 \end{bmatrix}\right)$.

Q: Following this exact same process, how might we generalize this result?

A: If $A = \begin{bmatrix} a & 0 \\ 0 & d \end{bmatrix}$, then $E_a = \text{span}(x\text{-axis})$ and $E_d = \text{span}(y\text{-axis})$.

19. From the remarks prior to Example 4.4, we have the following key insight:
x is an eigenvector of A if and only if A transforms **x** to a parallel vector.
So, lines that do *not* bend at the unit circle represent eigenvectors.
The extension beyond the circle tells us if the vector has been stretched.

Since the lines do not bend on the x-axis and the y-axis, we consider vectors **v** parallel to the x-axis, parallel to the y-axis, and not parallel to either axis.

- x-axis: On the x-axis, the lines do not bend and extend precisely one unit beyond it. So:
  If **v** is parallel to the x-axis, S transforms **v** to itself. That is, $S(\mathbf{v}) = \mathbf{v}$.
  So, all nonzero vectors parallel to the x-axis are eigenvectors of A corresponding to 1.
- y-axis: On the y-axis, the lines do not bend and extend precisely 2 units beyond it. So:
  If **v** is parallel to the y-axis, S transforms **v** to twice itself. That is, $S(\mathbf{v}) = 2\mathbf{v}$.
  So, all nonzero vectors parallel to the y-axis are eigenvectors of A corresponding to 2.
- *neither*: Off the x-axis and y-axis, the lines *do* bend at the unit circle. So:
  If **v** is not parallel to either axis, S transforms **v** to a nonparallel vector.
  So, all vectors not parallel to either axis are not eigenvectors of A.

So $E_1 = \text{span}(x\text{-axis}) = \text{span}\left(\begin{bmatrix} 1 \\ 0 \end{bmatrix}\right)$ and $E_2 = \text{span}(y\text{-axis}) = \text{span}\left(\begin{bmatrix} 0 \\ 1 \end{bmatrix}\right)$.

21. From the remarks prior to Example 4.4, we have the following key insight:
    x is an eigenvector of A if and only if A transforms x to a parallel vector.
    So, lines that do *not* bend at the unit circle represent eigenvectors.
    The extension beyond the circle tells us if the vector has been stretched.

    Since the lines do not bend on the line ℓ $y = -x$ with direction vector $\mathbf{d} = \begin{bmatrix} 1 \\ 1 \end{bmatrix}$,

    So, we consider vectors **v** parallel to **d**, perpendicular to **d**, and neither.

    parallel: On the line $y = x$, the lines do not bend and extend precisely 2 units beyond it. So:
    If **v** is parallel to the **d**, S transforms **v** to thrice itself. That is, $S(\mathbf{v}) = 2\mathbf{v}$.
    So, all nonzero vectors parallel to **d** are eigenvectors of A corresponding to 2.

    perpendicular: On the line $y = -x$, the lines extend precisely 0 units beyond the unit circle. So:
    If **v** is perpendicular to the **d**, S transforms **v** to **0**. That is, $S(\mathbf{v}) = 0\mathbf{v}$.
    So, all nonzero vectors perpendicular to **d** are eigenvectors of A corresponding to 0.

    neither: Off the lines $y = x$ and $y = -x$, the lines *do* bend at the unit circle. So:
    If **v** is not parallel or perpendicular to **d**, S transforms **v** to a nonparallel vector.
    So, all vectors not parallel or perpendicular to **d** are not eigenvectors of A.

    So $E_2 = \text{span}(\mathbf{d}) = \text{span}\left(\begin{bmatrix} 1 \\ 1 \end{bmatrix}\right)$ and $E_0 = \text{span}\left(\begin{bmatrix} 1 \\ -1 \end{bmatrix}\right)$.

23. As in Example 4.5, we find all solutions λ of the equation $\det(A - \lambda I) = 0$.

$$\det(A - \lambda I) = \det \begin{bmatrix} 4 - \lambda & -1 \\ 2 & 1 - \lambda \end{bmatrix} = \lambda^2 - 5\lambda + 6$$

Since $\lambda^2 - 5\lambda + 6 = (\lambda - 2)(\lambda - 3) = 0$, the solutions are $\lambda = 2$ and $\lambda = 3$.

$\lambda = 2$: $A - 2I = \begin{bmatrix} 4 - 2 & -1 \\ 2 & 1 - 2 \end{bmatrix} = \begin{bmatrix} 2 & -1 \\ 2 & -1 \end{bmatrix} \longrightarrow \begin{bmatrix} 2 & -1 \\ 0 & 0 \end{bmatrix}$

So, if $\mathbf{x} = \begin{bmatrix} x_1 \\ x_2 \end{bmatrix}$ is an eigenvector corresponding to the eigenvalue 2, then $x_2 = 2x_1$.

These eigenvectors are of the form $\mathbf{x} = \begin{bmatrix} x_1 \\ 2x_1 \end{bmatrix}$. That is nonzero multiples of $\begin{bmatrix} 1 \\ 2 \end{bmatrix}$.

So, $E_2 = \text{null}(A - 2I) = \text{span}\left(\begin{bmatrix} 1 \\ 2 \end{bmatrix}\right)$.

$\lambda = 3$: $A - 3I = \begin{bmatrix} 4 - 3 & -1 \\ 2 & 1 - 3 \end{bmatrix} \longrightarrow \begin{bmatrix} 1 & -1 \\ 0 & 0 \end{bmatrix}$

So, if $\mathbf{x} = \begin{bmatrix} x_1 \\ x_2 \end{bmatrix}$ is an eigenvector corresponding to the eigenvalue 3, then $x_2 = x_1$.

These eigenvectors are of the form $\mathbf{x} = \begin{bmatrix} x_1 \\ x_1 \end{bmatrix}$. That is nonzero multiples of $\begin{bmatrix} 1 \\ 1 \end{bmatrix}$.

So, $E_3 = \text{null}(A - 3I) = \text{span}\left(\begin{bmatrix} 1 \\ 1 \end{bmatrix}\right)$.

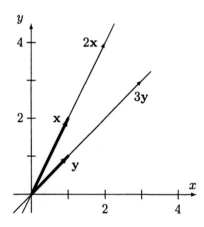

25. As in Example 4.5, we find all solutions λ of the equation $\det(A - \lambda I) = 0$.

$$\det(A - \lambda I) = \det \begin{bmatrix} 2 - \lambda & 5 \\ 0 & 2 - \lambda \end{bmatrix} = (2 - \lambda)^2$$

Since $(2 - \lambda)^2 = (\lambda - 2)(\lambda - 2) = 0$, the solution is $\lambda = 2$.

$\lambda = 2$: $A - 2I = \begin{bmatrix} 2-2 & 5 \\ 0 & 2-2 \end{bmatrix} = \begin{bmatrix} 0 & 5 \\ 0 & 0 \end{bmatrix} \longrightarrow \begin{bmatrix} 0 & 1 \\ 0 & 0 \end{bmatrix}$

So, if $\mathbf{x} = \begin{bmatrix} x_1 \\ x_2 \end{bmatrix}$ is an eigenvector corresponding to 2, then $x_1 = t$, $x_2 = 0$.

These eigenvectors are of the form $\mathbf{x} = \begin{bmatrix} t \\ 0 \end{bmatrix}$. That is nonzero multiples of $\begin{bmatrix} 1 \\ 0 \end{bmatrix}$.

So, $E_2 = \text{null}(A - 2I) = \text{span}\left(\begin{bmatrix} 1 \\ 0 \end{bmatrix}\right)$.

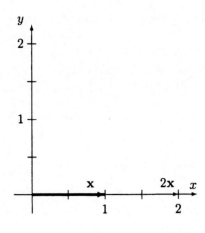

4.1 Introduction to Eigenvalues and Eigenvectors

27. As in Example 4.7, we find all solutions λ of the equation $\det(A - \lambda I) = 0$.

$$\det(A - \lambda I) = \det \begin{bmatrix} 1 - \lambda & 1 \\ -1 & 1 - \lambda \end{bmatrix} = \lambda^2 - 2\lambda + 2$$

Since $\lambda^2 - 2\lambda + 2 = 0$, the solutions are $\lambda = 1 + i, 1 - i$.

$1 + i$: $A - (1+i)I = \begin{bmatrix} 1 - (1+i) & 1 \\ -1 & 1 - (1+i) \end{bmatrix} = \begin{bmatrix} -i & 1 \\ -1 & -i \end{bmatrix} \longrightarrow \begin{bmatrix} 1 & i \\ 0 & 0 \end{bmatrix}$

So, if $\mathbf{x} = \begin{bmatrix} x_1 \\ x_2 \end{bmatrix}$ is an eigenvector corresponding to $1 + i$, then $x_1 = -ix_2 = -it$.

These eigenvectors are of the form $\mathbf{x} = \begin{bmatrix} -it \\ t \end{bmatrix}$. That is nonzero multiples of $\begin{bmatrix} -i \\ 1 \end{bmatrix}$.

So, $E_{1+i} = \text{null}(A - (1+i)I) = \text{span}\left(\begin{bmatrix} -i \\ 1 \end{bmatrix} \right)$.

$1 - i$: $A - (1-i)I = \begin{bmatrix} 1 - (1-i) & 1 \\ -1 & 1 - (1-i) \end{bmatrix} = \begin{bmatrix} i & 1 \\ -1 & i \end{bmatrix} \longrightarrow \begin{bmatrix} 1 & -i \\ 0 & 0 \end{bmatrix}$

So, if $\mathbf{x} = \begin{bmatrix} x_1 \\ x_2 \end{bmatrix}$ is an eigenvector corresponding to $1 - i$, then $x_1 = ix_2 = it$.

These eigenvectors are of the form $\mathbf{x} = \begin{bmatrix} it \\ t \end{bmatrix}$. That is nonzero multiples of $\begin{bmatrix} i \\ 1 \end{bmatrix}$.

So, $E_{1-i} = \text{null}(A - (1-i)I) = \text{span}\left(\begin{bmatrix} i \\ 1 \end{bmatrix} \right)$.

29. As in Example 4.7, we find all solutions λ of the equation $\det(A - \lambda I) = 0$.

$$\det(A - \lambda I) = \det \begin{bmatrix} 1 - \lambda & i \\ i & 1 - \lambda \end{bmatrix} = \lambda^2 - 2\lambda + 2$$

Since $\lambda^2 - 2\lambda + 2 = 0$, the solutions are $\lambda = 1 + i, 1 - i$.

$1 + i$: $A - (1+i)I = \begin{bmatrix} 1 - (1+i) & i \\ i & 1 - (1+i) \end{bmatrix} = \begin{bmatrix} -i & i \\ i & -i \end{bmatrix} \longrightarrow \begin{bmatrix} 1 & -1 \\ 0 & 0 \end{bmatrix}$

So, if $\mathbf{x} = \begin{bmatrix} x_1 \\ x_2 \end{bmatrix}$ is an eigenvector corresponding to $1 + i$, then $x_1 = x_2 = t$.

These eigenvectors are of the form $\mathbf{x} = \begin{bmatrix} t \\ t \end{bmatrix}$. That is nonzero multiples of $\begin{bmatrix} 1 \\ 1 \end{bmatrix}$.

So, $E_{1+i} = \text{null}(A - (1+i)I) = \text{span}\left(\begin{bmatrix} 1 \\ 1 \end{bmatrix} \right)$.

$1 - i$: $A - (1-i)I = \begin{bmatrix} 1 - (1-i) & i \\ i & 1 - (1-i) \end{bmatrix} = \begin{bmatrix} i & i \\ i & i \end{bmatrix} \longrightarrow \begin{bmatrix} 1 & 1 \\ 0 & 0 \end{bmatrix}$

So, if $\mathbf{x} = \begin{bmatrix} x_1 \\ x_2 \end{bmatrix}$ is an eigenvector corresponding to $1 - i$, then $x_1 = -x_2 = t$.

These eigenvectors are of the form $\mathbf{x} = \begin{bmatrix} t \\ -t \end{bmatrix}$. That is nonzero multiples of $\begin{bmatrix} 1 \\ -1 \end{bmatrix}$.

So, $E_{1-i} = \text{null}(A - (1-i)I) = \text{span}\left(\begin{bmatrix} 1 \\ -1 \end{bmatrix} \right)$.

31. As in Example 4.6, we find all solutions λ of the equation $\det(A - \lambda I) = 0$.

$$\det(A - \lambda I) = \det \begin{bmatrix} 1-\lambda & 0 \\ 1 & 2-\lambda \end{bmatrix} = \lambda^2 - 3\lambda + 2$$

Since $\lambda^2 - 3\lambda + 2 = (\lambda - 1)(\lambda - 2) = 0$ in $\mathbb{Z}_3$, the eigenvalues of A are $\lambda = 1, 2$.

33. As in Example 4.6, we find all solutions λ of the equation $\det(A - \lambda I) = 0$.

$$\det(A - \lambda I) = \det \begin{bmatrix} 3-\lambda & 1 \\ 4 & 0-\lambda \end{bmatrix} = \lambda^2 - 3\lambda - 4$$

Since $\lambda^2 - 3\lambda - 4 = \lambda^2 + 2\lambda + 1 = (\lambda + 1)^2 = 0$ in $\mathbb{Z}_5$, the eigenvalues of A are $\lambda = 4$.

35. (a) To find the eigenvalues of $A = \begin{bmatrix} a & b \\ c & d \end{bmatrix}$, we solve $\det(A - \lambda I) = 0 \Leftrightarrow$

$$\det \begin{bmatrix} a-\lambda & b \\ c & d-\lambda \end{bmatrix} = \lambda^2 - (a+d)\lambda + (ad - bc) = \lambda^2 - \operatorname{tr}(A)\lambda + \det A = 0.$$

(b) Using the quadratic formula, the solutions to the equation in part (a) are

$$\lambda = \frac{(a+d) \pm \sqrt{(a+d)^2 - 4(ad - bc)}}{2} = \frac{a+d \pm \sqrt{a^2 + d^2 + 2ad - 4ad + 4bc}}{2}$$

$$= \frac{1}{2}\left(a + d \pm \sqrt{(a-d)^2 + 4bc}\right).$$

(c) Let $\lambda_1 = \frac{1}{2}(a+d) + \sqrt{(a-d)^2 + 4bc}$ and $\lambda_2 = \frac{1}{2}(a+d) - \sqrt{(a-d)^2 + 4bc}$.
So, $\lambda_1 + \lambda_2 = \frac{1}{2}(a+d) + \frac{1}{2}(a+d) = a + d = \operatorname{tr}(A)$.
Also, $\lambda_1 \lambda_2 = \frac{1}{4}[(a+d)^2 - ((a-d)^2 + 4bc)] = \frac{1}{4}[4ad - 4bc] = ad - bc = \det A$.

4.1 Introduction to Eigenvalues and Eigenvectors

37. As in Example 4.5, we find all solutions λ of the equation $\det(A - \lambda I) = 0$.

$$\det(A - \lambda I) = \det \begin{bmatrix} a - \lambda & b \\ 0 & d - \lambda \end{bmatrix} = \lambda^2 - (a+d)\lambda + ad$$

Since $\lambda^2 - (a+d)\lambda + ad = (\lambda - a)(\lambda - d) = 0$, the solutions are $\lambda = a, d$.

$\lambda = a$: $A - aI = \begin{bmatrix} a - a & b \\ 0 & a - d \end{bmatrix} = \begin{bmatrix} 0 & b \\ 0 & a - d \end{bmatrix} \longrightarrow \begin{bmatrix} 0 & 1 \\ 0 & 0 \end{bmatrix}$

If $\mathbf{x} = \begin{bmatrix} x_1 \\ x_2 \end{bmatrix}$ is an eigenvector corresponding to a, $x_1 = t$, $x_2 = 0$.

These eigenvectors are of the form $\mathbf{x} = \begin{bmatrix} t \\ 0 \end{bmatrix}$, nonzero multiples of $\begin{bmatrix} 1 \\ 0 \end{bmatrix}$.

So, $E_a = \text{null}(A - aI) = \text{span}\left(\begin{bmatrix} 1 \\ 0 \end{bmatrix} \right)$.

$\lambda = d$: $A - dI = \begin{bmatrix} a - d & b \\ 0 & d - d \end{bmatrix} = \begin{bmatrix} a - d & b \\ 0 & 0 \end{bmatrix} \longrightarrow \begin{bmatrix} a - d & b \\ 0 & 0 \end{bmatrix}$

If $\mathbf{x} = \begin{bmatrix} x_1 \\ x_2 \end{bmatrix}$ is an eigenvector corresponding to d, $(d-a)x_1 = bx_2 = (d-a)bt$.

These eigenvectors are of the form $\mathbf{x} = \begin{bmatrix} bt \\ (d-a)t \end{bmatrix}$, nonzero multiples of $\begin{bmatrix} b \\ d-a \end{bmatrix}$.

So, $E_b = \text{null}(A - bI) = \text{span}\left(\begin{bmatrix} b \\ d-a \end{bmatrix} \right)$.

4.2 Determinants

Q: What are our main goals for Section 4.2?

A: To take a close at determinants and how they relate to a matrix being invertible. In addition, we will learn how to follow a series of theorems to a large conclusion. Finally, we will work on developing our own skills in explaining and writing math.

Key Definitions and Concepts

$\sum_{i=1}^{n} a_{ij} C_{ij}$ p.263   4.2 $\sum_{j=1}^{n} a_{ij} C_{ij} = a_{i1} C_{i1} + a_{i2} C_{i2} + \cdots + a_{in} C_{in}$

Shorthand for sums like $\prod_{i=1}^{n} A_i$ is for products

$\prod_{i=1}^{n} A_i$ p.263   4.2 $\prod_{i=1}^{n} A_i = A_1 A_2 \cdots A_n$

Shorthand for products like $\sum_{j=1}^{n} a_{ij} C_{ij}$ is for sums

A_{ij} p.263   4.2   submatrix of A obtained by deleting row i and column j

$\det A_{ij}$ p.263   4.2 $\det A_{ij}$, the (i,j)-minor of A

$A_i(\mathbf{b})$ p.273   4.2 $A_i(\mathbf{b})$ obtained by replacing column i of A by $\mathbf{b}$
That is, $A_i(\mathbf{b}) = \begin{bmatrix} \mathbf{a}_1 & \cdots & \mathbf{b} & \cdots & \mathbf{a}_n \end{bmatrix}$ (see Cramer's Rule)

adj A p.275   4.2   adj $A = [C_{ij}]^T$, the adjoint of A

$\det A$ p.264   4.2 $\det A = \sum_{j=1}^{n} a_{1j}(-1)^{i+1} \det A_{1j}$ or $\det A = \sum_{j=1}^{n} a_{1j} C_{1j}$

C_{ij} p.265   4.2 $C_{ij} = (-1)^{i+j} \det A_{ij}$, the (i,j)-cofactor of A

adjoint   p.275   4.2   adj $A = [C_{ij}]^T$, the adjoint of A, the matrix of cofactors

cofactor   p.265   4.2 $C_{ij} = (-1)^{i+j} \det A_{ij}$, the (i,j)-cofactor of A

determinant   p.264   4.2 $\det A = \sum_{j=1}^{n} a_{1j}(-1)^{i+1} \det A_{1j}$ or $\det A = \sum_{j=1}^{n} a_{1j} C_{1j}$

expansion   p.265   4.2 $\det A = \sum_{j=1}^{n} a_{ij} C_{ij}$ along the ith row

$\det A = \sum_{i=1}^{n} a_{ij} C_{ij}$ along the jth column

minor   p.263   4.2 $\det A_{ij}$, the (i,j)-minor of A

submatrix   p.263   4.2   submatrix of A obtained by deleting row i and column j

Theorems

Laplace	p.265	4.2	$\det A = \sum_{j=1}^{n} a_{ij}C_{ij}$ (any row) or $\sum_{i=1}^{n} a_{ij}C_{ij}$ (any column)
Cramer	p.274	4.2	If $A\mathbf{x} = \mathbf{b}$, then $x_i = \frac{\det(A_i(\mathbf{b}))}{\det A}$ (see Thm 4.11)
Thm 4.1	p.265	4.2	$\det A = \sum_{j=1}^{n} a_{ij}C_{ij}$ (any row) or $\sum_{i=1}^{n} a_{ij}C_{ij}$ (any column)
Thm 4.2	p.268	4.2	If A is triangular, then $\det A = a_{11}a_{22}\cdots a_{nn}$

Thm 4.3   p.268   4.2   a. through f. detail row (column) operations effects on $\det A$
  a. $\mathbf{A}_i = \mathbf{0} \Rightarrow \det A = 0$
  b. $A \xrightarrow{R_i \leftrightarrow R_j} B \Rightarrow \det B = -\det A$
  c. $\mathbf{A}_i = \mathbf{A}_j \Rightarrow \det A = 0$
  d. $A \xrightarrow{kR_i} B \Rightarrow \det B = k \det A$
  e. $\mathbf{C}_i = \mathbf{A}_i + \mathbf{B}_i \Rightarrow \det C = \det A + \det B$
  f. $A \xrightarrow{R_i + kR_j} B \Rightarrow \det B = \det A$

Thm 4.4   p.270   4.2   a. through c. detail row (column) operations effects on $\det E$
  a. $I \xrightarrow{R_i \leftrightarrow R_j} E \Rightarrow \det E = -1$
  b. $I \xrightarrow{kR_i} E \Rightarrow \det E = k$
  c. $I \xrightarrow{R_i + kR_j} E \Rightarrow \det E = 1$

Lem 4.5	p.271	4.2	E, elementary, then $\det(EB) = (\det E)(\det B)$
Thm 4.6	p.271	4.2	A is invertible if and only if $\det A \neq 0$
Thm 4.7	p.271	4.2	$\det(kA) = k^n \det A$
Thm 4.8	p.272	4.2	$\det(AB) = (\det A)(\det B)$
Thm 4.9	p.273	4.2	$\det(A^{-1}) = \frac{1}{\det A}$
Thm 4.10	p.273	4.2	$\det A = \det(A^T)$
Thm 4.11	p.274	4.2	If $A\mathbf{x} = \mathbf{b}$, then $x_i = \frac{\det(A_i(\mathbf{b}))}{\det A}$ (Cramer's Rule)
Thm 4.12	p.276	4.2	$A^{-1} = \frac{1}{\det A}\operatorname{adj} A$
Lem 4.13	p.276	4.2	(row 1) $\det A = \sum_{j=1}^{n} a_{1j}C_{1j} = \sum_{i=1}^{n} a_{i1}C_{i1}$ (column 1)
Lem 4.14	p.276	4.2	$A \xrightarrow{R_i \leftrightarrow R_j} B \Rightarrow \det B = -\det A$ (see Thm 4.3(f))

4.2 Determinants

Discussion of Definitions and Theorems

Statements for rows also hold true for columns

The statements in this discussion are often about rows, but they also hold true for columns.

det A is a scalar

To help us remember that det A is a scalar, see Exercise 55:

We use the fact that $A^2 = A \overset{\text{Ex }47}{\Rightarrow} (\det A)^2 = \det A$ to find all possible values of det A.

$$(\det A)^2 = \det A \Rightarrow (\det A)^2 - \det A = 0 \Rightarrow \det A(\det A - 1) = 0 \Rightarrow \det A = 0, 1$$

Q: If we let $k = \det A$, what does the above calculation look like?
A: With $k = \det A$: $k^2 = k \Rightarrow k^2 - k = 0 \Rightarrow k(k-1) = 0 \Rightarrow k = 0, 1$.

This clarifies the algebra and helps us remember that det A is a scalar.

adj A is the matrix of cofactors

By definition, adj A is the matrix of cofactors. So adj A is a matrix. See Exercise 61:

We compute $\det A = \begin{vmatrix} 1 & 1 \\ 1 & -1 \end{vmatrix} = -2$ and the four cofactors $C_{11} = +(-1) = -1$,

$C_{12} = -1$, $C_{21} = -1$, and $C_{22} = 1$. The adjoint is the transpose of the matrix of cofactors:

$$\text{adj } A = \begin{bmatrix} -1 & -1 \\ -1 & 1 \end{bmatrix}^T = \begin{bmatrix} -1 & -1 \\ -1 & 1 \end{bmatrix}. \text{ Then } A^{-1} = \frac{1}{\det A} \text{adj } A = -\frac{1}{2}\begin{bmatrix} -1 & -1 \\ -1 & 1 \end{bmatrix} = \begin{bmatrix} \frac{1}{2} & \frac{1}{2} \\ \frac{1}{2} & -\frac{1}{2} \end{bmatrix}.$$

Q: Is adj A a matrix or a scalar? What about det A?
A: By definition, adj A is the matrix of cofactors. So adj A is a matrix.
On the other hand, det A is a scalar. The names have a similar form. Be careful!

Practical consequences of the Laplace Expansion Theorem

The Laplace Expansion Theorem is extremely profound. This is evident in two separate ways:
1) It makes the proof of most of the results in this section straightforward and simple.
2) Its own proof is subtle and complex, requiring much care, insight, and perseverance.

Q: What does the Laplace Expansion Theorem tell us about computing det A?
A: When we compute det A, we can choose to expand along *any* row or column of A.

Q: What should we look for when choosing a row or column to expand along?
A: A row or column with the maximum number of zeroes. Why?

The maximum number of zeroes minimizes the number of cofactors we have to compute. See Exercises 1 through 15.

Determination of the coefficients and signs of $\det A_{ij}$ *and* C_{ij}

There are lots of subscripts and exponents in these formulas.
We should examine a few examples to see what those subscripts and formulas tell us.

Q: What does $C_{ij} = (-1)^{i+j} \det A_{ij}$ tell us about the sign of C_{ij}?
A: If $i+j$ is *even*, the sign of C_{ij} is 1. If $i+j$ is *odd*, the sign of C_{ij} is -1.

Q: Where does the sign of the coefficient of $\det A_{ij}$ come from?
A: From the sign of C_{ij} times the sign of a_{ij}. Why?
Because the entire coefficient of $\det A_{21}$ is $a_{ij}(-1)^{i+j}$.

Q: In Exercise 8, why is the coefficient of $\det A_{21}$ equal to -2 instead of 2?
A: Because the cofactor $C_{21} = (-1)^{2+1} \det A_{21} = -\det A_{21}$.
The entire coefficient of $\det A_{21}$ is $a_{21}(-1)^{2+1} = -2$.

Discussion of Theorem 4.6: *A is invertible if and only if* $\det A \neq 0$

Q: What is the contrapositive of Theorem 4.6?
A: A is *not* invertible if and only if $\det A = 0$.

Q: In Exercise 11, we find $\det A = ab(a+b)$. So, if $a = -b$ what do we know about A?
A: If $a = -b$, then A is *not* invertible (because $\det A = 0$).

Q: Since $\det A = ab(a+b)$, if $a = 0$ what do we know about A?

Q: What are the conditions on a and b that will guarantee A is invertible?
A: Both a and b are nonzero and $a \neq -b$. Why is this enough?

Connecting $\det A = 0$ *to linear dependence of rows and columns*

Q: In Exercise 12, we find $\det A = 0$. So, what do we know about A?
A: We know that A is *not* invertible.

Q: Why is it obvious that A is *not* invertible even *before* we compute $\det A$?
A: Since row 3 is a multiple of row 1, these rows are linearly dependent. Why is this enough?

Q: Is it equally obvious that $\det A = 0$ even *before* computing it?
A: Yes, because once we know A is *not* invertible, we know $\det A$ must equal zero.

Q: In Exercise 13, we find $\det A = 4$. What does that tell us about the rows of A?
A: The rows of A are linearly independent (because $\det A \neq 0$. Why is that enough?).

Q: What does $\det A = 4$ tell us about the columns of A?
Q: What does $\det B \neq 0$ tell us about the rows and columns of B?

4.2 Determinants

Paying attention to details in proofs

We use the solution of Exercise 21 as an example of this principle. All the details given after the proof support the proof given below. Investigate carefully. In general, we should be critical of our reasoning and actively seek out oversights.

21. We use induction to prove Theorem 4.2: if A is triangular then $\det A = a_{11}a_{22}\cdots a_{nn}$. See Appendix B for discussion and examples of *Mathematical Induction*. Since Theorem 4.10 asserts $\det A = \det A^T$, we can assume A is upper triangular.

   1: If A is 1×1, then $A = [a_{11}]$ so $\det A = \det([a_{11}]) = a_{11}$.
   This is obvious, so there is nothing to show.

   n: If A is $n \times n$ and upper triangular, then $\det A = a_{11}a_{22}\cdots a_{nn}$.
   This is the induction hypothesis.

   $n+1$: If A is $(n+1) \times (n+1)$ and upper triangular, then $\det A = a_{11}a_{22}\cdots a_{nn}a_{n+1\,n+1}$.
   This is the statement we must prove using the induction hypothesis.

   $$\det A \underset{\substack{A \text{ is upper} \\ \text{triangular}}}{=} a_{n+1\,n+1} \det A_{n+1\,n+1} \underset{\substack{\text{by} \\ \text{induction}}}{=} a_{n+1\,n+1}(a_{11}a_{22}\cdots a_{nn}) = a_{11}a_{22}\cdots a_{n+1\,n+1}$$

   We have proven (by induction) that if A is $n \times n$ and triangular, then $\det A = a_{11}a_{22}\cdots a_{nn}$.

   Q: Why does the fact that $\det A = \det A^T$ allow us to assume A is upper triangular?
   A: Because if A is lower triangular, then A^T is upper triangular.
   So, $\det A = \det A^T = a_{11}a_{22}\cdots a_{nn}$ because $[A]_{ii} = [A^T]_{ii}$.
   That is, the diagonal entries of A and A^T are equal.

   Q: Why does A being upper triangular imply $\det A = a_{n+1\,n+1} \det A_{n+1\,n+1}$?
   A: Because if A is upper triangular, then row $n+1 = \mathbf{A}_{n+1} = \begin{bmatrix} 0 & 0 & \cdots & a_{n+1\,n+1} \end{bmatrix}$.
   So when we expand along this row we have: $\det A = a_{n+1\,n+1} \det A_{n+1\,n+1}$.

   Q: Why is the coefficient of $\det A_{n+1\,n+1}$ equal to $a_{n+1\,n+1}$ instead of $-a_{n+1\,n+1}$?
   A: The cofactor $C_{n+1\,n+1} = (-1)^{2n+2} \det A_{n+1\,n+1} = \det A_{n+1\,n+1}$ because $2n+2$ is even.

   Q: Why do we get to apply the induction hypothesis to the matrix $A_{n+1\,n+1}$?
   A: Since $A_{n+1\,n+1}$ is created by removing the $n+1$ row and the $n+1$ column of A, $A_{n+1\,n+1}$ is $n \times n$ and upper triangular. Why is this obvious?
   Since A is $(n+1) \times (n+1)$, $A_{n+1\,n+1}$ (created by removing a row and column) is $n \times n$.
   Since A is upper triangular, $[A]_{ij} = 0 = [A_{n+1\,n+1}]_{ij}$ for $i > j$ when $i, j \leq n$.
   That is, $A_{n+1\,n+1}$ is upper triangular.

Loose explanations that elucidate proofs versus hand waving

Hand waving refers to faulty proofs that fail to be convincing because they omit detail.
On the other hand, loose explanations give an overview of the proof that elucidate the ideas.

Q: How might we state the essentials of our proof of Exercise 21 *loosely* in words?
A: Our goal is to elucidate the reasons that the stated conclusion is true.
For instance, we might point out that when computing the determinant
the zeros below the diagonal are paired with the entries above the diagonal.
So the entries on the diagonal are the only ones that *count*, so to speak.
From there, we could proceed to elucidate the details how that pairing takes place.
For a 2×2 matrix this pairing is obvious since a_{12} is multiplied by a_{21}.
This is the key fact that we use to prove the first step of induction.
Then as we expand along the columns from left to right, we continue to pair zeros with minors and expand until all the matrices involved are 2×2, so the conclusion holds.
This is essentially the process underlying the application of the induction hypothesis.

Again the essential difference between loose explanations and hand waving is this:
Hand waving attempts to obscure the weaknesses in our argument while
loose explanations attempt to elucidate the strengths of our argument.

The process of trying to elucidate our argument can often reveal weaknesses it to ourselves.

As an example of *hand waving* consider the following solution to Exercise 69:

We use induction on the number of rows of P. If P is 1×1,

then $\det A = \begin{vmatrix} P & Q \\ O & S \end{vmatrix} = (\det P)(\det S)$ by expansion along the first column.

Now we assume that the result holds for a $k \times k$ matrix P'.

Now if $P = \begin{bmatrix} p & Q' \\ O & P' \end{bmatrix}$, then expanding along the first column gives $\det P = p \det P'$.

Now, calculate the determinant of $A = \begin{bmatrix} P & Q \\ O & S \end{bmatrix} = \begin{bmatrix} p & Q' & Q \\ O & P' & q \\ O & 0 & S \end{bmatrix}$,

where q is an arbitrary entry.

Expanding along the first column, we have:
$\det A = p (\det P')(\det S) = (\det P)(\det S)$, as required.

Q: Is this proof satisfying? Can we explain in words why the conclusion is true?
Q: Compare the solution above to the one given among the solutions below.
Are the omitted details trivial or obvious?

4.2 Determinants

Loose descriptions of theorems

It is useful to state the conclusions of a theorem in a way we can understand intuitively. These descriptions do not capture all the details, but they do convey the essential conclusion.

Q: How might we state Theorem 4.1 (Laplace Expansion) *loosely* in words?
A: When we compute det A, we can choose to expand along *any* row or column of A.

Q: How might we state Theorem 4.2 (determinants of triangular matrices)?
A: The determinant of a triangular matrix is just the product of the entries on its diagonal.

Q: How might we state Theorem 4.3(a) (how matrix operations impact det A)?
A: The determinant of any matrix with a zero row or column is zero.

Q: Why is it obvious that det $O = 0$?
A: Because O has a zero row. In fact, all its rows are zero rows.

Q: How might we state Theorem 4.3(b) (how matrix operations impact det A)?
A: When we interchange rows in a matrix, we change the sign of the determinant.

Q: How might we state Theorem 4.3(d) (how matrix operations impact det A)?
A: The determinant of any matrix with duplicate rows or columns is zero.

Q: How might we state Theorem 4.8 ($\det(AB) = (\det A)(\det B)$)?
A: The determinant of the product equals the product of the determinants.

Drawing useful conclusions from theorems

Putting theorems into words allows us to draw useful conclusions not explicitly stated.

Q: What is an implied conclusion of Theorem 4.6 (A invertible $\Leftrightarrow \det A \neq 0$)?
A: The determinant of any matrix with linearly independent rows or columns cannot be zero. For a full discussion see questions and answers following Exercise 26 below.

Q: What about the contrapositive of Theorem 4.6 (A *not* invertible $\Leftrightarrow \det A = 0$)?
A: The determinant of any matrix with linearly dependent rows or columns must be zero.

Q: In Exercise 9, what does det $A = -12$ tell us about the rows and columns of A and B?

Q: What is an implied conclusion of Theorem 4.10 ($\det A = \det(A^T)$)?
A: Any statement about determinants that is true for rows is also true for columns. See how the key insight comes from the Laplace Expansion Theorem?

Generalizing theorems with insight, combination, and induction

As demonstrated by the proof of the Laplace Expansion Theorem, it is often useful to build slowly to a profound conclusion through a sequence of smaller, more easily understood steps.

Q: By combining Theorem 4.3(b) and Theorem 4.3(d), what new statement can we create?
A: If $A \xrightarrow{R_i \leftrightarrow kR_j} B$, then $\det B = -k \det A$.

Q: By combining Theorem 4.3(c) and Theorem 4.3(d), what new statement can we create?
A: If $\mathbf{A}_i = k\mathbf{A}_j$, then $\det A = 0$.

Q: How might we state this new conclusion in words?
A: If one row or column of a matrix is a multiple of another, then the determinant is zero.

Q: How are Theorem 4.3 and Theorem 4.4 related?
A: As noted in the text, Thm 4.4 comes directly from Thm 4.3 with $B = I$ since $\det I = 1$.

Q: By combining Theorem 4.8 and Lemma 4.14, what new statement can we create?
A: If B is obtained from A by n row interchanges, $\det B = (-1)^n \det A$.
   Try to prove this using induction.

Q: By using induction on Theorem 4.8, what new statement can we create?
A: If $B = A_1 A_2 \cdots A_n = \prod_{i=1}^{n} A_i$, then $\det B = (\det A_1)(\det A_2) \cdots (\det A_n) = \prod_{i=1}^{n} \det A_i$.

Q: By using induction on Theorem 4.4(d), what new statement can we create?
A: If $A \xrightarrow{k_i R_i} B$ for any n rows of A, then $\det B = (k_1 k_2 \cdots k_n) \det A = \left(\prod_{i=1}^{n} k_i \right) \det A$.

Q: A good proof should elucidate the reasons that it is true.
   In Exercise 44, for example, we see that we get a k^n
   because we have multiplied each of the n rows by k.
   Does the same result hold for columns? How might we prove that easily?
A: Use A^T. What is the advantage of proving the result for columns directly?

Q: In Exercise 44, we prove: If the first m rows of A
   have been multiplied by k to create B then $\det B = k^m \det A$.
   What are some ways in which we might generalize this result?
A: Prove this is true for *any* m rows not just the ones that come first.
   Prove this is true for *any* m columns not just rows.
   Prove this is true for *all* k_i not just one fixed k.

4.2 Determinants

Varying forms of solution and proof

As our knowledge and skills increase, the form and content of our solutions will vary.
It is possible to have two completely different solutions that are both correct.
Furthermore, as our knowledge increases not every detail will be spelled out.
Instead, we will be called upon to follow implications by ourselves as we read the solution.
In higher level math courses we joke that complex theorems have one-word proofs: *obvious*.

This joke points out our challenge in constructing a convincing argument.
We want to include sufficient detail to make our argument easy-to-follow,
but not so much that the idea of the proof is obscured by technical details.

As an example, consider two alternate solutions to Exercise 54:
1: We use Exercise 53 and $MM^{-1} = I$ to prove $\det(B^{-1}AB) = \det A$.

$$\det(B^{-1}AB) \stackrel{\text{Thm 4.8}}{=} (\det(B^{-1}))(\det A)(\det B) \stackrel{\text{scalars commute}}{=} (\det(B^{-1}))(\det B)(\det A)$$

$$\stackrel{\text{Thm 4.10}}{=} \left(\tfrac{1}{\det B}\right)(\det B)(\det A) \stackrel{\text{multiplicative inverse}}{=} \det A$$

2: We use Exercise 53, associativity, and $MM^{-1} = I$ to prove $\det(B^{-1}AB) = \det A$.

$$\det(B^{-1}AB) \stackrel{\text{assoc}}{=} \det(B^{-1}(AB)) \stackrel{\text{Ex 53}}{=} \det((AB)B^{-1}) \stackrel{\text{assoc}}{=} \det(A(BB^{-1})) \stackrel{MM^{-1}=I}{=} \det A$$

Q: We could have reworked solution 1 to be more direct as well. How?
A: Use Thms 4.8, 4.10, and $\left(\tfrac{1}{k}\right)k = 1$ to prove $\det(B^{-1}AB) = \det A$.

$$\det(B^{-1}AB) \stackrel{\text{Thm 4.8}}{=} (\det(B^{-1}))(\det A)(\det B) \stackrel{\text{Thm 4.10}}{=} \left(\tfrac{1}{\det B}\right)(\det A)(\det B) \stackrel{(1/k)k=1}{=} \det A$$

Q: What detail do we need to prove to complete the proof just given?
A: We need to show $\det(ABC) = (\det A)(\det B)(\det C)$.

Q: What are the strengths and weaknesses of the proofs given?

Q: Could we construct a proof based on the elementary matrices?
A: See Section 3.3 where B and B^{-1} are constructed from elementary matrices.

Q: Below we offer a proof using elementary matrices. Is it *hand waving*?

3: Consider a sequence of elementary $n \times n$ matrices E_i
whose ith row has been multiplied by k, so $\det E_i = k$ for all i.
Given an $n \times n$ matrix A, we have $kA = E_1 E_2 \cdots E_n A$.

Then repeated use of lemma (5) gives:

$\det(kA) = (\det E_1)(\det E_2) \cdots (\det E_n) \det A = k^n \det A$.

Q: Is one of these proofs more convincing than the others? Is one more complete?
Q: What are the strengths and weaknesses of each? Which one is preferable?

Solutions to odd-numbered exercises from Section 4.2

1. As in Example 4.8, we compute det A by expanding along the first *row* and the first *column*.

row: $\begin{vmatrix} 1 & 0 & 3 \\ 5 & 1 & 1 \\ 0 & 1 & 2 \end{vmatrix} = 1 \begin{vmatrix} 1 & 1 \\ 1 & 2 \end{vmatrix} - 0 \begin{vmatrix} 5 & 1 \\ 0 & 2 \end{vmatrix} + 3 \begin{vmatrix} 5 & 1 \\ 0 & 1 \end{vmatrix} = 1(1) + 3(5) = 16$

column: $\begin{vmatrix} 1 & 0 & 3 \\ 5 & 1 & 1 \\ 0 & 1 & 2 \end{vmatrix} = 1 \begin{vmatrix} 1 & 1 \\ 1 & 2 \end{vmatrix} - 5 \begin{vmatrix} 0 & 3 \\ 1 & 2 \end{vmatrix} + 0 \begin{vmatrix} 0 & 3 \\ 1 & 1 \end{vmatrix} = 1(1) - 5(-3) = 16$

3. As in Example 4.8, we compute det A by expanding along the first *row* and the first *column*.

row: $\begin{vmatrix} 1 & -1 & 0 \\ -1 & 0 & 1 \\ 0 & 1 & -1 \end{vmatrix} = 1 \begin{vmatrix} 0 & 1 \\ 1 & -1 \end{vmatrix} - (-1) \begin{vmatrix} -1 & 1 \\ 0 & -1 \end{vmatrix} = 1(-1) + 1(1) = 0$

column: $\begin{vmatrix} 1 & -1 & 0 \\ -1 & 0 & 1 \\ 0 & 1 & -1 \end{vmatrix} = 1 \begin{vmatrix} 0 & 1 \\ 1 & -1 \end{vmatrix} - (-1) \begin{vmatrix} -1 & 0 \\ 1 & -1 \end{vmatrix} = 1(-1) + 1(1) = 0$

5. As in Example 4.8, we compute det A by expanding along the first *row* and the first *column*.

row: $\begin{vmatrix} 1 & 2 & 3 \\ 2 & 3 & 1 \\ 3 & 1 & 2 \end{vmatrix} = 1 \begin{vmatrix} 3 & 1 \\ 1 & 2 \end{vmatrix} - 2 \begin{vmatrix} 2 & 1 \\ 3 & 2 \end{vmatrix} + 3 \begin{vmatrix} 2 & 3 \\ 3 & 1 \end{vmatrix} = 1(5) - 2(1) + 3(-7) = -18$

column: $\begin{vmatrix} 1 & 2 & 3 \\ 2 & 3 & 1 \\ 3 & 1 & 2 \end{vmatrix} = 1 \begin{vmatrix} 3 & 1 \\ 1 & 2 \end{vmatrix} - 2 \begin{vmatrix} 2 & 3 \\ 1 & 2 \end{vmatrix} + 3 \begin{vmatrix} 2 & 3 \\ 3 & 1 \end{vmatrix} = 1(5) - 2(1) + 3(-7) = -18$

7. As in Example 4.10, we choose a row or column that minimizes the number of calculations. Since $\mathbf{A}_3 = \begin{bmatrix} 3 & 0 & 0 \end{bmatrix}$ contains two zeroes, det $A = a_{31} C_{31} = a_{31}(-1)^{3+1} \det A_{31} = 3 \det A_{31}$.

row 3: $\begin{vmatrix} 5 & 2 & 2 \\ -1 & 1 & 2 \\ 3 & 0 & 0 \end{vmatrix} = 3 \begin{vmatrix} 2 & 2 \\ 1 & 2 \end{vmatrix} = 3(2) = 6$

Q: What should we look for when choosing a row or column to expand along?
A: A row or column with the maximum number of zeroes. Why?
The maximum number of zeroes minimizes the number of cofactors we have to compute.

9. As in Example 4.10, we choose a row or column that minimizes the number of calculations. Since $\mathbf{A}_3 = \begin{bmatrix} 1 & -1 & 0 \end{bmatrix}$ contains one zero, det $A = 1 \det A_{31} - (-1) \det A_{32}$.

row 3: $\begin{vmatrix} -4 & 1 & 3 \\ 2 & -2 & 4 \\ 1 & -1 & 0 \end{vmatrix} = 1 \begin{vmatrix} 1 & 3 \\ -2 & 4 \end{vmatrix} - (-1) \begin{vmatrix} -4 & 3 \\ 2 & 4 \end{vmatrix} = 1(10) + 1(-22) = -12$

Q: Since both row 3 and column 3 contain one zero, what makes row 3 a better choice?
A: Because the nonzero entries of row 3 are 1 and -1, but in column 3 they are 3 and 4.

4.2 Determinants

11. As in Example 4.10, we choose a row or column that minimizes the number of calculations. Since $\mathbf{A}_1 = \begin{bmatrix} a & b & 0 \end{bmatrix}$ contains one zero, $\det A = a \det A_{11} - b \det A_{12}$.

$$\text{row 1:} \quad \begin{vmatrix} a & b & 0 \\ 0 & a & b \\ a & 0 & b \end{vmatrix} = a \begin{vmatrix} a & b \\ 0 & b \end{vmatrix} - b \begin{vmatrix} 0 & b \\ a & b \end{vmatrix} = a(ab) - b(-ab) = ab(a+b)$$

Q: Since $\det A = ab(a+b)$, if $a = -b$ what do we know about A?
A: If $a = -b$, then A is *not* invertible (because $\det A = 0$).

Q: Since $\det A = ab(a+b)$, if $a = 0$ what do we know about A?

Q: What are the conditions on a and b that will guarantee A is invertible?
A: Both a and b are nonzero and $a \neq -b$. Why is this enough?

13. Since $\mathbf{A}_3 = \begin{bmatrix} 0 & 1 & 0 & 0 \end{bmatrix}$ contains three zeroes, $\det A = -1 \det A_{32}$.
If we let $B = A_{32}$ and expand along $\mathbf{B}_1$, $\det A = -\det B = -(1 \det B_{11} + 3 \det B_{13})$.

$$\text{row 3:} \quad \begin{vmatrix} 1 & -1 & 0 & 3 \\ 2 & 5 & 2 & 6 \\ 0 & 1 & 0 & 0 \\ 1 & 4 & 2 & 1 \end{vmatrix} = - \begin{vmatrix} 1 & 0 & 3 \\ 2 & 2 & 6 \\ 1 & 2 & 1 \end{vmatrix} = - \left(\begin{vmatrix} 2 & 6 \\ 2 & 1 \end{vmatrix} + 3 \begin{vmatrix} 2 & 2 \\ 1 & 2 \end{vmatrix} \right) = -(-10 + 3 \cdot 2) = 4.$$

Q: What does $\det A = 4$ tell us about the rows of A?
A: The rows of A are linearly independent (because $\det A \neq 0$. Why is that enough?).

Q: What does $\det A = 4$ tell us about the columns of A?
Q: What does $\det B \neq 0$ tell us about the rows and columns of B?

15. Since $\mathbf{A}_1 = \begin{bmatrix} 0 & 0 & 0 & a \end{bmatrix}$ contains three zeroes, $\det A = -a \det A_{14}$.
If we let $B = A_{14}$ and expand along $\mathbf{B}_1$, $\det A = -a \det B = -a(b \det B_{13})$.

$$\text{row 1:} \quad \begin{vmatrix} 0 & 0 & 0 & a \\ 0 & 0 & b & c \\ 0 & d & e & f \\ g & h & i & j \end{vmatrix} = -a \begin{vmatrix} 0 & 0 & b \\ 0 & d & e \\ g & h & i \end{vmatrix} = -ab \begin{vmatrix} 0 & d \\ g & h \end{vmatrix} = abdg.$$

Q: Since $\det A = abdg$, if a, b, d, g are nonzero then A is invertible. Why is this obvious?
A: Since if a, b, d, g are nonzero then A is *anti*-lower triangular. Why is that enough?

17. Following the method of Example 4.9, we have:

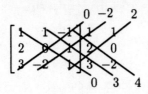

Adding the three products at the bottom and subtracting the three products at the top gives $\det A = 0 + 3 + 4 - 0 - (-2) - 2 = 7$.

19. Suppose that $A = \begin{bmatrix} a_{11} & a_{12} & a_{13} \\ a_{21} & a_{22} & a_{23} \\ a_{31} & a_{32} & a_{33} \end{bmatrix}$. Following the method of Example 4.9, we have:

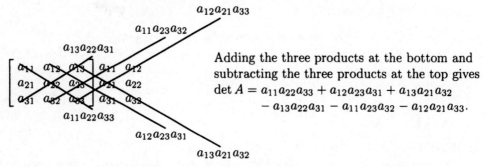

Adding the three products at the bottom and subtracting the three products at the top gives
$\det A = a_{11}a_{22}a_{33} + a_{12}a_{23}a_{31} + a_{13}a_{21}a_{32}$
$\quad - a_{13}a_{22}a_{31} - a_{11}a_{23}a_{32} - a_{12}a_{21}a_{33}.$

Now note, as above, expanding definition (1) gives:

$$\det A = a_{11}\begin{vmatrix} a_{22} & a_{23} \\ a_{32} & a_{33} \end{vmatrix} - a_{12}\begin{vmatrix} a_{21} & a_{23} \\ a_{31} & a_{33} \end{vmatrix} + a_{13}\begin{vmatrix} a_{21} & a_{22} \\ a_{31} & a_{32} \end{vmatrix}$$
$$= a_{11}a_{22}a_{33} - a_{11}a_{23}a_{32} - a_{12}a_{21}a_{33} + a_{12}a_{23}a_{31} + a_{13}a_{21}a_{32} - a_{13}a_{22}a_{31}.$$

4.2 Determinants

21. We use induction to prove Theorem 4.2: if A is triangular then $\det A = a_{11}a_{22}\cdots a_{nn}$. See Appendix B for discussion and examples of *Mathematical Induction*.
Since Theorem 4.10 asserts $\det A = \det A^T$, we can assume A is upper triangular.

   1: If A is 1×1, then $A = [a_{11}]$ so $\det A = \det([a_{11}]) = a_{11}$.
   This is obvious, so there is nothing to show.

   n: If A is $n \times n$ and upper triangular, then $\det A = a_{11}a_{22}\cdots a_{nn}$.
   This is the induction hypothesis.

   $n+1$: If A is $(n+1) \times (n+1)$ and upper triangular, then $\det A = a_{11}a_{22}\cdots a_{nn}a_{n+1n+1}$.
   This is the statement we must prove using the induction hypothesis.

   $$\det A \;\underset{\text{A is upper triangular}}{=}\; a_{n+1n+1}\det A_{n+1n+1} \;\underset{\text{by induction}}{=}\; a_{n+1n+1}(a_{11}a_{22}\cdots a_{nn}) = a_{11}a_{22}\cdots a_{n+1n+1}$$

   We have proven (by induction) that if A is $n \times n$ and triangular, then $\det A = a_{11}a_{22}\cdots a_{nn}$.

   Q: Why does the fact that $\det A = \det A^T$ allow us to assume A is upper triangular?
   A: Because if A is lower triangular, then A^T is upper triangular.
   So, $\det A = \det A^T = a_{11}a_{22}\cdots a_{nn}$ because $[A]_{ii} = [A^T]_{ii}$.
   That is, the diagonal entries of A and A^T are equal.

   Q: Why does A being upper triangular imply $\det A = a_{n+1n+1}\det A_{n+1n+1}$?
   A: Because if A is upper triangular, then row $n+1 = \mathbf{A}_{n+1} = \begin{bmatrix} 0 & 0 & \cdots & a_{n+1n+1} \end{bmatrix}$.
   So when we expand along this row we have: $\det A = a_{n+1n+1}\det A_{n+1n+1}$.

   Q: Why is the coefficient of $\det A_{n+1n+1}$ equal to a_{n+1n+1} instead of $-a_{n+1n+1}$?
   A: The cofactor $C_{n+1n+1} = (-1)^{2n+2}\det A_{n+1n+1} = \det A_{n+1n+1}$ because $2n+2$ is even.

   Q: Why do we get to apply the induction hypothesis to the matrix A_{n+1n+1}?
   A: Since A_{n+1n+1} is created by removing the $n+1$ row and the $n+1$ column of A, A_{n+1n+1} is $n \times n$ and upper triangular. Why is this obvious?
   Since A is $(n+1) \times (n+1)$, A_{n+1n+1} (created by removing a row and column) is $n \times n$.
   Since A is upper triangular, $[A]_{ij} = 0 = [A_{n+1n+1}]_{ij}$ for $i > j$ when $i,j \leq n$.
   That is, A_{n+1n+1} is upper triangular.

   All these details support the proof given above. We should carefully investigate them all. In general, we should be critical of our reasoning and actively seek out oversights.

23. As in Example 4.13, we use Theorem 4.3 to track adjustments to $\det A$ required $A \longrightarrow U$.

   $$\det A = \begin{vmatrix} -4 & 1 & 3 \\ 2 & -2 & 4 \\ 1 & -1 & 0 \end{vmatrix} \underset{\substack{R_1 \leftrightarrow R_3 \\ R_2 \leftrightarrow R_3}}{=} \begin{vmatrix} 1 & -1 & 0 \\ -4 & 1 & 3 \\ 2 & -2 & 4 \end{vmatrix} \underset{\substack{R_2+4R_1 \\ R_3-2R_1}}{=} \begin{vmatrix} 1 & -1 & 0 \\ 0 & -3 & 3 \\ 0 & 0 & 4 \end{vmatrix} = -12 = \det U$$

   Q: Though row interchanges were used in $A \longrightarrow U$, we still have $\det A = \det U$. Why?
   A: There were *two* row interchanges. So what does Theorem 4.3 say?
   Part b. asserts a row interchange changes the sign of the determinant, $\det U = -\det A$.
   If there are two interchanges the sign is changed twice, back to what it was originally.
   That is, $\det U = -(-\det A) = \det A$. Does this hold for even numbers of row interchanges?

25. As in Example 4.13, we use Theorem 4.3 to track adjustments to det A required $A \longrightarrow U$. We use a combination of row and column operations below to gain experience in doing so.

$$\begin{vmatrix} 2 & 0 & 3 & -1 \\ 1 & 0 & 2 & 2 \\ 0 & -1 & 1 & 4 \\ 2 & 0 & 1 & -3 \end{vmatrix} \begin{matrix} C_3 \leftrightarrow C_4 \\ C_3 \leftrightarrow C_2 \\ C_1 \leftrightarrow C_2 \\ = \end{matrix} - \begin{vmatrix} -1 & 2 & 0 & 3 \\ 2 & 1 & 0 & 2 \\ 4 & 0 & -1 & 1 \\ -3 & 2 & 0 & 1 \end{vmatrix} \begin{matrix} C_2 + 2C_1 \\ C_4 + 3C_1 + 13C_3 \\ \underline{} \end{matrix} - \begin{vmatrix} -1 & 0 & 0 & 0 \\ 2 & 5 & 0 & 8 \\ 4 & 8 & -1 & 0 \\ -3 & -4 & 0 & -8 \end{vmatrix} \begin{matrix} -R_1 \\ R_2 + R_4 \\ \underline{} \end{matrix} \begin{vmatrix} 1 & 0 & 0 & 0 \\ 2 & 1 & 0 & 0 \\ 4 & 8 & -1 & 0 \\ 3 & -4 & 0 & -8 \end{vmatrix} = 8$$

Q: Strictly speaking, we did not reduce A to row echelon form. Why is this still sufficient?
A: The resulting matrix L is lower triangular, so det L is the product of the diagonal entries.

Q: Why did we have to introduce a negative sign after step one?
A: Because we performed 3 column interchanges. Is this true for any odd number?

Q: Why did we remove the negative sign after step three?
A: Since we multiplied row 1 by -1, we had to multiply the determinant by -1.
Which part of Theorem 4.3 tells us we have to do this multiplication? Part d.

27. Since A is triangular, we have det $A = a_{11} a_{22} a_{33} = (3)(-2)(4) = -24$.

Q: Which Theorem from this Section did we have employ in reaching this conclusion?
A: Theorem 4.2.

Q: How does the proof of this Theorem in Exercise 21 suggest solving this problem directly?
A: By expanding along row 3. So:

$$\text{row 3: } \begin{vmatrix} 3 & 1 & 0 \\ 0 & -2 & 5 \\ 0 & 0 & 4 \end{vmatrix} = 4 \begin{vmatrix} 3 & 1 \\ 0 & -2 \end{vmatrix} = 4(3)(-2) = -24$$

Note: We should apply our proofs to specific examples to see if they make sense and work. Also: A_{33} is 2×2 and upper triangular as it should be according to proof in Exercise 21.

Q: Since $A \longrightarrow I$ (obviously), why is det $A = -24 \neq 1 = $ det I?
A: For $A \longrightarrow I$ we have to multiply row 1 by $\frac{1}{3}$, row 2 by $-\frac{1}{2}$, and row 3 by $\frac{1}{4}$. What does that tell us? We have to do the same thing to det A.
Therefore det $I = \left(\frac{1}{3}\right)\left(-\frac{1}{2}\right)\left(\frac{1}{4}\right)$ det $A = -\frac{1}{24}(-24) = 1$.

29. Since $\mathbf{a}_3 = \begin{bmatrix} -4 \\ -2 \\ 2 \end{bmatrix} = -2 \begin{bmatrix} 2 \\ 1 \\ -1 \end{bmatrix} = -2\mathbf{a}_1$, we have det $A = 0$.

Q: Does Theorem 4.3 imply if $\mathbf{a}_i = k\mathbf{a}_j$ then det $A = 0$?
A: Yes. Thm 4.6 implies if there is *any* dependence relation among the columns, det $A = 0$. For a full discussion see questions and answers following Exercise 26 above.

31. Since $\mathbf{a}_3 = \begin{bmatrix} 3 \\ -2 \\ 1 \end{bmatrix} = \begin{bmatrix} 4 \\ -2 \\ 5 \end{bmatrix} - \begin{bmatrix} 1 \\ 0 \\ 4 \end{bmatrix} = \mathbf{a}_1 - \mathbf{a}_2$, det $A = 0$.

Q: Does Theorem 4.3 imply if $\mathbf{a}_i = c_j \mathbf{a}_j + c_k \mathbf{a}_k$ then det $A = 0$?
A: Yes. Thm 4.6 implies if there is *any* dependence relation among the columns, det $A = 0$. For a full discussion see questions and answers following Exercise 26 above.

4.2 Determinants

33. Since $A \xrightarrow[R_3 \leftrightarrow R_4]{R_1 \leftrightarrow R_2} B$, $\det A = \det B = (-3)(2)(1)(4) = -24$ because B is triangular.

Q: In $A \longrightarrow B$ we used row interchanges, but $\det A = \det B$. Why?
A: Because we performed 2 row interchanges. Is this true for any even number? Yes.

35. Let A be the matrix given at the beginning of Exercises 35 – 40 with $\det A = 4$.
Let B be the matrix given in this Exercise which is derived from A. So:
Since $A \xrightarrow{2R_1} B$, $\det B = 2 \det A = 2(4) = 8$.

Q: What Theorem from this section supports our conclusion above that $\det B = 2 \det A$?
A: Theorem 4.3 part d. which asserts if $A \xrightarrow{kR_i} B$, then $\det B = k \det A$.

37. Let A be the matrix given at the beginning of Exercises 35 – 40 with $\det A = 4$.
Let B be the matrix given in this Exercise which is derived from A in 1 step.
Since $A \xrightarrow{R_1 \leftrightarrow R_2} B$, we conclude $\det B = -\det A = -4$.

Q: How might we generalize this result?
A: If $A \xrightarrow{R_i \leftrightarrow R_j} B$, $\det B = -\det A$. This is the assertion of Theorem 4.3, part b.

Q: Does this result hold for columns? If $A \xrightarrow{C_i \leftrightarrow C_j} B$, does $\det B = -\det A$?
A: Yes. Since $\det A = \det A^T$ this result holds for both rows and columns.

39. Let A be the matrix given at the beginning of Exercises 35 – 40 with $\det A = 4$.
Let C be the matrix given in this Exercise which is derived from A in 2 steps.
Since $A \xrightarrow{C_1 \leftrightarrow C_3} B \xrightarrow{2C_1} C$, we have $\det C = 2 \det B = 2(-\det A) = 2(-4) = -8$.

Q: How might we generalize this result?
A: If $A \xrightarrow{C_i \leftrightarrow C_j} B \xrightarrow{cC_k} C$, $\det C = -c \det A$. This is precisely Theorem 4.3, parts b. and d.

Q: Does this result hold for rows? If $A \xrightarrow{R_i \leftrightarrow R_j} B \xrightarrow{cR_k} C$, does $\det C = -c \det A$?
A: Yes. Since $\det A = \det A^T$ this result holds for both rows and columns.

41. We will first prove Theorem 4.3 part a. for rows and then for columns.

row: If A has a zero row, then $\det A = 0$. That is, if $\mathbf{A}_i = \mathbf{0}$, then $\det A = 0$.

By Theorem 4.1, $\det A = \sum_{j=1}^{n} a_{ij} C_{ij}$, where $\mathbf{A}_i = \begin{bmatrix} a_{i1} & a_{i2} & \cdots & a_{in} \end{bmatrix}$.

If $\mathbf{A}_i = \mathbf{0} = \begin{bmatrix} 0 & 0 & \cdots & 0 \end{bmatrix}$, then $a_{ij} = 0$ for all j. So, $\det A = \sum_{j=1}^{n} 0(C_{ij}) = 0$.

col: If A has a zero column, then $\det A = 0$. That is, if $\mathbf{a}_j = \mathbf{0}$, then $\det A = 0$.

By Theorem 4.1, $\det A = \sum_{i=1}^{n} a_{ij} C_{ij}$, where $\mathbf{a}_j = \begin{bmatrix} a_{1j} \\ a_{2j} \\ \vdots \\ a_{nj} \end{bmatrix}$.

If $\mathbf{a}_j = \mathbf{0} = \begin{bmatrix} 0 \\ 0 \\ \vdots \\ 0 \end{bmatrix}$, then $a_{ij} = 0$ for all i. So, $\det A = \sum_{i=1}^{n} 0(C_{ij}) = 0$.

Note: Our proof is trivial because the Laplace Expansion does all the work.

Q: How might we state our proof and conclusions about rows in words?

A: When we expand along a zero row, the coefficient of every cofactor is zero so $\det A = 0$. This illustrates an important point: symbols are essential to an elegant proof, but stating conclusions (loosely) in language is essential to solid understanding.

Q: Can we create a similar statement for our proof and conclusions about columns?

4.2 Determinants

43. We will apply Theorems 4.3 and 4.4 to prove Lemma 4.5: $\det(EB) = (\det E)(\det B)$. We consider each type of elementary matrix separately.

$R_i \leftrightarrow R_j$: Given $I \xrightarrow{R_i \leftrightarrow R_j} E$, we need to show $\det(EB) = (\det E)(\det B)$.
Since $B \xrightarrow{R_i \leftrightarrow R_j} EB$, Theorem 4.3(b) implies $\det(EB) = -\det B$.
Since $I \xrightarrow{R_i \leftrightarrow R_j} E$, Theorem 4.4(a) implies $\det E = -\det I = -1$.
So, $\det(EB) = -\det B = (-1)(\det B) = (\det E)(\det B)$.

Briefly in symbols: $\det(EB) \underset{\text{Thm 4.3(b)}}{\overset{R_i \leftrightarrow R_j}{=}} -\det B \underset{\text{Thm 4.4(a)}}{\overset{\det E = -1}{=}} (\det E)(\det B)$

kR_i: Given $I \xrightarrow{kR_i} E$, we need to show $\det(EB) = (\det E)(\det B)$.
Since $B \xrightarrow{kR_i} EB$, Theorem 4.3(d) implies $\det(EB) = k \det B$.
Since $I \xrightarrow{kR_i} E$, Theorem 4.4(b) implies $\det E = k \det I = k$.
So, $\det(EB) = k \det B = (k)(\det B) = (\det E)(\det B)$.

Briefly in symbols: $\det(EB) \underset{\text{Thm 4.3(d)}}{\overset{R_i \to kR_i}{=}} k \det B \underset{\text{Thm 4.4(b)}}{\overset{\det E = k}{=}} (\det E)(\det B)$

$R_i + kR_j$: Given $I \xrightarrow{R_i + kR_j} E$, we need to show $\det(EB) = (\det E)(\det B)$.
Since $B \xrightarrow{R_i + kR_j} EB$, Theorem 4.3(f) implies $\det(EB) = \det B$.
Since $I \xrightarrow{R_i + kR_j} E$, Theorem 4.4(c) implies $\det E = \det I = 1$.
So, $\det(EB) = 1 \det B = (1)(\det B) = (\det E)(\det B)$.

Briefly in symbols: $\det(EB) \underset{\text{Thm 4.3(f)}}{\overset{R_i \to R_i + kR_j}{=}} \det B \underset{\text{Thm 4.4(c)}}{\overset{\det E = 1}{=}} (\det E)(\det B)$

Q: How might we describe our result for $R_i \leftrightarrow R_j$ more completely in words?
A: If E results from interchanging two rows of I, then EB is derived from B by interchanging two rows of B. So by Theorem 4.3(b), $\det(EB) = -\det B$. But by Theorem 4.4(a), $\det E = -1$, so $\det(EB) = (\det E)(\det B)$ as required. Repeat this exercise for kR_i and $R_i + kR_j$.

45. Theorem 4.6 asserts A is invertible if and only if $\det A \neq 0$.
So, the contrapositive of Theorem 4.6 is: A is *not* invertible if and only if $\det A = 0$.
So we solve $\det A = 0$ to find the values of k we need to exclude in order for A to be invertible.

$$\det A = \begin{vmatrix} k & -k & 3 \\ 0 & k+1 & 1 \\ k & -8 & k-1 \end{vmatrix} \overset{R_3 - R_1}{=} \begin{vmatrix} k & -k & 3 \\ 0 & k+1 & 1 \\ 0 & k-8 & k-4 \end{vmatrix} \overset{\text{along } C_1}{=} k \begin{vmatrix} k+1 & 1 \\ k-8 & k-4 \end{vmatrix}$$

$= k\left(k^2 - 4k + 4\right) = k(k-2)^2 = 0 \Rightarrow k = 0, 2$.

So, the values of k we need to exclude in order for A to be invertible are $k = 0, 2$.
That is, A is invertible if and only if $k \neq 0, 2$.

47. We use Theorem 4.8, $\det(AB) = (\det A)(\det B)$, and the given values to compute $\det(AB)$.

$$\det(AB) \stackrel{\text{Thm 4.8}}{=} (\det A)(\det B) \stackrel{\text{givens}}{=} (3)(-2) = -6$$

Q: How might we state the conclusion $\det(AB) = (\det A)(\det B)$ in words?
A: The determinant of the product equals the product of the determinant.

49. We use Theorems 4.8, 4.9 and $\det A = 3$, $\det B = -2$ to compute $\det(B^{-1}A)$.

$$\det(B^{-1}A) \stackrel{\text{Thm 4.8}}{=} (\det(B^{-1}))(\det A) \stackrel{\text{Thm 4.9}}{=} \left(\frac{1}{\det B}\right)(\det A) \stackrel{\text{givens}}{=} \left(-\tfrac{1}{2}\right)(3) = -\tfrac{3}{2}$$

51. We use Theorem 4.7 with $k = 3$, Theorem 4.10, and the givens to compute $\det(3B^T)$.

$$\det(3B^T) \stackrel[\text{Thm 4.7}]{k=3}{=} 3^n \det(B^T) \stackrel{\text{Thm 4.10}}{=} 3^n \det B \stackrel{\text{givens}}{=} 3^n(-2) = -2 \cdot 3^n$$

53. We use Theorem 4.8 to prove $\det(AB) = \det(BA)$.

$$\det(AB) \stackrel{\text{Thm 4.8}}{=} (\det A)(\det B) \stackrel{\substack{\det M \text{ is a}\\ \text{scalar}}}{=} (\det B)(\det A) \stackrel{\text{Thm 4.8}}{=} \det(BA)$$

Q: What is the key insight to take away from this exercise?
A: That $\det A$ is a scalar (not a matrix), so it commutes.

55. We use the fact that $A^2 = A \stackrel{\text{Ex 47}}{\Rightarrow} (\det A)^2 = \det A$ to find all possible values of $\det A$.

$$(\det A)^2 = \det A \Rightarrow (\det A)^2 - \det A = 0 \Rightarrow \det A(\det A - 1) = 0 \Rightarrow \det A = 0, 1$$

Q: If we let $x = \det A$, what does the above calculation look like?
A: With $x = \det A$: $x^2 = x \Rightarrow x^2 - x = 0 \Rightarrow x(x-1) = 0 \Rightarrow x = 0, 1$.
This clarifies the algebra and helps us remember that $\det A$ is a scalar.

57. We solve the system using Theorem 4.11 (Cramer's Rule). So we have:

$$\det A = \begin{vmatrix} 1 & 1 \\ 1 & -1 \end{vmatrix} = -2,\ \det(A_1(\mathbf{b})) = \begin{vmatrix} 1 & 1 \\ 2 & -1 \end{vmatrix} = -3,\ \det(A_2(\mathbf{b})) = \begin{vmatrix} 1 & 1 \\ 1 & 2 \end{vmatrix} = 1.$$

By Cramer's Rule, $x = \dfrac{\det(A_1(\mathbf{b}))}{\det A} = \dfrac{-3}{-2} = \dfrac{3}{2}$ and $y = \dfrac{\det(A_2(\mathbf{b}))}{\det A} = \dfrac{1}{-2} = -\dfrac{1}{2}$.

59. We solve the system using Theorem 4.11 (Cramer's Rule). So we have:

$$\det A = \begin{vmatrix} 2 & 1 & 3 \\ 0 & 1 & 1 \\ 0 & 0 & 1 \end{vmatrix} = 2,\ \det(A_1(\mathbf{b})) = \begin{vmatrix} 1 & 1 & 3 \\ 1 & 1 & 1 \\ 1 & 0 & 1 \end{vmatrix} = 1 - 0 + 3(-1) = -2,$$

$$\det(A_2(\mathbf{b})) = \begin{vmatrix} 2 & 1 & 3 \\ 0 & 1 & 1 \\ 0 & 1 & 1 \end{vmatrix} = 0,\text{ and } \det(A_3(\mathbf{b})) = \begin{vmatrix} 2 & 1 & 1 \\ 0 & 1 & 1 \\ 0 & 0 & 1 \end{vmatrix} = 2.\text{ By Cramer's Rule,}$$

$$x = \frac{\det(A_1(\mathbf{b}))}{\det A} = \frac{-2}{2} = -1,\ y = \frac{\det(A_2(\mathbf{b}))}{\det A} = 0,\text{ and } z = \frac{\det(A_3(\mathbf{b}))}{\det A} = \frac{2}{2} = 1.$$

4.2 Determinants

61. We compute $\det A = \begin{vmatrix} 1 & 1 \\ 1 & -1 \end{vmatrix} = -2$ and the four cofactors $C_{11} = +(-1) = -1$,

$C_{12} = -1$, $C_{21} = -1$, and $C_{22} = 1$. The adjoint is the transpose of the matrix of cofactors:

$$\text{adj } A = \begin{bmatrix} -1 & -1 \\ -1 & 1 \end{bmatrix}^T = \begin{bmatrix} -1 & -1 \\ -1 & 1 \end{bmatrix}. \text{ Then } A^{-1} = \frac{1}{\det A}\text{adj } A = -\frac{1}{2}\begin{bmatrix} -1 & -1 \\ -1 & 1 \end{bmatrix} = \begin{bmatrix} \frac{1}{2} & \frac{1}{2} \\ \frac{1}{2} & -\frac{1}{2} \end{bmatrix}.$$

Q: Is adj A a matrix or a scalar? What about $\det A$?
A: By definition, adj A is the matrix of cofactors. So adj A is a matrix.
   On the other hand, $\det A$ is a scalar. The names have a similar form. Be careful!

63. $\det A = \begin{vmatrix} 2 & 1 & 3 \\ 0 & 1 & 1 \\ 0 & 0 & 1 \end{vmatrix} = 2$, $C_{11} = \begin{vmatrix} 1 & 1 \\ 0 & 1 \end{vmatrix} = 1$, $C_{12} = -\begin{vmatrix} 0 & 1 \\ 0 & 1 \end{vmatrix} = 0$

$C_{13} = 0$, $C_{21} = -1$, $C_{22} = 2$, $C_{23} = 0$, $C_{31} = -2$, $C_{32} = -2$, and $C_{33} = 2$.

$$\text{Thus adj } A = \begin{bmatrix} 1 & -1 & -2 \\ 0 & 2 & -2 \\ 0 & 0 & 2 \end{bmatrix} \text{ and } A^{-1} = \frac{1}{\det A}\text{adj } A = \frac{1}{2}\begin{bmatrix} 1 & -1 & -2 \\ 0 & 2 & -2 \\ 0 & 0 & 2 \end{bmatrix} = \begin{bmatrix} \frac{1}{2} & -\frac{1}{2} & -1 \\ 0 & 1 & -1 \\ 0 & 0 & 1 \end{bmatrix}.$$

65. We must show adj A is invertible, $(\text{adj } A)^{-1} = \frac{1}{\det A}A$, and $\text{adj}(A^{-1}) = \frac{1}{\det A}A$.

adj A: We will first show if A is invertible, then $B = kA$ ($k \neq 0$) is invertible.
We will then use this fact to prove that adj $A = (\det A)A^{-1}$ is invertible.
Since A is invertible implies $\det A \neq 0$, if $B = kA$, then $\det B = k^n \det A \neq 0$.
So, Theorem 4.6 implies that B is invertible.
Now by Theorem 4.12, $A^{-1} = \frac{1}{\det A}\text{adj } A$, so adj $A = (\det A)A^{-1}$.
Since A^{-1} is invertible, $\det A \neq 0$. Therefore, adj $A = (\det A)A^{-1}$ is invertible.
Q: How might we convey the above argument (loosely) in words?
A: Since adj A is a (nonzero) scalar multiple of an invertible matrix, adj A is invertible.

$(\text{adj } A)^{-1}$: As in Section 3.3, we prove X is the inverse of adj A by showing $(\text{adj } A)X = I$.
We will show $(\text{adj } A)(\frac{1}{\det A}A) = I$ which will imply $(\text{adj } A)^{-1} = \frac{1}{\det A}A$.
$(\text{adj } A)(\frac{1}{\det A}A) = ((\det A)A^{-1})(\frac{1}{\det A}A) = (\det A \frac{1}{\det A})(A^{-1}A) = A^{-1}A = I$

adj (A^{-1}): Note that Theorem 4.12 asserts $A^{-1} = \frac{1}{\det A}\text{adj } A$, so adj $A = (\det A)A^{-1}$.
So $(A^{-1})^{-1} = A = \frac{1}{\det(A^{-1})}\text{adj}(A^{-1})$, so adj $(A^{-1}) = (\det(A^{-1}))A$.
Now recall that $\det(A^{-1}) = \frac{1}{\det A}$, so adj $(A^{-1}) = (\det(A^{-1}))A = \frac{1}{\det A}A$ as claimed.

67. We use induction to show it requires n adjacent interchanges to move row s to row $s - n$. Then we prove it requires $2(s - r) - 1$ adjacent interchanges to interchange rows r and s.

- 1: Since $R_{s-1} \leftrightarrow R_s$ moves row s to row $s - 1$, only 1 adjacent interchange is required.

- k: It requires k adjacent interchanges to move row s to row $s - k$.
   This is the induction hypothesis, so there is nothing to show.

- $k+1$: It requires $k+1$ adjacent interchanges to move row s to row $s - (k+1)$.
   This is the step we must prove using the induction hypothesis.
   By induction, to move row s to row $s - k$ requires k adjacent interchanges.
   But to move row $s - k$ to row $s - (k+1)$ requires only 1 additional adjacent interchange.
   So, it requires $k+1$ adjacent interchanges to move row s to row $s - (k+1)$.

Now let $r = s - n$, then it takes $n = s - r$ adjacent interchanges to move row s to row r.
Since this process is symmetric, it also $s - r$ adjacent interchanges to move row r to row s.
However, the $R_r \leftrightarrow R_s$ adjacent exchange involves both row r and and row s simultaneously.
Therefore, we must subtract 1 in order not to double-count this shared interchange.
So, the total number of adjacent interchanges required is $(s - r) + (s - r) - 1 = 2(s - r) - 1$.

Note: This argument explains the presence of the -1 in the formula.

Q: What is the basic insight underlying the justification of this conclusion?
A: To move a row one row up or down requires one adjacent interchange.

4.2 Determinants

69. The result for *block* triangular matrices is neither trivial nor obvious.
It requires careful application of the definitions and use of all the given conditions.

As suggested in the hint, we proceed by induction on the number of rows of P.
Note that A, P, and S *must* be square but O and Q may or may not be square.
However, because A is square O *must* have the same number of columns as P.
Also, because A is square, Q *must* have the same number of rows as P.

1: If P has 1 row, we must show $\det A = (\det P)(\det S)$.

If $P = [p_{11}]$ and A is $m \times m$, then $\det P = p_{11}$ and $A = \begin{bmatrix} p_{11} & Q \\ O & S \end{bmatrix}$.

Compute $\det A$ by expansion along column 1: $\det A = \sum_{i=1}^{m} a_{i1}((-1)^{i+1} \det A_{i1})$.

Recall O is the matrix whose entries are all zero. So, $a_{11} = p_{11}$ and $a_{i1} = 0$ for $i > 1$.

So, $\det A = p_{11}((-1)^{1+1} \det A_{11}) + \sum_{i=2}^{m} 0((-1)^{i+1} \det A_{i1}) = p_{11} \det A_{11}$.

Recall, A_{11} is the submatrix of A obtained by deleting row 1 and column 1.
Since P is 1×1, O has 1 column and Q has 1 row. Therefore, $A_{11} = S$.

So, $\det A = p_{11} \det A_{11} \stackrel{A_{11}=S}{=} p_{11}(\det S) \stackrel{p_{11}=\det P}{=} (\det P)(\det S)$ as we were to show.

k: If P has k rows, then $\det A = (\det P)(\det S)$.
This is the induction hypothesis, so there is nothing to show.

$k+1$: If P has $k+1$ rows, then $\det A = (\det P)(\det S)$.
This is the step we must prove using the induction hypothesis.

As in step 1, we compute $\det A$ using column 1: $\det A = \sum_{i=1}^{m} a_{i1}((-1)^{i+1} \det A_{i1})$.

Similar to step 1, $a_{i1} = p_{i1}$ for $i \le k+1$ and $a_{i1} = 0$ for $i > k+1$.

So, $\det A = \sum_{i=1}^{k+1} p_{i1}((-1)^{i+1} \det A_{i1}) + \sum_{i=k+2}^{m} 0((-1)^{i+1} \det A_{i1}) = \sum_{i=1}^{k+1} p_{i1}((-1)^{i+1} \det A_{i1})$.

Recall, A_{i1} is the submatrix of A obtained from by deleting row i and column 1.
So, for $i \le k+1$ this process also removes row i and column 1 of P.
Since Q has $k+1$ rows, for $i \le k+1$ this process does not alter S.

Since $A_{ij} = \begin{bmatrix} P_{i1} & Q_i \\ O_1 & S \end{bmatrix}$, where P_{i1} has k rows, the induction hypothesis applies.

By induction, therefore, we have $\det A_{ij} = (\det P_{i1})(\det S)$ for $i \le k+1$.

So, $\det A = \sum_{i=1}^{k+1} p_{i1}((-1)^{i+1} \det A_{i1}) = \sum_{i=1}^{k+1} p_{i1}((-1)^{i+1}(\det P_{i1})(\det S))$.

Since every term in the sum is multiplied by $\det S$, we can factor it out.

So, $\det A = \sum_{i=1}^{k+1} p_{i1}((-1)^{i+1}(\det P_{i1})(\det S)) = (\det S) \sum_{i=1}^{k+1} p_{i1}((-1)^{i+1} \det P_{i1})$.

Recall, by expansion along column 1, $\det P = \sum_{i=1}^{k+1} p_{i1}((-1)^{i+1} \det P_{i1})$.

So, $\det A = (\det S) \left(\sum_{i=1}^{k+1} p_{i1}((-1)^{i+1} \det P_{i1}) \right) = (\det S)(\det P) = (\det P)(\det S)$.

Exploration: Geometric Applications of Determinants

Explorations are self-contained, so only odd-numbered solutions will be provided.

1. (a) $\mathbf{u} \times \mathbf{v} = \det \begin{bmatrix} \mathbf{e}_1 & u_1 & v_1 \\ \mathbf{e}_2 & u_2 & v_2 \\ \mathbf{e}_3 & u_3 & v_3 \end{bmatrix} = \begin{vmatrix} \mathbf{e}_1 & 0 & 3 \\ \mathbf{e}_2 & 1 & -1 \\ \mathbf{e}_3 & 1 & 2 \end{vmatrix}$

$= \mathbf{e}_1 (2 - (-1)) - \mathbf{e}_2 (0 - 3) + \mathbf{e}_3 (0 - 3) = 3\mathbf{e}_1 + 3\mathbf{e}_2 - 3\mathbf{e}_3$ or $\begin{bmatrix} 3 \\ 3 \\ -3 \end{bmatrix}$.

(b) $\mathbf{u} \times \mathbf{v} = \det \begin{bmatrix} \mathbf{e}_1 & u_1 & v_1 \\ \mathbf{e}_2 & u_2 & v_2 \\ \mathbf{e}_3 & u_3 & v_3 \end{bmatrix} = \begin{vmatrix} \mathbf{e}_1 & 3 & 0 \\ \mathbf{e}_2 & -1 & 1 \\ \mathbf{e}_3 & 2 & 1 \end{vmatrix}$

$= \mathbf{e}_1 (-1 - 2) - \mathbf{e}_2 (3 - 0) + \mathbf{e}_3 (3 - 0) = \begin{bmatrix} -3 \\ -3 \\ 3 \end{bmatrix}$.

(c) $\mathbf{u} \times \mathbf{v} = \begin{vmatrix} \mathbf{e}_1 & -1 & 2 \\ \mathbf{e}_2 & 2 & -4 \\ \mathbf{e}_3 & 3 & -6 \end{vmatrix} = O.$

(d) $\mathbf{u} \times \mathbf{v} = \begin{vmatrix} \mathbf{e}_1 & 1 & 1 \\ \mathbf{e}_2 & 1 & 2 \\ \mathbf{e}_3 & 1 & 3 \end{vmatrix} = \begin{bmatrix} 1 \\ -2 \\ 1 \end{bmatrix}$.

3. (a) $\mathbf{v} \times \mathbf{u} = \det \begin{bmatrix} \mathbf{e}_1 & v_1 & u_1 \\ \mathbf{e}_2 & v_2 & u_2 \\ \mathbf{e}_3 & v_3 & u_3 \end{bmatrix} \stackrel{C_2 \leftrightarrow C_3}{=} -\det \begin{bmatrix} \mathbf{e}_1 & u_1 & v_1 \\ \mathbf{e}_2 & u_2 & v_2 \\ \mathbf{e}_3 & u_3 & v_3 \end{bmatrix} = -(\mathbf{u} \times \mathbf{v})$.

(b) $\mathbf{u} \times \mathbf{0} = \begin{vmatrix} \mathbf{e}_1 & u_1 & 0 \\ \mathbf{e}_2 & u_2 & 0 \\ \mathbf{e}_3 & u_3 & 0 \end{vmatrix} = 0$.

(c) $\mathbf{u} \times \mathbf{u} = \begin{vmatrix} \mathbf{e}_1 & u_1 & u_1 \\ \mathbf{e}_2 & u_2 & u_2 \\ \mathbf{e}_3 & u_3 & u_3 \end{vmatrix} = 0$ (col 2 = col = 3).

(d) $\mathbf{u} \times k\mathbf{v} = \det \begin{bmatrix} \mathbf{e}_1 & u_1 & kv_1 \\ \mathbf{e}_2 & u_2 & kv_2 \\ \mathbf{e}_3 & u_3 & kv_3 \end{bmatrix} \stackrel{C_3/k}{=} k \det \begin{bmatrix} \mathbf{e}_1 & u_1 & v_1 \\ \mathbf{e}_2 & u_2 & v_2 \\ \mathbf{e}_3 & u_3 & v_3 \end{bmatrix} = k (\mathbf{u} \times \mathbf{v})$.

(e) $\mathbf{u} \times (\mathbf{v} + \mathbf{w}) = \det \begin{bmatrix} \mathbf{e}_1 & u_1 & v_1 + w_1 \\ \mathbf{e}_2 & u_2 & v_2 + w_2 \\ \mathbf{e}_3 & u_3 & v_3 + w_3 \end{bmatrix} = \det \begin{bmatrix} \mathbf{e}_1 & u_1 & v_1 \\ \mathbf{e}_2 & u_2 & v_2 \\ \mathbf{e}_3 & u_3 & v_3 \end{bmatrix} + \det \begin{bmatrix} \mathbf{e}_1 & u_1 & w_1 \\ \mathbf{e}_2 & u_2 & w_2 \\ \mathbf{e}_3 & u_3 & w_3 \end{bmatrix}$.

$= \mathbf{u} \times \mathbf{v} + \mathbf{u} \times \mathbf{w}$

(f) By Problem 2, $\mathbf{u} \cdot (\mathbf{u} \times \mathbf{v}) = \det \begin{bmatrix} u_1 & u_1 & v_1 \\ u_2 & u_2 & v_2 \\ u_3 & u_3 & v_3 \end{bmatrix} = 0$ since columns 1 and 2 are identical.

Similarly, $\mathbf{v} \cdot (\mathbf{u} \times \mathbf{v}) = \det \begin{bmatrix} v_1 & u_1 & v_1 \\ v_2 & u_2 & v_2 \\ v_3 & u_3 & v_3 \end{bmatrix} = 0$.

(g) By Problem 2, $\mathbf{u} \cdot (\mathbf{v} \times \mathbf{w}) = \det \begin{bmatrix} u_1 & v_1 & w_1 \\ u_2 & v_2 & w_2 \\ u_3 & v_3 & w_3 \end{bmatrix}$.

But $(\mathbf{u} \times \mathbf{v}) \cdot \mathbf{w} = \mathbf{w} \cdot (\mathbf{u} \times \mathbf{v}) = \det \begin{bmatrix} w_1 & u_1 & v_1 \\ w_2 & u_2 & v_2 \\ w_3 & u_3 & v_3 \end{bmatrix} = \det \begin{bmatrix} u_1 & v_1 & w_1 \\ u_2 & v_2 & w_2 \\ u_3 & v_3 & w_3 \end{bmatrix}$.

5. The area of the large rectangle is $(a+c)(b+d)$, the area of each of the large triangles is $\frac{1}{2}ab$, $\frac{1}{2}cd$, and of each of the small rectangles is bc.

Thus, $\mathcal{A} = (a+c)(b+d) - 2\left(\frac{1}{2}ab\right) - 2\left(\frac{1}{2}cd\right) - 2bc = ad - bc = \det \begin{bmatrix} a & b \\ c & d \end{bmatrix}$.

We must add an absolute value sign in case the two vectors are interchanged, in which case $ad < bc$. So $\mathcal{A} = \left| \det \begin{bmatrix} a & b \\ c & d \end{bmatrix} \right|$.

7. Recall, the dot product of two vectors is $\mathbf{a} \cdot \mathbf{b} = \|\mathbf{a}\| \|\mathbf{b}\| \cos \theta$, θ the angle between $\mathbf{a}$ and $\mathbf{b}$.

In this case, the cross product $\mathbf{v} \times \mathbf{w}$ is perpendicular to both $\mathbf{v}$ and $\mathbf{w}$, so $\mathbf{u} \cdot (\mathbf{v} \times \mathbf{w}) = (\|\mathbf{u}\| \cos \theta) \|\mathbf{v} \times \mathbf{w}\|$.

But $\|\mathbf{v} \times \mathbf{w}\|$ is the area of the parallelogram lying in the xy-plane determined by $\mathbf{v}$ and $\mathbf{w}$, and $\|\mathbf{u}\| \cos \theta$ is the height h of the parallelepiped.

So the volume of the parallelepiped is:
$V = \|\mathbf{v} \times \mathbf{w}\| h = (\|\mathbf{u}\| \cos \theta) \|\mathbf{v} \times \mathbf{w}\| = |\mathbf{u} \cdot (\mathbf{v} \times \mathbf{w})| = |\det [\mathbf{u} \ \mathbf{v} \ \mathbf{w}]|$ as required.

Exploration: Geometric Applications of Determinants

9. By Problem 4, the area of the parallelogram determined by $A\mathbf{u}$ and $A\mathbf{v}$ is

$$\left|\det \begin{bmatrix} A\mathbf{u} & A\mathbf{v} \end{bmatrix}\right| = \left|\det\left(A\begin{bmatrix} \mathbf{u} & \mathbf{v} \end{bmatrix}\right)\right| = \left|(\det A)\left(\det \begin{bmatrix} \mathbf{u} & \mathbf{v} \end{bmatrix}\right)\right| = |\det A|\,(\text{area of } P).$$

11. (a) The equation of the line through $(2,3)$ and $(-1,0)$ is $\begin{vmatrix} x & y & 1 \\ 2 & 3 & 1 \\ -1 & 0 & 1 \end{vmatrix} = 3x - 3y + 3 = 0.$

    (b) The equation of the line through $(1,2)$ and $(4,3)$ is $\begin{vmatrix} x & y & 1 \\ 1 & 2 & 1 \\ 4 & 3 & 1 \end{vmatrix} = -x + 2y - 2 = 0.$

13. There is a unique plane passing through these points, $ax + by + cz + d = 0$.
    Since the three given points are on this plane, their coordinates satisfy the equation:

$$ax_1 + by_1 + cz_1 + d = 0$$
$$ax_2 + by_2 + cz_2 + d = 0$$
$$ax_3 + by_3 + cz_3 + d = 0$$

This system has a nontrivial solution,

so the coefficient matrix $X = \begin{bmatrix} x & y & z & 1 \\ x_1 & y_1 & z_1 & 1 \\ x_2 & y_2 & z_2 & 1 \\ x_3 & y_3 & z_3 & 1 \end{bmatrix}$ cannot be invertible.

Thus, $|X| = 0$.

If the three points are collinear,

then the vectors $\begin{bmatrix} x_2 - x_1 \\ y_2 - y_1 \\ z_2 - z_1 \end{bmatrix}$ and $\begin{bmatrix} x_3 - x_1 \\ y_3 - y_1 \\ z_3 - z_1 \end{bmatrix}$ are multiples of one another.

That is, $\begin{bmatrix} x_3 - x_1 \\ y_3 - y_1 \\ z_3 - z_1 \end{bmatrix} = k \begin{bmatrix} x_2 - x_1 \\ y_2 - y_1 \\ z_2 - z_1 \end{bmatrix}, k \neq 0.$

So we can row-reduce the 4×4 matrix as follows:

$$\begin{bmatrix} x & y & z & 1 \\ x_1 & y_1 & z_1 & 1 \\ x_2 & y_2 & z_2 & 1 \\ x_3 & y_3 & z_3 & 1 \end{bmatrix} \xrightarrow[R_4-R_2]{R_3-R_2} \begin{bmatrix} x & y & z & 1 \\ x_1 & y_1 & z_1 & 1 \\ x_2-x_1 & y_2-y_1 & z_2-z_1 & 0 \\ x_3-x_1 & y_3-y_1 & z_3-z_1 & 0 \end{bmatrix} \xrightarrow{R_4-kR_3} \begin{bmatrix} x & y & z & 1 \\ x_1 & y_1 & z_1 & 1 \\ x_2-x_1 & y_2-y_1 & z_2-z_1 & 0 \\ 0 & 0 & 0 & 0 \end{bmatrix}.$$

In this case $|X|$ is trivially 0 and calculating the determinant gives us no information.

15. The three equations are
$$a - b + c = 10$$
$$a = 5$$
$$a + 3b + 9c = 2$$

which we can write as $PX = Q$ where $P = \begin{bmatrix} 1 & -1 & 1 \\ 1 & 0 & 0 \\ 1 & 3 & 9 \end{bmatrix}$, $X = \begin{bmatrix} a \\ b \\ c \end{bmatrix}$, and $Q = \begin{bmatrix} 10 \\ 5 \\ 2 \end{bmatrix}$.

Since $|P| = 12 \neq 0$, P is invertible and the system has the unique solution $X = P^{-1}Q$.

Solving the system, we add three times the first equation to the third equation to get:
$4a + 12c = 32 \Rightarrow 12c = 32 - 4(5) \Rightarrow c = 1$, so $5 - b + 1 = 10 \Rightarrow b = -4$
from the first equation.

Thus, the equation of the parabola is $y = 5 - 4x + x^2$.

17. If all three points are on the parabola $y = a + bx + cx^2$, then the system is:
$$a + ba_1 + ca_1^2 = b_1$$
$$a + ba_2 + ca_2^2 = b_2 \Rightarrow \text{the coefficient matrix } P = \begin{bmatrix} 1 & a_1 & a_1^2 \\ 1 & a_2 & a_2^2 \\ 1 & a_3 & a_3^2 \end{bmatrix},$$
$$a + ba_3 + ca_3^2 = b_3$$

and since the a_i are distinct:

$|P| \stackrel{R_2-R_1}{\underset{R_3-R_1}{=}} \begin{vmatrix} 1 & a_1 & a_1^2 \\ 0 & a_2-a_1 & a_2^2-a_1^2 \\ 0 & a_3-a_1 & a_3^2-a_1^2 \end{vmatrix} \stackrel{R_2/(a_2-a_1)}{\underset{R_3/(a_3-a_1)}{=}} (a_2-a_1)(a_3-a_1) \begin{vmatrix} 1 & a_1 & a_1^2 \\ 0 & 1 & a_1+a_2 \\ 0 & 1 & a_1+a_3 \end{vmatrix}$

$\stackrel{R_3-R_2}{=} (a_2-a_1)(a_3-a_1) \begin{vmatrix} 1 & a_1 & a_1^2 \\ 0 & 1 & a_1+a_2 \\ 0 & 0 & a_3-a_2 \end{vmatrix} = (a_2-a_1)(a_3-a_1)(a_3-a_2) \neq 0.$

Exploration: Geometric Applications of Determinants

19. For the $n \times n$ determinant given, we proceed as we did in Problem 18.
At the ith step, we subtract row 1 from all subsequent rows, then divide row j by $(a_j - a_i)$ for $2 \le j \le n$, then expand along the first column. So at the ith step we obtain a factor of $\prod_{j=i+1}^{n} (a_j - a_i)$.

The first step is as follows:

$$D = \begin{vmatrix} 1 & a_1 & a_1^2 & \cdots & a_1^{n-1} \\ 1 & a_2 & a_2^2 & \cdots & a_2^{n-1} \\ 1 & a_3 & a_3^2 & \cdots & a_3^{n-1} \\ \vdots & \vdots & \vdots & \ddots & \vdots \\ 1 & a_n & a_n^2 & \cdots & a_n^{n-1} \end{vmatrix} = \prod_{k=2}^{n} (a_k - a_1) \begin{vmatrix} 1 & a_1 & a_1^2 & \cdots & a_1^{n-1} \\ 0 & 1 & a_1 + a_2 & \cdots & \dfrac{a_2^{n-1} - a_1^{n-1}}{a_2 - a_1} \\ 0 & 1 & a_1 + a_3 & \cdots & \dfrac{a_3^{n-1} - a_1^{n-1}}{a_3 - a_1} \\ \vdots & \vdots & \vdots & \ddots & \vdots \\ 0 & 1 & a_1 + a_n & \cdots & \dfrac{a_n^{n-1} - a_1^{n-1}}{a_n - a_1} \end{vmatrix}$$

$$= \prod_{k=2}^{n} (a_k - a_1) \begin{vmatrix} 1 & a_1 + a_2 & \cdots & \dfrac{a_2^{n-1} - a_1^{n-1}}{a_2 - a_1} \\ 1 & a_1 + a_3 & \cdots & \dfrac{a_3^{n-1} - a_1^{n-1}}{a_3 - a_1} \\ \vdots & \vdots & \ddots & \vdots \\ 1 & a_1 + a_n & \cdots & \dfrac{a_n^{n-1} - a_1^{n-1}}{a_n - a_1} \end{vmatrix}.$$

We see that after $n-1$ iterations, we have

$$D = \left[\prod_{k=2}^{n}(a_k - a_1)\right]\left[\prod_{k=3}^{n}(a_k - a_2)\right] \cdots \left[\prod_{k=n-1}^{n}(a_k - a_{n-2})\right]\left[\prod_{k=n}^{n}(a_k - a_{n-1})\right]$$

$$= \prod_{1 \le i < j \le n} (a_j - a_i).$$

This means that for distinct $\{a_i\}$ and any $\{b_i\}$, $1 \le i \le n$, there is a unique polynomial of degree $n-1$ whose graph passes through the points (a_i, b_i).

4.3 Eigenvalues and Eigenvectors of $n \times n$ Matrices

Q: What are our main goals for Section 4.3?

A: To take our first in depth look at eigenvalues and eigenvectors.
The key equation an eigenvector satisfies is $A\mathbf{x} = \lambda\mathbf{x}$ with eigenvalue λ.
The associated characteristic equation is $\det(A - \lambda I) = c_A(\lambda) = \prod_{i=1}^{n}(\lambda_i - \lambda) = 0$.
A review of Appendix D on Polynomials is strongly recommended.

Key Definitions and Concepts

$c_A(\lambda)$	p.289	4.3	The *characteristic* polynomial is $c_A(\lambda) = \det(A - \lambda I)$. It is important to note that $c_A(\lambda)$ is a polynomial in λ. That is, λ is *not* a fixed eigenvalue, but a variable like x.
$\prod_{i=1}^{n}(\lambda_i - \lambda)$	p.289	4.3	$c_A(\lambda)$ using the eigenvalues of A, λ_i.
$\prod_{i=1}^{m}(\lambda_i - \lambda)^{k_i}$	p.289	4.3	$c_A(\lambda)$ emphasizing the algebraic multiplicity of λ_i, k_i.
algebraic multiplicity	p.291	4.3	The multiplicity of λ as a root of $\det(A - \lambda I) = 0$. That is, if $c_A(\lambda) = \prod_{i=1}^{m}(\lambda_i - \lambda)^{k_i}$ where λ_i are distinct, then the algebraic multiplicity of λ_i is k_i. That is, the algebraic multiplicity of an eigenvalue is equal to the exponent of its associated factor in the characteristic equation.
eigenvalue	p.289	4.3	The solutions of $\det(A - \lambda I) = 0$.
geometric multiplicity	p.291	4.3	The dimension of the eigenspace associated with λ, $\dim E_\lambda$, where $\dim E_\lambda$ is the number of vectors in a basis. That is, *geometric multiplicity* $= \dim E_\lambda$.
characteristic equation	p.289	4.3	The *characteristic* equation is $\det(A - \lambda I) = 0$. The solutions of $\det(A - \lambda I) = 0$ are the *eigenvalues* of A.
characteristic polynomial	p.289	4.3	The *characteristic* polynomial is $c_A(\lambda) = \det(A - \lambda I)$. The solutions of $\det(A - \lambda I) = 0$ are the *eigenvalues* of A.

Theorems

Thm 4.15	p.292	4.3	If A is triangular, then its eigenvalues are its diagonal entries.
Thm 4.16	p.292	4.3	A is invertible if and only if all $\lambda \neq 0$.
Thm 4.17	p.293	4.3	a. through o. give equivalent conditions for A to be invertible.
Thm 4.18	p.293	4.3	a. through c. relate eigenvalues, A^n and A^{-1}: For any integer n if $A\mathbf{x} = \lambda\mathbf{x}$, then $A^n\mathbf{x} = \lambda^n\mathbf{x}$.
Thm 4.19	p.294	4.3	If $\mathbf{x} \in \text{span}(\mathbf{v}_i)$, then $A^k\mathbf{x} = c_1\lambda_1^k\mathbf{v}_1 + c_2\lambda_2^k\mathbf{v}_2 + \cdots + c_m\lambda_m^k\mathbf{v}_m$ where $\mathbf{v}_i$ is the eigenvector corresponding to λ_i.
Thm 4.20	p.294	4.3	If λ_i are distinct, then $\{\mathbf{v}_i\}$ are linearly independent.

Discussion of Definitions and Theorems
The characteristic polynomial $c_A(\lambda) = \det(A - \lambda I)$

Q: What is the primary reason for defining $c_A(\lambda)$ as $\det(A - \lambda I)$?
A: Because we realize the roots of $\det(A - \lambda I)$ are the eigenvalues of A.

Q: What does it mean to say that x_i is a root of the polynomial $p(x)$?
A: It means x_i is a solution of the equation $p(x) = 0$. That is, $p(x_i) = 0$.
See Appendix D on Polynomials for a full discussion and development of these ideas.

Q: What does it mean to say that λ_i is a root of the polynomial $c_A(\lambda)$?
A: It means λ_i is a solution of $c_A(\lambda) = \det(A - \lambda I) = 0$. That is, $c_A(\lambda_i) = \det(A - \lambda_i I) = 0$.

Q: If λ is an eigenvalue of A, why does that imply that λ is a root of $\det(A - \lambda I)$?
A: This is argued in the text, but it is valuable to go through the details for ourselves.

If λ is an eigenvalue of A, then there exists $\mathbf{x} \neq \mathbf{0}$ such that $A\mathbf{x} = \lambda \mathbf{x}$.
So, $A\mathbf{x} - \lambda \mathbf{x} = \mathbf{0}$ which implies $(A - \lambda I)\mathbf{x} = \mathbf{0}$. What does this tell us?
The columns of $A - \lambda I$ are linearly dependent. Therefore, $\det(A - \lambda I) = 0$.

Q: We claim $(A - \lambda I)\mathbf{x} = \mathbf{0}$ implies the columns of $A - \lambda I$ are linearly dependent. Why?

A: Recall, $\mathbf{x} \neq \mathbf{0}$, so we have: $\mathbf{x} = \begin{bmatrix} c_1 \\ c_2 \\ \vdots \\ c_n \end{bmatrix}$, where at least one $c_i \neq 0$.

Now let $A - \lambda I = B = \begin{bmatrix} \mathbf{b}_1 & \mathbf{b}_2 & \cdots & \mathbf{b}_n \end{bmatrix}$, so $(A - \lambda I)\mathbf{x} = \sum_{i=1}^{n} c_i \mathbf{b}_i = \mathbf{0}$, at least one $c_i \neq 0$.

This is exactly what is required for the columns of $A - \lambda I$ to be linearly dependent.
The reasoning in this proof is useful in proving similar statements.

Q: If λ is a root of $\det(A - \lambda I)$, why does that imply that λ is an eigenvalue of A?
A: Hint: The argument is very similar to the one above.

Q: Why does the fact that λ_i are the roots of $c_A(\lambda) = \det(A - \lambda I)$, imply $c_A(\lambda) = \prod_{i=1}^{n}(\lambda_i - \lambda)$?
A: For a complete discussion, see the Factor Theorem in Appendix D on Polynomials.

We can note, however, $c_A(\lambda_k) = \prod_{i=1}^{n}(\lambda_i - \lambda_k) = 0$ because $\lambda_i - \lambda_k = 0$ when $i = k$.

4.3 Eigenvalues and Eigenvectors of $n \times n$ Matrices

Discussion of Definitions and Theorems

Algebraic and geometric multiplicity

Q: What is the definition of *algebraic multiplicity*?
A: The multiplicity of λ as a root of $\det(A - \lambda I) = 0$.

Q: What is the multiplicity of λ as a root of $\det(A - \lambda I) = 0$?
A: The multiplicity of λ as a root of $\det(A - \lambda I) = 0$ can be understood as follows:

If $c_A(\lambda) = \prod_{i=1}^{m} (\lambda_i - \lambda)^{k_i}$ where λ_i are distinct, then the algebraic multiplicity of λ_i is k_i.
That is, the algebraic multiplicity of an eigenvalue is equal to
the exponent of its associated factor in the characteristic equation.

Q: Consider: $\det(A - \lambda I) = \begin{vmatrix} 2-\lambda & 1 \\ -1 & 0-\lambda \end{vmatrix} = \lambda^2 - 2\lambda + 1 = (\lambda - 1)^2$.

Why does the eigenvalue 1 have an *algebraic multiplicity* of 2?
A: Since $c_A(\lambda) = (\lambda - 1)^2 = (1 - \lambda)^2$, the factor associated with 1 has an exponent of 2.

Q: What is the definition of *geometric multiplicity*?
A: The geometric multiplicity of $\lambda_i = \dim(E_{\lambda_i})$.
So, geometric multiplicity of $\lambda_i = \dim(E_{\lambda_i}) =$ number of vectors in a basis for E_{λ_i}.

Q: Consider: $A - 1I = \begin{bmatrix} 2-1 & 1 \\ -1 & 0-1 \end{bmatrix} \longrightarrow \begin{bmatrix} 1 & 1 \\ 0 & 0 \end{bmatrix}$, so $E_1 = \text{span}\left(\begin{bmatrix} 1 \\ -1 \end{bmatrix} \right)$.

Q: What does this tell us about the *geometric multiplicity* of the eigenvalue 1?
A: The geometric multiplicity of $1 = \dim(E_1) = 1$.

Q: Given A, how do we compute the algebraic and geometric multiplicities of λ_i?
A: The process is outlined in the *Box* prior to Example 4.18. See Exercises 1 through 12.

Q: In those exercises, compare the values of the algebraic and geometric multiplicities. What pattern emerges?
A: The geometric multiplicity is less than or equal to the algebraic multiplicity.

So, if $c_A(\lambda) = \prod_{i=1}^{m} (\lambda_i - \lambda)^{k_i}$, then $\dim(E_{\lambda_i}) \leq k_i$.

Continuing to improve the quality of our proofs

Q: What are two simple but important ways to improve the quality of our proofs?
A: State clearly *what* we are going to prove and *how* we are going to prove it.

Q: What is one effective way to test the success of our proof?
A: Test it out on our peers. If they cannot follow it, it may be lacking in detail.
We should ask them to be critical. To tell us if our proofs are compelling or confusing.

Theorem 4.16: *A is invertible if and only if all $\lambda \neq 0$*

Q: What is the contrapositive of Theorem 4.16?
A: A is *not* invertible if and only if 0 *is* an eigenvalue of A.

Q: What does the contrapositive of Theorem 4.16 imply?
A: $\det A = 0$ if and only if 0 *is* an eigenvalue of A. Why? Because the contrapositive of Theorem 4.6 in Section 4.2 implies $\det A = 0$ if and only if A is *not* invertible.

Theorem 4.17: *A is invertible if and only if all $\lambda \neq 0$*

Q: Why does an eigenvalue of 0 imply that A is not invertible?
If 0 is an eigenvalue of A, then there exists $\mathbf{x} \neq \mathbf{0}$ such that $A\mathbf{x} = 0\mathbf{x} = \mathbf{0}$.

Recall, $\mathbf{x} \neq \mathbf{0}$, so we have: $\mathbf{x} = \begin{bmatrix} c_1 \\ c_2 \\ \vdots \\ c_n \end{bmatrix}$, where at least one $c_i \neq 0$.

Now let $A = \begin{bmatrix} \mathbf{a}_1 & \mathbf{a}_2 & \cdots & \mathbf{a}_n \end{bmatrix}$, so $A\mathbf{x} = \sum_{i=1}^{n} c_i \mathbf{a}_i = \mathbf{0}$, at least one $c_i \neq 0$.

This is exactly what is required for the columns of A to be linearly dependent. Therefore, A is not invertible.

Why $A \longrightarrow B$ does not imply the eigenvalues of A and B are the same

Q: If the conjecture of Exercise 25 were true, what would that imply?
A: Since the only eigenvalue of I is 1, all invertible matrices would only have eigenvalue 1. This is clearly nonsense. However, they may be *related*.

Q: Let $\mathbf{x}$ be an eigenvector of A corresponding to eigenvalue λ.
If $A \xrightarrow{R_i \leftrightarrow R_j} B$, that is $B = E_{ij}A$, what goes wrong?
A: Since $B = E_{ij}A$, we have $B\mathbf{x} = E_{ij}(A\mathbf{x}) = \lambda(E_{ij}\mathbf{x})$.
So the components of $\mathbf{x}$ are interchanged and $\mathbf{x}$ fails to be an eigenvector for B.

Q: If $A \xrightarrow{kR_i} B$, what goes wrong?

Q: If $A \xrightarrow{R_i + kR_j} B$, what goes wrong?

Q: So, if $A \longrightarrow B$, we have seen their eigenvalues are not necessarily equal. However:
If $A \longrightarrow B$, is there a relationship among the eigenvalues and eigenvectors?
A: Hint: Consider the fact that $2I$ has eigenvalue 2. Can this process be generalized?

4.3 Eigenvalues and Eigenvectors of $n \times n$ Matrices

Solutions to odd-numbered exercises from Section 4.3

1. We follow the procedure outlined before Example 1.

   (a) The characteristic polynomial is $\det(A - \lambda I) = 0$, so we have:
   $$\det(A - \lambda I) = \begin{vmatrix} 1-\lambda & 3 \\ -2 & 6-\lambda \end{vmatrix} = (1-\lambda)(6-\lambda) - 3(-2) = \lambda^2 - 7\lambda + 12 = (\lambda - 3)(\lambda - 4).$$

   (b) The characteristic equation is $(\lambda - 3)(\lambda - 4) = 0$, which has solutions $\lambda_1 = 3$ and $\lambda_2 = 4$.

   (c) To find the eigenvectors corresponding to λ_1, we find the null space of $A - 3I = \begin{bmatrix} -2 & 3 \\ -2 & 3 \end{bmatrix}$.

   Row reduction produces $\left[\begin{array}{cc|c} -2 & 3 & 0 \\ -2 & 3 & 0 \end{array}\right] \longrightarrow \left[\begin{array}{cc|c} 1 & -\frac{3}{2} & 0 \\ 0 & 0 & 0 \end{array}\right]$.

   Thus, $\mathbf{x} = \begin{bmatrix} x_1 \\ x_2 \end{bmatrix}$ is in the eigenspace E_3 if and only if $x_1 - \frac{3}{2}x_2 = 0 \Leftrightarrow x_2 = \frac{2}{3}x_1$.

   Thus, $E_3 = \text{span}\left(\begin{bmatrix} 1 \\ \frac{2}{3} \end{bmatrix}\right)$.

   Similarly, $A - 4I = \begin{bmatrix} -3 & 3 \\ -2 & 2 \end{bmatrix} \longrightarrow \begin{bmatrix} 1 & -1 \\ 0 & 0 \end{bmatrix}$, so $E_4 = \text{span}\left(\begin{bmatrix} 1 \\ 1 \end{bmatrix}\right)$.

   (d) Each eigenvalue has algebraic and geometric multiplicity 1.

3. (a) $\det(A - \lambda I) = \begin{vmatrix} 1-\lambda & 1 & 0 \\ 0 & -2-\lambda & 1 \\ 0 & 0 & 3-\lambda \end{vmatrix} = (1-\lambda)(-2-\lambda)(3-\lambda).$

   (b) $(1-\lambda)(-2-\lambda)(3-\lambda) = 0 \Leftrightarrow \lambda_1 = -2, \lambda_2 = 1, \text{ or } \lambda_3 = 3.$

   (c) $A + 2I = \begin{bmatrix} 3 & 1 & 0 \\ 0 & 0 & 1 \\ 0 & 0 & 5 \end{bmatrix} \longrightarrow \begin{bmatrix} 1 & \frac{1}{3} & 0 \\ 0 & 0 & 1 \\ 0 & 0 & 0 \end{bmatrix}$, so $E_{-2} = \text{span}\left(\begin{bmatrix} 1 \\ -3 \\ 0 \end{bmatrix}\right)$.

   $A - I = \begin{bmatrix} 0 & 1 & 0 \\ 0 & -3 & 1 \\ 0 & 0 & 2 \end{bmatrix} \longrightarrow \begin{bmatrix} 0 & 1 & 0 \\ 0 & 0 & 1 \\ 0 & 0 & 0 \end{bmatrix}$, so $E_1 = \text{span}\left(\begin{bmatrix} 1 \\ 0 \\ 0 \end{bmatrix}\right)$.

   $A - 3I = \begin{bmatrix} -2 & 1 & 0 \\ 0 & -5 & 1 \\ 0 & 0 & 0 \end{bmatrix} \longrightarrow \begin{bmatrix} 1 & 0 & -\frac{1}{10} \\ 0 & 1 & -\frac{1}{5} \\ 0 & 0 & 0 \end{bmatrix}$, so $E_3 = \text{span}\left(\begin{bmatrix} 1 \\ 2 \\ 10 \end{bmatrix}\right)$.

   (d) Each eigenvalue has algebraic and geometric multiplicity 1.

5. (a) $\det(A - \lambda I) = \begin{vmatrix} 1-\lambda & 2 & 0 \\ -1 & -1-\lambda & 1 \\ 0 & 1 & 1-\lambda \end{vmatrix} = -(1-\lambda) + (1-\lambda)[(1-\lambda)(-1-\lambda) - (-2)]$
$= \lambda - 1 + (1 - \lambda + \lambda^2 - \lambda^3) = \lambda^2 - \lambda^3 = \lambda^2(1-\lambda)$.

(b) $\lambda^2(1-\lambda) = 0 \Leftrightarrow \lambda_1 = \lambda_2 = 0, \lambda_3 = 1$.

(c) $A = \begin{bmatrix} 1 & 2 & 0 \\ -1 & -1 & 1 \\ 0 & 1 & 1 \end{bmatrix} \longrightarrow \begin{bmatrix} 1 & 0 & -2 \\ 0 & 1 & 1 \\ 0 & 0 & 0 \end{bmatrix}$, so $E_0 = \text{span}\left(\begin{bmatrix} 1 \\ -\frac{1}{2} \\ \frac{1}{2} \end{bmatrix}\right)$.

$A - I = \begin{bmatrix} 0 & 2 & 0 \\ -1 & -2 & 1 \\ 0 & 1 & 0 \end{bmatrix} \longrightarrow \begin{bmatrix} 1 & 0 & -1 \\ 0 & 1 & 0 \\ 0 & 0 & 0 \end{bmatrix}$, so $E_1 = \text{span}\left(\begin{bmatrix} 1 \\ 0 \\ 1 \end{bmatrix}\right)$.

(d) 0 has algebraic multiplicity 2 and geometric multiplicity 1, while 1 has algebraic and geometric multiplicity 1.

7. (a) $\det(A - \lambda I) = \begin{vmatrix} 4-\lambda & 0 & 1 \\ 2 & 3-\lambda & 2 \\ -1 & 0 & 2-\lambda \end{vmatrix} = (3-\lambda)[(4-\lambda)(2-\lambda) - (-1)]$
$= (3-\lambda)(\lambda^2 - 6\lambda + 9) = (3-\lambda)^3$.

(b) $(3-\lambda)^3 = 0 \Leftrightarrow \lambda_1 = \lambda_2 = \lambda_3 = 3$.

(c) $A - 3I = \begin{bmatrix} 1 & 0 & 1 \\ 2 & 0 & 2 \\ -1 & 0 & -1 \end{bmatrix} \longrightarrow \begin{bmatrix} 1 & 0 & 1 \\ 0 & 0 & 0 \\ 0 & 0 & 0 \end{bmatrix}$, so $E_3 = \text{span}\left(\begin{bmatrix} 1 \\ 0 \\ -1 \end{bmatrix}, \begin{bmatrix} 0 \\ 1 \\ 0 \end{bmatrix}\right)$.

(d) 3 has algebraic multiplicity 3 and geometric multiplicity 2.

4.3 Eigenvalues and Eigenvectors of $n \times n$ Matrices

9. (a) $\det(A - \lambda I) = \begin{vmatrix} 3-\lambda & 1 & 0 & 0 \\ -1 & 1-\lambda & 0 & 0 \\ 0 & 0 & 1-\lambda & 4 \\ 0 & 0 & 1 & 1-\lambda \end{vmatrix} = \begin{vmatrix} 3-\lambda & 1 \\ -1 & 1-\lambda \end{vmatrix} \begin{vmatrix} 1-\lambda & 4 \\ 1 & 1-\lambda \end{vmatrix}$

$= [(3-\lambda)(1-\lambda) + 1]\left[(\lambda-1)^2 - 4\right] = (\lambda - 2)^2(\lambda - 3)(\lambda + 1)$.

(b) $(\lambda - 2)^2(\lambda - 3)(\lambda + 1) = 0 \Leftrightarrow \lambda_1 = -1, \lambda_2 = \lambda_3 = 2, \lambda_4 = 3$.

(c) $A + I = \begin{bmatrix} 4 & 1 & 0 & 0 \\ -1 & 2 & 0 & 0 \\ 0 & 0 & 2 & 4 \\ 0 & 0 & 1 & 2 \end{bmatrix} \longrightarrow \begin{bmatrix} 1 & 0 & 0 & 0 \\ 0 & 1 & 0 & 0 \\ 0 & 0 & 1 & 2 \\ 0 & 0 & 0 & 0 \end{bmatrix}$, so $E_{-1} = \text{span}\left(\begin{bmatrix} 0 \\ 0 \\ 1 \\ -\frac{1}{2} \end{bmatrix}\right)$.

$A - 2I = \begin{bmatrix} 1 & 1 & 0 & 0 \\ -1 & -1 & 0 & 0 \\ 0 & 0 & -1 & 4 \\ 0 & 0 & 1 & -1 \end{bmatrix} \longrightarrow \begin{bmatrix} 1 & 1 & 0 & 0 \\ 0 & 0 & 1 & 0 \\ 0 & 0 & 0 & 1 \\ 0 & 0 & 0 & 0 \end{bmatrix}$, so $E_2 = \text{span}\left(\begin{bmatrix} 1 \\ -1 \\ 0 \\ 0 \end{bmatrix}\right)$.

$A - 3I = \begin{bmatrix} 0 & 1 & 0 & 0 \\ -1 & -2 & 0 & 0 \\ 0 & 0 & -2 & 4 \\ 0 & 0 & 1 & -2 \end{bmatrix} \longrightarrow \begin{bmatrix} 1 & 0 & 0 & 0 \\ 0 & 1 & 0 & 0 \\ 0 & 0 & 1 & -2 \\ 0 & 0 & 0 & 0 \end{bmatrix}$, so $E_3 = \text{span}\left(\begin{bmatrix} 0 \\ 0 \\ 1 \\ \frac{1}{2} \end{bmatrix}\right)$.

(d) -1 and 3 have algebraic and geometric multiplicity 1, while 2 has algebraic multiplicity 2 and geometric multiplicity 1.

11. (a) $\det(A - \lambda I) = \begin{vmatrix} 1-\lambda & 0 & 0 & 0 \\ 0 & 1-\lambda & 0 & 0 \\ 1 & 1 & 3-\lambda & 0 \\ -2 & 1 & 2 & -1-\lambda \end{vmatrix} = (-1-\lambda)(1-\lambda)^2(3-\lambda)$.

(b) $(-1-\lambda)(1-\lambda)^2(3-\lambda) = 0 \Leftrightarrow \lambda_1 = -1, \lambda_2 = \lambda_3 = 1, \lambda_4 = 3$.

(c) $A + I = \begin{bmatrix} 2 & 0 & 0 & 0 \\ 0 & 2 & 0 & 0 \\ 1 & 1 & 4 & 0 \\ -2 & 1 & 2 & 0 \end{bmatrix} \longrightarrow \begin{bmatrix} 1 & 0 & 0 & 0 \\ 0 & 1 & 0 & 0 \\ 0 & 0 & 1 & 0 \\ 0 & 0 & 0 & 0 \end{bmatrix}$, so $E_{-1} = \text{span}\left(\begin{bmatrix} 0 \\ 0 \\ 0 \\ 1 \end{bmatrix}\right)$.

$A - I = \begin{bmatrix} 0 & 0 & 0 & 0 \\ 0 & 0 & 0 & 0 \\ 1 & 1 & 2 & 0 \\ -2 & 1 & 2 & -2 \end{bmatrix} \longrightarrow \begin{bmatrix} 1 & 0 & 0 & \frac{2}{3} \\ 0 & 1 & 2 & -\frac{2}{3} \\ 0 & 0 & 0 & 0 \\ 0 & 0 & 0 & 0 \end{bmatrix}$, so $E_1 = \text{span}\left(\begin{bmatrix} 1 \\ -1 \\ 0 \\ -\frac{3}{2} \end{bmatrix}, \begin{bmatrix} 1 \\ 0 \\ -\frac{1}{2} \\ -\frac{3}{2} \end{bmatrix}\right)$.

$A - 3I = \begin{bmatrix} -2 & 0 & 0 & 0 \\ 0 & -2 & 0 & 0 \\ 1 & 1 & 0 & 0 \\ -2 & 1 & 2 & -4 \end{bmatrix} \longrightarrow \begin{bmatrix} 1 & 0 & 0 & 0 \\ 0 & 1 & 0 & 0 \\ 0 & 0 & 1 & -2 \\ 0 & 0 & 0 & 0 \end{bmatrix}$, so $E_3 = \text{span}\left(\begin{bmatrix} 0 \\ 0 \\ 1 \\ \frac{1}{2} \end{bmatrix}\right)$.

(d) -1 and 3 have algebraic and geometric multiplicity 1, while 1 has algebraic and geometric multiplicity 2.

13. We need to show if $A\mathbf{x} = \lambda\mathbf{x}$, then $A^{-1}\mathbf{x} = \frac{1}{\lambda}\mathbf{x} = \lambda^{-1}\mathbf{x}$.
    Since $A\mathbf{x} = \lambda\mathbf{x}$, we have $A^{-1}(A\mathbf{x}) = A^{-1}(\lambda\mathbf{x}) = \lambda(A^{-1}\mathbf{x})$.
    So, $\lambda(A^{-1}\mathbf{x}) = (A^{-1}A)\mathbf{x} = \mathbf{x}$ which implies $A^{-1}\mathbf{x} = \frac{1}{\lambda}\mathbf{x} = \lambda^{-1}\mathbf{x}$ as required.

15. Since $\mathbf{x} = 2\mathbf{v}_1 + 3\mathbf{v}_2$ we have $A^{10}\mathbf{x} = A^{10}(2\mathbf{v}_1 + 3\mathbf{v}_2) = 2A^{10}\mathbf{v}_1 + 3A^{10}\mathbf{v}_2$.
    But by Theorem 4.4(a), $\mathbf{v}_1$ and $\mathbf{v}_2$ are eigenvectors of A^{10} with eigenvalues λ_1^{10} and λ_2^{10}.
    So, $2\lambda_1^{10}\mathbf{v}_1 + 3\lambda_2^{10}\mathbf{v}_2 = 2\left(\frac{1}{2}\right)^{10}\mathbf{v}_1 + 3(2)^{10}\mathbf{v}_2 = \frac{1}{512}\mathbf{v}_1 + 3072\mathbf{v}_2 = \begin{bmatrix} 3072 + \frac{1}{512} \\ 3072 - \frac{1}{512} \end{bmatrix}$.

17. We must find $\mathbf{x}$ as a linear combination $a_1\mathbf{v}_1 + a_2\mathbf{v}_2 + a_3\mathbf{v}_3$ of the eigenvectors. So:
    $$\begin{bmatrix} 2 \\ 1 \\ 2 \end{bmatrix} = a_1 \begin{bmatrix} 1 \\ 0 \\ 0 \end{bmatrix} + a_2 \begin{bmatrix} 1 \\ 1 \\ 0 \end{bmatrix} + a_3 \begin{bmatrix} 1 \\ 1 \\ 1 \end{bmatrix}$$
    We must have $a_3 = 2$, so this reduces to
    $$\begin{bmatrix} 0 \\ -1 \\ 0 \end{bmatrix} = a_1 \begin{bmatrix} 1 \\ 0 \\ 0 \end{bmatrix} + a_2 \begin{bmatrix} 1 \\ 1 \\ 0 \end{bmatrix}$$
    which has solution $a_1 = 1$, $a_2 = -1$. Thus,
    $$A^{20}\mathbf{x} = \lambda_1^{20}\mathbf{v}_1 - \lambda_2^{20}\mathbf{v}_2 + 2\cdot\lambda_3^{20}\mathbf{v}_3 = \begin{bmatrix} -\frac{1}{320} - \frac{1}{320} + 2 \\ -\frac{1}{320} + 2 \\ 2 \end{bmatrix} = \begin{bmatrix} 2 \\ 2 - \frac{1}{320} \\ 2 \end{bmatrix}$$

19. (a) The key observation is that $A^T - \lambda I = A^T - (\lambda I)^T = (A - \lambda I)^T$.
    Thus, using Theorem 4.10, the characteristic polynomial of A^T is
    $$\det(A^T - \lambda I) = \det(A - \lambda I)^T = \det(A - \lambda I)$$
    But $\det(A - \lambda I)$ is the characteristic polynomial of A.
    Therefore, A and A^T have the same eigenvalues as we were to show.

    (b) $A = \begin{bmatrix} 1 & 0 \\ 1 & 2 \end{bmatrix}$ has eigenspaces $E_1 = \text{span}\left(\begin{bmatrix} 1 \\ -1 \end{bmatrix}\right)$ and $E_2 = \text{span}\left(\begin{bmatrix} 0 \\ 1 \end{bmatrix}\right)$,
    while $A^T = \begin{bmatrix} 1 & 1 \\ 0 & 2 \end{bmatrix}$ has eigenspaces $E_1 = \text{span}\left(\begin{bmatrix} 1 \\ 0 \end{bmatrix}\right)$ and $E_2 = \text{span}\left(\begin{bmatrix} 1 \\ 1 \end{bmatrix}\right)$.

21. Suppose A is idempotent with eigenvector $\mathbf{x}$ corresponding to λ.
    Then $\lambda\mathbf{x} = A\mathbf{x} = A^2\mathbf{x} = A(A\mathbf{x}) = A(\lambda\mathbf{x}) = \lambda(A\mathbf{x}) = \lambda(\lambda\mathbf{x}) = \lambda^2\mathbf{x}$.
    So, we get $\lambda\mathbf{x} = \lambda^2\mathbf{x} \Rightarrow \lambda = \lambda^2$ (because $\mathbf{x} \neq \mathbf{0}$) $\Rightarrow \lambda^2 - \lambda = \lambda(\lambda - 1) = 0 \Rightarrow \lambda = 0$ or 1.

4.3 Eigenvalues and Eigenvectors of $n \times n$ Matrices

23. (a) $\det(A - \lambda I) = \begin{vmatrix} 3 - \lambda & 2 \\ 5 & 0 - \lambda \end{vmatrix} = \lambda^2 - 3\lambda - 10 = (\lambda + 2)(\lambda - 5) = 0 \Leftrightarrow \lambda = -2$ or 5.

$A + 2I = \begin{bmatrix} 5 & 2 \\ 5 & 2 \end{bmatrix}$, so $E_{-2} = \operatorname{span}\left(\begin{bmatrix} 1 \\ -\frac{5}{2} \end{bmatrix}\right)$, and $A - 5I = \begin{bmatrix} -2 & 2 \\ 5 & -5 \end{bmatrix}$.

So, $E_5 = \operatorname{span}\left(\begin{bmatrix} 1 \\ 1 \end{bmatrix}\right)$.

(b) By Theorem 4.4(b), A^{-1} has eigenvalues $-\frac{1}{2}$ and $\frac{1}{5}$ with

$E_{-1/2} = \operatorname{span}\left(\begin{bmatrix} 1 \\ -\frac{5}{2} \end{bmatrix}\right)$ and $E_{1/5} = \operatorname{span}\left(\begin{bmatrix} 1 \\ 1 \end{bmatrix}\right)$.

By Exercise 22, $A - 2I$ has eigenvalues -4 and 3 with

$E_{-4} = \operatorname{span}\left(\begin{bmatrix} 1 \\ -\frac{5}{2} \end{bmatrix}\right)$ and $E_3 = \operatorname{span}\left(\begin{bmatrix} 1 \\ 1 \end{bmatrix}\right)$, and

$A + 2I$ has eigenvalues 0 and 7 with

$E_0 = \operatorname{span}\left(\begin{bmatrix} 1 \\ -\frac{5}{2} \end{bmatrix}\right)$ and $E_7 = \operatorname{span}\left(\begin{bmatrix} 1 \\ 1 \end{bmatrix}\right)$.

25. As noted in Theorem 4.17(d), $A \longrightarrow I$ if and only if A is invertible.

    Q: If the conjecture of this exercise were true, what would that imply?
    A: Since the only eigenvalue of I is 1, all invertible matrices would only have eigenvalue 1. This is clearly nonsense. However, they may be *related*.

    Q: Let **x** be an eigenvector of A corresponding to eigenvalue λ.
    If $A \xrightarrow{R_i \leftrightarrow R_j} B$, that is $B = E_{ij}A$, what goes wrong?
    A: Since $B = E_{ij}A$, we have $B\mathbf{x} = E_{ij}(A\mathbf{x}) = \lambda(E_{ij}\mathbf{x})$.
    So the components of **x** are interchanged and **x** fails to be an eigenvector for B.

    Q: If $A \xrightarrow{kR_i} B$, what goes wrong?
    Q: If $A \xrightarrow{R_i + kR_j} B$, what goes wrong?

    Q: So, if $A \longrightarrow B$, we have seen their eigenvalues are not necessarily equal. However:
    If $A \longrightarrow B$, is there a relationship among the eigenvalues and eigenvectors?
    A: Hint: $2I$ has eigenvalue 2. Can this process be generalized? See Exercise 41.

27. The companion matrix of $p(x) = x^3 + 3x^2 - 4x + 12$ is $C(p) = \begin{bmatrix} -3 & 4 & -12 \\ 1 & 0 & 0 \\ 0 & 1 & 0 \end{bmatrix}$,

and the characteristic polynomial of $C(p)$ is

$\det \begin{bmatrix} -3 - \lambda & 4 & -12 \\ 1 & -\lambda & 0 \\ 0 & 1 & -\lambda \end{bmatrix} = -12 \begin{vmatrix} 1 & -\lambda \\ 0 & 1 \end{vmatrix} - \lambda \begin{vmatrix} -3 - \lambda & 4 \\ 1 & -\lambda \end{vmatrix} = -12 - \lambda[\lambda(3 + \lambda) - 4]$

$= -\lambda^3 - 3\lambda^2 + 4\lambda - 12.$

29. (a) $C(p) = \begin{bmatrix} -a & -b & -c \\ 1 & 0 & 0 \\ 0 & 1 & 0 \end{bmatrix}$, so the characteristic polynomial of $C(p)$ is

$$\begin{vmatrix} -a-\lambda & -b & -c \\ 1 & -\lambda & 0 \\ 0 & 1 & -\lambda \end{vmatrix} = -c \begin{vmatrix} 1 & -\lambda \\ 0 & 1 \end{vmatrix} - \lambda \begin{vmatrix} -a-\lambda & -b \\ 1 & -\lambda \end{vmatrix}$$

$$= -c - \lambda^3 - a\lambda^2 - b\lambda = -\left(\lambda^3 + a\lambda^2 + b\lambda + c\right).$$

(b) Suppose that λ is an eigenvalue of $C(p)$, with eigenvector $\mathbf{x} = \begin{bmatrix} x_1 \\ x_2 \\ x_3 \end{bmatrix}$. Then

$$\lambda \begin{bmatrix} x_1 \\ x_2 \\ x_3 \end{bmatrix} = C(p) \begin{bmatrix} x_1 \\ x_2 \\ x_3 \end{bmatrix} \Leftrightarrow \begin{bmatrix} \lambda x_1 \\ \lambda x_2 \\ \lambda x_3 \end{bmatrix} = \begin{bmatrix} -a & -b & -c \\ 1 & 0 & 0 \\ 0 & 1 & 0 \end{bmatrix} \begin{bmatrix} x_1 \\ x_2 \\ x_3 \end{bmatrix} = \begin{bmatrix} -ax_1 - bx_2 - cx_3 \\ x_1 \\ x_2 \end{bmatrix} \Leftrightarrow$$

$x_2 = \lambda x_3$ and $x_1 = \lambda x_2$, so $\begin{bmatrix} \lambda^2 \\ \lambda \\ 1 \end{bmatrix}$ is a corresponding eigenvector.

31. According to Exercise 29, the characteristic polynomial of the non-diagonal matrix

$C(p) = \begin{bmatrix} -a & -b & -c \\ 1 & 0 & 0 \\ 0 & 1 & 0 \end{bmatrix}$ is $-\left(\lambda^3 + a\lambda^2 + b\lambda + c\right)$. So we require that

$-\left(\lambda^3 + a\lambda^2 + b\lambda + c\right) = -\left(\lambda + 2\right)\left(\lambda - 1\right)\left(\lambda - 3\right) = -\left(\lambda^3 - 2\lambda^2 - 5\lambda + 6\right) \Leftrightarrow$
$a = -2$, $b = -5$, and $c = 6$.

Clearly, the matrix $\begin{bmatrix} 2 & 5 & -6 \\ 1 & 0 & 0 \\ 0 & 1 & 0 \end{bmatrix}$ is such a matrix.

33. The characteristic polynomial $c_A(\lambda)$ of A is $\det(A - \lambda I) = \begin{vmatrix} 1-\lambda & -1 \\ 2 & 3-\lambda \end{vmatrix} = \lambda^2 - 4\lambda + 5$.
We verify that

$$A^2 - 4A + 5I = \begin{bmatrix} 1 & -1 \\ 2 & 3 \end{bmatrix}^2 - 4 \begin{bmatrix} 1 & -1 \\ 2 & 3 \end{bmatrix} + 5 \begin{bmatrix} 1 & 0 \\ 0 & 1 \end{bmatrix}$$

$$= \begin{bmatrix} -1 & -4 \\ 8 & 7 \end{bmatrix} - \begin{bmatrix} 4 & -4 \\ 8 & 12 \end{bmatrix} + \begin{bmatrix} 5 & 0 \\ 0 & 5 \end{bmatrix} = O.$$

4.3 Eigenvalues and Eigenvectors of $n \times n$ Matrices

35. In Exercise 33, $c_A(\lambda) = \lambda^2 - 4\lambda + 5$, so $a = -4$ and $b = 5$.

Thus, $A^2 = -aA - bI = 4A - 5I = 4\begin{bmatrix} 1 & -1 \\ 2 & 3 \end{bmatrix} - 5\begin{bmatrix} 1 & 0 \\ 0 & 1 \end{bmatrix} = \begin{bmatrix} -1 & -4 \\ 8 & 7 \end{bmatrix}$.

Similarly, $A^3 = (a^2 - b)A + abI = 11A - 20I = 11\begin{bmatrix} 1 & -1 \\ 2 & 3 \end{bmatrix} - 20\begin{bmatrix} 1 & 0 \\ 0 & 1 \end{bmatrix} = \begin{bmatrix} -9 & -11 \\ 22 & 13 \end{bmatrix}$.

The corresponding formula for A^4 is given by

$A^4 = AA^3 = A[(a^2 - b)A + abI] = (a^2 - b)A^2 + abA = (a^2 - b)[-aA - bI] + abA$
$= (-a^3 + 2ab)A + (b^2 - a^2 b)I$.

So, $A^4 = 24A - 55I = 24\begin{bmatrix} 1 & -1 \\ 2 & 3 \end{bmatrix} - 55\begin{bmatrix} 1 & 0 \\ 0 & 1 \end{bmatrix} = \begin{bmatrix} -31 & -24 \\ 48 & 17 \end{bmatrix}$.

37. $A^{-1} = -\frac{1}{b}A - \frac{a}{b}I = -\frac{1}{5}\begin{bmatrix} 1 & -1 \\ 2 & 3 \end{bmatrix} + \frac{4}{5}\begin{bmatrix} 1 & 0 \\ 0 & 1 \end{bmatrix} = \begin{bmatrix} \frac{3}{5} & \frac{1}{5} \\ -\frac{2}{5} & \frac{1}{5} \end{bmatrix}$. Also,

$A^{-2} = (A^{-1})^2 = \left(-\frac{1}{b}A - \frac{a}{b}I\right)^2 = \frac{1}{b^2}A^2 + \frac{2a}{b^2}A + \frac{a^2}{b^2}I = \frac{1}{b^2}(-aA - bI) + \frac{2a}{b^2}A + \frac{a^2}{b^2}I$

$= \frac{a}{b^2}A + \frac{a^2 - b}{b^2}I = -\frac{4}{25}\begin{bmatrix} 1 & -1 \\ 2 & 3 \end{bmatrix} + \frac{11}{25}\begin{bmatrix} 1 & 0 \\ 0 & 1 \end{bmatrix} = \begin{bmatrix} \frac{7}{25} & \frac{4}{25} \\ -\frac{8}{25} & -\frac{1}{25} \end{bmatrix}$.

39. According to Exercise 69 in Section 4.2, $\det A = \det\begin{bmatrix} P & Q \\ O & S \end{bmatrix} = (\det P)(\det S)$.

We apply this result to $A - \lambda I$ instead of A below.

Note $c_A(\lambda) = \det(A - \lambda I)$, $c_P(\lambda) = \det(P - \lambda I)$, and $c_S(\lambda) = \det(S - \lambda I)$.

So, $c_A(\lambda) = \det(A - \lambda I) = \det(P - \lambda I)\det(S - \lambda I) = c_P(\lambda)c_S(\lambda)$.

Q: Why can we apply the result of Exercise 69 to $A - \lambda I$?
A: Because P becomes $P - \lambda I$ and S becomes $S - \lambda I$ so the proof holds.

That is, we have $A - \lambda I = \begin{bmatrix} P - \lambda I & Q \\ O & S - \lambda I \end{bmatrix}$.

Q: Neither O nor Q are affected. Why not?
A: Because P and S are square and I has all zero entries off the diagonal.

41. Given A has eigenvalues α_i and B has eigenvalues β_i, we need to show the following:
   1) If $C = A + B$ has eigenvalues γ_i, then $\sum_{i=1}^{n} \gamma_i = \sum_{i=1}^{n}(\alpha_i + \beta_i)$ and
   2) If $C = AB$ has eigenvalues γ_i, then $\prod_{i=1}^{n} \gamma_i = \prod_{i=1}^{n}(\alpha_i \beta_i)$.

   Q: How does this differ from the statement of Exercise 24?
   A: In Exercise 24, we show that if $C = A + B$ it does *not* follow that $\gamma_i = \alpha_i + \beta_i$.
   In this Exercise, we consider the sum (and product) of *all* the eigenvalues.

tr(C): If $C = A + B$ has eigenvalues γ_i, then $\sum_{i=1}^{n} \gamma_i = \sum_{i=1}^{n}(\alpha_i + \beta_i)$.

Since $[C]_{ii} = [A]_{ii} + [B]_{ii}$, we have $\operatorname{tr}(C) = \operatorname{tr}(A) + \operatorname{tr}(B)$.

Furthermore, Exercise 40 asserts for any matrix A, $\operatorname{tr}(A) = \sum_{i=1}^{n} \lambda_i$.

So, $\operatorname{tr}(C) = \operatorname{tr}(A) + \operatorname{tr}(B)$ implies $\sum_{i=1}^{n} \gamma_i = \sum_{i=1}^{n} \alpha_i + \sum_{i=1}^{n} \beta_i = \sum_{i=1}^{n}(\alpha_i + \beta_i)$.

Q: How might we state $[C]_{ii} = [A]_{ii} + [B]_{ii}$ in words?
A: A diagonal entry of C equals the sum of the corresponding entries in A and B.

Q: How might we state $\sum_{i=1}^{n} \gamma_i = \sum_{i=1}^{n}(\alpha_i + \beta_i)$ in words?
A: The sum of the eigenvalues of C equals the sum of all the eigenvalues of both A and B.

det(C): If $C = AB$ has eigenvalues γ_i, then $\prod_{i=1}^{n} \gamma_i = \prod_{i=1}^{n}(\alpha_i \beta_i)$.

By Theorem 4.8 in Section 4.2, $C = AB$ implies $\det C = (\det A)(\det B)$.

Furthermore, Exercise 40 asserts for any matrix A, $\det(A) = \prod_{i=1}^{n} \lambda_i$.

So, $\det C = (\det A)(\det B)$ implies $\prod_{i=1}^{n} \gamma_i = \left(\prod_{i=1}^{n} \alpha_i\right)\left(\prod_{i=1}^{n} \beta_i\right) = \prod_{i=1}^{n}(\alpha_i \beta_i)$.

Q: How might we state $\prod_{i=1}^{n} \gamma_i = \prod_{i=1}^{n}(\alpha_i \beta_i)$ in words?
A: The product of the eigenvalues of C equals the product of all eigenvalues of A, B.

4.4 Similarity and Diagonalization

Q: What are our main goals for Section 4.4?

A: To become familiar with the concept of similarity, $A \sim B$. That is, $P^{-1}AP = B$.
If A is diagonalizable, $P^{-1}AP = D$, realize the columns of P are the eigenvectors of A.
Furthermore, realize that the eigenvalues of A are the diagonal entries of D.

Key Definitions and Concepts

$c_A(\lambda)$	p.289	4.3	The *characteristic* polynomial is $c_A(\lambda) = \det(A - \lambda I)$.
$\prod_{i=1}^{m}(\lambda_i - \lambda)^{k_i}$	p.289	4.3	$c_A(\lambda)$ emphasizing the algebraic multiplicity of λ_i, k_i.
$A \sim B$	p.298	4.4	$A \sim B$ (A is *similar* to B) if $P^{-1}AP = B$, P invertible
algebraic multiplicity	p.291	4.3	The multiplicity of λ as a root of $\det(A - \lambda I) = 0$. If $c_A(\lambda) = \prod_{i=1}^{m}(\lambda_i - \lambda)^{k_i}$, k_i is algebraic multiplicity.
characteristic equation	p.289	4.3	The *characteristic* equation is $\det(A - \lambda I) = 0$. The solutions of $\det(A - \lambda I) = 0$ are the *eigenvalues* of A.
diagonalizable	p.300	4.4	$A \sim D$ if $P^{-1}AP = D$ where D is diagonal
geometric multiplicity	p.291	4.3	The dimension of the eigenspace associated with λ, $\dim E_\lambda$, where $\dim E_\lambda$ is the number of vectors in a basis.
similar	p.298	4.4	$A \sim B$ if $P^{-1}AP = B$ where P is invertible

Theorems

Thm 4.21	p.291	4.4	a. $A \sim A$, b. $A \sim B \Rightarrow B \sim A$, c. $A \sim B$, $B \sim C \Rightarrow A \sim C$
Thm 4.22	p.299	4.4	parts a. through e. list implications of $A \sim B$ a. $\det A = \det B$ b. A is invertible if and only if B is invertible c. $\text{rank}(A) = \text{rank}(B)$ d. $\det(A - \lambda I) = \det(B - \lambda I)$ e. λ is an eigenvalue of A if and only if λ is an eigenvalue of B
Thm 4.23	p.300	4.4	$P^{-1}AP = D \Leftrightarrow [D]_{ii} = \lambda_i$ and $\mathbf{p}_i = \mathbf{v}_i$ (the eigenvectors)
Thm 4.24	p.302	4.4	If λ_i are distinct, their basis vectors are linearly independent.
Thm 4.25	p.303	4.4	If all the λ_i are distinct, then A is diagonalizable.
Lem 4.26	p.303	4.4	geometric multiplicity $\leq$ algebraic multiplicity: That is, $\dim(E_{\lambda_i}) \leq k_i$ where $c_A(\lambda) = \prod_{i=1}^{m}(\lambda_i - \lambda)^{k_i}$.
Thm 4.27	p.304	4.4	parts a. through c. list equivalences to $A = P^{-1}DP$ a. A is diagonalizable, $A = P^{-1}DP$ b. bases of the eigenvectors contain n vectors c. geometric multiplicity = algebraic multiplicity, $\dim(E_{\lambda_i}) = k_i$

Diagonal and diagonalizable matrices

Diagonalizable matrices can be thought of as a *generalization* of diagonal matrices.
We can see that in the key condition to be diagonalizable, $P^{-1}AP = D$.
Diagonal matrices are simple and have lots of desirable properties.
Let's review those properties and compare them to the properties of diagonalizable matrices.

Q: The eigenvalues of a diagonal matrix are its diagonal entries.
   How does this property get generalized for diagonalizable matrices?
A: If $P^{-1}AP = D$, then the eigenvalues of A are the diagonal entries of D.

Q: Diagonal matrices commute.
   How does this property get generalized for diagonalizable matrices?
A: In order to understand this generalization, it is useful to begin by looking at Exercise 44.

Q: In Exercise 44, we show the following:
   In order for two matrices to commute, they must have the same eigenvectors.
   Using this principle, explain why all diagonal matrices commute.
A: Because $\mathbf{e}_i$ are the eigenvectors for all diagonal matrices.

Q: Why are $\mathbf{e}_i$ the eigenvectors for all diagonal matrices?
A: Since $D = \begin{bmatrix} \lambda_1 \mathbf{e}_1 & \lambda_2 \mathbf{e}_2 & \cdots & \lambda_n \mathbf{e}_n \end{bmatrix}$, $D\mathbf{e}_i = \lambda_i \mathbf{e}_i$.

Q: In the case of diagonalizable matrices, what additional condition do we need?
A: In Exercise 45, the matrices were required to have distinct eigenvalues. Why?
   Because a matrix with distinct eigenvalues has distinct eigenvectors that span $\mathbb{R}^n$.

Q: Diagonal matrices are invertible.
   How does this property get generalized for diagonalizable matrices?
A: A diagonalizable matrix is invertible if its eigenvectors span $\mathbb{R}^n$.
   This is essentially the assertion of Theorem 4.27(b).

Q: If D is diagonal, then D^{-1} is diagonal. How does this generalize?
A: If A is diagonalizable, so is A^{-1}?
   See Exercise 42. Note: the fact that D^{-1} is diagonal is critical to that proof.

   Proof that D^{-1} is diagonal:

$$DD^{-1} = D(D^{-1})^T = \begin{bmatrix} \lambda_1 \mathbf{e}_1 & \lambda_2 \mathbf{e}_2 & \cdots & \lambda_n \mathbf{e}_n \end{bmatrix} \begin{bmatrix} \frac{1}{\lambda_1}\mathbf{e}_1 \\ \frac{1}{\lambda_2}\mathbf{e}_2 \\ \vdots \\ \frac{1}{\lambda_n}\mathbf{e}_n \end{bmatrix} =$$

$$\begin{bmatrix} \lambda_1 \frac{1}{\lambda_1}\mathbf{e}_1 \cdot \mathbf{e}_1 & \lambda_2 \frac{1}{\lambda_2}\mathbf{e}_2 \cdot \mathbf{e}_2 & \cdots & \lambda_n \frac{1}{\lambda_n}\mathbf{e}_n \cdot \mathbf{e}_n \end{bmatrix} = \begin{bmatrix} \mathbf{e}_1 & \mathbf{e}_2 & \cdots & \mathbf{e}_n \end{bmatrix} = I$$

4.4 Similarity and Diagonalization

Similarity, $A \sim B$

Q: How are $A \sim B$ and $A \longrightarrow B$ alike?
A: Both of these properties are both transitive.

Q: What does that tell us?
A: $A \sim B$: If two matrices are similar to a third matrix, they are similar to each other.
$A \longrightarrow B$: If two matrices reduce to a third matrix, they reduce to each other.

Q: How are $A \sim B$ and $A \longrightarrow B$ different?
A: If $A \sim B$, then A and B *do* have the same eigenvalues.
This is precisely the assertion of Theorem 4.24(e).
However, as we saw in the last section:
If $A \longrightarrow B$, then A and B do not necessarily have the same eigenvalues.

Discussion of proofs

We should continue to develop our understanding and ability to put our proofs into words.

Q: How might we summarize the result of our proof to Exercise 41 in words?
A: Not only is A^T diagonalizable, its diagonal matrix is the same as that of A.
Furthermore, its matrix of diagonalization is the inverse tranpose of the one for A.

Q: How might we summarize the result of our proof to Exercise 42 in words?
A: Not only is A^{-1} diagonalizable, its diagonal matrix is the inverse of that for A.
Furthermore, its matrix of diagonalization is the *same* as the one for A.

Q: Below are two proofs of Exercise 33. Compare and contrast them.

1: A and B have the same characteristic polynomial, by Theorem 4.22(d).
It follows immediately that they have the same eigenvalues.

2: Since we are proving Theorem 4.22(e), we need to prove the following two statements:
1) Given $A \sim B$, if α is an eigenvalue of A then α is an eigenvalue of B and
2) Given $A \sim B$, if β is an eigenvalue of B then β is an eigenvalue of A.

Since λ if an eigenvalue of M if and only if there exists an eigenvector $\mathbf{v} \neq \mathbf{0}$
such that $M\mathbf{v} = \lambda\mathbf{v}$, it suffices to show the following:
1) If $A\mathbf{x} = \alpha\mathbf{x}$, $\mathbf{x} \neq \mathbf{0}$, then $B(P^{-1}\mathbf{x}) = \alpha(P^{-1}\mathbf{x})$ and
2) If $B\mathbf{y} = \beta\mathbf{y}$, $\mathbf{y} \neq \mathbf{0}$, then $A(P\mathbf{y}) = \beta(P\mathbf{y})$.

Note, $P^{-1}AP = B$ implies $A = PBP^{-1}$, where P is invertible and so is P^{-1}. Therefore:
If $A\mathbf{x} = \alpha\mathbf{x}$, $\mathbf{x} \neq \mathbf{0}$, then $(PBP^{-1})\mathbf{x} = \alpha\mathbf{x}$.
So, left multiplying both sides by P^{-1} implies $B(P^{-1}\mathbf{x}) = P^{-1}(\alpha\mathbf{x}) = \alpha(P^{-1}\mathbf{x})$.

Likewise, if $B\mathbf{y} = \beta\mathbf{y}$, $\mathbf{y} \neq \mathbf{0}$, then $(P^{-1}AP)\mathbf{y} = \beta\mathbf{y}$.
So, left multiplying both sides by P implies $A(P\mathbf{y}) = P(\beta\mathbf{y}) = \beta(P\mathbf{y})$.

Detailed proof of why distinct eigenvalues have distinct eigenvectors

The ideas and techniques in this proof have broad application. See Exercise 44.

Q: Since the eigenvectors of A are *distinct* and $A(B\mathbf{p}_i) = \alpha_i(B\mathbf{p}_i)$,
in Exercise 44 we claim that $B\mathbf{p}_i$ must be a multiple of $\mathbf{p}_i$. Why? Prove this.

A: Since A has n distinct eigenvalues, Theorem 4.20 in Section 4.3 implies
the corresponding eigenvectors $\mathbf{p}_i$ are linearly independent and therefore span $\mathbb{R}^n$.

So, we can express $\mathbf{x}$ as a linear combination of the $\mathbf{p}_i$: $\mathbf{x} = \sum_{k=1}^{n} \beta_k \mathbf{p}_k$.

Now let $\mathbf{x}$ be *any* nonzero vector such that $A\mathbf{x} = \alpha_i \mathbf{x}$.

On the *right* hand side, we have $A\mathbf{x} = A\left(\sum_{k=1}^{n} \beta_k \mathbf{p}_k\right) = \sum_{k=1}^{n} \beta_k (A\mathbf{p}_k) = \sum_{k=1}^{n} \beta_k \alpha_k \mathbf{p}_k$.

On the *left* hand side, we have $\alpha_i \mathbf{x} = \alpha_i \left(\sum_{k=1}^{n} \beta_k \mathbf{p}_k\right) = \sum_{k=1}^{n} \beta_k \alpha_i \mathbf{p}_k$.

Setting these two sides equal yields: $\sum_{k=1}^{n} \beta_k \alpha_k \mathbf{p}_k = \sum_{k=1}^{n} \beta_k \alpha_i \mathbf{p}_k$.

It is important to note that α_i is the one *fixed* eigenvalue we are considering.
However, the α_k are changing with the index value in the summation.

Now since the $\mathbf{p}_i$ are linearly independent, every scalar in the above summation must be zero.

Therefore, $\beta_k(\alpha_k - \alpha_i) = 0$ for all k.

Now since the eigenvalues of A are distinct, $\alpha_k \neq \alpha_i$ for all $k \neq i$.
So, $\alpha_k - \alpha_i \neq 0$ for all $k \neq i$.
Therefore, $\beta_k(\alpha_k - \alpha_i) = 0$ for all k implies $\beta_k = 0$ for all $k \neq i$.

Furthermore, since $\mathbf{x}$ is nonzero, $\beta_i \neq 0$.

Therefore, as claimed: $\mathbf{x} = \sum_{k=1}^{n} \beta_k \mathbf{p}_k = \beta_i \mathbf{p}_i + \sum_{k \neq i}^{n} \beta_k \mathbf{p}_k = \beta_i \mathbf{p}_i + \sum_{k \neq i}^{n} 0 \mathbf{p}_k = \beta_i \mathbf{p}_i$.

Q: According to this proof, what is the eigenvalue of $\mathbf{p}_i$ for B?

4.4 Similarity and Diagonalization

Solutions to odd-numbered exercises from Section 4.4

1. $\det(A - \lambda I) = \begin{vmatrix} 4-\lambda & 1 \\ 3 & 1-\lambda \end{vmatrix} = \lambda^2 - 5\lambda + 1$, while

   $\det(B - \lambda I) = \begin{vmatrix} 1-\lambda & 0 \\ 0 & 1-\lambda \end{vmatrix} = \lambda^2 - 2\lambda + 1$. Thus, by Theorem 4.22(d), $A \not\sim B$.

3. A has eigenvalues 2, 2, 4; B has eigenvalues 1, 4, and $4 \Rightarrow A \not\sim B$.

5. Since $\begin{bmatrix} 2 & -1 \\ -1 & 1 \end{bmatrix} \begin{bmatrix} 5 & -1 \\ 2 & 2 \end{bmatrix} \begin{bmatrix} 1 & 1 \\ 1 & 2 \end{bmatrix} = \begin{bmatrix} 4 & 0 \\ 0 & 3 \end{bmatrix}$ is of the form $P^{-1}AP = D$,

   we see that the eigenvalues of A are $\lambda_1 = 4$ and $\lambda_2 = 3$.

   So, E_4, E_3 have bases $\mathbf{p}_1 = \begin{bmatrix} 1 \\ 1 \end{bmatrix}$ and $\mathbf{p}_2 = \begin{bmatrix} 1 \\ 2 \end{bmatrix}$.

7. The eigenvalues of A are 6 and $-2 \Rightarrow E_6, E_{-2}$ have bases $\begin{bmatrix} 3 \\ 2 \\ 3 \end{bmatrix}$ and $\left\{ \begin{bmatrix} 0 \\ 1 \\ -1 \end{bmatrix}, \begin{bmatrix} 1 \\ 0 \\ -1 \end{bmatrix} \right\}$.

9. $\det(A - \lambda I) = \begin{vmatrix} -3-\lambda & 4 \\ -1 & 1-\lambda \end{vmatrix} = \lambda^2 + 2\lambda + 1 = (\lambda + 1)^2$,

   so A has eigenvalue -1 with algebraic multiplicity 2.
   To find the corresponding eigenspace, we calculate

   $A + I = \begin{bmatrix} -2 & 4 \\ -1 & 2 \end{bmatrix} \longrightarrow \begin{bmatrix} 1 & -2 \\ 0 & 0 \end{bmatrix}$, so the eigenvalue -1 has geometric multiplicity 1,

   and thus A is not diagonalizable, by the Diagonalization Theorem.

11. Expanding along the first row,

$$\det(A - \lambda I) = \begin{vmatrix} 1-\lambda & 0 & 1 \\ 0 & 1-\lambda & 1 \\ 1 & 1 & -\lambda \end{vmatrix} = (1-\lambda)[(1-\lambda)(-\lambda) - 1] - (1-\lambda)$$

$$= (1-\lambda)(\lambda^2 - \lambda - 2) = -(\lambda+1)(\lambda-1)(\lambda-2).$$

So, A has eigenvalues -1, 1, and 2, and Theorem 4.25 tells us that that A is diagonalizable. We find bases for the eigenspaces:

$$A + I = \begin{bmatrix} 2 & 0 & 1 \\ 0 & 2 & 1 \\ 1 & 1 & 1 \end{bmatrix} \longrightarrow \begin{bmatrix} 1 & 0 & \frac{1}{2} \\ 0 & 1 & \frac{1}{2} \\ 0 & 0 & 0 \end{bmatrix}, \text{ so } E_{-1} \text{ has basis } \mathbf{p}_1 = \begin{bmatrix} 1 \\ 1 \\ -2 \end{bmatrix}.$$

$$A - I = \begin{bmatrix} 0 & 0 & 1 \\ 0 & 0 & 1 \\ 1 & 1 & -1 \end{bmatrix} \longrightarrow \begin{bmatrix} 1 & 1 & 0 \\ 0 & 0 & 1 \\ 0 & 0 & 0 \end{bmatrix}, \text{ so } E_1 \text{ has basis } \mathbf{p}_2 = \begin{bmatrix} 1 \\ -1 \\ 0 \end{bmatrix}.$$

$$A - 2I = \begin{bmatrix} -1 & 0 & 1 \\ 0 & -1 & 1 \\ 1 & 1 & -2 \end{bmatrix} \longrightarrow \begin{bmatrix} 1 & 0 & -1 \\ 0 & 1 & -1 \\ 0 & 0 & 0 \end{bmatrix}, \text{ so } E_2 \text{ has basis } \mathbf{p}_3 = \begin{bmatrix} 1 \\ 1 \\ 1 \end{bmatrix}.$$

Thus, $P = \begin{bmatrix} 1 & 1 & 1 \\ 1 & -1 & 1 \\ -2 & 0 & 1 \end{bmatrix}$ and $D = \begin{bmatrix} -1 & 0 & 0 \\ 0 & 1 & 0 \\ 0 & 0 & 2 \end{bmatrix}$ satisfy $P^{-1}AP = D$.

13. $\det(A - \lambda I) = \begin{vmatrix} 1-\lambda & 2 & 1 \\ -1 & -\lambda & 1 \\ 1 & 1 & -\lambda \end{vmatrix} = \lambda^2 - \lambda^3 = -\lambda^2(\lambda - 1)$. Thus, A has eigenvalues 0 and 1.

But $A - 0I = \begin{bmatrix} 1 & 2 & 1 \\ -1 & 0 & 1 \\ 1 & 1 & 0 \end{bmatrix} \longrightarrow \begin{bmatrix} 1 & 0 & -1 \\ 0 & 1 & 1 \\ 0 & 0 & 0 \end{bmatrix}$, so 0 has geometric multiplicity 1,

and thus A is not diagonalizable, by the Diagonalization Theorem.

15. A has eigenvalues -2 and 2. We check

$$A + 2I = \begin{bmatrix} 4 & 0 & 0 & 4 \\ 0 & 4 & 0 & 0 \\ 0 & 0 & 0 & 0 \\ 0 & 0 & 0 & 0 \end{bmatrix} \longrightarrow \begin{bmatrix} 1 & 0 & 0 & 1 \\ 0 & 1 & 0 & 0 \\ 0 & 0 & 0 & 0 \\ 0 & 0 & 0 & 0 \end{bmatrix}, \text{ so } E_{-2} \text{ has basis } \left\{ \begin{bmatrix} 1 \\ 0 \\ 0 \\ -1 \end{bmatrix}, \begin{bmatrix} 0 \\ 0 \\ 1 \\ 0 \end{bmatrix} \right\}.$$

$$A - 2I = \begin{bmatrix} 0 & 0 & 0 & 4 \\ 0 & 0 & 0 & 0 \\ 0 & 0 & -4 & 0 \\ 0 & 0 & 0 & -4 \end{bmatrix} \longrightarrow \begin{bmatrix} 0 & 0 & 1 & 0 \\ 0 & 0 & 0 & 1 \\ 0 & 0 & 0 & 0 \\ 0 & 0 & 0 & 0 \end{bmatrix}, \text{ so } E_2 \text{ has basis } \left\{ \begin{bmatrix} 1 \\ 0 \\ 0 \\ 0 \end{bmatrix}, \begin{bmatrix} 0 \\ 1 \\ 0 \\ 0 \end{bmatrix} \right\}.$$

Thus $P = \begin{bmatrix} 1 & 0 & 1 & 0 \\ 0 & 0 & 0 & 1 \\ 0 & 1 & 0 & 0 \\ -1 & 0 & 0 & 0 \end{bmatrix}$ and $D = \begin{bmatrix} -2 & 0 & 0 & 0 \\ 0 & -2 & 0 & 0 \\ 0 & 0 & 2 & 0 \\ 0 & 0 & 0 & 2 \end{bmatrix}$ satisfy $P^{-1}AP = D$.

4.4 Similarity and Diagonalization

17. We diagonalize A, first finding its eigenvalues and eigenvectors:

$$\det(A - \lambda I) = \begin{vmatrix} -1-\lambda & 6 \\ 1 & -\lambda \end{vmatrix} = \lambda^2 + \lambda - 6 = (\lambda+3)(\lambda-2), \text{ so } A \text{ has eigenvalues } -3 \text{ and } 2.$$

$A + 3I = \begin{bmatrix} 2 & 6 \\ 1 & 3 \end{bmatrix} \longrightarrow \begin{bmatrix} 1 & 3 \\ 0 & 0 \end{bmatrix}$, so E_{-3} has basis $\begin{bmatrix} 3 \\ -1 \end{bmatrix}$.

$A - 2I = \begin{bmatrix} -3 & 6 \\ 1 & -2 \end{bmatrix} \longrightarrow \begin{bmatrix} 1 & -2 \\ 0 & 0 \end{bmatrix}$, so E_2 has basis $\begin{bmatrix} 2 \\ 1 \end{bmatrix}$.

Therefore, $P^{-1}AP = D$ is satisfied if $P = \begin{bmatrix} 3 & 2 \\ -1 & 1 \end{bmatrix}$ and $D = \begin{bmatrix} -3 & 0 \\ 0 & 2 \end{bmatrix}$. Therefore,

$$\begin{bmatrix} -1 & 6 \\ 1 & 0 \end{bmatrix}^{10} = (PDP^{-1})^{10} = PD^{10}P^{-1} = \begin{bmatrix} 3 & 2 \\ -1 & 1 \end{bmatrix} \begin{bmatrix} -3 & 0 \\ 0 & 2 \end{bmatrix}^{10} \begin{bmatrix} 3 & 2 \\ -1 & 1 \end{bmatrix}^{-1}$$

$$= \begin{bmatrix} 3 & 2 \\ -1 & 1 \end{bmatrix} \begin{bmatrix} (-3)^{10} & 0 \\ 0 & 2^{10} \end{bmatrix} \begin{bmatrix} \frac{1}{5} & -\frac{2}{5} \\ \frac{1}{5} & \frac{3}{5} \end{bmatrix} = \begin{bmatrix} 35,839 & -69,630 \\ -11,605 & 24,234 \end{bmatrix}.$$

19. $\det(A - \lambda I) = \begin{vmatrix} -\lambda & 3 \\ 1 & 2-\lambda \end{vmatrix} = \lambda^2 - 2\lambda - 3 = (\lambda+1)(\lambda-3).$

$A + I = \begin{bmatrix} 1 & 3 \\ 1 & 3 \end{bmatrix} \Rightarrow E_{-1} = \text{span}\left(\begin{bmatrix} 3 \\ -1 \end{bmatrix}\right)$ and $A - 3I = \begin{bmatrix} -3 & 3 \\ 1 & -1 \end{bmatrix} \Rightarrow E_3 = \text{span}\left(\begin{bmatrix} 1 \\ 1 \end{bmatrix}\right).$

Therefore

$$\begin{bmatrix} 0 & 3 \\ 1 & 2 \end{bmatrix}^k = (PDP^{-1})^k = \begin{bmatrix} 3 & 1 \\ -1 & 1 \end{bmatrix} \begin{bmatrix} -1 & 0 \\ 0 & 3 \end{bmatrix}^k \begin{bmatrix} 3 & 1 \\ -1 & 1 \end{bmatrix}^{-1}$$

$$= \begin{bmatrix} 3 & 1 \\ -1 & 1 \end{bmatrix} \begin{bmatrix} (-1)^k & 0 \\ 0 & 3^k \end{bmatrix} \frac{1}{4}\begin{bmatrix} 1 & -1 \\ 1 & 3 \end{bmatrix} = \frac{1}{4}\begin{bmatrix} 3(-1)^k + 3^k & 3(-1)^{k+1} + 3^{k+1} \\ (-1)^{k+1} + 3^k & (-1)^k + 3^{k+1} \end{bmatrix}.$$

21. Since $D = \begin{bmatrix} 1 & 0 & 0 \\ 0 & -1 & 0 \\ 0 & 0 & -1 \end{bmatrix}$, we have

$$A^{2002} = (PDP^{-1})^{2002} = P\begin{bmatrix} 1 & 0 & 0 \\ 0 & -1 & 0 \\ 0 & 0 & -1 \end{bmatrix}^{2002} P^{-1} = PP^{-1} = I_3.$$

23. $\det(A - \lambda I) = \begin{vmatrix} 1-\lambda & 1 & 0 \\ 2 & -2-\lambda & 2 \\ 0 & 1 & 1-\lambda \end{vmatrix} = -\lambda^3 + 7\lambda - 6 = -(\lambda+3)(\lambda-1)(\lambda-2)$.

$A + 3I = \begin{bmatrix} 4 & 1 & 0 \\ 2 & 1 & 2 \\ 0 & 1 & 4 \end{bmatrix} \longrightarrow \begin{bmatrix} 1 & 0 & -1 \\ 0 & 1 & 4 \\ 0 & 0 & 0 \end{bmatrix} \Rightarrow E_{-3} = \text{span}\left(\begin{bmatrix} 1 \\ -4 \\ 1 \end{bmatrix}\right)$.

$A - I = \begin{bmatrix} 0 & 1 & 0 \\ 2 & -3 & 2 \\ 0 & 1 & 0 \end{bmatrix} \longrightarrow \begin{bmatrix} 1 & 0 & 1 \\ 0 & 1 & 0 \\ 0 & 0 & 0 \end{bmatrix} \Rightarrow E_1 = \text{span}\left(\begin{bmatrix} 1 \\ 0 \\ -1 \end{bmatrix}\right)$.

$A - 2I = \begin{bmatrix} -1 & 1 & 0 \\ 2 & -4 & 2 \\ 0 & 1 & -1 \end{bmatrix} \longrightarrow \begin{bmatrix} 1 & 0 & -1 \\ 0 & 1 & -1 \\ 0 & 0 & 0 \end{bmatrix}$, so $E_2 = \text{span}\left(\begin{bmatrix} 1 \\ 1 \\ 1 \end{bmatrix}\right)$. Therefore

$\begin{bmatrix} 1 & 1 & 0 \\ 2 & -2 & 2 \\ 0 & 1 & 1 \end{bmatrix}^k = \begin{bmatrix} 1 & 1 & 1 \\ -4 & 0 & 1 \\ 1 & -1 & 1 \end{bmatrix} \begin{bmatrix} -3 & 0 & 0 \\ 0 & 1 & 0 \\ 0 & 0 & 2 \end{bmatrix}^k \begin{bmatrix} 1 & 1 & 1 \\ -4 & 0 & 1 \\ 1 & -1 & 1 \end{bmatrix}^{-1}$

$= \dfrac{1}{10} \begin{bmatrix} 1 & 1 & 1 \\ -4 & 0 & 1 \\ 1 & -1 & 1 \end{bmatrix} \begin{bmatrix} (-3)^k & 0 & 0 \\ 0 & 1 & 0 \\ 0 & 0 & 2^k \end{bmatrix} \begin{bmatrix} 1 & -2 & 1 \\ 5 & 0 & -5 \\ 4 & 2 & 4 \end{bmatrix}$

$= \dfrac{1}{10} \begin{bmatrix} 4(2)^k + (-3)^k + 5 & 2(2)^k - 2(-3)^k & 4(2)^k + (-3)^k - 5 \\ 4(2)^k - 4(-3)^k & 2(2)^k + 8(-3)^k & 4(2)^k - 4(-3)^k \\ 4(2)^k + (-3)^k - 5 & 2(2)^k - 2(-3)^k & 4(2)^k + (-3)^k + 5 \end{bmatrix}$.

25. $\det(A - \lambda I) = \begin{vmatrix} 1-\lambda & k \\ 0 & 1-\lambda \end{vmatrix}$. The only eigenvalue is 1, and $A - I = \begin{bmatrix} 0 & k \\ 0 & 0 \end{bmatrix}$.

If $k = 0$, then A is already diagonal;
otherwise 1 has geometric multiplicity 1 and thus A is not diagonalizable.
Thus, A is diagonalizable only if $k = 0$.

27. A has eigenvalue 1 with algebraic multiplicity 3. $A - I = \begin{bmatrix} 0 & 0 & k \\ 0 & 0 & 0 \\ 0 & 0 & 0 \end{bmatrix}$.

If $k = 0$ then A is already diagonal;
otherwise 1 has geometric multiplicity 2 and thus A is not diagonalizable.
Thus, A is diagonalizable only if $k = 0$.

4.4 Similarity and Diagonalization

29. $\det(A - \lambda I) = \begin{vmatrix} 1-\lambda & 1 & k \\ 1 & 1-\lambda & k \\ 1 & 1 & k-\lambda \end{vmatrix}$

$= (1-\lambda)[(1-\lambda)(k-\lambda) - k] - [(k-\lambda) - k] + k[1 - (1-\lambda)]$

$= (1-\lambda)(\lambda^2 - k\lambda - \lambda) + \lambda + k\lambda = -\lambda^3 + 2\lambda^2 + k\lambda^2 = \lambda^2(k - \lambda + 2)$.

Now if $k = -2$, then A has only 0 as an eigenvalue (with algebraic multiplicity 3) and since

$A = \begin{bmatrix} 1 & 1 & -2 \\ 1 & 1 & -2 \\ 1 & 1 & -2 \end{bmatrix} \longrightarrow \begin{bmatrix} 1 & 1 & -2 \\ 0 & 0 & 0 \\ 0 & 0 & 0 \end{bmatrix}$, 0 has geometric multiplicity 2 and A is not diagonalizable.

If $k \neq -2$, then 0 has algebraic multiplicity 2 and $A = \begin{bmatrix} 1 & 1 & k \\ 1 & 1 & k \\ 1 & 1 & k \end{bmatrix} \longrightarrow \begin{bmatrix} 1 & 1 & k \\ 0 & 0 & 0 \\ 0 & 0 & 0 \end{bmatrix}$,

so 0 has geometric multiplicity 2 and A is diagonalizable.

Thus, A is diagonalizable provided $k \neq -2$.

31. $AP = PB \Rightarrow \det(AP) = \det(PB) \Rightarrow \det(A)\det(P) = \det(P)\det(B) \Rightarrow \det(A) = \det(B)$.
So A is invertible if and only if B is invertible.

33. Since we are proving Theorem 4.22(e), we need to prove the following two statements:
1) Given $A \sim B$, if α is an eigenvalue of A then α is an eigenvalue of B and
2) Given $A \sim B$, if β is an eigenvalue of B then β is an eigenvalue of A.

Since λ if an eigenvalue of M if and only if there exists an eigenvector $\mathbf{v} \neq \mathbf{0}$ such that $M\mathbf{v} = \lambda\mathbf{v}$, it suffices to show the following:

1) If $A\mathbf{x} = \alpha\mathbf{x}$, $\mathbf{x} \neq \mathbf{0}$, then $B(P^{-1}\mathbf{x}) = \alpha(P^{-1}\mathbf{x})$ and
2) If $B\mathbf{y} = \beta\mathbf{y}$, $\mathbf{y} \neq \mathbf{0}$, then $A(P\mathbf{y}) = \beta(P\mathbf{y})$.

Note, $P^{-1}AP = B$ implies $A = PBP^{-1}$, where P is invertible and so is P^{-1}. Therefore:
If $A\mathbf{x} = \alpha\mathbf{x}$, $\mathbf{x} \neq \mathbf{0}$, then $(PBP^{-1})\mathbf{x} = \alpha\mathbf{x}$.
So, left multiplying both sides by P^{-1} implies $B(P^{-1}\mathbf{x}) = P^{-1}(\alpha\mathbf{x}) = \alpha(P^{-1}\mathbf{x})$.

Likewise, if $B\mathbf{y} = \beta\mathbf{y}$, $\mathbf{y} \neq \mathbf{0}$, then $(P^{-1}AP)\mathbf{y} = \beta\mathbf{y}$.
So, left multiplying both sides by P implies $A(P\mathbf{y}) = P(\beta\mathbf{y}) = \beta(P\mathbf{y})$.

Q: Why do we *not* have to specify that $P^{-1}\mathbf{x}$ and $P\mathbf{y}$ are not the zero vector?
A: Because P is invertible and so P^{-1} is invertible, too. Therefore:
If $\mathbf{u} \neq \mathbf{0}$, then $P\mathbf{u} \neq \mathbf{0}$ and $P^{-1}\mathbf{u} \neq \mathbf{0}$. Why?

Q: How might we state our stronger result in words?
A: Not only is α an eigenvalue for B, one associated eigenvector of α is $P^{-1}\mathbf{x}$.
Not only is β an eigenvalue for A, one associated eigenvector of β is $P\mathbf{y}$.

35. We need to show if $A \sim B$, then $\text{tr}(A) = \text{tr}(B)$.
That is, if $P^{-1}AP = B$ or equivalently $A = PBP^{-1}$, then $\text{tr}(A) = \text{tr}(B)$.

The key insight is: $\text{tr}(MN) \overset{(1)}{=} \text{tr}(NM)$.
This result is not trivial. It was proven in Exercise 45 of Section 3.2.

$$\text{tr}(A) \overset{\text{by def}}{=} \text{tr}(PBP^{-1}) \overset{\text{by assoc}}{=} \text{tr}(P(BP^{-1})) \overset{\text{by (1)}}{=} \text{tr}((BP^{-1})P) \overset{\text{by assoc}}{=} \text{tr}(B(P^{-1}P)) = \text{tr}(B)$$

Alternate: We know that A and B have the same eigenvalues $\lambda_1, \lambda_2, \ldots, \lambda_n$.
So by Exercise 40 from Section 4.3, $\text{tr}(A) = \lambda_1 + \lambda_2 + \cdots + \lambda_n = \text{tr}(B)$.

37. $\det(A - \lambda I) = \begin{vmatrix} 5-\lambda & -3 \\ 4 & -2-\lambda \end{vmatrix} = \lambda^2 - 3\lambda + 2 = (\lambda - 1)(\lambda - 2)$, so A has eigenvalues 1, 2.

$\det(B - \lambda I) = \begin{vmatrix} -1-\lambda & 1 \\ -6 & 4-\lambda \end{vmatrix} = \lambda^2 - 3\lambda + 2$, so B has eigenvalues 1, 2.

Therefore, A and B are similar to $\begin{bmatrix} 1 & 0 \\ 0 & 2 \end{bmatrix}$.

For A, we have $E_1 = \text{span}\left(\begin{bmatrix} 3 \\ 4 \end{bmatrix}\right)$ and $E_2 = \text{span}\left(\begin{bmatrix} 1 \\ 1 \end{bmatrix}\right)$.

For B, $E_1 = \text{span}\left(\begin{bmatrix} 1 \\ 2 \end{bmatrix}\right)$ and $E_2 = \text{span}\left(\begin{bmatrix} 1 \\ 3 \end{bmatrix}\right)$.

So $Q^{-1}AQ = D = R^{-1}BR$, where $Q = \begin{bmatrix} 3 & 1 \\ 4 & 1 \end{bmatrix}$, $D = \begin{bmatrix} 1 & 0 \\ 0 & 2 \end{bmatrix}$, and $R = \begin{bmatrix} 1 & 1 \\ 2 & 3 \end{bmatrix}$.

Thus, $RQ^{-1}AQR^{-1} = (QR^{-1})^{-1} A (QR^{-1}) = B \Rightarrow$

$P^{-1}AP = B$ with $P = QR^{-1} = \begin{bmatrix} 3 & 1 \\ 4 & 1 \end{bmatrix} \begin{bmatrix} 1 & 1 \\ 2 & 3 \end{bmatrix}^{-1} = \begin{bmatrix} 3 & 1 \\ 4 & 1 \end{bmatrix} \begin{bmatrix} 3 & -1 \\ -2 & 1 \end{bmatrix} = \begin{bmatrix} 7 & -2 \\ 10 & -3 \end{bmatrix}$.

4.4 Similarity and Diagonalization

39. $\det(A - \lambda I) = \begin{vmatrix} 1-\lambda & 0 & 2 \\ 1 & -1-\lambda & 1 \\ 2 & 0 & 1-\lambda \end{vmatrix} = (-1-\lambda)\left[(1-\lambda)^2 - 4\right] = -(\lambda+1)^2(\lambda-3)$, while

$\det(B - \lambda I) = \begin{vmatrix} -3-\lambda & -2 & 0 \\ 6 & 5-\lambda & 0 \\ 4 & 4 & -1-\lambda \end{vmatrix} = -(\lambda+1)[(-3-\lambda)(5-\lambda) + 12] = -(\lambda+1)^2(\lambda-3)$.

So, A and B both have eigenvalues -1 and 3.

For A, $E_{-1} = \text{span}\left(\begin{bmatrix} 1 \\ 0 \\ -1 \end{bmatrix}, \begin{bmatrix} 0 \\ 1 \\ 0 \end{bmatrix}\right)$ and $E_3 = \text{span}\left(\begin{bmatrix} 2 \\ 1 \\ 2 \end{bmatrix}\right)$.

For B, $E_{-1} = \left(\begin{bmatrix} 1 \\ -1 \\ 0 \end{bmatrix}, \begin{bmatrix} 0 \\ 0 \\ 1 \end{bmatrix}\right)$ and $E_3 = \text{span}\left(\begin{bmatrix} 1 \\ -3 \\ -2 \end{bmatrix}\right)$.

Thus, as in Exercises 36 and 37, we have

$P^{-1}AP = B$ where $P = QR^{-1} = \begin{bmatrix} 1 & 0 & 2 \\ 0 & 1 & 1 \\ -1 & 0 & 2 \end{bmatrix} \begin{bmatrix} 1 & 0 & 1 \\ -1 & 0 & -3 \\ 0 & 1 & -2 \end{bmatrix}^{-1} = \begin{bmatrix} \frac{1}{2} & -\frac{1}{2} & 0 \\ -\frac{3}{2} & -\frac{3}{2} & 1 \\ -\frac{5}{2} & -\frac{3}{2} & 0 \end{bmatrix}$.

41. Given A is diagonalizable, we need to show A^T is diagonalizable.
It suffices to show if $P^{-1}AP = D$, then $P^T A^T (P^T)^{-1} = D$.

This is obvious since $D \underset{\text{diagonal}}{=} D^T \underset{\text{def}}{=} (P^{-1}AP)^T \underset{\substack{\text{Thm 3.9} \\ \text{Sect 3.3}}}{=} P^T A^T (P^T)^{-1}$

Q: Why is this sufficient?
A: Because P is invertible, so is P^T. Why? $\det(P^T) = \det P$.

Q: How might we summarize this stronger result in words?
A: Not only is A^T diagonalizable, its diagonal matrix is the same as that of A.
Furthermore, its matrix of diagonalization is the inverse tranpose of the one for A.

43. Suppose $P^{-1}AP = D$. Now each entry on the diagonal of D is λ (the only eigenvalue of A), So $D = \lambda I$. Thus, $A = PDP^{-1} = P(\lambda I)P^{-1} = \lambda PP^{-1}I = \lambda I$.

45. A and B are similar and thus have the same characteristic polynomial, by Theorem 4.22(d). It follows that the algebraic multiplicities of the eigenvalues of A and B are the same. Why?

47. We need to show if $A = P^{-1}DP$ is diagonalizable and $\lambda_i = 0$ or 1, then $A^2 = A$.
The key observation is: if $\lambda_i = 0$ or 1, then $\lambda_i^2 = \lambda_i$. Prove this.

So, since $\lambda_i^2 = \lambda_i$ and $D = [\lambda_1 e_1 \ \lambda_2 e_2 \ \cdots \ \lambda_n e_n]$, $D^2 \overset{(1)}{=} D$. Prove this. So:

$A^2 = (P^{-1}DP)^2 = (P^{-1}DP)(P^{-1}DP) = P^{-1}D^2P \overset{\text{by}\ (1)}{=} P^{-1}DP = A$

49. We note that Lemma 4.26 is the key to (a) and Theorem 4.27(c) is the key to (b).

(a) Since $A\mathbf{v}_1 = \mathbf{v}_1$, $A\mathbf{v}_2 = \mathbf{v}_2$, $A\mathbf{v}_3 = \mathbf{v}_3$, implies $\mathbf{v}_1, \mathbf{v}_2, \mathbf{v}_3$ are in E_1,
it suffices to show that if $c_A(\lambda) = (1+\lambda)(1-\lambda)^2(2-\lambda)^3$, then $\dim(E_1) \leq 2$.
Q: Why is this sufficient?
A: If the dimension of a subspace is ≤ 2, any 3 vectors in it must be linearly dependent.
The key observation is: if $c_A(\lambda) = \prod_{i=1}^{m}(\lambda_i - \lambda)^{k_i}$, then $\dim E_{\lambda_i} \leq k_i$. That is:

The dimension of the eigenspace associated with an eigenvector is less than or equal to the exponent of the factor of its eigenvalue in the characteristic equation.

Since $c_A(\lambda) = (1+\lambda)(1-\lambda)^2(2-\lambda)^3$, $\dim(E_1) \leq 2$.
Therefore any three vectors in E_1 are linearly dependent. So:
If $A\mathbf{v}_1 = \mathbf{v}_1$, $A\mathbf{v}_2 = \mathbf{v}_2$, $A\mathbf{v}_3 = \mathbf{v}_3$, then $\mathbf{v}_1, \mathbf{v}_2, \mathbf{v}_3$ must be linearly dependent.

(b) Since A is diagonalizable and $c_A(\lambda) = (1+\lambda)(1-\lambda)^2(2-\lambda)^3$,
Theorem 4.27(c) implies $\dim(E_{-1}) = 1$, $\dim(E_1) = 2$, and $\dim(E_2) = 3$.

4.5 Iterative Methods for Computing Eigenvalues

Q: What are our main goals for Section 4.5?

A: To practice the various power methods for determining eigenvalues.
To explore what Gerschgorin disks tell us about eigenvalues and their associated matrices.

Key Definitions and Concepts

λ_1	p.308	4.5	*dominant eigenvalue*, $\|\lambda_1\| > \|\lambda_i\|$
$\mathbf{v}_1$	p.308	4.5	*dominant eigenvector*, associated with dominant eigenvalue
diagonalizable	p.300	4.4	$A \sim D$ if $P^{-1}AP = D$ where D is diagonal
dominant eigenvalue	p.308	4.5	$\|\lambda_1\| > \|\lambda_i\|$, where λ_i are the *other* eigenvalues of A
dominant eigenvector	p.308	4.5	The eigenvector associated with $\|\lambda_1\| > \|\lambda_i\|$, that is, $A\mathbf{x}_1 = \lambda_1 \mathbf{x}_1$, where λ_1 is dominant
Gerschgorin disk	p.316	4.5	$D_i = \{z \text{ in } \mathbb{C} : \|z - a_{ii}\| \leq r_i\}$ where $r_i = \sum_{j \neq i} \|a_{ij}\|$
Power Method	p.312	4.5	This two-step iterative method is described in detail on p. 312 and illustrated in Example 4.31
Inverse	p.314	4.5	Inverse Power Method: See Example 4.33
Shifted	p.313	4.5	Shifted Power Method: See Example 4.32
Shifted Inverse	p.315	4.5	Shifted Inverse Power Method: See Example 4.34

Theorems

Thm 4.28	p.309	4.5	If A is diagonalizable with dominant eigenvalue λ_1, then $\mathbf{x}_1 = A\mathbf{x}_0$, $\mathbf{x}_2 = A\mathbf{x}_1$, $\mathbf{x}_k = A\mathbf{x}_{k-1}$ where the $\mathbf{x}_k$ are approaching the dominant eigenvector of A
Thm 4.29	p.318	4.5	Every eigenvalue of A is contained in a Gerschgorin Disk

Discussion of Definitions and Theorems
The discussion is included with the solutions below.

Solutions to odd-numbered exercises from Section 4.5

1. (a) $\mathbf{x}_5 = \begin{bmatrix} 4443 \\ 11,109 \end{bmatrix}$, so we estimate a dominant eigenvector of A to be $\begin{bmatrix} 1 \\ \frac{11,109}{4443} \end{bmatrix} \approx \begin{bmatrix} 1 \\ 2.500 \end{bmatrix}$.

To find the dominant eigenvalue λ_1, we calculate

$\mathbf{x}_6 = A\mathbf{x}_5 = \begin{bmatrix} 1 & 2 \\ 5 & 4 \end{bmatrix} \begin{bmatrix} 4443 \\ 11,109 \end{bmatrix} = \begin{bmatrix} 26,661 \\ 66,651 \end{bmatrix} \approx \lambda_1 \begin{bmatrix} 4443 \\ 11,109 \end{bmatrix} \Rightarrow \lambda_1 \approx \frac{26,661}{4443} \approx 6.001$.

(b) $\det(A - \lambda I) = \begin{vmatrix} 1-\lambda & 2 \\ 5 & 4-\lambda \end{vmatrix} = \lambda^2 - 5\lambda - 6 = (\lambda+1)(\lambda-6)$.

So, 6 is the dominant eigenvalue.

3. (a) $\mathbf{x}_5 = \begin{bmatrix} 144 \\ 89 \end{bmatrix}$, so we estimate a dominant eigenvector of A to be $\begin{bmatrix} 1 \\ \frac{89}{144} \end{bmatrix} \approx \begin{bmatrix} 1 \\ 0.618 \end{bmatrix}$.

To find the dominant eigenvalue λ_1, we calculate

$\mathbf{x}_6 = A\mathbf{x}_5 = \begin{bmatrix} 2 & 1 \\ 1 & 1 \end{bmatrix} \begin{bmatrix} 144 \\ 89 \end{bmatrix} = \begin{bmatrix} 377 \\ 233 \end{bmatrix} \Rightarrow \lambda_1 \approx \frac{377}{144} = 2.618$.

(b) $\det(A - \lambda I) = \begin{vmatrix} 2-\lambda & 1 \\ 1 & 1-\lambda \end{vmatrix} = \lambda^2 - 3\lambda + 1 = 0 \Rightarrow \lambda = \frac{3 \pm \sqrt{5}}{2}$.

So, $\frac{3+\sqrt{5}}{2}$ is the dominant eigenvalue.

5. (a) $\mathbf{x}_5 = \begin{bmatrix} -3.667 \\ 11.001 \end{bmatrix} \Rightarrow m_5 = 11.001$ and $\mathbf{y}_5 = \begin{bmatrix} -\frac{3.667}{11.001} \\ 1 \end{bmatrix} = \begin{bmatrix} -0.333 \\ 1 \end{bmatrix}$.

(b) $A\mathbf{y}_5 = \begin{bmatrix} 2 & -3 \\ -3 & 10 \end{bmatrix} \begin{bmatrix} -0.333 \\ 1 \end{bmatrix} = \begin{bmatrix} -3.666 \\ 10.999 \end{bmatrix}$ while $\lambda_1 \mathbf{y}_5 = 11.001 \begin{bmatrix} -0.333 \\ 1 \end{bmatrix} = \begin{bmatrix} -3.663 \\ 11.001 \end{bmatrix}$.

So we have indeed approximated an eigenvalue and an eigenvector of A.

7. (a) $\mathbf{x}_8 = \begin{bmatrix} 10.000 \\ 0.001 \\ 10.000 \end{bmatrix} \Rightarrow m_8 = 10.000$ and $\mathbf{y}_8 = \begin{bmatrix} 1 \\ \frac{0.001}{10.000} \\ 1 \end{bmatrix} = \begin{bmatrix} 1 \\ 0.0001 \\ 1 \end{bmatrix}$.

(b) $A\mathbf{y}_8 = \begin{bmatrix} 4 & 0 & 6 \\ -1 & 3 & 1 \\ 6 & 0 & 4 \end{bmatrix} \begin{bmatrix} 1 \\ 0.0001 \\ 1 \end{bmatrix} = \begin{bmatrix} 10 \\ 0.0003 \\ 10 \end{bmatrix}$ while $\lambda_1 \mathbf{y}_8 = 10.000 \begin{bmatrix} 1 \\ 0.0001 \\ 1 \end{bmatrix} = \begin{bmatrix} 10 \\ 0.001 \\ 10 \end{bmatrix}$.

9. With $A = \begin{bmatrix} 14 & 12 \\ 5 & 3 \end{bmatrix}$ and $\mathbf{x}_0 = \begin{bmatrix} 1 \\ 1 \end{bmatrix}$ as the initial vector, we get the values in the table.

k	0	1	2	3	4	5
$\mathbf{x}_k$	$\begin{bmatrix} 1 \\ 1 \end{bmatrix}$	$\begin{bmatrix} 26 \\ 8 \end{bmatrix}$	$\begin{bmatrix} 17.696 \\ 5.924 \end{bmatrix}$	$\begin{bmatrix} 18.017 \\ 6.004 \end{bmatrix}$	$\begin{bmatrix} 17.999 \\ 6.000 \end{bmatrix}$	$\begin{bmatrix} 18.000 \\ 6.000 \end{bmatrix}$
$\mathbf{y}_k$	$\begin{bmatrix} 1 \\ 1 \end{bmatrix}$	$\begin{bmatrix} 1 \\ 0.308 \end{bmatrix}$	$\begin{bmatrix} 1 \\ 0.335 \end{bmatrix}$	$\begin{bmatrix} 1 \\ 0.333 \end{bmatrix}$	$\begin{bmatrix} 1 \\ 0.333 \end{bmatrix}$	$\begin{bmatrix} 1 \\ 0.333 \end{bmatrix}$
m_k	1	26	17.696	18.017	17.999	18.000

We deduce that a dominant eigenvector of A is $\begin{bmatrix} 3 \\ 1 \end{bmatrix}$ with eigenvalue 18.

4.5 Iterative Methods for Computing Eigenvalues

11. With $A = \begin{bmatrix} 7 & 2 \\ 2 & 3 \end{bmatrix}$ and $\mathbf{x}_0 = \begin{bmatrix} 1 \\ 0 \end{bmatrix}$ as the initial vector, we get the values in the table.

k	0	1	2	3	4	5	6
$\mathbf{x}_k$	$\begin{bmatrix} 1 \\ 0 \end{bmatrix}$	$\begin{bmatrix} 7 \\ 2 \end{bmatrix}$	$\begin{bmatrix} 7.572 \\ 2.858 \end{bmatrix}$	$\begin{bmatrix} 7.754 \\ 3.131 \end{bmatrix}$	$\begin{bmatrix} 7.808 \\ 3.212 \end{bmatrix}$	$\begin{bmatrix} 7.822 \\ 3.233 \end{bmatrix}$	$\begin{bmatrix} 7.826 \\ 3.239 \end{bmatrix}$
$\mathbf{y}_k$	$\begin{bmatrix} 1 \\ 0 \end{bmatrix}$	$\begin{bmatrix} 1 \\ 0.286 \end{bmatrix}$	$\begin{bmatrix} 1 \\ 0.377 \end{bmatrix}$	$\begin{bmatrix} 1 \\ 0.404 \end{bmatrix}$	$\begin{bmatrix} 1 \\ 0.411 \end{bmatrix}$	$\begin{bmatrix} 1 \\ 0.413 \end{bmatrix}$	$\begin{bmatrix} 1 \\ 0.414 \end{bmatrix}$
m_k	1	7	7.572	7.754	7.808	7.822	7.826

We deduce that a dominant eigenvector of A is approximately $\begin{bmatrix} 1 \\ 0.414 \end{bmatrix}$ and the corresponding eigenvalue is about 7.83.

(In fact, a dominant eigenvector is $\begin{bmatrix} 1 \\ \sqrt{2}-1 \end{bmatrix}$ and the corresponding eigenvalue is $5 + 2\sqrt{2}$.)

13. With $A = \begin{bmatrix} 9 & 4 & 8 \\ 4 & 15 & -4 \\ 8 & -4 & 9 \end{bmatrix}$ and $\mathbf{x}_0 = \begin{bmatrix} 1 \\ 1 \\ 1 \end{bmatrix}$ as the initial vector, we get the values in the table.

k	0	1	2	3	4	5
$\mathbf{x}_k$	$\begin{bmatrix} 1 \\ 1 \\ 1 \end{bmatrix}$	$\begin{bmatrix} 21 \\ 15 \\ 13 \end{bmatrix}$	$\begin{bmatrix} 16.808 \\ 12.234 \\ 10.715 \end{bmatrix}$	$\begin{bmatrix} 17.008 \\ 12.372 \\ 10.821 \end{bmatrix}$	$\begin{bmatrix} 16.996 \\ 12.361 \\ 10.816 \end{bmatrix}$	$\begin{bmatrix} 16.996 \\ 12.361 \\ 10.816 \end{bmatrix}$
$\mathbf{y}_k$	$\begin{bmatrix} 1 \\ 1 \\ 1 \end{bmatrix}$	$\begin{bmatrix} 1 \\ 0.714 \\ 0.619 \end{bmatrix}$	$\begin{bmatrix} 1 \\ 0.728 \\ 0.637 \end{bmatrix}$	$\begin{bmatrix} 1 \\ 0.727 \\ 0.636 \end{bmatrix}$	$\begin{bmatrix} 1 \\ 0.727 \\ 0.636 \end{bmatrix}$	$\begin{bmatrix} 1 \\ 0.727 \\ 0.636 \end{bmatrix}$
m_k	1	21	16.808	17.008	16.996	16.996

We deduce that a dominant eigenvector of $A \approx \begin{bmatrix} 1 \\ 0.727 \\ 0.636 \end{bmatrix}$ with eigenvalue 17.

(In fact, this eigenvalue corresponds to the eigenspace spanned by $\begin{bmatrix} 1 \\ 2 \\ 0 \end{bmatrix}$ and $\begin{bmatrix} 0 \\ -2 \\ 1 \end{bmatrix}$.)

15. With $A = \begin{bmatrix} 4 & 1 & 3 \\ 0 & 2 & 0 \\ 1 & 1 & 2 \end{bmatrix}$ and $\mathbf{x}_0 = \begin{bmatrix} 1 \\ 1 \\ 1 \end{bmatrix}$ as the initial vector, we get the values in the table.

k	0	1	2	4	6	7	8
$\mathbf{x}_k$	$\begin{bmatrix} 1 \\ 1 \\ 1 \end{bmatrix}$	$\begin{bmatrix} 8 \\ 2 \\ 4 \end{bmatrix}$	$\begin{bmatrix} 5.75 \\ 0.5 \\ 2.25 \end{bmatrix}$	$\begin{bmatrix} 5.11 \\ 0.06 \\ 1.75 \end{bmatrix}$	$\begin{bmatrix} 5.02 \\ 0 \\ 1.68 \end{bmatrix}$	$\begin{bmatrix} 4.99 \\ 0 \\ 1.66 \end{bmatrix}$	$\begin{bmatrix} 4.99 \\ 0 \\ 1.66 \end{bmatrix}$
$\mathbf{y}_k$	$\begin{bmatrix} 1 \\ 1 \\ 1 \end{bmatrix}$	$\begin{bmatrix} 1 \\ 0.25 \\ 0.5 \end{bmatrix}$	$\begin{bmatrix} 1 \\ 0.09 \\ 0.39 \end{bmatrix}$	$\begin{bmatrix} 1 \\ 0.01 \\ 0.34 \end{bmatrix}$	$\begin{bmatrix} 1 \\ 0 \\ 0.33 \end{bmatrix}$	$\begin{bmatrix} 1 \\ 0 \\ 0.33 \end{bmatrix}$	$\begin{bmatrix} 1 \\ 0 \\ 0.33 \end{bmatrix}$
m_k	1	8	5.75	5.11	5.02	4.99	4.99

We deduce that a dominant eigenvector of A is $\begin{bmatrix} 3 \\ 0 \\ 1 \end{bmatrix}$ with eigenvalue 5.

17. With $A = \begin{bmatrix} 7 & 2 \\ 2 & 3 \end{bmatrix}$ and $\mathbf{x}_0 = \begin{bmatrix} 1 \\ 0 \end{bmatrix}$ as the initial vector, we get the values in the table.

k	0	1	2	3
$\mathbf{x}_k$	$\begin{bmatrix} 1 \\ 0 \end{bmatrix}$	$\begin{bmatrix} 7 \\ 2 \end{bmatrix}$	$\begin{bmatrix} 7.572 \\ 2.858 \end{bmatrix}$	$\begin{bmatrix} 7.754 \\ 3.131 \end{bmatrix}$
$R(\mathbf{x}_k)$	7	7.755	7.823	7.828

19. With $A = \begin{bmatrix} 9 & 4 & 8 \\ 4 & 15 & -4 \\ 8 & -4 & 9 \end{bmatrix}$ and $\mathbf{x}_0 = \begin{bmatrix} 1 \\ 1 \\ 1 \end{bmatrix}$ as the initial vector, we get the values in the table.

k	0	1	2
$\mathbf{x}_k$	$\begin{bmatrix} 1 \\ 1 \\ 1 \end{bmatrix}$	$\begin{bmatrix} 21 \\ 15 \\ 13 \end{bmatrix}$	$\begin{bmatrix} 16.808 \\ 12.234 \\ 10.715 \end{bmatrix}$
$R(\mathbf{x}_k)$	16.333	16.998	17.000

4.5 Iterative Methods for Computing Eigenvalues

21. With $A = \begin{bmatrix} 4 & 1 \\ 0 & 4 \end{bmatrix}$ and $\mathbf{x}_0 = \begin{bmatrix} 1 \\ 1 \end{bmatrix}$ as the initial vector, we get the values in the table.

k	0	1	2	3	4	5	6	7	8
$\mathbf{x}_k$	$\begin{bmatrix} 1 \\ 1 \end{bmatrix}$	$\begin{bmatrix} 5 \\ 4 \end{bmatrix}$	$\begin{bmatrix} 4.8 \\ 3.2 \end{bmatrix}$	$\begin{bmatrix} 4.67 \\ 2.68 \end{bmatrix}$	$\begin{bmatrix} 4.57 \\ 2.28 \end{bmatrix}$	$\begin{bmatrix} 4.5 \\ 2.0 \end{bmatrix}$	$\begin{bmatrix} 4.44 \\ 1.76 \end{bmatrix}$	$\begin{bmatrix} 4.4 \\ 1.6 \end{bmatrix}$	$\begin{bmatrix} 4.36 \\ 1.44 \end{bmatrix}$
$\mathbf{y}_k$	$\begin{bmatrix} 1 \\ 1 \end{bmatrix}$	$\begin{bmatrix} 1 \\ 0.8 \end{bmatrix}$	$\begin{bmatrix} 1 \\ 0.67 \end{bmatrix}$	$\begin{bmatrix} 1 \\ 0.57 \end{bmatrix}$	$\begin{bmatrix} 1 \\ 0.50 \end{bmatrix}$	$\begin{bmatrix} 1 \\ 0.44 \end{bmatrix}$	$\begin{bmatrix} 1 \\ 0.40 \end{bmatrix}$	$\begin{bmatrix} 1 \\ 0.36 \end{bmatrix}$	$\begin{bmatrix} 1 \\ 0.33 \end{bmatrix}$
m_k	1	5	4.8	4.67	4.57	4.5	4.44	4.4	4.36

A has the double eigenvalue 4 with corresponding eigenvector $\begin{bmatrix} 1 \\ 0 \end{bmatrix}$.

It seems that the power method is converging, but very slowly.

23. With $A = \begin{bmatrix} 4 & 0 & 1 \\ 0 & 4 & 0 \\ 0 & 0 & 1 \end{bmatrix}$ and $\mathbf{x}_0 = \begin{bmatrix} 1 \\ 1 \\ 1 \end{bmatrix}$ as the initial vector, we get the values in the table.

k	0	1	2	3	4	5	6
$\mathbf{x}_k$	$\begin{bmatrix} 1 \\ 1 \\ 1 \end{bmatrix}$	$\begin{bmatrix} 5 \\ 4 \\ 1 \end{bmatrix}$	$\begin{bmatrix} 4.2 \\ 3.2 \\ 0.2 \end{bmatrix}$	$\begin{bmatrix} 4.05 \\ 3.04 \\ 0.05 \end{bmatrix}$	$\begin{bmatrix} 4.01 \\ 3 \\ 0.01 \end{bmatrix}$	$\begin{bmatrix} 4 \\ 3 \\ 0 \end{bmatrix}$	$\begin{bmatrix} 4 \\ 3 \\ 0 \end{bmatrix}$
$\mathbf{y}_k$	$\begin{bmatrix} 1 \\ 1 \\ 1 \end{bmatrix}$	$\begin{bmatrix} 1 \\ 0.8 \\ 0.2 \end{bmatrix}$	$\begin{bmatrix} 1 \\ 0.76 \\ 0.05 \end{bmatrix}$	$\begin{bmatrix} 1 \\ 0.75 \\ 0.01 \end{bmatrix}$	$\begin{bmatrix} 1 \\ 0.75 \\ 0 \end{bmatrix}$	$\begin{bmatrix} 1 \\ 0.75 \\ 0 \end{bmatrix}$	$\begin{bmatrix} 1 \\ 0.75 \\ 0 \end{bmatrix}$
m_k	1	5	4.2	4.05	4.01	4	4

It seems that an eigenvector of A is $\begin{bmatrix} 4 \\ 3 \\ 0 \end{bmatrix}$ with eigenvalue 4.

In fact, the eigenspace of 4 is span $\left(\begin{bmatrix} 1 \\ 0 \\ 0 \end{bmatrix}, \begin{bmatrix} 0 \\ 1 \\ 0 \end{bmatrix} \right)$.

25. With $A = \begin{bmatrix} -1 & 2 \\ -1 & 1 \end{bmatrix}$ and $\mathbf{x}_0 = \begin{bmatrix} 1 \\ 1 \end{bmatrix}$ $\Rightarrow$ A has no real eigenvalue:

k	0	1	2
$\mathbf{x}_k$	$\begin{bmatrix} 1 \\ 1 \end{bmatrix}$	$\begin{bmatrix} 1 \\ 0 \end{bmatrix}$	$\begin{bmatrix} -1 \\ -1 \end{bmatrix}$
$\mathbf{y}_k$	$\begin{bmatrix} 1 \\ 1 \end{bmatrix}$	$\begin{bmatrix} 1 \\ 0 \end{bmatrix}$	$\begin{bmatrix} 1 \\ 1 \end{bmatrix}$
m_k	1	1	-1

27. With $A = \begin{bmatrix} -5 & 1 & 7 \\ 0 & 4 & 0 \\ 7 & 1 & -5 \end{bmatrix}$ and $\mathbf{x}_0 = \begin{bmatrix} 1 \\ 1 \\ 1 \end{bmatrix}$ as the initial vector, we get:

k	0	1	2	3	4	5	6
$\mathbf{x}_k$	$\begin{bmatrix} 1 \\ 1 \\ 1 \end{bmatrix}$	$\begin{bmatrix} 3 \\ 4 \\ 3 \end{bmatrix}$	$\begin{bmatrix} 2.5 \\ 4 \\ 2.5 \end{bmatrix}$	$\begin{bmatrix} 2.26 \\ 4 \\ 2.26 \end{bmatrix}$	$\begin{bmatrix} 2.14 \\ 4 \\ 2.14 \end{bmatrix}$	$\begin{bmatrix} 2.08 \\ 4 \\ 2.08 \end{bmatrix}$	$\begin{bmatrix} 2.04 \\ 4 \\ 2.04 \end{bmatrix}$
$\mathbf{y}_k$	$\begin{bmatrix} 1 \\ 1 \\ 1 \end{bmatrix}$	$\begin{bmatrix} 0.75 \\ 1 \\ 0.75 \end{bmatrix}$	$\begin{bmatrix} 0.63 \\ 1 \\ 0.63 \end{bmatrix}$	$\begin{bmatrix} 0.57 \\ 1 \\ 0.57 \end{bmatrix}$	$\begin{bmatrix} 0.54 \\ 1 \\ 0.54 \end{bmatrix}$	$\begin{bmatrix} 0.52 \\ 1 \\ 0.52 \end{bmatrix}$	$\begin{bmatrix} 0.51 \\ 1 \\ 0.51 \end{bmatrix}$
m_k	1	4	4	4	4	4	4

The power method is converging to $\begin{bmatrix} 1 \\ 2 \\ 1 \end{bmatrix}$ and 4,

which are indeed an eigenvector and eigenvalue of A.

But, in fact, the dominant eigenvalue of A is -12, with corresponding eigenvector $\begin{bmatrix} 1 \\ 0 \\ -1 \end{bmatrix}$.

We did not find these values because our initial vector $\mathbf{x}_0$ is orthogonal to $\begin{bmatrix} 1 \\ 0 \\ -1 \end{bmatrix}$.

29. In Exercise 9 we found that $\lambda_1 = 18$.

To find λ_2, we apply the power method to $A - 18I = \begin{bmatrix} -4 & 12 \\ 5 & -15 \end{bmatrix}$ with $\mathbf{x}_0 = \begin{bmatrix} 1 \\ 1 \end{bmatrix}$:

k	0	1	2	3
$\mathbf{x}_k$	$\begin{bmatrix} 1 \\ 1 \end{bmatrix}$	$\begin{bmatrix} 8 \\ -10 \end{bmatrix}$	$\begin{bmatrix} 15.2 \\ -19.0 \end{bmatrix}$	$\begin{bmatrix} 15.2 \\ -19.0 \end{bmatrix}$
$\mathbf{y}_k$	$\begin{bmatrix} 1 \\ 1 \end{bmatrix}$	$\begin{bmatrix} -0.8 \\ 1 \end{bmatrix}$	$\begin{bmatrix} -0.8 \\ 1 \end{bmatrix}$	$\begin{bmatrix} -0.8 \\ 1 \end{bmatrix}$
m_k	1	-10	-19	-19

We deduce that $\lambda_2 - \lambda_1 = -19$, so $\lambda_2 = -1$ is the second eigenvalue of A.

4.5 Iterative Methods for Computing Eigenvalues

31. $\lambda_1 = 17$, so we apply the power method to $A - 17I = \begin{bmatrix} -8 & 4 & 8 \\ 4 & -2 & -4 \\ 8 & -4 & -8 \end{bmatrix}$ with $\mathbf{x}_0 = \begin{bmatrix} 1 \\ 1 \\ 1 \end{bmatrix}$:

k	0	1	2	3
$\mathbf{x}_k$	$\begin{bmatrix} 1 \\ 1 \\ 1 \end{bmatrix}$	$\begin{bmatrix} 4 \\ -2 \\ -4 \end{bmatrix}$	$\begin{bmatrix} -18 \\ 9 \\ 18 \end{bmatrix}$	$\begin{bmatrix} -18 \\ 9 \\ 18 \end{bmatrix}$
$\mathbf{y}_k$	$\begin{bmatrix} 1 \\ 1 \\ 1 \end{bmatrix}$	$\begin{bmatrix} 1 \\ -0.5 \\ -1 \end{bmatrix}$	$\begin{bmatrix} 1 \\ -0.5 \\ -1 \end{bmatrix}$	$\begin{bmatrix} 1 \\ -0.5 \\ -1 \end{bmatrix}$
m_k	1	4	-18	-18

We deduce that $\lambda_2 - \lambda_1 = -18 \Rightarrow \lambda_2 = -1$.

33. We solve $A\mathbf{x}_1 = \mathbf{y}_0$ using row reduction: $\begin{bmatrix} A & | & \mathbf{y}_0 \end{bmatrix} = \begin{bmatrix} 14 & 12 & | & 1 \\ 5 & 3 & | & 1 \end{bmatrix} \longrightarrow \begin{bmatrix} 1 & 0 & | & 0.5 \\ 0 & 1 & | & -0.5 \end{bmatrix}$.
Thus, $\mathbf{x}_1 = \begin{bmatrix} 0.5 \\ -0.5 \end{bmatrix}$, so $\mathbf{y}_1 = \begin{bmatrix} 1 \\ -1 \end{bmatrix}$. Continuing in this way we calculate:

k	0	1	2	3	4	5
$\mathbf{x}_k$	$\begin{bmatrix} 1 \\ 1 \end{bmatrix}$	$\begin{bmatrix} 0.5 \\ -0.5 \end{bmatrix}$	$\begin{bmatrix} -0.833 \\ 1.056 \end{bmatrix}$	$\begin{bmatrix} 0.798 \\ -0.997 \end{bmatrix}$	$\begin{bmatrix} 0.8 \\ -1 \end{bmatrix}$	$\begin{bmatrix} 0.8 \\ -1 \end{bmatrix}$
$\mathbf{y}_k$	$\begin{bmatrix} 1 \\ 1 \end{bmatrix}$	$\begin{bmatrix} 1 \\ -1 \end{bmatrix}$	$\begin{bmatrix} -0.789 \\ 1 \end{bmatrix}$	$\begin{bmatrix} -0.800 \\ 1 \end{bmatrix}$	$\begin{bmatrix} -0.8 \\ 1 \end{bmatrix}$	$\begin{bmatrix} -0.8 \\ 1 \end{bmatrix}$
m_k	1	0.5	1.056	-0.997	-1	-1

We deduce that the eigenvalue of A smallest in magnitude is $\frac{1}{-1} = -1$.

35. We solve $A\mathbf{x}_1 = \mathbf{y}_0$ using row reduction: $\begin{bmatrix} A \mid \mathbf{y}_0 \end{bmatrix} = \begin{bmatrix} 4 & 0 & 6 & \mid & 1 \\ -1 & 3 & 1 & \mid & 1 \\ 6 & 0 & 4 & \mid & -1 \end{bmatrix} \longrightarrow \begin{bmatrix} 1 & 0 & 0 & \mid & -0.5 \\ 0 & 1 & 0 & \mid & 0 \\ 0 & 0 & 1 & \mid & 0.5 \end{bmatrix}$.

Thus, $\mathbf{x}_1 = \begin{bmatrix} -0.5 \\ 0 \\ 0.5 \end{bmatrix}$, so $\mathbf{y}_1 = \begin{bmatrix} 1 \\ 0 \\ -1 \end{bmatrix}$. Continuing in this way we calculate:

k	0	1	2	3	4	5
$\mathbf{x}_k$	$\begin{bmatrix} 1 \\ 1 \\ -1 \end{bmatrix}$	$\begin{bmatrix} -0.5 \\ 0 \\ 0.5 \end{bmatrix}$	$\begin{bmatrix} -0.5 \\ -0.333 \\ 0.5 \end{bmatrix}$	$\begin{bmatrix} -0.5 \\ -0.111 \\ 0.5 \end{bmatrix}$	$\begin{bmatrix} -0.5 \\ -0.259 \\ 0.5 \end{bmatrix}$	$\begin{bmatrix} -0.5 \\ -0.161 \\ 0.5 \end{bmatrix}$
$\mathbf{y}_k$	$\begin{bmatrix} 1 \\ 1 \\ -1 \end{bmatrix}$	$\begin{bmatrix} 1 \\ 0 \\ -1 \end{bmatrix}$	$\begin{bmatrix} 1 \\ 0.666 \\ -1 \end{bmatrix}$	$\begin{bmatrix} 1 \\ 0.222 \\ -1 \end{bmatrix}$	$\begin{bmatrix} 1 \\ 0.518 \\ -1 \end{bmatrix}$	$\begin{bmatrix} 1 \\ 0.322 \\ -1 \end{bmatrix}$
m_k	1	-0.5	-0.5	-0.5	-0.5	-0.5

We deduce that the eigenvalue of A that is smallest in magnitude is $\frac{1}{-0.5} = -2$.

37. Since $\alpha = 0$, this is the inverse power method. So, the solution to Exercise 33 applies.

39. We first shift: $A - 5I = \begin{bmatrix} -1 & 0 & 6 \\ -1 & -2 & 1 \\ 6 & 0 & -1 \end{bmatrix}$. Now, taking $\mathbf{x}_0 = \mathbf{y}_0 = \begin{bmatrix} 1 \\ 1 \\ 1 \end{bmatrix}$,

we solve $\begin{bmatrix} A - 5I \mid \mathbf{y}_0 \end{bmatrix} = \begin{bmatrix} -1 & 0 & 6 & \mid & 1 \\ -1 & -2 & 1 & \mid & 1 \\ 6 & 0 & -1 & \mid & 1 \end{bmatrix} \longrightarrow \begin{bmatrix} 1 & 0 & 0 & \mid & 0.2 \\ 0 & 1 & 0 & \mid & -0.5 \\ 0 & 0 & 1 & \mid & 0.2 \end{bmatrix}$.

Thus, $\mathbf{x}_1 = \begin{bmatrix} 0.2 \\ -0.5 \\ 0.2 \end{bmatrix}$ and $\mathbf{y}_1 = \begin{bmatrix} -0.4 \\ 1 \\ -0.4 \end{bmatrix}$. Continuing, we get the values shown in the table.

k	0	1	2	3
$\mathbf{x}_k$	$\begin{bmatrix} 1 \\ 1 \\ 1 \end{bmatrix}$	$\begin{bmatrix} 0.2 \\ -0.5 \\ 0.2 \end{bmatrix}$	$\begin{bmatrix} -0.08 \\ -0.5 \\ -0.08 \end{bmatrix}$	$\begin{bmatrix} 0.032 \\ -0.5 \\ 0.032 \end{bmatrix}$
$\mathbf{y}_k$	$\begin{bmatrix} 1 \\ 1 \\ 1 \end{bmatrix}$	$\begin{bmatrix} -0.4 \\ 1 \\ -0.4 \end{bmatrix}$	$\begin{bmatrix} 0.16 \\ 1 \\ 0.16 \end{bmatrix}$	$\begin{bmatrix} -0.064 \\ 1 \\ -0.064 \end{bmatrix}$
m_k	1	-0.5	-0.5	-0.5

It appears that the eigenvalue of A closest to 5 is $5 + \frac{1}{-0.5} = 5 - 2 = 3$.

4.5 Iterative Methods for Computing Eigenvalues

41. The companion matrix of $p(x) = x^2 + 2x - 2$ is $C(p) = \begin{bmatrix} -2 & 2 \\ 1 & 0 \end{bmatrix}$.

So, we apply the inverse power method with $\mathbf{x}_0 = \mathbf{y}_0 = \begin{bmatrix} 1 \\ 1 \end{bmatrix}$.

We first row-reduce $[\,C(p)\,|\,\mathbf{y}_0\,] = \begin{bmatrix} -2 & 2 & | & 1 \\ 1 & 0 & | & 1 \end{bmatrix} \longrightarrow \begin{bmatrix} 1 & 0 & | & 1 \\ 0 & 1 & | & 1.5 \end{bmatrix}$.

Thus, $\mathbf{x}_1 = \begin{bmatrix} 1 \\ 1.5 \end{bmatrix}$. Continuing, we get:

k	0	1	2	3	4	5	6
$\mathbf{x}_k$	$\begin{bmatrix} 1 \\ 1 \end{bmatrix}$	$\begin{bmatrix} 1 \\ 1.5 \end{bmatrix}$	$\begin{bmatrix} 1 \\ 1.333 \end{bmatrix}$	$\begin{bmatrix} 1 \\ 1.375 \end{bmatrix}$	$\begin{bmatrix} 1 \\ 1.364 \end{bmatrix}$	$\begin{bmatrix} 1 \\ 1.367 \end{bmatrix}$	$\begin{bmatrix} 1 \\ 1.366 \end{bmatrix}$
$\mathbf{y}_k$	$\begin{bmatrix} 1 \\ 1 \end{bmatrix}$	$\begin{bmatrix} 0.666 \\ 1 \end{bmatrix}$	$\begin{bmatrix} 0.750 \\ 1 \end{bmatrix}$	$\begin{bmatrix} 0.727 \\ 1 \end{bmatrix}$	$\begin{bmatrix} 0.733 \\ 1 \end{bmatrix}$	$\begin{bmatrix} 0.732 \\ 1 \end{bmatrix}$	$\begin{bmatrix} 0.732 \\ 1 \end{bmatrix}$
m_k	1	1.5	1.333	1.375	1.364	1.367	1.366

Thus, we estimate that the root of $p(x)$ closest to $\alpha = 0$ is approximately $\frac{1}{1.366} \approx 0.732$.

43. $C(p) = \begin{bmatrix} 2 & 0 & -1 \\ 1 & 0 & 0 \\ 0 & 1 & 0 \end{bmatrix}$, so we apply the inverse power method with $\mathbf{x}_0 = \mathbf{y}_0 = \begin{bmatrix} 1 \\ 1 \\ 1 \end{bmatrix}$.

Row-reducing $[\,C(p)\,|\,\mathbf{y}_0\,] = \begin{bmatrix} 2 & 0 & -1 & | & 1 \\ 1 & 0 & 0 & | & 1 \\ 0 & 1 & 0 & | & 1 \end{bmatrix} \longrightarrow \begin{bmatrix} 1 & 0 & 0 & | & 1 \\ 0 & 1 & 0 & | & 1 \\ 0 & 0 & 1 & | & 1 \end{bmatrix}$.

We see that $\mathbf{y}_0$ is an eigenvector, but this is undesirable since the root $x = 1$ is not necessarily the closest to 0.

So we start again with $\mathbf{x}_0 = \mathbf{y}_0 = \begin{bmatrix} 1 \\ 0 \\ 1 \end{bmatrix}$, obtaining the values:

k	0	1	2	4	6	8	9
$\mathbf{x}_k$	$\begin{bmatrix} 1 \\ 0 \\ 1 \end{bmatrix}$	$\begin{bmatrix} 0 \\ 1 \\ -1 \end{bmatrix}$	$\begin{bmatrix} 1 \\ -1 \\ 2 \end{bmatrix}$	$\begin{bmatrix} -0.666 \\ 1 \\ -1.665 \end{bmatrix}$	$\begin{bmatrix} -0.625 \\ 1 \\ -1.625 \end{bmatrix}$	$\begin{bmatrix} -0.619 \\ 1 \\ -1.619 \end{bmatrix}$	$\begin{bmatrix} -0.618 \\ 1 \\ -1.618 \end{bmatrix}$
$\mathbf{y}_k$	$\begin{bmatrix} 1 \\ 0 \\ 1 \end{bmatrix}$	$\begin{bmatrix} 0 \\ 1 \\ -1 \end{bmatrix}$	$\begin{bmatrix} 0.5 \\ -0.5 \\ 1 \end{bmatrix}$	$\begin{bmatrix} 0.4 \\ -0.6 \\ 1 \end{bmatrix}$	$\begin{bmatrix} 0.385 \\ -0.615 \\ 1 \end{bmatrix}$	$\begin{bmatrix} 0.382 \\ -0.618 \\ 1 \end{bmatrix}$	$\begin{bmatrix} 0.382 \\ -0.618 \\ 1 \end{bmatrix}$
m_k	1	1	2	−1.665	−1.625	−1.619	−1.618

Thus, we conclude that the root of $p(x)$ closest to 0 is $\frac{1}{-1.618} \approx -0.618$.

45. To prove $\frac{1}{\lambda-\alpha}$ is an eigenvalue of $(A - \alpha I)^{-1}$ we must show:
There exists $\mathbf{x} \neq \mathbf{0}$ such that $(A - \alpha I)^{-1}\mathbf{x} = \frac{1}{\lambda-\alpha}\mathbf{x}$.

Since λ is an eigenvalue of A, there exists $\mathbf{x} \neq \mathbf{0}$ such that $A\mathbf{x} = \lambda \mathbf{x}$.
Subtracting $\alpha \mathbf{x}$ from both sides yields: $A\mathbf{x} - \alpha \mathbf{x} = \lambda \mathbf{x} - \alpha \mathbf{x}$. So, $(A - \alpha I)\mathbf{x} = (\lambda - \alpha)\mathbf{x}$.
Left multiplying both sides by $(A - \alpha I)^{-1}$ gives us: $\mathbf{x} = (A - \alpha I)^{-1}(\lambda - \alpha)\mathbf{x}$.
Dividing both sides by $\lambda - \alpha$ yields: $(A - \alpha I)^{-1}\mathbf{x} = \frac{1}{\lambda-\alpha}\mathbf{x}$ as required.

Q: Why does the fact that α is *not* an eigenvalue of A imply $A - \alpha I$ is invertible?
A: If $\det(A - \alpha I) = 0$, then α *is* an eigenvalue of A.

Q: What is the contrapositive of this statement?
A: If α is *not* an eigenvalue of A, then $\det(A - \alpha I) \neq 0$. What does that tell us?
If $\det(A - \alpha I) \neq 0$, then $A - \alpha I$ is invertible.

47. We find the Gerschgorin's Disk Theorem for rows and columns separately.

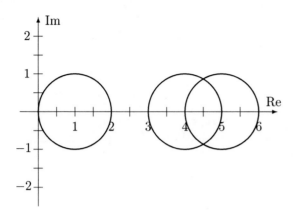

Gerschgorin's Disk Theorem for rows:
$$\begin{bmatrix} 1 & 1 & 0 \end{bmatrix} \Rightarrow c = a_{11} = 1,\ r = a_{12} + a_{13} = 1$$
$$\begin{bmatrix} \frac{1}{2} & 4 & \frac{1}{2} \end{bmatrix} \Rightarrow c = 4,\ r = \frac{1}{2} + \frac{1}{2} = 1$$
$$\begin{bmatrix} 1 & 0 & 5 \end{bmatrix} \Rightarrow c = 5,\ r = 1 + 0 = 1$$

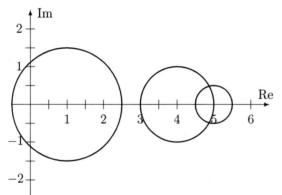

Gerschgorin's Disk Theorem for columns:
$$\begin{bmatrix} 1 \\ \frac{1}{2} \\ 1 \end{bmatrix} \Rightarrow c = a_{11} = 1,\ r = a_{21} + a_{31} = \frac{3}{2}$$
$$\mathbf{a_2} \Rightarrow c = 4,\ r = 1 + 0 = 1$$
$$\mathbf{a_3} \Rightarrow c = 5,\ r = 0 + \frac{1}{2} = \frac{1}{2}$$

Q: The centers of the disks do not change for the rows and columns. Why?
A: Because they come from the diagonal entries.

Q: The radii of the disks *do* change for the rows and columns. Why?
A: Because they are created from the rows and columns respectively.

Q: What does the fact that the disk of radius 1 centered at 1 is disjoint tell us?
A: There is a real eigenvalue of A in the interval $[0, 2]$.

49. We find the Gerschgorin's Disk Theorem for rows and columns separately.

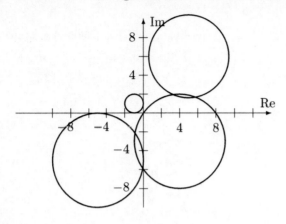

Gerschgorin's Disk Theorem for rows:
$\mathbf{A}_1 \Rightarrow c = 4 - 3i, r = 5$
$\mathbf{A}_2 \Rightarrow c = -1 + i, r = 1$
$\mathbf{A}_3 \Rightarrow c = 5 + 6i, r = 3 + \sqrt{2}$
$\mathbf{A}_4 \Rightarrow c = -5 - 5i, r = 5$

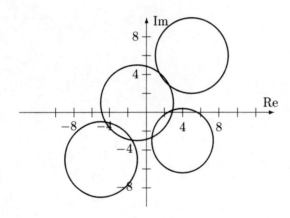

Gerschgorin's Disk Theorem for columns:
$\mathbf{a}_1 \Rightarrow c = 4 - 3i, r = 2 + \sqrt{2}$
$\mathbf{a}_2 \Rightarrow c = -1 + i, r = 4$
$\mathbf{a}_3 \Rightarrow c = 5 + 6i, r = 4$
$\mathbf{a}_4 \Rightarrow c = -5 - 5i, r = 4$

4.5 Iterative Methods for Computing Eigenvalues

51. Given A *strictly diagonally dominant*, $|a_{ii}| > \sum_{j \neq i} |a_{ij}|$, we must show A is invertible.

We will use $|a_{ii}| > \sum_{j \neq i} |a_{ij}|$, to prove that all the eigenvalues of A are nonzero.

By Theorem 4.17(o) in Section 4.3, if 0 is not an eigenvalue of A then A is invertible.

Let λ be an eigenvalue of A. Then by the Gerschgorin's Disk Theorem, Theorem 4.29, we know λ is contained in a Gerschgorin disk for some row i. So: $r_i = \sum_{j \neq i} |a_{ij}| \geq |a_{ii} - \lambda|$.

Since A is *strictly diagonally dominant*, we have: $|a_{ii}| > \sum_{j \neq i} |a_{ij}| \geq |a_{ii} - \lambda|$.

That is, $|a_{ii}| > |a_{ii} - \lambda|$. So, by the Triangle Inequality for complex numbers, we have $|a_{ii}| > |a_{ii} - \lambda| \geq |a_{ii}| - |\lambda|$. That is, $|a_{ii}| > |a_{ii}| - |\lambda|$.
Subtracting $|a_{ii}|$ from both sides and adding $|\lambda|$ to both sides, yields: $|\lambda| > 0$.
But $|\lambda| > 0$ implies $\lambda \neq 0$, as we were to show.
Therefore, all the eigenvalues of A are nonzero which implies A is invertible.

Q: How might we generalize this result?
A: Hint: If the dominant entry is *not* on the diagonal, can we permute the rows?

Q: What does the Triangle Inequality for complex numbers state?
A: It states $|z| + |w| \geq |z + w|$. See Appendix C, *Complex Numbers*.

Q: Why does $|z| + |w| \geq |z + w|$ imply $|a_{ii} - \lambda| \geq |a_{ii}| - |\lambda|$?
A: Hint: Let $z = a_{ii} - \lambda$ and $w = \lambda$.

53. We need to show $|\lambda| \leq 1$, where λ is the eigenvalue of a *stochastic* matrix A.
We will show this using the result of Exercise 52 applied to A^T, after showing $\|A^T\| = 1$.
From Section 3.7, a *stochastic* matrix is a square matrix whose columns are probability vectors.
That is, the sum of all the entries in any column of A is 1: $\sum_{i=1}^{n} |a_{ij}| = 1$.

Therefore, the sum of all the entries in any row of A^T is 1: $\sum_{i=1}^{n} |a_{ij}| = 1$.

So by Exercise 52, we have $|\lambda| \leq \|A^T\| = 1$,

where $\|A^T\| = \max_{1 \leq j \leq n} \left(\sum_{i=1}^{n} |a_{ij}| \right) = 1$ and λ is an eigenvalue of A^T.

Recall from Exercise 19 of Section 4.3, that the eigenvalues of A^T and A are the same.
Therefore $|\lambda| \leq 1$, where λ is the eigenvalue of a *stochastic* matrix A as claimed.

4.6 Applications and the Perron-Frobenius Theorem

Q: What are our main goals for Section 4.6?

A: To explore various applications of methods involving iteration and sequences.

Key Definitions and Concepts

attractor	p.347	4.6	See definition on p.347 and Example 4.48.
conditions	p.333	4.6	*initial conditions.* See definition p. 333.
recurrence	p.333	4.6	*linear recurrence.* See definition p. 333.
repeller	p.349	4.6	See definition on p.349 and Example 4.50.
saddle point	p.349	4.6	See definition on p.349 and Example 4.50.
trajectory	p.346	4.6	See definition on p.346 prior to Example 4.48.

Theorems

Perron's	p.330	4.6	The Perron's Theorem. See Theorem 4.36.
Perron-Fro	p.332	4.6	The Perron-Frobenius Theorem. See Theorem 4.37.
Thm 4.30	p.322	4.6	If P is the $n \times n$ transition matrix of a Markov chain, then 1 is an eigenvalue of P
Thm 4.31	p.322	4.6	Let P be an $n \times n$ transition matrix with eigenvalue λ. a. $|\lambda| \leq 1$ b. If P is regular and $\lambda \neq 1$, then $|\lambda| < 1$.
Lem 4.32	p.324	4.6	Let P be an $n \times n$ transition matrix. If P is diagonalizable, then the dominant eigenvalue $\lambda_1 = 1$ has algebraic multiplicity 1.
Thm 4.33	p.325	4.6	Let P be an $n \times n$ transition matrix. Then as $k \longrightarrow \infty$, P^k approaches an $n \times n$ matrix L whose columns are $\mathbf{x}$, the steady state probability vector for P.
Thm 4.34	p.326	4.6	Let P be an $n \times n$ transition matrix ... as in Theorem 4.33. Then for any probability vector $\mathbf{x}_0$, $\mathbf{x}_k$ approaches $\mathbf{x}$.
Thm 4.35	p.328	4.6	Every Leslie matrix has a unique positive eigenvalue and a corresponding eigenvector with positive components.
Thm 4.36	p.330	4.6	Let A be a positive $n \times n$ matrix. A has a real eigenvalue λ_1: a. $\lambda_1 > 0$ b. λ_1 has a corresponding positive eigenvector. c. If λ is any other eigenvalue of A, then $|\lambda| \leq \lambda_1$.
Thm 4.37	p.332	4.6	Let A be an irreducible nonnegative $n \times n$ matrix. Then: a. $\lambda_1 > 0$ b. λ_1 has a corresponding positive eigenvector. c. If λ is any other eigenvalue of A, then $|\lambda| \leq \lambda_1$. If A is primitive, then this inequality is strict. d. If λ is an eigenvalue of A such that $|\lambda| = \lambda_1$, then λ is a (complex) root of the equation $\lambda^n - \lambda_1^n = 0$. e. λ_1 has algebraic multiplicity 1.
Thm 4.38	p.336	4.6	See statement of Theorem on p. 336.
Thm 4.39	p.337	4.6	See statement of Theorem on p. 337.
Thm 4.40	p.339	4.6	See statement of Theorem on p. 339.
Thm 4.41	p.344	4.6	See statement of Theorem on p. 344.
Thm 4.42	p.350	4.6	See statement of Theorem on p. 350.
Thm 4.43	p.351	4.6	See statement of Theorem on p. 351.

Discussion of Definitions and Theorems

Discussion is included with the solutions below.

4.6 Applications and the Perron-Frobenius Theorem

Solutions to odd-numbered exercises from Section 4.6

1. $\begin{bmatrix} 0 & 1 \\ 1 & 0 \end{bmatrix}^2 = \begin{bmatrix} 1 & 0 \\ 0 & 1 \end{bmatrix}$, so any power of $\begin{bmatrix} 0 & 1 \\ 1 & 0 \end{bmatrix}$ contains zeros. Thus $\begin{bmatrix} 0 & 1 \\ 1 & 0 \end{bmatrix}$ is not regular.

3. $\begin{bmatrix} \frac{1}{3} & 1 \\ \frac{2}{3} & 0 \end{bmatrix}^2 = \begin{bmatrix} \frac{7}{9} & \frac{1}{3} \\ \frac{2}{9} & \frac{2}{3} \end{bmatrix}$, which is positive, so $\begin{bmatrix} \frac{1}{3} & 1 \\ \frac{2}{3} & 0 \end{bmatrix}$ is regular.

5. $\begin{bmatrix} 0.1 & 0 & 0.5 \\ 0.5 & 1 & 0 \\ 0.4 & 0 & 0.5 \end{bmatrix}^n$ has 0s as its $(1,2)$ and $(3,2)$ entries for all n, so $\begin{bmatrix} 0.1 & 0 & 0.5 \\ 0.5 & 1 & 0 \\ 0.4 & 0 & 0.5 \end{bmatrix}$ is not regular.

7. The transition matrix $P = \begin{bmatrix} \frac{1}{3} & \frac{1}{6} \\ \frac{2}{3} & \frac{5}{6} \end{bmatrix}$ has characteristic equation

$$\det(P - \lambda I) = \begin{vmatrix} \frac{1}{3} - \lambda & \frac{1}{6} \\ \frac{2}{3} & \frac{5}{6} - \lambda \end{vmatrix} = \frac{1}{6}(\lambda - 1)(6\lambda - 1)$$ so its eigenvalues are $1, \frac{1}{6}$.

The eigenspaces are $E_1 = \text{span}\left(\begin{bmatrix} 1 \\ 4 \end{bmatrix}\right)$ and $E_{1/6} = \text{span}\left(\begin{bmatrix} 1 \\ -1 \end{bmatrix}\right)$.

So, taking $Q = \begin{bmatrix} 1 & 1 \\ 4 & -1 \end{bmatrix}$, we know that $Q^{-1}PQ = \begin{bmatrix} 1 & 0 \\ 0 & \frac{1}{6} \end{bmatrix} = D$, and

$$P^k = QD^kQ^{-1} = \begin{bmatrix} 1 & 1 \\ 4 & -1 \end{bmatrix} \begin{bmatrix} 1^k & 0 \\ 0 & (\frac{1}{6})^k \end{bmatrix} \begin{bmatrix} 1 & 1 \\ 4 & -1 \end{bmatrix}^{-1}.$$

As $k \to \infty$, $(\frac{1}{6})^k \to 0$, so $D^k \to \begin{bmatrix} 1 & 0 \\ 0 & 0 \end{bmatrix}$, $P^k \to \begin{bmatrix} 1 & 1 \\ 4 & -1 \end{bmatrix} \begin{bmatrix} 1 & 0 \\ 0 & 0 \end{bmatrix} \begin{bmatrix} 1 & 1 \\ 4 & -1 \end{bmatrix}^{-1} = \begin{bmatrix} \frac{1}{5} & \frac{1}{5} \\ \frac{4}{5} & \frac{4}{5} \end{bmatrix} = L.$

9. The eigenspace of $P = \begin{bmatrix} 0.2 & 0.3 & 0.4 \\ 0.6 & 0.1 & 0.4 \\ 0.2 & 0.6 & 0.2 \end{bmatrix}$ corresponding to the eigenvalue 1 is

$E_1 = \text{span}\left(\begin{bmatrix} 24 \\ 28 \\ 27 \end{bmatrix}\right)$, so $L = \frac{1}{79}\begin{bmatrix} 24 & 24 & 24 \\ 28 & 28 & 28 \\ 27 & 27 & 27 \end{bmatrix}$.

11. $\det(L - \lambda I) = \begin{vmatrix} -\lambda & 2 \\ 0.5 & -\lambda \end{vmatrix} = (\lambda - 1)(\lambda + 1)$, so the positive eigenvalue of L is 1.

$[L - I \mid 0] = \begin{bmatrix} -1 & 2 & \mid & 0 \\ 0.5 & -1 & \mid & 0 \end{bmatrix} \to \begin{bmatrix} 1 & -2 & \mid & 0 \\ 0 & 0 & \mid & 0 \end{bmatrix}$, so $E_1 = \text{span}\left(\begin{bmatrix} 2 \\ 1 \end{bmatrix}\right)$.

13. $\det(L - \lambda I) = \begin{vmatrix} -\lambda & 7 & 4 \\ 0.5 & -\lambda & 0 \\ 0 & 0.5 & -\lambda \end{vmatrix} = -\lambda^3 + 3.5\lambda + 1 = -\frac{1}{2}(\lambda - 2)(2\lambda^2 + 4\lambda + 1).$

So, the positive eigenvalue of L is 2.

$[\, L - 2I \mid \mathbf{0} \,] = \begin{bmatrix} -2 & 7 & 4 & 0 \\ 0.5 & -2 & 0 & 0 \\ 0 & 0.5 & -2 & 0 \end{bmatrix} \longrightarrow \begin{bmatrix} 1 & 0 & -16 & 0 \\ 0 & 1 & -4 & 0 \\ 0 & 0 & 0 & 0 \end{bmatrix}$, so $E_2 = \text{span}\left(\begin{bmatrix} 16 \\ 4 \\ 1 \end{bmatrix}\right).$

15. If $\lambda_1 > 1$, then the population will increase without limit.
If $\lambda_1 < 1$, then the population will decline and eventually vanish.
If $\lambda_1 = 1$, then the population will be stable.

17. $P^{-1} = \begin{bmatrix} 1 & 0 & 0 & \cdots & 0 \\ 0 & \frac{1}{s_1} & 0 & \cdots & 0 \\ 0 & 0 & \frac{1}{s_1 s_2} & \cdots & 0 \\ \vdots & \vdots & \vdots & \ddots & \vdots \\ 0 & 0 & 0 & \cdots & \frac{1}{s_1 s_2 \cdots s_{n-1}} \end{bmatrix}$ and $L = \begin{bmatrix} b_1 & b_2 & b_3 & \cdots & b_{n-1} & b_n \\ s_1 & 0 & 0 & \cdots & 0 & 0 \\ 0 & s_2 & 0 & \cdots & 0 & 0 \\ 0 & 0 & s_3 & \cdots & 0 & 0 \\ \vdots & \vdots & \vdots & \ddots & \vdots & \vdots \\ 0 & 0 & 0 & \cdots & s_{n-1} & 0 \end{bmatrix}.$

So, $P^{-1}L = \begin{bmatrix} b_1 & b_2 & b_3 & \cdots & b_{n-1} & b_n \\ 1 & 0 & 0 & \cdots & 0 & 0 \\ 0 & \frac{1}{s_1} & 0 & \cdots & 0 & 0 \\ 0 & 0 & \frac{1}{s_1 s_2} & \cdots & 0 & 0 \\ \vdots & \vdots & \vdots & \ddots & \vdots & \vdots \\ 0 & 0 & 0 & \cdots & \frac{1}{s_1 s_2 \cdots s_{n-2}} & 0 \end{bmatrix}$ and

$P^{-1}LP = \begin{bmatrix} b_1 & b_2 s_1 & b_3 s_1 s_2 & \cdots & b_{n-1} s_1 s_2 \cdots s_{n-2} & b_n s_1 s_2 \cdots s_{n-1} \\ 1 & 0 & 0 & \cdots & 0 & 0 \\ 0 & 1 & 0 & \cdots & 0 & 0 \\ 0 & 0 & 1 & \cdots & 0 & 0 \\ \vdots & \vdots & \vdots & \ddots & \vdots & \vdots \\ 0 & 0 & 0 & \cdots & 1 & 0 \end{bmatrix}.$

This is the companion matrix of the polynomial:
$p(x) = x^n - b_1 x^{n-1} - b_2 s_1 x^{n-2} - \cdots - b_{n-1} s_1 s_2 \cdots s_{n-2} x - b_n s_1 s_2 \cdots s_{n-1}.$
So, by Exercise 32 in Section 4.3, the characteristic polynomial of $P^{-1}LP$ is $(-1)^n p(\lambda)$.

So, the characteristic polynomial of L is:

$\det(L - \lambda I) = \det(P^{-1}LP - \lambda I) = (-1)^n p(\lambda)$
$= (-1)^n \left(\lambda^n - b_1 \lambda^{n-1} - b_2 s_1 \lambda^{n-2} - \cdots - b_n s_1 s_2 \cdots s_{n-1}\right).$

4.6 Applications and the Perron-Frobenius Theorem

19. Using a CAS, we find the positive eigenvalue of $L = \begin{bmatrix} 1 & 1 & 3 \\ 0.7 & 0 & 0 \\ 0 & 0.5 & 0 \end{bmatrix}$ is ≈ 1.7456.

    So, this is the steady state growth rate of the population with this Leslie matrix. Now according to Exercise 18, a corresponding eigenvector is
    $\begin{bmatrix} 1 & 0.7/1.7456 & (0.7 \cdot 0.5)/(1.7456)^2 \end{bmatrix}^T \approx \begin{bmatrix} 1 & 0.401 & 0.115 \end{bmatrix}^T$.
    So, the age classes are in the ratio $1 : 0.401 : 0.115$.

21. The positive eigenvalue of L is approximately 1.0924, so this is the steady state growth rate.

    A corresponding eigenvector is $\begin{bmatrix} 1 \\ \frac{0.3}{1.0924} \\ \frac{0.3 \cdot 0.7}{(1.0924)^2} \\ \frac{0.3 \cdot 0.7 \cdot 0.9}{(1.0924)^3} \\ \frac{0.3 \cdot 0.7 \cdot 0.9 \cdot 0.9}{(1.0924)^4} \\ \frac{0.3 \cdot 0.7 \cdot 0.9 \cdot 0.9 \cdot 0.9}{(1.0924)^5} \\ \frac{0.3 \cdot 0.7 \cdot 0.9 \cdot 0.9 \cdot 0.9 \cdot 0.6}{(1.0924)^6} \end{bmatrix} \approx \begin{bmatrix} 1 \\ 0.275 \\ 0.176 \\ 0.145 \\ 0.119 \\ 0.098 \\ 0.054 \end{bmatrix}$.

23. (a) Each term in r corresponds to the probability of a given female having a daughter while she is a member of a particular age class. Thus, the sum of the terms gives the average number of daughters born to a single female over her lifetime.

    (b) We define $g(\lambda)$ as suggested. Then $g(\lambda) = 1 \Leftrightarrow$
    $\frac{b_1}{\lambda} + \frac{b_2 s_1}{\lambda^2} + \cdots + \frac{b_n s_1 s_2 \cdots s_{n-1}}{\lambda^n} = 1 \Leftrightarrow \lambda^n - b_1 \lambda^{n-1} - b_2 s_1 \lambda^{n-2} - \cdots - b_n s_1 s_2 \cdots s_{n-1} = 0$.
    But from Exercise 16, this is true if and only if λ is a zero of the characteristic polynomial of the Leslie matrix, indicating that λ is an eigenvalue of L.
    Now $r = b_1 + b_2 s_1 + \cdots + b_n s_1 s_2 \cdots s_{n-1} = 1 \Leftrightarrow \lambda = 1$, as desired.

    (c) If $r < 1$, each female averages less than 1 daughter $\Rightarrow$ decreasing population.
    If $r > 1$, each female averages more than 1 daughter $\Rightarrow$ increasing population.

25. (a) In Exercise 21 we found the positive eigenvalue of L to be $\lambda_1 \approx 1.0924$, so from Exercise 24, $h = 1 - 1/\lambda_1 \approx 1 - 1/1.0924 \approx 0.0846$.

    (b) We reduce the initial population levels by a factor of 0.0846, so:
    $\mathbf{x}'_0 = (1 - 0.0846) \begin{bmatrix} 10 & 2 & 8 & 5 & 12 & 0 & 1 \end{bmatrix}^T = \begin{bmatrix} 9.15 & 1.831 & 7.323 & 4.577 & 10.985 & 0 & 0.915 \end{bmatrix}^T$.

    Then $\mathbf{x}'_1 = L\mathbf{x}'_0 = \begin{bmatrix} 0 & 0.4 & 1.8 & 1.8 & 1.8 & 1.6 & 0.6 \\ 0.3 & 0 & 0 & 0 & 0 & 0 & 0 \\ 0 & 0.7 & 0 & 0 & 0 & 0 & 0 \\ 0 & 0 & 0.9 & 0 & 0 & 0 & 0 \\ 0 & 0 & 0 & 0.9 & 0 & 0 & 0 \\ 0 & 0 & 0 & 0 & 0.9 & 0 & 0 \\ 0 & 0 & 0 & 0 & 0 & 0.6 & 0 \end{bmatrix} \begin{bmatrix} 9.15 \\ 1.831 \\ 7.323 \\ 4.577 \\ 10.985 \\ 0 \\ 0.915 \end{bmatrix} = \begin{bmatrix} 42.47 \\ 2.75 \\ 1.28 \\ 6.59 \\ 4.12 \\ 9.89 \\ 0 \end{bmatrix}$.

    In fact, the population increases substantially.

27. $\lambda_1 = r_1 (\cos 0 + i \sin 0) = r_1$ since it is positive.

Let $\lambda = r (\cos \theta + i \sin \theta)$ be some other eigenvector with $\theta \neq 0$.

Then equating the two expressions for $g(\lambda)$, applying De Moivre's Theorem, and taking the real part of each side, we have:

$$\frac{b_1}{r_1} + \frac{b_2 s_1}{r_1^2} + \frac{b_3 s_1 s_2}{r_1^3} + \cdots + \frac{b_n s_1 s_2 \cdots s_{n-1}}{r_1^n}$$

$$= \frac{b_1}{r} \cos \theta + \frac{b_2 s_1}{r^2} \cos 2\theta + \frac{b_3 s_1 s_2}{r^3} \cos 3\theta + \cdots + \frac{b_n s_1 s_2 \cdots s_{n-1}}{r^n} \cos n\theta$$

But $\cos k\theta \leq 1$ for all k, so

$$\frac{b_1}{r_1} + \frac{b_2 s_1}{r_1^2} + \frac{b_3 s_1 s_2}{r_1^3} + \cdots + \frac{b_n s_1 s_2 \cdots s_{n-1}}{r_1^n} \leq \frac{b_1}{r} + \frac{b_2 s_1}{r^2} + \frac{b_3 s_1 s_2}{r^3} + \cdots + \frac{b_n s_1 s_2 \cdots s_{n-1}}{r^n}$$

showing that $r_1 \geq r$.

29. $\det(A - \lambda I) = \begin{vmatrix} 1-\lambda & 3 \\ 2 & -\lambda \end{vmatrix} = \lambda^2 - \lambda - 6 = (\lambda + 2)(\lambda - 3)$, so $\lambda_1 = 3$ is the Perron root.

$A - 3I = \begin{bmatrix} -2 & 3 \\ 2 & -3 \end{bmatrix} \longrightarrow \begin{bmatrix} 2 & -3 \\ 0 & 0 \end{bmatrix}$, so the Perron eigenvector is $\begin{bmatrix} \frac{3}{5} \\ \frac{2}{5} \end{bmatrix}$.

31. $\det(A - \lambda I) = \begin{vmatrix} 2-\lambda & 1 & 1 \\ 1 & 1-\lambda & 0 \\ 1 & 0 & 1-\lambda \end{vmatrix} = -\lambda(\lambda-1)(\lambda-3)$, so $\lambda_1 = 3$ is the Perron root.

$A - 3I = \begin{bmatrix} -1 & 1 & 1 \\ 1 & -2 & 0 \\ 1 & 0 & -2 \end{bmatrix} \longrightarrow \begin{bmatrix} 1 & 0 & -2 \\ 0 & 1 & -1 \\ 0 & 0 & 0 \end{bmatrix}$, so the Perron eigenvector is $\begin{bmatrix} \frac{1}{2} \\ \frac{1}{4} \\ \frac{1}{4} \end{bmatrix}$.

33. $(I + A)^{n-1} = \begin{bmatrix} 1 & 0 & 1 & 0 \\ 0 & 1 & 1 & 1 \\ 1 & 0 & 1 & 0 \\ 1 & 1 & 0 & 1 \end{bmatrix}^3 = \begin{bmatrix} 4 & 0 & 4 & 0 \\ 6 & 4 & 6 & 4 \\ 4 & 0 & 4 & 0 \\ 6 & 4 & 6 & 4 \end{bmatrix}$, so A is reducible.

In particular, the permutations $R_1 \leftrightarrow R_3$ and $R_2 \leftrightarrow R_4$ put A into the desired block form.

35. $(I + A)^{n-1} = \begin{bmatrix} 1 & 1 & 0 & 0 & 0 \\ 0 & 1 & 0 & 0 & 1 \\ 1 & 0 & 1 & 0 & 1 \\ 0 & 0 & 1 & 1 & 0 \\ 0 & 0 & 0 & 1 & 2 \end{bmatrix}^4 = \begin{bmatrix} 1 & 4 & 1 & 5 & 11 \\ 1 & 1 & 5 & 11 & 16 \\ 5 & 6 & 6 & 12 & 21 \\ 6 & 4 & 5 & 6 & 12 \\ 5 & 1 & 11 & 16 & 22 \end{bmatrix}$, so A is irreducible.

4.6 Applications and the Perron-Frobenius Theorem

37. (a) Suppose G is bipartite and let v and w be vertices which are connected. Then all paths between v and w have odd length, whereas all paths between v and itself have even length. So, any power of adjacency matrix A contains zeros, showing A is not primitive.

    (b) We can write the adjacency matrix of G as $A = \begin{bmatrix} O & B \\ B^T & O \end{bmatrix}$ and partition the eigenvector $\mathbf{v}$ of λ as $\begin{bmatrix} \mathbf{v}_1 \\ \mathbf{v}_2 \end{bmatrix}$, where $\mathbf{v}_1, \mathbf{v}_2$ are column vectors and $\mathbf{v}_2$ has the same number of rows as B does columns. Then $A\mathbf{v} = \lambda\mathbf{v} \Leftrightarrow \begin{bmatrix} B\mathbf{v}_2 \\ B^T\mathbf{v}_1 \end{bmatrix} = \begin{bmatrix} \lambda\mathbf{v}_1 \\ \lambda\mathbf{v}_2 \end{bmatrix}$, so $B\mathbf{v}_2 = \lambda\mathbf{v}_1$ and $B^T\mathbf{v}_1 = \lambda\mathbf{v}_2$. Then $-\lambda \begin{bmatrix} \mathbf{v}_1 \\ -\mathbf{v}_2 \end{bmatrix} = \begin{bmatrix} -B\mathbf{v}_2 \\ B^T\mathbf{v}_1 \end{bmatrix} = A \begin{bmatrix} \mathbf{v}_1 \\ -\mathbf{v}_2 \end{bmatrix}$, showing that $-\lambda$ is an eigenvalue of A with corresponding eigenvector $\begin{bmatrix} \mathbf{v}_1 \\ -\mathbf{v}_2 \end{bmatrix}$.

39. This exploration and the subsequent explanation are left to the reader.

41. $x_0 = 1$, $x_1 = 2$, $x_2 = 4$, $x_3 = 8$, $x_4 = 16$.

43. $y_0 = 0$, $y_1 = 1$, $y_2 = 1$, $y_3 = 0$, $y_4 = -1$.

45. We compute the eigenvalues of $A = \begin{bmatrix} 3 & 4 \\ 1 & 0 \end{bmatrix}$ to be -1 and 4.

    So, we can immediately write $x_n = c_1(-1)^n + c_2 4^n$, where c_1 and c_2 are to be determined. Using the initial conditions, we have:
    $0 = x_0 = c_1(-1)^0 + c_2 4^0 = c_1 + c_2$ and $5 = x_1 = c_1(-1)^1 + c_2 4^1 = -c_1 + 4c_2$.
    We solve the system to find that $c_1 = -1$ and $c_2 = 1$, so $x_n = -(-1)^n + 4^n$.

47. The recurrence is $y_n - 4y_{n-1} + 4y_{n-2} = 0$, so the only eigenvalue is 2.
    Thus, by Theorem 4.38(b), $y_n = c_1(2)^n + c_2 n 2^n$.
    We solve $y_1 = 1 = 2c_1 + 2c_2$ and $y_2 = 6 = 4c_1 + 8c_2$ to get $c_1 = -\frac{1}{2}$ and $c_2 = 1$.
    Thus, $y_n = -\frac{1}{2}(2)^n + n 2^n$.

49. The recurrence is $b_n - 2b_{n-1} - 2b_{n-2} = 0$, so the eigenvalues are $1 \pm \sqrt{3}$. We solve:
    $b_0 = 0 = c_1 + c_2$ and $b_1 = 1 = (1 - \sqrt{3})c_1 + (1 + \sqrt{3})c_2$ to get $c_1 = -\frac{\sqrt{3}}{6}$, $c_2 = \frac{\sqrt{3}}{6}$.
    Thus, $b_n = -\frac{\sqrt{3}}{6}(1 - \sqrt{3})^n + \frac{\sqrt{3}}{6}(1 + \sqrt{3})^n$.

51. Since A is diagonalizable, we have $P^{-1}AP = D = \begin{bmatrix} \lambda_1 & 0 \\ 0 & \lambda_2 \end{bmatrix}$.

Suppose $P = \begin{bmatrix} a & b \\ c & d \end{bmatrix}$, so $P^{-1} = \frac{1}{ad-bc}\begin{bmatrix} d & -b \\ -c & a \end{bmatrix}$. Then:

$A^k = PD^kP^{-1} = \frac{1}{ad-bc}\begin{bmatrix} a & b \\ c & d \end{bmatrix}\begin{bmatrix} \lambda_1^k & 0 \\ 0 & \lambda_2^k \end{bmatrix}\begin{bmatrix} d & -b \\ -c & a \end{bmatrix} = \frac{1}{ad-bc}\begin{bmatrix} ad\lambda_1^k - bc\lambda_2^k & -ab\lambda_1^k + ab\lambda_2^k \\ cd\lambda_1^k - cd\lambda_2^k & -bc\lambda_1^k + ad\lambda_2^k \end{bmatrix}$.

If the two given terms of the sequence are x_{j-1} and x_j, then for any n we have

$$\mathbf{x}_n = A^{n-j}\mathbf{x}_j = \frac{1}{ad-bc}\begin{bmatrix} ad\lambda_1^{n-j} - bc\lambda_2^{n-j} & -ab\lambda_1^{n-j} + ab\lambda_2^{n-j} \\ cd\lambda_1^{n-j} - cd\lambda_2^{n-j} & -bc\lambda_1^{n-j} + ad\lambda_2^{n-j} \end{bmatrix}\begin{bmatrix} x_j \\ x_{j-1} \end{bmatrix}$$

$$= \frac{1}{ad-bc}\begin{bmatrix} \left(ad\lambda_1^{n-j} - bc\lambda_2^{n-j}\right)x_j + \left(-ab\lambda_1^{n-j} + ab\lambda_2^{n-j}\right)x_{j-1} \\ \left(cd\lambda_1^{n-j} - cd\lambda_2^{n-j}\right)x_j + \left(-bc\lambda_1^{n-j} + ad\lambda_2^{n-j}\right)x_{j-1} \end{bmatrix}$$

$$= \frac{1}{ad-bc}\begin{bmatrix} \left[(adx_j - abx_{j-1})\lambda_1^{-j}\right]\lambda_1^n + \left[(-bcx_j + abx_{j-1})\lambda_2^{-j}\right]\lambda_2^n \\ \left[(cdx_j - bcx_{j-1})\lambda_1^{-j}\right]\lambda_1^n + \left[(-cdx_j + adx_{j-1})\lambda_2^{-j}\right]\lambda_2^n \end{bmatrix}.$$

Thus, for constants $c_1 = \frac{adx_j - abx_{j-1}}{ad-bc}\lambda_1^{-j}$ and $c_2 = -\frac{(bcx_j - abx_{j-1})}{ad-bc}\lambda_2^{-j}$, we have $x_n = c_1\lambda_1^n + c_2\lambda_2^n$ as required.

53. (a) For $n = 1$, $A = \begin{bmatrix} 1 & 1 \\ 1 & 0 \end{bmatrix} = \begin{bmatrix} f_2 & f_1 \\ f_1 & f_0 \end{bmatrix}$. So, assume that $A^k = \begin{bmatrix} f_{k+1} & f_k \\ f_k & f_{k-1} \end{bmatrix}$. Then:

$A^{k+1} = A^k A = \begin{bmatrix} f_{k+1} & f_k \\ f_k & f_{k-1} \end{bmatrix}\begin{bmatrix} 1 & 1 \\ 1 & 0 \end{bmatrix} = \begin{bmatrix} f_{k+1} + f_k & f_{k+1} \\ f_k + f_{k-1} & f_k \end{bmatrix} = \begin{bmatrix} f_{k+2} & f_{k+1} \\ f_{k+1} & f_k \end{bmatrix}$.

(b) $\det A = -1$, so $\det(A^n) = (\det A)^n \Leftrightarrow \begin{vmatrix} f_{n+1} & f_n \\ f_n & f_{n-1} \end{vmatrix} = f_{n+1}f_{n-1} - f_n^2 = (-1)^n$.

(c) If we interpret the 5×13 "rectangle" as being composed of two "triangles," we see that, in fact, they are not triangles.
The "diagonal" of the "rectangle" is not straight because the 8×3 triangular pieces are not similar to the entire triangles (in particular, $\frac{5}{13} \neq \frac{3}{8}$).
So there is empty space along the diagonal of the figure (1 square unit of space, in fact).

55. (a) $d_1 = 1$, $d_2 = 2$, $d_3 = 3$, $d_4 = 5$, $d_5 = 8$. As in Exercise 54, $d_0 = 1$.

(b) If we lengthen an $(n-1) \times 2$ rectangle by 1, we can complete the tiling in only one way. If we lengthen an $(n-2) \times 2$ rectangle by 2, there is only one way to complete the tiling that is distinct from the $(n-1) \times 2$ case. Thus, $d_n = d_{n-1} + d_{n-2}$.

(c) This is simply the Fibonacci sequence shifted by 1, so we know that $d_n = \frac{1}{\sqrt{5}}\left(\frac{1+\sqrt{5}}{2}\right)^{n+1} - \frac{1}{\sqrt{5}}\left(\frac{1-\sqrt{5}}{2}\right)^{n+1}$.
This formula agrees with the data in part (a), including our value for d_0.

4.6 Applications and the Perron-Frobenius Theorem

57. The coefficient matrix is $\begin{bmatrix} 1 & 3 \\ 2 & 2 \end{bmatrix}$, which has eigenvalues $\lambda_1 = 4$ and $\lambda_2 = -1$ with corresponding eigenvectors $\mathbf{v}_1 = \begin{bmatrix} 1 \\ 1 \end{bmatrix}$ and $\mathbf{v}_2 = \begin{bmatrix} 3 \\ -2 \end{bmatrix}$.

Thus, $\mathbf{y}' = P^{-1}AP\mathbf{y} = D\mathbf{y}$ is the system $y_1' = 4y_1$, $y_2' = -y_2$ which has the general solution
$$\mathbf{y} = \begin{bmatrix} C_1 e^{4t} \\ C_2 e^{-t} \end{bmatrix} \text{ and } \mathbf{x} = P\mathbf{y} = \begin{bmatrix} 1 & 3 \\ 1 & -2 \end{bmatrix} \begin{bmatrix} C_1 e^{4t} \\ C_2 e^{-t} \end{bmatrix} = \begin{bmatrix} C_1 e^{4t} + 3C_2 e^{-t} \\ C_1 e^{4t} - 2C_2 e^{-t} \end{bmatrix}.$$

We substitute the initial conditions $x(0) = 0$ and $y(0) = 5$ to find that $C_1 = 3$ and $C_2 = -1$. So, the solution to the initial value problem is $x = 3e^{4t} - 3e^{-t}$, $y = 3e^{4t} + 2e^{-t}$.

59. The coefficient matrix is $\begin{bmatrix} 1 & 1 \\ 1 & -1 \end{bmatrix}$, which has eigenvalues $\lambda_1 = \sqrt{2}$ and $\lambda_2 = -\sqrt{2}$ with corresponding eigenvectors $\mathbf{v}_1 = \begin{bmatrix} 1 \\ -1 + \sqrt{2} \end{bmatrix}$ and $\mathbf{v}_2 = \begin{bmatrix} 1 \\ -1 - \sqrt{2} \end{bmatrix}$.

Thus, $\mathbf{y} = \begin{bmatrix} C_1 e^{\sqrt{2}t} \\ C_2 e^{-\sqrt{2}t} \end{bmatrix}$ and $\mathbf{x} = P\mathbf{y} = \begin{bmatrix} C_1 e^{\sqrt{2}t} + C_2 e^{-\sqrt{2}t} \\ (-1+\sqrt{2}) C_1 e^{\sqrt{2}t} + (-1-\sqrt{2}) C_2 e^{-\sqrt{2}t} \end{bmatrix}$.

We substitute $x_1(0) = 1$ and $x_2(0) = 0$ to find that $C_1 = \frac{2+\sqrt{2}}{4}$ and $C_2 = \frac{2-\sqrt{2}}{4}$. So, the solution to the initial value problem is

$x_1 = \frac{2+\sqrt{2}}{4} e^{\sqrt{2}t} + \frac{2-\sqrt{2}}{4} e^{-\sqrt{2}t}$,

$x_2 = (-1+\sqrt{2}) \left(\frac{2+\sqrt{2}}{4}\right) e^{\sqrt{2}t} + (-1-\sqrt{2}) \left(\frac{2-\sqrt{2}}{4}\right) e^{-\sqrt{2}t} = \frac{\sqrt{2}}{4} e^{\sqrt{2}t} - \frac{\sqrt{2}}{4} e^{-\sqrt{2}t}$.

61. The coefficient matrix $\begin{bmatrix} 0 & 1 & -1 \\ 1 & 0 & 1 \\ 1 & 1 & 0 \end{bmatrix}$ has eigenvalues $\lambda_1 = 1$, $\lambda_2 = 0$, $\lambda_3 = -1$ with corresponding eigenvectors $\mathbf{v}_1 = \begin{bmatrix} 0 \\ 1 \\ 1 \end{bmatrix}$, $\mathbf{v}_2 = \begin{bmatrix} 1 \\ -1 \\ -1 \end{bmatrix}$, and $\mathbf{v}_3 = \begin{bmatrix} 1 \\ -1 \\ 0 \end{bmatrix}$.

Thus, $\mathbf{y} = \begin{bmatrix} C_1 e^t \\ C_2 \\ C_3 e^{-t} \end{bmatrix}$ and $\mathbf{x} = P\mathbf{y} = \begin{bmatrix} C_2 + C_3 e^{-t} \\ C_1 e^t - C_2 - C_3 e^{-t} \\ C_1 e^t - C_2 \end{bmatrix}$.

We substitute $x(0) = 1$, $y(0) = 0$, and $z(0) = -1$ to find that $C_1 = 1$, $C_2 = 2$, and $C_3 = -1$. So, the solution to the initial value problem is $x = 2 - e^{-t}$, $y = e^t - 2 + e^{-t}$, and $z = e^t - 2$.

63. (a) The coefficient matrix is $\begin{bmatrix} 1.2 & -0.2 \\ -0.2 & 1.5 \end{bmatrix}$, which has eigenvalues $\lambda_1 = 1.6$ and $\lambda_2 = 1.1$ with corresponding eigenvectors $\mathbf{v}_1 = \begin{bmatrix} 1 \\ -2 \end{bmatrix}$ and $\mathbf{v}_2 = \begin{bmatrix} 2 \\ 1 \end{bmatrix}$.

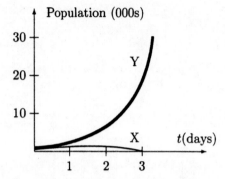

Thus, $\mathbf{y} = \begin{bmatrix} C_1 e^{1.6t} \\ C_2 e^{1.1t} \end{bmatrix}$

and

$\mathbf{x} = P\mathbf{y} = \begin{bmatrix} C_1 e^{1.6t} + 2C_2 e^{1.1t} \\ -2C_1 e^{1.6t} + C_2 e^{1.1t} \end{bmatrix}.$

We substitute $x(0) = 400$ and $y(0) = 500$ to find that $C_1 = -120$ and $C_2 = 260$.
So, the solution is $x = -120e^{1.6t} + 520e^{1.1t}$, $y = 240e^{1.6t} + 260e^{1.1t}$.
We see from the graph:
Bacteria X dies out after about 3 days, while bacteria Y increases indefinitely.

(b) Solving the general case, we find that $C_1 = \frac{1}{5}a - \frac{2}{5}b$ and $C_2 = \frac{2}{5}a + \frac{1}{5}b$.
Since $e^{1.6t}$ term is dominant, the fate of the populations depends on the positivity of C_1.

Thus, if $a = 2b$, then both strains of bacteria will thrive.
Otherwise, only one strain of bacteria will survive indefinitely.
$a > 2b \Rightarrow$, bacteria X survives, $a < 2b \Rightarrow$ bacteria Y survives.

4.6 Applications and the Perron-Frobenius Theorem

65. $\mathbf{x}' = A\mathbf{x} + \mathbf{b} = \begin{bmatrix} x+y-30 \\ -x+y-10 \end{bmatrix}$. We wish to have $\mathbf{x}' = A\mathbf{u}$ where $\mathbf{u} = \begin{bmatrix} u \\ v \end{bmatrix} = \mathbf{x} + \begin{bmatrix} a \\ b \end{bmatrix}$, so we solve $\begin{bmatrix} x+y-30 \\ -x+y-10 \end{bmatrix} = \begin{bmatrix} (x+a)+(y+b) \\ -(x+a)+(y+b) \end{bmatrix}$ to get $a = -10$, $b = -20$.

Our new initial conditions are $u(0) = x(0) - 10 = 10$ and $v(0) = y(0) - 20 = 10$.

A has eigenvalues $1+i$ and $1-i$ corresponding to eigenvectors $\begin{bmatrix} 1 \\ i \end{bmatrix}$ and $\begin{bmatrix} 1 \\ -i \end{bmatrix}$.

So, our solution has the form $\mathbf{u}(t) = C_1 e^{(1+i)t} \begin{bmatrix} 1 \\ i \end{bmatrix} + C_2 e^{(1-i)t} \begin{bmatrix} 1 \\ -i \end{bmatrix}$.

From $\mathbf{u}(0) = \begin{bmatrix} 10 \\ 10 \end{bmatrix}$ we get $C_1 = 5 - 5i$ and $C_2 = 5 + 5i$.

So applying Euler's Formula and expanding, we get:

$$\mathbf{u}(t) = (5-5i)e^{(1+i)t}\begin{bmatrix} 1 \\ i \end{bmatrix} + (5+5i)e^{(1-i)t}\begin{bmatrix} 1 \\ -i \end{bmatrix}$$

$$= (5-5i)e^t(\cos t + i\sin t)\begin{bmatrix} 1 \\ i \end{bmatrix} + (5+5i)e^t(\cos(-t) + i\sin(-t))\begin{bmatrix} 1 \\ -i \end{bmatrix}$$

$$= \begin{bmatrix} 10e^t \cos t + 10e^t \sin t \\ 10e^t \cos t - 10e^t \sin t \end{bmatrix}.$$

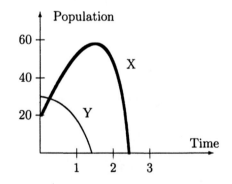

Rewriting in terms of the original variables, we get:

$$\mathbf{x}(t) = \mathbf{u}(t) - \begin{bmatrix} -10 \\ -20 \end{bmatrix}$$

$$= \begin{bmatrix} 10e^t \cos t + 10e^t \sin t + 10 \\ 10e^t \cos t - 10e^t \sin t + 20 \end{bmatrix}.$$

From the graph, we see that both populations die out.
Species Y after about 1.2 units of time and species X after about 2.4 units.

67. (a) Making the change of variables $y = x'$ and $z = x$, we have $x'' + ax' + bx = 0 \Leftrightarrow y' + ay + bz = 0$ and $y' + az' + bz = 0 \Leftrightarrow y' = -ay - bz$ and $z' = y$.

(b) The coefficient matrix is $\begin{bmatrix} -a & -b \\ 1 & 0 \end{bmatrix}$,

whose characteristic polynomial is $\begin{vmatrix} -a-\lambda & -b \\ 1 & -\lambda \end{vmatrix} = \lambda^2 + \lambda a + b$.

69. From Exercise 67, we know that the characteristic polynomial is $\lambda^2 - 5\lambda + 6$, so the eigenvalues are 2 and 3. The general solution is thus $x(t) = C_1 e^{2t} + C_2 e^{3t}$.

71. We know that $A = \begin{bmatrix} 1 & 3 \\ 2 & 2 \end{bmatrix}$ has eigenvalues $\lambda_1 = 4$ and $\lambda_2 = -1$

with corresponding eigenvectors $\mathbf{v}_1 = \begin{bmatrix} 1 \\ 1 \end{bmatrix}$ and $\mathbf{v}_2 = \begin{bmatrix} 3 \\ -2 \end{bmatrix}$.

So by Theorem 4.41, the system $\mathbf{x}' = A\mathbf{x}$ has solution $\mathbf{x} = e^{At}\mathbf{x}(0)$.

In this case $\mathbf{x}(0) = \begin{bmatrix} 0 \\ 5 \end{bmatrix}$ and we calculate

$$e^{At} = P(e^t)^D P^{-1} = \begin{bmatrix} 1 & 3 \\ 1 & -2 \end{bmatrix} \begin{bmatrix} e^{4t} & 0 \\ 0 & e^{-t} \end{bmatrix} \begin{bmatrix} 1 & 3 \\ 1 & -2 \end{bmatrix}^{-1} = \begin{bmatrix} \frac{2}{5}e^{4t} + \frac{3}{5}e^{-t} & \frac{3}{5}e^{4t} - \frac{3}{5}e^{-t} \\ \frac{2}{5}e^{4t} - \frac{2}{5}e^{-t} & \frac{3}{5}e^{4t} + \frac{2}{5}e^{-t} \end{bmatrix}.$$

So, the solution is $\begin{bmatrix} \frac{2}{5}e^{4t} + \frac{3}{5}e^{-t} & \frac{3}{5}e^{4t} - \frac{3}{5}e^{-t} \\ \frac{2}{5}e^{4t} - \frac{2}{5}e^{-t} & \frac{3}{5}e^{4t} + \frac{2}{5}e^{-t} \end{bmatrix} \begin{bmatrix} 0 \\ 5 \end{bmatrix} = \begin{bmatrix} 3e^{-4t} - 3e^{-t} \\ 3e^{-t} + 2e^{-t} \end{bmatrix}$ as before.

73. $A = \begin{bmatrix} 0 & 1 & -1 \\ 1 & 0 & 1 \\ 1 & 1 & 0 \end{bmatrix}$ has eigenvalues $\lambda_1 = 1$, $\lambda_2 = 0$, and $\lambda_3 = -1$

with corresponding eigenvectors $\mathbf{v}_1 = \begin{bmatrix} 0 \\ 1 \\ 1 \end{bmatrix}$, $\mathbf{v}_2 = \begin{bmatrix} 1 \\ -1 \\ -1 \end{bmatrix}$, and $\mathbf{v}_3 = \begin{bmatrix} 1 \\ -1 \\ 0 \end{bmatrix}$.

Thus, by Theorem 4.41, the solution is

$$\mathbf{x} = e^{At}\mathbf{x}(0) = \begin{bmatrix} 0 & 1 & 1 \\ 1 & -1 & -1 \\ 1 & -1 & 0 \end{bmatrix} \begin{bmatrix} e^t & 0 & 0 \\ 0 & 1 & 0 \\ 0 & 0 & e^{-t} \end{bmatrix} \begin{bmatrix} 0 & 1 & 1 \\ 1 & -1 & -1 \\ 1 & -1 & 0 \end{bmatrix}^{-1} \begin{bmatrix} 1 \\ 0 \\ -1 \end{bmatrix} = \begin{bmatrix} 2 - e^{-t} \\ e^t - 2 + e^{-t} \\ e^t - 2 \end{bmatrix}.$$

75. We follow the insights and methods outlined in Examples 4.48 through 4.51.

(a) Since $A = \begin{bmatrix} 2 & 1 \\ 0 & 3 \end{bmatrix}$, we have $\mathbf{x}_1 = A\mathbf{x}_0 = \begin{bmatrix} 3 \\ 3 \end{bmatrix}$, $\mathbf{x}_2 = A\mathbf{x}_1 = \begin{bmatrix} 9 \\ 9 \end{bmatrix}$, $\mathbf{x}_3 = A\mathbf{x}_2 = \begin{bmatrix} 27 \\ 27 \end{bmatrix}$.

(b) Since $A = \begin{bmatrix} 2 & 1 \\ 0 & 3 \end{bmatrix}$, we have $\mathbf{x}_1 = A\mathbf{x}_0 = \begin{bmatrix} 2 \\ 0 \end{bmatrix}$, $\mathbf{x}_2 = A\mathbf{x}_1 = \begin{bmatrix} 4 \\ 0 \end{bmatrix}$, $\mathbf{x}_3 = A\mathbf{x}_2 = \begin{bmatrix} 8 \\ 0 \end{bmatrix}$.

(c) Since $\lambda = 2, 3$, we see 0 is a repeller.

(d) Dashed lines are other trajectories.

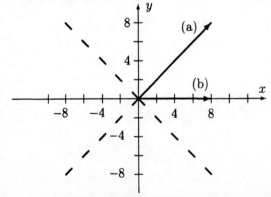

4.6 Applications and the Perron-Frobenius Theorem

77. We follow the insights and methods outlined in Examples 4.48 through 4.51.

(a) Since $A = \begin{bmatrix} 2 & -1 \\ -1 & 2 \end{bmatrix}$, we have $\mathbf{x}_1 = A\mathbf{x}_0 = \begin{bmatrix} 1 \\ 1 \end{bmatrix}$, $\mathbf{x}_2 = A\mathbf{x}_1 = \begin{bmatrix} 1 \\ 1 \end{bmatrix}$, $\mathbf{x}_3 = A\mathbf{x}_2 = \begin{bmatrix} 1 \\ 1 \end{bmatrix}$.

(b) $\mathbf{x}_1 = A\mathbf{x}_0 = \begin{bmatrix} 2 \\ -1 \end{bmatrix}$, $\mathbf{x}_2 = A\mathbf{x}_1 = \begin{bmatrix} 5 \\ -4 \end{bmatrix}$, $\mathbf{x}_3 = A\mathbf{x}_2 = \begin{bmatrix} 14 \\ -13 \end{bmatrix}$.

(c) Since $\lambda = 1, 3$, we see 0 is a saddle point.

(d) Sketches are left to the reader.

79. We follow the insights and methods outlined in Examples 4.48 through 4.51.

(a) $\mathbf{x}_1 = A\mathbf{x}_0 = \begin{bmatrix} 0.5 \\ -1.0 \end{bmatrix}$, $\mathbf{x}_2 = A\mathbf{x}_1 = \begin{bmatrix} 1.75 \\ -0.50 \end{bmatrix}$, $\mathbf{x}_3 = A\mathbf{x}_2 = \begin{bmatrix} 3.125 \\ -1.750 \end{bmatrix}$.

(b) $\mathbf{x}_1 = A\mathbf{x}_0 = \begin{bmatrix} 1.5 \\ -1.0 \end{bmatrix}$, $\mathbf{x}_2 = A\mathbf{x}_1 = \begin{bmatrix} 3.25 \\ -1.50 \end{bmatrix}$, $\mathbf{x}_3 = A\mathbf{x}_2 = \begin{bmatrix} 6.375 \\ -3.250 \end{bmatrix}$.

(c) Since $|\lambda| = 0.5, 2$, we see 0 is a saddle point.

(d) Sketches are left to the reader.

81. We follow the insights and methods outlined in Examples 4.48 through 4.51.

(a) $\mathbf{x}_1 = A\mathbf{x}_0 = \begin{bmatrix} 0.6 \\ 0.6 \end{bmatrix}$, $\mathbf{x}_2 = A\mathbf{x}_1 = \begin{bmatrix} 0.36 \\ 0.36 \end{bmatrix}$, $\mathbf{x}_3 = A\mathbf{x}_2 = \begin{bmatrix} 0.216 \\ 0.216 \end{bmatrix}$.

(b) $\mathbf{x}_1 = A\mathbf{x}_0 = \begin{bmatrix} 0.2 \\ -0.2 \end{bmatrix}$, $\mathbf{x}_2 = A\mathbf{x}_1 = \begin{bmatrix} -0.04 \\ -0.20 \end{bmatrix}$, $\mathbf{x}_3 = A\mathbf{x}_2 = \begin{bmatrix} -0.088 \\ -0.152 \end{bmatrix}$.

(c) Since $|\lambda| = 0.4, 0.6$, we see 0 is an attractor (every trajectory converges to 0).

(d) Sketches are left to the reader.

83. We follow the insights in the proof of Theorem 4.42 and the remarks following it.

Since $A = \begin{bmatrix} 1 & -1 \\ 1 & 1 \end{bmatrix}$ and $r = \sqrt{a^2 + b^2}$, the scaling factor $r = \sqrt{1^2 + 1^2} = \sqrt{2}$.

So, we factor A as $A = \begin{bmatrix} r & 0 \\ 0 & r \end{bmatrix} \begin{bmatrix} \cos\theta & -\sin\theta \\ \sin\theta & \cos\theta \end{bmatrix} = \begin{bmatrix} \sqrt{2} & 0 \\ 0 & \sqrt{2} \end{bmatrix} \begin{bmatrix} \frac{1}{\sqrt{2}} & -\frac{1}{\sqrt{2}} \\ \frac{1}{\sqrt{2}} & \frac{1}{\sqrt{2}} \end{bmatrix}$.

Since $\cos\theta = \frac{1}{\sqrt{2}}$ and $\sin\theta = \frac{1}{\sqrt{2}}$, the angle of rotation $\theta = 45°$.

Finally, since $r = |\lambda| = \sqrt{2} > 1$, the origin is a spiral repeller.

Plots and sketches are left to the reader.

85. We follow the insights in the proof of Theorem 4.42 and the remarks following it.

Since $A = \begin{bmatrix} 1 & \sqrt{3} \\ -\sqrt{3} & 1 \end{bmatrix}$ and $r = \sqrt{a^2 + b^2}$, the scaling factor $r = \sqrt{1^2 + (\sqrt{3})^2} = 2$.

So, we factor A as $A = \begin{bmatrix} r & 0 \\ 0 & r \end{bmatrix} \begin{bmatrix} \cos\theta & -\sin\theta \\ \sin\theta & \cos\theta \end{bmatrix} = \begin{bmatrix} 2 & 0 \\ 0 & 2 \end{bmatrix} \begin{bmatrix} \frac{1}{2} & \frac{\sqrt{3}}{2} \\ -\frac{\sqrt{3}}{2} & \frac{1}{2} \end{bmatrix}$.

Since $\cos\theta = \frac{1}{2}$ and $\sin\theta = -\frac{\sqrt{3}}{2}$, the angle of rotation $\theta = 300°$.

Finally, since $r = |\lambda| = 2 > 1$, the origin is a spiral repeller.

Plots and sketches are left to the reader.

87. We follow the insights in the proof of Theorem 4.43 and the remarks following it.

Since $A = \begin{bmatrix} 0.1 & -0.2 \\ 0.1 & 0.3 \end{bmatrix}$ has eigenvalues $\lambda = 0.2 \pm 0.1i$ with eigenvector $\mathbf{x} = \begin{bmatrix} -1-i \\ 1+0i \end{bmatrix}$.

So $P = [\text{Re } \mathbf{x} \ \text{Im } \mathbf{x}] = \begin{bmatrix} -1 & -1 \\ 1 & 0 \end{bmatrix}$.

Therefore $C = P^{-1}AP = \begin{bmatrix} 0 & 1 \\ -1 & -1 \end{bmatrix} \begin{bmatrix} 0.1 & -0.2 \\ 0.1 & 0.3 \end{bmatrix} \begin{bmatrix} -1 & -1 \\ 1 & 0 \end{bmatrix} = \begin{bmatrix} 0.2 & -0.1 \\ 0.1 & 0.2 \end{bmatrix}$.

Since $|\lambda| = 0.2 < 1$, the origin is a spiral attractor. Plots and sketches are left to the reader.

89. We follow the insights in the proof of Theorem 4.43 and the remarks following it.

Since $A = \begin{bmatrix} 1 & -1 \\ 1 & 0 \end{bmatrix}$ has eigenvalues $\lambda = \frac{1}{2} \pm \frac{\sqrt{3}}{2}i$ with eigenvector $\mathbf{x} = \begin{bmatrix} \frac{1}{2} - \frac{\sqrt{3}}{2}i \\ 1 + 0i \end{bmatrix}$.

So $P = [\text{Re } \mathbf{x} \ \text{Im } \mathbf{x}] = \begin{bmatrix} \frac{1}{2} & -\frac{\sqrt{3}}{2} \\ 1 & 0 \end{bmatrix}$.

Therefore $C = P^{-1}AP = \begin{bmatrix} 0 & 1 \\ -\frac{2}{\sqrt{3}} & \frac{1}{\sqrt{3}} \end{bmatrix} \begin{bmatrix} 1 & -1 \\ 1 & 0 \end{bmatrix} \begin{bmatrix} \frac{1}{2} & -\frac{\sqrt{3}}{2} \\ 1 & 0 \end{bmatrix} = \begin{bmatrix} \frac{1}{2} & -\frac{\sqrt{3}}{2} \\ \frac{\sqrt{3}}{2} & \frac{1}{2} \end{bmatrix}$.

Since $|\lambda| = 1$, the origin is an orbital center. Plots and sketches are left to the reader.

Chapter 4 Review

Key Definitions and Concepts

This list includes most but not all of the definitions listed at the end of each section. We begin the list with key symbols and symbol-based definitions.

λ	p.253	4.1	λ often used to denote an eigenvalue of A
E_λ	p.255	4.1	$E_\lambda = \text{null}(A - \lambda I) = \{\text{eigenvectors of } \lambda\} \cup \{\text{the zero vector, } \mathbf{0}\}$
$\sum_{i=1}^{n} a_{ij}C_{ij}$	p.263	4.2	$\sum_{j=1}^{n} a_{ij}C_{ij} = a_{i1}C_{i1} + a_{i2}C_{i2} + \cdots + a_{in}C_{in}$
$\prod_{i=1}^{n} A_i$	p.263	4.2	$\prod_{i=1}^{n} A_i = A_1 A_2 \cdots A_n$
			Shorthand for products like $\sum_{j=1}^{n} a_{ij}C_{ij}$ is for sums
A_{ij}	p.263	4.2	submatrix of A obtained by deleting row i and column j
$\det A_{ij}$	p.263	4.2	$\det A_{ij}$, the (i,j)-minor of A
$A_i(\mathbf{b})$	p.273	4.2	$A_i(\mathbf{b})$ obtained by replacing column i of A by $\mathbf{b}$
			That is, $A_i(\mathbf{b}) = \begin{bmatrix} \mathbf{a}_1 & \cdots & \mathbf{b} & \cdots & \mathbf{a}_n \end{bmatrix}$ (see Cramer's Rule)
$\text{adj } A$	p.275	4.2	$\text{adj } A = [C_{ij}]^T$, the adjoint of A
$\det A$	p.264	4.2	$\det A = \sum_{j=1}^{n} a_{1j}(-1)^{i+1} \det A_{1j}$ or $\det A = \sum_{j=1}^{n} a_{1j}C_{1j}$
C_{ij}	p.265	4.2	$C_{ij} = (-1)^{i+j} \det A_{ij}$, the (i,j)-cofactor of A
$c_A(\lambda)$	p.289	4.3	The *characteristic* polynomial is $c_A(\lambda) = \det(A - \lambda I)$. It is important to note that $c_A(\lambda)$ is a polynomial in λ. That is, λ is *not* a fixed eigenvalue, but a variable like x.
$\prod_{i=1}^{n}(\lambda_i - \lambda)$	p.289	4.3	$c_A(\lambda)$ using the eigenvalues of A, λ_i.
$\prod_{i=1}^{m}(\lambda_i - \lambda)^{k_i}$	p.289	4.3	$c_A(\lambda)$ emphasizing the algebraic multiplicity of λ_i, k_i.
$A \sim B$	p.298	4.4	$A \sim B$ (A is *similar* to B) if $P^{-1}AP = B$, P invertible
λ_1	p.308	4.5	*dominant eigenvalue*, $|\lambda_1| > |\lambda_i|$
$\mathbf{v}_1$	p.308	4.5	*dominant eigenvector*, associated with dominant eigenvalue

Key Definitions and Concepts, continued

adjoint	p.275	4.2	adj $A = [C_{ij}]^T$, the adjoint of A, the matrix of cofactors
algebraic multiplicity	p.291	4.3	The multiplicity of λ as a root of $\det(A - \lambda I) = 0$. That is, if $c_A(\lambda) = \prod_{i=1}^{m}(\lambda_i - \lambda)^{k_i}$ where λ_i are distinct, then the algebraic multiplicity of λ_i is k_i. That is, the algebraic multiplicity of an eigenvalue is equal to the exponent of its associated factor in the characteristic equation.
attractor	p.347	4.6	See definition on p.347 and Example 4.48.
characteristic equation	p.289	4.3	The *characteristic* equation is $\det(A - \lambda I) = 0$. The solutions of $\det(A - \lambda I) = 0$ are the *eigenvalues* of A.
characteristic polynomial	p.289	4.3	The *characteristic* polynomial is $c_A(\lambda) = \det(A - \lambda I)$. The solutions of $\det(A - \lambda I) = 0$ are the *eigenvalues* of A.
cofactor	p.265	4.2	$C_{ij} = (-1)^{i+j} \det A_{ij}$, the (i,j)-cofactor of A
conditions	p.333	4.6	*initial conditions*. See definition p. 333.
determinant	p.264	4.2	$\det A = \sum_{j=1}^{n} a_{1j}(-1)^{i+1} \det A_{1j}$ or $\det A = \sum_{j=1}^{n} a_{1j}C_{1j}$
diagonalizable	p.300	4.4	$A \sim D$ if $P^{-1}AP = D$ where D is diagonal
dominant eigenvalue	p.308	4.5	$|\lambda_1| > |\lambda_i|$, where λ_i are the *other* eigenvalues of A
dominant eigenvector	p.308	4.5	The eigenvector associated with $|\lambda_1| > |\lambda_i|$, that is, $A\mathbf{x}_1 = \lambda_1 \mathbf{x}_1$, where λ_1 is dominant
eigenvalue	p.253	4.1	if $A\mathbf{x} = \lambda\mathbf{x}$, then A has eigenvalue λ
eigenvalue	p.289	4.3	The solutions of $\det(A - \lambda I) = 0$.
eigenvector	p.253	4.1	if $A\mathbf{x} = \lambda\mathbf{x}$, then $\mathbf{x}$ is an eigenvector of A
eigenspace	p.255	4.1	$E_\lambda = \text{null}(A - \lambda I) = \{\text{eigenvectors of } \lambda\} \cup \{\text{the zero vector, } \mathbf{0}\}$. That is all vectors such that $(A - \lambda I)\mathbf{x} = \mathbf{0}$ or $A\mathbf{x} = \lambda\mathbf{x}$
expansion	p.265	4.2	$\det A = \sum_{j=1}^{n} a_{ij}C_{ij}$ along the ith row $\det A = \sum_{i=1}^{n} a_{ij}C_{ij}$ along the jth column

Chapter 4 Review

Key Definitions and Concepts, continued

geometric multiplicity	p.291	4.3	The dimension of the eigenspace associated with λ, $\dim E_\lambda$, where $\dim E_\lambda$ is the number of vectors in a basis. That is, *geometric multiplicity* $= \dim E_\lambda$.
Gerschgorin disk	p.316	4.5	$D_i = \{z \text{ in } \mathbb{C} : |z - a_{ii}| \leq r_i\}$ where $r_i = \sum_{j \neq i} |a_{ij}|$
minor	p.263	4.2	$\det A_{ij}$, the (i,j)-minor of A
Power Method	p.312	4.5	This two-step iterative method is described in detail on p. 312 and illustrated in Example 4.31
Inverse	p.314	4.5	Inverse Power Method: See Example 4.33
Shifted	p.313	4.5	Shifted Power Method: See Example 4.32
Shifted Inverse	p.315	4.5	Shifted Inverse Power Method: See Example 4.34
recurrence	p.333	4.6	*linear recurrence.* See definition p. 333.
repeller	p.349	4.6	See definition on p.349 and Example 4.50.
saddle point	p.349	4.6	See definition on p.349 and Example 4.50.
similar	p.298	4.4	$A \sim B$ if $P^{-1}AP = B$ where P is invertible
submatrix	p.263	4.2	submatrix of A obtained by deleting row i and column j
trajectory	p.346	4.6	See definition on p.346 prior to Example 4.48.

Theorems

Cramer	p.274	4.2	If $Ax = b$, then $x_i = \frac{\det(A_i(b))}{\det A}$ (see Thm 4.11)
Laplace	p.265	4.2	$\det A = \sum_{j=1}^{n} a_{ij}C_{ij}$ (any row) or $\sum_{i=1}^{n} a_{ij}C_{ij}$ (any column)
Thm 4.1	p.265	4.2	$\det A = \sum_{j=1}^{n} a_{ij}C_{ij}$ (any row) or $\sum_{i=1}^{n} a_{ij}C_{ij}$ (any column)
Thm 4.2	p.268	4.2	If A is triangular, then $\det A = a_{11}a_{22}\cdots a_{nn}$
Thm 4.3	p.268	4.2	a. through f. detail row (column) operations effects on $\det A$ a. $\mathbf{A}_i = \mathbf{0} \Rightarrow \det A = 0$ b. $A \xrightarrow{R_i \leftrightarrow R_j} B \Rightarrow \det B = -\det A$ c. $\mathbf{A}_i = \mathbf{A}_j \Rightarrow \det A = 0$ d. $A \xrightarrow{kR_i} B \Rightarrow \det B = k\det A$ e. $\mathbf{C}_i = \mathbf{A}_i + \mathbf{B}_i \Rightarrow \det C = \det A + \det B$ f. $A \xrightarrow{R_i + kR_j} B \Rightarrow \det B = \det A$
Thm 4.4	p.270	4.2	a. through c. detail row (column) operations effects on $\det E$ a. $I \xrightarrow{R_i \leftrightarrow R_j} E \Rightarrow \det E = -1$ b. $I \xrightarrow{kR_i} E \Rightarrow \det E = k$ c. $I \xrightarrow{R_i + kR_j} E \Rightarrow \det E = 1$
Lem 4.5	p.271	4.2	E, elementary, then $\det(EB) = (\det E)(\det B)$
Thm 4.6	p.271	4.2	A is invertible if and only if $\det A \neq 0$
Thm 4.7	p.271	4.2	$\det(kA) = k^n \det A$
Thm 4.8	p.272	4.2	$\det(AB) = (\det A)(\det B)$
Thm 4.9	p.273	4.2	$\det(A^{-1}) = \frac{1}{\det A}$
Thm 4.10	p.273	4.2	$\det A = \det(A^T)$
Thm 4.11	p.274	4.2	If $Ax = b$, then $x_i = \frac{\det(A_i(b))}{\det A}$ (Cramer's Rule)
Thm 4.12	p.276	4.2	$A^{-1} = \frac{1}{\det A}\operatorname{adj} A$
Lem 4.13	p.276	4.2	(row 1) $\det A = \sum_{j=1}^{n} a_{1j}C_{1j} = \sum_{i=1}^{n} a_{i1}C_{i1}$ (column 1)
Lem 4.14	p.276	4.2	$A \xrightarrow{R_i \leftrightarrow R_j} B \Rightarrow \det B = -\det A$ (see Thm 4.3(f))

Theorems, continued

Thm 4.15	p.292	4.3	If A is triangular, then its eigenvalues are its diagonal entries.
Thm 4.16	p.292	4.3	A is invertible if and only if all $\lambda \neq 0$.
Thm 4.17	p.293	4.3	a. through o. give equivalent conditions for A to be invertible.
Thm 4.18	p.293	4.3	a. through c. relate eigenvalues, A^n and A^{-1}: For any integer n if $A\mathbf{x} = \lambda\mathbf{x}$, then $A^n\mathbf{x} = \lambda^n\mathbf{x}$.
Thm 4.19	p.294	4.3	If $\mathbf{x} \in \text{span}(\mathbf{v}_i)$, then $A^k\mathbf{x} = c_1\lambda_1^k\mathbf{v}_1 + c_2\lambda_2^k\mathbf{v}_2 + \cdots + c_m\lambda_m^k\mathbf{v}_m$ where $\mathbf{v}_i$ is the eigenvector corresponding to λ_i.
Thm 4.20	p.294	4.3	If λ_i are distinct, then $\{\mathbf{v}_i\}$ are linearly independent.
Thm 4.21	p.291	4.4	a. $A \sim A$, b. $A \sim B \Rightarrow B \sim A$, c. $A \sim B, B \sim C \Rightarrow A \sim C$
Thm 4.22	p.299	4.4	parts a. through e. list implications of $A \sim B$ a. $\det A = \det B$ b. A is invertible if and only if B is invertible c. $\text{rank}(A) = \text{rank}(B)$ d. $\det(A - \lambda I) = \det(B - \lambda I)$ e. λ is an eigenvalue of A if and only if λ is an eigenvalue of B
Thm 4.23	p.300	4.4	$P^{-1}AP = D \Leftrightarrow [D]_{ii} = \lambda_i$ and $\mathbf{p}_i = \mathbf{v}_i$ (the eigenvectors)
Thm 4.24	p.302	4.4	If λ_i are distinct, their basis vectors are linearly independent.
Thm 4.25	p.303	4.4	If all the λ_i are distinct, then A is diagonalizable.
Lem 4.26	p.303	4.4	geometric multiplicity $\leq$ algebraic multiplicity: That is, $\dim(E_{\lambda_i}) \leq k_i$ where $c_A(\lambda) = \prod_{i=1}^{m}(\lambda_i - \lambda)^{k_i}$.
Thm 4.27	p.304	4.4	parts a. through c. list equivalences to $A = P^{-1}DP$ a. A is diagonalizable, $A = P^{-1}DP$ b. bases of the eigenvectors contain n vectors c. geometric multiplicity = algebraic multiplicity, $\dim(E_{\lambda_i}) = k_i$
Thm 4.28	p.309	4.5	If A is diagonalizable with dominant eigenvalue λ_1, then $\mathbf{x}_1 = A\mathbf{x}_0$, $\mathbf{x}_2 = A\mathbf{x}_1$, $\mathbf{x}_k = A\mathbf{x}_{k-1}$ where the $\mathbf{x}_k$ are approaching the dominant eigenvector of A
Thm 4.29	p.318	4.5	Every eigenvalue of A is contained in a Gerschgorin Disk

Theorems, continued

Perron's	p.330	4.6	The Perron's Theorem. See Theorem 4.36.
Perron-Fro	p.332	4.6	The Perron-Frobenius Theorem. See Theorem 4.37.
Thm 4.30	p.322	4.6	If P is the $n \times n$ transition matrix of a Markov chain, then 1 is an eigenvalue of P
Thm 4.31	p.322	4.6	Let P be an $n \times n$ transition matrix with eigenvalue λ. a. $\|\lambda\| \leq 1$ b. If P is regular and $\lambda \neq 1$, then $\|\lambda\| < 1$.
Lem 4.32	p.324	4.6	Let P be an $n \times n$ transition matrix. If P is diagonalizable, then the dominant eigenvalue $\lambda_1 = 1$ has algebraic multiplicity 1.
Thm 4.33	p.325	4.6	Let P be an $n \times n$ transition matrix. Then as $k \longrightarrow \infty$, P^k approaches an $n \times n$ matrix L whose columns are **x**, the steady state probability vector for P.
Thm 4.34	p.326	4.6	Let P be an $n \times n$ transition matrix ... as in Theorem 4.33. Then for any probability vector $\mathbf{x}_0$, $\mathbf{x}_k$ approaches **x**.
Thm 4.35	p.328	4.6	Every Leslie matrix has a unique positive eigenvalue and a corresponding eigenvector with positive components.
Thm 4.36	p.330	4.6	Let A be a positive $n \times n$ matrix. A has a real eigenvalue λ_1: a. $\lambda_1 > 0$ b. λ_1 has a corresponding positive eigenvector. c. If λ is any other eigenvalue of A, then $\|\lambda\| \leq \lambda_1$.
Thm 4.37	p.332	4.6	Let A be an irreducible nonnegative $n \times n$ matrix. Then: a. $\lambda_1 > 0$ b. λ_1 has a corresponding positive eigenvector. c. If λ is any other eigenvalue of A, then $\|\lambda\| \leq \lambda_1$. If A is primitive, then this inequality is strict. d. If λ is an eigenvalue of A such that $\|\lambda\| = \lambda_1$, then λ is a (complex) root of the equation $\lambda^n - \lambda_1^n = 0$. e. λ_1 has algebraic multiplicity 1.
Thm 4.38	p.336	4.6	See statement of Theorem on p. 336.
Thm 4.39	p.337	4.6	See statement of Theorem on p. 337.
Thm 4.40	p.339	4.6	See statement of Theorem on p. 339.
Thm 4.41	p.344	4.6	See statement of Theorem on p. 344.
Thm 4.42	p.349	4.6	See statement of Theorem on p. 349.

Chapter 4 Review

Solutions to odd-numbered exercises from Chapter 4 *Review*

1. We will explain and give counter examples to justify our answers below.

   (a) **False.** See Theorem 4.7 of Section 4.2.
   Not quite accurate. Theorem 4.7 of Section 4.2 states $\det(kA) = k^n A$.
   So letting $k = -1$, we see $\det(-A) = (-1)^n \det A$, where A is $n \times n$.

   (b) **True.** See Theorem 4.8 in Section 4.2.
   This is precisely the statement of Theorem 4.8 in Section 4.2.

   (c) **False.** See Theorem 4.3(b) in Section 4.2.
   Not necessarily. Theorem 4.3(b) states: $A \xrightarrow{R_i \leftrightarrow R_j} B \Rightarrow \det B = -\det A$.
   So if B is obtained from A by one interchange of two columns, then $\det B = -\det A$.
   However, if we perform a second interchange, we get $\det B = -(-\det A) = \det A$.
   So, if $A \longrightarrow B$ by an *odd* number of column exchanges, then $\det B = -\det A$.
   But, if $A \longrightarrow B$ by an *even* number of column exchanges, then $\det B = \det A$.

   (d) **False.** See Theorems 4.9 and 4.10 in Section 4.2.
   Theorem 4.9 asserts $\det(A^{-1}) = \frac{1}{\det A}$ while Theorem 4.10 asserts $\det A = \det(A^T)$.
   Therefore, $\det(A^{-1}) = \frac{1}{\det(A^T)} = [\det(A^T)]^{-1}$.

   (e) **False.** Consider the counter example, $\begin{bmatrix} 0 & 1 \\ 0 & 0 \end{bmatrix}$.
   Q: Why is it obvious that 0 is the only eigenvalue of this matrix?
   Q: What is the general pattern underlying this specific counterexample?
   Q: If all the eigenvalues of a matrix A are 0, what can we conclude about $\det A$?

   (f) **False.** See Exercise 7 in Section 4.3.
   A simple but constructive counter example is the matrix I.
   Since $I\mathbf{e}_i = \mathbf{e}_i$, all nonzero vectors correspond to the eigenvalue $\lambda = 1$.
   That is, $E_1 = \mathbb{R}^n$ for I, the identity matrix.
   So, there are n linearly independent vectors corresponding to the eigenvalue $\lambda = 1$.

   (g) **True.** See Theorem 4.25 in Section 4.4.

   (h) **False.** A simple counterexample is the identity matrix, I.
   Is every diagonal matrix diagonalizable? Why or why not?

   (i) **False.** See Theorem 4.22 and Exercise 33 in Section 4.4.
   Theorem 4.22(e) states λ is an eigenvalue of A if and only if λ is an eigenvalue of B.
   So if A is similar to B, then they have the same eigen*values* not eigen*vectors*.
   In Exercise 33, we see how the eigenvectors are related given $A = PBP^{-1}$.
   If $A\mathbf{x} = \alpha\mathbf{x}$, $\mathbf{x} \neq \mathbf{0}$, then $(PBP^{-1})\mathbf{x} = \alpha\mathbf{x}$.
   So, left multiplying both sides by P^{-1} implies $B(P^{-1}\mathbf{x}) = P^{-1}(\alpha\mathbf{x}) = \alpha(P^{-1}\mathbf{x})$.
   That is, not only is α an eigenvalue for B, one associated eigenvector of α is $P^{-1}\mathbf{x}$.

   (j) **False.** An instructive counter example is all invertible matrices.
   If A is invertible, then $A \longrightarrow I$.
   But $A \sim I$ implies the only eigenvalue of A is $\lambda = 1$. Why?
   We would be forced to conclude the only eigenvalue of any invertible matrix is 1.
   That conclusion is clearly nonsense.

3. See Exercises 35 through 40 in Section 4.2.

   Let A be the first matrix given at the beginning of this exercise with $\det A = 3$.
   Let E be the second matrix given in this Exercise which is derived from A in 4 steps.
   $$A \xrightarrow{R_1 \leftrightarrow R_2} B \xrightarrow{3C_1} C \xrightarrow{2C_2} D \xrightarrow{C_2 - 4C_3} E$$
   So, $\det E = \det D = 2(\det C) = 2((3\det B)) = 2((3(-\det A))) = -6\det A = -18$.

   We show the steps and matrices created below:
   $$A = \begin{bmatrix} a & b & c \\ d & e & f \\ g & h & i \end{bmatrix} \xrightarrow{R_1 \leftrightarrow R_2} \begin{bmatrix} d & e & f \\ a & b & c \\ g & h & i \end{bmatrix} \xrightarrow{3C_1} \begin{bmatrix} 3d & e & f \\ 3a & b & c \\ 3g & h & i \end{bmatrix} \xrightarrow{2C_2} \begin{bmatrix} 3d & 2e & f \\ 3a & 2b & c \\ 3g & 2h & i \end{bmatrix}$$
   $$\xrightarrow{C_2 - 4C_3} \begin{bmatrix} 3d & 2e-4f & f \\ 3a & 2b-4c & c \\ 3g & 2h-4i & i \end{bmatrix} = E$$

5. See Exercises 37 to 43 in Section 3.2 including the definition of skew-symmetric, $A^T = -A$.

   Since $A^T = -A$, $\det A = \det(A^T) = \det(-A) = (-1)^n \det A$. So, $\det A = (-1)^n \det A$.
   So when n is odd, $(-1)^n = -1$ which implies $\det A = -1 \det A$.
   Therefore when n is odd, $2\det A = 0$ which implies $\det A = 0$.

7. See Exercises 1 through 5 and Example 4.1 in Section 4.1.

   If $A\mathbf{x} = \lambda\mathbf{x}$, then $\mathbf{x}$ is an eigenvector of A corresponding to λ.

   So, as in Example 4.1, since $A\mathbf{v} = \begin{bmatrix} 3 & 1 \\ 4 & 3 \end{bmatrix} \begin{bmatrix} 1 \\ 2 \end{bmatrix} = \begin{bmatrix} 5 \\ 10 \end{bmatrix} = 5\begin{bmatrix} 1 \\ 2 \end{bmatrix} = 5\mathbf{v}$,

   we see $\mathbf{v}$ is an eigenvector of A corresponding to (the eigenvalue) 5.

Chapter 4 Review

9. See Examples 4.18, 4.19 and Exercises 1 through 12 in Section 4.3.
   See the box outlining this procedure in Section 4.3 on p.289.

   (a) See the definition of the characteristic polynomial given in Section 4.3 on p.289.
   Recall the characteristic polynomial of A is defined as $c_A(\lambda) = \det(A - \lambda I)$.
   We expand along the third row because that row contains two zeroes.
   $$c_A(\lambda) = \det(A - \lambda I) = \begin{vmatrix} -5-\lambda & -6 & 3 \\ 3 & 4-\lambda & -3 \\ 0 & 0 & -2-\lambda \end{vmatrix}$$
   $$= (-2-\lambda)\begin{vmatrix} -5-\lambda & -6 \\ 3 & 4-\lambda \end{vmatrix} = (-2-\lambda)(\lambda^2 + \lambda - 2) = -(\lambda+2)^2(\lambda-1)$$

   (b) See the definition of eigenvalues given in Section 4.3 on p.289.
   Recall the eigenvalues of A are the solutions of $\det(A - \lambda I) = 0$.
   We expand along the third row because that row contains two zeroes.
   $\det(A - \lambda I) = -(\lambda+2)^2(\lambda-1) = 0$ which implies the eigenvalues of A are $\lambda = -2, 1$.

   (c) The eigenspace $E_\lambda = \text{null}(A - \lambda I)$, so $E_{-2} = \text{null}(A + 2I)$ and $E_1 = \text{null}(A - I)$.
   $$[\,A+2I\,|\,0\,] = \begin{bmatrix} -3 & -6 & 3 & | & 0 \\ 3 & 6 & -3 & | & 0 \\ 0 & 0 & 0 & | & 0 \end{bmatrix} \xrightarrow{R_2+R_1} \begin{bmatrix} -3 & -6 & 3 & | & 0 \\ 0 & 0 & 0 & | & 0 \\ 0 & 0 & 0 & | & 0 \end{bmatrix} \xrightarrow{\frac{1}{3}R_1} \begin{bmatrix} -1 & -2 & 1 & | & 0 \\ 0 & 0 & 0 & | & 0 \\ 0 & 0 & 0 & | & 0 \end{bmatrix}$$
   So $x_1 = -2x_2 + x_3$, from which it follows that $E_{-2} = \text{span}\left(\begin{bmatrix} 2 \\ -1 \\ 0 \end{bmatrix}, \begin{bmatrix} 1 \\ 0 \\ 1 \end{bmatrix}\right)$.

   $$[\,A-I\,|\,0\,] = \begin{bmatrix} -6 & -6 & 3 & | & 0 \\ 3 & 3 & -3 & | & 0 \\ 0 & 0 & -3 & | & 0 \end{bmatrix} \xrightarrow{2R_3} \begin{bmatrix} -6 & -6 & 3 & | & 0 \\ 6 & 6 & -6 & | & 0 \\ 0 & 0 & -3 & | & 0 \end{bmatrix} \xrightarrow{R_2+R_1} \begin{bmatrix} -6 & -6 & 3 & | & 0 \\ 0 & 0 & -3 & | & 0 \\ 0 & 0 & -3 & | & 0 \end{bmatrix}$$
   So $x_3 = 0$ which implies $x_1 = -x_2$, from which it follows that $E_1 = \text{span}\left(\begin{bmatrix} 1 \\ -1 \\ 0 \end{bmatrix}\right)$.

11. This is a straightforward application of Theorem 4.19 on p.294 in Section 4.3.
   Since $\begin{bmatrix} 3 \\ 7 \end{bmatrix} = 5\begin{bmatrix} 1 \\ 1 \end{bmatrix} - 2\begin{bmatrix} 1 \\ -1 \end{bmatrix}$, $A^{-5}\begin{bmatrix} 3 \\ 7 \end{bmatrix} = 5\left(\frac{1}{2}\right)^{-5}\begin{bmatrix} 1 \\ 1 \end{bmatrix} - 2(-1)^{-5}\begin{bmatrix} 1 \\ -1 \end{bmatrix} = \begin{bmatrix} 162 \\ 158 \end{bmatrix}$.

13. Since the characteristic polynomial of A is not equal to the characteristic polynomial of B, A is not similar to B ($A \nsim B$). That is:
    Since $c_A(\lambda) = (4-\lambda)(1-\lambda) - 6 \neq (2-\lambda)(2-\lambda) - 6 = c_B(\lambda)$, we have $A \nsim B$.

15. A and B are *not* similar. Why?
    This is an example of a pair of matrices that share the characteristics
    of Theorem 4.22 in Section 4.4 but are still *not* similar.
    See the remarks on p.300 in Section 4.4.

17. If **x** is an eigenvector of A, $A^3\mathbf{x} = \lambda^3\mathbf{x}$. What theorem asserts this?

Since $A^3 = A$, we have $\lambda\mathbf{x} = A\mathbf{x} = A^3\mathbf{x} = \lambda^3\mathbf{x}$.

So, $\lambda\mathbf{x} = \lambda^3\mathbf{x}$. Since $\mathbf{x} \neq \mathbf{0}$, we have:
$\lambda = \lambda^3 \Rightarrow \lambda^3 - \lambda = 0 \Rightarrow \lambda(\lambda^2 - 1) = 0 \Rightarrow \lambda(\lambda - 1)(\lambda + 1) = 0 \Rightarrow \lambda = -1, 0, 1$.

19. Given **x** is an eigenvector and $B = A^2 - 5A + 2$, we compute $A^2\mathbf{x} - 5A\mathbf{x} + 2\mathbf{x}$ directly.

We get: $B\mathbf{x} = A^2\mathbf{x} - 5A\mathbf{x} + 2\mathbf{x} = 3^2\mathbf{x} - 5 \cdot 3\mathbf{x} + 2\mathbf{x} = -4\mathbf{x}$.

This shows that **x** is an eigenvector of B with eigenvalue $\lambda = -4$. Why?

Chapter 5

Orthogonality

5.1 Orthogonality in $\mathbb{R}^n$

Q: What are our main goals for Section 5.1?
A: To take our first look at *orthornormal* bases and the closely-related *orthogonal* matrices.

Key Definitions and Concepts

$\mathcal{B}$	p.369	5.1	$\mathcal{B}$ often used to denote a set of basis vectors
$\mathbf{q}_i$	p.369	5.1	$\mathbf{q}_i$ often used to denote a vector in an *orthonormal* basis
Q	p.371	5.1	Q often used to denote an *orthogonal* matrix

orthogonal   5.1   This adjective is used for sets, bases, and matrices:
- set   p.366 $\mathbf{v}_i \cdot \mathbf{v}_j = 0$ when $i \neq j$
- basis   p.367   a basis $\mathcal{B}$ that is an orthogonal set
- matrix   p.371   a *square* matrix whose columns form an orthonormal set
  That is, Q where columns $\{\mathbf{q}_i\}$ are orthonormal
  Matrix is ortho*gonal* when columns are ortho*normal*

orthonormal   5.1   This adjective is used for sets and bases:
     p.369   For a set to be orthonormal it must have two key properties:
$$\mathbf{q}_i \cdot \mathbf{q}_j = \begin{cases} 0 & \text{if } i \neq j \quad \text{Property 1} \\ 1 & \text{if } i = j \quad \text{Property 2} \end{cases}$$
- set   p.369   an orthogonal set of *unit* vectors
- basis   p.369   a basis $\mathcal{B}$ that is an orthonormal set

Theorems

Thm 5.1   p.366   5.1 $\mathcal{B} = \{\mathbf{v}_1, \mathbf{v}_2, \ldots, \mathbf{v}_n\}$ orthogonal $\Rightarrow \mathcal{B}$ is linearly independent

Thm 5.2   p.368   5.1   If $\mathcal{B} = \{\mathbf{v}_1, \mathbf{v}_2, \ldots, \mathbf{v}_n\}$ orthogonal and $\mathbf{w} = \sum_{i=1}^{n} c_i \mathbf{v}_i$, then $c_i = \frac{\mathbf{w} \cdot \mathbf{v}_i}{\mathbf{v}_i \cdot \mathbf{v}_i}$.

Thm 5.3   p.370   5.1 $\mathbf{w} = \sum_{i=1}^{n} (\mathbf{w} \cdot \mathbf{q}_1) \mathbf{q}_i$ (this representation is unique)

Thm 5.4   p.371   5.1 $Q^T Q = I$ if and only if $\{\mathbf{q}_i\}$ is an orthonormal set.

Thm 5.5   p.371   5.1 Q is orthogonal if and only if $Q^{-1} = Q^T$

Thm 5.6   p.372   5.1   a. through c. give equivalent conditions for Q to be orthogonal:
     a. Q is orthogonal
     b. $\|Q\mathbf{x}\| = \|\mathbf{x}\|$ for every $\mathbf{x}$ in $\mathbb{R}^n$
     c. $Q\mathbf{x} \cdot Q\mathbf{x} = \mathbf{x} \cdot \mathbf{y}$ for every $\mathbf{x}, \mathbf{y}$ in $\mathbb{R}^n$

Thm 5.7   p.373   5.1 Q is orthogonal, then its *rows* form an orthonormal set.

Thm 5.8   p.372   5.1   a. through d. list implications of Q being orthogonal:
     a. Q^{-1} is orthogonal
     b. $\det Q = \pm 1$
     c. If λ is an eigenvalue of Q, then $|\lambda| = 1$.
     d. If Q_1 and Q_2 are orthogonal $n \times n$ matrices, then so is $Q_1 Q_2$.

Discussion of Key Definitions and Concepts

Orthogonal sets versus Orthonormal sets

Q: What does it mean to say a set of vectors is *orthogonal*?
A: $\mathbf{v}_i \cdot \mathbf{v}_j = 0$ when $i \neq j$.

Q: Are we familiar with a set of vectors that is *orthogonal*?
A: Yes, the standard basis vectors, $\mathcal{E} = \{\mathbf{e}_i\}$.
The realization that $\mathcal{E}$ has many desirable properties motivates this definition.

Q: Using $\mathcal{E}$, can we easily we construct other *orthogonal* sets?
A: Yes, any set $\mathcal{E}_\alpha = \{\alpha_i \mathbf{e}_i\}$ is *orthogonal* provided $\alpha_i \neq 0$.

Q: How might we describe this condition in words?
A: Every vector in the set is orthogonal to every other vector in the set.
Using words helps clarify the strength of this condition.

Q: Geometrically, why should we expect an orthogonal set to be linearly independent?
A: In $\mathbb{R}^2$ and $\mathbb{R}^3$, vectors parallel to the axes are orthogonal sets.
Pairwise, we realize that being orthogonal is a very special way of *not* being parallel.
In particular, orthogonal vectors form a right angle between them. A right angle
is the only angle at which the projection of one vector onto another is the zero vector.
In this sense, the directions of the two vectors are completely independent.

Q: What does it mean to say a set of vectors is *orthonormal*?
A: $\mathbf{v}_i \cdot \mathbf{v}_j = 0$ when $i \neq j$ and $\mathbf{v}_i \cdot \mathbf{v}_i = 1$.

Q: How might we describe this additional condition, $\mathbf{v}_i \cdot \mathbf{v}_i = 1$, in words?
A: Every vector in the set has length 1.

Q: Why is this set referred to as ortho*normal*?
A: Normalizing a vector means adjusting its length to 1.
That is, finding a vector in the same direction of the given vector but of length 1.
See the definition of *normalizing* on p. 18 in Section 1.2.

Q: Using $\mathcal{E}$, can we easily we construct lots of other *orthonormal* sets?
A: No. Why not? The additional condition of all lengths being 1 makes $\mathcal{E}$ almost unique.
That is, $\mathcal{E}_\alpha = \{\alpha_i \mathbf{e}_i\}$ is *not* orthonormal when $|\alpha_i| \neq 1$.

Q: Using $\mathcal{E}$, what are the only other *orthonormal* sets we can construct?
A: $\mathcal{E}_\alpha = \{\alpha_i \mathbf{e}_i\}$ is *orthonormal* when $|\alpha_i| = 1$.
For example, $\mathcal{E}_{-1} = \{-\mathbf{e}_i\}$. Verify that all such sets are orthonormal.

Q: Why are we interested in generalizing $\{\mathbf{e}_i\}$ by identifying two of its key properties?
A: Given these two properties, we want to see how many of the familiar implications hold.
This broadens our understanding of what is key and gives us a greater freedom
in choosing bases that work well in a given situation or application.

Q: Why would we want a basis $\mathcal{B}$ to be not just an orthogonal set but an *orthonormal* set?
A: Hint: Consider the coordinate vector $[\mathbf{x}]_\mathcal{B}$ when $\mathcal{B}$ is *not* orthonormal.

5.1 Orthogonality in $\mathbb{R}^n$

Discussion of Key Definitions and Concepts, continued

Coordinate vectors over an orthonormal basis

Q: Given $\mathcal{B} = \mathbf{b}_1 = \begin{bmatrix} 2 \\ 0 \end{bmatrix}$, and $\mathbf{b}_2 = \begin{bmatrix} 0 \\ 2 \end{bmatrix}$, find $[\mathbf{x}]_\mathcal{B}$ when $\mathbf{x} = \begin{bmatrix} 2 \\ 2 \end{bmatrix}$.

How do the components of this coordinate vector, $[\mathbf{x}]_\mathcal{B}$, relate to $\mathbf{x} \cdot \mathbf{b}_i$?

A: Since $\mathbf{x} = 1\mathbf{b}_1 + 1\mathbf{b}_2$, $[\mathbf{x}]_\mathcal{B} = \begin{bmatrix} 1 \\ 1 \end{bmatrix}$ but $\mathbf{x} \cdot \mathbf{b}_1 = 4$ and $\mathbf{x} \cdot \mathbf{b}_2 = 4$.

The i-th component of $[\mathbf{x}]_\mathcal{B}$ does *not* equal the dot product with the i-th basis vector.

Q: Given $\mathcal{E} = \mathbf{e}_1 = \begin{bmatrix} 1 \\ 0 \end{bmatrix}$, and $\mathbf{e}_2 = \begin{bmatrix} 0 \\ 1 \end{bmatrix}$, find $[\mathbf{x}]_\mathcal{B}$ when $\mathbf{x} = \begin{bmatrix} 1 \\ 1 \end{bmatrix}$.

How do the components of this coordinate vector, $[\mathbf{x}]_\mathcal{B}$, relate to $\mathbf{x} \cdot \mathbf{b}_i$?

A: Since $\mathbf{x} = 1\mathbf{e}_1 + 1\mathbf{e}_2$, $[\mathbf{x}]_\mathcal{E} = \begin{bmatrix} 1 \\ 1 \end{bmatrix}$ and $\mathbf{x} \cdot \mathbf{e}_1 = 1$ and $\mathbf{x} \cdot \mathbf{e}_2 = 1$.

The i-th component of $[\mathbf{x}]_\mathcal{E}$ *does* equal the dot product with the i-th basis vector.

Q: Given $\mathcal{B}$ is an orthonormal basis, consider the following question:
Does the i-th component of $[\mathbf{x}]_\mathcal{B}$ equal the dot product with the i-th basis vector?
That is, does $[x_i]_\mathcal{B} = \mathbf{x} \cdot \mathbf{b}_i$? If this is true, prove it.

A: Yes, this is one of the many properties that makes orthonormal bases desirable.
This is one the key implications of Theorems 5.2 and 5.3 in this Section.
For another desirable property, see *Parseval's Identity* in Exercise 37 below.

Q: How might we state the above observation using symbols?

A: If $\mathcal{B} = \{\mathbf{b}_i\}$ is an orthonormal basis and $\mathbf{x} = \sum_{k=1}^{n} \beta_k \mathbf{b}_k$, then $\mathbf{x} \cdot \mathbf{b}_i = \beta_i$.

By definition, β_i is the i-th component in the coordinate vector $[\mathbf{x}]_\mathcal{B}$. Why?

Q: How might we restate these observations using all the symbols referenced above?

A: If $\mathcal{B} = \{\mathbf{b}_i\}$ is an orthonormal basis, then $\mathbf{x} = \sum_{k=1}^{n} ([x_k]_\mathcal{B}) \mathbf{b}_k = \sum_{k=1}^{n} (\mathbf{x} \cdot \mathbf{b}_k) \mathbf{b}_k$.

Discussion of Key Definitions and Concepts, continued

Orthonormal columns imply an Orthogonal matrix

Q: If the columns $\{\mathbf{a}_i\}$ of A are *orthogonal*, is A *orthogonal*?
A: Unfortunately, no. What is required is the following:
   If the columns $\{\mathbf{a}_i\}$ of A are ortho*normal*, then A is *orthogonal*.

Q: If the rows $\{\mathbf{A}_i\}$ of A are ortho*normal*, is A is *orthogonal*?
A: Yes, this is precisely the assertion of Theorem 5.7.

Q: Why is it obvious that $A = \begin{bmatrix} 2 & 0 \\ 0 & 2 \end{bmatrix}$ is *not* orthogonal?

A: Because the columns of A are *orthogonal* but *not* ortho*normal*.

Q: Why is it obvious that the columns of A are not ortho*normal*?
A: A vector has been *normalized* if and only if its length is 1.
   The length of each column of A as a vector is obviously 2, not 1.

Q: Why is it obvious that the columns of A *are* orthogonal?
A: Because $\mathbf{a}_1 \cdot \mathbf{a}_2 = (2)(0) + (0)(2) = 0$.

Q: But, is there an easy way to create a orthogonal matrix from A?
A: Yes, since the columns of A are orthogonal, we simply normalize the columns to create a new matrix B whose columns are orthonormal. Whence, B is orthogonal.

Q: What does it mean to *normalize* the columns of A?
A: To *normalize*, we find a vector in the same direction as the given vector but of length one. This is very simply accomplished by computing the length of the given vector and then dividing each component by that length. The resulting vector will have length 1. So in the case of A, we would divide each component in both columns by 2 to obtain I.

5.1 Orthogonality in $\mathbb{R}^n$

Discussion of Key Definitions and Concepts, continued

Identifying Orthogonal matrices

Q: Why is it obvious that $B = \begin{bmatrix} 0.5 & 0.5 \\ 0.5 & -0.5 \end{bmatrix}$ is *not* orthogonal?

A: Because the columns of B do *not* have length 1.
Be careful. The length of a vector is not simply the sum of its components.
The length of a vector $\mathbf{x}$, $\|\mathbf{x}\|$, is $\|\mathbf{x}\| = \sqrt{\sum_{k=1}^{n} x_k^2}$.

Q: Why is it obvious that the columns of B *are* orthogonal?
A: Because they follow the pattern $\mathbf{b}_1 \cdot \mathbf{b}_2 = (b)(b) + (b)(-b) = b^2 - b^2 = 0$.

Q: Why is it obvious that $C = \begin{bmatrix} 1 & 2 \\ 3 & 4 \end{bmatrix}$ is *not* orthogonal?

A: Because the columns of C are obviously *not* orthogonal.

Q: Why is it obvious that the columns of C are *not* orthogonal?
A: Because all the entries of C are positive. Why is that enough?
Consider the fact that $\mathbf{c}_1 \cdot \mathbf{c}_2 = c_{11}c_{12} + c_{21}c_{22}$.
Can a summation of products ever sum to zero if all of the products are positive?

Q: Why is it obvious that $D = \begin{bmatrix} 1 & 2 \\ 2 & 4 \end{bmatrix}$ is *not* orthogonal?

A: Because the columns of D are obviously linearly dependent, $\mathbf{d}_2 = 2\mathbf{d}_1$.
Vectors that are linearly dependent cannot be orthogonal. Why not?
Being orthogonal is a *stronger* condition than being linearly independent.
That is, if two vectors are orthogonal, then they are linearly independent.

Q: What is the contrapositive of this last statement?
A: If two vectors are linearly dependent, then they cannot be orthogonal.

Discussion of Key Definitions and Concepts, continued

Identifying Orthogonal matrices, continued

Q: Can a matrix with linearly dependent columns (or rows) be orthogonal?
A: No. Why not? Because its columns (or rows) cannot be orthogonal.

Q: Can a matrix with nullity$(A) \neq 0$ be orthogonal?
A: No. Because nullity$(A) \neq 0$ implies the columns of A are linearly dependent.

Q: Can a matrix with null$(A) \neq \mathbf{0}$ be orthogonal?
Can a matrix with rank(A) less than the number of columns be orthogonal?
A: The answer to all these types of questions is, *No*. Why?
Because any condition that implies linear dependence among the rows or columns of A implies that A cannot be orthogonal. Why?
If A is orthogonal, its columns (and rows) must be linearly independent.
Again, this is a much *weaker* condition than what is actually required.
That is, that the columns (and rows) be orthonormal.
However, we should take care not to lose sight of this useful underlying condition.

Q: Can a matrix with an eigenvalue of 2 be orthogonal?
A: No. Why not? This is the assertion of Theorem 5.8(c).

Q: Prove that a matrix with an eigenvalue of 2 cannot be orthogonal using Theorem 5.6(b).
A: Since there exists an eigenvector $\mathbf{x}$ such that $A\mathbf{x} = 2\mathbf{x}$, $\|A\mathbf{x}\| = \|2\mathbf{x}\| = 2\|\mathbf{x}\| \neq \|\mathbf{x}\|$.
This shows that A is *not* orthogonal. Why?

Q: The above argument requires that $\|\mathbf{x}\| \neq 0$. How do we know that is true?
A: Because $\mathbf{x}$ is an eigenvector, so it is nonzero. Therefore, $\|\mathbf{x}\| \neq 0$.

Q: All we have shown is that $\|A\mathbf{x}\| \neq \|\mathbf{x}\|$ for one single vector $\mathbf{x}$.
Why is that enough?
A: Because Thm 5.6(b) asserts if A is orthogonal then $\|A\mathbf{x}\| = \|\mathbf{x}\|$ for *every* $\mathbf{x}$.
This example should help us appreciate what a strong condition Thm 5.6(b) really is.

5.1 Orthogonality in $\mathbb{R}^n$

Solutions to odd-numbered exercises from Section 5.1

1. $\mathbf{v}_1 \cdot \mathbf{v}_2 = (-3)(2) + 1(4) + 2(1) = 0$, $\mathbf{v}_2 \cdot \mathbf{v}_3 = 2(1) + 4(-1) + 1(2) = 0$,
   $\mathbf{v}_1 \cdot \mathbf{v}_3 = (-3) + 1(-1) + 2(2) = 0 \Rightarrow$ This set of vectors is orthogonal.

3. $\mathbf{v}_1 \cdot \mathbf{v}_2 = 3(-1) + 1(2) + (-1)(1) \neq 0 \Rightarrow$ not orthogonal.

5. $\mathbf{v}_1 \cdot \mathbf{v}_2 = \mathbf{v}_2 \cdot \mathbf{v}_3 = \mathbf{v}_1 \cdot \mathbf{v}_3 = 0 \Rightarrow$ This set of vectors is orthogonal.

7. $c_1 = \frac{4+6}{16+4} = \frac{1}{2}$, $c_2 = \frac{1-6}{1+4} = -1 \Rightarrow [\mathbf{w}]_{\mathcal{B}} = \begin{bmatrix} 1/2 \\ -1 \end{bmatrix}$. $\mathbf{w} = \frac{1}{2}\begin{bmatrix} 4 \\ -2 \end{bmatrix} + (-1)\begin{bmatrix} 1 \\ -2 \end{bmatrix} = \begin{bmatrix} 1 \\ -3 \end{bmatrix}$.

9. $c_1 = \frac{1+0-1}{1+0+1} = 0$, $c_2 = \frac{1+2+1}{1+4+1} = \frac{2}{3}$, $c_3 = \frac{1-1+1}{1+1+1} = \frac{1}{3} \Rightarrow [\mathbf{w}]_{\mathcal{B}} = \begin{bmatrix} 0 \\ 2/3 \\ 1/3 \end{bmatrix}$.

11. $\|\mathbf{v}_1\| = \sqrt{(\frac{3}{5})^2 + (\frac{4}{5})^2} = 1$, $\|\mathbf{v}_2\| = \sqrt{(-\frac{4}{5})^2 + (\frac{3}{5})^2} = 1 \Rightarrow$ This set is orthonormal.

13. $\|\mathbf{v}_1\| = \sqrt{(\frac{1}{3})^2 + 2(\frac{2}{3})^2} = 1$, $\|\mathbf{v}_2\| = \sqrt{(\frac{2}{3})^2 + (-\frac{1}{3})^2} = \frac{\sqrt{5}}{3}$, $\|\mathbf{v}_3\| = \sqrt{(1^2 + 2^2 + (-\frac{5}{2})^2} = \frac{3\sqrt{5}}{2}$
    $\Rightarrow \left\{ \begin{bmatrix} 1/3 \\ 2/3 \\ 2/3 \end{bmatrix}, \frac{3}{\sqrt{5}}\begin{bmatrix} 2/3 \\ -1/3 \\ 0 \end{bmatrix} = \begin{bmatrix} 2/\sqrt{5} \\ -1/\sqrt{5} \\ 0 \end{bmatrix}, \frac{2}{3\sqrt{5}}\begin{bmatrix} 1 \\ 2 \\ -5/2 \end{bmatrix} = \begin{bmatrix} 2/3\sqrt{5} \\ 4/3\sqrt{5} \\ -5/3\sqrt{5} \end{bmatrix} \right\}$.

15. $\|\mathbf{v}_1\| = \|\mathbf{v}_2\| = \|\mathbf{v}_3\| = \|\mathbf{v}_4\| = 1 \Rightarrow$ orthonormal.

17. $QQ^T = I \Rightarrow Q$ is orthogonal and $Q^{-1} = Q^T = \begin{bmatrix} 1/\sqrt{2} & -1/\sqrt{2} \\ 1/\sqrt{2} & 1/\sqrt{2} \end{bmatrix}$.

19. $QQ^T = I \Rightarrow Q$ is orthogonal and $Q^{-1} = Q^T = \begin{bmatrix} \cos\theta\sin\theta & \cos^2\theta & \sin\theta \\ -\cos\theta & \sin\theta & 0 \\ -\sin^2\theta & -\cos\theta\sin\theta & \cos\theta \end{bmatrix}$.

21. $\mathbf{q}_1 \cdot \mathbf{q}_4 = 1(\frac{1}{\sqrt{6}}) + 0(\frac{1}{\sqrt{6}}) + 0(-\frac{1}{\sqrt{6}}) + 0(-\frac{1}{\sqrt{2}}) \neq 0 \Rightarrow$ not orthogonal (columns must be $\perp$).

23. We will show this using the fact that $\det I = 1$, $QQ^T = I$, and $\det Q = \det Q^T$.
    $1 = \det I = \det(QQ^T) = \det Q \det Q^T = \det Q \det Q \Rightarrow \sqrt{(\det Q)^2} = \sqrt{1} \Rightarrow \det Q = \pm 1$.

25. Induction on Exercise 24 $\Rightarrow Q_1, Q_2, \ldots, Q_n$ orthogonal $\Rightarrow Q = Q_n \ldots Q_2 Q_1$ orthogonal.
    Let $P = P_n \ldots P_2 P_1$, where P_k is an elementary matrix corresponding to a row interchange.
    Then $P_k P_k^T = I$ and Theorem 5.5 $\Rightarrow P_k$ is orthogonal $\Rightarrow P = P_n \ldots P_2 P_1$ is orthogonal.

27. Let θ be the angle between $\mathbf{x}$ and $\mathbf{y}$ and ϕ be the angle between $Q\mathbf{x}$ and $Q\mathbf{y}$.
    We need to show that $\theta = \phi$. We will use Theorem 6(b,c): $\|Q\mathbf{x}\| = \|\mathbf{x}\|$, $Q\mathbf{x} \cdot Q\mathbf{y} = \mathbf{x} \cdot \mathbf{y}$.
    $0 \leq \theta, \phi \leq \pi \Rightarrow \cos\theta = \cos\phi \Rightarrow \theta = \phi$. So $\cos\theta = \frac{\mathbf{x} \cdot \mathbf{y}}{\|\mathbf{x}\|\|\mathbf{y}\|} = \frac{Q\mathbf{x} \cdot Q\mathbf{y}}{\|Q\mathbf{x}\|\|Q\mathbf{y}\|} = \cos\phi \Rightarrow \theta = \phi$.

29. By Exercise 28(d), $\det Q = 1 \Rightarrow$ rotation. So, since $\cos\theta = \frac{1}{\sqrt{2}}$, $\sin\theta = \frac{1}{\sqrt{2}}$, $\theta = \frac{\pi}{4}$ or $45°$.

31. $\det Q = -1 \Rightarrow$ reflection. Line of reflection is $y = -\tan\theta \, x = -\frac{\sin\theta}{\cos\theta}x = -(\frac{\sqrt{3}/2}{-1/2})x = \sqrt{3}\,x$.

33. (a) $\Rightarrow AA^T = I$, $B^TB = I \Rightarrow A(A^T + B^T)B = AA^TB + AB^TB = IB + AI = A + B$.

(b) Note: B orthogonal and Theorem 5.8(b) $\Rightarrow \det B = \pm 1 \Rightarrow \det B \det B = 1$.
If $\det A + \det B = 0$ also, then $\det A = -\det B \Rightarrow \det A \det B = -\det B \det B = -1$.
Also, recall: $\det(AB) = \det A \det B$ and $\det(A^T + B^T) = \det(A + B)^T = \det(A + B)$.
In order to prove $A + B$ is not invertible, we need only show that $\det(A + B) = 0$.
$A + B = A(A^T + B^T)B \Rightarrow \det(A+B) = \det(A(A^T+B^T)B) = \det A \det B \det(A^T + B^T)$.
Now use the fact that A, B orthogonal and $\det A + \det B = 0 \Rightarrow \det A \det B = -1$.
So, $\det(A + B) = \det A \det B \det(A^T + B^T) = -\det(A^T + B^T) = -\det(A + B)$.
Therefore, $\det(A + B) = -\det(A + B) \Rightarrow 2\det(A + B) = 0 \Rightarrow \det(A+B) = 0 \Rightarrow A + B$ is not invertible.

35. If Q is orthogonal and upper triangular, we must show Q is diagonal.
That is, if Q is orthogonal and $q_{ij} = 0$ for $j > i$, we must show $q_{ij} = 0$ for $j < i$.
Note: Since Q is orthogonal, $Q^TQ = I$. That is, as remarked in the proof of Theorem 5.4:

$$\mathbf{q}_i \cdot \mathbf{q}_j = \sum_{k=1}^{n} q_{ik}q_{jk} \overset{Q \text{ is upper triangular}}{=} \sum_{k=1}^{j} q_{ik}q_{jk} = \begin{cases} 0 & \text{if } i \neq j \\ 1 & \text{if } i = j \end{cases}$$

We proceed to show $q_{ij} = 0$ for $j < i$ by induction.

1: $q_{i1} = 0$ for $1 < i$.

$$\mathbf{q}_i \cdot \mathbf{q}_1 = \sum_{k=1}^{n} q_{ik}q_{1k} \overset{Q \text{ is upper triangular}}{=} \sum_{k=1}^{1} q_{ik}q_{1k} = q_{i1}q_{11} = 0.$$

Therefore, since $q_{11} \neq 0$, we have $q_{i1} = 0$ for $1 < i$ as required.

j: $q_{ij} = 0$ for $j < i$.
This is the induction hypothesis.

$j+1$: $q_{i(j+1)} = 0$ for $j + 1 < i$.
This is the statement we must prove using the induction hypothesis.

$$\mathbf{q}_i \cdot \mathbf{q}_{j+1} = \sum_{k=1}^{n} q_{ik}q_{(j+1)k} \overset{Q \text{ is upper triangular}}{=} \sum_{k=1}^{j+1} q_{ik}q_{(j+1)k} \overset{\text{by induction}}{=} q_{i(j+1)}q_{(j+1)(j+1)} = 0.$$

Therefore, since $q_{(j+1)(j+1)} \neq 0$, we have $q_{i(j+1)} = 0$ for $j + 1 < i$ as required.

We have proven (by induction) that $q_{ij} = 0$ for $j < i$. That is, Q is diagonal.

Q: Why does Q being upper triangular imply $\sum_{k=1}^{n} q_{ik}q_{jk} \overset{Q \text{ is upper triangular}}{=} \sum_{k=1}^{j} q_{ik}q_{jk}$?

A: Because $q_{jk} = 0$ for all $k > j$ by definition of upper triangular.

Q: Why does the induction hypothesis imply $\sum_{k=1}^{j+1} q_{ik}q_{(j+1)k} \overset{\text{by induction}}{=} q_{i(j+1)}q_{(j+1)(j+1)} = 0$?

A: Because $q_{ik} = 0$ for all $k \leq j$ by the induction hypothesis.

5.1 Orthogonality in $\mathbb{R}^n$

37. We are given $\mathcal{B} = \{\mathbf{v}_1, \mathbf{v}_2, \ldots, \mathbf{v}_n\}$ is an orthonormal basis for $\mathbb{R}^n$.

(a) We need to show $\mathbf{x} \cdot \mathbf{y} = (\mathbf{x} \cdot \mathbf{v}_1)(\mathbf{y} \cdot \mathbf{v}_1) + \cdots + (\mathbf{x} \cdot \mathbf{v}_n)(\mathbf{y} \cdot \mathbf{v}_n) = \sum_{k=1}^{n}(\mathbf{x} \cdot \mathbf{v}_k)(\mathbf{y} \cdot \mathbf{v}_k)$.

Since $\mathcal{B}$ is orthonormal, we have the following two key properties:
$$\mathbf{v}_i \cdot \mathbf{v}_j = \begin{cases} 0 & \text{if } i \neq j \quad \text{Property 1} \\ 1 & \text{if } i = j \quad \text{Property 2} \end{cases}$$

Since $\mathcal{B}$ is a basis, there exist α_k and β_k such that $\mathbf{x} = \sum_{k=1}^{n} \alpha_k \mathbf{v}_k$, $\mathbf{y} = \sum_{k=1}^{n} \beta_k \mathbf{v}_k$.

The key fact is $\mathbf{x} \cdot \mathbf{v}_i = \left(\sum_{k=1}^{n} \alpha_k \mathbf{v}_k\right) \cdot \mathbf{v}_i = \alpha_i$ and $\mathbf{y} \cdot \mathbf{v}_i = \left(\sum_{k=1}^{n} \beta_k \mathbf{v}_k\right) \cdot \mathbf{v}_i = \beta_i$.

So, $\mathbf{x} \cdot \mathbf{y} = \left(\sum_{k=1}^{n} \alpha_k \mathbf{v}_k\right) \cdot \left(\sum_{k=1}^{n} \beta_k \mathbf{v}_k\right) \stackrel{\text{Prop 1}}{=} \left(\sum_{k=1}^{n} (\alpha_k \mathbf{v}_k) \cdot (\beta_k \mathbf{v}_k)\right) \stackrel{\text{Prop 2}}{=} \left(\sum_{k=1}^{n} \alpha_k \beta_k\right)$

$\stackrel{\text{key fact}}{=} \sum_{k=1}^{n}(\mathbf{x} \cdot \mathbf{v}_k)(\mathbf{y} \cdot \mathbf{v}_k)$ as we were to show.

(b) What does our proof in part (a) suggest is the relationship between $\mathbf{x} \cdot \mathbf{y}$ and $[\mathbf{x}]_\mathcal{B} \cdot [\mathbf{y}]_\mathcal{B}$?

Since $[\mathbf{x}]_\mathcal{B} = [\alpha_k]$, $[\mathbf{y}]_\mathcal{B} = [\beta_k]$, as shown above $\mathbf{x} \cdot \mathbf{y} = \left(\sum_{k=1}^{n} \alpha_k \beta_k\right) = [\mathbf{x}]_\mathcal{B} \cdot [\mathbf{y}]_\mathcal{B}$.

That is, these two dot products are exactly the same.

Q: How might we summarize this finding in words?
A: The dot product is invariant under representation by orthonormal bases.

Q: The *key fact* cited above is implicit in what Theorem from this Section?
A: Theorem 5.3.

5.2 Orthogonal Complements and Projections

Q: What are our main goals for Section 5.2?
A: To take our first look at orthogonal *complements* and *projections*.

Key Definitions and Concepts

$W^\perp$	p.375	5.2	$W^\perp = \{\mathbf{v} \text{ in } \mathbb{R}^n : \mathbf{v} \cdot \mathbf{w} = 0 \text{ for all } \mathbf{w} \text{ in } W\}$
$\text{perp}_W(\mathbf{v})$ or $\mathbf{w}^\perp$	p.379	5.2	component of $\mathbf{v}$ orthogonal to W, $\mathbf{v} - \text{proj}_W(\mathbf{v})$
$\text{proj}_W(\mathbf{v})$	p.379	5.2	*orthogonal projection* of $\mathbf{v}$ onto W, $\sum_{k=1}^{n} \left(\frac{\mathbf{u}_k \cdot \mathbf{v}}{\mathbf{u}_k \cdot \mathbf{u}_k} \right) \mathbf{u}_k$
fundamental subspaces	p.377	5.2	There are four *fundamental subspaces* (two pair): $\text{row}(A), \text{null}(A)$ and $\text{col}(A), \text{null}(A^T)$
orthogonal		5.2	This adjective applies to *complements* and *projections*:
complement	p.375		$W^\perp = \{\mathbf{v} \text{ in } \mathbb{R}^n : \mathbf{v} \cdot \mathbf{w} = 0 \text{ for all } \mathbf{w} \text{ in } W\}$
component	p.379		$\text{perp}_W(\mathbf{v}) = \mathbf{w}^\perp = \mathbf{v} - \text{proj}_W(\mathbf{v})$
projection	p.379		$\text{proj}_W(\mathbf{v}) = \sum_{k=1}^{n} \left(\frac{\mathbf{u}_k \cdot \mathbf{v}}{\mathbf{u}_k \cdot \mathbf{u}_k} \right) \mathbf{u}_k = \sum_{k=1}^{n} \text{proj}_{\mathbf{u}_k}(\mathbf{v})$

Theorems

Thm 5.9	p.376	5.2	a. through d. give properties of $W^\perp$: a. $W^\perp$ is a subspace of $\mathbb{R}^n$ b. $(W^\perp)^\perp = W$ c. $W \cap W^\perp = \{\mathbf{0}\}$ d. If $W = \text{span}(\mathbf{w}_1, \ldots, \mathbf{w}_k)$, then $\mathbf{v}$ is in $W^\perp$ if and only if $\mathbf{v} \cdot \mathbf{w}_i = 0$ for all i
Thm 5.10	p.376	5.2	$(\text{row}(A))^\perp = \text{null}(A)$ and $(\text{col}(A))^\perp = \text{null}(A^T)$
Thm 5.11	p.381	5.2	**The Orthogonal Decomposition Theorem:** $\mathbf{v} = \mathbf{w} + \mathbf{w}^\perp$
Cor 5.12	p.382	5.2	$(W^\perp)^\perp = W$
Thm 5.13	p.383	5.2	$\dim W + \dim W^\perp = n$
Cor 5.14	p.381	5.2	**The Rank Theorem:** Given A is $m \times n$ then ... $\text{rank}(A) + \text{nullity}(A) = n$ and $\text{rank}(A) + \text{nullity}(A^T) = m$

Discussion of Theorems

Q: Can we explain why $W \cap W^\perp = \mathbf{0}$?
A: First of all, both W and $W^\perp$ are subspaces and $\mathbf{0}$ is in *every* subspace. Furthermore, $\mathbf{0}$ is the only vector that is orthogonal to *every* vector.

Q: How do the facts $\mathbf{v} = \mathbf{w} + \mathbf{w}^\perp$ and $\dim W + \dim W^\perp = n$ go together?
A: Every vector breaks into two pieces and therefore so does all of $\mathbb{R}^n$.

Q: Theorem 5.10 asserts $(\text{row}(A))^\perp = \text{null}(A)$. What does this prove?
A: Since $(\text{row}(A))^\perp = \text{null}(A)$, this shows $(\text{row}(A))^\perp$ is a subspace. Why? Because we know that $\text{null}(A)$ is a subspace.

Q: Can we show in detail why $(\text{row}(A))^\perp \subseteq \text{null}(A)$?
A: We need to show if $\mathbf{x}$ is in $(\text{row}(A))^\perp$, then $\mathbf{x}$ is in $\text{null}(A)$.
Let $\mathcal{B} = \{\mathbf{b}_i\}$ be a basis for $(\text{row}(A))^\perp$.
Then $\mathbf{x} = \sum_{i=1}^{k} \alpha_i \mathbf{b}_i$, where $\mathbf{A}_j \cdot \mathbf{b}_i = 0$ for all i and for all j.

So, $A\mathbf{x} = \begin{bmatrix} \mathbf{A}_1 \\ \vdots \\ \mathbf{A}_m \end{bmatrix} \left(\sum_{i=1}^{k} \alpha_i \mathbf{b}_i \right) = \sum_{i=1}^{k} \alpha_i \left(\sum_{j=1}^{m} \mathbf{A}_j \cdot \mathbf{b}_i \right) = \mathbf{0}.$

So, by definition, $\mathbf{x}$ is in $\text{null}(A)$.

Q: Can we show in detail why $\text{null}(A) \subseteq (\text{row}(A))^\perp$?
A: Hint: Is the argument used above reversible?

Q: Can we show in detail why $(\text{col}(A))^\perp \subseteq \text{null}(A^T)$?

Q: Given any $W^\perp$, is it possible to find a matrix A such that $W^\perp = \text{null}(A)$?
A: Hint: Let $\mathcal{B} = \{\mathbf{b}_i\}$ be a basis for W.
What vectors might we use as the columns of A?

Q: In $\mathbb{R}^3$, if W is a plane with normal $\mathbf{n}$, what is $W^\perp$?
A: A line that has direction vector $\mathbf{n}$. See Exercises 1 through 5.

Q: In $\mathbb{R}^3$, if W is a line with direction vector $\mathbf{d}$, what is $W^\perp$?
A: A plane with normal $\mathbf{d}$.

5.2 Orthogonal Complements and Projections

Solutions to odd-numbered exercises from Section 5.2

1. Since $\mathbf{w} = \begin{bmatrix} 1 \\ 2 \end{bmatrix}$ is a basis for W, for all $\mathbf{v} = \begin{bmatrix} x \\ y \end{bmatrix}$ in $W^\perp$ we have $\mathbf{v} \cdot \mathbf{w} = x + 2y = 0$.

   So, $W^\perp = \left\{ \begin{bmatrix} x \\ y \end{bmatrix} : x + 2y = 0 \right\}$ which implies $x = -2y$. Therefore, $\begin{bmatrix} -2 \\ 1 \end{bmatrix}$ is a basis.

   Note the following general pattern:
   The lines that describe the necessary condition to be in W and $W^\perp$ are necessarily perpendicular.
   So, in this case, $2x - y = 0$ and $x + 2y = 0$ are necessarily perpendicular.

3. W consists of $\begin{bmatrix} x \\ y \\ x+y \end{bmatrix} = x \begin{bmatrix} 1 \\ 0 \\ 1 \end{bmatrix} + y \begin{bmatrix} 0 \\ 1 \\ 1 \end{bmatrix} \Rightarrow \left\{ \begin{bmatrix} 1 \\ 0 \\ 1 \end{bmatrix}, \begin{bmatrix} 0 \\ 1 \\ 1 \end{bmatrix} \right\}$ is a basis for W.

   So, for all $\mathbf{v} = \begin{bmatrix} x \\ y \\ z \end{bmatrix}$ in $W^\perp$ we have $\begin{array}{l} \mathbf{v} \cdot \mathbf{w_1} = x + z = 0 \\ \mathbf{v} \cdot \mathbf{w_2} = y + z = 0 \end{array}$ which imply $\begin{array}{l} x = y \\ z = -x \end{array}$.

   Therefore, $W^\perp = \left\{ \begin{bmatrix} x \\ y \\ z \end{bmatrix} : x = t, y = t, z = -t \right\}$ which has basis $\begin{bmatrix} 1 \\ 1 \\ -1 \end{bmatrix}$.

   Note the following general pattern:
   If W is a plane with normal $\mathbf{n}$, then $W^\perp$ is a line that has direction vector $\mathbf{n}$.
   Likewise, if W is a line that has direction vector $\mathbf{n}$, then $W^\perp$ is a plane with normal $\mathbf{n}$.

5. Since $\mathbf{w} = \begin{bmatrix} 1 \\ -1 \\ 3 \end{bmatrix}$ is a basis for W, for $\mathbf{v} = \begin{bmatrix} x \\ y \\ z \end{bmatrix}$ in $W^\perp$, $\mathbf{v} \cdot \mathbf{w} = x - y + 3z = 0$.

   Therefore, $W^\perp = \left\{ \begin{bmatrix} x \\ y \\ z \end{bmatrix} : x - y + 3z = 0 \right\}$, which implies $y = x + 3z$.

   So, $W^\perp$ consists of $\begin{bmatrix} x \\ x+3z \\ z \end{bmatrix} = x \begin{bmatrix} 1 \\ 1 \\ 0 \end{bmatrix} + z \begin{bmatrix} 0 \\ 3 \\ 1 \end{bmatrix} \Rightarrow \left\{ \begin{bmatrix} 1 \\ 1 \\ 0 \end{bmatrix}, \begin{bmatrix} 0 \\ 3 \\ 1 \end{bmatrix} \right\}$ is a basis for $W^\perp$.

7. Since $A \longrightarrow \begin{bmatrix} 1 & 0 & 1 \\ 0 & 1 & -2 \\ 0 & 0 & 0 \\ 0 & 0 & 0 \end{bmatrix}$, $\{[\,1\ 0\ 1\,], [\,0\ 1\ -2\,]\}$ is a basis for row(A).

Since $\begin{bmatrix} 1 & 0 & 1 & | & 0 \\ 0 & 1 & -2 & | & 0 \\ 0 & 0 & 0 & | & 0 \\ 0 & 0 & 0 & | & 0 \end{bmatrix} \Rightarrow$ x in null(A) $= \begin{bmatrix} -x_3 \\ 2x_3 \\ x_3 \end{bmatrix}$, $\left\{ \begin{bmatrix} -1 \\ 2 \\ 1 \end{bmatrix} \right\}$ is a basis for null(A).

$[\,1\ 0\ 1\,]\begin{bmatrix} -1 \\ 2 \\ 1 \end{bmatrix} = 0, [\,0\ 1\ -2\,]\begin{bmatrix} -1 \\ 2 \\ 1 \end{bmatrix} = 0 \Rightarrow$ v in row(A) $\perp$ x in null(A).

9. Note: We can stop row reduction at any step to keep calculations simple.

$A^T \longrightarrow \begin{bmatrix} 1 & 5 & 0 & -1 \\ -1 & 2 & 1 & -1 \\ 0 & 0 & 0 & 0 \end{bmatrix}$, $\left\{ \begin{bmatrix} 1 \\ 5 \\ 0 \\ -1 \end{bmatrix}, \begin{bmatrix} -1 \\ 2 \\ 1 \\ -1 \end{bmatrix} \right\}$ is a basis for row(A^T) = col(A).

$\begin{bmatrix} 1 & 5 & 0 & -1 & | & 0 \\ -1 & 2 & 1 & -1 & | & 0 \\ 0 & 0 & 0 & 0 & | & 0 \end{bmatrix} \Rightarrow$ y $= \begin{bmatrix} y_4 - 5y_2 \\ y_2 \\ 2y_4 - 7y_2 \\ y_4 \end{bmatrix}$, $\left\{ \begin{bmatrix} 1 \\ 0 \\ 2 \\ 1 \end{bmatrix}, \begin{bmatrix} -5 \\ 1 \\ -7 \\ 0 \end{bmatrix} \right\}$ is a basis for null(A^T).

We now verify that the basis vectors for col(A) and null(A^T) are orthogonal:

$[\,1\ 5\ 0\ -1\,]\begin{bmatrix} 1 \\ 0 \\ 2 \\ 1 \end{bmatrix} = 1 - 1 = 0.$ $[\,1\ 5\ 0\ -1\,]\begin{bmatrix} -5 \\ 1 \\ -7 \\ 0 \end{bmatrix} = -5 + 5 = 0.$

$[\,-1\ 2\ 1\ -1\,]\begin{bmatrix} 1 \\ 0 \\ 2 \\ 1 \end{bmatrix} = -1 + 2 - 1 = 0.$ $[\,-1\ 2\ 1\ -1\,]\begin{bmatrix} -5 \\ 1 \\ -7 \\ 0 \end{bmatrix} = 5 + 2 - 7 = 0.$

11. $[A\,|\,\mathbf{0}] = \begin{bmatrix} 2 & 1 & -2 & | & 0 \\ 4 & 0 & 1 & | & 0 \end{bmatrix} \Rightarrow$ x $= \begin{bmatrix} x_1 \\ -10x_1 \\ -4x_1 \end{bmatrix} \Rightarrow \left\{ \begin{bmatrix} 1 \\ -10 \\ -4 \end{bmatrix} \right\}$ is a basis for $W^\perp$.

13. $\begin{bmatrix} 2 & -1 & 6 & 3 & | & 0 \\ 0 & 3 & 0 & -1 & | & 0 \\ 0 & 0 & 0 & 0 & | & 0 \end{bmatrix} \Rightarrow \begin{matrix} 2x_1 - x_2 + 6x_3 + 3x_4 = 0 \\ 3x_2 - x_4 = 0 \end{matrix} \Rightarrow \begin{matrix} x_1 = -4x_2 - 3x_3 \\ x_4 = 3x_2 \end{matrix} \Rightarrow$

Every x in null(A) $= \begin{bmatrix} -4x_2 - 3x_3 \\ x_2 \\ x_3 \\ 3x_2 \end{bmatrix} \Rightarrow \left\{ \begin{bmatrix} -4 \\ 1 \\ 0 \\ 3 \end{bmatrix}, \begin{bmatrix} -3 \\ 0 \\ 1 \\ 0 \end{bmatrix} \right\}$ is a basis for $W^\perp$.

15. So, $\mathbf{u}_1 \cdot \mathbf{v} = 3$, $\mathbf{u}_1 \cdot \mathbf{u}_1 = 2$, and $\mathbf{u}_1 = \begin{bmatrix} 1 \\ 1 \end{bmatrix} \Rightarrow \text{proj}_W(\mathbf{v}) = \frac{3}{2}\begin{bmatrix} 1 \\ 1 \end{bmatrix} = \begin{bmatrix} 3/2 \\ 3/2 \end{bmatrix}$.

5.2 Orthogonal Complements and Projections

17. $\text{proj}_W(\mathbf{v}) = \frac{1}{9}\begin{bmatrix} 2 \\ -2 \\ 1 \end{bmatrix} + \frac{13}{18}\begin{bmatrix} -1 \\ 1 \\ 4 \end{bmatrix} = \begin{bmatrix} -1/2 \\ 1/2 \\ 3 \end{bmatrix}$.

19. $\text{proj}_W(\mathbf{v}) + (\mathbf{v} - \text{proj}_W(\mathbf{v})) = -\frac{4}{10}\begin{bmatrix} 1 \\ 3 \end{bmatrix} + \left(\begin{bmatrix} 2 \\ -2 \end{bmatrix} - \frac{-4}{10}\begin{bmatrix} 1 \\ 3 \end{bmatrix}\right) = \begin{bmatrix} -2/5 \\ -6/5 \end{bmatrix} + \begin{bmatrix} 12/5 \\ -4/5 \end{bmatrix}$.

21. $\left(\frac{3}{6}\begin{bmatrix} 1 \\ 2 \\ 1 \end{bmatrix} + \frac{9}{3}\begin{bmatrix} 1 \\ -1 \\ 1 \end{bmatrix}\right) + \left(\begin{bmatrix} 4 \\ -2 \\ 3 \end{bmatrix} - \left(\frac{3}{6}\begin{bmatrix} 1 \\ 2 \\ 1 \end{bmatrix} + \frac{9}{3}\begin{bmatrix} 1 \\ -1 \\ 1 \end{bmatrix}\right)\right) = \begin{bmatrix} 7/2 \\ -2 \\ 7/2 \end{bmatrix} + \begin{bmatrix} 1/2 \\ 0 \\ -1/2 \end{bmatrix}$.

23. Need to show if $\mathbf{w}$ in W and $W^\perp$, then $\mathbf{w} = \mathbf{0}$: $\mathbf{w}$ in W and $W^\perp \Rightarrow \mathbf{w} \cdot \mathbf{w} = 0 \Rightarrow \mathbf{w} = \mathbf{0}$.

25. Orthogonality to $\mathbf{w}$ in subspace W does not guarantee orthogonality to W. Example:

Let $\mathbf{w}' = \begin{bmatrix} -1 \\ 1 \\ 0 \end{bmatrix}$ and $W = \text{span}\left(\mathbf{w} = \begin{bmatrix} 1 \\ 1 \\ 0 \end{bmatrix}, \mathbf{v}' = \begin{bmatrix} 0 \\ 1 \\ 1 \end{bmatrix}\right)$, then $\mathbf{w}' \cdot \mathbf{w} = 0$, but $\mathbf{w}' \cdot \mathbf{v}' \neq 0$.

27. Since this is an if and only if statement, there are two statements to prove.
Let $\{\mathbf{u}_k\}$ be an orthogonal basis for W.

if: If $\mathbf{x} = \text{proj}_W(\mathbf{x})$ then $\mathbf{x}$ is in W.

If $\mathbf{x} = \text{proj}_W(\mathbf{x}) = \sum_{k=1}^{n}\left(\frac{\mathbf{u}_k \cdot \mathbf{x}}{\mathbf{u}_k \cdot \mathbf{u}_k}\right)\mathbf{u}_k \stackrel{\mathbf{u}_k \cdot \mathbf{u}_k = 1}{=} \sum_{k=1}^{n}(\mathbf{u}_k \cdot \mathbf{x})\mathbf{u}_k$, then $\mathbf{x}$ is in $\text{span}(\mathbf{u}_k) = W$.

only if: If $\mathbf{x}$ is in W then $\mathbf{x} = \text{proj}_W(\mathbf{x})$.

If $\mathbf{x}$ is in W then $\mathbf{x} = \sum_{k=1}^{n} \alpha_k \mathbf{u}_k$, but $\mathbf{u}_i \cdot \mathbf{x} = \mathbf{u}_i \cdot \left(\sum_{k=1}^{n} \alpha_k \mathbf{u}_k\right) = \alpha_i$. Why?

Because $\{\mathbf{u}_k\}$ is an orthogonal basis for W.
For a basis to be orthonormal it must have two key properties:

$$\mathbf{u}_i \cdot \mathbf{u}_j = \begin{cases} 0 & \text{if } i \neq j \quad \text{Property 1} \\ 1 & \text{if } i = j \quad \text{Property 2} \end{cases}$$

Therefore, $\mathbf{x} = \sum_{k=1}^{n} \alpha_k \mathbf{u}_k = \sum_{k=1}^{n} (\mathbf{u}_k \cdot \mathbf{x})\mathbf{u}_k = \text{proj}_W(\mathbf{x})$.

29. We will use our insights from Exercise 27 to prove this assertion.
Note that $\text{proj}_W(\mathbf{x})$ is in W. Why?

Because $\text{proj}_W(\mathbf{x}) = \sum_{k=1}^{n}(\mathbf{u}_k \cdot \mathbf{x})\mathbf{u}_k$, so $\text{proj}_W(\mathbf{x})$ is in $\text{span}(\mathbf{u}_k) = W$.

By Exercise 27, if $\mathbf{v}$ is in W then $\text{proj}_W(\mathbf{v}) = \mathbf{v}$.
Since $\mathbf{v} = \text{proj}_W(\mathbf{x})$ is in W, $\text{proj}_W(\text{proj}_W(\mathbf{x})) = \text{proj}_W(\mathbf{x})$.

5.3 The Gram-Schmidt Process and the QR Factorization

Q: What are our main goals for Section 5.3?

A: To practice the Gram-Schmidt Process.

Theorems

Thm 5.15 p.386 5.3 **The Gram-Schmidt Process:**
Let $\{x_1, \ldots, w_k\}$ be a basis for W then ...
$v_1 = x_1$, $W_1 = \text{span}(x_1)$
$v_2 = x_2 - \left(\frac{v_1 \cdot x_2}{v_1 \cdot v_1}\right) v_1$, $W_2 = \text{span}(x_1, x_2)$
$v_3 = x_3 - \left(\frac{v_1 \cdot x_3}{v_1 \cdot v_1}\right) v_1 - \left(\frac{v_2 \cdot x_3}{v_2 \cdot v_2}\right) v_2$, $W_3 = \text{span}(x_1, x_2, x_3)$
...
$v_k = x_k - \sum_{i=1}^{k-1} \frac{v_i \cdot x_k}{v_i \cdot v_i}$, $W_k = \text{span}(\{x_k\})$

... $\{v_1, \ldots, v_k\}$ is an orthogonal basis for W.

Thm 5.16 p.390 5.3 **The QR Factorization:**
Let A be an $m \times n$ matrix with linearly independent columns.
Then A can be factored as $A = QR$, where
Q is an $m \times n$ matrix with orthonormal columns and
R is an invertible upper triangular matrix.

Discussion of Key Definitions and Concepts

Since this section focuses on computation, the discussion accompanies the exercises below.

Solutions to odd-numbered exercises from Section 5.3

1. Applying Gram-Schmidt $\Rightarrow \mathbf{v}_1 = \mathbf{x}_1$, $\mathbf{v}_2 = \begin{bmatrix} 1 \\ 2 \end{bmatrix} - \frac{3}{2}\begin{bmatrix} 1 \\ 1 \end{bmatrix} = \begin{bmatrix} -1/2 \\ 1/2 \end{bmatrix}$.

$\|\mathbf{v}_1\| = \sqrt{2}$, $\|\mathbf{v}_2\| = \frac{\sqrt{2}}{2} \Rightarrow \mathbf{q}_1 = \frac{1}{\sqrt{2}}\begin{bmatrix} 1 \\ 1 \end{bmatrix} = \begin{bmatrix} 1/\sqrt{2} \\ 1/\sqrt{2} \end{bmatrix}$, $\mathbf{q}_2 = \frac{2}{\sqrt{2}}\begin{bmatrix} -1/2 \\ 1/2 \end{bmatrix} = \begin{bmatrix} -1/\sqrt{2} \\ 1/\sqrt{2} \end{bmatrix}$.

3. G-S $\Rightarrow \mathbf{v}_2 = \begin{bmatrix} 0 \\ 3 \\ 3 \end{bmatrix} + 2\begin{bmatrix} 1 \\ -1 \\ -1 \end{bmatrix} = \begin{bmatrix} 2 \\ 1 \\ 1 \end{bmatrix}$, $\mathbf{v}_3 = \begin{bmatrix} 3 \\ 2 \\ 4 \end{bmatrix} + 1\begin{bmatrix} 1 \\ -1 \\ -1 \end{bmatrix} - 2\begin{bmatrix} 2 \\ 1 \\ 1 \end{bmatrix} = \begin{bmatrix} 0 \\ -1 \\ 1 \end{bmatrix}$.

Therefore, $\|\mathbf{v}_1\| = \sqrt{3}$, $\|\mathbf{v}_2\| = \sqrt{6}$, $\|\mathbf{v}_3\| = \sqrt{2} \Rightarrow$

$\mathbf{q}_1 = \begin{bmatrix} 1/\sqrt{3} \\ -1/\sqrt{3} \\ -1/\sqrt{3} \end{bmatrix}$, $\mathbf{q}_2 = \begin{bmatrix} 2/\sqrt{6} \\ 1/\sqrt{6} \\ 1/\sqrt{6} \end{bmatrix}$, $\mathbf{q}_3 = \begin{bmatrix} 0 \\ -1/\sqrt{2} \\ 1/\sqrt{2} \end{bmatrix}$.

5. $\mathbf{v}_2 = \begin{bmatrix} 3 \\ 4 \\ 2 \end{bmatrix} - \frac{7}{2}\begin{bmatrix} 1 \\ 1 \\ 0 \end{bmatrix} = \begin{bmatrix} -1/2 \\ 1/2 \\ 2 \end{bmatrix} \Rightarrow$ Orthogonal basis $= \left\{ \begin{bmatrix} 1 \\ 1 \\ 0 \end{bmatrix}, \begin{bmatrix} -1/2 \\ 1/2 \\ 2 \end{bmatrix} \right\}$.

7. $\left(0\begin{bmatrix} 1 \\ 1 \\ 0 \end{bmatrix} + \frac{4}{9}\begin{bmatrix} -1/2 \\ 1/2 \\ 2 \end{bmatrix} \right) + \left(\begin{bmatrix} 4 \\ -4 \\ 3 \end{bmatrix} - \left(0\begin{bmatrix} 1 \\ 1 \\ 0 \end{bmatrix} + \frac{4}{9}\begin{bmatrix} -1/2 \\ 1/2 \\ 2 \end{bmatrix} \right) \right) = \begin{bmatrix} -2/9 \\ 2/9 \\ 8/9 \end{bmatrix} + \begin{bmatrix} 38/9 \\ -38/9 \\ 19/9 \end{bmatrix}$.

9. $\mathbf{v}_2 = \begin{bmatrix} 1 \\ 0 \\ 1 \end{bmatrix} - \frac{1}{2}\begin{bmatrix} 0 \\ 1 \\ 1 \end{bmatrix} = \begin{bmatrix} 1 \\ -1/2 \\ 1/2 \end{bmatrix}$, $\mathbf{v}_3 = \begin{bmatrix} 1 \\ 1 \\ 0 \end{bmatrix} - \frac{1}{2}\begin{bmatrix} 0 \\ 1 \\ 1 \end{bmatrix} - \frac{1}{3}\begin{bmatrix} 1 \\ -1/2 \\ 1/2 \end{bmatrix} = \begin{bmatrix} 2/3 \\ 2/3 \\ -2/3 \end{bmatrix}$.

Therefore, $\left\{ \begin{bmatrix} 0 \\ 1 \\ 1 \end{bmatrix}, \begin{bmatrix} 1 \\ -1/2 \\ 1/2 \end{bmatrix}, \begin{bmatrix} 2/3 \\ 2/3 \\ -2/3 \end{bmatrix} \right\}$ is an orthogonal basis for col(A).

11. $\mathbf{v}_2 = \begin{bmatrix} 0 \\ 1 \\ 0 \end{bmatrix} - \frac{1}{35}\begin{bmatrix} 3 \\ 1 \\ 5 \end{bmatrix} = \begin{bmatrix} -3/35 \\ 34/35 \\ -1/7 \end{bmatrix}$, $\mathbf{v}_3 = \begin{bmatrix} 0 \\ 0 \\ 1 \end{bmatrix} - \frac{1}{7}\begin{bmatrix} 3 \\ 1 \\ 5 \end{bmatrix} + \frac{5}{34}\begin{bmatrix} -3/35 \\ 34/35 \\ -1/7 \end{bmatrix} = \begin{bmatrix} -15/34 \\ 0 \\ 9/34 \end{bmatrix}$.

Therefore, $\left\{ \begin{bmatrix} 3 \\ 1 \\ 5 \end{bmatrix}, \begin{bmatrix} -3/35 \\ 34/35 \\ -1/7 \end{bmatrix}, \begin{bmatrix} -15/34 \\ 0 \\ 9/34 \end{bmatrix} \right\}$ is an orthogonal basis.

5.3 The Gram-Schmidt Process and the QR Factorization

13. We need to choose an $\mathbf{x}_3$ and then use the Gram-Schmidt Process.

Since $\mathbf{q}_1 = \begin{bmatrix} \frac{1}{\sqrt{2}} \\ 0 \\ -\frac{1}{\sqrt{2}} \end{bmatrix}$ and $\mathbf{q}_2 = \begin{bmatrix} \frac{1}{\sqrt{3}} \\ \frac{1}{\sqrt{3}} \\ \frac{1}{\sqrt{3}} \end{bmatrix}$, we can let $\mathbf{x}_3 = \begin{bmatrix} 0 \\ 0 \\ 1 \end{bmatrix}$. Why?

Q: How can we see that this is a good choice for $\mathbf{x}_3$?
A: It may help to temporarily ignore the fact that the columns of Q have been normalized.

That is, consider $P = \begin{bmatrix} 1 & 1 & * \\ 0 & 1 & * \\ -1 & 1 & * \end{bmatrix}$ then let $\mathbf{p}_3 = \mathbf{x}_3$. So $P = \begin{bmatrix} 1 & 1 & 0 \\ 0 & 1 & 0 \\ -1 & 1 & 1 \end{bmatrix}$.

Then $P \longrightarrow I$. That is, the columns of P are linearly independent.

Applying the Gram-Schmidt Process, we have:

$$\mathbf{v}_3 = \mathbf{x}_3 - \left(\frac{\mathbf{q}_1 \cdot \mathbf{x}_3}{\mathbf{q}_1 \cdot \mathbf{q}_1}\right)\mathbf{q}_1 - \left(\frac{\mathbf{q}_2 \cdot \mathbf{x}_3}{\mathbf{q}_2 \cdot \mathbf{q}_2}\right)\mathbf{q}_2 = \begin{bmatrix} 0 \\ 0 \\ 1 \end{bmatrix} + \frac{1}{\sqrt{2}}\begin{bmatrix} \frac{1}{\sqrt{2}} \\ 0 \\ -\frac{1}{\sqrt{2}} \end{bmatrix} - \frac{1}{\sqrt{3}}\begin{bmatrix} \frac{1}{\sqrt{3}} \\ \frac{1}{\sqrt{3}} \\ \frac{1}{\sqrt{3}} \end{bmatrix} = \begin{bmatrix} -\frac{1}{2} \\ 1 \\ \frac{1}{2} \end{bmatrix}.$$

In order for Q to be an orthogonal matrix, its columns must be orthonormal.
Therefore, we must compute the length of $\mathbf{x}_3$ and divide each of the components by it.

Clearly, $\|\mathbf{x}_3\| = \sqrt{\left(\frac{1}{2}\right)^2 + 1^2 + \left(-\frac{1}{2}\right)^2} = \frac{\sqrt{6}}{2}$.

Therefore, the completed matrix $Q = \begin{bmatrix} \frac{1}{\sqrt{2}} & \frac{1}{\sqrt{3}} & -\frac{1}{\sqrt{6}} \\ 0 & \frac{1}{\sqrt{3}} & \frac{2}{\sqrt{6}} \\ -\frac{1}{\sqrt{2}} & \frac{1}{\sqrt{3}} & \frac{1}{\sqrt{6}} \end{bmatrix}$.

Q: Why is Q an orthogonal matrix?
A: Because it has orthonormal columns.

Q: Therefore, what two conditions do the columns of Q satisfy?
A: For a set to be orthonormal it must have two key properties:

$$\mathbf{q}_i \cdot \mathbf{q}_j = \begin{cases} 0 & \text{if } i \neq j \quad \text{Property 1} \\ 1 & \text{if } i = j \quad \text{Property 2} \end{cases}$$

Verify this.

Q: Do the rows of Q also form an orthonormal set?
A: According to Theorem 5.7 of Section 5.1, they should. Verify this.
This verification also provides a quick check on our final result.

Q: If we have an orthogonal set can we divide by the lengths to create an orthonormal set?
A: Yes. Why?
If $\mathbf{v} \cdot c\mathbf{w} = 0$ ($c \neq 0$), then $\mathbf{v} \cdot \mathbf{w} = 0$.
Why is this sufficient? Verify that this claim is true.

15. From 9, $\|\mathbf{v}_1\| = \sqrt{2}$, $\|\mathbf{v}_2\| = \frac{\sqrt{6}}{2}$, $\|\mathbf{v}_3\| = \frac{2\sqrt{3}}{3} \Rightarrow Q = \begin{bmatrix} 0 & 2/\sqrt{6} & 1/\sqrt{3} \\ 1/\sqrt{2} & -1/\sqrt{6} & 1/\sqrt{3} \\ 1/\sqrt{2} & 1/\sqrt{6} & -1/\sqrt{3} \end{bmatrix}$.

Therefore, $R = Q^T A = \begin{bmatrix} 0 & 1/\sqrt{2} & 1/\sqrt{2} \\ 2/\sqrt{6} & -1/\sqrt{6} & 1/\sqrt{6} \\ 1/\sqrt{3} & 1/\sqrt{3} & -1/\sqrt{3} \end{bmatrix} \begin{bmatrix} 0 & 1 & 1 \\ 1 & 0 & 1 \\ 1 & 1 & 0 \end{bmatrix} = \begin{bmatrix} \sqrt{2} & 1/\sqrt{2} & 1/\sqrt{2} \\ 0 & 3/\sqrt{6} & 1/\sqrt{6} \\ 0 & 0 & 2/\sqrt{3} \end{bmatrix}$.

Verify that $A = QR = \begin{bmatrix} 0 & 2/\sqrt{6} & 1/\sqrt{3} \\ 1/\sqrt{2} & -1/\sqrt{6} & 1/\sqrt{3} \\ 1/\sqrt{2} & 1/\sqrt{6} & -1/\sqrt{3} \end{bmatrix} \begin{bmatrix} \sqrt{2} & 1/\sqrt{2} & 1/\sqrt{2} \\ 0 & 3/\sqrt{6} & 1/\sqrt{6} \\ 0 & 0 & 2/\sqrt{3} \end{bmatrix} = \begin{bmatrix} 0 & 1 & 1 \\ 1 & 0 & 1 \\ 1 & 1 & 0 \end{bmatrix}$.

17. $R = Q^T A = \begin{bmatrix} 2/3 & 1/3 & -2/3 \\ 1/3 & 2/3 & 2/3 \\ 2/3 & -2/3 & 1/3 \end{bmatrix} \begin{bmatrix} 2 & 8 & 2 \\ 1 & 7 & -1 \\ -2 & -2 & 1 \end{bmatrix} = \begin{bmatrix} 3 & 9 & 1/3 \\ 0 & 6 & 2/3 \\ 0 & 0 & 7/3 \end{bmatrix}$.

Verify that $A = QR = \begin{bmatrix} 2/3 & 1/3 & 2/3 \\ 1/3 & 2/3 & -2/3 \\ -2/3 & 2/3 & 1/3 \end{bmatrix} \begin{bmatrix} 3 & 9 & 1/3 \\ 0 & 6 & 2/3 \\ 0 & 0 & 7/3 \end{bmatrix} = \begin{bmatrix} 2 & 8 & 2 \\ 1 & 7 & -1 \\ -2 & -2 & 1 \end{bmatrix}$.

19. Since A is orthogonal, simply let $A = Q \Rightarrow A = AI$, where I is obviously upper triangular.

21. $A^{-1} = R^{-1}Q^T = \begin{bmatrix} 1/\sqrt{2} & -1/\sqrt{6} & -1/2\sqrt{3} \\ 0 & 2/\sqrt{6} & -1/2\sqrt{3} \\ 0 & 0 & 3/2\sqrt{3} \end{bmatrix} \begin{bmatrix} 0 & 1/\sqrt{2} & 1/\sqrt{2} \\ 2/\sqrt{6} & -1/\sqrt{6} & 1/\sqrt{6} \\ 1/\sqrt{3} & 1/\sqrt{3} & -1/\sqrt{3} \end{bmatrix}$

$= \begin{bmatrix} -1/2 & 1/2 & 1/2 \\ 1/2 & -1/2 & 1/2 \\ 1/2 & 1/2 & -1/2 \end{bmatrix}$, so $AA^{-1} = \begin{bmatrix} 0 & 1 & 1 \\ 1 & 0 & 1 \\ 1 & 1 & 0 \end{bmatrix} \begin{bmatrix} -1/2 & 1/2 & 1/2 \\ 1/2 & -1/2 & 1/2 \\ 1/2 & 1/2 & -1/2 \end{bmatrix} = \begin{bmatrix} 1 & 0 & 0 \\ 0 & 1 & 0 \\ 0 & 0 & 1 \end{bmatrix}$.

23. Recall, Q orthogonal $\Rightarrow Q^T$ orthogonal $\Rightarrow \|Q^T\mathbf{x}\| = \|\mathbf{x}\|$.

We need to show $R\mathbf{x} = \mathbf{0} \Rightarrow \mathbf{x} = \mathbf{0}$, where $A = QR \Rightarrow R = Q^T A$.

So, $R\mathbf{x} = \mathbf{0} \Rightarrow \|R\mathbf{x}\| = 0 \Rightarrow \|(Q^T A)\mathbf{x}\| = \|Q^T(A\mathbf{x})\| = \|A\mathbf{x}\| = 0 \Rightarrow$
$\sum \mathbf{a}_i x_i = 0$ (since $A\mathbf{x}$ is just a linear combination of the columns of A) $\Rightarrow$
$x_i = 0$ because the columns of A, $\mathbf{a}_i$, are linearly independent $\Rightarrow$
$A\mathbf{x} = \mathbf{0} \Rightarrow \mathbf{x} = \mathbf{0}$.

Therefore, R is invertible by property (c) of the Fundamental Theorem.

Exploration: The Modified QR Factorization

Explorations are self-contained, so only odd-numbered solutions will be provided.

1. $Q = \begin{bmatrix} 1 - 2d_1^2 & -2d_1 d_2 \\ -2d_1 d_2 & 1 - 2d_2^2 \end{bmatrix} = \begin{bmatrix} 1 & 0 \\ 0 & 1 \end{bmatrix} - 2 \begin{bmatrix} d_1^2 & d_1 d_2 \\ d_1 d_2 & d_2^2 \end{bmatrix} = I - 2\mathbf{u}\mathbf{u}^T.$

3. (a) $Q^T = (I - 2\mathbf{u}\mathbf{u}^T)^T = I - 2(\mathbf{u}^T)^T \mathbf{u}^T = I - 2\mathbf{u}\mathbf{u}^T.$
   (b) Show $Q^T = Q^{-1}$, then apply Section 5.1, Theorem 5.5.
   (c) Follows from (a) and (b) since $I = QQ^T = QQ = Q^2 = I$.
   In particular, note that $\mathbf{u}^T\mathbf{u} = 1$ is key to this property.
   $QQ^T = Q^2 = (I - 2\mathbf{u}\mathbf{u}^T)(I - 2\mathbf{u}\mathbf{u}^T) = I^2 - 4\mathbf{u}\mathbf{u}^T - (2\mathbf{u}\mathbf{u}^T)(2\mathbf{u}\mathbf{u}^T)$
   $= I - 4\mathbf{u}\mathbf{u}^T - 4\mathbf{u}(\mathbf{u}^T\mathbf{u})\mathbf{u}^T = I - 4\mathbf{u}\mathbf{u}^T - 4\mathbf{u}\mathbf{u}^T = I.$

5. Correction: Note that we have to normalize the given $\mathbf{u}$ to get:

$$\mathbf{u} = \begin{bmatrix} \frac{1}{\sqrt{6}} \\ -\frac{1}{\sqrt{6}} \\ \frac{2}{\sqrt{6}} \end{bmatrix} \Rightarrow \mathbf{u}\mathbf{u}^T = \begin{bmatrix} \frac{1}{6} & -\frac{1}{6} & \frac{1}{3} \\ -\frac{1}{6} & \frac{1}{6} & -\frac{1}{3} \\ \frac{1}{3} & -\frac{1}{3} & \frac{2}{3} \end{bmatrix} \Rightarrow Q = I - 2\mathbf{u}\mathbf{u}^T = \begin{bmatrix} \frac{2}{3} & \frac{1}{3} & -\frac{2}{3} \\ \frac{1}{3} & \frac{2}{3} & \frac{2}{3} \\ -\frac{2}{3} & \frac{2}{3} & -\frac{1}{3} \end{bmatrix}.$$

Verification that $Q^T = Q$ and $QQ^T = Q^2 = I$.

$$Q^T = \begin{bmatrix} \frac{2}{3} & \frac{1}{3} & -\frac{2}{3} \\ \frac{1}{3} & \frac{2}{3} & \frac{2}{3} \\ -\frac{2}{3} & \frac{2}{3} & -\frac{1}{3} \end{bmatrix} = Q. \qquad QQ^T = Q^2 = \begin{bmatrix} \frac{2}{3} & \frac{1}{3} & -\frac{2}{3} \\ \frac{1}{3} & \frac{2}{3} & \frac{2}{3} \\ -\frac{2}{3} & \frac{2}{3} & -\frac{1}{3} \end{bmatrix} \begin{bmatrix} \frac{2}{3} & \frac{1}{3} & -\frac{2}{3} \\ \frac{1}{3} & \frac{2}{3} & \frac{2}{3} \\ -\frac{2}{3} & \frac{2}{3} & -\frac{1}{3} \end{bmatrix} = I.$$

Verification that if $\mathbf{v} = c\mathbf{u}$, then $Q\mathbf{v} = -\mathbf{v}$.

$$Q\mathbf{v} = Q(c\mathbf{u}) = cQ\mathbf{u} = c \begin{bmatrix} \frac{2}{3} & \frac{1}{3} & -\frac{2}{3} \\ \frac{1}{3} & \frac{2}{3} & \frac{2}{3} \\ -\frac{2}{3} & \frac{2}{3} & -\frac{1}{3} \end{bmatrix} \begin{bmatrix} \frac{1}{\sqrt{6}} \\ -\frac{1}{\sqrt{6}} \\ \frac{2}{\sqrt{6}} \end{bmatrix} = c \begin{bmatrix} -\frac{1}{\sqrt{6}} \\ \frac{1}{\sqrt{6}} \\ -\frac{2}{\sqrt{6}} \end{bmatrix} = -c\mathbf{u} = -\mathbf{v}.$$

Verification that if $\mathbf{v} \cdot \mathbf{u} = 0$, then $Q\mathbf{v} = \mathbf{v}$.

Note, $\mathbf{v} \cdot \mathbf{u} = \mathbf{u}^T \mathbf{v} = 0 \Rightarrow \begin{bmatrix} \frac{1}{\sqrt{6}} & -\frac{1}{\sqrt{6}} & \frac{2}{\sqrt{6}} \end{bmatrix} \begin{bmatrix} x_1 \\ x_2 \\ x_3 \end{bmatrix} = \frac{1}{\sqrt{6}} x_1 - \frac{1}{\sqrt{6}} x_2 + \frac{2}{\sqrt{6}} x_3 = 0 \Rightarrow$

$x_1 - x_2 + 2x_3 = 0 \Rightarrow x_2 = x_1 + 2x_3 \Rightarrow$ If $\mathbf{v} \cdot \mathbf{u} = 0$ then $\mathbf{v} = s \begin{bmatrix} 1 \\ 1 \\ 0 \end{bmatrix} + t \begin{bmatrix} 0 \\ 2 \\ 1 \end{bmatrix}.$

So, we have:

$$Q\mathbf{v} = Q\left(s \begin{bmatrix} 1 \\ 1 \\ 0 \end{bmatrix} + t \begin{bmatrix} 0 \\ 2 \\ 1 \end{bmatrix}\right) = sQ\left(\begin{bmatrix} 1 \\ 1 \\ 0 \end{bmatrix}\right) + tQ\left(\begin{bmatrix} 0 \\ 2 \\ 1 \end{bmatrix}\right)$$

$$= s \begin{bmatrix} \frac{2}{3} & \frac{1}{3} & -\frac{2}{3} \\ \frac{1}{3} & \frac{2}{3} & \frac{2}{3} \\ -\frac{2}{3} & \frac{2}{3} & -\frac{1}{3} \end{bmatrix} \begin{bmatrix} 1 \\ 1 \\ 0 \end{bmatrix} + t \begin{bmatrix} \frac{2}{3} & \frac{1}{3} & -\frac{2}{3} \\ \frac{1}{3} & \frac{2}{3} & \frac{2}{3} \\ -\frac{2}{3} & \frac{2}{3} & -\frac{1}{3} \end{bmatrix} \begin{bmatrix} 0 \\ 2 \\ 1 \end{bmatrix} = s \begin{bmatrix} 1 \\ 1 \\ 0 \end{bmatrix} + t \begin{bmatrix} 0 \\ 2 \\ 1 \end{bmatrix} = \mathbf{v}.$$

7. Note, $\mathbf{u} = \frac{1}{2\sqrt{3}}\begin{bmatrix} -2 \\ 2 \\ 2 \end{bmatrix} \Rightarrow \mathbf{uu}^T = \begin{bmatrix} \frac{1}{3} & -\frac{1}{3} & -\frac{1}{3} \\ -\frac{1}{3} & \frac{1}{3} & \frac{1}{3} \\ -\frac{1}{3} & \frac{1}{3} & \frac{1}{3} \end{bmatrix} \Rightarrow Q = I - 2\mathbf{uu}^T = \begin{bmatrix} \frac{1}{3} & \frac{2}{3} & \frac{2}{3} \\ \frac{2}{3} & \frac{1}{3} & -\frac{2}{3} \\ \frac{2}{3} & -\frac{2}{3} & \frac{1}{3} \end{bmatrix}.$

So, we verify: $Q\mathbf{x} = \begin{bmatrix} \frac{1}{3} & \frac{2}{3} & \frac{2}{3} \\ \frac{2}{3} & \frac{1}{3} & -\frac{2}{3} \\ \frac{2}{3} & -\frac{2}{3} & \frac{1}{3} \end{bmatrix} \begin{bmatrix} 1 \\ 2 \\ 2 \end{bmatrix} = \begin{bmatrix} 3 \\ 0 \\ 0 \end{bmatrix} = \mathbf{y}.$

9. Show $Q_2 Q_2^T = I$, using the fact that $P_2 P_2^T = I$.

Then: $Q_2 Q_1 A = \begin{bmatrix} 1 & 0 \\ 0 & P_2 \end{bmatrix} \begin{bmatrix} * & * \\ 0 & A_1 \end{bmatrix} = \begin{bmatrix} * & * \\ 0 & P_2 A_1 \end{bmatrix} = \begin{bmatrix} * & * & * \\ 0 & * & * \\ 0 & 0 & A_2 \end{bmatrix}.$

11. Since $Q_{m-1} \cdots Q_2 Q_1 A = R$, take $Q^T = Q_{m-1} \cdots Q_2 Q_1$, so we have:
$Q^T A = R \Rightarrow A = QR$, where $Q = (Q_{m-1} \cdots Q_2 Q_1)^T = Q_1^T Q_2^T \cdots Q_{m-1}^T$ orthogonal.

Exploration: Approximating Eigenvalues with the QR Algorithm

Explorations are self-contained, so only odd-numbered solutions will be provided.

1. The definition of similar in Section 4.4 states: $B = P^{-1}AP$, where P is invertible $\Rightarrow A \sim B$. So, when Q is orthogonal this definition becomes $B = QAQ^T \Rightarrow A \sim B$.
   Now in this case, we have $A_1 = RQ$, where $A = QR$.
   So, $A = QR = QR(QQ^T) = Q(RQ)Q^T = QA_1Q^T \Rightarrow A \sim A_1$.
   Since eigenvalues are the solutions of $\det(A - \lambda I) = 0$, we need only show $\det(A - \lambda I) = \det(A_1 - \lambda I)$, that is A and A_1 have the same characteristic polynomial.
   $\det(A - \lambda I) = \det(QA_1Q^T - \lambda I) = \det(QA_1Q^T - \lambda QIQ^T) = \det Q \det(A_1 - \lambda I)\det Q^T$
   $= \det Q \det Q^T \det(A_1 - \lambda I) = \det QQ^T \det(A_1 - \lambda I) = \det I \det(A_1 - \lambda I) = \det(A_1 - \lambda I)$.

3. We will proceed by induction.
   $n = 1$: This case was proven in Exercise 1.
   $n = k - 1$: Assume $A_n \sim A$, for $n \leq k - 1$. We need to show $A_k \sim A$.
   We have $A_k = R_{k-1}Q_{k-1}$, where $A_{k-1} = Q_{k-1}R_{k-1}$.
   So, $A_k = Q_{k-1}R_{k-1} = Q_{k-1}R_{k-1}(Q_{k-1}Q_{k-1}^T) = Q_{k-1}(R_{k-1}Q_{k-1})Q_{k-1}^T = Q_{k-1}A_kQ_{k-1}^T \Rightarrow A_k \sim A_{k-1}$. Therefore, $A_k \sim A$, by the transitive property of similarity.

5. It's clear that the diagonal entries of U will be the eigenvalues of A because the eigenvalues of U and A are the same.

7. We follow the same process as in Exercise 2.

$$A = QR = \begin{bmatrix} \frac{2}{\sqrt{5}} & -\frac{1}{\sqrt{5}} \\ -\frac{1}{\sqrt{5}} & -\frac{2}{\sqrt{5}} \end{bmatrix} \begin{bmatrix} \sqrt{5} & \frac{8}{\sqrt{5}} \\ 0 & \frac{1}{\sqrt{5}} \end{bmatrix} \Rightarrow A_1 = RQ = \begin{bmatrix} \sqrt{5} & \frac{8}{\sqrt{5}} \\ 0 & \frac{1}{\sqrt{5}} \end{bmatrix} \begin{bmatrix} \frac{2}{\sqrt{5}} & -\frac{1}{\sqrt{5}} \\ -\frac{1}{\sqrt{5}} & -\frac{2}{\sqrt{5}} \end{bmatrix} = \begin{bmatrix} \frac{2}{5} & -\frac{21}{5} \\ -\frac{1}{5} & -\frac{2}{5} \end{bmatrix}.$$

$$A_1 = Q_1R_1 = \begin{bmatrix} \frac{2}{\sqrt{5}} & -\frac{1}{\sqrt{5}} \\ -\frac{1}{\sqrt{5}} & -\frac{2}{\sqrt{5}} \end{bmatrix} \begin{bmatrix} \frac{1}{\sqrt{5}} & -\frac{8}{\sqrt{5}} \\ 0 & \sqrt{5} \end{bmatrix} \Rightarrow A_2 = R_1Q_2 = \begin{bmatrix} 2 & 3 \\ -1 & -2 \end{bmatrix} = A.$$

In this case, the QR algorithm returns to A because $Q_1 = Q^{-1}$ (and $R_1 = R^{-1}$).
In general, $A_1 = Q^{-1}AQ$ and $A_2 = Q_1^{-1}A_1Q_1$, so we have:
$A_2 = Q_1^{-1}A_1Q_1 = Q_1^{-1}(Q^{-1}AQ)Q_1 = (Q_1Q)^{-1}A(Q_1Q) = A.$

9. Note in this problem, unlike above, we take $A = A_1 = Q_1 R_1$. We proceed by induction.

$n = 2$: $Q_1 A_2 = Q_1(R_1 Q_1) = (Q_1 R_1) Q_1 = A Q_1$.

$n = k - 1$: Assume $Q_1 Q_2 \cdots Q_{k-2} A_{k-1} = A Q_1 Q_2 \cdots Q_{k-2}$, then we must show
$$Q_1 Q_2 \cdots Q_{k-1} A_k = A Q_1 Q_2 \cdots Q_{k-1}.$$

Note, $A_k = R_{k-1} Q_{k-1}$. We use this fact in the second step below.

$Q_1 Q_2 \cdots Q_{k-1} A_k = Q_1 Q_2 \cdots Q_{k-2}(Q_{k-1} A_k) = Q_1 Q_2 \cdots Q_{k-2}(Q_{k-1}(R_{k-1} Q_{k-1}))$
$= Q_1 Q_2 \cdots Q_{k-2}(Q_{k-1} R_{k-1}) Q_{k-1} = (Q_1 Q_2 \cdots Q_{k-2} A_{k-1}) Q_{k-1}$
$= (A Q_1 Q_2 \cdots Q_{k-2}) Q_{k-1}$ (by the $k-1$th assumption) $= A Q_1 Q_2 \cdots Q_{k-1}$.

So, we have shown: $Q_1 Q_2 \cdots Q_{k-1} A_k = A Q_1 Q_2 \cdots Q_{k-1}$.

Next, we need to show $(Q_1 Q_2 \cdots Q_k)(R_k R_{k-1} \cdots R_1) = A(Q_1 Q_2 \cdots Q_{k-1})(R_{k-1} \cdots R_2 R_1)$.
This follows immediately from the above by multiplying both sides by $R_{k-1} \cdots R_2 R_1$.
We need only note that $A_k R_{k-1} = Q_k R_k R_{k-1}$.

$Q_1 Q_2 \cdots Q_{k-1} A_k (R_{k-1} \cdots R_2 R_1) = A Q_1 Q_2 \cdots Q_{k-1}(R_{k-1} \cdots R_2 R_1) \Rightarrow$
$(Q_1 Q_2 \cdots Q_k)(R_k R_{k-1} \cdots R_1) = A(Q_1 Q_2 \cdots Q_{k-1})(R_{k-1} \cdots R_2 R_1)$.

Finally, we need to show $A^k = (Q_1 Q_2 \cdots Q_k)(R_k R_{k-1} \cdots R_1)$.
We proceed by induction.

$n = 1$: $A = Q_1 R_1$.

$n = k - 1$: Assume $A^{k-1} = (Q_1 Q_2 \cdots Q_{k-1})(R_{k-1} R_{k-2} \cdots R_1)$, then we must show
$$A^k = (Q_1 Q_2 \cdots Q_k)(R_k R_{k-1} \cdots R_1).$$

Note, from the first equation we proved above, we have: $A_k = (Q_1 Q_2 \cdots Q_{k-1})^{-1} A Q_1 Q_2 \cdots Q_{k-1}$.
We will use this fact in step 3 below.

$(Q_1 Q_2 \cdots Q_k)(R_k R_{k-1} \cdots R_1) = (Q_1 Q_2 \cdots Q_{k-1})(Q_k R_k)(R_{k-1} R_{k-2} \cdots R_1)$
$= (Q_1 Q_2 \cdots Q_{k-1})(A_k)(R_{k-1} R_{k-2} \cdots R_1)$
$= (Q_1 Q_2 \cdots Q_{k-1})((Q_1 Q_2 \cdots Q_{k-1})^{-1} A Q_1 Q_2 \cdots Q_{k-1})(R_{k-1} R_{k-2} \cdots R_1)$
$= A(Q_1 Q_2 \cdots Q_{k-1})(R_{k-1} R_{k-2} \cdots R_1) = A A^{k-1}$ (by the $k-1$th assumption) $= A^k$.

5.4 Orthogonal Diagonalization of Symmetric Matrices

Q: What are our main goals for Section 5.4?
A: To explore symmetric matrices. That is, learn how to compute
Q (orthogonal) and D (diagonal) such that $Q^T A Q = D$ or $A = Q D Q^T$.

Key Definitions and Concepts

orthogonally diagonalizable	p.397	5.4	If there exists an orthogonal Q and diagonal D such that $Q^T A Q = D$, then A is *orthogonally diagonalizable*.
projection form	p.402	5.4	If A is a real symmetric matrix with $A = QDQ^T$, then the *projection form of the Spectral Theorem* is $A = \lambda_1 \mathbf{q}_1 \mathbf{q}_1^T + \cdots + \lambda_n \mathbf{q}_n \mathbf{q}_n^T = \sum_{i=1}^{n} \lambda_i \mathbf{q}_i \mathbf{q}_i^T.$
spectral decomposition	p.402	5.4	If A is a real symmetric matrix with $A = QDQ^T$, then $A = \lambda_1 \mathbf{q}_1 \mathbf{q}_1^T + \cdots + \lambda_n \mathbf{q}_n \mathbf{q}_n^T = \sum_{i=1}^{n} \lambda_i \mathbf{q}_i \mathbf{q}_i^T.$
spectrum	p.400	5.4	The eigenvalues of an *orthogonally diagonalizable* matrix.
symmetric	p.149	3.1	A matrix A is *symmetric* if and only if $A^T = A$.

Theorems

Thm 5.17 p.398 5.4 If A is *orthogonally diagonalizable*,
then A is *symmetric*.

Thm 5.18 p.398 5.4 If A is a *real* symmetric matrix,
then the eigenvalues of A are *real*.

Thm 5.19 p.399 5.4 If A is a *symmetric* matrix, then any two eigenvectors
corresponding to distinct eigenvalues of A are *orthogonal*.

Thm 5.20 p.400 5.4 **The Spectral Theorem:**
Let A be an $n \times n$ *real* matrix. Then A is *symmetric*
if and only if A is *orthogonally diagonalizable*.

Discussion of Key Definitions and Concepts

It is very instructive to compare this Section to Section 4.4, *Similarity and Diagonalization*. In Section 4.4, we observed that in the key condition to be diagonalizable, $P^{-1}AP = D$. So, diagonalizable matrices are a kind of *generalization* of diagonal matrices.

Q: What is the stronger condition on A we have in this section?
A: In this Section, we require that A be real symmetric as well.

Q: What does this stronger condition on A imply about Q?
A: When A is real symmetric, we get Q is orthogonal.

Q: In Section 4.4, we observed that the columns of P were eigenvectors of A. Is that also true of Q?
A: Yes, in fact this is precisely why Q is orthogonal.

Q: What theorem from this Section coupled with the above implies Q is orthogonal?
A: Theorem 5.19 asserts that the eigenvectors of two distinct eigenvalues of A are orthogonal. Using Gram-Schmidt and normalizing allows us to create a matrix Q that is orthogonal. Q is orthogonal means its columns form an orthonormal set. Recall:

For a set to be orthonormal it must have two key properties:

$$\mathbf{q}_i \cdot \mathbf{q}_j = \begin{cases} 0 & \text{if } i \neq j \text{ Property 1} \\ 1 & \text{if } i = j \text{ Property 2} \end{cases}$$

Q: The eigenvalues of a diagonal matrix are its diagonal entries. How does this property get generalized for *orthogonally diagonalizable* matrices?
A: If $Q^T A Q = D$, then the eigenvalues of A are the diagonal entries of D.

Q: Diagonal matrices commute. How does this property get generalized for *orthogonally diagonalizable* matrices?
A: In order to understand this generalization, see the solution to Exercise 15.

Q: Diagonal matrices are invertible. How does this property get generalized for diagonalizable matrices?
A: A diagonalizable matrix is invertible if its eigenvectors span $\mathbb{R}^n$. This is essentially the assertion of Theorem 4.27(b) of Section 4.4.

Q: So, is an *orthogonally diagonalizable* matrice invertible?
A: Yes, because its eigenvectors span $\mathbb{R}^n$. More directly, we observe Q and D are invertible and $A = QDQ^T$.

Q: If D is diagonal, then D^{-1} is diagonal. How does this generalize? If A is *orthogonally diagonalizable*, so is A^{-1}?
A: See Exercise 14. Is the fact that D^{-1} is diagonal is critical to the proof as it was in Section 4.4?

5.4 Orthogonal Diagonalization of Symmetric Matrices

Solutions to odd-numbered exercises from Section 5.4

1. $\begin{vmatrix} 4-\lambda & 1 \\ 1 & 4-\lambda \end{vmatrix} = 0 \Rightarrow \lambda^2 - 8\lambda + 15 = (\lambda-5)(\lambda-3) = 0 \Rightarrow \lambda_1 = 5, \lambda_2 = 3 \Rightarrow D = \begin{bmatrix} 5 & 0 \\ 0 & 3 \end{bmatrix}$.

As a reminder, we show how to find $\mathbf{v}_1$ (make sure $\mathbf{v}_2$ is orthogonal to $\mathbf{v}_1$).

$(A - 5I)\mathbf{v}_1 = \begin{bmatrix} -1 & 1 \\ 1 & -1 \end{bmatrix} \begin{bmatrix} x_1 \\ x_2 \end{bmatrix} = \begin{bmatrix} 0 \\ 0 \end{bmatrix} \Rightarrow x_1 = x_2 \Rightarrow \mathbf{v}_1 = \begin{bmatrix} 1 \\ 1 \end{bmatrix}$ is a basis.

So, $\mathbf{v}_1 = \begin{bmatrix} 1 \\ 1 \end{bmatrix}, \mathbf{v}_2 = \begin{bmatrix} 1 \\ -1 \end{bmatrix}$. Normalizing $\Rightarrow Q = \begin{bmatrix} 1/\sqrt{2} & 1/\sqrt{2} \\ 1/\sqrt{2} & -1/\sqrt{2} \end{bmatrix}$.

Verify $Q^T A Q = D = \begin{bmatrix} 1/\sqrt{2} & 1/\sqrt{2} \\ 1/\sqrt{2} & -1/\sqrt{2} \end{bmatrix} \begin{bmatrix} 4 & 1 \\ 1 & 4 \end{bmatrix} \begin{bmatrix} 1/\sqrt{2} & 1/\sqrt{2} \\ 1/\sqrt{2} & -1/\sqrt{2} \end{bmatrix} = \begin{bmatrix} 5 & 0 \\ 0 & 3 \end{bmatrix}$.

3. $\begin{vmatrix} 1-\lambda & \sqrt{2} \\ \sqrt{2} & 0-\lambda \end{vmatrix} = 0 \Rightarrow \lambda^2 - \lambda - 2 = (\lambda-2)(\lambda+1) = 0 \Rightarrow \lambda_1 = 2, \lambda_2 = -1 \Rightarrow D = \begin{bmatrix} 2 & 0 \\ 0 & -1 \end{bmatrix}$.

Furthermore, $\mathbf{v}_1 = \begin{bmatrix} \sqrt{2} \\ 1 \end{bmatrix}, \mathbf{v}_2 = \begin{bmatrix} 1 \\ -\sqrt{2} \end{bmatrix}$. Normalizing $\Rightarrow Q = \begin{bmatrix} 2/\sqrt{6} & 1/\sqrt{3} \\ 1/\sqrt{3} & -2/\sqrt{6} \end{bmatrix}$.

5. $\begin{vmatrix} 5-\lambda & 0 & 0 \\ 0 & 1-\lambda & 3 \\ 0 & 3 & 1-\lambda \end{vmatrix} = 0 \Rightarrow (5-\lambda)(\lambda^2 - 2\lambda - 8) = 0 \Rightarrow D = \begin{bmatrix} 5 & 0 & 0 \\ 0 & 4 & 0 \\ 0 & 0 & -2 \end{bmatrix}$.

As a reminder, in this case we show how to find $\mathbf{v}_1$ ($\mathbf{v}_2$ and $\mathbf{v}_3$ are extremely similar).

$(A - 5I)\mathbf{v}_1 = \begin{bmatrix} 0 & 0 & 0 \\ 0 & -4 & 3 \\ 0 & 3 & -4 \end{bmatrix} \begin{bmatrix} x_1 \\ x_2 \\ x_3 \end{bmatrix} = \begin{bmatrix} 0 \\ 0 \\ 0 \end{bmatrix} \Rightarrow \begin{array}{c} x_1 \text{ is free} \\ x_2 = x_3 = 0 \end{array} \Rightarrow \mathbf{v}_1 = \begin{bmatrix} 1 \\ 0 \\ 0 \end{bmatrix}$ is a basis.

So, $\mathbf{v}_1 = \begin{bmatrix} 1 \\ 0 \\ 0 \end{bmatrix}, \mathbf{v}_2 = \begin{bmatrix} 0 \\ 1 \\ 1 \end{bmatrix}, \mathbf{v}_3 = \begin{bmatrix} 0 \\ -1 \\ 1 \end{bmatrix}$. Normalizing $\Rightarrow Q = \begin{bmatrix} 1 & 0 & 0 \\ 0 & 1/\sqrt{2} & -1/\sqrt{2} \\ 0 & 1/\sqrt{2} & 1/\sqrt{2} \end{bmatrix}$.

7. $\begin{vmatrix} 1-\lambda & 0 & -1 \\ 0 & 1-\lambda & 0 \\ -1 & 0 & 1-\lambda \end{vmatrix} = 0 \Rightarrow (1-\lambda)(\lambda^2 - 2\lambda) = 0 \Rightarrow D = \begin{bmatrix} 2 & 0 & 0 \\ 0 & 1 & 0 \\ 0 & 0 & 0 \end{bmatrix}$.

So, $\mathbf{v}_1 = \begin{bmatrix} -1 \\ 0 \\ 1 \end{bmatrix}, \mathbf{v}_2 = \begin{bmatrix} 0 \\ 1 \\ 0 \end{bmatrix}, \mathbf{v}_3 = \begin{bmatrix} 1 \\ 0 \\ 1 \end{bmatrix}$. Normalizing $\Rightarrow Q = \begin{bmatrix} -1/\sqrt{2} & 0 & 1/\sqrt{2} \\ 0 & 1 & 0 \\ 1/\sqrt{2} & 0 & 1/\sqrt{2} \end{bmatrix}$.

9. $\begin{vmatrix} 1-\lambda & 1 & 0 & 0 \\ 1 & 1-\lambda & 0 & 0 \\ 0 & 0 & 1-\lambda & 1 \\ 0 & 0 & 1 & 1-\lambda \end{vmatrix} = 0 \Rightarrow \lambda^2(\lambda-2)^2 = 0 \Rightarrow D = \begin{bmatrix} 2 & 0 & 0 & 0 \\ 0 & 2 & 0 & 0 \\ 0 & 0 & 0 & 0 \\ 0 & 0 & 0 & 0 \end{bmatrix}.$

$\lambda_1 = 2$ of multiplicity $2 \Rightarrow (A - 2I)\mathbf{v} = \mathbf{0} \Rightarrow \begin{matrix} x_2 = x_1 \\ x_4 = x_3 \end{matrix} \Rightarrow \mathbf{v}_1 = \begin{bmatrix} 1 \\ 1 \\ 0 \\ 0 \end{bmatrix}, \mathbf{v}_2 = \begin{bmatrix} 0 \\ 0 \\ 1 \\ 1 \end{bmatrix}.$

$\lambda_2 = 0$ of multiplicity $2 \Rightarrow (A - 0I)\mathbf{v} = \mathbf{0} \Rightarrow \begin{matrix} x_2 = -x_1 \\ x_4 = -x_3 \end{matrix} \Rightarrow \mathbf{v}_3 = \begin{bmatrix} 1 \\ -1 \\ 0 \\ 0 \end{bmatrix}, \mathbf{v}_4 = \begin{bmatrix} 0 \\ 0 \\ 1 \\ -1 \end{bmatrix}.$

These choices make the $\mathbf{v}_i$ orthogonal. Normalizing $\Rightarrow Q = \begin{bmatrix} 1/\sqrt{2} & 0 & 1/\sqrt{2} & 0 \\ 1/\sqrt{2} & 0 & -1/\sqrt{2} & 0 \\ 0 & 1/\sqrt{2} & 0 & 1/\sqrt{2} \\ 0 & 1/\sqrt{2} & 0 & -1/\sqrt{2} \end{bmatrix}.$

11. $\begin{vmatrix} a-\lambda & b \\ b & a-\lambda \end{vmatrix} = 0 \Rightarrow (a-\lambda)^2 - b^2 = 0 \Rightarrow \lambda = a \pm b, \Rightarrow D = \begin{bmatrix} a+b & 0 \\ 0 & a-b \end{bmatrix}.$

We show how to find $\mathbf{v}_1$ (make sure $\mathbf{v}_2$ is orthogonal to $\mathbf{v}_1$).

$(A - (a+b)I)\mathbf{v}_1 = \begin{bmatrix} -b & b \\ b & -b \end{bmatrix}\begin{bmatrix} x_1 \\ x_2 \end{bmatrix} = \begin{bmatrix} 0 \\ 0 \end{bmatrix} \Rightarrow x_1 = x_2 \Rightarrow \mathbf{v}_1 = \begin{bmatrix} 1 \\ 1 \end{bmatrix}$ is a basis.

So, $\mathbf{v}_1 = \begin{bmatrix} 1 \\ 1 \end{bmatrix}, \mathbf{v}_2 = \begin{bmatrix} 1 \\ -1 \end{bmatrix}$. Normalizing $\Rightarrow Q = \begin{bmatrix} 1/\sqrt{2} & 1/\sqrt{2} \\ 1/\sqrt{2} & -1/\sqrt{2} \end{bmatrix}.$

Verify $Q^T A Q = D = \begin{bmatrix} 1/\sqrt{2} & 1/\sqrt{2} \\ 1/\sqrt{2} & -1/\sqrt{2} \end{bmatrix}\begin{bmatrix} a & b \\ b & a \end{bmatrix}\begin{bmatrix} 1/\sqrt{2} & 1/\sqrt{2} \\ 1/\sqrt{2} & -1/\sqrt{2} \end{bmatrix} = \begin{bmatrix} a+b & 0 \\ 0 & a-b \end{bmatrix}.$

13. (a) A, B orthogonally diagonalizable $\Rightarrow A, B$ symmetric $\Rightarrow A + B$ symmetric $\Rightarrow Q^T(A+B)Q = D$.

(b) A orthogonally diagonalizable $\Rightarrow A$ symmetric $\Rightarrow cA$ symmetric $\Rightarrow Q^T(cA)Q = D$.

(c) AA^T is symmetric $\Rightarrow$ if A symmetric, then $AA^T = AA = A^2$ symmetric.
So, A orthogonally diagonalizable $\Rightarrow A$ symmetric $\Rightarrow A^2$ symmetric $\Rightarrow Q^T(A^2)Q = D$.

15. By Exercise 36 in Section 3.2, A, B symmetric and $AB = BA \Rightarrow AB$ symmetric.
So, A, B symmetric and $AB = BA$ (given) $\Rightarrow AB$ symmetric $\Rightarrow Q^T(AB)Q = D$.

17. $A = \lambda_1 \mathbf{q}_1 \mathbf{q}_1^T + \lambda_2 \mathbf{q}_2 \mathbf{q}_2^T$

$= 5 \begin{bmatrix} 1/2 & 1/2 \\ 1/2 & 1/2 \end{bmatrix} + 3 \begin{bmatrix} 1/2 & -1/2 \\ -1/2 & 1/2 \end{bmatrix} = \begin{bmatrix} 5/2 & 5/2 \\ 5/2 & 5/2 \end{bmatrix} + \begin{bmatrix} 3/2 & -3/2 \\ -3/2 & 3/2 \end{bmatrix}.$

5.4 Orthogonal Diagonalization of Symmetric Matrices

19. $A = 5 \begin{bmatrix} 1 & 0 & 0 \\ 0 & 0 & 0 \\ 0 & 0 & 0 \end{bmatrix} + 4 \begin{bmatrix} 0 & 0 & 0 \\ 0 & 1/2 & 1/2 \\ 0 & 1/2 & 1/2 \end{bmatrix} - 2 \begin{bmatrix} 0 & 0 & 0 \\ 0 & 1/2 & -1/2 \\ 0 & -1/2 & 1/2 \end{bmatrix}$

$= \begin{bmatrix} 5 & 0 & 0 \\ 0 & 0 & 0 \\ 0 & 0 & 0 \end{bmatrix} + \begin{bmatrix} 0 & 0 & 0 \\ 0 & 2 & 2 \\ 0 & 2 & 2 \end{bmatrix} + \begin{bmatrix} 0 & 0 & 0 \\ 0 & -1 & 1 \\ 0 & 1 & -1 \end{bmatrix}.$

21. $A = \lambda_1 q_1 q_1^T + \lambda_2 q_2 q_2^T = (-1) \begin{bmatrix} 1/2 & 1/2 \\ 1/2 & 1/2 \end{bmatrix} + 2 \begin{bmatrix} 1/2 & -1/2 \\ -1/2 & 1/2 \end{bmatrix} = \begin{bmatrix} 1/2 & -3/2 \\ -3/2 & 1/2 \end{bmatrix}.$

23. $A = 1 \begin{bmatrix} 1/2 & 1/2 & 0 \\ 1/2 & 1/2 & 0 \\ 0 & 0 & 0 \end{bmatrix} + 2 \begin{bmatrix} 1/3 & -1/3 & 1/3 \\ -1/3 & 1/3 & -1/3 \\ 1/3 & -1/3 & 1/3 \end{bmatrix} + 3 \begin{bmatrix} 1/6 & -1/6 & -1/3 \\ -1/6 & 1/6 & 1/3 \\ -1/3 & 1/3 & 2/3 \end{bmatrix}$

$= \begin{bmatrix} 5/3 & -2/3 & -1/3 \\ -2/3 & 5/3 & 1/3 \\ -1/3 & 1/3 & 8/3 \end{bmatrix}.$

25. $\text{proj}_W(\mathbf{v}) = (\frac{\mathbf{q} \cdot \mathbf{v}}{\mathbf{q} \cdot \mathbf{q}})\mathbf{q} = (\mathbf{q}^T \mathbf{v})\mathbf{q} = \mathbf{q}(\mathbf{q}^T \mathbf{v}) = (\mathbf{q}\mathbf{q}^T)\mathbf{v}.$

   Note: $(\mathbf{q}^T \mathbf{v})\mathbf{q} = \mathbf{q}(\mathbf{q}^T \mathbf{v})$ because $\mathbf{q}^T \mathbf{v}$ is a scalar.

27. Let $\mathbf{v}_1$ be the normalized eigenvector for λ_1 then extend $\mathbf{v}_1$ to orthonormal basis $\{\mathbf{v}_1, \ldots, \mathbf{v}_n\} \Rightarrow$

$Q_1^T A Q_1 = \begin{bmatrix} \mathbf{v}_1^T \\ \vdots \\ \mathbf{v}_n^T \end{bmatrix} A \begin{bmatrix} \mathbf{v}_1 & \cdots & \mathbf{v}_n \end{bmatrix} = \begin{bmatrix} \mathbf{v}_1^T \\ \vdots \\ \mathbf{v}_n^T \end{bmatrix} \begin{bmatrix} A\mathbf{v}_1 & \cdots & A\mathbf{v}_n \end{bmatrix}$

$= \begin{bmatrix} \mathbf{v}_1^T \\ \vdots \\ \mathbf{v}_n^T \end{bmatrix} \begin{bmatrix} \lambda_1 \mathbf{v}_1 & \cdots & \lambda_n \mathbf{v}_n \end{bmatrix} = \left[\begin{array}{c|c} \lambda_1 & * \\ \hline 0 & A_1 \end{array} \right] = T.$

Note that T is upper triangular and that the result follows by induction.

5.5 Applications

Q: What are our main goals for Section 5.5?
A: To explore applications of $W^\perp$ (dual codes) and decomposition (quadratic forms).
To better understand dual codes, it may be useful to review Section 3.7 starting on p. 240.

Key Definitions and Concepts

1	p.411	5.5	**1** (the vector in $\mathbb{Z}_2^n$ all of whose entries are 1)
C	p.408	5.5	C (a set of code vectors in $\mathbb{Z}_2^n$)
$C^\perp$	p.408	5.5	$C^\perp = \{\mathbf{x} \text{ in } \mathbb{Z}_2^n : \mathbf{c} \cdot \mathbf{x} = 0 \text{ for all } \mathbf{c} \text{ in } C\}$
G	p.405	5.5	$G = \left[\dfrac{I_m}{A}\right]$ (standard generator matrix for a code)
P	p.405	5.5	$G = [\,B\,|\,I_{n-m}\,]$ (standard parity check matrix for a code)
$w(\mathbf{x})$	p.411	5.5	$w(\mathbf{x})$ (*weight* of $\mathbf{x}$), the number of 1s in $\mathbf{x}$ (in $\mathbb{Z}_2^n$)
$f(\mathbf{x})$	p.412	5.5	$f(\mathbf{x}) = \mathbf{x}^T A \mathbf{x}$ (*quadratic form*), (A, the matrix associated with f, is symmetric)
$\mathbf{x}^T A \mathbf{x}$	p.412	5.5	$\mathbf{x}^T A \mathbf{x} = a_{11} x_1^2 + a_{22} x_2^2 + \cdots + a_{nn} x_n^2 + \displaystyle\sum_{i<j} 2 a_{ij} x_i x_j$
definite	p.415	5.5	A quadratic form $f(\mathbf{x}) = \mathbf{x}^T A \mathbf{x}$ is classified as follows: 1. *positive definite* $f(\mathbf{x}) > 0$ for all $\mathbf{x} \neq \mathbf{0}$ 2. *positive semidefinite* $f(\mathbf{x}) \geq 0$ for all $\mathbf{x}$ 3. *negative definite* $f(\mathbf{x}) < 0$ for all $\mathbf{x} \neq \mathbf{0}$ 4. *negative semidefinite* $f(\mathbf{x}) \leq 0$ for all $\mathbf{x}$ 5. *indefinite* if $f(\mathbf{x})$ takes on both positive and negative values The associated matrix A is classified in exactly the same way.
generator matrix	p.406	5.5	For $n > m$, an $n \times m$ matrix G and an $(n-k) \times n$ matrix P (with entries in $\mathbb{Z}_2$) are a *generator matrix* and a *parity check matrix* respectively for an (n, k) binary code C if: 1. The columns of G are linearly independent. 2. The rows of P are linearly independent. 3. $PG = O$.
parity check	p.406	5.5	See *generator matrix* for details: the definitions and conditions are associated.
quadratic form	p.412	5.5	$f(\mathbf{x}) = \mathbf{x}^T A \mathbf{x}$ (*quadratic form*), (A, the matrix associated with f, is symmetric)
weight	p.411	5.5	$w(\mathbf{x})$ (*weight* of $\mathbf{x}$), the number of 1s in $\mathbf{x}$ (in $\mathbb{Z}_2^n$)

Theorems

Thm 5.21   p.409   5.5   If C is an (n, k) binary code with generator matrix G and parity check matrix P, then $C^\perp$ is an $(n, n-k)$ binary code such that
a. G^T is parity check matrix for $C^\perp$.
b. P^T is generator matrix for $C^\perp$.

Thm 5.22   p.411   5.5   If C is a self dual code, then
a. Every vector in C has even weight.
b. $\mathbf{1}$ is in C.

Thm 5.23   p.411   5.5   **The Principal Axes Theorem:**
Every quadratic form can be diagonalized:
$\mathbf{x}^T A \mathbf{x} = \mathbf{y}^T D \mathbf{y} = \lambda_1 y_1^2 + \cdots + \lambda_n y_n^2$
(see statement of Theorem in the text for details)

Thm 5.24   p.416   5.5   The quadratic form $f(\mathbf{x}) = \mathbf{x}^T A \mathbf{x}$ is:
a. *positive definite* $\Leftrightarrow \lambda_i > 0$
b. *positive semidefinite* $\Leftrightarrow \lambda_i \geq 0$
c. *negative definite* $\Leftrightarrow \lambda_i < 0$
d. *negative semidefinite* $\Leftrightarrow \lambda_i \leq 0$
e. *indefinite* $\Leftrightarrow \lambda_i$ are both positive and negative

Thm 5.25   p.417   5.5   Given $f(\mathbf{x}) = \mathbf{x}^T A \mathbf{x}$,
$\lambda_1 \geq \lambda_2 \geq \cdots \geq \lambda_n$, and $\|\mathbf{x}\| = 1$:
a. $\lambda_1 \geq f(\mathbf{x}) \geq \lambda_n$
b. $\max f(\mathbf{x}) = \lambda_1$ occurs for a unit eigenvector
c. $\min f(\mathbf{x}) = \lambda_n$ occurs for a unit eigenvector

Discussion of Key Definitions and Concepts

Since these applications are fully explained in the text, we simply present solutions here.

5.5   Applications

Solutions to odd-numbered exercises from Section 5.5

1. $C_1' = C_1 + C_2 \Rightarrow G' \longrightarrow \begin{bmatrix} 1 & 0 \\ 0 & 1 \\ 1 & 0 \end{bmatrix}$ (standard form, $\begin{bmatrix} I \\ \hline A \end{bmatrix}$) $\Rightarrow C' = C$ (R1 not required).

3. $R_1 = 0 \Rightarrow$ R1 required $\Rightarrow C' \neq C$.

$\begin{array}{l} R_1' = R_4 \\ R_4' = R_1 \end{array} \Rightarrow G \longrightarrow \begin{bmatrix} 1 & 1 & 1 \\ 1 & 0 & 1 \\ 0 & 1 & 1 \\ 0 & 0 & 0 \end{bmatrix}$, then $\begin{array}{l} C_1' = C_1 + C_2 + C_3 \\ C_2' = C_2 + C_3 \\ C_3' = C_1 + C_3 \end{array} \Rightarrow G' \longrightarrow \begin{bmatrix} 1 & 0 & 0 \\ 0 & 1 & 0 \\ 0 & 0 & 1 \\ 0 & 0 & 0 \end{bmatrix}$.

5. Only one row $\Rightarrow$ C1 required $\Rightarrow C' \neq C$.

$\begin{array}{l} C_1' = C_1 \\ C_2' = C_1 + C_2 \\ C_3' = C_1 \end{array} \Rightarrow P' \longrightarrow \begin{bmatrix} 1 & 0 & 1 \end{bmatrix}$.

7. $C_3 = C_4 \Rightarrow$ C1 required $\Rightarrow C' \neq C$.

$\begin{array}{l} C_1' = C_4 \\ C_4' = C_1 \end{array} \Rightarrow P \longrightarrow \begin{bmatrix} 1 & 1 & 1 & 0 & 0 \\ 0 & 1 & 0 & 1 & 1 \\ 1 & 0 & 1 & 0 & 1 \end{bmatrix}$, then $\begin{array}{l} R_1' = R_1 \\ R_2' = R_1 + R_2 + R_3 \\ R_3' = R_1 + R_3 \end{array} \Rightarrow P' \longrightarrow \begin{bmatrix} 1 & 1 & 1 & 0 & 0 \\ 0 & 0 & 0 & 1 & 0 \\ 0 & 1 & 0 & 0 & 1 \end{bmatrix}$.

9. $C = \left\{ \begin{bmatrix} 0 \\ 0 \\ 0 \end{bmatrix}, \begin{bmatrix} 0 \\ 1 \\ 0 \end{bmatrix} \right\} \Rightarrow A = \begin{bmatrix} 0 & 0 & 0 \\ 0 & 1 & 0 \end{bmatrix}$. Therefore, to find $null(A)$ we compute $A\mathbf{x} = 0$.

$\Rightarrow \begin{bmatrix} 0 & 0 & 0 \\ 0 & 1 & 0 \end{bmatrix} \begin{bmatrix} x_1 \\ x_2 \\ x_3 \end{bmatrix} = \begin{bmatrix} 0 \\ 0 \\ 0 \end{bmatrix} \Rightarrow \begin{array}{l} x_1, x_3 = 0 \text{ or } 1 \\ x_2 = 0 \end{array} \Rightarrow C^\perp = \left\{ \begin{bmatrix} 0 \\ 0 \\ 0 \end{bmatrix}, \begin{bmatrix} 0 \\ 0 \\ 1 \end{bmatrix}, \begin{bmatrix} 1 \\ 0 \\ 0 \end{bmatrix}, \begin{bmatrix} 1 \\ 0 \\ 1 \end{bmatrix} \right\}$.

11. $\begin{bmatrix} 0 & 0 & 0 & 0 \\ 0 & 1 & 0 & 0 \\ 0 & 1 & 0 & 1 \\ 0 & 0 & 0 & 1 \end{bmatrix} \begin{bmatrix} x_1 \\ x_2 \\ x_3 \\ x_4 \end{bmatrix} = 0 \Rightarrow \begin{array}{l} x_1, x_3 = 0 \text{ or } 1 \\ x_2 = x_4 = 0 \end{array} \Rightarrow C^\perp = \left\{ \begin{bmatrix} 0 \\ 0 \\ 0 \\ 0 \end{bmatrix}, \begin{bmatrix} 0 \\ 0 \\ 1 \\ 0 \end{bmatrix}, \begin{bmatrix} 1 \\ 0 \\ 0 \\ 0 \end{bmatrix}, \begin{bmatrix} 1 \\ 0 \\ 1 \\ 0 \end{bmatrix} \right\}$.

13. $P^\perp = G^T = \begin{bmatrix} 1 & 1 & 1 & 0 \\ 1 & 1 & 0 & 1 \end{bmatrix}$ (standard) $\Rightarrow A = \begin{bmatrix} 1 & 1 \\ 1 & 1 \end{bmatrix} \Rightarrow G^\perp = \begin{bmatrix} I \\ \hline A \end{bmatrix} = \begin{bmatrix} 1 & 0 \\ 0 & 1 \\ 1 & 1 \\ 1 & 1 \end{bmatrix}$.

15. $\begin{array}{l} C_1' = C_1 + C_2 \\ C_2' = C_2 \end{array} \Rightarrow G^\perp \longrightarrow \begin{bmatrix} 1 & 0 \\ 0 & 1 \\ 1 & 0 \\ 1 & 1 \end{bmatrix} \Rightarrow A = \begin{bmatrix} 1 & 0 \\ 1 & 1 \end{bmatrix} \Rightarrow P^\perp = \begin{bmatrix} 1 & 0 & 1 & 0 \\ 1 & 1 & 0 & 1 \end{bmatrix}$.

17. $P^\perp = G^T = \begin{bmatrix} 1 & 0 & 0 & 0 & 1 & 1 & 0 \\ 0 & 1 & 0 & 0 & 1 & 0 & 1 \\ 0 & 0 & 1 & 0 & 0 & 1 & 1 \\ 0 & 0 & 0 & 1 & 1 & 1 & 1 \end{bmatrix}$ and $G^\perp = P^T = \begin{bmatrix} 1 & 1 & 0 \\ 1 & 0 & 1 \\ 0 & 1 & 1 \\ 1 & 1 & 1 \\ 1 & 0 & 0 \\ 0 & 1 & 0 \\ 0 & 0 & 1 \end{bmatrix}$.

19. We will show $Rep_n \subseteq E_n^\perp$ and $E_n^\perp \subseteq Rep_n \Rightarrow Rep_n = E_n^\perp$.

    $\mathbf{x}$ in $E_n \Rightarrow w(\mathbf{x}) = 0$ because vectors in E_n have even weight.
    Therefore, $\mathbf{x} \cdot \mathbf{0} = 0$ and $\mathbf{x} \cdot \mathbf{1} = w(\mathbf{x}) = 0 \Rightarrow Rep_n \subseteq E_n^\perp$.

    To show $E_n^\perp \subseteq Rep_n$ we will show that any vector $\mathbf{v}$ that is *not* in Rep_n is *not* in $E_n^\perp$.

    $\mathbf{v}$ not in $Rep_n \Rightarrow \mathbf{v}$ is not all 0s, 1s $\Rightarrow$ there is an ith position $= 0$ and a jth position $= 1$.
    $\mathbf{e}_{ij}$ with a 1 in the ith and j^{th} positions and 0s elsewhere has even weight $\Rightarrow \mathbf{e}_{ij}$ is in E_n.
    However, $\mathbf{e}_{ij} \cdot \mathbf{v} = 1 \neq 0 \Rightarrow \mathbf{v}$ is not in $E_n^\perp \Rightarrow E_n^\perp \subseteq Rep_n \Rightarrow Rep_n = E_n^\perp$.

21. We need only show that $(C^\perp)^\perp$ and C have the same generator matrix G.

    $G = \begin{bmatrix} I \\ A \end{bmatrix} \Rightarrow P = [\, A \,|\, I \,] \Rightarrow G^\perp = P^T \Rightarrow (C^\perp)^\perp$ has $(P^\perp)^\perp = (G^\perp)^T = (P^T)^T = P$

    So, $P = [\, A \,|\, I \,] \Rightarrow G = \begin{bmatrix} I \\ A \end{bmatrix} \Rightarrow (C^\perp)^\perp$ has generator matrix $G \Rightarrow (C^\perp)^\perp = C$.

23. $f(\mathbf{x}) = \mathbf{x}^T A \mathbf{x} = 2x^2 + 4y^2 + 6xy$.

25. $f(\mathbf{x}) = \mathbf{x}^T A \mathbf{x} = 3(1)^2 + 4(6)^2 - 4(1)(6) = 123$.

27. $f(\mathbf{x}) = \mathbf{x}^T A \mathbf{x} = 1(2)^2 + 2(-1)^2 + 3(1)^2 + 0(2)(-1) - 6(2)(1) + 2(-1)(1) = -5$.

29. $x_1^2 + 2x_2^2 + 6x_1x_2 \Rightarrow$ diagonal $= 1, 2$ and corners $= \frac{1}{2}6 = 3 \Rightarrow A = \begin{bmatrix} 1 & 3 \\ 3 & 2 \end{bmatrix}$.

31. $3x^2 - 3xy - y^2 \Rightarrow$ diagonal $= 3, -1$ and corners $= \frac{1}{2}(-3) = -\frac{3}{2} \Rightarrow A = \begin{bmatrix} 3 & -3/2 \\ -3/2 & -1 \end{bmatrix}$.

33. $\Rightarrow$ diagonal $= 5, -1, 2$ and off diagonal $= \frac{1}{2}2 = 1, \frac{1}{2}(-4) = -2, \frac{1}{2}4 = 2 \Rightarrow A = \begin{bmatrix} 5 & 1 & -2 \\ 1 & -1 & 2 \\ -2 & 2 & 2 \end{bmatrix}$.

35. $A = \begin{bmatrix} 2 & -2 \\ -2 & 5 \end{bmatrix} \Rightarrow \begin{vmatrix} 2-\lambda & -2 \\ -2 & 5-\lambda \end{vmatrix} = \lambda^2 - 7\lambda + 6 = 0 \Rightarrow D = \begin{bmatrix} 1 & 0 \\ 0 & 6 \end{bmatrix}$.

    Furthermore, $\mathbf{v}_1 = \begin{bmatrix} 2 \\ 1 \end{bmatrix}$, $\mathbf{v}_2 = \begin{bmatrix} 1 \\ -2 \end{bmatrix}$. Normalizing $\Rightarrow Q = \begin{bmatrix} 2/\sqrt{5} & 1/\sqrt{5} \\ 1/\sqrt{5} & -2/\sqrt{5} \end{bmatrix}$.

    So, the new quadratic form is $f(\mathbf{y}) = \mathbf{y}^T D \mathbf{y} = \begin{bmatrix} y_1 & y_2 \end{bmatrix} \begin{bmatrix} 1 & 0 \\ 0 & 6 \end{bmatrix} \begin{bmatrix} y_1 \\ y_2 \end{bmatrix} = y_1^2 + 6y_2^2$.

5.5 Applications

37. $A = \begin{bmatrix} 7 & 4 & 4 \\ 4 & 1 & -8 \\ 4 & -8 & 1 \end{bmatrix} \Rightarrow \begin{vmatrix} 7-\lambda & 4 & 4 \\ 4 & 1-\lambda & -8 \\ 4 & -8 & 1-\lambda \end{vmatrix} = (\lambda-9)^2(\lambda+9) = 0 \Rightarrow D = \begin{bmatrix} 9 & 0 & 0 \\ 0 & 9 & 0 \\ 0 & 0 & -9 \end{bmatrix}$.

$(A - 9I)\mathbf{v} = \mathbf{0}$ (of multiplicity 2) $\Rightarrow x_1 = 2(x_2 + x_3) \Rightarrow \mathbf{v}_1 = \begin{bmatrix} 2 \\ 0 \\ 1 \end{bmatrix}, \mathbf{v}_2 = \begin{bmatrix} 2 \\ 1 \\ 0 \end{bmatrix}$.

$(A - (-9)I)\mathbf{v} = \mathbf{0}$ (of multiplicity 1) $\Rightarrow x_1 = -\frac{1}{4}(x_2 + x_3) \Rightarrow \mathbf{v}_3 = \begin{bmatrix} -1 \\ 2 \\ 2 \end{bmatrix}$.

$\Rightarrow Q = \begin{bmatrix} 2/\sqrt{5} & 2/\sqrt{5} & -1/3 \\ 0 & 1/\sqrt{5} & 2/3 \\ 1/\sqrt{5} & 0 & 2/3 \end{bmatrix} \Rightarrow f(\mathbf{y}) = \mathbf{y}^T D \mathbf{y} = 9y_1^2 + 9y_2^2 - 9y_3^2$.

39. $A = \begin{bmatrix} 1 & -1 & 0 \\ -1 & 0 & 1 \\ 0 & 1 & 1 \end{bmatrix} \Rightarrow \begin{vmatrix} 1-\lambda & -1 & 0 \\ -1 & 0-\lambda & 1 \\ 0 & 1 & 1-\lambda \end{vmatrix} = (\lambda - 2)(\lambda^2 - 1) = 0 \Rightarrow D = \begin{bmatrix} 2 & 0 & 0 \\ 0 & 1 & 0 \\ 0 & 0 & -1 \end{bmatrix}$.

$(A - \lambda I)\mathbf{v} = \mathbf{0} \Rightarrow \mathbf{v}_1 = \begin{bmatrix} 1 \\ -1 \\ -1 \end{bmatrix}, \mathbf{v}_2 = \begin{bmatrix} 1 \\ 0 \\ 1 \end{bmatrix}, \mathbf{v}_3 = \begin{bmatrix} 1 \\ 2 \\ -1 \end{bmatrix}$.

$\Rightarrow Q = \begin{bmatrix} 1/\sqrt{3} & 1/\sqrt{2} & 1/\sqrt{6} \\ -1/\sqrt{3} & 0 & 2/\sqrt{6} \\ -1/\sqrt{3} & 1/\sqrt{2} & -1/\sqrt{6} \end{bmatrix} \Rightarrow f(\mathbf{x}') = (\mathbf{x}')^T D(\mathbf{x}') = 2(x')^2 + (y')^2 - (z')^2$.

41. $A = \begin{bmatrix} 1 & 0 \\ 0 & 2 \end{bmatrix} \Rightarrow \begin{vmatrix} 1-\lambda & 0 \\ 0 & 2-\lambda \end{vmatrix} = 0 \Rightarrow \lambda = 1, 2 > 0 \Rightarrow f$ is positive definite.

43. $A = \begin{bmatrix} -2 & 1 \\ 1 & -2 \end{bmatrix} \Rightarrow \begin{vmatrix} -2-\lambda & 1 \\ 1 & -2-\lambda \end{vmatrix} = 0 \Rightarrow \lambda = -3, -1 < 0 \Rightarrow f$ is negative definite.

45. $A = \begin{bmatrix} 2 & 1 & 1 \\ 1 & 2 & 1 \\ 1 & 1 & 2 \end{bmatrix} \Rightarrow \begin{vmatrix} 2-\lambda & 1 & 1 \\ 1 & 2-\lambda & 1 \\ 1 & 1 & 2-\lambda \end{vmatrix} = 0 \Rightarrow \lambda = 1, 4 > 0 \Rightarrow f$ is positive definite.

47. $\Rightarrow \begin{vmatrix} 1-\lambda & 2 & 0 \\ 2 & 1-\lambda & 0 \\ 0 & 0 & -1-\lambda \end{vmatrix} = 0 \Rightarrow \lambda = -1, 3 \Rightarrow f$ is indefinite.

49. For 5.24a., Theorem 5.23 $\Rightarrow f(\mathbf{x}) = \lambda_1 y_1^2 + \ldots + \lambda_n y_n^2 > 0 \Leftrightarrow \mathbf{x} \neq \mathbf{0}$ and $\lambda_i > 0$.
    Likewise 5.24b. to 5.24e.

51. Note $\mathbf{x}^T A \mathbf{x} = \mathbf{x}^T B^T B \mathbf{x} = (B\mathbf{x})^T (B\mathbf{x}) = \|B\mathbf{x}\|^2 \geq 0$.
    So, if $\mathbf{x}^T A \mathbf{x} = 0$, then $\|B\mathbf{x}\|^2 = 0 \Rightarrow \mathbf{x} = \mathbf{0}$ because B is invertible
    $\Rightarrow \mathbf{x}^T A \mathbf{x} > 0$ for all $\mathbf{x} \neq \mathbf{0} \Rightarrow A = B^T B$ is positive definite.

53. (a) $\mathbf{x}^T(cA)\mathbf{x} = c(\mathbf{x}^T A\mathbf{x}) > 0$ because $c > 0$ and $\mathbf{x}^T A\mathbf{x} > 0$ when $\mathbf{x} \neq 0$.
    (b) $\mathbf{x}^T A^2 \mathbf{x} = (\mathbf{x}^T A\mathbf{x})(\mathbf{x}^T A\mathbf{x}) = (\mathbf{x}^T A\mathbf{x})^2 > 0$.
    (c) $\mathbf{x}^T(A+B)\mathbf{x} = (\mathbf{x}^T A\mathbf{x}) + (\mathbf{x}^T B\mathbf{x}) > 0$ because $\mathbf{x}^T A\mathbf{x} > 0$ and $\mathbf{x}^T B\mathbf{x} > 0$.
    (d) Note, A is invertible by Exercise 52 since $A = (CQ^T)^T CQ^T$, where CQ^T is invertible. Therefore, the eigenvalues of A^{-1} are $\frac{1}{\lambda_i} > 0 \Rightarrow A^{-1}$ is positive definite by Theorem 4a.

55. $\lambda = 0, 2 \Rightarrow \max = 2$, $\min = 0$. So, $(A - 2I)\mathbf{v} = 0 \Rightarrow \mathbf{v}_1 = \begin{bmatrix} 1 \\ -1 \end{bmatrix}$, $A\mathbf{v} = 0 \Rightarrow \mathbf{v}_2 = \begin{bmatrix} 1 \\ 1 \end{bmatrix}$.

    For $\|\mathbf{x}\| = 1$ max occurs at $\mathbf{x} = \pm \begin{bmatrix} 1/\sqrt{2} \\ -1/\sqrt{2} \end{bmatrix}$ and min occurs at $\mathbf{x} = \pm \begin{bmatrix} 1/\sqrt{2} \\ 1/\sqrt{2} \end{bmatrix}$.

57. $\lambda = 1, 4 \Rightarrow \max = 4$, $\min = 1$. So, $(A - \lambda I)\mathbf{v} = 0 \Rightarrow \mathbf{v}_1 = \begin{bmatrix} 1 \\ 1 \\ 1 \end{bmatrix}$, $\mathbf{v}_2 = \begin{bmatrix} 1 \\ 0 \\ -1 \end{bmatrix}$, $\mathbf{v}_3 = \begin{bmatrix} -1 \\ 1 \\ 0 \end{bmatrix}$.

    So, max occurs at $\mathbf{x} = \pm \begin{bmatrix} 1/\sqrt{3} \\ 1/\sqrt{3} \\ 1/\sqrt{3} \end{bmatrix}$ and min occurs at $\mathbf{x} = \pm \begin{bmatrix} 1/\sqrt{2} \\ 0 \\ -1/\sqrt{2} \end{bmatrix}$ or $\pm \begin{bmatrix} -1/\sqrt{2} \\ 1/\sqrt{2} \\ 0 \end{bmatrix}$.

59. $f(x) = \lambda_1 y_1^2 + \lambda_2 y_2^2 + \ldots + \lambda_n y_n^2 \geq \lambda_n(y_1^2 + y_2^2 + \ldots + y_n^2) = \lambda_n \|\mathbf{y}\| = \lambda_n$ because $\lambda_1 \geq \ldots \geq \lambda_n$.

61. $x^2 + 5y^2 = 25 \Rightarrow \frac{x^2}{25} + \frac{y^2}{5} = 1 \Rightarrow$ ellipse.

63. $x^2 - y - 1 = 0 \Rightarrow y = x^2 - 1 \Rightarrow$ parabola.

65. $3x^2 = y^2 - 1 \Rightarrow \frac{y^2}{1} - \frac{x^2}{1/3} = 1 \Rightarrow$ hyperbola.

67.

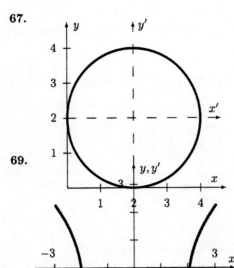

$x^2 + y^2 - 4x - 4y + 4 = 0 \Rightarrow$
$(x-2)^2 + (y-2)^2 = 4 \Rightarrow$
circle $(x')^2 + (y')^2 = 4$.

69.

$9x^2 - 4y^2 - 4y = 37 \Rightarrow$
$\frac{(x-0)^2}{4} - \frac{(y+1/2)^2}{9} = 1 \Rightarrow$
hyperbola $\frac{(x')^2}{4} - \frac{(y')^2}{9} = 1$.

71.

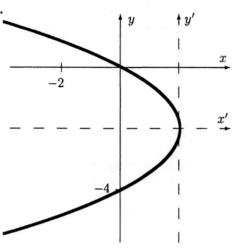

$2y^2 + 8y + 4x = 0 \Rightarrow$
$x = -\frac{1}{2}(y+2)^2 + 2 \Rightarrow$
parabola $x' = -\frac{1}{2}(y')^2$.

73. $x^2 + xy + y^2 = 6 \Rightarrow A = \begin{bmatrix} 1 & \frac{1}{2} \\ \frac{1}{2} & 1 \end{bmatrix} \Rightarrow \lambda = \frac{3}{2}, \frac{1}{2} \Rightarrow \frac{3}{2}(x')^2 + \frac{1}{2}(y')^2 = 6 \Rightarrow \frac{(x')^2}{4} + \frac{(y')^2}{12} = 1.$

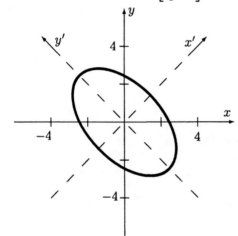

$Q = \begin{bmatrix} \frac{1}{\sqrt{2}} & -\frac{1}{\sqrt{2}} \\ \frac{1}{\sqrt{2}} & \frac{1}{\sqrt{2}} \end{bmatrix} \Rightarrow$

$\mathbf{e}_1 = \begin{bmatrix} 1 \\ 0 \end{bmatrix} \longrightarrow \begin{bmatrix} \frac{1}{\sqrt{2}} \\ \frac{1}{\sqrt{2}} \end{bmatrix}$ and

$\mathbf{e}_2 = \begin{bmatrix} 0 \\ 1 \end{bmatrix} \longrightarrow \begin{bmatrix} -\frac{1}{\sqrt{2}} \\ \frac{1}{\sqrt{2}} \end{bmatrix} \Rightarrow$

45° rotation

Ellipse

75. $4x^2 + 6xy - 4y^2 = 5 \Rightarrow A = \begin{bmatrix} 4 & 3 \\ 3 & -4 \end{bmatrix} \Rightarrow \lambda = 5, -5 \Rightarrow 5(x')^2 - 5(y')^2 = 5 \Rightarrow (x')^2 - (y')^2 = 1.$

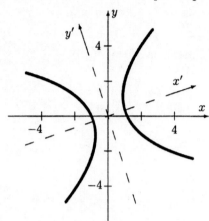

$Q = \begin{bmatrix} \frac{3}{\sqrt{10}} & -\frac{1}{\sqrt{10}} \\ \frac{1}{\sqrt{10}} & \frac{3}{\sqrt{10}} \end{bmatrix} \Rightarrow$

$\mathbf{e}_1 = \begin{bmatrix} 1 \\ 0 \end{bmatrix} \longrightarrow \begin{bmatrix} \frac{3}{\sqrt{10}} \\ \frac{1}{\sqrt{10}} \end{bmatrix}$ and

$\mathbf{e}_2 = \begin{bmatrix} 0 \\ 1 \end{bmatrix} \longrightarrow \begin{bmatrix} -\frac{1}{\sqrt{10}} \\ \frac{3}{\sqrt{10}} \end{bmatrix} \Rightarrow$

$\approx 18.435°$ rotation

Hyperbola

77. $3x^2 - 4xy + 3y^2 - 28\sqrt{2}\,x + 22\sqrt{2}\,y + 84 = 0 \Rightarrow$

$A = \begin{bmatrix} 3 & -2 \\ -2 & 3 \end{bmatrix}, B = \begin{bmatrix} -28\sqrt{2} & 22\sqrt{2} \end{bmatrix} \Rightarrow \lambda = 1, 5$ and $Q = \begin{bmatrix} \frac{1}{\sqrt{2}} & \frac{1}{\sqrt{2}} \\ -\frac{1}{\sqrt{2}} & \frac{1}{\sqrt{2}} \end{bmatrix} \Rightarrow$

$BQ = \begin{bmatrix} -6 & -50 \end{bmatrix} \Rightarrow (x')^2 - 6x' + 5(y')^2 - 50y' = -84 \Rightarrow \frac{(x'-3)^2}{50} + \frac{(y'-5)^2}{10} = 1 \Rightarrow$

$\frac{(x'')^2}{50} + \frac{(y'')^2}{10} = 1$, an ellipse.

79. $2xy + 2\sqrt{2}\,x - 1 = 0 \Rightarrow$

$A = \begin{bmatrix} 0 & 1 \\ 1 & 0 \end{bmatrix}, B = \begin{bmatrix} 2\sqrt{2} & 0 \end{bmatrix} \Rightarrow \lambda = 1, -1$ and $Q = \begin{bmatrix} \frac{1}{\sqrt{2}} & \frac{1}{\sqrt{2}} \\ \frac{1}{\sqrt{2}} & -\frac{1}{\sqrt{2}} \end{bmatrix} \Rightarrow$

$BQ = \begin{bmatrix} 2 & 2 \end{bmatrix} \Rightarrow (x')^2 + 2x' - (y')^2 + 2y' = 1 \Rightarrow (x'+1)^2 - (y'-1)^2 = 1 \Rightarrow$

$(x'')^2 - (y'')^2 = 1$, a hyperbola.

81.

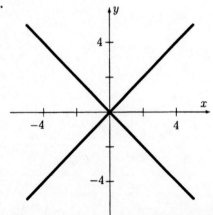

$x^2 - y^2 = 0 \Rightarrow$

$y = \pm x$

Degenerate

Two lines

83.

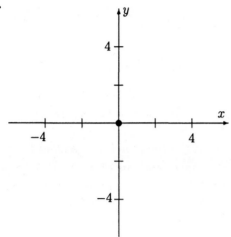

$3x^2 + y^2 = 0 \Rightarrow$
$y^2 = -3x^2 \Rightarrow$
$y = x = 0 \Rightarrow$
Degenerate
A point $(0, 0)$

85. $x^2 - 2xy + y^2 + 2\sqrt{2}\,x - 2\sqrt{2}\,y = 0 \Rightarrow$

$A = \begin{bmatrix} 1 & -1 \\ -1 & 1 \end{bmatrix}$, $B = \begin{bmatrix} 2\sqrt{2} & 2\sqrt{2} \end{bmatrix} \Rightarrow \lambda = 2, 0$ and $Q = \begin{bmatrix} \frac{1}{\sqrt{2}} & \frac{1}{\sqrt{2}} \\ -\frac{1}{\sqrt{2}} & \frac{1}{\sqrt{2}} \end{bmatrix} \Rightarrow$

$BQ = \begin{bmatrix} 4 & 0 \end{bmatrix} \Rightarrow 2(x')^2 + 2x' + 0(y')^2 + 4x' = 0 \Rightarrow x' = 0, -2 \Rightarrow$ degenerate, two lines $\Rightarrow$

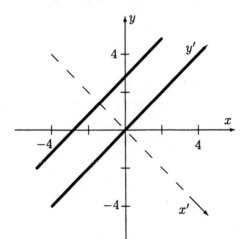

$Q = \begin{bmatrix} \frac{1}{\sqrt{2}} & \frac{1}{\sqrt{2}} \\ -\frac{1}{\sqrt{2}} & \frac{1}{\sqrt{2}} \end{bmatrix} \Rightarrow$

$\mathbf{e}_1 = \begin{bmatrix} 1 \\ 0 \end{bmatrix} \longrightarrow \begin{bmatrix} \frac{1}{\sqrt{2}} \\ -\frac{1}{\sqrt{2}} \end{bmatrix}$ and

$\mathbf{e}_2 = \begin{bmatrix} 0 \\ 1 \end{bmatrix} \longrightarrow \begin{bmatrix} \frac{1}{\sqrt{2}} \\ \frac{1}{\sqrt{2}} \end{bmatrix} \Rightarrow$

$-45°$ rotation

87. Recall, $Q^T A Q = D \Rightarrow A = QDQ^T$, where $D = \begin{bmatrix} \lambda_1 & 0 \\ 0 & \lambda_2 \end{bmatrix}$.

So, $\det A = (\det Q)(\det D)(\det Q^T) = \det D = \lambda_1 \lambda_2$.

Furthermore, The Principal Axes Theorem $\Rightarrow \mathbf{x}^T A \mathbf{x} = \mathbf{x}'^T D \mathbf{x}' = \lambda_1 (x')^2 + \lambda_2 (y')^2 = k$.

(a) We are given $k \neq 0$ and $\det A = \lambda_1 \lambda_2 < 0 \Rightarrow$ The coefficients of $(x')^2$ and $(y')^2$ in $\mathbf{x}^T A \mathbf{x} = \lambda_1 (x')^2 + \lambda_2 (y')^2 = k$, that is $\frac{\lambda_1}{k}(x')^2 + \frac{\lambda_2}{k}(y')^2 = 1$, are opposite in sign $\Rightarrow$ The graph of $\mathbf{x}^T A \mathbf{x}$ is a hyperbola.

Note, we use $k \neq 0$ when we divide by k to put the equation into standard form.

(b) We are given $k \neq 0$ and $\det A = \lambda_1 \lambda_2 > 0 \Rightarrow$ The coefficients of $(x')^2$ and $(y')^2$ in $\mathbf{x}^T A \mathbf{x} = \lambda_1 (x')^2 + \lambda_2 (y')^2 = k$, that is $\frac{\lambda_1}{k}(x')^2 + \frac{\lambda_2}{k}(y')^2 = 1$, have the same sign $\Rightarrow$
If $k < 0$, then the graph of $\mathbf{x}^T A \mathbf{x}$ is an imaginary conic.
If $k > 0$ and $\lambda_1 = \lambda_2$, then the graph of $\mathbf{x}^T A \mathbf{x}$ is a circle.
If $k > 0$ and $\lambda_1 \neq \lambda_2$, then the graph of $\mathbf{x}^T A \mathbf{x}$ is an ellipse.

(c) We are given $k \neq 0$ and $\det A = \lambda_1 \lambda_2 = 0 \Rightarrow$ One of the coefficients of $(x')^2$ and $(y')^2$ in $\mathbf{x}^T A \mathbf{x} = \lambda_1 (x')^2 + \lambda_2 (y')^2 = k$, that is $\frac{\lambda_1}{k}(x')^2 + \frac{\lambda_2}{k}(y')^2 = 1$, is zero $\Rightarrow$
If $k < 0$, then the graph of $\mathbf{x}^T A \mathbf{x}$ is an imaginary conic.
If $k > 0$ and $\lambda_1 = 0$, then we have $y' = \pm\sqrt{\frac{k}{\lambda_2}}$, so the graph is two straight lines.
If $k > 0$ and $\lambda_2 = 0$, then we have $x' = \pm\sqrt{\frac{k}{\lambda_1}}$, so the graph is two straight lines.

(d) We are given $k = 0$ and $\det A = \lambda_1 \lambda_2 \neq 0 \Rightarrow$ The coefficients of $(x')^2$ and $(y')^2$ in $\mathbf{x}^T A \mathbf{x} = \lambda_1 (x')^2 + \lambda_2 (y')^2 = 0$, are nonzero $\Rightarrow y' = \pm\sqrt{\frac{-\lambda_1}{\lambda_2}}\, x'$.

If λ_1 and λ_2 have the same sign, then $x' = y' = 0$, the graph is a single point (0,0).
If λ_1 and λ_2 have opposite signs, the graph is two straight lines.

(e) We are given $k = 0$ and $\det A = \lambda_1 \lambda_2 = 0 \Rightarrow$ One of the coefficients of $(x')^2$ and $(y')^2$ in $\mathbf{x}^T A \mathbf{x} = \lambda_1 (x')^2 + \lambda_2 (y')^2 = 0$, is zero $\Rightarrow$
If $\lambda_1 = 0$, then $y' = 0$, so the graph of $\mathbf{x}^T A \mathbf{x}$ is a straight line.
If $\lambda_2 = 0$, then $x' = 0$, so the graph of $\mathbf{x}^T A \mathbf{x}$ is a straight line.

89. $x^2 + y^2 + z^2 - 4yz = 1 \Rightarrow$

$A = \begin{bmatrix} 1 & 0 & 0 \\ 0 & 1 & -2 \\ 0 & -2 & 1 \end{bmatrix} \Rightarrow \lambda = 1, -1, 3 \Rightarrow (x')^2 - (y')^2 + 3(z')^2 = 1 \Rightarrow$

The graph is a hyperboloid of one sheet.

91. $2xy + z = 0 \Rightarrow$

$A = \begin{bmatrix} 0 & 1 & 0 \\ 1 & 0 & 0 \\ 0 & 0 & 0 \end{bmatrix}$ and $B = \begin{bmatrix} 0 & 0 & 1 \end{bmatrix} \Rightarrow \lambda = 1, -1, 0 \Rightarrow Q = \begin{bmatrix} \frac{1}{\sqrt{2}} & \frac{1}{\sqrt{2}} & 0 \\ \frac{1}{\sqrt{2}} & -\frac{1}{\sqrt{2}} & 0 \\ 0 & 0 & 1 \end{bmatrix} \Rightarrow$

$BQ = \begin{bmatrix} 0 & 0 & 1 \end{bmatrix} \Rightarrow z' = z$ and $(x')^2 - (y')^2 + 0(z)^2 + z = 0 \Rightarrow$
$z = -(x')^2 + (y')^2 \Rightarrow$ The graph is a hyperbolic paraboloid.

5.5 Applications

93. $x^2 + y^2 - 2z^2 + 4xy - 2xz + 2yz - x + y + z = 0 \Rightarrow$

$A = \begin{bmatrix} 1 & 2 & -1 \\ 2 & 1 & 1 \\ -1 & 1 & -2 \end{bmatrix}$ and $B = \begin{bmatrix} -1 & 1 & 1 \end{bmatrix} \Rightarrow \lambda = 0, 3, -3 \Rightarrow Q = \begin{bmatrix} -\frac{1}{\sqrt{3}} & \frac{1}{\sqrt{2}} & \frac{1}{\sqrt{6}} \\ \frac{1}{\sqrt{3}} & \frac{1}{\sqrt{2}} & -\frac{1}{\sqrt{6}} \\ \frac{1}{\sqrt{3}} & 0 & \frac{2}{\sqrt{6}} \end{bmatrix} \Rightarrow$

$BQ = \begin{bmatrix} \sqrt{3} & 0 & 0 \end{bmatrix} \Rightarrow 0(x')^2 + 3(y')^2 - 3(z)^2 + \sqrt{3}\,x' = 0 \Rightarrow$

$x' = -\sqrt{3}(y')^2 + \sqrt{3}(z')^2 \Rightarrow$ The graph is a hyperbolic paraboloid.

95. $11x^2 + 11y^2 + 14z^2 + 2xy + 8xz - 8yz - 12x + 12y + 12z = 6 \Rightarrow$

$A = \begin{bmatrix} 11 & 1 & 4 \\ 1 & 11 & -4 \\ 4 & -4 & 14 \end{bmatrix}$ and $B = \begin{bmatrix} -12 & 12 & 12 \end{bmatrix} \Rightarrow \lambda = 18, 6, 12 \Rightarrow$

$Q = \begin{bmatrix} \frac{1}{\sqrt{6}} & -\frac{1}{\sqrt{3}} & \frac{1}{\sqrt{2}} \\ -\frac{1}{\sqrt{6}} & \frac{1}{\sqrt{3}} & \frac{1}{\sqrt{2}} \\ \frac{2}{\sqrt{6}} & \frac{1}{\sqrt{3}} & 0 \end{bmatrix} \Rightarrow BQ = \begin{bmatrix} 0 & 12\sqrt{3} & 0 \end{bmatrix} \Rightarrow 18(x')^2 + 6(y')^2 + 12(z')^2 + 12\sqrt{3}\,y' = 6 \Rightarrow$

$18(x')^2 + 6(y' + \sqrt{3})^2 + 12(z')^2 = 24 \Rightarrow 3(x'')^2 + (y'')^2 + 2(z'')^2 = 4 \Rightarrow$

The graph is an ellipsoid.

Note that $x'' = x'$ and $z'' = z'$.

Chapter 5 Review

Key Definitions and Concepts

This list includes most but not all of the definitions listed at the end of each section. We begin the list with key symbols and symbol-based definitions.

$\mathcal{B}$	p.369	5.1	$\mathcal{B}$ often used to denote a set of basis vectors
$\mathbf{q}_i$	p.369	5.1	$\mathbf{q}_i$ often used to denote a vector in an *orthonormal* basis
Q	p.371	5.1	Q often used to denote an *orthogonal* matrix
$\mathbf{w}^\perp$	p.379	5.2	component of $\mathbf{v}$ orthogonal to W, $\mathbf{v} - \text{proj}_W(\mathbf{v})$
$W^\perp$	p.375	5.2	$W^\perp = \{\mathbf{v} \text{ in } \mathbb{R}^n : \mathbf{v} \cdot \mathbf{w} = 0 \text{ for all } \mathbf{w} \text{ in } W\}$
$\text{perp}_W(\mathbf{v})$	p.379	5.2	component of $\mathbf{v}$ orthogonal to W, $\mathbf{v} - \text{proj}_W(\mathbf{v})$
$\text{proj}_W(\mathbf{v})$	p.379	5.2	*orthogonal projection* of $\mathbf{v}$ onto W, $\sum_{k=1}^{n} \left(\frac{\mathbf{u}_k \cdot \mathbf{v}}{\mathbf{u}_k \cdot \mathbf{u}_k} \right) \mathbf{u}_k$
$\mathbf{1}$	p.411	5.5	$\mathbf{1}$ (the vector in $\mathbb{Z}_2^n$ all of whose entries are 1)
C	p.408	5.5	C (a set of code vectors in $\mathbb{Z}_2^n$)
$C^\perp$	p.408	5.5	$C^\perp = \{\mathbf{x} \text{ in } \mathbb{Z}_2^n : \mathbf{c} \cdot \mathbf{x} = 0 \text{ for all } \mathbf{c} \text{ in } C\}$
G	p.405	5.5	$G = \begin{bmatrix} I_m \\ A \end{bmatrix}$ (standard generator matrix for a code)
P	p.405	5.5	$G = [\, B \mid I_{n-m} \,]$ (standard parity check matrix for a code)
$w(\mathbf{x})$	p.411	5.5	$w(\mathbf{x})$ (*weight* of $\mathbf{x}$), the number of 1s in $\mathbf{x}$ (in $\mathbb{Z}_2^n$)
$f(\mathbf{x})$	p.412	5.5	$f(\mathbf{x}) = \mathbf{x}^T A \mathbf{x}$ (*quadratic form*), (A, the matrix associated with f, is symmetric)
$\mathbf{x}^T A \mathbf{x}$	p.412	5.5	$\mathbf{x}^T A \mathbf{x} = a_{11} x_1^2 + a_{22} x_2^2 + \cdots + a_{nn} x_n^2 + \sum_{i<j} 2 a_{ij} x_i x_j$

Key Definitions and Concepts, continued

definite	p.415	5.5	A quadratic form $f(\mathbf{x}) = \mathbf{x}^T A \mathbf{x}$ is classified as follows:

1. *positive definite* $f(\mathbf{x}) > 0$ for all $\mathbf{x} \neq \mathbf{0}$
2. *positive semidefinite* $f(\mathbf{x}) \geq 0$ for all $\mathbf{x}$
3. *negative definite* $f(\mathbf{x}) < 0$ for all $\mathbf{x} \neq \mathbf{0}$
4. *negative semidefinite* $f(\mathbf{x}) \leq 0$ for all $\mathbf{x}$
5. *indefinite* if $f(\mathbf{x})$ takes on both positive and negative values

The associated matrix A is classified in exactly the same way.

fundamental subspaces	p.377	5.2	There are four *fundamental subspaces* (two pair): $\text{row}(A), \text{null}(A)$ and $\text{col}(A), \text{null}(A^T)$
generator matrix	p.406	5.5	For $n > m$, an $n \times m$ matrix G and an $(n-k) \times n$ matrix P (with entries in $\mathbb{Z}_2$) are a *generator matrix* and a *parity check matrix* respectively for an (n, k) binary code C if:

1. The columns of G are linearly independent.
2. The rows of P are linearly independent.
3. $PG = O$.

orthogonal		5.1	This adjective is used for sets, bases, and matrices:
set	p.366		$\mathbf{v}_i \cdot \mathbf{v}_j = 0$ when $i \neq j$
basis	p.367		a basis $\mathcal{B}$ that is an orthogonal set
matrix	p.371		a *square* matrix whose columns form an orthonormal set
			That is, Q where columns $\{\mathbf{q}_i\}$ are orthonormal
			Matrix is ortho*gonal* when columns are ortho*normal*
orthogonal		5.2	This adjective applies to *complements* and *projections*:
complement	p.375		$W^\perp = \{\mathbf{v} \text{ in } \mathbb{R}^n : \mathbf{v} \cdot \mathbf{w} = 0 \text{ for all } \mathbf{w} \text{ in } W\}$
component	p.379		$\text{perp}_W(\mathbf{v}) = \mathbf{w}^\perp = \mathbf{v} - \text{proj}_W(\mathbf{v})$
projection	p.379		$\text{proj}_W(\mathbf{v}) = \sum_{k=1}^{n} \left(\frac{\mathbf{u}_k \cdot \mathbf{v}}{\mathbf{u}_k \cdot \mathbf{u}_k} \right) \mathbf{u}_k = \sum_{k=1}^{n} \text{proj}_{\mathbf{u}_k}(\mathbf{v})$
orthogonally diagonalizable	p.397	5.4	If there exists an orthogonal Q and diagonal D such that $Q^T A Q = D$, then A is *orthogonally diagonalizable*.
orthonormal		5.1	This adjective is used for sets and bases:
	p.369		For a set to be orthonormal it must have two key properties:
			$\mathbf{q}_i \cdot \mathbf{q}_j = \begin{cases} 0 & \text{if } i \neq j \quad \text{Property 1} \\ 1 & \text{if } i = j \quad \text{Property 2} \end{cases}$
set	p.369		an orthogonal set of *unit* vectors
basis	p.369		a basis $\mathcal{B}$ that is an orthonormal set

Key Definitions and Concepts, continued

parity check	p.406	5.5	See *generator matrix* for details: the definitions and conditions are associated.
$\text{perp}_W(\mathbf{v})$	p.379	5.2	component of $\mathbf{v}$ orthogonal to W, $\mathbf{v} - \text{proj}_W(\mathbf{v})$
$\text{proj}_W(\mathbf{v})$	p.379	5.2	*orthogonal projection* of $\mathbf{v}$ onto W, $\sum_{k=1}^{n} \left(\frac{\mathbf{u}_k \cdot \mathbf{v}}{\mathbf{u}_k \cdot \mathbf{u}_k} \right) \mathbf{u}_k$
projection form	p.402	5.4	If A is a real symmetric matrix with $A = QDQ^T$, then the *projection form of the Spectral Theorem* is $A = \lambda_1 \mathbf{q}_1 \mathbf{q}_1^T + \cdots + \lambda_n \mathbf{q}_n \mathbf{q}_n^T = \sum_{i=1}^{n} \lambda_i \mathbf{q}_i \mathbf{q}_i^T.$
quadratic form	p.412	5.5	$f(\mathbf{x}) = \mathbf{x}^T A \mathbf{x}$ (*quadratic form*), (A, the matrix associated with f, is symmetric)
spectral decomposition	p.402	5.4	If A is a real symmetric matrix with $A = QDQ^T$, then $A = \lambda_1 \mathbf{q}_1 \mathbf{q}_1^T + \cdots + \lambda_n \mathbf{q}_n \mathbf{q}_n^T = \sum_{i=1}^{n} \lambda_i \mathbf{q}_i \mathbf{q}_i^T.$
spectrum	p.400	5.4	The eigenvalues of an *orthogonally diagonalizable* matrix.
symmetric	p.149	3.1	A matrix A is *symmetric* if and only if $A^T = A$.
weight	p.411	5.5	$w(\mathbf{x})$ (*weight* of $\mathbf{x}$), the number of 1s in $\mathbf{x}$ (in $\mathbb{Z}_2^n$)

Theorems with names

Thm 5.15 p.386 5.3 **The Gram-Schmidt Process:**
Let $\{\mathbf{x}_1,\ldots,\mathbf{w}_k\}$ be a basis for W then ...
$\mathbf{v}_1 = \mathbf{x}_1$, $W_1 = \mathrm{span}(\mathbf{x}_1)$
$\mathbf{v}_2 = \mathbf{x}_2 - \left(\frac{\mathbf{v}_1 \cdot \mathbf{x}_2}{\mathbf{v}_1 \cdot \mathbf{v}_1}\right)\mathbf{v}_1$, $W_2 = \mathrm{span}(\mathbf{x}_1, \mathbf{x}_2)$
$\mathbf{v}_3 = \mathbf{x}_3 - \left(\frac{\mathbf{v}_1 \cdot \mathbf{x}_3}{\mathbf{v}_1 \cdot \mathbf{v}_1}\right)\mathbf{v}_1 - \left(\frac{\mathbf{v}_2 \cdot \mathbf{x}_3}{\mathbf{v}_2 \cdot \mathbf{v}_2}\right)\mathbf{v}_2$, $W_3 = \mathrm{span}(\mathbf{x}_1, \mathbf{x}_2, \mathbf{x}_3)$
...
$\mathbf{v}_k = \mathbf{x}_k - \sum_{i=1}^{k-1} \frac{\mathbf{v}_i \cdot \mathbf{x}_k}{\mathbf{v}_i \cdot \mathbf{v}_i}$, $W_k = \mathrm{span}(\{\mathbf{x}_k\})$
... $\{\mathbf{v}_1,\ldots,\mathbf{v}_k\}$ is an orthogonal basis for W.

Thm 5.11 p.381 5.2 **The Orthogonal Decomposition Theorem:** $\mathbf{v} = \mathbf{w} + \mathbf{w}^\perp$

Thm 5.23 p.411 5.5 **The Principal Axes Theorem:**
Every quadratic form can be diagonalized:
$\mathbf{x}^T A \mathbf{x} = \mathbf{y}^T D \mathbf{y} = \lambda_1 y_1^2 + \cdots + \lambda_n y_n^2$
(see statement of Theorem in the text for details)

Thm 5.16 p.390 5.3 **The QR Factorization:**
Let A be an $m \times n$ matrix with linearly independent columns.
Then A can be factored as $A = QR$, where
Q is an $m \times n$ matrix with orthonormal columns and
R is an invertible upper triangular matrix.

Cor 5.14 p.381 5.2 **The Rank Theorem:** Given A is $m \times n$ then ...
$\mathrm{rank}(A) + \mathrm{nullity}(A) = n$ and $\mathrm{rank}(A) + \mathrm{nullity}(A^T) = m$

Thm 5.20 p.400 5.4 **The Spectral Theorem:**
Let A be an $n \times n$ *real* matrix. Then A is *symmetric*
if and only if A is *orthogonally diagonalizable*.

Theorems

Thm 5.1	p.366	5.1	$\mathcal{B} = \{\mathbf{v}_1, \mathbf{v}_2, \ldots, \mathbf{v}_n\}$ orthogonal $\Rightarrow$ $\mathcal{B}$ is linearly independent
Thm 5.2	p.368	5.1	If $\mathcal{B} = \{\mathbf{v}_1, \mathbf{v}_2, \ldots, \mathbf{v}_n\}$ is orthogonal and $\mathbf{w} = \sum_{i=1}^{n} c_i \mathbf{v}_i$, then $c_i = \frac{\mathbf{w} \cdot \mathbf{v}_i}{\mathbf{v}_i \cdot \mathbf{v}_i}$.
Thm 5.3	p.370	5.1	$\mathbf{w} = \sum_{i=1}^{n} (\mathbf{w} \cdot \mathbf{q}_1) \mathbf{q}_i$ (this representation is unique)
Thm 5.4	p.371	5.1	$Q^T Q = I$ if and only if $\{\mathbf{q}_i\}$ is an orthonormal set.
Thm 5.5	p.371	5.1	Q is orthogonal if and only if $Q^{-1} = Q^T$
Thm 5.6	p.372	5.1	a. through c. give equivalent conditions for Q to be orthogonal: a. Q is orthogonal b. $\|Q\mathbf{x}\| = \|\mathbf{x}\|$ for every $\mathbf{x}$ in $\mathbb{R}^n$ c. $Q\mathbf{x} \cdot Q\mathbf{x} = \mathbf{x} \cdot \mathbf{y}$ for every $\mathbf{x}, \mathbf{y}$ in $\mathbb{R}^n$
Thm 5.7	p.373	5.1	Q is orthogonal, then its *rows* form an orthonormal set.
Thm 5.8	p.372	5.1	a. through d. list implications of Q being orthogonal: a. Q^{-1} is orthogonal b. $\det Q = \pm 1$ c. If λ is an eigenvalue of Q, then $|\lambda| = 1$. d. If Q_1 and Q_2 are orthogonal $n \times n$ matrices, then so is $Q_1 Q_2$.
Thm 5.9	p.376	5.2	a. through d. give properties of $W^\perp$: a. $W^\perp$ is a subspace of $\mathbb{R}^n$ b. $(W^\perp)^\perp = W$ c. $W \cap W^\perp = \{\mathbf{0}\}$ d. If $W = \text{span}(\mathbf{w}_1, \ldots, \mathbf{w}_k)$, then $\mathbf{v}$ is in $W^\perp$ if and only if $\mathbf{v} \cdot \mathbf{w}_i = 0$ for all i
Thm 5.10	p.376	5.2	$(\text{row}(A))^\perp = \text{null}(A)$ and $(\text{col}(A))^\perp = \text{null}(A^T)$
Thm 5.11	p.381	5.2	The *Orthogonal Decomposition* Theorem, $\mathbf{v} = \mathbf{w} + \mathbf{w}^\perp$
Cor 5.12	p.382	5.2	$(W^\perp)^\perp = W$
Thm 5.13	p.383	5.2	$\dim W + \dim W^\perp = n$
Cor 5.14	p.381	5.2	The *Rank* Theorem: Given A is $m \times n$ then ... $\text{rank}(A) + \text{nullity}(A) = n$ and $\text{rank}(A) + \text{nullity}(A^T) = m$

Theorems, continued

Thm 5.15   p.389   5.3   The Gram-Schmidt Process:
Let $\{\mathbf{x}_1,\ldots,\mathbf{w}_k\}$ be a basis for W then ...
$\mathbf{v}_1 = \mathbf{x}_1$, $W_1 = \text{span}(\mathbf{x}_1)$
$\mathbf{v}_2 = \mathbf{x}_2 - \left(\frac{\mathbf{v}_1 \cdot \mathbf{x}_2}{\mathbf{v}_1 \cdot \mathbf{v}_1}\right)\mathbf{v}_1$, $W_2 = \text{span}(\mathbf{x}_1, \mathbf{x}_2)$
$\mathbf{v}_3 = \mathbf{x}_3 - \left(\frac{\mathbf{v}_1 \cdot \mathbf{x}_3}{\mathbf{v}_1 \cdot \mathbf{v}_1}\right)\mathbf{v}_1 - \left(\frac{\mathbf{v}_2 \cdot \mathbf{x}_3}{\mathbf{v}_2 \cdot \mathbf{v}_2}\right)\mathbf{v}_2$, $W_3 = \text{span}(\mathbf{x}_1, \mathbf{x}_2, \mathbf{x}_3)$
...
$\mathbf{v}_k = \mathbf{x}_k - \sum_{i=1}^{k-1} \frac{\mathbf{v}_i \cdot \mathbf{x}_k}{\mathbf{v}_i \cdot \mathbf{v}_i}$, $W_k = \text{span}(\{\mathbf{x}_k\})$
... $\{\mathbf{v}_1, \ldots, \mathbf{v}_k\}$ is an orthogonal basis for W.

Thm 5.16   p.390   5.3   The QR Factorization:
Let A be an $m \times n$ matrix with linearly independent columns.
Then A can be factored as $A = QR$, where
Q is an $m \times n$ matrix with orthonormal columns and
R is an invertible upper triangular matrix.

Thm 5.17   p.398   5.4   If A is *orthogonally diagonalizable*,
then A is *symmetric*.

Thm 5.18   p.398   5.4   If A is a *real* symmetric matrix,
then the eigenvalues of A are *real*.

Thm 5.19   p.399   5.4   If A is a *symmetric* matrix, then any two eigenvectors
corresponding to distinct eigenvalues of A are *orthogonal*.

Thm 5.20   p.400   5.4   The Spectral Theorem:
Let A be an $n \times n$ *real* matrix. Then A is *symmetric*
if and only if A is *orthogonally diagonalizable*.

Theorems, continued

Thm 5.21   p.409   5.5   If C is an (n, k) binary code with generator matrix G and parity check matrix P, then $C^\perp$ is an $(n, n-k)$ binary code such that
    a. G^T is parity check matrix for $C^\perp$.
    b. P^T is generator matrix for $C^\perp$.

Thm 5.22   p.411   5.5   If C is a self dual code, then
    a. Every vector in C has even weight.
    b. **1** is in C.

Thm 5.23   p.411   5.5   The Principal Axes Theorem:
Every quadratic form can be diagonalized:
$\mathbf{x}^T A \mathbf{x} = \mathbf{y}^T D \mathbf{y} = \lambda_1 y_1^2 + \cdots + \lambda_n y_n^2$
(see statement of Theorem in the text for details)

Thm 5.24   p.416   5.5   The quadratic form $f(\mathbf{x}) = \mathbf{x}^T A \mathbf{x}$ is:
    a. *positive definite* $\Leftrightarrow \lambda_i > 0$
    b. *positive semidefinite* $\Leftrightarrow \lambda_i \geq 0$
    c. *negative definite* $\Leftrightarrow \lambda_i < 0$
    d. *negative semidefinite* $\Leftrightarrow \lambda_i \leq 0$
    e. *indefinite* $\Leftrightarrow \lambda_i$ are both positive and negative

Thm 5.25   p.417   5.5   Given $f(\mathbf{x}) = \mathbf{x}^T A \mathbf{x}$, $\lambda_1 \geq \lambda_2 \geq \cdots \geq \lambda_n$, and $\|\mathbf{x}\| = 1$:
    a. $\lambda_1 \geq f(\mathbf{x}) \geq \lambda_n$
    b. $\max f(\mathbf{x}) = \lambda_1$ occurs for a unit eigenvector
    c. $\min f(\mathbf{x}) = \lambda_n$ occurs for a unit eigenvector

Solutions to odd-numbered exercises from Chapter 5 Review

1. We will explain and give counter examples to justify our answers below.

   (a) **True**. See definition on p.369 of Section 5.1.

   It is important to note the following hierarchy of definitions:
   ortho*normal* $\Rightarrow$ orthogonal $\Rightarrow$ linearly independent.
   Why?
   For a set of vectors $\{\mathbf{q}_i\}$ to be orthonormal it must satisfy two properties:
   $$\mathbf{q}_i \cdot \mathbf{q}_j = \begin{cases} 0 & \text{if } i \neq j \quad \text{Property 1} \\ 1 & \text{if } i = j \quad \text{Property 2} \end{cases}$$
   A set of vectors $\{\mathbf{q}_i\}$ that satisfies only Property 1, $\mathbf{q}_i \cdot \mathbf{q}_j = 0$ if $i \neq j$, is only orthogonal, not orthonormal.
   That is, we can make an orthogonal set orthonormal by normalizing all of the vectors. What does that mean? We have to make the length of each of the vectors 1.
   That is, an orthogonal set of *unit* vectors is an ortho*normal* set.

   Q: Why is it obvious that an orthogonal set of vectors is linearly independent?
   A: A set of vectors is linearly independent if $\sum_{i=1}^{n} c_i \mathbf{v}_i = \mathbf{0}$ implies all the $c_i = 0$.
   If that set of vectors is *orthogonal* with each $\mathbf{v}_j \neq \mathbf{0}$, then
   $$\left(\sum_{i=1}^{n} c_i \mathbf{v}_i\right) \mathbf{v}_j = \sum_{i \neq j} c_i 0 + c_j \mathbf{v}_j \cdot \mathbf{v}_j = c_j \|\mathbf{v}_j\|^2 = 0 \text{ implies } c_j = 0.$$
   So, both orthogonality and orthonormality imply linear independence.

   (b) **True**. See Theorem 5.17 in Section 5.3.

   This is what the Gram-Schmidt Process of Theorem 5.17 in Section 5.3 asserts:
   Given any basis for a subspace W, we can construct an orthogonal basis.

   (c) **True**. See definition on p.371 in Section 5.1.

   A *square* matrix Q whose columns $\{\mathbf{q}_i\}$ form an orthonormal set.
   It is important that Q be square to satisfy the definition.

   (d) **True**. See definition on p.371 in Section 5.1.

   A *square* matrix Q whose columns $\{\mathbf{q}_i\}$ form an orthonormal set.
   Since its columns are orthonormal, they are linearly independent.
   Therefore, Q is invertible.

   Q: Theorem 5.8(b) of Section 5.1 asserts if Q is orthogonal, then $|\det Q| = 1$.
   Does this also imply if Q is orthogonal then Q is invertible? Why or why not?
   A: Hint: See Theorem 4.6 in Section 4.2.

Chapter 5 Review

1. We will explain and give counter examples to justify our answers below. (continued)

    (e) **False**. See Theorem 5.8(b) in Section 5.1 and Theorem 4.2 in Section 4.2.
    Theorem 5.8(b) of Section 5.1 asserts if Q is orthogonal, then $|\det Q| = 1$.
    However, Theorem 4.2 of Section 4.2 asserts if A is triangular, then $\det A = a_{11}a_{22}\cdots a_{nn}$.
    This tells us how to construct many counter examples. For instance, consider:
    $$A = \begin{bmatrix} 1 & 1 \\ 0 & 1 \end{bmatrix}, B = \begin{bmatrix} 1 & 2 \\ 0 & 1 \end{bmatrix}, C = \begin{bmatrix} 1 & 3 \\ 0 & 1 \end{bmatrix}, \text{ and } D = \begin{bmatrix} 1 & 4 \\ 0 & 1 \end{bmatrix}.$$
    Q: Why is it obvious that that none of A, B, C, or D are orthogonal?
    A: Hint: Are the columns orthogonal? Do they have length 1?
    Again, we should note what a strong condition orthogonality in matrices really is.

    (f) **True**. See Theorem 5.10 in Section 5.2.
    Theorem 5.10 in Section 5.2 asserts that $(\text{row}(A))^\perp = \text{null}(A)$.
    So if $(\text{row}(A))^\perp = \text{null}(A) = \mathbb{R}$, then $A\mathbf{x} = \mathbf{0}$ for every $\mathbf{x}$.
    Q: Why does $A\mathbf{x} = \mathbf{0}$ for every $\mathbf{x}$ imply that $A = O$?
    A: Hint: What does $A\mathbf{e}_i$ equal?

    (g) **False**. See Theorems 5.11 to 5.14 in Section 5.2 and Exercises 1 to 5 in Section 5.2.
    As a counterexample, consider: if $W = \text{span}(\mathbf{e}_1)$, then $\text{proj}_W(\mathbf{e}_2) = \mathbf{0}$.
    As another counterexample, see the solution to Exercise 3 of Section 5.2:
    More generally, what is going on? $\mathbb{R}^n$ breaks into two pieces: W and $W^\perp$.
    One of the underlying assertions of Theorems 5.11 to 5.14 is the following:
    $$W^\perp = \{\mathbf{v} : \text{proj}_W(\mathbf{v}) = \mathbf{0}\}.$$
    See Exercise 28 in Section 5.2.
    Q: Given the above, if $\text{proj}_W(\mathbf{v}) = \mathbf{0} \Rightarrow \mathbf{v} = \mathbf{0}$, what is W?
    A: Hint: If $\mathbf{0}$ is the only vector in $W^\perp$, what is $\dim W^\perp$?

    (h) **True**. See Theorem 5.5 in Section 5.1 and *symmetric* defined on p.149 in Section 3.1.
    By Thm 5.5, if A is orthogonal then $A^{-1} = A^T$. If A is also symmetric, then $A^T = A$.
    Therefore, if A is symmetric and orthogonal, then $A = A^T = A^{-1}$.
    So, if A is symmetric and orthogonal, then $A^2 = AA = AA^T = AA^{-1} = I$.

    (i) **False**. See definition on p.397 in Section 5.4. Consider diagonal matrices.
    If A is *orthogonally diagonalizable*, then there exists an orthogonal matrix Q and a diagonal matrix D such that $Q^T A Q = D$. Therefore, $A = QDQ^T$.
    So if the diagonal matrix D is not invertible, neither is A.

    (j) **True**. It is left to the reader to research and prove this assertion.

3. See Example 5.4, Exercises 7 to 10, and Theorem 5.2 in Section 5.1.

   As in Example 5.4, we apply the formula from Theorem 5.2, $c_i = \frac{\mathbf{v} \cdot \mathbf{u}_i}{\mathbf{u}_i \cdot \mathbf{u}_i}$.

   Let $\mathbf{u}_1 = \begin{bmatrix} 1 \\ 0 \\ 1 \end{bmatrix}$, $\mathbf{u}_2 = \begin{bmatrix} 1 \\ 1 \\ -1 \end{bmatrix}$, $\mathbf{u}_3 = \begin{bmatrix} -1 \\ 2 \\ 1 \end{bmatrix}$ and $\mathbf{v} = \begin{bmatrix} 7 \\ -3 \\ 2 \end{bmatrix}$.

   So $c_1 = \frac{7+2}{1+1} = \frac{9}{2}$, $c_2 = \frac{7-3-2}{1+1+1} = \frac{2}{3}$, and $c_3 = \frac{-7-6+2}{1+4+1} = -\frac{11}{6}$.

   Therefore, $[\mathbf{v}]_\mathbf{B} = \begin{bmatrix} \frac{9}{2} \\ \frac{2}{3} \\ -\frac{11}{6} \end{bmatrix}$.

   Q: What is the basic definition of $[\mathbf{v}]_\mathbf{B}$?

   A: If $\mathbf{B} = \{\mathbf{u}_1, \ldots, \mathbf{u}_n\}$ and $\mathbf{v} = \sum_{i=1}^{n} c_i \mathbf{u}_i$, then $[\mathbf{v}]_\mathbf{B} = \begin{bmatrix} c_1 \\ \vdots \\ c_n \end{bmatrix}$.

   Q: How might we put this basic definition of $[\mathbf{v}]_\mathbf{B}$ into words?
   A: $[\mathbf{v}]_\mathbf{B}$ is the coefficients in the linear combination equal to $\mathbf{v}$.

   Q: Given this basic definition, what is another method we might use to find $[\mathbf{v}]_\mathbf{B}$?
   A: Simply row reduce $[\,\mathbf{u}_1\ \mathbf{u}_2\ \mathbf{u}_3\,|\,\mathbf{v}\,] \longrightarrow [\,I\,|\,[\mathbf{v}]_\mathbf{B}\,]$. Verify this.

   See Examples 3.53 and 3.54 in Section 3.4 for further detail.

5. As in Theorem 5.5 of Section 5.1, we will show Q is orthogonal by showing that $QQ^T = I$.

$$QQ^T = \begin{bmatrix} \frac{6}{7} & \frac{2}{7} & \frac{3}{7} \\ -\frac{1}{\sqrt{5}} & 0 & \frac{2}{\sqrt{5}} \\ \frac{4}{7\sqrt{5}} & -\frac{15}{7\sqrt{5}} & \frac{2}{7\sqrt{5}} \end{bmatrix} \begin{bmatrix} \frac{6}{7} & -\frac{1}{\sqrt{5}} & \frac{4}{7\sqrt{5}} \\ \frac{2}{7} & 0 & -\frac{15}{7\sqrt{5}} \\ \frac{3}{7} & \frac{2}{\sqrt{5}} & \frac{2}{7\sqrt{5}} \end{bmatrix} = \begin{bmatrix} 1 & 0 & 0 \\ 0 & 1 & 0 \\ 0 & 0 & 1 \end{bmatrix} = I$$

   Q: Why is the entry in the first row and first column of the resulting matrix equal to 1?
   A: Because $\frac{36+4+9}{49} = \frac{49}{49} = 1$. Verify all the entries in the product matrix.

7. See the definition of orthonormal given in Section 5.1 on p.369.

   Given Q is orthogonal and $\mathbf{v}_i \cdot \mathbf{v}_j = 0$ and $\mathbf{v}_i \cdot \mathbf{v}_i = 1$,
   we need to show $Q\mathbf{v}_i \cdot Q\mathbf{v}_j = 0$ and $Q\mathbf{v}_i \cdot Q\mathbf{v}_i = 1$.

   First: $Q\mathbf{v}_i \cdot Q\mathbf{v}_j = (Q\mathbf{v}_i)^T Q\mathbf{v}_j = \mathbf{v}_i^T(Q^TQ)\mathbf{v}_j \stackrel{Q\text{ is orthogonal}}{=} \mathbf{v}_i^T I \mathbf{v}_j = \mathbf{v}_i^T \mathbf{v}_j = \mathbf{v}_i \cdot \mathbf{v}_j = 0$.
   Likewise: $Q\mathbf{v}_i \cdot Q\mathbf{v}_i = \mathbf{v}_i \cdot \mathbf{v}_i = 1$.

   Q: What does the fact that $Q\mathbf{v}_i \cdot Q\mathbf{v}_i = \mathbf{v}_i \cdot \mathbf{v}_i = 1$ tell us?
   A: Multiplication by an orthogonal matrix preserves length.
   Q: What does the fact that $Q\mathbf{v}_i \cdot Q\mathbf{v}_j = \mathbf{v}_i \cdot \mathbf{v}_j = 0$ tell us?
   A: Multiplication by an orthogonal matrix preserves orthogonality.
   Q: Does multiplication by an orthogonal matrix preserve all angles at which vectors meet?

Chapter 5 Review

9. See Exercises 1 through 6 and Example 5.8 in Section 5.2.

$\mathbf{v} \cdot \mathbf{w} = 5x + 2y = 0 \Rightarrow W^\perp = \left\{ \begin{bmatrix} x \\ y \end{bmatrix} : 5x + 2y = 0 \right\} \Rightarrow 5x = -2y \Rightarrow \begin{bmatrix} -2 \\ 5 \end{bmatrix}$ is a basis.

Note the following general pattern:
The lines that describe the necessary condition to be in W and $W^\perp$ are necessarily perpendicular.
So, in this case, $2x - 5y = 0$ and $5x + 2y = 0$ are necessarily perpendicular.

11. We use the *cross product* to find a vector orthogonal to both of the given vectors. See *Exploration: The Cross Product* following Section 1.3.

Since $\mathbf{x} = \mathbf{u} \times \mathbf{v} = \begin{bmatrix} u_2v_3 - u_3v_2 \\ u_3v_1 - u_1v_3 \\ u_1v_2 - u_2v_1 \end{bmatrix} = \begin{bmatrix} (-1)(-3) - (4)(1) \\ (4)(0) - (1)(-3) \\ (1)(1) - (-1)(0) \end{bmatrix} = \begin{bmatrix} -1 \\ 3 \\ 1 \end{bmatrix}$.

Verify that $\mathbf{x} \cdot \mathbf{u} = \mathbf{x} \cdot \mathbf{v} = 0$. So $W^\perp = \left\{ \begin{bmatrix} -1 \\ 3 \\ 1 \end{bmatrix} \right\}$ is a basis for $W^\perp$.

13. See Example 5.9 and Exercises 7 through 10 in Section 5.2.

$A \longrightarrow \begin{bmatrix} 1 & 0 & 2 & 3 & 4 \\ 0 & 1 & 0 & 2 & 1 \\ 0 & 0 & 0 & 0 & 0 \\ 0 & 0 & 0 & 0 & 0 \end{bmatrix}, \{[1\ 0\ 2\ 3\ 4], [0\ 1\ 0\ 2\ 1]\}$ is a basis for row(A).

$\begin{bmatrix} 1 & 0 & 2 & 3 & 4 & | & 0 \\ 0 & 1 & 0 & 2 & 1 & | & 0 \\ 0 & 0 & 0 & 0 & 0 & | & 0 \\ 0 & 0 & 0 & 0 & 0 & | & 0 \end{bmatrix} \Rightarrow \mathbf{x} = \begin{bmatrix} -2x_3 - 3x_4 - 4x_5 \\ -2x_4 - x_5 \\ x_3 \\ x_4 \\ x_5 \end{bmatrix} \Rightarrow \left\{ \begin{bmatrix} -2 \\ 0 \\ 1 \\ 0 \\ 0 \end{bmatrix}, \begin{bmatrix} -3 \\ -2 \\ 0 \\ 1 \\ 0 \end{bmatrix}, \begin{bmatrix} -4 \\ -1 \\ 0 \\ 0 \\ 1 \end{bmatrix} \right\}$ is a basis for null(A).

$A^T \longrightarrow \begin{bmatrix} 1 & 0 & 5 & 1 \\ 0 & 1 & 3 & -2 \\ 0 & 0 & 0 & 0 \\ 0 & 0 & 0 & 0 \\ 0 & 0 & 0 & 0 \end{bmatrix}, \left\{ \begin{bmatrix} 1 \\ -1 \\ 2 \\ 3 \end{bmatrix}, \begin{bmatrix} -1 \\ 2 \\ 1 \\ -5 \end{bmatrix} \right\}$ is a basis for row(A^T) = col(A).

$\begin{bmatrix} 1 & 0 & 5 & 1 & | & 0 \\ 0 & 1 & 3 & -2 & | & 0 \\ 0 & 0 & 0 & 0 & | & 0 \\ 0 & 0 & 0 & 0 & | & 0 \\ 0 & 0 & 0 & 0 & | & 0 \end{bmatrix} \Rightarrow \mathbf{y} = \begin{bmatrix} -5y_3 - y_4 \\ -3y_3 + 2y_4 \\ y_3 \\ y_4 \end{bmatrix} \Rightarrow \left\{ \begin{bmatrix} -5 \\ -3 \\ 1 \\ 0 \end{bmatrix}, \begin{bmatrix} -1 \\ 2 \\ 0 \\ 1 \end{bmatrix} \right\}$ is a basis for null(A^T).

15. For (a), see Example 5.13 and Exercise 12 in Section 5.3.
For (b), see Example 5.15 and Exercises 15 and 16 in Section 5.3.

(a) We compute the orthogonal basis vectors $\mathbf{v}_1$, $\mathbf{v}_2$, and $\mathbf{v}_3$ in sequence:

$$\mathbf{v}_1 = \mathbf{x}_1 = \begin{bmatrix} 1 \\ 1 \\ 1 \\ 1 \end{bmatrix}, \quad \mathbf{v}_2 = \mathbf{x}_2 - \left(\frac{\mathbf{v}_1 \cdot \mathbf{x}_2}{\mathbf{v}_1 \cdot \mathbf{v}_1}\right)\mathbf{v}_1 = \begin{bmatrix} 1 \\ 1 \\ 1 \\ 0 \end{bmatrix} - \frac{3}{4}\begin{bmatrix} 1 \\ 1 \\ 1 \\ 1 \end{bmatrix} = \begin{bmatrix} \frac{1}{4} \\ \frac{1}{4} \\ \frac{1}{4} \\ -\frac{3}{4} \end{bmatrix},$$

$$\mathbf{v}_3 = \mathbf{x}_3 - \left(\frac{\mathbf{v}_1 \cdot \mathbf{x}_3}{\mathbf{v}_1 \cdot \mathbf{v}_1}\right)\mathbf{v}_1 - \left(\frac{\mathbf{v}_2 \cdot \mathbf{x}_3}{\mathbf{v}_2 \cdot \mathbf{v}_2}\right)\mathbf{v}_2 = \begin{bmatrix} 0 \\ 1 \\ 1 \\ 1 \end{bmatrix} - \frac{3}{4}\begin{bmatrix} 1 \\ 1 \\ 1 \\ 1 \end{bmatrix} + \frac{1}{3}\begin{bmatrix} \frac{1}{4} \\ \frac{1}{4} \\ \frac{1}{4} \\ -\frac{3}{4} \end{bmatrix} = \begin{bmatrix} -\frac{2}{3} \\ \frac{1}{3} \\ \frac{1}{3} \\ 0 \end{bmatrix}.$$

(b) To find $A = QR$ we normalize the vectors from (a) to find Q and then set $R = Q^T A$:

Normalizing $\mathbf{v}_1$, $\mathbf{v}_2$, and $\mathbf{v}_3$ yields: $\mathbf{q}_1 = \dfrac{\mathbf{v}_1}{\|\mathbf{v}_1\|} = \dfrac{1}{2}\begin{bmatrix} 1 \\ 1 \\ 1 \\ 1 \end{bmatrix} = \begin{bmatrix} \frac{1}{2} \\ \frac{1}{2} \\ \frac{1}{2} \\ \frac{1}{2} \end{bmatrix},$

$$\mathbf{q}_2 = \frac{\mathbf{v}_2}{\|\mathbf{v}_2\|} = \frac{2\sqrt{3}}{3}\begin{bmatrix} \frac{1}{4} \\ \frac{1}{4} \\ \frac{1}{4} \\ -\frac{3}{4} \end{bmatrix} = \begin{bmatrix} \frac{\sqrt{3}}{6} \\ \frac{\sqrt{3}}{6} \\ \frac{\sqrt{3}}{6} \\ -\frac{\sqrt{3}}{2} \end{bmatrix}, \quad \mathbf{q}_3 = \frac{\mathbf{v}_3}{\|\mathbf{v}_3\|} = \frac{\sqrt{6}}{2}\begin{bmatrix} -\frac{2}{3} \\ \frac{1}{3} \\ \frac{1}{3} \\ 0 \end{bmatrix} = \begin{bmatrix} -\frac{\sqrt{6}}{3} \\ \frac{\sqrt{6}}{6} \\ \frac{\sqrt{6}}{6} \\ 0 \end{bmatrix}.$$

So $R = Q^T A = \begin{bmatrix} \frac{1}{2} & \frac{1}{2} & \frac{1}{2} & \frac{1}{2} \\ \frac{\sqrt{3}}{6} & \frac{\sqrt{3}}{6} & \frac{\sqrt{3}}{6} & -\frac{\sqrt{3}}{2} \\ -\frac{\sqrt{6}}{3} & \frac{\sqrt{6}}{6} & \frac{\sqrt{6}}{6} & 0 \end{bmatrix}\begin{bmatrix} 1 & 1 & 0 \\ 1 & 1 & 1 \\ 1 & 1 & 1 \\ 1 & 0 & 1 \end{bmatrix} = \begin{bmatrix} 2 & \frac{3}{2} & \frac{3}{2} \\ 0 & \frac{\sqrt{3}}{2} & -\frac{\sqrt{3}}{6} \\ 0 & 0 & \frac{\sqrt{6}}{3} \end{bmatrix}.$

Verify that $A = QR$.

That is: $\begin{bmatrix} 1 & 1 & 0 \\ 1 & 1 & 1 \\ 1 & 1 & 1 \\ 1 & 0 & 1 \end{bmatrix} = \begin{bmatrix} \frac{1}{2} & \frac{\sqrt{3}}{6} & -\frac{\sqrt{6}}{3} \\ \frac{1}{2} & \frac{\sqrt{3}}{6} & \frac{\sqrt{6}}{6} \\ \frac{1}{2} & \frac{\sqrt{3}}{6} & \frac{\sqrt{6}}{6} \\ \frac{1}{2} & -\frac{\sqrt{3}}{2} & 0 \end{bmatrix}\begin{bmatrix} 2 & \frac{3}{2} & \frac{3}{2} \\ 0 & \frac{\sqrt{3}}{2} & -\frac{\sqrt{3}}{6} \\ 0 & 0 & \frac{\sqrt{6}}{3} \end{bmatrix}.$

Chapter 5 Review

17. Similar to Exercise 16 above. However, we begin by finding a basis for W.

Since $x_1 + x_2 + x_3 + x_4 = 0 \Rightarrow x_4 = -x_1 - x_2 - x_3$,

W has basis $\mathcal{B} = \left\{ \begin{bmatrix} 1 \\ 0 \\ 0 \\ -1 \end{bmatrix}, \begin{bmatrix} 0 \\ 1 \\ 0 \\ -1 \end{bmatrix}, \begin{bmatrix} 0 \\ 0 \\ 1 \\ -1 \end{bmatrix} \right\}$. Why is this obvious?

Now we compute orthogonal basis vectors $\mathbf{v}_1$, $\mathbf{v}_2$, and $\mathbf{v}_3$ in sequence:

$$\mathbf{v}_1 = \mathbf{x}_1 = \begin{bmatrix} 1 \\ 0 \\ 0 \\ -1 \end{bmatrix}, \quad \mathbf{v}_2 = \mathbf{x}_2 - \left(\frac{\mathbf{v}_1 \cdot \mathbf{x}_2}{\mathbf{v}_1 \cdot \mathbf{v}_1} \right) \mathbf{v}_1 = \begin{bmatrix} 0 \\ 1 \\ 0 \\ -1 \end{bmatrix} - \frac{1}{2} \begin{bmatrix} 1 \\ 0 \\ 0 \\ -1 \end{bmatrix} = \begin{bmatrix} -\frac{1}{2} \\ 1 \\ 0 \\ -\frac{1}{2} \end{bmatrix},$$

$$\mathbf{v}_3 = \mathbf{x}_3 - \left(\frac{\mathbf{v}_1 \cdot \mathbf{x}_3}{\mathbf{v}_1 \cdot \mathbf{v}_1} \right) \mathbf{v}_1 - \left(\frac{\mathbf{v}_2 \cdot \mathbf{x}_3}{\mathbf{v}_2 \cdot \mathbf{v}_2} \right) \mathbf{v}_2 = \begin{bmatrix} 0 \\ 0 \\ 1 \\ -1 \end{bmatrix} - \frac{1}{2} \begin{bmatrix} 1 \\ 0 \\ 0 \\ -1 \end{bmatrix} - \frac{1}{3} \begin{bmatrix} -\frac{1}{2} \\ 1 \\ 0 \\ -\frac{1}{2} \end{bmatrix} = \begin{bmatrix} -\frac{1}{3} \\ -\frac{1}{3} \\ 1 \\ -\frac{1}{3} \end{bmatrix}.$$

Q: Is it reasonable to think of W as a hyperplane in $\mathbb{R}^4$? Why or why not?
A: Consider $ax + by + cz = 0$ in $\mathbb{R}^3$. Is this familiar? If so, what is it?

19. See Example 5.20 and Exercises 23 and 24 in Section 5.4.

We are given $\mathbf{x}_1 = \begin{bmatrix} 1 \\ 1 \\ 0 \end{bmatrix}$, $\mathbf{x}_2 = \begin{bmatrix} 1 \\ 1 \\ 1 \end{bmatrix}$ (from E_1), and $\mathbf{x}_3 = \begin{bmatrix} 1 \\ -1 \\ 0 \end{bmatrix}$ (from E_{-2}).

Now we compute orthogonal basis vectors $\mathbf{v}_1$, $\mathbf{v}_2$, and $\mathbf{v}_3$ in sequence:

$$\mathbf{v}_1 = \mathbf{x}_1 = \begin{bmatrix} 1 \\ 1 \\ 0 \end{bmatrix}, \quad \mathbf{v}_2 = \mathbf{x}_2 - \left(\frac{\mathbf{v}_1 \cdot \mathbf{x}_2}{\mathbf{v}_1 \cdot \mathbf{v}_1}\right) \mathbf{v}_1 = \begin{bmatrix} 1 \\ 1 \\ 1 \end{bmatrix} - \frac{2}{2} \begin{bmatrix} 1 \\ 1 \\ 0 \end{bmatrix} = \begin{bmatrix} 0 \\ 0 \\ 1 \end{bmatrix},$$

$$\mathbf{v}_3 = \mathbf{x}_3 - \left(\frac{\mathbf{v}_1 \cdot \mathbf{x}_3}{\mathbf{v}_1 \cdot \mathbf{v}_1}\right) \mathbf{v}_1 - \left(\frac{\mathbf{v}_2 \cdot \mathbf{x}_3}{\mathbf{v}_2 \cdot \mathbf{v}_2}\right) \mathbf{v}_2 = \begin{bmatrix} 1 \\ -1 \\ 0 \end{bmatrix} - \frac{0}{2} \begin{bmatrix} 1 \\ 1 \\ 0 \end{bmatrix} - \frac{0}{1} \begin{bmatrix} 0 \\ 0 \\ 1 \end{bmatrix} = \begin{bmatrix} 1 \\ -1 \\ 0 \end{bmatrix}.$$

Normalizing $\Rightarrow \mathbf{q}_1 = \begin{bmatrix} \frac{\sqrt{2}}{2} \\ \frac{\sqrt{2}}{2} \\ 0 \end{bmatrix}$, $\mathbf{q}_2 = \begin{bmatrix} 0 \\ 0 \\ 1 \end{bmatrix}$, and $\mathbf{q}_3 = \begin{bmatrix} \frac{\sqrt{2}}{2} \\ -\frac{\sqrt{2}}{2} \\ 0 \end{bmatrix}$.

Therefore, $\mathbf{q}_1 \mathbf{q}_1^T = \begin{bmatrix} \frac{\sqrt{2}}{2} \\ \frac{\sqrt{2}}{2} \\ 0 \end{bmatrix} \begin{bmatrix} \frac{\sqrt{2}}{2} & \frac{\sqrt{2}}{2} & 0 \end{bmatrix} = \begin{bmatrix} \frac{1}{2} & \frac{1}{2} & 0 \\ \frac{1}{2} & \frac{1}{2} & 0 \\ 0 & 0 & 0 \end{bmatrix}.$

Likewise, $\mathbf{q}_2 \mathbf{q}_2^T = \begin{bmatrix} 0 \\ 0 \\ 1 \end{bmatrix} \begin{bmatrix} 0 & 0 & 1 \end{bmatrix} = \begin{bmatrix} 0 & 0 & 0 \\ 0 & 0 & 0 \\ 0 & 0 & 1 \end{bmatrix}.$

Finally, $\mathbf{q}_3 \mathbf{q}_3^T = \begin{bmatrix} \frac{\sqrt{2}}{2} \\ -\frac{\sqrt{2}}{2} \\ 0 \end{bmatrix} \begin{bmatrix} \frac{\sqrt{2}}{2} & -\frac{\sqrt{2}}{2} & 0 \end{bmatrix} = \begin{bmatrix} \frac{1}{2} & -\frac{1}{2} & 0 \\ -\frac{1}{2} & \frac{1}{2} & 0 \\ 0 & 0 & 0 \end{bmatrix}.$

Now let $A = 1\mathbf{q}_1 \mathbf{q}_1^T + 1\mathbf{q}_2 \mathbf{q}_2^T - 2\mathbf{q}_3 \mathbf{q}_3^T$

$$= \begin{bmatrix} \frac{1}{2} & \frac{1}{2} & 0 \\ \frac{1}{2} & \frac{1}{2} & 0 \\ 0 & 0 & 0 \end{bmatrix} + \begin{bmatrix} 0 & 0 & 0 \\ 0 & 0 & 0 \\ 0 & 0 & 1 \end{bmatrix} - 2 \begin{bmatrix} \frac{1}{2} & -\frac{1}{2} & 0 \\ -\frac{1}{2} & \frac{1}{2} & 0 \\ 0 & 0 & 0 \end{bmatrix} = \begin{bmatrix} -\frac{1}{2} & \frac{3}{2} & 0 \\ \frac{3}{2} & -\frac{1}{2} & 0 \\ 0 & 0 & 1 \end{bmatrix}.$$

Verify that the eigenvalues of A are 1 and -2 with the given eigenspaces E_1 and E_{-2}.

Q: How do we verify the eigenvalues and eigenspaces E_1 and E_{-2}?
A: Hint: Is $A\mathbf{x}_1 = \mathbf{x}_1$? Should it be? What about $A\mathbf{x}_2$ and $A\mathbf{x}_3$?

Chapter 6

Vector Spaces

6.1 Vector Spaces and Subspaces

Q: What are our main goals for Section 6.1?
A: To take our second and much more in-depth look at *vector spaces* and *subspaces*.
See Section 3.5, *Subspaces, Basis, Dimension, and Rank*.
Recall, the opening line of that section:
This section introduces perhaps the most important ideas in the entire book.
Throughout the rest of the book, we will see exactly why that is true.

Key Definitions and Concepts

V	p.433	6.1	V often used to denote a *vector space*
W	p.438	6.1	W often used to denote a *subspace*
span(S)	p.442	6.1	If $S = \{\mathbf{v}_i\}$, then span(S) equals all linear combinations of $\mathbf{v}_i$.
subspace	p.438	6.1	$W \subseteq V$ and W satisfies the axioms of a vector space
trivial subspaces	p.441	6.1	The subspaces $\{\mathbf{0}\}$ and V are called the *trivial subspaces* of V. Therefore, all other subspaces are nontrivial.
vector space	p.433	6.1	If the following axioms hold, then V is a *vector space*:

1. $\mathbf{u} + \mathbf{v}$ is in V (*Closure under addition*)
2. $\mathbf{u} + \mathbf{v} = \mathbf{v} + \mathbf{u}$ (*Commutativity*)
3. $(\mathbf{u} + \mathbf{v}) + \mathbf{w} = \mathbf{u} + (\mathbf{v} + \mathbf{w})$ (*Associativity*)
4. $\mathbf{u} + \mathbf{0} = \mathbf{u}$ (*Zero vector*)
5. $\mathbf{u} + (-\mathbf{u}) = \mathbf{0}$
6. $c\mathbf{u}$ is in V (*Closure under scalar multiplication*)
7. $c(\mathbf{u} + \mathbf{v}) = c\mathbf{u} + c\mathbf{v}$ (*Distributivity*)
8. $(c + d)\mathbf{u} = c\mathbf{u} + d\mathbf{u}$ (*Distributivity*)
9. $c(d\mathbf{u}) = (cd)\mathbf{u}$
10. $1\mathbf{u} = \mathbf{u}$

Theorems

Thm 6.1 p.437 6.1 Let V be a vector space, $\mathbf{u}$ a vector, c a scalar:
a. $0\mathbf{u} = \mathbf{0}$
b. $c\mathbf{0} = \mathbf{0}$
c. $(-1)\mathbf{u} = -\mathbf{u}$
d. If $c\mathbf{u} = \mathbf{0}$, then $c = 0$ or $\mathbf{u} = \mathbf{0}$.

Thm 6.2 p.438 6.1 Let $W(\neq \emptyset) \subseteq V$, then W is a subspace if:
a. $\mathbf{u}, \mathbf{v} \in W \Rightarrow \mathbf{u} + \mathbf{v} \in W$
b. $\mathbf{u} \in W \Rightarrow c\mathbf{u} \in W$

Thm 6.3 p.445 6.1 Let $\mathbf{v}_1, \mathbf{v}_2, \ldots, \mathbf{v}_k$ be vectors in a vector space V.
a. span($\mathbf{v}_1, \mathbf{v}_2, \ldots, \mathbf{v}_k$) is a subspace of V
b. span($\mathbf{v}_1, \mathbf{v}_2, \ldots, \mathbf{v}_k$) is smallest subspace with $\mathbf{v}_1, \mathbf{v}_2, \ldots, \mathbf{v}_k$

Discussion of Key Definitions and Concepts

Q: What is the easiest way to determine if W is a subspace?
A: By applying Theorem 6.2. That is, verifying that it is closed under both operations. That is the main point of all the exercises in this section.

Q: What are the two technical points we need to be aware of when proving W is a subspace?
A: Be sure to verify that $W \neq \emptyset$. Also, as noted in the text, $\mathbf{0}$ of V must be in W.

Q: What is the key issue that causes W to fail to be a subspace in Exercise 27?
A: The presence of *absolute value*. Why? It is not additive. See solution below:

27. $\begin{bmatrix} a \\ b \\ |a| \end{bmatrix} + \begin{bmatrix} a' \\ b' \\ |a'| \end{bmatrix} \begin{bmatrix} a+a' \\ b+b' \\ |a|+|a'| \end{bmatrix} \neq \begin{bmatrix} a+a' \\ b+b' \\ |a+a'| \end{bmatrix} \Rightarrow W$ is not a subspace.

Q: Exercises 49 and 50 explore the vector space $U \times W$.
What is this? Why should we already be very familiar with $U \times W$?
A: $U \times W$ is the generalization of $\mathbb{R}^2$. Verify this.
Hint: Is $\mathbb{R} \times \mathbb{R} = \mathbb{R}^2$? Why or why not?

Q: How can we create $\mathbb{R}^3$ using $U \times W$?
That is, what should we chose U and V to be to make $U \times W = \mathbb{R}^3$?
See the discussions below.

We will show that $U \times W$ satisfies axioms 1 and 6 $\Rightarrow U \times W$ is a subspace.

1. $\mathbf{x}, \mathbf{x}' \in U \times W \Rightarrow \mathbf{x} + \mathbf{x}' = (\mathbf{u}, \mathbf{v}) + (\mathbf{u}', \mathbf{v}') = (\mathbf{u} + \mathbf{u}', \mathbf{v} + \mathbf{v}') \Rightarrow \mathbf{x} + \mathbf{x}' \in U \times W$.
6. $\mathbf{x} \in U \times W \Rightarrow c\mathbf{x} = (c\mathbf{u}, c\mathbf{v}) \Rightarrow c\mathbf{x} \in U \times W$.

Also, note that $U \times W$ is not empty. Why not?

We will show that Δ satisfies axioms 1 and 6 $\Rightarrow \Delta$ is a subspace of $V \times V$.

1. $\mathbf{x}, \mathbf{x}' \in \Delta \Rightarrow \mathbf{x} + \mathbf{x}' = (\mathbf{w}, \mathbf{w}) + (\mathbf{w}', \mathbf{w}') = (\mathbf{w} + \mathbf{w}', \mathbf{w} + \mathbf{w}') \Rightarrow \mathbf{x} + \mathbf{x}' \in \Delta$.
6. $\mathbf{x} \in \Delta \Rightarrow c\mathbf{x} = (c\mathbf{w}, c\mathbf{w}) \Rightarrow c\mathbf{x} \in \Delta$.

Finally, since $W \subseteq V$, Δ is a subspace of $V \times V$.

Q: In the proof above, where did we use the fact that W is a subspace of V?
A: When asserting (indirectly) that both $\mathbf{w} + \mathbf{w}'$ and $c\mathbf{w}$ are in W.

Q: Is Δ that same thing as $W \times W$? Why or why not?
A: Hint: Does $W \times W$ require that the components in the ordered pair be identical?

Q: Is $W \times W$ also a subspace of $V \times V$? Why or why not?

Q: Is $\Delta_V = \{(\mathbf{v}, \mathbf{v}) : \mathbf{v} \text{ is in } V\}$ a subspace of $V \times V$?
A: Can V be thought of as a subspace of itself?

6.1 Vector Spaces and Subspaces

Solutions to odd-numbered exercises from Section 6.1

1. Let $V = \left\{ \begin{bmatrix} x \\ x \end{bmatrix} \right\}$. We will show V satisfies all 10 axioms required of a vector space.

   1. $\begin{bmatrix} x \\ x \end{bmatrix}, \begin{bmatrix} y \\ y \end{bmatrix} \in V \Rightarrow \begin{bmatrix} x+y \\ x+y \end{bmatrix} \in V.$

   2. $\begin{bmatrix} x \\ x \end{bmatrix} + \begin{bmatrix} y \\ y \end{bmatrix} = \begin{bmatrix} x+y \\ x+y \end{bmatrix} = \begin{bmatrix} y \\ y \end{bmatrix} + \begin{bmatrix} x \\ x \end{bmatrix}.$

   3. $\left(\begin{bmatrix} x \\ x \end{bmatrix} + \begin{bmatrix} y \\ y \end{bmatrix} \right) + \begin{bmatrix} z \\ z \end{bmatrix} = \begin{bmatrix} x+y+z \\ x+y+z \end{bmatrix} = \begin{bmatrix} x \\ x \end{bmatrix} + \left(\begin{bmatrix} y \\ y \end{bmatrix} + \begin{bmatrix} z \\ z \end{bmatrix} \right).$

   4. $\begin{bmatrix} 0 \\ 0 \end{bmatrix} \in V$ and $\begin{bmatrix} x \\ x \end{bmatrix} + \begin{bmatrix} 0 \\ 0 \end{bmatrix} = \begin{bmatrix} x \\ x \end{bmatrix}.$

   5. $\begin{bmatrix} x \\ x \end{bmatrix} \in V \Rightarrow \begin{bmatrix} -x \\ -x \end{bmatrix} \in V$ and $\begin{bmatrix} x \\ x \end{bmatrix} + \begin{bmatrix} -x \\ -x \end{bmatrix} = \begin{bmatrix} 0 \\ 0 \end{bmatrix}.$

   6. $\begin{bmatrix} x \\ x \end{bmatrix} \in V \Rightarrow \begin{bmatrix} cx \\ cx \end{bmatrix} \in V.$

   7. $c \left(\begin{bmatrix} x \\ x \end{bmatrix} + \begin{bmatrix} y \\ y \end{bmatrix} \right) = \begin{bmatrix} cx+cy \\ cx+cy \end{bmatrix} = c \begin{bmatrix} x \\ x \end{bmatrix} + c \begin{bmatrix} y \\ y \end{bmatrix}.$

   8. $(c+d) \begin{bmatrix} x \\ x \end{bmatrix} = \begin{bmatrix} cx+dx \\ cx+dx \end{bmatrix} = c \begin{bmatrix} x \\ x \end{bmatrix} + d \begin{bmatrix} x \\ x \end{bmatrix}.$

   9. $c \left(d \begin{bmatrix} x \\ x \end{bmatrix} \right) = \begin{bmatrix} cdx \\ cdx \end{bmatrix} = (cd) \begin{bmatrix} x \\ x \end{bmatrix}.$

   10. $1 \begin{bmatrix} x \\ x \end{bmatrix} = \begin{bmatrix} 1x \\ 1x \end{bmatrix} = \begin{bmatrix} x \\ x \end{bmatrix}.$

3. Since $V = \left\{\begin{bmatrix} x \\ y \end{bmatrix} : xy \geq 0\right\}$ fails to satisfy axiom 1, V is not a vector space.

   1. $x \neq y$, $\begin{bmatrix} x \\ y \end{bmatrix}, \begin{bmatrix} -y \\ -x \end{bmatrix} \in V \Rightarrow \begin{bmatrix} x \\ y \end{bmatrix} + \begin{bmatrix} -y \\ -x \end{bmatrix} = \begin{bmatrix} x-y \\ y-x \end{bmatrix} \notin V$ since $(x-y)(y-x) < 0$.

5. $\begin{bmatrix} cx \\ y \end{bmatrix} + \begin{bmatrix} dx \\ y \end{bmatrix} \neq \begin{bmatrix} (c+d)x \\ y \end{bmatrix} \Rightarrow$ axiom 8 fails $\Rightarrow V$ is not a vector space.

7. All axioms apply $\Rightarrow V$ is a vector space.
   1. $x > 0, y > 0 \Rightarrow xy > 0$ 2. $xy = yx$ 3. $(xy)z = x(yz)$
   4. $x1 = x$ (so $1 = \mathbf{0}$!) 5. $x\frac{1}{x} = 1$ 6. $x > 0 x^c > 0$
   7. $(xy)^c = x^c y^c$ 8. $x^{(c+d)} = x^c x^d$ 9. $(x^d)^c = x^{dc}$
   10. $x^1 = x$

9. All axioms apply $\Rightarrow V$ is a vector space. We will show only the key axioms of 1 and 6.
   1. $\begin{bmatrix} a & b \\ 0 & c \end{bmatrix} + \begin{bmatrix} a' & b' \\ 0 & c' \end{bmatrix} = \begin{bmatrix} a+a' & b+b' \\ 0 & c+c' \end{bmatrix}$ 6. $d\begin{bmatrix} a & b \\ 0 & c \end{bmatrix} = \begin{bmatrix} da & db \\ 0 & dc \end{bmatrix}$

11. All axioms apply $\Rightarrow V$ is a vector space. We will show only the key axioms of 1 and 6.
    1. $(A+B)^T = B^T + A^T = -B + (-A) = -(A+B)$ 6. $(cA)^T = cA^T = c(-A) = -(cA)$

13. Need only show axioms 7, 8, 9 and 10 are satisfied.
    7. $c(f+g)(x) = cf(x) + cg(x) = (cf)(x) + (cg)(x)$.
    8. $(c+d)((f(x)) = cf(x) + df(x) = (cf)(x) + (df)(x)$.
    9. $c((df)(x)) = c(d(f(x))) = (cd)f(x)$.
    10. $1(f(x)) = 1\,f(x) = f(x)$.

15. All axioms apply $\Rightarrow V$ is a complex vector space.

17. $c \in \mathbb{C}$ and $c \notin \mathbb{R} \Rightarrow c\mathbf{x} \notin \mathbb{R}^n \Rightarrow$ axiom 6 fails $\Rightarrow V$ is not a complex vector space.

19. Axioms 1, 4, and 6 fail $\Rightarrow V$ is not a vector space.
    1. $w(\mathbf{x}+\mathbf{y}) = w(\mathbf{x}) + w(\mathbf{y}) - 2\,\text{overlap} \Rightarrow w(\mathbf{x}+\mathbf{y})$ is even $\Rightarrow \mathbf{x}+\mathbf{y} \notin V$.
    4. $\mathbf{0} \notin V$ because it has an even number (0) of 1s.
    6. Given $\mathbf{x} \in V$, $0\mathbf{x} = \mathbf{0} \notin V$.

21. Addition, multiplication not well-defined (do not match) $\Rightarrow V$ is not a vector space.

23. Will show $\mathbf{u} + (-1)\mathbf{u} = \mathbf{0} \Rightarrow (-1)\mathbf{u} = -\mathbf{u}$: $\mathbf{u} + (-1)\mathbf{u} = 1\mathbf{u} + (-1)\mathbf{u} = (1+(-1))\mathbf{u} = 0\mathbf{u} = \mathbf{0}$.

25. $\begin{bmatrix} a \\ -a \\ 2a \end{bmatrix} + \begin{bmatrix} b \\ -b \\ 2b \end{bmatrix} = \begin{bmatrix} a+b \\ -(a+b) \\ 2(a+b) \end{bmatrix}, c\begin{bmatrix} a \\ -a \\ 2a \end{bmatrix} = \begin{bmatrix} ca \\ -ca \\ 2ca \end{bmatrix} \Rightarrow W$ is a subspace.

27. $\begin{bmatrix} a \\ b \\ |a| \end{bmatrix} + \begin{bmatrix} a' \\ b' \\ |a'| \end{bmatrix} \begin{bmatrix} a+a' \\ b+b' \\ |a|+|a'| \end{bmatrix} \neq \begin{bmatrix} a+a' \\ b+b' \\ |a+a'| \end{bmatrix} \Rightarrow W$ is not a subspace.

6.1 Vector Spaces and Subspaces

29. $a, b \neq 0$, $\begin{bmatrix} a & 0 \\ 0 & b \end{bmatrix}, \begin{bmatrix} -a & a \\ b & -b \end{bmatrix} \in W$ but $\begin{bmatrix} a & 0 \\ 0 & b \end{bmatrix} + \begin{bmatrix} -a & a \\ b & -b \end{bmatrix} = \begin{bmatrix} 0 & a \\ b & 0 \end{bmatrix} \notin W \Rightarrow$ no.

31. Note, $D \in W \Leftrightarrow D = \sum \lambda_i I$. We will show W satisfies axioms 1 and 6 $\Rightarrow W$ is a subspace.
    1. $D + D' = \sum \lambda_i I + \sum \lambda_i' I = \sum (\lambda_i + \lambda_i') I \Rightarrow D + D' \in W$.
    6. $cD = c \sum \lambda_i I \sum_c \lambda_i I \Rightarrow cD \in W$.

33. We will show that W satisfies axioms 1 and 6 $\Rightarrow W$ is a subspace.
    1. $(A + C)B = AB + CB = BA + BC = B(A + C) \Rightarrow A + C \in W$.
    6. $(cA)B = c(AB) = c(BA) = B(cA) \Rightarrow cA \in W$.

35. We will show that W satisfies axioms 1 and 6 $\Rightarrow W$ is a subspace.
    1. Let $\mathbf{v} = a + b + cx^2$, $\mathbf{v}' = a' + b' + c'x^2 \in W$, that is $a + b + c = 0$ and $a' + b' + c' = 0$.
       Then $(a + a') + (b + b') + (c + c') = 0 \Rightarrow \mathbf{v} + \mathbf{v}' \in W$.
    6. Let $\mathbf{v} = a + b + cx^2 \in W$, that is $a + b + c = 0$.
       Then $c(a + b + c) = 0 \Rightarrow c\mathbf{v} \in W$.

37. $\mathbf{v} = x^3, \mathbf{w} = -x^3 \in W \Rightarrow \mathbf{v} + \mathbf{w} = x^3 + (-x^3) = 0 \notin W$ (because degree $= 0$) $\Rightarrow$ no.

39. We will show that W satisfies axioms 1 and 6 $\Rightarrow W$ is a subspace.
    1. $(f + g)(-x) = f(-x) + g(-x) = -(f(x) + g(x)) = -(f + g)(x) \Rightarrow f + g \in W$.
    6. $(cf)(-x) = cf(-x) = -cf(x) = -(cf)(x) \Rightarrow cf \in W$.

41. $(f + g)(x) = f(x) + g(x) = 0 + 0 = 0$ and $(cf)(x) = cf(x) = c0 = 0 \Rightarrow W$ is a subspace.

43. $f(x) = x \in W$ but $-f(x) = -x \notin W$ $(-f'(x) = -1) \Rightarrow W$ is not a subspace.

45. $f(x) = \frac{1}{x} \in W$ but $-f(x) = -\frac{1}{x} \notin W$ (since $\lim(-f) = -\lim f = -\infty$) $\Rightarrow$ W is not a subspace.

47. Let $U = \left\{ \begin{bmatrix} x \\ 0 \end{bmatrix} \right\}$ (the x-axis) and $W = \left\{ \begin{bmatrix} 0 \\ y \end{bmatrix} \right\}$ (the y-axis).
    $x, y \neq 0$, $\begin{bmatrix} x \\ 0 \end{bmatrix} \in U$, $\begin{bmatrix} 0 \\ y \end{bmatrix} \in W$, but $\begin{bmatrix} x \\ 0 \end{bmatrix} + \begin{bmatrix} 0 \\ y \end{bmatrix} = \begin{bmatrix} x \\ y \end{bmatrix} \notin U \cup W$ ($\begin{bmatrix} x \\ y \end{bmatrix}$ not on axes).

49. We will show that $U \times W$ satisfies the axioms of a vector space.
    1. $\mathbf{x}, \mathbf{x}' \in U \times W \Rightarrow \mathbf{x} + \mathbf{x}' = (\mathbf{u}, \mathbf{v}) + (\mathbf{u}', \mathbf{v}') = (\mathbf{u} + \mathbf{u}', \mathbf{v} + \mathbf{v}') \Rightarrow \mathbf{x} + \mathbf{x}' \in U \times W$.
    2. $\mathbf{x}, \mathbf{x}' \in U \times W \Rightarrow \mathbf{x} + \mathbf{x}' = (\mathbf{u} + \mathbf{u}', \mathbf{v} + \mathbf{v}') = (\mathbf{u}' + \mathbf{u}, \mathbf{v}' + \mathbf{v}) = \mathbf{x}' + \mathbf{x}$.
    6. $\mathbf{x} \in U \times W \Rightarrow c\mathbf{x} = (c\mathbf{u}, c\mathbf{v}) \Rightarrow c\mathbf{x} \in U \times W$.

    The remaining axioms are proved similarly. The details are left to the reader.

51. $aA + bB = C \Rightarrow \begin{bmatrix} a+b & a-b \\ b-a & a \end{bmatrix} = \begin{bmatrix} 1 & 2 \\ 3 & 4 \end{bmatrix} \Rightarrow \begin{array}{l} a+b=1 \\ b-a=3 \end{array} \Rightarrow a = -1 \neq 4 \Rightarrow$ no.

53. $a(1 - 2x) + b(x - x^2) + c(-2 + 3x + x^2) = 3 - 5x - x^2 \Rightarrow \begin{array}{l} a - 2c = 3 \\ -2a + b + 3c = -5 \\ -b + c = -1 \end{array} \Rightarrow$

    $a = -2, b = -1, c = 0 \Rightarrow s(x) = -2p(x) - q(x) \Rightarrow s(x) \in \text{span}(p(x), q(x), r(x))$.

55. $h(x) = 1 = \cos^2 x + \sin^2 x = f(x) + g(x) \Rightarrow h(x) \in \text{span}(f(x), g(x))$.

57. $h(x) = \sin 2x = a f(x) + b g(x) = a\sin^2 x + b\cos^2 x$ for all $x \Rightarrow$
    $x = 0 \Rightarrow a(0)^2 + b(1)^2 = 0 \Rightarrow b = 0$
    $x = \frac{\pi}{2} \Rightarrow a(1)^2 + b(0)^2 = 0 \Rightarrow a = 0$ $\Rightarrow h(x) = \sin 2x$ would be the zero function $\Rightarrow$
    This system has no solution $\Rightarrow h(x) \notin \text{span}(f(x), g(x))$.
    To see the absurdity of $h(x) = \sin 2x =$ the zero function consider $\sin\left(2\frac{\pi}{4}\right) = 1 \Rightarrow 1 \neq 0$.

59. Let $V_1 = \begin{bmatrix} 1 & 1 \\ 0 & 1 \end{bmatrix}, V_2 = \begin{bmatrix} 0 & 1 \\ 1 & 0 \end{bmatrix}, V_3 = \begin{bmatrix} 1 & 0 \\ 1 & 1 \end{bmatrix}, V_4 = \begin{bmatrix} 0 & -1 \\ 1 & 0 \end{bmatrix}$.
    Then since $V_4 = V_3 - V_1$, $\text{span}(V_1, V_2, V_3, V_4) = \text{span}(V_1, V_2, V_3)$.
    To show $\text{span}(V_1, V_2, V_3) \neq M_{22}$ we will show $E_{11} \notin \text{span}(V_1, V_2, V_3)$.
    $aV_1 + bV_2 + cV_3 = E_{11} \Rightarrow \begin{bmatrix} a+c=1 & a+b=0 \\ b+c=0 & a+c=0 \end{bmatrix} \Rightarrow 0 = 1 \Rightarrow$
    This system has no solution $\Rightarrow E_{11} \notin \text{span}(V_1, V_2, V_3) \Rightarrow \text{span}(V_1, V_2, V_3) \neq M_{22}$.

61. Let $p(x) = 1 + x$, $q(x) = x + x^2$, $r(x) = 1 + x^2$. Then note, $1 = a p(x) + b q(x) + c r(x) \Rightarrow$
    $a + c = 1$
    $a + b = 0 \Rightarrow a = c = \frac{1}{2}, b = -\frac{1}{2} \Rightarrow 1 = \frac{1}{2} p(x) - \frac{1}{2} q(x) + \frac{1}{2} r(x)$.
    $b + c = 0$
    Likewise, $x = \frac{1}{2} p(x) + \frac{1}{2} q(x) - \frac{1}{2} r(x)$, $x^2 = -\frac{1}{2} p(x) + \frac{1}{2} q(x) + \frac{1}{2} r(x) \Rightarrow$
    $\text{span}(p(x), q(x), r(x)) = \text{span}(1, x, x^2) = \mathscr{P}_2$. Explicitly, $s(x) \in \mathscr{P}_2 \Rightarrow$
    $s(x) = a + bx + cx^2$
    $= a(\frac{1}{2} p(x) - \frac{1}{2} q(x) + \frac{1}{2} r(x)) + b(\frac{1}{2} p(x) + \frac{1}{2} q(x) - \frac{1}{2} r(x)) + c(-\frac{1}{2} p(x) + \frac{1}{2} q(x) + \frac{1}{2} r(x)) \Rightarrow$
    $s(x) \in \text{span}(p(x), q(x), r(x)) = \text{span}(1, x, x^2) = \mathscr{P}_2$.

63. Let $\mathbf{0}$ and $\mathbf{0'}$ satisfy axiom 4. We will show $\mathbf{0'} = \mathbf{0}$. Axiom 4 $\Rightarrow \mathbf{0'} = \mathbf{0'} + \mathbf{0} = \mathbf{0}$.

6.2 Linear Independence, Basis, and Dimension

Q: What are our main goals for Section 6.2?
A: To take our second and much more in-depth look at *basis* and *dimension*.
Again, we explore the themes of Section 3.5, *Subspaces, Basis, Dimension, and Rank*.

Key Definitions and Concepts

$\{\mathbf{e}_i\}$	p.451	6.2	$\{\mathbf{e}_i\}$, standard basis for $\mathbb{R}^n$ (see Example 6.29)
$\{x^i\}$	p.451	6.2	$\{x^i\}$, standard basis for $\mathscr{P}_n$ (see Example 6.30)
$\{E_{ij}\}$	p.451	6.2	$\mathcal{E} = \{E_{ij}\}$, standard basis for M_{mn} (see Example 6.31)
$[\mathbf{v}]_{\mathcal{B}}$	p.453	6.2	If $\mathcal{B} = \{\mathbf{u}_i\}$ is a basis and $\mathbf{x} = \sum_{i=1}^{n} c_i \mathbf{u}_i$, then $[\mathbf{v}]_{\mathcal{B}} = \begin{bmatrix} c_1 \\ \vdots \\ c_n \end{bmatrix}$.
basis	p.450	6.2	A subset $\mathcal{B}$ of a vector space V is a *basis* for V if 1. $\mathcal{B}$ spans V, that is span$(\mathcal{B}) = V$ 2. $\mathcal{B}$ is linearly independent
coordinate vector	p.453	6.2	If $\mathcal{B} = \{\mathbf{u}_i\}$ is a basis and $\mathbf{x} = \sum_{i=1}^{n} c_i \mathbf{u}_i$, then $[\mathbf{v}]_{\mathcal{B}} = \begin{bmatrix} c_1 \\ \vdots \\ c_n \end{bmatrix}$.
dimensional finite-dim infinite-dim zero-dim	p.457	6.2	A vector space is *blank*-dimensional if: A basis has finitely many vectors A basis has infinitely many vectors The dimension of the subspace $\mathbf{0}$ is defined to be 0
linearly dependent	p.447	6.2	vector can be written as linear combination of others $c_1 \mathbf{v}_1 + \cdots + c_n \mathbf{v}_n = \mathbf{0}$ with at least one $c_i \neq 0$
linearly independent	p.447	6.2	no vector can be written as linear combination of others or $c_1 \mathbf{v}_1 + \cdots + c_n \mathbf{v}_n = \mathbf{0} \Leftrightarrow$ all the $c_i = 0$

Theorems

Thm 6.4   p.448   6.2   vector can be written as linear combination of others
                                  $\Leftrightarrow$ *linearly dependent*

Thm 6.5   p.452   6.2   If $\mathcal{B} = \{\mathbf{u}_i\}$ is a basis and $\mathbf{x} = \sum_{i=1}^{n} c_i \mathbf{u}_i$,
                                  then c_i are unique.

Thm 6.6   p.455   6.2   Let $\mathcal{B} = \{\mathbf{u}_i\}$ be a basis for V then:
                                  a. $[\mathbf{u} + \mathbf{v}]_\mathcal{B} = [\mathbf{u}]_\mathcal{B} + [\mathbf{v}]_\mathcal{B}$
                                  b. $[c\mathbf{u}]_\mathcal{B} = c[\mathbf{u}]_\mathcal{B}$

Thm 6.7   p.456   6.2   Let $\mathcal{B} = \{\mathbf{u}_i\}$ be a basis for V then:
                                  $\{\mathbf{u}_i\}$ are linearly independent in V if and only if
                                  $[\mathbf{u}_i]_\mathcal{B}$ are linearly independent in $\mathbb{R}^n$

Thm 6.8   p.456   6.2   Let $\mathcal{B} = \{\mathbf{u}_i\}$ be a basis for V then:
                                  a. Any set of more than n vectors in V must be linearly dependent
                                  b. Any set of fewer than n vectors cannot span V

Thm 6.9   p.457   6.2   Every basis has exactly the same number of vectors

Thm 6.10   p.458   6.2   Let V be a vector space with $\dim V = n$. Then:
                                  a. Any linearly independent set in V contains at most n vectors.
                                  b. Any spanning set for V contains at least n vectors.
                                  c. Any lin indpt set of exactly n vectors in V is a basis for V.
                                  d. Any spanning set for V of exactly n vectors is a basis for V.
                                  e. Any lin indpt set in V can be extended to a basis for V.
                                  f. Any spanning set for V can be reduced to a basis for V.

Thm 6.11   p.460   6.2   Let W be a subspace of a finite-dimensional vector space V.
                                  a. W is finite-dimensional and $\dim W \leq \dim V$.
                                  b. $\dim W = \dim V$ if and only if $W = V$.

6.2 Linear Independence, Basis, and Dimension

Discussion of Key Definitions and Concepts

How to prove or disprove B is a linearly independent set

Q: Why are we interested in proving (or disproving) that B is linearly independent?
A: Because we are often interested in proving B is a basis.
To be a basis for V, B must span V (span$(B) = V$) and be linearly independent.

Q: What is one method of showing that $B = \{\mathbf{u}_i\}$ is linearly independent?
A: By showing that $c_1 \mathbf{v}_1 + \cdots + c_n \mathbf{v}_n = \mathbf{0}$ implies all the $c_i = 0$.
For example, see the solution to Exercise 19 below:

19. Let $V_1 = \begin{bmatrix} 1 & 0 \\ 0 & 1 \end{bmatrix}$, $V_2 = \begin{bmatrix} 0 & -1 \\ 1 & 0 \end{bmatrix}$, $V_3 = \begin{bmatrix} 1 & 1 \\ 1 & 1 \end{bmatrix}$, $V_4 = \begin{bmatrix} 1 & 1 \\ 1 & -1 \end{bmatrix}$. So,

$$aV_1 + bV_2 + cV_3 + dV_4 = 0 \Rightarrow \begin{bmatrix} a+b+d = 0 & -b+c+d = 0 \\ b+c+d = 0 & a-c-d = 0 \end{bmatrix} \Rightarrow a = b = c = d = 0 \Rightarrow$$

V_1, V_2, V_3, V_4 linearly independent $\Rightarrow$
The number of linearly independent vectors in $B = 4 = \dim M_{22} \Rightarrow B$ is a basis for M_{22}.

Q: What is one method of showing that $B = \{\mathbf{u}_i\}$ is *not* linearly independent?
A: By showing that one of the vectors can be written as a combination of the others.
For example, see the solution to Exercise 11 below:

11. $1 = \sin^2 x + \cos^2 x \Rightarrow \{1, \sin^2 x, \cos^2 x\}$ linearly dependent.

Proving B is not a basis

Q: What is one method of showing that $B = \{\mathbf{u}_i\}$ is *not* a basis?
A: One method is to show that B has too few *linearly independent* vectors in it.
For example, see the solution to Exercise 23 below:

23. $x - x^2 = -(1-x) + (1-x^2) \Rightarrow$ number of linearly independent vectors in $B \leq 2 < \dim \mathcal{P}_2 \Rightarrow$
B is not a basis for $\mathcal{P}_2$.

Q: What is another method of showing that $B = \{\mathbf{u}_i\}$ is *not* a basis?
A: Another method is to show that B has too *many* vectors in it.
For example, see the solution to Exercise 21 below:

21. The number of vectors in $B = 5 > 4 = \dim M_{22} \Rightarrow B$ is not a basis for M_{22}.

Comparison of $U \times V$ to $\mathbb{R}^n$ and $W \times W$ to Δ

Q: In Exercise 43, we show $\dim(U \times V) = \dim U + \dim V$.
Does this agree with the dimensions of $\mathbb{R}^2$, $\mathbb{R}^3$, and $\mathbb{R}^n$?
If so, construct $\mathbb{R}^2$, $\mathbb{R}^3$, and $\mathbb{R}^n$ using $U \times V$.

Q: Does this make it easier to prove or disprove that $W \times W = \Delta$?
A: Hint: Compare the dimension of each of these subspaces.

Discussion of Theorems

Putting the key ideas of theorems into words

Q: How might we put the key conclusion of Thm 6.5 into words?
A: There is only *one* way to write a vector as a linear combination of vectors in a basis.

Q: How might we summarize Thm 6.4 and Thm 6.5 together as a unit?
A: The only way to write $\mathbf{b}_j$ as a linear combination of the vectors in $\mathcal{B}$ is $\mathbf{b}_j$. Why?

Q: What is the essence of Thm 6.6?
A: Coordinate vectors for any basis behave like the coordinate vectors for a standard basis. That is, like the vectors in $\mathbb{R}^n$ with which we are comfortable and familiar.

Q: What is the essence of Thm 6.7?
A: Vectors that are linearly independent are really and truly linearly independent. That is, it does not matter what basis we use to write them down, they remain linearly independent.
This is an extremely important conclusion, idea, and insight.

Q: How do Thms 6.8, 6.9, and 6.10 differ?
A: In many different ways, obviously. But perhaps the key is statement of Thm 6.9:
Every basis has exactly the same number of vectors.
If we understand this claim, the conclusions of Thm 6.8 and 6.10 should not surprise us.

Q: What are some of the major points of Thm 6.10?
A: There are several key themes. One of the most important is that of *extending to a basis*. See Exercise 42 for an interesting application of this idea to $U + V$.

Q: The parallel claim is that a spanning set can be *reduced* to a basis. How?
A: By removing the *redundant* vectors, so to speak. That is, those vectors that can be written as a linear combination of the remaining vectors.

Q: What are three major conclusions of Thm 6.11?
A: The dimension of any subspace must be less than or equal to the dimension of the space.
If the subspace is not the *whole* space, its dimension is *less* than that of the whole space.
The only subspace whose dimension is equal to that of the whole space *is* the whole space.

6.2 Linear Independence, Basis, and Dimension

Solutions to odd-numbered exercises from Section 6.2

1. Let $V_1 = \begin{bmatrix} 1 & 1 \\ 0 & -1 \end{bmatrix}, V_2 = \begin{bmatrix} 1 & -1 \\ 1 & 0 \end{bmatrix}, V_3 = \begin{bmatrix} 1 & 0 \\ 3 & 2 \end{bmatrix}$.

   Then, $aV_1 + bV_2 + cV_3 = \begin{bmatrix} 0 & 0 \\ 0 & 0 \end{bmatrix} \Rightarrow \begin{bmatrix} a+b+c=0 & a-b=0 \\ b+3c=0 & -a+2c=0 \end{bmatrix} \Rightarrow a=b=c=0 \Rightarrow$

   V_1, V_2, V_3 linearly independent by definition.

3. $aV_1 + bV_2 + cV_3 + dV_4 = \begin{bmatrix} 0 & 0 \\ 0 & 0 \end{bmatrix} \Rightarrow \begin{bmatrix} -a+3b-d=0 & a+2c=0 \\ -2a+b+-3c-d=0 & 2a+b+c+7d=0 \end{bmatrix} \Rightarrow$

   $a=4, b=1, c=-2, d=-1 \Rightarrow V_4 = 4V_1 + V_2 - 2V_3$.

5. Let $p(x) = x$, $q(x) = 1+x$, then $a\,p(x) + b\,q(x) = a(x) + b(1+x) = b + (a+b)x = 0 \Rightarrow$
   $b=0$ $a+b=0 \Rightarrow a=b=0 \Rightarrow p(x), q(x)$ linearly independent.

7. $a\,p(x) + b\,q(x) + c\,r(x) = (a+2b+3c)x + (-b+2c)x^2 = 0 \Rightarrow$
   $0=0, \; a+2b+3c=0, \; -b+2c=0 \Rightarrow a=7, b=-2, c=-1 \Rightarrow r(x) = 7p(x) - 2q(x)$.
   Since we had only 2 equations and 3 unknowns a was free so there are infinitely many solutions.

9. $a\,p(x)+b\,q(x)+c\,r(x)+d\,s(x) = (a+c+3d)+(-2a+3b+2)x+(b+c)x^2+(-b+2c+3d)x^3 = 0 \Rightarrow$
   $\begin{array}{l} a+c+3d=0 \quad -2a+3b+2=0 \\ b+c=0 \quad\quad\quad -b+2c+3d=0 \end{array} \Rightarrow a=b=c=d=0 \Rightarrow$ linearly independent.

11. $1 = \sin^2 x + \cos^2 x \Rightarrow \{1, \sin^2 x, \cos^2 x\}$ linearly dependent.

13. $\ln(x^2) = -2\ln 2 \cdot 1 + 2 \cdot \ln(2x) \Rightarrow \{1, \ln(2x), \ln(x^2)\}$ linearly dependent.

15. $W(x) \neq 0$ for some $x \Rightarrow f(x), g(x)$ linearly independent $\Leftrightarrow$
    $f(x), g(x)$ linearly dependent $\Rightarrow W(x) = g'(x)f(x) - f'(x)g(x) = 0$ for all x.
    This is what we will show.

    $f(x), g(x)$ linearly dependent $\Rightarrow f(x)+ag(x) = 0, a \neq 0 \Rightarrow\Rightarrow f'(x)+a'g(x)=0 \Rightarrow a = -\frac{f'(x)}{g'(x)}$.

    Substituting this value into $f(x) + ag(x) = 0 \Rightarrow f(x) - \frac{f'(x)}{g'(x)}g(x) = 0 \Rightarrow$
    $W(x) = g'(x)f(x) - f'(x)g(x) = 0$ for all x.

17. (a) $a(\mathbf{u}+\mathbf{v}) + b(\mathbf{v}+\mathbf{w}) + c(\mathbf{u}+\mathbf{w}) = (a+c)\mathbf{u} + (a+b)\mathbf{v} + (b+c)\mathbf{w} = \mathbf{0} = 0 \Rightarrow$
    $a+c=0, \; a+b=0, \; b+c=0 \Rightarrow a=b=c=0 \Rightarrow$
    $\mathbf{u}+\mathbf{v}, \mathbf{v}+\mathbf{w}, \mathbf{u}+\mathbf{w}$ are linearly independent.

    (b) $a(\mathbf{u}-\mathbf{v}) + b(\mathbf{v}-\mathbf{w}) + c(\mathbf{u}-\mathbf{w}) = (a+c)\mathbf{u} + (-a+b)\mathbf{v} + (-b-c)\mathbf{w} = \mathbf{0} = 0 \Rightarrow$
    $a+c=0, \; -a+b=0, \; -b-c=0 \Rightarrow a=b=1, c=-1$ is a solution
    $\Rightarrow \mathbf{u}-\mathbf{v}, \mathbf{v}-\mathbf{w}, \mathbf{u}-\mathbf{w}$ are linearly dependent.
    Note, a was free so there are infinitely many solutions.
    Eg.: Let $\mathbf{u}=1, \mathbf{v}=x, \mathbf{w}=x^2$, then $\mathbf{u}-\mathbf{v}=1-x, \mathbf{v}-\mathbf{w}=x-x^2, \mathbf{u}-\mathbf{w}=1-x^2$.
    Since $1-x^2 = 1(1-x) + 1(x-x^2)$, $\mathbf{u}-\mathbf{v}, \mathbf{v}-\mathbf{w}, \mathbf{u}-\mathbf{w}$ are clearly linearly dependent.

19. Let $V_1 = \begin{bmatrix} 1 & 0 \\ 0 & 1 \end{bmatrix}, V_2 = \begin{bmatrix} 0 & -1 \\ 1 & 0 \end{bmatrix}, V_3 = \begin{bmatrix} 1 & 1 \\ 1 & 1 \end{bmatrix}, V_4 = \begin{bmatrix} 1 & 1 \\ 1 & -1 \end{bmatrix}$. So,

$aV_1 + bV_2 + cV_3 + dV_4 = 0 \Rightarrow \begin{bmatrix} a+c+d=0 & -b+c+d=0 \\ b+c+d=0 & a+c-d=0 \end{bmatrix} \Rightarrow a=b=c=d=0 \Rightarrow$

V_1, V_2, V_3, V_4 linearly independent $\Rightarrow$
The number of linearly independent vectors in $B = 4 = \dim M_{22} \Rightarrow B$ is a basis for M_{22}.

21. The number of vectors in $B = 5 > 4 = \dim M_{22} \Rightarrow B$ is not a basis for M_{22}.

23. $x - x^2 = -(1-x) + (1-x^2) \Rightarrow$ number of linearly independent vectors in $B \leq 2 < \dim \mathscr{P}_2 \Rightarrow$
B is not a basis for $\mathscr{P}_2$.

25. The number of vectors in $B = 4 > 3 = \dim \mathscr{P}_2 \Rightarrow B$ is not a basis for $\mathscr{P}_2$.

27. Let $V_1 = \begin{bmatrix} 1 & 0 \\ 0 & 0 \end{bmatrix}, V_2 = \begin{bmatrix} 1 & 1 \\ 0 & 0 \end{bmatrix}, V_3 = \begin{bmatrix} 1 & 1 \\ 1 & 0 \end{bmatrix}, V_4 = \begin{bmatrix} 1 & 1 \\ 1 & 1 \end{bmatrix}$.

Then $A = \begin{bmatrix} 1 & 2 \\ 3 & 4 \end{bmatrix} = aV_1 + bV_2 + cV_3 + dV_4 \Rightarrow \begin{bmatrix} a+b+c+d=1 & b+c+d=2 \\ c+d=3 & d=4 \end{bmatrix} \Rightarrow$

$A = -V_1 - V_2 - V_3 + 4V_4 \Rightarrow [A]_B = \begin{bmatrix} -1 \\ -1 \\ -1 \\ 4 \end{bmatrix}$.

29. $p(x) = 2 - x + 3x^2 = a(1) + b(1+x) + c(-1+x^2) = (a+b-c) + bx + cx^2 \Rightarrow$

$a+b-c=2, b=-1, c=3 \Rightarrow a=6, b=-1, c=3 \Rightarrow [p(x)]_B = \begin{bmatrix} 6 \\ -1 \\ 3 \end{bmatrix}$.

31. $[c_1\mathbf{u}_1 + \ldots + c_n\mathbf{u}_n]_B = [c_1\mathbf{u}_1]_B + \ldots + [c_n\mathbf{u}_n]_B = c_1[\mathbf{u}_1]_B + \ldots + c_n[\mathbf{u}_n]_B$.

33. Let $\text{span}(S) = \mathbb{R}^n$ and let $\mathbf{v} \in V \Rightarrow \mathbf{x} = [\mathbf{v}]_B = \sum c_i[\mathbf{u}_i]_B = [\sum c_i\mathbf{u}_i]_B \Rightarrow \mathbf{v} = \sum c_i\mathbf{u}_i$
because the representation of $\mathbf{v}$ with respect to basis B is unique.
Therefore, $V = \text{span}(\mathbf{u}_i)$.

Let $V = \text{span}(\mathbf{u}_i)$ and let $\mathbf{x} = \begin{bmatrix} 1 \\ c_2 \\ \vdots \\ c_n \end{bmatrix} \in \mathbb{R}^n \Rightarrow \sum c_i\mathbf{u}_i = \mathbf{v} \in V$ because $V = \text{span}(\mathbf{u}_i)$.

Therefore, $\mathbf{x} = [\mathbf{v}]_B = \sum[c_i\mathbf{u}_i]_B = \sum c_i[\mathbf{u}_i]_B \Rightarrow \mathbf{x} \in \text{span}(S)$.
Therefore, $\text{span}(S) = \mathbb{R}^n$.

35. $p(1) = a(1)^2 + b(1) + c = 0 \Rightarrow a = -(b+c)$, b, c free $\Rightarrow$
$B = \{1 - x, 1 - x^2\}$ is a basis $\Rightarrow \dim V = 2$.

37. $B = \{E_{11}, E_{12}, E_{22}\}$ is a basis $\Rightarrow \dim V = 3$.

6.2 Linear Independence, Basis, and Dimension

39. $AB = BA \Rightarrow \begin{bmatrix} a & a+b \\ c & c+d \end{bmatrix} = \begin{bmatrix} a+c & b+d \\ c & d \end{bmatrix} \Rightarrow$

    $c = 0, a = d, a, b$ free $\Rightarrow B = \left\{ \begin{bmatrix} 1 & 0 \\ 0 & 1 \end{bmatrix}, \begin{bmatrix} 0 & 1 \\ 0 & 0 \end{bmatrix} \right\}$ is a basis $\Rightarrow \dim V = 2$.

41. We should begin by noting that we need to count the number of free variables. We start with n^2 free variables from M_{nn}, then reduce that number by applying the conditions. $A^T = -A \Rightarrow a_{ii} = 0$ which reduces the number of free variables by n to $n^2 - n$. Then $A^T = -A \Rightarrow a_{ij} = -a_{ji}$ cuts the remaining number in half, so $\dim V = \frac{n^2-n}{2}$.

43. We will find bases, then use those bases to compute the dimension. Let $\dim U = m$ with basis $\mathcal{B} = \{\mathbf{u}_1, \ldots, \mathbf{u}_m\}$ and $\dim V = n$ with basis $\mathcal{C} = \{\mathbf{v}_1, \ldots, \mathbf{v}_n\}$.

    (a) We will show that $\mathcal{D} = \{(\mathbf{u}_i, \mathbf{0}), (\mathbf{0}, \mathbf{v}_j)\}$ is a basis for $U \times V$. This will imply that $\dim(U \times V) = \dim U + \dim V$.

    spans: We will show that $\text{span}(\mathcal{D}) = U \times V$. Let $\mathbf{x} = (\mathbf{u}, \mathbf{v}) \in U \times V$ then:
    $$\mathbf{x} = (\mathbf{u}, \mathbf{v}) = (\sum_{i=1}^{m} c_i \mathbf{u}_i, \sum_{j=1}^{n} d_j \mathbf{v}_j) = \sum_{i=1}^{m} c_i(\mathbf{u}_i, \mathbf{0}) + \sum_{j=1}^{n} d_j(\mathbf{0}, \mathbf{v}_j).$$

    lin ind: We will show that $\mathcal{D}$ is linearly independent. That is, we will show if $\mathbf{0} = (\mathbf{0}, \mathbf{0}) = \sum_{i=1}^{m} c_i(\mathbf{u}_i, \mathbf{0}) + \sum_{j=1}^{n} d_j(\mathbf{0}, \mathbf{v}_j)$, then $c_i = d_j = 0$.
    $$\mathbf{0} = (\mathbf{0}, \mathbf{0}) = \sum_{i=1}^{m} c_i(\mathbf{u}_i, \mathbf{0}) + \sum_{j=1}^{n} d_j(\mathbf{0}, \mathbf{v}_j) = (\sum_{i=1}^{m} c_i \mathbf{u}_i = \mathbf{0}, \sum_{j=1}^{n} d_j \mathbf{v}_j = \mathbf{0})$$
    Since $\mathbf{u}_i$ and $\mathbf{v}_j$ are linearly independent, $\sum_{i=1}^{m} c_i \mathbf{u}_i = \mathbf{0}, \sum_{j=1}^{n} d_j \mathbf{v}_j = \mathbf{0}$ implies $c_i = d_j = 0$, as we were to show.

    (b) Let $\dim W = k$ with basis $\mathcal{B} = \{\mathbf{w}_1, \ldots, \mathbf{w}_k\}$. Then: Similar to (a), $\mathcal{C} = \{(\mathbf{w}_i, \mathbf{w}_i)\}$ is a basis for Δ, so $\dim \Delta = \dim W$.

45. $a(1 + x) + b(1 + x + x^2) + c(1) = (a + b + c) + (a + b)x + bx^2 = 0 \Rightarrow$
    $a + b + c = 0, a + b = 0, b = 0 \Rightarrow$
    $a = b = c = 0 \Rightarrow \mathcal{B} = \{1, 1 + x, 1 + x + x^2\}$ linearly independent $\Rightarrow$
    $\mathcal{B}$ is a basis for $\mathcal{P}_2$ because $\dim \mathcal{P}_2 = 3 =$ the number of vectors in $\mathcal{B}$.

47. Let $V_1 = \begin{bmatrix} 1 & 0 \\ 0 & 1 \end{bmatrix}, V_2 = \begin{bmatrix} 0 & 1 \\ 1 & 0 \end{bmatrix}, V_3 = \begin{bmatrix} 0 & -1 \\ 1 & 0 \end{bmatrix}$.

    Then, $aV_1 + bV_2 + cV_3 + dE_{11} = 0 \Rightarrow \begin{bmatrix} a+d = 0 & b-c = 0 \\ b+c = 0 & a = 0 \end{bmatrix} \Rightarrow$

    $a = b = c = d = 0 \Rightarrow \mathcal{B} = \{V_1, V_2, V_3, E_{11}\}$ linearly independent $\Rightarrow$
    $\mathcal{B}$ is a basis for M_{22} because $\dim M_{22} = 4 =$ the number of vectors in $\mathcal{B}$.

49. We will show $\text{span}(1, 1 + x, 2x) = \text{span}(1, 1 + x)$ and $\{1, 1 + x\}$ linearly independent which will imply $\{1, 1 + x\}$ is a basis for $\text{span}(1, 1 + x, 2x)$.
    $2x = -2(1) + 2(1 + x) \Rightarrow \text{span}(1, 1 + x, 2x) = \text{span}(1, 1 + x)$.
    $a(1) + b(1 + x) = (a + b) + bx = 0 \Rightarrow a + b = 0, b = 0 \Rightarrow a = b = 0 \Rightarrow$
    $\{1, 1 + x\}$ linearly independent. Therefore, $\{1, 1 + x\}$ is a basis for $\text{span}(1, 1 + x, 2x)$.

51. We will show $\text{span}(1-x, x-x^2, 1-x^2, 1-2x+x^2) = \text{span}(1-x, x-x^2)$ and $\{1-x, x-x^2\}$ linearly independent which will imply
$\{1-x, x-x^2\}$ is a basis for $\text{span}(1-x, x-x^2, 1-x^2, 1-2x+x^2)$.
$1-x^2 = 1(1-x) + 1(x-x^2)$ and $1-2x+x^2 = 1(1-x) - (x-x^2) \Rightarrow$
$\text{span}(1-x, x-x^2, 1-x^2, 1-2x+x^2) = \text{span}(1-x, x-x^2)$.
$a(1-x) + b(x-x^2) = a + (-a+b)x + (-b)x^2 = 0 \Rightarrow$
$\Rightarrow a = b = 0 \Rightarrow \{1-x, x-x^2\}$ linearly independent.

Therefore, $\{1-x, x-x^2\}$ is a basis for $\text{span}(1-x, x-x^2, 1-x^2, 1-2x+x^2)$.

53. We will show $\text{span}(\sin^2 x, \cos^2 x, \cos 2x) = \text{span}(\sin^2 x, \cos^2 x)$ and $\{\sin^2 x, \cos^2 x\}$ linearly independent which will imply
$\{\sin^2 x, \cos^2 x\}$ is a basis for $\text{span}(\sin^2 x, \cos^2 x, \cos 2x)$.
$\cos 2x = \cos^2 x - \sin^2 x \Rightarrow \text{span}(\sin^2 x, \cos^2 x, \cos 2x) = \text{span}(\sin^2 x, \cos^2 x)$.
$a \sin^2 x + b \cos^2 x = 0 \Rightarrow x = 0 \Rightarrow b = 0$ and $x = \frac{\pi}{2} \Rightarrow a = 0 \Rightarrow a = b = 0 \Rightarrow$
$\{\sin^2 x, \cos^2 x\}$ linearly independent $\Rightarrow \{\sin^2 x, \cos^2 x\}$ is a basis for $\text{span}(\sin^2 x, \cos^2 x, \cos 2x)$.

55. Must show $\mathbf{v} \in V \Leftrightarrow$ there exist c_i such that $\mathbf{v} = c_1 \mathbf{v}_1 + \ldots + c_{n-1} \mathbf{v}_{n-1}$.
Note this will prove that $V = \text{span}(S) = \text{span}(S')$.

$V = \text{span}(S) \Rightarrow \mathbf{v} = c_1 \mathbf{v}_1 + \ldots + c_{n-1} \mathbf{v}_{n-1} + c_n \mathbf{v}_n$.
$\mathbf{v}_n \in \text{span}(S') \Rightarrow \mathbf{v}_n = d_1 \mathbf{v}_1 + \ldots + d_{n-1} \mathbf{v}_{n-1}$. Therefore, substitution gives us:
$\mathbf{v} = c_1 \mathbf{v}_1 + \ldots + c_{n-1} \mathbf{v}_{n-1} + c_n(d_n \mathbf{v}_1 + \ldots + d_{n-1} \mathbf{v}_{n-1})$
$= (c_1 + c_n d_1) \mathbf{v}_1 + \ldots + (c_{n-1} + c_n d_{n-1}) \mathbf{v}_{n-1} \Rightarrow \mathbf{v} \in \text{span}(S')$ as required.

57. Note $c_i \neq 0 \Rightarrow$ we can divide by every c_i. We will show $V = \text{span}(c_1 \mathbf{v}_1, \ldots, c_n \mathbf{v}_n)$ and $\{c_1 \mathbf{v}_1, \ldots, c_n \mathbf{v}_n\}$ linearly independent which will imply $\{c_1 \mathbf{v}_1, \ldots, c_n \mathbf{v}_n\}$ is a basis for V.
$\{\mathbf{v}_1, \ldots, \mathbf{v}_n\}$ is a basis for $V \Rightarrow \mathbf{v} = d_1 \mathbf{v}_1 + \ldots + d_n \mathbf{v}_n = \frac{d_1}{c_1} c_1 \mathbf{v}_1 + \ldots + \frac{d_n}{c_n} c_n \mathbf{v}_n \Rightarrow$
$V = \text{span}(c_1 \mathbf{v}_1, \ldots, c_n \mathbf{v}_n)$.
$0 = d_1 c_1 \mathbf{v}_1 + \ldots + d_1 c_n \mathbf{v}_n \Rightarrow d_i c_i = 0$ for all i because $\mathbf{v}_i$ are linearly independent $\Rightarrow$
$d_i = 0$ because c_i are all nonzero $\Rightarrow \{c_1 \mathbf{v}_1, \ldots, c_n \mathbf{v}_n\}$ linearly independent.

Therefore, $\{c_1 \mathbf{v}_1, \ldots, c_n \mathbf{v}_n\}$ is a basis for V.

59. (a) $p_0(x) = \frac{(x-2)(x-3)}{(1-2)(1-3)} = \frac{1}{2} x^2 - \frac{5}{2} x + 3$.

$p_1(x) = \frac{(x-1)(x-3)}{(2-1)(2-3)} = -x^2 + 4x - 3$.

$p_2(x) = \frac{(x-1)(x-2)}{(3-1)(3-2)} = \frac{1}{2} x^2 - \frac{3}{2} x + 1$.

(b) $p_i(a_i) = \frac{(a_i-a_0)\ldots(a_i-a_{i-1})(a_i-a_{i+1})\ldots(a_i-a_n)}{(a_i-a_0)\ldots(a_i-a_{i-1})(a_i-a_{i+1})\ldots(a_i-a_n)} = 1$.

$p_i(a_j) = \frac{(a_j-a_0)\ldots(a_j-a_j)\ldots(a_j-a_n)}{(a_i-a_0)\ldots(a_i-a_j)\ldots(a_i-a_n)} = 0$.

6.2 Linear Independence, Basis, and Dimension

61. (a) Exercise 59(b) $\Rightarrow q(a_i) = c_0\, p_0(a_i) + \ldots + c_n\, p_n(a_i) = c_i p_i(a_i) = c_i$ for all $i \Rightarrow$
    $q(x) = q(a_0)\, p_0(a_i) + \ldots + q(a_n)\, p_n(a_i)$.

    (b) Exercise 59(b) $\Rightarrow q(a_i) = c_0\, p_0(a_i) + \ldots + c_n\, p_n(a_i) = c_i \Rightarrow$
    q passes through $(a_0, c_0), \ldots, (a_n, c_n)$. So, q is unique because $\mathcal{B}$ is a basis for $\mathcal{P}_n$.
    Note that a_i must be distinct to avoid division by zero.

    (c) (i) $a_0 = 1$, $a_1 = 2$, $a_3 = 3 \Rightarrow$ we can use $p_0(x)$, $p_1(x)$, and $p_2(x)$ from Exercise 59(a).

    Therefore, (c) $\Rightarrow q(x) = 6\, p_0(x) - p_1(x) - 2\, p_2(x) = 3x^2 - 16x + 19$.

    (ii) With $a_0 = -1$, $a_1 = 0$, and $a_3 = 3$, we find:

    $p_0(x) = \frac{(x-0)(x-3)}{(-1-0)(-1-3)} = \frac{1}{4} x^2 - \frac{3}{4} x$.

    $p_1(x) = \frac{(x+1)(x-3)}{(0+1)(0-3)} = -\frac{1}{3} x^2 + \frac{2}{3} x + 1$.

    $p_2(x) = \frac{(x+1)(x-0)}{(3+1)(3-0)} = \frac{1}{12} x^2 + \frac{1}{12} x$.

    Therefore, (c) $\Rightarrow q(x) = 10\, p_0(x) + 5\, p_1(x) + 2\, p_2(x) = x^2 - 4x + 5$.

63. Recall, a $n \times n$ matrix A is invertible $\Leftrightarrow$ its columns form a basis for col(I).
    Therefore, a matrix in $M_{nn}(\mathbb{Z}_p)$ is invertible $\Leftrightarrow$ its columns form a basis for col(I) = $\mathbb{Z}_p^n$.

    As suggested in the hint, we will count the number of ways to construct a basis for $\mathbb{Z}_p^n$.

    Note, when constructing a basis, we can start with any vector except for $\mathbf{0}$.
    Therefore, since there are p^n vectors in $\mathbb{Z}_p^n$, there are $p^n - 1$ ways to choose the first vector.
    Call this vector $\mathbf{v}_1$.

    For our second vector, we can choose any vector not in span($\mathbf{v}_1$) = $\{a_1 \mathbf{v}_1 : a_1 = 1, \ldots, p\}$.
    Since there are p vectors in this span, therefore there are $p^n - p$ ways to choose $\mathbf{v}_2$.

    Likewise, choose $\mathbf{v}_3 \notin \mathrm{span}(\mathbf{v}_1, \mathbf{v}_2) = \{a_1 \mathbf{v}_1 + a_2 \mathbf{v}_2 : a_1, a_2 = 1, \ldots, p\}$.
    Since there are p^2 vectors in this span, therefore there are $p^n - p^2$ ways to choose $\mathbf{v}_3$.

    In general, therefore, there are $p^n - p^{i-1}$ ways to choose the ith vector in our basis.
    Taking the product of these to find the total number of ways possible, we find:

    the number of different bases for $\mathbb{Z}_p^n$ = the number of invertible matrices in $M_{nn}(\mathbb{Z}_p)$
    $= (p^n - 1)(p^n - p)(p^n - p^2) \ldots (p^n - p^{n-1})$.

Exploration: Magic Squares

Since Explorations are self-contained, only odd-numbered solutions will be provided.

1. We want to show: $\text{wt}(M) = \frac{n(n^2+1)}{2}$.

   A *classical magic square* contains each of entries $1, 2, \cdots, n^2$ exactly once.
   Since $\text{wt}(M)$ is the common sum of any row or column we have the following:
   $$\text{wt}(M) = \tfrac{1}{n}(1 + 2 + \cdots + n^2)$$
   Using induction, we will prove $1 + 2 + \cdots + n^2 = \frac{n^2(n^2+1)}{2}$ for all $n \geq 1$.

   1: $1 = \frac{1^2(1+1)}{2}$
   This is obvious, so there is nothing to show.

   k: $1 + 2 + \cdots + k^2 = \frac{k^2(k^2+1)}{2}$
   This is the induction hypothesis, so there is nothing to show.

   $k+1$: $1 + 2 + \cdots + k^2 + (k+1)^2 = \frac{(k+1)^2((k+1)^2+1)}{2}$
   This is the statement we must prove using the induction hypothesis.

   $$1 + 2 + \cdots + k^2 + (k+1)^2 \stackrel{\text{by induction}}{=} \frac{k^2(k^2+1)}{2} + (k^2+1) + (k^2+2) + \cdots + (k^2+2k+1).$$

   By induction $1 + 2 + \cdots + (2k+1) = \frac{(2k+1)(2k+2)}{2} = (2k+1)(k+1)$.
   See Appendix B for further discussion of *Mathematical Induction*.

   We also have $2k+1$ copies of k^2. So:
   $$1 + 2 + \cdots + k^2 + (k+1)^2 = \frac{k^2(k^2+1)}{2} + (k^2+1) + (k^2+2) + \cdots + (k^2+2k+1)$$
   $$= \frac{k^2(k^2+1)}{2} + (2k+1)k^2 + (2k+1)(k+1)$$
   $$= \frac{k^4 + 4k^3 + 7k^2 + 6k + 2}{2} = \frac{(k+1)^2((k+1)^2+1)}{2}$$

   We have proven (by induction) that
   $$1 + 2 + \cdots + n^2 = \frac{n^2(n^2+1)}{2} \text{ for all } n \geq 1.$$

   So we have $\text{wt}(M)$ is the the common sum of any row or column we have the following:
   $$\text{wt}(M) = \tfrac{1}{n}(1 + 2 + \cdots + n^2) = \tfrac{1}{n}\left(\frac{n^2(n^2+1)}{2}\right) = \frac{n(n^2+1)}{2}.$$

   Q: Why do we multiply the sum by the factor $\frac{1}{n}$?
   A: Because each of the n columns of M sum to $\text{wt}(M)$.
   So the sum of all the entries equals n times the $\text{wt}(M)$.
   That is, we have $\text{total} = n \times \text{wt}(M)$, so $\text{wt}(M) = \tfrac{1}{n} \times \text{total}$.

3. We want to construct a 3×3 magic square of $\text{wt}(N) = 1$ with all different entries.
   From Exercise 2, we have M with $\text{wt}(M) = 15$, so we can let $N = \frac{1}{15}M$.

$$N = \tfrac{1}{15}M = \tfrac{1}{15}\begin{bmatrix} 8 & 3 & 4 \\ 1 & 5 & 9 \\ 6 & 7 & 2 \end{bmatrix} = \begin{bmatrix} \tfrac{8}{15} & \tfrac{3}{15} & \tfrac{4}{15} \\ \tfrac{1}{15} & \tfrac{5}{15} & \tfrac{9}{15} \\ \tfrac{6}{15} & \tfrac{7}{15} & \tfrac{2}{15} \end{bmatrix} = \begin{bmatrix} \tfrac{8}{15} & \tfrac{1}{5} & \tfrac{4}{15} \\ \tfrac{1}{15} & \tfrac{1}{3} & \tfrac{3}{5} \\ \tfrac{2}{5} & \tfrac{7}{15} & \tfrac{2}{15} \end{bmatrix}.$$

   Given any weight w, we can construct the classical magic square M. Then let $N = \frac{w}{\text{wt}(M)}M$.

5. We need to show if M is in Mag_3 with $\text{wt}(M) = w$ then $M = M_0 + kJ$,
   where M_0 has weight zero and J is the matrix whose entries are all ones.

   Following the hint, we will show that $M - kJ$ is in Mag_3^0.

   Let $k = \frac{w}{3} = \frac{\text{wt}(M)}{3}$ and $M_0 = M - kJ = M - \frac{\text{wt}(M)}{3}J = M - \frac{w}{3}J$.

   With this value of k, note that $\text{wt}(\frac{w}{3}J) = w$.

   Why?

   Because, for example, the first row of $\frac{w}{3}J$ is $\begin{bmatrix} \frac{w}{3} & \frac{w}{3} & \frac{w}{3} \end{bmatrix}$.

   When we sum these three entries, we get: $\frac{w}{3} + \frac{w}{3} + \frac{w}{3} = w$.

   So $\text{wt}(M_0) = \text{wt}(M) - \text{wt}\left(\frac{w}{3}J\right) = w - w = 0$.

7. Since there are two unknowns s and t, it is clear that $\dim Mag_3^0 = 2$. Why?

   From the general description of M given in the text,
   it is obvious that the following set is a basis for Mag_3^0:

$$\left\{ M_s = \begin{bmatrix} 1 & -1 & 0 \\ -1 & 0 & 1 \\ 0 & 1 & -1 \end{bmatrix}, M_t = \begin{bmatrix} 0 & -1 & 1 \\ 1 & 0 & -1 \\ -1 & 1 & 0 \end{bmatrix}, \right\}$$

   Q: Think of these matrices as basis vectors for Mag_3^0.
   Is it obvious that these matrices are linearly independent? Why or why not?
   A: Consider $kM_s + jM_t = O$. What do k and j have to be?

   Q: Define the dot product between these matrices as $M \cdot N = \sum m_{ij} \cdot n_{ij}$.
   Using this definition, are these matrices orthogonal? Is this a good definition?

9. Using the hint we note the following: If we add the diagonals and the middle column,
   we get row 1 and row 2 and central $(2,2)$ entry 3 times.
   So, if we subtract row 1 and row 2 we are left with three times the central entry, c.

   That is: $(d_1 + d_2 + c_2) - (r_1 + r_2) = 3c$.

   Also, since we subtracted two rows from two diagonals and a column we get the weight, w.

   That is: $w = (d_1 + d_2 + c_2) - (r_1 + r_2) = 3c$.

   Therefore, the central entry $(2,2)$ must be $c = \frac{w}{3}$ as we were to show.

6.3 Change of Basis

Q: What are our main goals for Section 6.3?
A: In this section, we extend and formalize the notion of *change-of-basis*.
Again, this idea was originally introduced in Section 3.5.
In this Section, we look at how to perform this change by matrix multiplication.

Key Definitions and Concepts

$P_{C \leftarrow B}$ p.469   6.3   *change-of-basis* matrix: $P_{C \leftarrow B} = \begin{bmatrix} [\mathbf{u}_1]_C & \cdots & [\mathbf{u}_n]_C \end{bmatrix}$

change-of-basis   p.469   6.3   *change-of-basis* matrix: $P_{C \leftarrow B} = \begin{bmatrix} [\mathbf{u}_1]_C & \cdots & [\mathbf{u}_n]_C \end{bmatrix}$

Theorems

Thm 6.12   p.469   6.3   Let $\mathcal{B} = \{\mathbf{u}_1, \ldots, \mathbf{u}_n\}$, $\mathcal{C} = \{\mathbf{v}_1, \ldots, \mathbf{v}_n\}$ be bases for V.
let $P_{C \leftarrow B}$ be a *change-of-basis* matrix from $\mathcal{B}$ to $\mathcal{C}$. Then:
a. $P_{C \leftarrow B}[\mathbf{x}]_B = [\mathbf{x}]_C$ for all $\mathbf{x}$ in V.
b. $P_{C \leftarrow B}$ is the unique matrix P with property (a).
c. $P_{C \leftarrow B}$ is invertible and $(P_{C \leftarrow B})^{-1} = P_{B \leftarrow C}$.

Thm 6.13   p.474   6.3   Let $\mathcal{B} = \{\mathbf{u}_1, \ldots, \mathbf{u}_n\}$, $\mathcal{C} = \{\mathbf{v}_1, \ldots, \mathbf{v}_n\}$ be bases for V.
Let $B = \begin{bmatrix} [\mathbf{u}_1]_\mathcal{E} & \cdots & [\mathbf{u}_n]_\mathcal{E} \end{bmatrix}$, $C = \begin{bmatrix} [\mathbf{v}_1]_\mathcal{E} & \cdots & [\mathbf{v}_n]_\mathcal{E} \end{bmatrix}$.
Then: $\begin{bmatrix} C | B \end{bmatrix} \longrightarrow \begin{bmatrix} I | P_{C \leftarrow B} \end{bmatrix}$.

Discussion of Key Definitions and Concepts

Putting the description of the change-of-basis matrix into words

Q: How might we put the construction of the *change-of-basis* matrix into words?
A: Let us look at it, piece-by-piece. First of all, we have the following:
$P_{C \leftarrow B} = \begin{bmatrix} [\mathbf{u}_1]_C & \cdots & [\mathbf{u}_n]_C \end{bmatrix}$. What does this mean?
The columns of $P_{C \leftarrow B}$ are coordinate vectors in the *new* basis C.
However, those coordinate vectors are the basis vectors of the *old* basis, B.
To see that, we need to keep in mind that $B = \{\mathbf{u}_i\}$.

There is lots of nice symbolism here, but it takes time for its meaning to sink in. Make sure to make and take that time up front. It will ease the confusion greatly.

Q: What does $P_{C \leftarrow B}[\mathbf{x}]_B = [\mathbf{x}]_C$ say in words?
A: To find the coordinate vector in the *new* basis, we have to multiply the coordinate vector in the *old* basis by the *change-of-basis* matrix.
This is precisely why $P_{C \leftarrow B}$ is called the *change-of-basis* matrix.

See Exercises 1 through 10 for detailed examples of this process.

Discussion of Theorems

Q: What does Thm 6.12a. claim?
A: It simply asserts that $P_{C \leftarrow B}[\mathbf{x}]_B = [\mathbf{x}]_C$ works.
That is, $P_{C \leftarrow B}$ as defined really does work as the *change-of-basis* matrix.

Q: What does Thm 6.12b. claim?
A: $P_{C \leftarrow B}$ is the only matrix that works as the *change-of-basis* matrix.

Q: What does Thm 6.12c. claim?
A: $P_{B \leftarrow C}$ is the inverse of $P_{C \leftarrow B}$.

Q: Why should we expect $P_{C \leftarrow B}$ to be invertible?
A: Since the set $\{\mathbf{u}_i\}$ are linearly independent, so is the set $\{[\mathbf{u}_i]_B\}$. Why?
Since the columns of $P_{C \leftarrow B}$ are linearly independent, $P_{C \leftarrow B}$ is invertible.

Q: Intuitively, why should we expect the inverse of $P_{C \leftarrow B}$ to be $P_{B \leftarrow C}$?
A: To get back the *old* coordinate vectors we have to reverse all the operations that created the *new* ones.

Q: What is the main point of Thm 6.13?
A: It gives us a way to compute any change-of-basis matrix $P_{C \leftarrow B}$ using row reduction.

Q: How might we put the process of Thm 6.13 into words?
A: We create two matrices. How?
By finding the coordinate vectors of both bases involved in the change.
All those coordinate vectors are relative to the *standard basis*, $\mathcal{E}$.
Then we simply row reduce to find the change-of-basis matrix.

See the solution of Exercise 17 below for a detailed example of this process.

6.3 Change of Basis

Solutions to odd-numbered exercises from Section 6.3

1. (a) $\begin{bmatrix} 2 \\ 3 \end{bmatrix} = a_1 \begin{bmatrix} 1 \\ 0 \end{bmatrix} + a_2 \begin{bmatrix} 0 \\ 1 \end{bmatrix} \Rightarrow \begin{bmatrix} a_1 = 2 \\ a_2 = 3 \end{bmatrix} \Rightarrow [\mathbf{x}]_B = \begin{bmatrix} 2 \\ 3 \end{bmatrix}.$

$\begin{bmatrix} 2 \\ 3 \end{bmatrix} = b_1 \begin{bmatrix} 1 \\ 1 \end{bmatrix} + b_2 \begin{bmatrix} 1 \\ -1 \end{bmatrix} \Rightarrow \begin{bmatrix} b_1 + b_2 = 2 \\ b_1 - b_2 = 3 \end{bmatrix} \Rightarrow \begin{matrix} b_1 = \frac{5}{2} \\ b_2 = -\frac{1}{2} \end{matrix} \Rightarrow [\mathbf{x}]_C = \begin{bmatrix} \frac{5}{2} \\ -\frac{1}{2} \end{bmatrix}.$

(b) Let $\mathcal{E} = \left\{ \begin{bmatrix} 1 \\ 0 \end{bmatrix}, \begin{bmatrix} 0 \\ 1 \end{bmatrix} \right\}$. Then Theorem 6.13, which states $[C|B] \longrightarrow [I|P_{C \leftarrow B}]$, implies

$\begin{bmatrix} 1 & 1 & | & 1 & 0 \\ 1 & -1 & | & 0 & 1 \end{bmatrix} \longrightarrow \begin{bmatrix} 1 & 0 & | & \frac{1}{2} & \frac{1}{2} \\ 0 & 1 & | & \frac{1}{2} & -\frac{1}{2} \end{bmatrix} \Rightarrow P_{C \leftarrow B} = \begin{bmatrix} \frac{1}{2} & \frac{1}{2} \\ \frac{1}{2} & -\frac{1}{2} \end{bmatrix}.$

(c) $[\mathbf{x}]_C = P_{C \leftarrow B} [\mathbf{x}]_B = \begin{bmatrix} \frac{1}{2} & \frac{1}{2} \\ \frac{1}{2} & -\frac{1}{2} \end{bmatrix} \begin{bmatrix} 2 \\ 3 \end{bmatrix} = \begin{bmatrix} \frac{5}{2} \\ -\frac{1}{2} \end{bmatrix}.$

(d) Let $\mathcal{E} = \left\{ \begin{bmatrix} 1 \\ 0 \end{bmatrix}, \begin{bmatrix} 0 \\ 1 \end{bmatrix} \right\}$. Then Theorem 6.13, which states $[B|C] \longrightarrow [I|P_{B \leftarrow C}]$, implies

$\begin{bmatrix} 1 & 0 & | & 1 & 1 \\ 0 & 1 & | & 1 & -1 \end{bmatrix} = \begin{bmatrix} 1 & 0 & | & 1 & 1 \\ 0 & 1 & | & 1 & -1 \end{bmatrix} \Rightarrow P_{B \leftarrow C} = \begin{bmatrix} 1 & 1 \\ 1 & -1 \end{bmatrix}.$

(e) $[\mathbf{x}]_B = P_{B \leftarrow C} [\mathbf{x}]_C = \begin{bmatrix} 1 & 1 \\ 1 & -1 \end{bmatrix} \begin{bmatrix} \frac{5}{2} \\ -\frac{1}{2} \end{bmatrix} = \begin{bmatrix} 2 \\ 3 \end{bmatrix}.$

3. (a) $\begin{bmatrix} 1 \\ 0 \\ -1 \end{bmatrix} = a_1 \begin{bmatrix} 1 \\ 0 \\ 0 \end{bmatrix} + a_2 \begin{bmatrix} 0 \\ 1 \\ 0 \end{bmatrix} + a_3 \begin{bmatrix} 0 \\ 0 \\ 1 \end{bmatrix} \Rightarrow \begin{matrix} a_1 = 1 \\ a_2 = 0 \\ a_3 = -1 \end{matrix} \Rightarrow [\mathbf{x}]_\mathcal{B} = \begin{bmatrix} 1 \\ 0 \\ -1 \end{bmatrix}.$

$\begin{bmatrix} 1 \\ 0 \\ -1 \end{bmatrix} = b_1 \begin{bmatrix} 1 \\ 1 \\ 1 \end{bmatrix} + b_2 \begin{bmatrix} 0 \\ 1 \\ 1 \end{bmatrix} + b_3 \begin{bmatrix} 0 \\ 0 \\ 1 \end{bmatrix} \Rightarrow \begin{matrix} b_1 = 1 \\ b_2 = -1 \\ b_3 = -1 \end{matrix} \Rightarrow [\mathbf{x}]_\mathcal{C} = \begin{bmatrix} 1 \\ -1 \\ -1 \end{bmatrix}.$

(b) $\left[\begin{array}{ccc|ccc} 1 & 0 & 0 & 1 & 0 & 0 \\ 1 & 1 & 0 & 0 & 1 & 0 \\ 1 & 1 & 1 & 0 & 0 & 1 \end{array}\right] \longrightarrow \left[\begin{array}{ccc|ccc} 1 & 0 & 0 & 1 & 0 & 0 \\ 0 & 1 & 0 & -1 & 1 & 0 \\ 0 & 0 & 1 & 0 & -1 & 1 \end{array}\right] \Rightarrow P_{\mathcal{C}\leftarrow\mathcal{B}} = \begin{bmatrix} 1 & 0 & 0 \\ -1 & 1 & 0 \\ 0 & -1 & 1 \end{bmatrix}.$

(c) $[\mathbf{x}]_\mathcal{C} = P_{\mathcal{C}\leftarrow\mathcal{B}} [\mathbf{x}]_\mathcal{B} = \begin{bmatrix} 1 & 0 & 0 \\ -1 & 1 & 0 \\ 0 & -1 & 1 \end{bmatrix} \begin{bmatrix} 1 \\ 0 \\ -1 \end{bmatrix} = \begin{bmatrix} 1 \\ -1 \\ -1 \end{bmatrix}.$

(d) $\left[\begin{array}{ccc|ccc} 1 & 0 & 0 & 1 & 0 & 0 \\ 0 & 1 & 0 & 1 & 1 & 0 \\ 0 & 0 & 1 & 1 & 1 & 1 \end{array}\right] \longrightarrow \left[\begin{array}{ccc|ccc} 1 & 0 & 0 & 1 & 0 & 0 \\ 0 & 1 & 0 & 1 & 1 & 0 \\ 0 & 0 & 1 & 1 & 1 & 1 \end{array}\right] \Rightarrow P_{\mathcal{B}\leftarrow\mathcal{C}} = \begin{bmatrix} 1 & 0 & 0 \\ 1 & 1 & 0 \\ 1 & 1 & 1 \end{bmatrix}.$

(e) $[\mathbf{x}]_\mathcal{B} = P_{\mathcal{B}\leftarrow\mathcal{C}} [\mathbf{x}]_\mathcal{C} = \begin{bmatrix} 1 & 0 & 0 \\ 1 & 1 & 0 \\ 1 & 1 & 1 \end{bmatrix} \begin{bmatrix} 1 \\ -1 \\ -1 \end{bmatrix} = \begin{bmatrix} 1 \\ 0 \\ -1 \end{bmatrix}.$

5. (a) $\begin{bmatrix} 2 \\ -1 \end{bmatrix} = a_1 \begin{bmatrix} 1 \\ 0 \end{bmatrix} + a_2 \begin{bmatrix} 0 \\ 1 \end{bmatrix} \Rightarrow \begin{matrix} a_1 = 2 \\ a_2 = -1 \end{matrix} \Rightarrow [p(x)]_\mathcal{B} = \begin{bmatrix} 2 \\ -1 \end{bmatrix}.$

$\begin{bmatrix} 2 \\ -1 \end{bmatrix} = b_1 \begin{bmatrix} 0 \\ 1 \end{bmatrix} + b_2 \begin{bmatrix} 1 \\ 1 \end{bmatrix} \Rightarrow \begin{matrix} b_2 = 2 \\ b_1 + b_2 = -1 \end{matrix} \Rightarrow \begin{matrix} b_1 = -3 \\ b_2 = 2 \end{matrix} \Rightarrow [p(x)]_\mathcal{C} = \begin{bmatrix} -3 \\ 2 \end{bmatrix}.$

(b) $\left[\begin{array}{cc|cc} 0 & 1 & 1 & 0 \\ 1 & 1 & 0 & 1 \end{array}\right] \longrightarrow \left[\begin{array}{cc|cc} 1 & 0 & -1 & 1 \\ 0 & 1 & 1 & 0 \end{array}\right] \Rightarrow P_{\mathcal{C}\leftarrow\mathcal{B}} = \begin{bmatrix} -1 & 1 \\ 1 & 0 \end{bmatrix}.$

(c) $[p(x)]_\mathcal{C} = P_{\mathcal{C}\leftarrow\mathcal{B}} [p(x)]_\mathcal{B} = \begin{bmatrix} -1 & 1 \\ 1 & 0 \end{bmatrix} \begin{bmatrix} 2 \\ -1 \end{bmatrix} = \begin{bmatrix} -3 \\ 2 \end{bmatrix}.$

(d) $\left[\begin{array}{cc|cc} 1 & 0 & 0 & 1 \\ 0 & 1 & 1 & 1 \end{array}\right] = \left[\begin{array}{cc|cc} 1 & 0 & 0 & 1 \\ 0 & 1 & 1 & 1 \end{array}\right] \Rightarrow P_{\mathcal{B}\leftarrow\mathcal{C}} = \begin{bmatrix} 0 & 1 \\ 1 & 1 \end{bmatrix}.$

(e) $[p(x)]_\mathcal{B} = P_{\mathcal{B}\leftarrow\mathcal{C}} [p(x)]_\mathcal{C} = \begin{bmatrix} 0 & 1 \\ 1 & 1 \end{bmatrix} \begin{bmatrix} -3 \\ 2 \end{bmatrix} = \begin{bmatrix} 2 \\ -1 \end{bmatrix}.$

6.3 Change of Basis

7. (a) $\begin{bmatrix} 1 \\ 0 \\ 1 \end{bmatrix} = a_1 \begin{bmatrix} 1 \\ 1 \\ 1 \end{bmatrix} + a_2 \begin{bmatrix} 0 \\ 1 \\ 1 \end{bmatrix} + a_3 \begin{bmatrix} 0 \\ 0 \\ 1 \end{bmatrix} \Rightarrow \begin{matrix} a_1 = 1 \\ a_2 = -1 \\ a_3 = 1 \end{matrix} \Rightarrow [p(x)]_\mathcal{B} = \begin{bmatrix} 1 \\ -1 \\ 1 \end{bmatrix}.$

$\begin{bmatrix} 1 \\ 0 \\ 1 \end{bmatrix} = b_1 \begin{bmatrix} 1 \\ 0 \\ 0 \end{bmatrix} + b_2 \begin{bmatrix} 0 \\ 1 \\ 0 \end{bmatrix} + b_3 \begin{bmatrix} 0 \\ 0 \\ 1 \end{bmatrix} \Rightarrow \begin{matrix} b_1 = 1 \\ b_2 = 0 \\ b_3 = 1 \end{matrix} \Rightarrow [p(x)]_\mathcal{C} = \begin{bmatrix} 1 \\ 0 \\ 1 \end{bmatrix}.$

(b) $\left[\begin{array}{ccc|ccc} 1 & 0 & 0 & 1 & 0 & 0 \\ 0 & 1 & 0 & 1 & 1 & 0 \\ 0 & 0 & 1 & 1 & 1 & 1 \end{array}\right] \longrightarrow \left[\begin{array}{ccc|ccc} 1 & 0 & 0 & 1 & 0 & 0 \\ 0 & 1 & 0 & 1 & 1 & 0 \\ 0 & 0 & 1 & 1 & 1 & 1 \end{array}\right] \Rightarrow P_{\mathcal{C} \leftarrow \mathcal{B}} = \begin{bmatrix} 1 & 0 & 0 \\ 1 & 1 & 0 \\ 1 & 1 & 1 \end{bmatrix}.$

(c) $[p(x)]_\mathcal{C} = P_{\mathcal{C} \leftarrow \mathcal{B}} [p(x)]_\mathcal{B} = \begin{bmatrix} 1 & 0 & 0 \\ 1 & 1 & 0 \\ 1 & 1 & 1 \end{bmatrix} \begin{bmatrix} 1 \\ -1 \\ 1 \end{bmatrix} = \begin{bmatrix} 1 \\ 0 \\ 1 \end{bmatrix}.$

(d) $\left[\begin{array}{ccc|ccc} 1 & 0 & 0 & 1 & 0 & 0 \\ 1 & 1 & 0 & 0 & 1 & 0 \\ 1 & 1 & 1 & 0 & 0 & 1 \end{array}\right] \longrightarrow \left[\begin{array}{ccc|ccc} 1 & 0 & 0 & 1 & 0 & 0 \\ 0 & 1 & 0 & -1 & 1 & 0 \\ 0 & 0 & 1 & 0 & -1 & 1 \end{array}\right] \Rightarrow P_{\mathcal{B} \leftarrow \mathcal{C}} = \begin{bmatrix} 1 & 0 & 0 \\ -1 & 1 & 0 \\ 0 & -1 & 1 \end{bmatrix}.$

(e) $[p(x)]_\mathcal{B} = P_{\mathcal{B} \leftarrow \mathcal{C}} [p(x)]_\mathcal{C} = \begin{bmatrix} 1 & 0 & 0 \\ -1 & 1 & 0 \\ 1 & -1 & 1 \end{bmatrix} \begin{bmatrix} 1 \\ 0 \\ 1 \end{bmatrix} = \begin{bmatrix} 1 \\ -1 \\ 1 \end{bmatrix}.$

9. (a) $\begin{bmatrix} 4 \\ 2 \\ 0 \\ -1 \end{bmatrix} = a_1 \begin{bmatrix} 1 \\ 0 \\ 0 \\ 0 \end{bmatrix} + a_2 \begin{bmatrix} 0 \\ 1 \\ 0 \\ 0 \end{bmatrix} + a_3 \begin{bmatrix} 0 \\ 0 \\ 1 \\ 0 \end{bmatrix} + a_4 \begin{bmatrix} 0 \\ 0 \\ 0 \\ 1 \end{bmatrix} \Rightarrow \begin{matrix} a_1 = 4 \\ a_2 = 2 \\ a_3 = 0 \\ a_4 = -1 \end{matrix} \Rightarrow [A]_\mathcal{B} = \begin{bmatrix} 4 \\ 2 \\ 0 \\ -1 \end{bmatrix}.$

$\begin{bmatrix} 4 \\ 2 \\ 0 \\ -1 \end{bmatrix} = b_1 \begin{bmatrix} 1 \\ 2 \\ 0 \\ -1 \end{bmatrix} + b_2 \begin{bmatrix} 2 \\ 1 \\ 1 \\ 0 \end{bmatrix} + b_3 \begin{bmatrix} 1 \\ 1 \\ 0 \\ 1 \end{bmatrix} + b_4 \begin{bmatrix} 1 \\ 0 \\ 0 \\ 1 \end{bmatrix} \Rightarrow$

$\begin{matrix} b_1 = 5/2 \\ b_2 = 0 \\ b_3 = -3 \\ b_4 = 9/2 \end{matrix} \Rightarrow [A]_\mathcal{C} = \begin{bmatrix} 5/2 \\ 0 \\ -3 \\ 9/2 \end{bmatrix}.$

(b) $\begin{bmatrix} 1 & 2 & 1 & 1 & | & 1 & 0 & 0 & 0 \\ 2 & 1 & 1 & 0 & | & 0 & 1 & 0 & 0 \\ 0 & 1 & 0 & 0 & | & 0 & 0 & 1 & 0 \\ -1 & 0 & 1 & 1 & | & 0 & 0 & 0 & 1 \end{bmatrix} \longrightarrow \begin{bmatrix} 1 & 0 & 0 & 0 & | & 1/2 & 0 & -1 & -1/2 \\ 0 & 1 & 0 & 0 & | & 0 & 0 & 1 & 0 \\ 0 & 0 & 1 & 0 & | & -1 & 1 & 1 & 1 \\ 0 & 0 & 0 & 1 & | & 3/2 & -1 & -2 & -1/2 \end{bmatrix} \Rightarrow$

$P_{\mathcal{C}\leftarrow\mathcal{B}} = \begin{bmatrix} 1/2 & 0 & -1 & -1/2 \\ 0 & 0 & 1 & 0 \\ -1 & 1 & 1 & 1 \\ 3/2 & -1 & -2 & -1/2 \end{bmatrix}.$

(c) $[A]_\mathcal{C} = P_{\mathcal{C}\leftarrow\mathcal{B}} [A]_\mathcal{B} = \begin{bmatrix} 1/2 & 0 & -1 & -1/2 \\ 0 & 0 & 1 & 0 \\ -1 & 1 & 1 & 1 \\ 3/2 & -1 & -2 & -1/2 \end{bmatrix} \begin{bmatrix} 4 \\ 2 \\ 0 \\ -1 \end{bmatrix} = \begin{bmatrix} 5/2 \\ 0 \\ -3 \\ 9/2 \end{bmatrix}.$

(d) $\begin{bmatrix} 1 & 0 & 0 & 0 & | & 1 & 2 & 1 & 1 \\ 0 & 1 & 0 & 0 & | & 2 & 1 & 1 & 0 \\ 0 & 0 & 1 & 0 & | & 0 & 1 & 0 & 0 \\ 0 & 0 & 0 & 1 & | & -1 & 0 & 1 & 1 \end{bmatrix} \longrightarrow \begin{bmatrix} 1 & 0 & 0 & 0 & | & 1 & 2 & 1 & 1 \\ 0 & 1 & 0 & 0 & | & 2 & 1 & 1 & 0 \\ 0 & 0 & 1 & 0 & | & 0 & 1 & 0 & 0 \\ 0 & 0 & 0 & 1 & | & -1 & 0 & 1 & 1 \end{bmatrix} \Rightarrow P_{\mathcal{B}\leftarrow\mathcal{C}} = \begin{bmatrix} 1 & 2 & 1 & 1 \\ 2 & 1 & 1 & 0 \\ 0 & 1 & 0 & 0 \\ -1 & 0 & 1 & 1 \end{bmatrix}.$

(e) $[A]_\mathcal{B} = P_{\mathcal{B}\leftarrow\mathcal{C}} [A]_\mathcal{C} = \begin{bmatrix} 1 & 2 & 1 & 1 \\ 2 & 1 & 1 & 0 \\ 0 & 1 & 0 & 0 \\ -1 & 0 & 1 & 1 \end{bmatrix} \begin{bmatrix} 5/2 \\ 0 \\ -3 \\ 9/2 \end{bmatrix} = \begin{bmatrix} 4 \\ 2 \\ 0 \\ -1 \end{bmatrix}.$

6.3 Change of Basis

11. (a) To find $[f(x)]_B$, we need to solve the following function equation for a and b.
$f(x) = 2\sin x - 3\cos x = a(\sin x + \cos x) + b(\cos x) = a\sin x + (a+b)\cos x \Rightarrow$
$a = -2$ and $a + b = -3 \Rightarrow b = -5 \Rightarrow [f(x)]_B = \begin{bmatrix} 2 \\ -5 \end{bmatrix}$.

Using coordinate vectors, we would find $[f(x)]_B$ as follows:
$\begin{bmatrix} 2 \\ -3 \end{bmatrix} = a_1 \begin{bmatrix} 1 \\ 1 \end{bmatrix} + a_2 \begin{bmatrix} 0 \\ 1 \end{bmatrix} \Rightarrow \begin{matrix} a_1 = 2 \\ a_2 = -5 \end{matrix} \Rightarrow [f(x)]_B == \begin{bmatrix} 2 \\ -5 \end{bmatrix}$.

Likewise, to find $[f(x)]_C$, we need to solve the following for a and b.
$f(x) = 2\sin x - 3\cos x = a(\sin x) + b(\cos x) \Rightarrow a = 2$ and $b = -3 \Rightarrow [f(x)]_C = \begin{bmatrix} 2 \\ -3 \end{bmatrix}$.

Using coordinate vectors, we would find $[f(x)]_C$ as follows:
$\begin{bmatrix} 2 \\ -3 \end{bmatrix} = b_1 \begin{bmatrix} 1 \\ 0 \end{bmatrix} + b_2 \begin{bmatrix} 0 \\ 1 \end{bmatrix} \Rightarrow \begin{matrix} b_1 = 2 \\ b_2 = -3 \end{matrix} \Rightarrow [f(x)]_C = \begin{bmatrix} 2 \\ -3 \end{bmatrix}$.

(b) $\left[\begin{array}{cc|cc} 1 & 0 & 1 & 0 \\ 1 & 0 & 0 & 1 \end{array}\right] \longrightarrow \left[\begin{array}{cc|cc} 1 & 0 & 1 & 0 \\ 0 & 1 & 1 & 1 \end{array}\right] \Rightarrow P_{C \leftarrow B} = \begin{bmatrix} 1 & 0 \\ 1 & 1 \end{bmatrix} \Rightarrow$.

(c) $[f(x)]_C = P_{C \leftarrow B}[f(x)]_B = \begin{bmatrix} 1 & 0 \\ 1 & 1 \end{bmatrix}\begin{bmatrix} 2 \\ -5 \end{bmatrix} = \begin{bmatrix} 2 \\ -3 \end{bmatrix}$.

(d) $\left[\begin{array}{cc|cc} 1 & 0 & 1 & 0 \\ 0 & 1 & 1 & 1 \end{array}\right] \longrightarrow \left[\begin{array}{cc|cc} 1 & 0 & 1 & 0 \\ 0 & 1 & -1 & 1 \end{array}\right] \Rightarrow P_{B \leftarrow C} = \begin{bmatrix} 1 & 0 \\ -1 & 1 \end{bmatrix} \Rightarrow$.

(e) $[f(x)]_B = P_{B \leftarrow C}[f(x)]_C = \begin{bmatrix} 1 & 0 \\ -1 & 1 \end{bmatrix} = \begin{bmatrix} 2 \\ -3 \end{bmatrix} = \begin{bmatrix} 2 \\ -5 \end{bmatrix}$.

13. (a) Section 3.6, Example 3.58 $\Rightarrow P_{C \leftarrow B} = \begin{bmatrix} \cos 60° & \sin 60° \\ -\sin 60° & \cos 60° \end{bmatrix} = \begin{bmatrix} 1/2 & \sqrt{3}/2 \\ -\sqrt{3}/2 & 1/2 \end{bmatrix}$.

So, $[\mathbf{x}]_C = P_{C \leftarrow B}[\mathbf{x}]_B = \begin{bmatrix} 1/2 & \sqrt{3}/2 \\ -\sqrt{3}/2 & 1/2 \end{bmatrix}\begin{bmatrix} 3 \\ 2 \end{bmatrix} = \begin{bmatrix} (3 + 2\sqrt{3})/2 \\ (2 - 3\sqrt{3})/2 \end{bmatrix}$.

(b) We have: $[\mathbf{y}]_C = P_{C \leftarrow B}[\mathbf{y}]_B = \begin{bmatrix} 1/2 & -\sqrt{3}/2 \\ \sqrt{3}/2 & 1/2 \end{bmatrix}\begin{bmatrix} 4 \\ -4 \end{bmatrix} = \begin{bmatrix} 2 + 2\sqrt{3} \\ 2\sqrt{3} - 2 \end{bmatrix}$.

15. $P_{C \leftarrow B} = \begin{bmatrix} 1 & -1 \\ -1 & 2 \end{bmatrix} \Rightarrow [\mathbf{u}_1]_C = \begin{bmatrix} 1 \\ -1 \end{bmatrix}, [\mathbf{u}_2]_C = \begin{bmatrix} -1 \\ 2 \end{bmatrix} \Rightarrow \mathbf{u}_1 = \begin{bmatrix} 1 \\ 2 \end{bmatrix} - \begin{bmatrix} 2 \\ 3 \end{bmatrix} = \begin{bmatrix} -1 \\ -1 \end{bmatrix}$.

Likewise, $\mathbf{u}_2 = -\begin{bmatrix} 1 \\ 2 \end{bmatrix} + 2\begin{bmatrix} 2 \\ 3 \end{bmatrix} = \begin{bmatrix} 3 \\ 4 \end{bmatrix} \Rightarrow \mathcal{B} = \{\mathbf{u}_1, \mathbf{u}_2\} = \left\{\begin{bmatrix} -1 \\ -1 \end{bmatrix}, \begin{bmatrix} 3 \\ 4 \end{bmatrix}\right\}$.

17. $a = 1 \Rightarrow \mathcal{B} = \{1, x-1, (x-1)^2 = 1 - 2x + x^2\}$ is the Taylor Polynomial basis.

Let $\mathcal{C} = \{1, x, x^2\} \Rightarrow [p(x)]_\mathcal{C} = \begin{bmatrix} 1 \\ 2 \\ -5 \end{bmatrix}$.

$[\mathcal{B}\,|\,\mathcal{C}] \longrightarrow [I\,|\,P_{\mathcal{B}\leftarrow\mathcal{C}}] \Rightarrow \begin{bmatrix} 1 & -1 & 1 & 1 & 0 & 0 \\ 0 & 1 & -2 & 0 & 1 & 0 \\ 0 & 0 & 1 & 0 & 0 & 1 \end{bmatrix} \longrightarrow \begin{bmatrix} 1 & 0 & 0 & 1 & 1 & 1 \\ 0 & 1 & 0 & 0 & 1 & 2 \\ 0 & 0 & 1 & 0 & 0 & 1 \end{bmatrix} \Rightarrow$

$P_{\mathcal{B}\leftarrow\mathcal{C}} = \begin{bmatrix} 1 & 1 & 1 \\ 0 & 1 & 2 \\ 0 & 0 & 1 \end{bmatrix} \Rightarrow [p(x)]_\mathcal{B} = P_{\mathcal{B}\leftarrow\mathcal{C}}\,[p(x)]_\mathcal{C} = \begin{bmatrix} 1 & 1 & 1 \\ 0 & 1 & 2 \\ 0 & 0 & 1 \end{bmatrix} \begin{bmatrix} 1 \\ 2 \\ -5 \end{bmatrix} = \begin{bmatrix} -2 \\ -8 \\ -5 \end{bmatrix} \Rightarrow$

$p(x) = -2(1) - 8(x-1) - 5(x-1)^2$.

19. $a = -1 \Rightarrow \mathcal{B} = \{1, x+1, (x+1)^2 = 1 + 2x + x^2, (x+1)^3 = 1 + 3x + 3x^2 + x^3\}$
is the Taylor Polynomial basis. Let $\mathcal{C} = \{1, x, x^2, x^3\} \Rightarrow$

$[1]_\mathcal{C} = \begin{bmatrix} 1 \\ 0 \\ 0 \\ 0 \end{bmatrix}, [x+1]_\mathcal{C} = \begin{bmatrix} 1 \\ 1 \\ 0 \\ 0 \end{bmatrix}, [(x+1)^2]_\mathcal{C} = \begin{bmatrix} 1 \\ 2 \\ 1 \\ 0 \end{bmatrix}, [(x+1)^3]_\mathcal{C} = \begin{bmatrix} 1 \\ 3 \\ 3 \\ 1 \end{bmatrix} \Rightarrow$

$[\mathcal{B}\,|\,\mathcal{C}] \longrightarrow [I\,|\,P_{\mathcal{B}\leftarrow\mathcal{C}}] \Rightarrow$

$\begin{bmatrix} 1 & 1 & 1 & 1 & 1 & 0 & 0 & 0 \\ 0 & 1 & 2 & 3 & 0 & 1 & 0 & 0 \\ 0 & 0 & 1 & 3 & 0 & 0 & 1 & 0 \\ 0 & 0 & 0 & 1 & 0 & 0 & 0 & 1 \end{bmatrix} \longrightarrow \begin{bmatrix} 1 & 0 & 0 & 0 & 1 & -1 & 1 & -1 \\ 0 & 1 & 0 & 0 & 0 & 1 & -2 & 3 \\ 0 & 0 & 1 & 0 & 0 & 0 & 1 & -3 \\ 0 & 0 & 0 & 1 & 0 & 0 & 0 & 1 \end{bmatrix} \Rightarrow$

$P_{\mathcal{B}\leftarrow\mathcal{C}} = \begin{bmatrix} 1 & -1 & 1 & -1 \\ 0 & 1 & -2 & 3 \\ 0 & 0 & 1 & -3 \\ 0 & 0 & 0 & 1 \end{bmatrix} \Rightarrow$

$[p(x)]_\mathcal{B} = P_{\mathcal{B}\leftarrow\mathcal{C}}\,[p(x)]_\mathcal{C} = \begin{bmatrix} 1 & -1 & 1 & -1 \\ 0 & 1 & -2 & 3 \\ 0 & 0 & 1 & -3 \\ 0 & 0 & 0 & 1 \end{bmatrix} \begin{bmatrix} 0 \\ 0 \\ 0 \\ 1 \end{bmatrix} = \begin{bmatrix} -1 \\ 3 \\ -3 \\ 1 \end{bmatrix} \Rightarrow$

$p(x) = -(1) + 3(x+1) - 3(x+1)^2 + 1(x+1)^3$.

21. Need only show $[\mathbf{x}]_\mathcal{D} = P_{\mathcal{D}\leftarrow\mathcal{C}}\,P_{\mathcal{C}\leftarrow\mathcal{B}}\,[\mathbf{x}]_\mathcal{B}$ since this matrix with this property is unique.
By definition, $[\mathbf{x}]_\mathcal{D} = P_{\mathcal{D}\leftarrow\mathcal{C}}\,[\mathbf{x}]_\mathcal{C}$ (1) and $[\mathbf{x}]_\mathcal{C} = P_{\mathcal{C}\leftarrow\mathcal{B}}\,[\mathbf{x}]_\mathcal{B}$ (2).
Substituting (2) into (1) $\Rightarrow [\mathbf{x}]_\mathcal{D} = P_{\mathcal{D}\leftarrow\mathcal{C}}\,P_{\mathcal{C}\leftarrow\mathcal{B}}\,[\mathbf{x}]_\mathcal{B}$ which is what we needed to show.

6.4 Linear Transformations

Q: What are our main goals for Section 6.4?
A: In this section, we revisit linear transformations, especially in the context of *bases*.
Recall, that linear transformations were originally introduced in Section 3.6.
We explore everything from continuous functions to integration and apply *composition*.
So, one of our primary goals is to vastly expand what we think of as linear transformations.

Key Definitions and Concepts

$\mathscr{C}[a,b]$	p.477	6.4	The space of continuous functions (See Example 6.52).
$\int_a^b f(x)\,dx$	p.477	6.4	$S: \mathscr{C}[a,b] \to \mathbb{R}$, $S(f) = \int_a^b f(x)\,dx$ (See Example 6.52).
I	p.478	6.4	$I: V \to W$ maps every vector to itself. That is, $I(\mathbf{v}) = \mathbf{v}$ (See Example 6.54).
$S \circ T$	p.481	6.4	If $T: U \to V$ and $S: V \to W$ then the composition of S with T is $(S \circ T)(\mathbf{u}) = S(T(\mathbf{u}))$.
T_0	p.478	6.4	$T_0: V \to W$ maps every vector to the zero vector. That is, $T_0(\mathbf{v}) = \mathbf{0}$ (See Example 6.54).
T^{-1}	p.483	6.4	T^{-1}: $T^{-1} \circ T = I_V$, $T \circ T^{-1} = I_W$ (See *invertible*).
composition of S with T	p.481	6.4	If $T: U \to V$ and $S: V \to W$ then the *composition of S with T* is the mapping $S \circ T$: $(S \circ T)(\mathbf{u}) = S(T(\mathbf{u}))$.
differential operator	p.477	6.4	The *differential operator* is defined as follows: $D: \mathscr{D} \to \mathscr{F}$, $D(f) = f'$ (See Example 6.51).
identity transformation	p.478	6.4	$I: V \to W$ maps every vector to itself. That is, $I(\mathbf{v}) = \mathbf{v}$ (See Example 6.54).
inverse	p.483	6.4	T^{-1}: $T^{-1} \circ T = I_V$, $T \circ T^{-1} = I_W$ (See *invertible*).
invertible	p.482	6.4	A linear transformation $T: V \to W$ is *invertible* if there is a linear transformation $T': W \to V$ such that: $T' \circ T = I_V$ and $T \circ T' = I_W$ In this case, T' is called the *inverse* for T.
linear transformation	p.476	6.4	A *linear transformation* from a vector space V to a vector space W is a mapping $T: V \to W$ such that: 1. $T(\mathbf{u}+\mathbf{v}) = T(\mathbf{u}+\mathbf{v})$ 2. $T(c\mathbf{u}) = cT(\mathbf{u})$ or $T(c_1\mathbf{v}_1 + \cdots + c_k\mathbf{v}_k) = c_1T(\mathbf{v}_1) + \cdots + c_kT(\mathbf{v}_k)$
zero transformation	p.478	6.4	$T_0: V \to W$ maps every vector to the zero vector. That is, $T_0(\mathbf{v}) = \mathbf{0}$ (See Example 6.54).

Theorems

Thm 6.14 p.479 6.4 Let $T : V \to W$ be a linear transformation. Then:
a. $T(\mathbf{0}) = \mathbf{0}$.
b. $T(-\mathbf{v}) = -T(\mathbf{v})$ for all $\mathbf{v}$ in V.
c. $T(\mathbf{u} - \mathbf{v}) = T(\mathbf{u}) - T(\mathbf{v})$ for all $\mathbf{u}$, $\mathbf{v}$ in V.

Thm 6.15 p.480 6.4 Let $T : V \to W$ be a linear transformation.
Let $\mathcal{B} = \{\mathbf{v}_1, \ldots, \mathbf{v}_n\}$ be a spanning set for V. Then:
$T(\mathcal{B}) = \{T(\mathbf{v}_1), \ldots, T(\mathbf{v}_n)\}$ is a spanning set for the range of T.

Thm 6.16 p.481 6.4 If $T : U \to V$ and $S : V \to W$ then
$S \circ T : U \to W$, $(S \circ T)(\mathbf{u}) = S(T(\mathbf{u}))$, is a linear transformation.

Thm 6.17 p.483 6.4 If T is an *invertible* linear transformation, then its *inverse* is unique.
If $R \circ T = I_V$, $T \circ R = I_W$ and $S \circ T = I_V$, $T \circ S = I_W$, then $R = S$.

Two essential linear transformations

Q: What is the zero transformation T_0? What matrix does it correspond to?
A: As the definition states, T_0 maps every vector to the zero vector.
This is exactly what happens when we multiply any vector by the zero matrix, O.

Q: What is the range of T_0?
A: The zero vector but thought of as $\{\mathbf{0}\}$, the subspace of V.

Q: What is the identity matrix I? What matrix does it correspond to?
A: As the definition states, I maps every vector to itself.
This is exactly what happens when we multiply any vector by the identity matrix, I.

Q: What is the range of I?
A: The entire vector space.

Discussion of theorems

Q: What is the underlying point of Thm 6.14?
A: To lay the groundwork for showing that linear transformations preserve spanning sets.
The assertions here are obvious and easy to prove. Verify this.

Q: What is one possible extension or application of Thm 6.15 to the concept of bases?
A: We would like to see if $T : V \to W$ maps a basis for V to a basis for W.

Q: Does a linear transformation always map a basis for V to a basis for W?
A: Hint: Consider T_0 and I. Which one does map a basis to a basis? Which one does not?

Q: Can we generalize the property needed to insure that a basis is mapped to a basis?
A: Hint: Why does T_0 fail to map a basis to a basis? How can we fix that deficiency?

Q: Thm 6.17 asserts that the inverse of a linear transformation is unique.
Why does this come as no surprise to us?
A: Every linear transformation corresponds to a matrix and the inverse of a matrix is unique.

6.4 Linear Transformations

Solutions to odd-numbered exercises from Section 6.4

1. $T\left(\begin{bmatrix} a & b \\ c & d \end{bmatrix} + \begin{bmatrix} a' & b' \\ c' & d' \end{bmatrix}\right) = T\begin{bmatrix} a+a' & b+b' \\ c+c' & d+d' \end{bmatrix} = \begin{bmatrix} a+a'+b+b' & 0 \\ 0 & c+c'+d+d' \end{bmatrix}$

    $= \begin{bmatrix} a+b & 0 \\ 0 & c+d \end{bmatrix} + \begin{bmatrix} a'+b' & 0 \\ 0 & c'+d' \end{bmatrix} = T\begin{bmatrix} a & b \\ c & d \end{bmatrix} + T\begin{bmatrix} a' & b' \\ c' & d' \end{bmatrix}.$

    $T\left(\alpha \begin{bmatrix} a & b \\ c & d \end{bmatrix}\right) = T\begin{bmatrix} \alpha a & \alpha b \\ \alpha c & \alpha d \end{bmatrix} = \begin{bmatrix} \alpha a + \alpha b & 0 \\ 0 & \alpha c + \alpha d \end{bmatrix} = \alpha \begin{bmatrix} a+b & 0 \\ 0 & c+d \end{bmatrix} = \alpha T \begin{bmatrix} a & b \\ c & d \end{bmatrix} \Rightarrow$

    T is a linear transformation.

3. $T(A+C) = (A+C)B = AB + CB = T(A) + T(C)$ and
   $T(\alpha A) = (\alpha A)B = \alpha(AB) = \alpha T(A) \Rightarrow T$ is a linear transformation.

5. By Exercise 44, Section 3.2 $\text{tr}(A+C) = \text{tr}(A) + \text{tr}(C)$ and $\text{tr}(\alpha A) = \alpha \text{tr}(A) \Rightarrow$
   $T = \text{tr}$ is a linear transformation.

7. Let A be a matrix with $\text{rank}(A) \neq 0$, then $T(A) = T(-A) = \text{rank}(A) \Rightarrow$
   $T(-\mathbf{v}) \neq -T(\mathbf{v}) \Rightarrow T$ is *not* a linear transformation because it fails Theorem 6.14b.
   Note this is another application of the principle $p \to q \Leftrightarrow -q \to -p$.
   We specify $\text{rank}(A) \neq 0$ since zero is the only number such that $0 = -0$.

9. $T((a + bx + cx^2) + (a' + b'x + c'x^2)) = T((a+a') + (b+b')x + (c+c')x^2)$
   $= (a+a') + (b+b')(x+1) + (c+c')(x+1)^2 = (a+b(x+1)+c(x+1)^2) + (a'+b'(x+1)+c'(x+1)^2)$
   $= T(a + bx + cx^2) + T(a' + b'x + c'x^2)$ and
   $T(\alpha(a+bx+cx^2)) = T(\alpha a + \alpha bx + \alpha cx^2) = \alpha a + \alpha b(x+1) + \alpha c(x+1)^2$
   $= \alpha(a + b(x+1) + c(x+1)^2) = \alpha T(a+bx+cx^2) \Rightarrow T$ is a linear transformation.

11. $T(f) = (f(x))^2 = (-f(x))^2 = T(-f) \Rightarrow T(-\mathbf{v}) \neq -T(\mathbf{v}) \Rightarrow T$ is *not* linear.

13. $S(p(x) + q(x)) = x(p(x) + q(x)) = xp(x) + xq(x) = S(p(x)) + S(q(x))$ and
    $S(\lambda p(x)) = x(\lambda p(x)) = \lambda(xp(x)) = \lambda S(p(x)) \Rightarrow S$ is a linear transformation.

    $T\left(\begin{bmatrix} a \\ b \end{bmatrix} + \begin{bmatrix} a' \\ b' \end{bmatrix}\right) = T\begin{bmatrix} a+a' \\ b+b' \end{bmatrix}$

    $= (a+a') + ((a+a') + (b+b'))x = (a + (a+b)x) + (a' + (a'+b')x) = T\begin{bmatrix} a \\ b \end{bmatrix} + T\begin{bmatrix} a' \\ b' \end{bmatrix}.$

    $T\left(\alpha \begin{bmatrix} a \\ b \end{bmatrix}\right) = T\begin{bmatrix} \alpha a \\ \alpha b \end{bmatrix} = \alpha a + (\alpha a + \alpha b)x = \alpha(a + (a+b)x) = \alpha a T\begin{bmatrix} a \\ b \end{bmatrix} \Rightarrow T$ is linear.

15. First, we need to solve $\begin{bmatrix} -7 \\ 9 \end{bmatrix} = c_1 \begin{bmatrix} 1 \\ 1 \end{bmatrix} + c_2 \begin{bmatrix} 3 \\ -1 \end{bmatrix} \Rightarrow \begin{matrix} c_1 + 3c_2 = -7 \\ c_1 - c_2 = 9 \end{matrix} \Rightarrow \begin{matrix} c_1 = 5 \\ c_2 = -4 \end{matrix}$.

So, since $\begin{bmatrix} -7 \\ 9 \end{bmatrix} = 5\begin{bmatrix} 1 \\ 1 \end{bmatrix} - 4\begin{bmatrix} 3 \\ -1 \end{bmatrix}$, we have:

$T\begin{bmatrix} -7 \\ 9 \end{bmatrix} = 5T\begin{bmatrix} 1 \\ 1 \end{bmatrix} - 4T\begin{bmatrix} 3 \\ -1 \end{bmatrix} = 5(1-2x) - 4(x+2x^2) = 5 - 14x - 8x^2$.

Now, we need to solve $\begin{bmatrix} a \\ b \end{bmatrix} = c_1 \begin{bmatrix} 1 \\ 1 \end{bmatrix} + c_2 \begin{bmatrix} 3 \\ -1 \end{bmatrix} \Rightarrow \begin{matrix} c_1 + 3c_2 = a \\ c_1 - c_2 = b \end{matrix} \Rightarrow \begin{matrix} c_1 = \frac{a+3b}{4} \\ c_2 = \frac{a-b}{4} \end{matrix}$.

So, since $\begin{bmatrix} a \\ b \end{bmatrix} = \frac{a+3b}{4}\begin{bmatrix} 1 \\ 1 \end{bmatrix} + \frac{a-b}{4}\begin{bmatrix} 3 \\ -1 \end{bmatrix}$, we have:

$T\begin{bmatrix} a \\ b \end{bmatrix} = \frac{a+3b}{4} T\begin{bmatrix} 1 \\ 1 \end{bmatrix} + \frac{a-b}{4} T\begin{bmatrix} 3 \\ -1 \end{bmatrix} = \frac{a+3b}{4}(1-2x) + \frac{a-b}{4}(x+2x^2) = \frac{a+3b}{4} - \frac{a+7b}{4}x + \frac{a-b}{4}x^2$.

17. We need to solve $4 - x + 3x^2 = c_1(1+x) + c_2(x+x^2) + c_3(1+x^2)$

$= (c_1+c_3) + (c_1+c_2)x + (c_2+c_3)x^2 \Rightarrow \begin{matrix} c_1 + c_3 = 4 \\ c_1 + c_2 = -1 \\ c_2 + c_3 = 3 \end{matrix} \Rightarrow \begin{matrix} c_1 = 0 \\ c_2 = -1 \\ c_3 = 4 \end{matrix}$.

So, since $4 - x + 3x^2 = 0(1+x) - (x+x^2) + 4(1+x^2)$, we have:

$T(4-x+3x^2) = 0\,T(1+x) - T(x+x^2) + 4\,T(1+x^2) = 0(1+x^2) - (x-x^2) + 4(1+x+x^2)$
$= 4 + 3x + 5x^2$.

We need to solve $a + bx + cx^2 = c_1(1+x) + c_2(x+x^2) + c_3(1+x^2)$

$= (c_1+c_3) + (c_1+c_2)x + (c_2+c_3)x^2 \Rightarrow \begin{matrix} c_1 + c_3 = a \\ c_1 + c_2 = b \\ c_2 + c_3 = c \end{matrix} \Rightarrow \begin{matrix} c_1 = \frac{a+b-c}{2} \\ c_2 = \frac{-a+b+c}{2} \\ c_3 = \frac{a-b+c}{2} \end{matrix}$.

So, since $a + bx + cx^2 = \frac{a+b-c}{2}(1+x) + \frac{-a+b+c}{2}(x+x^2) + \frac{a-b+c}{2}(1+x^2)$, we have:

$T(a+bx+cx^2) = \frac{a+b-c}{2} T(1+x) + \frac{-a+b+c}{2} T(x+x^2) + \frac{a-b+c}{2} T(1+x^2)$
$= \frac{a+b-c}{2}(1+x^2) + \frac{-a+b+c}{2}(x-x^2) + \frac{a-b+c}{2}(1+x+x^2) = a + cx + \left(\frac{3a-b-c}{2}\right)x^2$.

19. Let $T(E_{11}) = a$, $T(E_{12}) = b$, $T(E_{21}) = c$, and $T(E_{22}) = d$.

Then note $\begin{bmatrix} w & x \\ y & z \end{bmatrix} = wE_{11} + xE_{12} + yE_{21} + zE_{22} \Rightarrow$

$T\begin{bmatrix} w & x \\ y & z \end{bmatrix} = T(wE_{11} + xE_{12} + yE_{21} + zE_{22}) = T(wE_{11}) + T(xE_{12}) + T(yE_{21}) + T(zE_{22})$

$= aw + bx + cy + dz$ as required.

21. Recall, by definition, in any vector space V, $\mathbf{v} + (-\mathbf{v}) = \mathbf{0}$. Then Theorem 6.14a $\Rightarrow$
$T(\mathbf{0}) = T(\mathbf{v} + (-\mathbf{v})) = T(\mathbf{v}) + T(-\mathbf{v}) = 0 \Rightarrow T(-\mathbf{v}) = -T(\mathbf{v})$ *by definition.*

23. Let $p(x) = \sum a_k x^k$, then $T(p(x)) = T(\sum a_k x^k) = \sum a_k T(x^k) = \sum a_k k x^{k-1} = p'(x)$.

6.4 Linear Transformations

25. $(S \circ T) \begin{bmatrix} 2 \\ 1 \end{bmatrix} = S \left(T \begin{bmatrix} 2 \\ 1 \end{bmatrix} \right) = S \begin{bmatrix} 5 \\ -1 \end{bmatrix} = \begin{bmatrix} 5 + (-1) & -1 \\ 0 & 5 - (-1) \end{bmatrix} = \begin{bmatrix} 4 & -1 \\ 0 & 6 \end{bmatrix}.$

$(S \circ T) \begin{bmatrix} x \\ y \end{bmatrix} = S \left(T \begin{bmatrix} x \\ y \end{bmatrix} \right) = S \begin{bmatrix} 2x + y \\ -y \end{bmatrix} = \begin{bmatrix} 2x & -y \\ 0 & 2x + 2y \end{bmatrix}.$

Domain of $T = \mathbb{R}^2 \neq M_{22} =$ codomain of $S \Rightarrow$ we cannot compute $T \circ S$.

27. The Chain Rule states $(p \circ g)'(x) = g'(x)p'(g(x))$. So, $g(x) = x + 1 \Rightarrow g'(x) = 1$.
    $(S \circ T)(p(x)) = S(T(p(x))) = S(p'(x)) = p'(x + 1)$.
    $(T \circ S)(p(x)) = T(S(p(x))) = T(p(x + 1)) = p'(x + 1)$ (because the derivative of $x + 1$ is 1).

29. We need to show $S \circ T = I_{\mathbb{R}^2}$ and $T \circ S = I_{\mathbb{R}^2}$.

    That is, we must show $S \circ T \begin{bmatrix} x \\ y \end{bmatrix} = \begin{bmatrix} x \\ y \end{bmatrix}$ and $T \circ S \begin{bmatrix} x \\ y \end{bmatrix} = \begin{bmatrix} x \\ y \end{bmatrix}$ for all $\begin{bmatrix} x \\ y \end{bmatrix}$.

    $S \circ T \begin{bmatrix} x \\ y \end{bmatrix} = S \left(T \begin{bmatrix} x \\ y \end{bmatrix} \right) = S \left(\begin{bmatrix} x - y \\ -3x + 4y \end{bmatrix} \right) = \begin{bmatrix} 4(x - y) + (-3x + 4y) \\ 3(x - y) + (-3x + 4y) \end{bmatrix} = \begin{bmatrix} x \\ y \end{bmatrix}.$

    $T \circ S \begin{bmatrix} x \\ y \end{bmatrix} = T \left(S \begin{bmatrix} x \\ y \end{bmatrix} \right) = T \left(\begin{bmatrix} 4x + y \\ 3x + y \end{bmatrix} \right) = \begin{bmatrix} (4x + y) - (3x + y) \\ -3(4x + y) + 4(3x + y) \end{bmatrix} = \begin{bmatrix} x \\ y \end{bmatrix}.$

31. Let T', T'' be such that $\begin{array}{l} T' \circ T = T'' \circ T = I_V \\ T \circ T' = T \circ T'' = I_W \end{array}$. Need to show $T'(\mathbf{w}) = T''(\mathbf{w})$ for all $\mathbf{w} \in W$.

    $T'(\mathbf{w}) = I_V \circ T'(\mathbf{w}) = T'' \circ T \circ T'(\mathbf{w}) = T'' \circ I_W(\mathbf{w}) = T''(\mathbf{w})$ for all $\mathbf{w} \in W$.

33. (a) $\{\mathbf{v}, T(\mathbf{v})\}$ linearly dependent $\Leftrightarrow$ there exists $c \neq 0$, $c\mathbf{v} + T(\mathbf{v}) = 0 \Rightarrow c = 0 \Leftrightarrow T(\mathbf{v}) = 0$.
    $c\mathbf{v} + T(\mathbf{v}) = 0$ (1) $\Leftrightarrow$ $T(c\mathbf{v} + T(\mathbf{v})) = 0$ $\Leftrightarrow$ $cT(\mathbf{v}) + T \circ T(\mathbf{v}) = 0 \Leftrightarrow$
    (Now use the fact that $T \circ T = T$.)
    $cT(\mathbf{v}) + T(\mathbf{v}) = 0$ (2)
    Combining (1) and (2) into a system, we have:
    $\begin{array}{l} c\mathbf{v} + T(\mathbf{v}) = 0 \\ cT(\mathbf{v}) + T(\mathbf{v}) = 0 \end{array} \Leftrightarrow c(\mathbf{v} - T(\mathbf{v})) = 0 \Leftrightarrow c = 0$ or $T(\mathbf{v}) = \mathbf{v} \Leftrightarrow T(\mathbf{v}) = 0$ or $T(\mathbf{v}) = \mathbf{v}$.

    (b) Example: $T \begin{bmatrix} x \\ y \end{bmatrix} = \begin{bmatrix} x \\ 0 \end{bmatrix}$.

    Then: $T \circ T \begin{bmatrix} x \\ y \end{bmatrix} = T(T \begin{bmatrix} x \\ y \end{bmatrix}) = T \begin{bmatrix} x \\ 0 \end{bmatrix} = \begin{bmatrix} x \\ 0 \end{bmatrix} = T \begin{bmatrix} x \\ y \end{bmatrix} \Rightarrow T \circ T = T.$

35. Below we are intrinsically using the fact that V and W are vector spaces. That tells us that $\mathbf{v}$ in V and $S(\mathbf{v})$ in W satisfy the ten axioms of a vector space.

Note: To prove $S = T$ below, we show $S(\mathbf{v}) = T(\mathbf{v})$ for all $\mathbf{v}$ in V.

1. Exercise 34 $\Rightarrow S, T \in \mathscr{L} \Rightarrow S + T \in \mathscr{L}$.
2. $(S+T)(\mathbf{v}) = S(\mathbf{v}) + T(\mathbf{v}) = T(\mathbf{v}) + S(\mathbf{v}) = (T+S)(\mathbf{v})$.
3. $((R+S)+T)(\mathbf{v}) = (R+S)(\mathbf{v}) + T(\mathbf{v}) = (R(\mathbf{v}) + S(\mathbf{v})) + T(\mathbf{v}) = R(\mathbf{v}) + (S(\mathbf{v}) + T(\mathbf{v}))$
$= (R + (S+T))(\mathbf{v})$.
4. Set $Z(\mathbf{v}) = \mathbf{0}$ for all $\mathbf{v} \Rightarrow (S+Z)(\mathbf{v}) = S(\mathbf{v}) + Z(\mathbf{v}) = S(\mathbf{v}) + \mathbf{0} = S(\mathbf{v})$.
5. Set $(-S)(\mathbf{v}) = -S(\mathbf{v})$ for all $\mathbf{v} \Rightarrow$
$(S + (-S))(\mathbf{v}) = S(\mathbf{v}) + (-S)(\mathbf{v}) = S(\mathbf{v}) + (-S(\mathbf{v})) = S(\mathbf{v}) + S(-\mathbf{v})$
$= S(\mathbf{v} + (-\mathbf{v})) = S(\mathbf{0}) = \mathbf{0}$.
6. Exercise 34 $\Rightarrow S \in \mathscr{L} \Rightarrow cS \in \mathscr{L}$.
7. $c(S+T)(\mathbf{v}) = c(S(\mathbf{v}) + T(\mathbf{v})) = cS(\mathbf{v}) + cT(\mathbf{v})$.
8. $(c+d)S(\mathbf{v}) = cS(\mathbf{v}) + dS(\mathbf{v})$.
9. $(cd)S(\mathbf{v}) = c(dS(\mathbf{v}))$.
10. Set $1(S(\mathbf{v})) = 1\,S(\mathbf{v})$ for all $\mathbf{v}$.

6.5 The Kernel and Range of a Linear Transformation

Q: What are our main goals for Section 6.5?
A: In this section, we revisit linear transformations, especially in the context of *subspaces*.
The two key subspaces associated with a linear transformation are: the *kernel* and *range*.
Recall, that linear transformations were originally introduced in Section 3.6.
Our goal is to deepen our understanding of linear transformations through these subspaces.

Key Definitions and Concepts

ker(T)	p.486	6.5	ker(T) = {**v** in V : $T(\mathbf{v}) = \mathbf{0}$} (See *kernel*.)
nullity(T)	p.488	6.5	nullity(T) = dim(ker(T)) (See *nullity*.)
range(T)	p.486	6.5	range(T) = {**w** in W : **w** = $T(\mathbf{v})$ for some **v** in V} (See *range*.)
rank(T)	p.488	6.5	rank(T) = dim(range(T)) (See *rank*.)
$V \cong W$	p.497	6.5	If $T : V \to W$ where T is an *isomorphism*. Then: V and W are *isomorphic*. This is written $V \cong W$.
isomorphic	p.497	6.5	If $T : V \to W$ where T is an *isomorphism*. Then: V and W are *isomorphic*. This is written $V \cong W$.
isomorphism	p.497	6.5	$T : V \to W$ is an *isomorphism* if T is *one-to-one* and *onto*.
kernel (of **T**)	p.486	6.5	Let $T : V \to W$ be a linear transformation. The *kernel* of T is the set of all vectors that T maps to zero. That is, ker(T) = {**v** in V : $T(\mathbf{v}) = \mathbf{0}$}.
nullity (of **T**)	p.488	6.5	Let $T : V \to W$ be a linear transformation. The *nullity* of T is the dimension of the kernel of T. That is, nullity(T) = dim(ker(T)).
one-to-one	p.492	6.5	T maps distinct vectors in V to distinct vectors in W. That is, $\mathbf{v} \neq \mathbf{u}$ implies $T(\mathbf{v}) \neq T(\mathbf{u})$, or $T(\mathbf{v}) = T(\mathbf{u})$ implies $\mathbf{v} = \mathbf{u}$.
onto	p.492	6.5	If range(T) = W, then T is called *onto*. For all **w**, there is a **v** such that $\mathbf{w} = T(\mathbf{v})$.
range (of **T**)	p.486	6.5	Let $T : V \to W$ be a linear transformation. The *range* of T is the set of all vectors that are images under T. That is, range(T) = {**w** in W : **w** = $T(\mathbf{v})$ for some **v** in V}.
rank (of **T**)	p.488	6.5	Let $T : V \to W$ be a linear transformation. The *rank* of T is the dimension of the range of T. That is, rank(T) = dim(range(T)).

Theorems

Thm 6.18   p.488   6.5   Let $T: V \to W$ be a linear transformation. Then:
    a. The *kernel* of T is a subspace of V
    b. The *range* of T is a subspace of W

Thm 6.19   p.490   6.5   Let $T: V \to W$ be a linear transformation. Then:
$\text{rank}(T) + \text{nullity}(T) = \dim V$

Thm 6.20   p.494   6.5   Let $T: V \to W$ be a linear transformation. Then:
T is *one-to-one* if and only if $\ker(T) = \{\mathbf{0}\}$.

Thm 6.21   p.494   6.5   Let $\dim V = \dim W = n$. Then:
$T: V \to W$ is *one-to-one* if and only if T is *onto*.

Thm 6.22   p.495   6.5   Let $T: V \to W$ be a *one-to-one* linear transformation.
If $S = \{\mathbf{v}_1, \ldots, \mathbf{v}_k\}$ is a linearly independent set in V. Then:
$T(S) = \{T(\mathbf{v}_1), \ldots, T(\mathbf{v}_k)\}$ is a linearly independent set in W.

Cor 6.23   p.495   6.5   Let $\dim V = \dim W = n$. Then:
If T is *one-to-one*, then T maps a basis for V to a basis for W.

Thm 6.24   p.495   6.5   Let $T: V \to W$ be a linear transformation. Then:
T is *invertible* if and only if T is *one-to-one* and *onto*.

Thm 6.25   p.498   6.5   Let V and W be two finite dimensional vector spaces. Then:
V is *isomorphic* to W ($V \cong W$) if and only if $\dim V = \dim W$.

6.5 The Kernel and Range of a Linear Transformation

Discussion of Key Definitions and Concepts

The kernel of a linear transformation

Q: What does the definition of the *kernel* of T, $\ker(T) = \{\mathbf{v} \text{ in } V : T(\mathbf{v}) = \mathbf{0}\}$, mean?
A: The *kernel* of T is all vectors that get mapped to the zero vector by T.

Q: What is the *kernel* of $T_0 : V \to W$? Explain in words.
A: Since T_0 maps all vectors to the zero vector, $\ker(T_0) = V$.

Q: What is the *kernel* of $I : V \to V$? Explain in words.
A: Since the only vector I maps to the zero vector *is* the zero vector, $\ker(I) = \{\mathbf{0}\}$.

Q: Is the *kernel* of $T : V \to W$ a subspace of V or of W? Explain.
A: Though the vectors in $\ker(T)$ are mapped to the zero vector in W, they actually lie in V.
   Therefore, the *kernel* of $T : V \to W$ a subspace of V.

Q: Recall that a linear transformation T can be associated with a matrix A.
   Which space of A is associated with $\ker(T)$?
A: Since $\ker(T)$ is all vectors mapped to the zero vector, $\ker(T)$ is associated with $\text{null}(A)$.

Q: Is $\ker(T) = \text{null}(A)$? Why or why not?

The range of a linear transformation

Q: What does $\text{range}(T) = \{\mathbf{w} \text{ in } W : \mathbf{w} = T(\mathbf{v}) \text{ for some } \mathbf{v} \text{ in } V\}$ mean?
A: The *range* of T is all vectors that are mapped *to* by any vector in V. See Thm 6.18.

Q: What is the *range* of $T_0 : V \to W$? Explain in words.
A: Since the only vector T_0 maps *to* is the zero vector, $\text{range}(T_0) = \{\mathbf{0}\}$.

Q: What is the *range* of $I : V \to V$? Explain in words.
A: Since I maps every vector to itself, $\text{range}(I) = V$.

Q: Is the *range* of $T : V \to W$ a subspace of V or of W? Explain.
A: Though it is vectors in V that are being mapped by T,
   it is their images that make up the *range* and those images lie in W.
   Therefore, the *range* of $T : V \to W$ a subspace of W.

Q: Recall that a linear transformation T can be associated with a matrix A.
   Which space of A is associated with $\text{range}(T)$?
A: Since $\text{range}(T)$ is all vectors we can create with T, $\text{range}(T)$ is associated with $\text{col}(A)$.

Q: Is $\text{range}(T) = \text{col}(A)$? Why or why not?

Discussion of Theorems

These theorems are extremely important and inter-related, but not difficult to prove. We should understand how to prove all of them to help us get a sense of how they go together.

Q: Thm 6.19 asserts rank(T)+nullity(T) = dim V. What is somewhat surprising about this?
A: Recall that rank(T) = dim(range(T)), but range(T) is a subspace of W, *not* of V. However, the theorem claims that the dimension of range(T) plus the dimension of ker(T) (which *is* a subspace of V) equals the dimension of V.

Q: So how can we make sense of what Thm 6.19 telling us?
A: T breaks V breaks into two pieces:
vectors mapped to zero and vectors *not* mapped to zero.
So, it is obvious that every vector in V must fall into one of these two categories.

Q: What are the two pieces V is broken into by T_0?

Q: If dim $V = n$, what is rank(T_0)?

Q: What are the two pieces V is broken into by I?

Q: If dim $V = n$, what is rank(I) (the linear transformation, not the matrix)?

Q: Half of what Thm 6.20 asserts is: if T is *one-to-one*, then ker$(T) = \{\mathbf{0}\}$. Explain why this is obvious in words.
A: Since T is *one-to-one* and it must map zero to zero (why?), the only vector T maps to zero is zero (because it maps exactly one vector to zero).

Q: So if T is *one-to-one*, what is nullity(T)?
A: Since T is *one-to-one*, ker$(T) = \{\mathbf{0}\}$ which implies nullity$(T) = 0$.

Q: If T is *one-to-one*, what is the relationship between dim V and dim(range(T))?
A: They are equal. Why?
Since dim(range(T)) = rank(T), we have:
dim(range(T)) = rank(T) = rank(T) + 0 = rank(T) + nullity(T) = dim V.

Q: If T is *one-to-one*, to what does T map a basis $\mathcal{B}$ for V? That is, what is $T(\mathcal{B})$?
A: T maps a basis for V to a basis for range(T). That is, $T(\mathcal{B})$ is a basis for range(T).

Q: If T is *one-to-one*, is $T(\mathcal{B})$ also a basis for W?
A: No. This is only true if range$(T) = W$. That is, if T is *onto*.

Q: If T is *not* one-to-one, is $T(\mathcal{B})$ still a basis for range(T)? Why or why not?

6.5 The Kernel and Range of a Linear Transformation

Discussion of Theorems, continued

Q: Half of Thm 6.21 is: if $\dim V = \dim W = n$ and T is *one-to-one* then T is *onto*.
Explain this conclusion in the light of our discussion of Thm 6.20.

A: When T is one-to-one, if $\mathcal{B}$ is a basis for V then $T(\mathcal{B})$ is a basis for range(T).
However, if $\dim V = \dim W = n$, then range$(T) = W$. That is, T is *onto* as claimed.

Q: What is the assertion of Cor 6.23?

A: A clear statement of what is implied in our argument above. Namely:
If $\dim V = \dim W = n$ and $\mathcal{B}$ is a basis for V,
then $T(\mathcal{B})$ is a basis for W because $W = $ range(T).
Keep in mind: if T is one-to-one, $T(\mathcal{B})$ is *automatically* a basis for range(T).

Q: Thm 6.22 is the backbone of many of the conclusions in this section.
Is Thm 6.22 true if T is *not* one-to-one? Why do we need to know that T is one-to-one?

A: If T is *not* one-to-one, linear independence is *not* preserved. Consider T_0.

Q: Compare Thm 6.21 and Thm 6.24. What is the difference between them?

A: Thm 6.21 tells us if T is one-to-one it is also onto provided $\dim V = \dim W$.
Thm 6.24 tells us that T has an inverse if and only if T is one-to-one and onto.
Note: T being one-to-one and onto is not *quite* enough to conclude $\dim V = \dim W$.

Q: Given our discussion Thm 6.21 and Thm 6.24, what is the point of Thm 6.25?

A: If we know that V and W are finite-dimensional,
then T being one-to-one and onto *is* enough to conclude $\dim V = \dim W$.

Q: Where in the statement of Thm 6.25 does it say that T is one-to-one and onto?

A: This is precisely what is implied when we say V is *isomorphic* to W ($V \cong W$).
That is, to say V is *isomorphic* to W ($V \cong W$) means:
There exists $T : V \to W$ such that T is one-to-one and onto.

Solutions to odd-numbered exercises from Section 6.5

1. (a) $\begin{bmatrix} a & b \\ c & d \end{bmatrix}$ in $\ker(T) \Leftrightarrow a = d = 0 \Rightarrow$ only (ii) $= \begin{bmatrix} 0 & 4 \\ 2 & 0 \end{bmatrix}$ in $\ker(T)$.

    (b) $\begin{bmatrix} a & b \\ c & d \end{bmatrix}$ in $\text{range}(T) \Leftrightarrow b = c = 0 \Rightarrow$ only (iii) $= \begin{bmatrix} 3 & 0 \\ 0 & -3 \end{bmatrix}$ in $\text{range}(T)$.

    (c) $\begin{bmatrix} a & b \\ c & d \end{bmatrix}$ in $\ker(T) \Leftrightarrow a = d = 0 \Rightarrow \ker(T) = \left\{ \begin{bmatrix} 0 & b \\ c & 0 \end{bmatrix} \right\}$.

    $\begin{bmatrix} a & b \\ c & d \end{bmatrix}$ in $\text{range}(T) \Leftrightarrow b = c = 0 \Rightarrow \text{range}(T) = \left\{ \begin{bmatrix} a & 0 \\ 0 & d \end{bmatrix} \right\}$.

3. (a) $p(x)$ in $\ker(T) \Leftrightarrow \begin{matrix} a - b = 0 \\ b + c = 0 \end{matrix} \Leftrightarrow a = b = -c \Rightarrow$ only (iii) in $\ker(T)$.

    (b) $p(x)$ in $\text{range}(T) \Leftrightarrow a, b, c \in \mathbb{R} \Rightarrow$ (i), (ii), (iii) in $\text{range}(T)$.

    (c) $p(x)$ in $\ker(T) \Leftrightarrow a = b = -c \Rightarrow \ker(T) = \{t + tx - tx^2\}$.
    $p(x)$ in $\text{range}(T) \Leftrightarrow a, b, c \in \mathbb{R} \Rightarrow \text{range}(T) = \mathbb{R}^2$.

5. $\ker(T) = \left\{ \begin{bmatrix} 0 & b \\ c & 0 \end{bmatrix} \right\} \Rightarrow \left\{ \begin{bmatrix} 0 & 1 \\ 0 & 0 \end{bmatrix}, \begin{bmatrix} 0 & 0 \\ 1 & 0 \end{bmatrix} \right\}$ is basis $\Rightarrow \text{nullity}(T) = \dim \ker(T) = 2$.

    Likewise, $\text{range}(T) = \left\{ \begin{bmatrix} 1 & 0 \\ 0 & 0 \end{bmatrix}, \begin{bmatrix} 0 & 0 \\ 0 & 1 \end{bmatrix} \right\}$ is a basis $\Rightarrow \text{rank}(T) = \dim \text{range}(T) = 2$.

7. $\ker(T) = \{t + tx - tx^2\} \Rightarrow \{1 + x - x^2\}$ is basis $\Rightarrow \text{nullity}(T) = 1$.
    $\text{range}(T) = \mathbb{R}^2 \Rightarrow \text{rank}(T) = \dim \text{range}(T) = \dim \mathbb{R}^2 = 2$.

9. $A \in \ker(T) \Rightarrow = \begin{matrix} a - b = 0 \\ c - d = 0 \end{matrix} \Rightarrow \begin{matrix} a = b \\ c = d \end{matrix} \Rightarrow$

    $\ker(T) = \left\{ \begin{bmatrix} a & a \\ c & c \end{bmatrix} \right\} \Rightarrow \ker(T) = \text{span} \left(\begin{bmatrix} 1 & 1 \\ 0 & 0 \end{bmatrix}, \begin{bmatrix} 0 & 0 \\ 1 & 1 \end{bmatrix} \right) \Rightarrow \text{nullity}(T) = \dim \ker(T) = 2$.
    Therefore, $\text{rank}(T) = \dim M_{22} - \text{nullity}(T) = 4 - 2 = 2$.

11. $T(E_{11}) = E_{11}B = \begin{bmatrix} 1 & 0 \\ 0 & 0 \end{bmatrix} \begin{bmatrix} 1 & -1 \\ -1 & 1 \end{bmatrix} = \begin{bmatrix} 1 & -1 \\ 0 & 0 \end{bmatrix}$. Likewise, $T(E_{12}) = E_{12}B = \begin{bmatrix} -1 & 1 \\ 0 & 0 \end{bmatrix}$,

    $T(E_{21}) = E_{21}B = \begin{bmatrix} 0 & 0 \\ 1 & -1 \end{bmatrix}$, $T(E_{22}) = E_{22}B = \begin{bmatrix} 0 & 0 \\ -1 & 1 \end{bmatrix} \Rightarrow$

    $\text{range}(T) = \text{span} \left(\begin{bmatrix} 1 & -1 \\ 0 & 0 \end{bmatrix}, \begin{bmatrix} 0 & 0 \\ 1 & -1 \end{bmatrix} \right) \Rightarrow \text{rank}(T) = 2$.

    Therefore, $\text{nullity}(T) = \dim M_{22} - \text{rank}(T) = 4 - 2 = 2$.

13. $p(x) \in \ker(T) \Leftrightarrow p'(0) = (b + 2c(0)) = 0 \Rightarrow b = 0, a, c$ free $\Rightarrow$
    $\ker(T) = \{a + cx^2\} = \text{span}(1, x^2) \Rightarrow \text{nullity}(T) = \dim \ker(T) = 2$.
    Therefore, $\text{rank}(T) = \dim \mathscr{P}_2 - \text{nullity}(T) = 3 - 2 = 1$.

6.5 The Kernel and Range of a Linear Transformation

15. (a) $\begin{bmatrix} x \\ y \end{bmatrix} \in \ker(T) \Rightarrow = \begin{matrix} 2x - y = 0 \\ x + 2y = 0 \end{matrix} \Rightarrow x = y = 0 \Rightarrow \ker(T) = \{\mathbf{0}\} \Rightarrow T$ is one-to-one.

    (b) Since (a) $\Rightarrow T$ is one-to-one and $V = W = \mathbb{R}^2$, Theorem 6.21 implies T is onto.

17. (a) Note, $T(1) = \begin{bmatrix} 2 \\ 1 \\ -1 \end{bmatrix}, T(x) = \begin{bmatrix} -1 \\ 1 \\ 0 \end{bmatrix}, T(x^2) = \begin{bmatrix} 0 \\ -3 \\ 1 \end{bmatrix} \Rightarrow$

    $-T(1) - 2T(x) = T(x^2) \Rightarrow T(1 + 2x + x^2) = 0 \Rightarrow \ker(T) = \text{span}(1 + 2x + x^2) \neq \{\mathbf{0}\} \Rightarrow$ T is *not* one-to-one.

    (b) Since (a) $\Rightarrow T$ is *not* one-to-one and $\dim \mathscr{P}_2 = \dim \mathbb{R}^3$, Theorem 6.21 implies no T can be onto.

19. (a) $\begin{bmatrix} a \\ b \\ c \end{bmatrix} \in \ker(T) \Rightarrow \begin{bmatrix} a-b & b-c \\ a+b & b+c \end{bmatrix} = 0 \Rightarrow \begin{matrix} a = 0 \\ b = 0 \\ c = 0 \end{matrix} \Rightarrow \ker(T) = \{\mathbf{0}\} \Rightarrow T$ is one-to-one.

    (b) (a) $\Rightarrow \text{rank}(T) = \dim \mathbb{R}^3 - \text{nullity}(T) = 3 - 0 = 3 \Rightarrow$
    $\dim \text{range}(T) = 3 < \dim M_{22} = 4 \Rightarrow T$ is *not* onto.
    Furthermore, $\dim \text{range}(T) \leq 3 < \dim M_{22} = 4 \Rightarrow$ no T can be onto.

21. $D_3 = \text{span}(E_{11}, E_{22}, E_{33}) \Rightarrow \dim D_3 = 3 = \dim \mathbb{R}^3 \Rightarrow D_3 \cong \mathbb{R}^3$.

    Define $T \begin{bmatrix} x & 0 & 0 \\ 0 & y & 0 \\ 0 & 0 & z \end{bmatrix} = \begin{bmatrix} x \\ y \\ z \end{bmatrix}$. Then $A \in \ker(T) \Rightarrow \begin{bmatrix} x & 0 & 0 \\ 0 & y & 0 \\ 0 & 0 & z \end{bmatrix} = 0 \Rightarrow x = y = z = 0 \Rightarrow$

    $\ker(T) = \{\mathbf{0}\} \Rightarrow T$ is one-to-one.
    Since T is one-to-one and $\dim D_3 = \dim \mathbb{R}^3$, Theorem 6.21 implies T is onto.

23. $A \in S'_3 \Rightarrow A = -A^T \Rightarrow \begin{matrix} a_{21} = a_{12} \\ a_{31} = a_{13} \\ a_{32} = a_{23} \end{matrix}$ and $a_{11} = a_{22} = a_{33} = 0 \Rightarrow \dim S'_3 = 3$.

    Exercise 22 $\Rightarrow \dim S_3 = 6 \neq \dim S'_3 = 3 \Rightarrow S_3 \not\cong S'_3$.

25. $\mathbb{C} = \text{span}(1, i) \Rightarrow \dim \mathbb{C} = 2$. Likewise, $\mathbb{R}^2 = \text{span}\left(\begin{bmatrix} 1 \\ 0 \end{bmatrix}, \begin{bmatrix} 0 \\ 1 \end{bmatrix}\right) \Rightarrow \dim \mathbb{R}^2 = 2$,

    Therefore, since $\mathbb{C} = \dim \mathbb{R}^2 = 2$, $\mathbb{C} \cong \mathbb{R}^2$.

    Define $T(a + ib) = \begin{bmatrix} a \\ b \end{bmatrix}$.

    Then $a + ib \in \ker(T) \Rightarrow a = b = 0 \Rightarrow \ker(T) = \{\mathbf{0}\} \Rightarrow T$ is one-to-one.
    Since T is one-to-one and $\dim \mathbb{C} = \dim \mathbb{R}^2$, Theorem 6.21 implies T is onto.

27. Need only show $p(x) + p'(x) = 0 \Rightarrow p(x) = 0 \Rightarrow \ker(T) = \{\mathbf{0}\} \Rightarrow T$ is one-to-one.
    Let $p(x) = a_0 + a_1 x + \ldots + a_n x^n$ be such that $p(x) + p'(x) = 0 \Rightarrow$
    $(a_0 + a_1 x + \ldots + a_n x^n) + (a_1 + 2a_2 x + \ldots + na_n x^{n-1})$
    $= (a_0 + a_1) + (a_1 + 2a_2)x + \ldots + (a_n + (n-1)a_{n-1} x^{n-1}) + a_n x^n = 0 \Rightarrow a_n = 0 \Rightarrow$
    $0 + (n-1)a_{n-1} = 0 \Rightarrow a_{n-1} = 0 \ldots a_0 = 0 \Rightarrow p(x) = 0$.

29. Need only show $x^n p(\frac{1}{x}) = 0 \Rightarrow p(x) = 0$.

    Let $p(x) = a_0 + a_1 x + \ldots + a_n x^n$ be such that $x^n p(\frac{1}{x}) = 0$. We will show $p(x) = 0$.
    $x^n p(\frac{1}{x}) = a_0 x^n + a_1 x^{n-1} + \ldots + a_n = 0 \Rightarrow a_k = 0$
    because x^k are linearly independent $\Rightarrow p(x) = 0$.

31. Define $T(f) = f(\frac{1}{2} x)$. Need only show that T is one-to-one and onto.

    One-to-one: $f(\frac{1}{2} x) = 0$ for all $x \in [0, 2] \Rightarrow$
    $$f(y) = 0 \text{ for all } y \in [0, 1] \text{ where } y = \tfrac{x}{2} \Rightarrow$$
    $$f(x) = 0 \text{ for all } x \in [0, 1].$$

    Onto: Given any $f(x) \in \mathscr{C}[0, 2]$, we can define $g \in \mathscr{C}[0, 1]$ as $g(x) = f(2x)$.

    Then simply note that $T(g) = f(2(\frac{1}{2} x)) = f(x)$ which shows T is onto.

33. (a) Recall L is one-to-one if and only if $L(\mathbf{v}) = \mathbf{0} \Leftrightarrow \mathbf{v} = \mathbf{0}$ because $\ker L = \{\mathbf{0}\}$.
    So, we need to show $(S \circ T)(\mathbf{u}) = \mathbf{0} \Leftrightarrow \mathbf{u} = \mathbf{0}$.
    $(S \circ T)(\mathbf{u}) = \mathbf{0} \Leftrightarrow S(T(\mathbf{u})) = \mathbf{0} \Leftrightarrow T(\mathbf{u}) = \mathbf{0} \Leftrightarrow \mathbf{u} = \mathbf{0}$ which shows $S \circ T$ is one-to-one.

    (b) Recall L is onto for every $\mathbf{w} \in W$ there exists $\mathbf{v} \in V$ such that $L(\mathbf{v}) = \mathbf{w}$.
    So, we need to show for every $\mathbf{w} \in W$ there exists $\mathbf{u} \in U$ such that $(S \circ T)(\mathbf{u}) = \mathbf{w}$.
    S onto $\Rightarrow$ for every $\mathbf{w} \in W$ there exists $\mathbf{v} \in V$ such that $S(\mathbf{v}) = \mathbf{w}$.
    Furthermore, T onto $\Rightarrow$ there exists $\mathbf{u} \in U$ such that $T(\mathbf{u}) = \mathbf{v} \Rightarrow$
    $(S \circ T)(\mathbf{u}) = S(T(\mathbf{u})) = S(\mathbf{v}) = \mathbf{w}$ which shows $S \circ T$ is onto.

35. (a) Recall Theorem 6.11b from Section 6.2 says if W is a subspace
    of a finite-dimensional vector space V, then $\dim W = \dim V \Leftrightarrow W = V$.
    Note, the contrapositive of this statement is $\dim W \neq \dim V \Leftrightarrow W \neq V$.
    Given $T: V \to W$, we will show $\dim V < \dim W \Rightarrow \dim \operatorname{range}(T) \neq \dim W$
    which will then imply $\operatorname{range}(T) \neq W$ and therefore T cannot be onto.
    Recall, Theorem 6.18b of this section asserts $\operatorname{range}(T)$ is a subspace of W.
    Now, Theorem 6.20, therefore implies $\dim \operatorname{range}(T) = \dim V - \operatorname{nullity}(T) \leq \dim V$.
    Therefore, $\dim \operatorname{range}(T) < \dim W \Rightarrow T$ cannot be onto.
    Need to show $T(\mathbf{u}) = \mathbf{0} \Leftrightarrow \mathbf{u} = \mathbf{0}$.
    $T(\mathbf{u}) = \mathbf{0} \Leftrightarrow (S \circ T)(\mathbf{u}) = \mathbf{0} \Leftrightarrow \mathbf{u} = \mathbf{0}$ (because $S \circ T$ is one-to-one).

    (b) Since Theorem 6.20 states T is one-to-one $\Leftrightarrow \ker(T) = \{\mathbf{0}\}$,
    we need to show $\dim V > \dim W \Rightarrow \ker(T) \neq \{\mathbf{0}\}$.
    Recall, Theorem 6.18b of this section asserts
    $\operatorname{range}(T)$ is a subspace of $W \Rightarrow \dim W \geq \operatorname{range}(T) = \operatorname{rank}(T)$.
    Therefore, $\dim W + \operatorname{nullity}(T) \geq \operatorname{rank}(T) + \operatorname{nullity}(T) = \dim V \Rightarrow$
    $\dim \ker(T) = \operatorname{nullity}(T) \geq \dim V - \dim W > 0 \Rightarrow \ker(T) \neq \{\mathbf{0}\}$.

6.5 The Kernel and Range of a Linear Transformation

37. **NOTE**: The *null spaces* in the text should be referred to as *kernels*.
Following the hint, we will use the Rank Theorem to show $\ker(T) = \ker(T^2)$.
The Rank Theorem implies $\operatorname{rank}(T) + \operatorname{nullity}(T) = \operatorname{rank}(T^2) + \operatorname{nullity}(T^2) = \dim V$.
So since $\operatorname{rank}(T) = \operatorname{rank}(T^2)$, we have $\operatorname{nullity}(T^2) = \operatorname{nullity}(T)$.

So we will show $\ker(T) \subseteq \ker(T^2)$ and conclude that $\ker(T) = \ker(T^2)$.
Let **v** be in $\ker(T)$, then $T^2(\mathbf{v}) = T(T(\mathbf{v})) = T(\mathbf{0}) = \mathbf{0}$ which implies $\ker(T) \subseteq \ker(T^2)$.
So $\ker(T) = \ker(T^2)$ as we were to show.

Now let **v** be in $\operatorname{range}(T) \cap \ker(T)$. We want to show $\mathbf{v} = \mathbf{0}$.
Then $\mathbf{v} = T(\mathbf{w})$ and $T(\mathbf{v}) = T(T(\mathbf{w})) = \mathbf{0}$ which implies **w** is in $\ker(T^2)$.
But $\ker(T) = \ker(T^2)$, so **w** is also in $\ker(T)$.
That is, $\mathbf{v} = T(\mathbf{w}) = \mathbf{0}$ as we were to show.

6.6 The Matrix of a Linear Transformation

Q: What are our main goals for Section 6.6?
A: To take another look at subspaces, especially the big three related to matrices: the row space, the column space and null space. These ideas are profound.
See Section 3.5 for a reminder of the first investigation of Linear Transformations.

Key Definitions and Concepts

$[T]_{C \leftarrow B}$ p.502   6.6   The matrix of T with respect to bases B and C.
By Thm 6.26 it satisfies: $[T]_{C \leftarrow B}[v]_B = [T(v)]_C$
See remarks following Thm 6.26.

$[T]_B$ p.502   6.6   Special case when $V = W$ and $B = C$.
By Thm 6.26 it satisfies: $[T]_B[v]_B = [T(v)]_B$
See remarks following Thm 6.26.

$[T(v)]_C$ p.502   6.6 $A[v]_B = [T(v)]_C$ We are asked to show this in Exercise 39.
See remarks following Thm 6.26.

commutative diagram   p.503   6.6   See detailed remarks following Thm 6.26.
The diagram referred to is on p.502.

diagonalizable   p.513   6.6   Let V be a finite dimensional vector space.
Let $T : V \to V$ be a linear transformation. Then:
T is called *diagonalizable* if there is a basis C such that $[T]_C$ is diagonal matrix.

Theorems

Thm 6.26 p.502 6.6 Let V and W be two finite dimensional vector spaces with bases $\mathcal{B}$ and $\mathcal{C}$ respectively, where $\mathcal{B} = \{\mathbf{v}_1, \ldots, \mathbf{v}_n\}$. Let $T : V \to W$ be a linear transformation, then the $n \times n$ matrix $A = \begin{bmatrix} [T(\mathbf{v}_1)]_\mathcal{C} & | & [T(\mathbf{v}_2)]_\mathcal{C} & | & \cdots & | & [T(\mathbf{v}_n)]_\mathcal{C} \end{bmatrix}$ satisfies $A[\mathbf{v}]_\mathcal{B} = [T(\mathbf{v})]_\mathcal{C}$ for every vector $\mathbf{v}$ in V.

Thm 6.27 p.508 6.6 Let U, V, and W be finite dimensional vector spaces with bases $\mathcal{B}$, $\mathcal{C}$, and $\mathcal{D}$ respectively. Let $T : U \to V$ and $S : V \to W$ be linear transformations. Then: $[S \circ T]_{\mathcal{D} \leftarrow \mathcal{B}} = [S]_{\mathcal{D} \leftarrow \mathcal{C}} [T]_{\mathcal{C} \leftarrow \mathcal{B}}$

Thm 6.28 p.509 6.6 Let V and W be finite dimensional vector spaces with bases $\mathcal{B}$ and $\mathcal{C}$, respectively. Let $T : V \to W$ be a linear transformation. Then: T is invertible if and only if the matrix $[T]_{\mathcal{C} \leftarrow \mathcal{B}}$ is invertible. In this case, $([T]_{\mathcal{C} \leftarrow \mathcal{B}})^{-1} = [T^{-1}]_{\mathcal{C} \leftarrow \mathcal{B}}$

Thm 6.29 p.512 6.6 Let V be a finite dimensional vector space with bases $\mathcal{B}$ and $\mathcal{C}$. Let $T : V \to V$ be a linear transformation. Then: $[T]_\mathcal{C} = P^{-1}[T]_\mathcal{B} P$ where P is the change-of-basis matrix from $\mathcal{C}$ to $\mathcal{B}$.

Thm 6.30 p.516 6.6 *Fundamental Theorem of Invertible Matrices*: Version 4
Let A be an $n \times n$ matrix and let $T : V \to W$ be a linear transformation whose matrix $([T]_{\mathcal{C} \leftarrow \mathcal{B}})$ with respect to bases $\mathcal{B}$ and $\mathcal{C}$ of V and W respectively, is A. Statements a. through t. are equivalent (see p.516).

p. T is invertible
q. T is one-to-one
r. T is onto
s. $\ker(T) = \mathbf{0}$
t. $\text{range}(T) = W$

6.6 The Matrix of a Linear Transformation

Discussion of Key Definitions and Concepts

Q: What is the easiest way to determine if W is a subspace?

A: By applying Theorem 6.2. That is, verifying that it is closed under both operations. That is the main point of all the exercises in this section.

Q: What are the two technical points we need to be aware of when proving W is a subspace?

A: Be sure to verify that $W \neq \emptyset$. Also, as noted in the text, $\mathbf{0}$ of V must be in W.

Solutions to odd-numbered exercises from Section 6.6

1. Directly, $T(4 + 2x) = 2 - 4x$.
$[T(1)]_C = [0 - x]_C = \begin{bmatrix} 0 \\ -1 \end{bmatrix}$, $[T(x)]_C = [1 - 0x]_C = \begin{bmatrix} 1 \\ 0 \end{bmatrix} \Rightarrow [T]_{C \leftarrow B} = \begin{bmatrix} 0 & 1 \\ -1 & 0 \end{bmatrix}$.

So, $[T]_{C \leftarrow B} [4 + 2x]_B = \begin{bmatrix} 0 & 1 \\ -1 & 0 \end{bmatrix} \begin{bmatrix} 4 \\ 2 \end{bmatrix} = \begin{bmatrix} 2 \\ -4 \end{bmatrix} = [2 - 4x]_C = [T(4 + 2x)]_C$.

3. Directly, $T(a + bx + cx^2) = a + b(x + 2) + c(x + 2)^2$.

$[T(1)]_C = [1]_C = \begin{bmatrix} 1 \\ 0 \\ 0 \end{bmatrix}$, $[T(x)]_C = [x + 2]_C = \begin{bmatrix} 0 \\ 1 \\ 0 \end{bmatrix}$, $[T(x^2)]_C = [(x + 2)^2]_C = \begin{bmatrix} 0 \\ 0 \\ 1 \end{bmatrix} \Rightarrow$

$[T]_{C \leftarrow B} = \begin{bmatrix} 1 & 0 & 0 \\ 0 & 1 & 0 \\ 0 & 0 & 1 \end{bmatrix}$. So:

$[T]_{C \leftarrow B} [4 + 2x]_B = \begin{bmatrix} 1 & 0 & 0 \\ 0 & 1 & 0 \\ 0 & 0 & 1 \end{bmatrix} \begin{bmatrix} a \\ b \\ c \end{bmatrix} = \begin{bmatrix} a \\ b \\ c \end{bmatrix} = [a + b(x + 2) + c(x + 2)^2]_C$

$= [T(a + bx + cx^2)]_C$.

5. Directly, $T(a + bx + cx^2) = \begin{bmatrix} a \\ a + b + c \end{bmatrix}$.

$[T(1)]_C = \begin{bmatrix} 1 \\ 1 \end{bmatrix}_C = \begin{bmatrix} 1 \\ 1 \end{bmatrix}$. Likewise, $[T(x)]_C = [T(x^2)]_C = \begin{bmatrix} 0 \\ 1 \end{bmatrix} \Rightarrow [T]_{C \leftarrow B} = \begin{bmatrix} 1 & 0 & 0 \\ 1 & 1 & 1 \end{bmatrix}$.

So, $[T]_{C \leftarrow B} [a + bx + cx^2]_B = \begin{bmatrix} 1 & 0 & 0 \\ 1 & 1 & 1 \end{bmatrix} \begin{bmatrix} a \\ b \\ c \end{bmatrix} = \begin{bmatrix} a \\ a + b + c \end{bmatrix}_C = [T(a + bx + cx^2)]_C$.

7. Directly, $T \begin{bmatrix} -7 \\ 7 \end{bmatrix} = \begin{bmatrix} 7 \\ 7 \\ 7 \end{bmatrix}$. Also, note since $\begin{bmatrix} -7 \\ 7 \end{bmatrix} = 2 \begin{bmatrix} 1 \\ 2 \end{bmatrix} - 3 \begin{bmatrix} 3 \\ -1 \end{bmatrix}$, $\begin{bmatrix} -7 \\ 7 \end{bmatrix}_B = \begin{bmatrix} 2 \\ -3 \end{bmatrix}$.

Likewise, $\left[T\begin{bmatrix} 1 \\ 2 \end{bmatrix}\right]_C = \begin{bmatrix} 5 \\ -1 \\ 2 \end{bmatrix}_C = \begin{bmatrix} 6 \\ -3 \\ 2 \end{bmatrix}$, $\left[T\begin{bmatrix} 3 \\ -1 \end{bmatrix}\right]_C = \begin{bmatrix} 1 \\ -3 \\ -1 \end{bmatrix}_C = \begin{bmatrix} 4 \\ -2 \\ -1 \end{bmatrix} \Rightarrow$

$[T]_{C \leftarrow B} = \begin{bmatrix} 6 & 4 \\ -3 & -2 \\ 2 & -1 \end{bmatrix}$.

So, $[T]_{C \leftarrow B} \begin{bmatrix} -7 \\ 7 \end{bmatrix}_B = \begin{bmatrix} 6 & 4 \\ -3 & -2 \\ 2 & -1 \end{bmatrix} \begin{bmatrix} 2 \\ -3 \end{bmatrix} = \begin{bmatrix} 0 \\ 0 \\ 7 \end{bmatrix} = \begin{bmatrix} 7 \\ 7 \\ 7 \end{bmatrix}_C = \left[T\begin{bmatrix} -7 \\ 7 \end{bmatrix}\right]_C$.

6.6 The Matrix of a Linear Transformation

9. Directly $T\begin{bmatrix} a & b \\ c & d \end{bmatrix} = \begin{bmatrix} a & b \\ c & d \end{bmatrix}^T = \begin{bmatrix} a & c \\ b & d \end{bmatrix}.$

$[T(E_{11})]_C = [E_{11}]_C = \begin{bmatrix} 1 \\ 0 \\ 0 \\ 0 \end{bmatrix}$, Likewise, $[T(E_{12})]_C = [E_{21}]_C = \begin{bmatrix} 0 \\ 0 \\ 1 \\ 0 \end{bmatrix}$,

$[T(E_{21})]_C = [E_{12}]_C = \begin{bmatrix} 0 \\ 1 \\ 0 \\ 0 \end{bmatrix}$, $[T(E_{22})]_C = \begin{bmatrix} 0 \\ 0 \\ 0 \\ 1 \end{bmatrix}$ $\Rightarrow [T]_{C \leftarrow B} = \begin{bmatrix} 1 & 0 & 0 & 0 \\ 0 & 0 & 1 & 0 \\ 0 & 1 & 0 & 0 \\ 0 & 0 & 0 & 1 \end{bmatrix}.$

So, $[T]_{C \leftarrow B}[A]_B = \begin{bmatrix} 1 & 0 & 0 & 0 \\ 0 & 0 & 1 & 0 \\ 0 & 1 & 0 & 0 \\ 0 & 0 & 0 & 1 \end{bmatrix} \begin{bmatrix} a \\ b \\ c \\ d \end{bmatrix} = \begin{bmatrix} a \\ c \\ b \\ d \end{bmatrix} = \left[\begin{bmatrix} a & c \\ b & d \end{bmatrix}\right]_C = [T(A)]_C.$

11. Directly, $T(A) = AB - BA = \begin{bmatrix} a-b & b-a \\ c-d & d-c \end{bmatrix} - \begin{bmatrix} a-c & b-d \\ c-a & d-b \end{bmatrix} = \begin{bmatrix} c-b & d-a \\ a-d & b-c \end{bmatrix}.$

$[T(E_{11})]_C = [E_{11}B - BE_{11}]_C = \begin{bmatrix} 0 & -1 \\ 1 & 0 \end{bmatrix}_C = \begin{bmatrix} 0 \\ -1 \\ 1 \\ 0 \end{bmatrix}.$

Likewise, $[T(E_{12})]_C = \begin{bmatrix} -1 & 0 \\ 0 & 1 \end{bmatrix}_C = \begin{bmatrix} -1 \\ 0 \\ 0 \\ 1 \end{bmatrix}$, $[T(E_{21})]_C = \begin{bmatrix} 1 & 0 \\ 0 & -1 \end{bmatrix}_C = \begin{bmatrix} 1 \\ 0 \\ 0 \\ -1 \end{bmatrix},$

$[T(E_{22})]_C = \begin{bmatrix} 0 & 1 \\ -1 & 0 \end{bmatrix}_C = \begin{bmatrix} 0 \\ 1 \\ -1 \\ 0 \end{bmatrix} \Rightarrow [T]_{C \leftarrow B} = \begin{bmatrix} 0 & -1 & 1 & 0 \\ -1 & 0 & 0 & 1 \\ 1 & 0 & 0 & -1 \\ 0 & 1 & -1 & 0 \end{bmatrix}.$ So:

$[T]_{C \leftarrow B}[A]_B = \begin{bmatrix} 0 & -1 & 1 & 0 \\ -1 & 0 & 0 & 1 \\ 1 & 0 & 0 & -1 \\ 0 & 1 & -1 & 0 \end{bmatrix} \begin{bmatrix} a \\ b \\ c \\ d \end{bmatrix} = \begin{bmatrix} c-b \\ d-a \\ a-d \\ b-c \end{bmatrix} = \left[\begin{bmatrix} c-b & d-a \\ a-d & b-c \end{bmatrix}\right]_C$

$= [AB - BA]_C = [T(A)]_C.$

13. (a) $D(\sin x) = \cos x, D(\cos x) = -\sin x \Rightarrow$
    range$(D) = \text{span}(\sin x, \cos x) = W$ as was to be shown.

(b) $\mathcal{B} = \{\sin x, \cos x\} \Rightarrow [\sin x]_\mathcal{B} = \begin{bmatrix} 1 \\ 0 \end{bmatrix}, [\cos x]_\mathcal{B} = \begin{bmatrix} 0 \\ 1 \end{bmatrix} \Rightarrow$

$[D(\sin x)]_\mathcal{B} = \begin{bmatrix} 0 \\ 1 \end{bmatrix}, [D(\cos x)]_\mathcal{B} = \begin{bmatrix} -1 \\ 0 \end{bmatrix} \Rightarrow [D]_\mathcal{B} = \begin{bmatrix} 0 & -1 \\ 1 & 0 \end{bmatrix}.$

(c) $[f(x)]_\mathcal{B} = [3\sin x - 5\cos x]_\mathcal{B} = \begin{bmatrix} 3 \\ -5 \end{bmatrix} \Rightarrow$

$[D(f(x))]_\mathcal{B} = [D]_\mathcal{B}[f(x)]_\mathcal{B} = \begin{bmatrix} 0 & -1 \\ 1 & 0 \end{bmatrix}\begin{bmatrix} 3 \\ -5 \end{bmatrix} = \begin{bmatrix} 5 \\ 3 \end{bmatrix} \Rightarrow$

$f'(x) = 5\sin x + 3\cos x.$

Directly, we have $f'(x) = 3(\cos x) - 5(-\sin x) = 5\sin x + 3\cos x.$
This shows the indirect and direct methods give the same answer.

15. (a) $\mathcal{B} = \{e^{2x}, e^{2x}\cos x, e^{2x}\sin x\} \Rightarrow$

$[D(e^{2x})]_\mathcal{B} = [2e^{2x}]_\mathcal{B} = \begin{bmatrix} 2 \\ 0 \\ 0 \end{bmatrix}, [D(e^{2x}\cos x)]_\mathcal{B} = [2e^{2x}\cos x - e^{2x}\sin x]_\mathcal{B} = \begin{bmatrix} 0 \\ 2 \\ -1 \end{bmatrix},$

$[D(e^{2x}\sin x)]_\mathcal{B} = [e^{2x}\cos x + 2e^{2x}\sin x]_\mathcal{B} = \begin{bmatrix} 0 \\ 1 \\ 2 \end{bmatrix} \Rightarrow [D]_\mathcal{B} = \begin{bmatrix} 2 & 0 & 0 \\ 0 & 2 & 1 \\ 0 & -1 & 2 \end{bmatrix}.$

(b) $[f(x)]_\mathcal{B} = [3e^{2x} - e^{2x}\cos x + 2e^{2x}\sin x]_\mathcal{B} = \begin{bmatrix} 3 \\ -1 \\ 2 \end{bmatrix} \Rightarrow$

$[D(f(x))]_\mathcal{B} = [D]_\mathcal{B}[f(x)]_\mathcal{B} = \begin{bmatrix} 2 & 0 & 0 \\ 0 & 2 & 1 \\ 0 & -1 & 2 \end{bmatrix}\begin{bmatrix} 3 \\ -1 \\ 2 \end{bmatrix} = \begin{bmatrix} 6 \\ 0 \\ 5 \end{bmatrix} \Rightarrow f'(x) = 6e^{2x} + 5e^{2x}\sin x.$

Directly, we have:
$f'(x) = 3(2e^{2x}) - (2e^{2x}\cos x - e^{2x}\sin x) + 2(e^{2x}\cos x + 2e^{2x}\sin x) = 6e^{2x} + 5e^{2x}\sin x.$
This shows the indirect and direct methods give the same answer.

6.6 The Matrix of a Linear Transformation

17. (a) Let $p(x) = c + dx$, then $T(p(x)) = T(c + dx) = \begin{bmatrix} p(0) \\ p(1) \end{bmatrix} = \begin{bmatrix} c \\ c+d \end{bmatrix} \Rightarrow$

$(S \circ T)(p(x)) = S(T(p(x))) = S(T(c+dx)) = S\begin{bmatrix} c \\ c+d \end{bmatrix} = \begin{bmatrix} c - 2(c+d) \\ 2c - (c+d) \end{bmatrix} = \begin{bmatrix} -c - 2d \\ c - d \end{bmatrix} \Rightarrow$

$[(S \circ T)(1)]_{D \leftarrow B} = \begin{bmatrix} -1 - 2(0) \\ 1 - 0 \end{bmatrix} = \begin{bmatrix} -1 \\ 1 \end{bmatrix}$ and

$[(S \circ T)(x)]_{D \leftarrow B} = \begin{bmatrix} -0 - 2(1) \\ 0 - 1 \end{bmatrix} = \begin{bmatrix} -2 \\ -1 \end{bmatrix} \Rightarrow [S \circ T]_{D \leftarrow B} = \begin{bmatrix} -1 & -2 \\ 1 & -1 \end{bmatrix}$.

(b) $[T(1)]_{C \leftarrow B} = \begin{bmatrix} p(0) = 1 \\ p(1) = 1 \end{bmatrix} = \begin{bmatrix} 1 \\ 1 \end{bmatrix}$ and $[T(x))]_{C \leftarrow B} = \begin{bmatrix} p(0) = 0 \\ p(1) = 1 \end{bmatrix} = \begin{bmatrix} 0 \\ 1 \end{bmatrix} \Rightarrow$

$[T]_{C \leftarrow B} = \begin{bmatrix} 1 & 0 \\ 1 & 1 \end{bmatrix}$.

$[S \begin{bmatrix} 1 \\ 0 \end{bmatrix}]_{D \leftarrow C} = \begin{bmatrix} 1 - 2(0) \\ 2(1) - 0 \end{bmatrix} = \begin{bmatrix} 1 \\ 2 \end{bmatrix}$ and $S\begin{bmatrix} 0 \\ 1 \end{bmatrix}]_{D \leftarrow C} = \begin{bmatrix} 0 - 2(1) \\ 2(0) - 1 \end{bmatrix} = \begin{bmatrix} -2 \\ -1 \end{bmatrix} \Rightarrow$

$[S]_{D \leftarrow C} = \begin{bmatrix} 1 & -2 \\ 2 & -1 \end{bmatrix}$.

Therefore, $[S]_{D \leftarrow B} = [S]_{D \leftarrow C}[T]_{C \leftarrow B} = \begin{bmatrix} 1 & -2 \\ 2 & -1 \end{bmatrix}\begin{bmatrix} 1 & 0 \\ 1 & 1 \end{bmatrix} = \begin{bmatrix} -1 & -2 \\ 1 & -1 \end{bmatrix}$ as in (a).

19. In Exercise 1, since both bases were already standard, we have:

$[T]_{\mathcal{E}' \leftarrow \mathcal{E}} = \begin{bmatrix} 0 & 1 \\ -1 & 0 \end{bmatrix}$ invertible $\Rightarrow T$ invertible and

$[T^{-1}(a + bx)]_{\mathcal{E}' \leftarrow \mathcal{E}} = ([T]_{\mathcal{E}' \leftarrow \mathcal{E}})^{-1}\begin{bmatrix} a \\ b \end{bmatrix} = \begin{bmatrix} 0 & 1 \\ -1 & 0 \end{bmatrix}^{-1}\begin{bmatrix} a \\ b \end{bmatrix} = \begin{bmatrix} 0 & -1 \\ 1 & 0 \end{bmatrix}\begin{bmatrix} a \\ b \end{bmatrix} = \begin{bmatrix} -b \\ a \end{bmatrix} \Rightarrow$

$T^{-1}(a + bx) = -b + ax$.

21. In Exercise 3, C was not the standard basis, so we need to compute $[T]_{\mathcal{E}' \leftarrow \mathcal{E}}$ before we begin.

$[T(1)]_{\mathcal{E}} = [1]_{\mathcal{E}} = \begin{bmatrix} 1 \\ 0 \\ 0 \end{bmatrix}$, $[T(x)]_{\mathcal{E}} = [x + 2]_{\mathcal{E}} = \begin{bmatrix} 2 \\ 1 \\ 0 \end{bmatrix}$, $[T(x^2)]_{\mathcal{E}} = [(x+2)^2]_{\mathcal{E}} = \begin{bmatrix} 4 \\ 4 \\ 1 \end{bmatrix} \Rightarrow$

$[T]_{\mathcal{E}' \leftarrow \mathcal{E}} = \begin{bmatrix} 1 & 2 & 4 \\ 0 & 1 & 4 \\ 0 & 0 & 1 \end{bmatrix}$.

Therefore, $[T]_{\mathcal{E}' \leftarrow \mathcal{E}}$ invertible $\Rightarrow T$ invertible and

$[T^{-1}(p(x))]_{\mathcal{E} \leftarrow \mathcal{E}'} = [T^{-1}(a + bx + cx^2)]_{\mathcal{E} \leftarrow \mathcal{E}'} = ([T]_{\mathcal{E}' \leftarrow \mathcal{E}})^{-1}[a + bx + cx^2]_{\mathcal{E}' \leftarrow \mathcal{E}}$

$= \begin{bmatrix} 1 & 2 & 4 \\ 0 & 1 & 4 \\ 0 & 0 & 1 \end{bmatrix}^{-1} \begin{bmatrix} a \\ b \\ c \end{bmatrix} = \begin{bmatrix} 1 & -2 & 4 \\ 0 & 1 & -4 \\ 0 & 0 & 1 \end{bmatrix}\begin{bmatrix} a \\ b \\ c \end{bmatrix} = \begin{bmatrix} a - 2b + 4c \\ b - 4c \\ c \end{bmatrix} \Rightarrow$

$T^{-1}(a + bx + cx^2) = (a - 2b + 4c) + (b - 4c)x + cx^2 = a + b(x - 2) + c(x - 2)^2 \Rightarrow T^{-1}(p(x)) = p(x - 2)$.

23. $[T(1)]_{\mathcal{E}} = [1+0]_{\mathcal{E}} = \begin{bmatrix} 1 \\ 0 \\ 0 \end{bmatrix}$, $[T(x)]_{\mathcal{E}} = [x+1]_{\mathcal{E}} = \begin{bmatrix} 1 \\ 1 \\ 0 \end{bmatrix}$, $[T(x^2)]_{\mathcal{E}} = [x^2+2x]_{\mathcal{E}} = \begin{bmatrix} 0 \\ 2 \\ 1 \end{bmatrix} \Rightarrow$

$[T]_{\mathcal{E}' \leftarrow \mathcal{E}} = \begin{bmatrix} 1 & 1 & 0 \\ 0 & 1 & 2 \\ 0 & 0 & 1 \end{bmatrix} \Rightarrow [T]_{\mathcal{E}' \leftarrow \mathcal{E}}$ is invertible$\Rightarrow T$ is invertible and

$[T^{-1}(p(x))]_{\mathcal{E} \leftarrow \mathcal{E}'} = [T^{-1}(a+bx+cx^2)]_{\mathcal{E} \leftarrow \mathcal{E}'} = ([T]_{\mathcal{E}' \leftarrow \mathcal{E}})^{-1}[a+bx+cx^2]_{\mathcal{E}' \leftarrow \mathcal{E}}$

$= \begin{bmatrix} 1 & 1 & 0 \\ 0 & 1 & 2 \\ 0 & 0 & 1 \end{bmatrix}^{-1} \begin{bmatrix} a \\ b \\ c \end{bmatrix} = \begin{bmatrix} 1 & -1 & 2 \\ 0 & 1 & -2 \\ 0 & 0 & 1 \end{bmatrix} \begin{bmatrix} a \\ b \\ c \end{bmatrix} = \begin{bmatrix} a-b+2c \\ b-2c \\ c \end{bmatrix} \Rightarrow$

$T^{-1}(a+bx+cx^2) = (a-b+2c) + (b-2c)x + cx^2 = (a+bx+cx^2) - (b+2cx) + (2c) \Rightarrow$
$T^{-1}(p(x)) = p(x) - p'(x) + p''(x)$.

25. $\det([T]_{\mathcal{E}' \leftarrow \mathcal{E}}) = 0 \Rightarrow [T]_{\mathcal{E}' \leftarrow \mathcal{E}}$ is not invertible$\Rightarrow T$ is not invertible.

27. Let $\mathcal{E}' = \mathcal{E} = \{\sin x, \cos x\}$. Then Example 6.83 $\Rightarrow [\int (a \sin x + b \cos x) dx]_{\mathcal{E}'} = [D]_{\mathcal{E}' \leftarrow \mathcal{E}}^{-1} \begin{bmatrix} a \\ b \end{bmatrix}$.

Recall from Exercise 13 that $[D]_{\mathcal{E}' \leftarrow \mathcal{E}} = \begin{bmatrix} 0 & -1 \\ 1 & 0 \end{bmatrix} \Rightarrow$

$[\int (\sin x - 3\cos x) dx]_{\mathcal{E}'} = [D]_{\mathcal{E}' \leftarrow \mathcal{E}}^{-1} \begin{bmatrix} 1 \\ -3 \end{bmatrix} = \begin{bmatrix} 0 & 1 \\ -1 & 0 \end{bmatrix} \begin{bmatrix} 1 \\ -3 \end{bmatrix} = \begin{bmatrix} -3 \\ -1 \end{bmatrix} \Rightarrow$

$\int (\sin x - 3\cos x) dx = -3\sin x - \cos x + C$.

Directly, we have $(-3\sin x - \cos x)' = -3(\cos x) - (-\sin x) = \sin x - 3\cos x$.

29. Let $\mathcal{E}' = \mathcal{E} = \{e^{2x}, e^{2x}\cos x, e^{2x}\sin x\}$.

Then Example 6.83 $\Rightarrow [\int (a e^{2x} + b e^{2x}\cos x + c e^{2x}\sin x) dx]_{\mathcal{E}'} = [D]_{\mathcal{E}' \leftarrow \mathcal{E}}^{-1} \begin{bmatrix} a \\ b \\ c \end{bmatrix}$.

Recall from Exercise 15 that $[D]_{\mathcal{E}' \leftarrow \mathcal{E}} = \begin{bmatrix} 2 & 0 & 0 \\ 0 & 2 & 1 \\ 0 & -1 & 2 \end{bmatrix} \Rightarrow$

$[\int (e^{2x}\cos x - 2e^{2x}\sin x) dx]_{\mathcal{E}'} = [D]_{\mathcal{E}' \leftarrow \mathcal{E}}^{-1} \begin{bmatrix} 0 \\ 1 \\ -2 \end{bmatrix} = \begin{bmatrix} 1/2 & 0 & 0 \\ 0 & 2/5 & -1/5 \\ 0 & 1/5 & 2/5 \end{bmatrix} \begin{bmatrix} 0 \\ 1 \\ -2 \end{bmatrix} = \begin{bmatrix} 0 \\ 4/5 \\ -3/5 \end{bmatrix} \Rightarrow$

$\int (e^{2x}\cos x - 2e^{2x}\sin x) dx = \frac{4}{5} e^{2x}\cos x - \frac{3}{5} e^{2x}\sin x + C$.

Directly we have

$(\frac{4}{5} e^{2x}\cos x - \frac{3}{5} e^{2x}\sin x)' = \frac{4}{5}(2e^{2x}\cos x - e^{2x}\sin x) - \frac{3}{5}(e^{2x}\cos x + 2e^{2x}\sin x)$
$= e^{2x}\cos x - 2e^{2x}\sin x$.

6.6 The Matrix of a Linear Transformation

31. Let $\mathcal{E}' = \mathcal{E} = \left\{ \begin{bmatrix} 1 \\ 0 \end{bmatrix}, \begin{bmatrix} 0 \\ 1 \end{bmatrix} \right\}$.

Then $[T \begin{bmatrix} 1 \\ 0 \end{bmatrix}]_\mathcal{E} = [\begin{bmatrix} -4(0) \\ 1+5(0) \end{bmatrix}] = \begin{bmatrix} 0 \\ 1 \end{bmatrix}$ and $[T \begin{bmatrix} 0 \\ 1 \end{bmatrix}]_\mathcal{E} = [\begin{bmatrix} -4(1) \\ 0+5(1) \end{bmatrix}]_\mathcal{E} = \begin{bmatrix} -4 \\ 5 \end{bmatrix} \Rightarrow$

$[T]_\mathcal{E} = \begin{bmatrix} 0 & -4 \\ 1 & 5 \end{bmatrix}$.

Example 6.86b $\Rightarrow$ the eigenvectors of $[T]_\mathcal{E}$ will give us a basis $\mathcal{C}$.

Section 4.3 $\Rightarrow$ the eigenvalues of $[T]_\mathcal{E}$ are 4, 1 with corresponding eigenvectors

$\mathcal{C} = \left\{ \begin{bmatrix} -1 \\ 1 \end{bmatrix}, \begin{bmatrix} -4 \\ 1 \end{bmatrix} \right\}$.

33. Let $\mathcal{E}' = \mathcal{E} = \{1, x\}$.

Then $[T(1)]_\mathcal{E} = [(4(1) + 2(0)) + (1 + 3(0))x]_\mathcal{E} = [4 + x]_\mathcal{E} = \begin{bmatrix} 4 \\ 1 \end{bmatrix}$ and

$[T(x)]_\mathcal{E} = [(4(0) + 2(1)) + (0 + 3(1))x]_\mathcal{E} = [2 + 3x]_\mathcal{E} = \begin{bmatrix} 2 \\ 3 \end{bmatrix} \Rightarrow [T]_\mathcal{E} = \begin{bmatrix} 4 & 2 \\ 1 & 3 \end{bmatrix}$.

Example 6.86b $\Rightarrow$ the eigenvectors of $[T]_\mathcal{E}$ will give us a basis $\mathcal{C}$.

Section 4.3 $\Rightarrow$ the eigenvalues of $[T]_\mathcal{E}$ are 2, 5 with corresponding eigenvectors

$\begin{bmatrix} 1 \\ -1 \end{bmatrix}$ and $\begin{bmatrix} 2 \\ 1 \end{bmatrix} \Rightarrow \mathcal{C} = \{1 - x, 2 + x\}$.

35. Let $\mathcal{E}' = \mathcal{E} = \{1, x\}$.

Then $[T(1)]_\mathcal{E} = [(1 + (0)]_\mathcal{E} = [1]_\mathcal{E} = \begin{bmatrix} 1 \\ 0 \end{bmatrix}$ and

$[T(x)]_\mathcal{E} = [x + x(1)]_\mathcal{E} = [2x]_\mathcal{E} = \begin{bmatrix} 0 \\ 2 \end{bmatrix} \Rightarrow [T]_\mathcal{E} = \begin{bmatrix} 1 & 0 \\ 0 & 2 \end{bmatrix}$.

$\Rightarrow$ the eigenvalues of $[T]_\mathcal{E}$ are 1, 2 with corresponding eigenvectors

$\begin{bmatrix} 1 \\ 0 \end{bmatrix}$ and $\begin{bmatrix} 0 \\ 1 \end{bmatrix} \Rightarrow \mathcal{C} = \{1, x\}$.

37. Let T be reflection in line and let $\mathbf{d} = \begin{bmatrix} d_1 \\ d_2 \end{bmatrix}$ and $\mathbf{d}' = \begin{bmatrix} -d_2 \\ d_1 \end{bmatrix}$ as in Example 6.85.

Then $\mathcal{D} = \{\mathbf{d}, \mathbf{d}'\}$ is a basis for $\mathbb{R}^2$, again as noted in Example 6.85.

In particular, Example 6.85 $\Rightarrow \begin{bmatrix} -d_2 \\ d_1 \end{bmatrix}$ is orthogonal to line $\Rightarrow T\begin{bmatrix} -d_2 \\ d_1 \end{bmatrix} = \begin{bmatrix} d_2 \\ -d_1 \end{bmatrix}$.

Therefore, $[T(\mathbf{d})]_\mathcal{D} = \begin{bmatrix} 1 \\ 0 \end{bmatrix}$ and $[T(\mathbf{d}')]_\mathcal{D} = \begin{bmatrix} 0 \\ -1 \end{bmatrix} \Rightarrow [T]_\mathcal{D} = \begin{bmatrix} 1 & 0 \\ 0 & -1 \end{bmatrix}$. Furthermore:

$P_{\mathcal{E}\leftarrow\mathcal{D}} = \begin{bmatrix} d_1 & d_2 \\ d_2 & -d_1 \end{bmatrix} \Rightarrow P_{\mathcal{D}\leftarrow\mathcal{E}} = (P_{\mathcal{E}\leftarrow\mathcal{D}})^{-1} = \begin{bmatrix} \frac{d_1}{d_1^2+d_2^2} & \frac{d_2}{d_1^2+d_2^2} \\ \frac{d_2}{d_1^2+d_2^2} & -\frac{d_1}{d_1^2+d_2^2} \end{bmatrix} \Rightarrow P_{\mathcal{E}\leftarrow\mathcal{D}} = \begin{bmatrix} d_1 & d_2 \\ d_2 & -d_1 \end{bmatrix} \Rightarrow$

$[T]_\mathcal{E} = P_{\mathcal{E}\leftarrow\mathcal{D}}[T]_\mathcal{D} P_{\mathcal{D}\leftarrow\mathcal{E}} = \begin{bmatrix} d_1 & d_2 \\ d_2 & -d_1 \end{bmatrix}\begin{bmatrix} 1 & 0 \\ 0 & -1 \end{bmatrix}\begin{bmatrix} \frac{d_1}{d_1^2+d_2^2} & \frac{d_2}{d_1^2+d_2^2} \\ \frac{d_2}{d_1^2+d_2^2} & -\frac{d_1}{d_1^2+d_2^2} \end{bmatrix} = \begin{bmatrix} \frac{d_1^2-d_2^2}{d_1^2+d_2^2} & \frac{2d_1d_2}{d_1^2+d_2^2} \\ \frac{2d_1d_2}{d_1^2+d_2^2} & \frac{d_2^2-d_1^2}{d_1^2+d_2^2} \end{bmatrix}$.

39. Theorem 6.26 $\Rightarrow [T]_{\mathcal{C}\leftarrow\mathcal{B}} = [\,[T(\mathbf{v}_1)]\,|\,[T(\mathbf{v}_2)]\,|\,\ldots\,|\,[T(\mathbf{v}_n)]\,]$ that is the ith column of $[T]_{\mathcal{C}\leftarrow\mathcal{B}} = [T(\mathbf{v}_i)]$ where $\mathcal{B} = \{\mathbf{v}_1, \ldots, \mathbf{v}_n\}$.

We will show: $A[\mathbf{v}]_\mathcal{B} = [T(\mathbf{v})]_\mathcal{C} \Rightarrow \mathbf{a}_i = [T(\mathbf{v}_i)]$.

That is, the ith column of $A = [T(\mathbf{v}_i)] \Rightarrow A = [T]_{\mathcal{C}\leftarrow\mathcal{B}}$.

Note, $[\mathbf{v}_i]_\mathcal{B} = \mathbf{e}_i$ for all i by definition.

$A[\mathbf{v}]_\mathcal{B} = [T(\mathbf{v})]_\mathcal{C} \Rightarrow A[\mathbf{v}_i]_\mathcal{B} A \mathbf{e}_i = \mathbf{a}_i = [T(\mathbf{v}_i)]_\mathcal{C} \Rightarrow$ the ith column of $A = [T(\mathbf{v}_i)] \Rightarrow A = [T]_{\mathcal{C}\leftarrow\mathcal{B}}$ as required.

41. By Exercise 39, $A = [T]_{\mathcal{C}\leftarrow\mathcal{B}} \Leftrightarrow A[\mathbf{b}_i]_\mathcal{B} = A\mathbf{e}_i = \mathbf{a}_i = [T(\mathbf{b}_i)]_\mathcal{C}$.

Also, $\text{rank}(A) = \dim \text{col}(A)$, where $\text{col}(A) = \text{span}(\mathbf{a}_i : \mathbf{a}_i \text{ linearly independent})$ and
$\text{rank}(T) = \dim \text{range}(T)$, where $\text{range}(T) = \text{span}(T(\mathbf{b}_i) : T(\mathbf{b}_i) \text{ linearly independent})$.

We will show $\{\mathbf{a}_1, \ldots, \mathbf{a}_m\}$ is a basis for $\text{col}(A) \Leftrightarrow \{T(\mathbf{b}_1), \ldots, T(\mathbf{b}_m)\}$ is a basis for $\text{range}(T)$.

$\mathbf{a}_i = A[\mathbf{b}_i]_\mathcal{B} = [T(\mathbf{b}_i)]_\mathcal{C}$ linearly independent $\Leftrightarrow T(\mathbf{b}_i)$ linearly independent $\Rightarrow$
$\text{col}(A) = \text{span}(\mathbf{a}_1, \ldots, \mathbf{a}_m) \Leftrightarrow \text{range}(T) = \text{span}(T(\mathbf{b}_1), \ldots, T(\mathbf{b}_m)) \Rightarrow \text{rank}(T) = \text{rank}(A)$.

43. Section 3.4, Theorem 6.19 $\Rightarrow \dim V = \text{rank}([T]_{\mathcal{C}\leftarrow\mathcal{B}}) + \text{nullity}([T]_{\mathcal{C}\leftarrow\mathcal{B}})$.
So, Exercises 40 and 41 $\Rightarrow \dim V = \text{rank}(T) + \text{nullity}(T)$.

45. For $T \in \mathscr{L}(V, W)$, define $\varphi(T) = [T]_{\mathcal{C}\leftarrow\mathcal{B}} \in M_{nn}$. We will show φ is a linear transformation.
Recall, $[T]_{\mathcal{C}\leftarrow\mathcal{B}} = [\,[T\mathbf{b}_i]_\mathcal{C}\,]$, where $[\,[T\mathbf{b}_i]_\mathcal{C}\,] = [\,[T\mathbf{b}_1]_\mathcal{C}\,|\,\ldots\,|\,[T\mathbf{b}_n]_\mathcal{C}\,]$.
Then, $\varphi(S + T) = [\,[(S + T)\mathbf{b}_i]_\mathcal{C}\,] = [\,[S\mathbf{b}_i]_\mathcal{C} + [T\mathbf{b}_i]_\mathcal{C}\,] = [\,[S\mathbf{b}_i]_\mathcal{C}\,] + [\,[T\mathbf{b}_i]_\mathcal{C}\,] = \varphi(S) + \varphi(T)$.
Likewise, $\varphi(cT) = [\,[cT\mathbf{b}_i]_\mathcal{C}\,] = [\,c[T\mathbf{b}_i]_\mathcal{C}\,] = c[\,[T\mathbf{b}_i]_\mathcal{C}\,] = c\varphi(T)$.

To show φ is an isomorphism, we need only show that $\ker(\varphi) = \mathbf{0} \Rightarrow \text{nullity}(T) = n$ since the Rank Theorem then implies $\ker(T) = V$, that is $T\mathbf{v} = \mathbf{0}$ for all $\mathbf{v}$ so $T = \mathbf{0}$, the zero linear transformation.

$\varphi(T) = \mathbf{0} \Rightarrow [T]_{\mathcal{C}\leftarrow\mathcal{B}} = [\mathbf{0}]_{nn} \Rightarrow \text{null}(T) = \mathbb{R}^n \Rightarrow \text{nullity}([T]_{\mathcal{C}\leftarrow\mathcal{B}}) = n = \text{nullity}(T)$.

Therefore, φ is one-to-one and hence onto since $\dim \mathscr{L}(V, W) = n^2 = \dim M_{nn} \Rightarrow$
φ is an isomorphism $\Rightarrow \mathscr{L}(V, W) \cong M_{nn}$.

Exploration: Tilings, Lattices, and Crystallographic Restriction

Explorations are self-contained, so only odd-numbered solutions will be provided.

1. Let $P =$ the pattern in Figure 6.15.
   We are given $P + \mathbf{u} = P$ and $P + \mathbf{v} = P \Rightarrow P + a\mathbf{u} = P$ and $P + b\mathbf{v} = P$.
   Therefore, $P + a\mathbf{u} + b\mathbf{v} = (P + a\mathbf{u}) + b\mathbf{v} = P + b\mathbf{v} = P$.

3. Let $\text{rot} P_O^\theta = P$ rotated by $\theta > 0$ through rotation center O.
   Then $\text{rot} P_O^{n\theta} = \text{rot} P_O^\theta \; n$ times $= P$ (because $\text{rot} P_O^\theta = P$ every time). Likewise for $-\theta$.
   We need to show that if $0 < \theta \leq 360°$, then $\frac{360}{\theta} = n$ an integer, i.e. $n\theta = 360°$.
   Clearly, $\text{rot} P_O^{360°} = P$ and $\text{rot} P_O^{360°+\varphi} = \text{rot} P_O^\varphi$.
   Let $\theta' = \min \{\varphi : \text{rot} P_O^\varphi = P, 0 < \varphi \leq 360°\}$ and $m = \min \{k : k\theta' \geq 360, k \text{ a positive integer}\}$.
   Now let $n\theta' = 360° + \psi$. We will show $\psi < \theta'$ which will imply $\psi = 0°$.
   $m\theta' = 360° + \psi \Rightarrow m\theta' - \psi = 360° > (m-1)\theta'$ (because m is the minimum) $\Rightarrow$
   $m\theta' - \psi > m\theta' - \theta' \Rightarrow -\psi > -\theta' \Rightarrow \psi < \theta'$.
   So, since $P = \text{rot} P_O^{m\theta'} = \text{rot} P_O^{360°+\psi} = \text{rot} P_O^\psi$, we have $\psi = 0°$ (because θ' is the minimum).
   Therefore, $m\theta' = 360° \Rightarrow \frac{360}{\theta'} = m$ an integer.
   Now if θ, $0 < \theta \leq 360°$ and $\text{rot} P_O^\theta = P$ we will show that $\theta = n\theta'$ to complete the proof.
   Let $n = \max \{k : \theta - k\theta' \geq 0, k \text{ a positive integer}\}$.
   Now let $\theta - n\theta' = \psi$. We will show $\psi < \theta'$ which will imply $\psi = 0°$.
   $\theta - n\theta' = \psi \Rightarrow \theta - n\theta' - \psi > \theta - (n+1)\theta'$ (because n is the maximum) $\Rightarrow$
   $\theta - n\theta' - \psi > \theta - n\theta' - \theta' \Rightarrow -\psi > -\theta' \Rightarrow \psi < \theta'$.
   So, since $P = \text{rot} P_O^{\theta-n\theta'} = \text{rot} P_O^\psi$, we have $\psi = 0°$ (because θ' is the minimum).
   Therefore, $\theta - n\theta' = 0 \Rightarrow \theta = n\theta' \Rightarrow \frac{\theta}{\theta'} = n \Rightarrow \frac{360}{\theta}$ is an integer.

5. It is impossible to draw a lattice with 8-fold symmetry as we will show below.

7. Since the lattice is invariant under R_θ, $\mathbf{u}$ and $\mathbf{v}$ must mapped to lattice points. So:
   $$[R_\theta(\mathbf{u})]_\mathcal{B} = [a\mathbf{u} + c\mathbf{v}]_\mathcal{B} = \begin{bmatrix} a \\ c \end{bmatrix} \text{ and } [R_\theta(\mathbf{v})]_\mathcal{B} = [b\mathbf{u} + d\mathbf{v}]_\mathcal{B} = \begin{bmatrix} b \\ d \end{bmatrix} \Rightarrow [R_\theta]_\mathcal{B} = \begin{bmatrix} a & b \\ c & d \end{bmatrix}.$$

9. From 8, $2\cos\theta =$ an integer $\Rightarrow \theta = 0°, 60°, 90°, 120°, 180°, 240°, 270°, 300°,$ and $360°$.
   Now relating these to the integers n, where $n = \frac{360}{\theta}$, we have:
   $n = \frac{360}{360} = 1$, $n = \frac{360}{180} = 2$, $n = \frac{360}{120} = 3$, $n = \frac{360}{90} = 4$, and $n = \frac{360}{60} = 6$.
   Note the other θ listed above do not give integer results when divided into 360.

6.7 Applications

Q: What are our main goals for Section 6.7?
A: To see how subspaces apply to *differential equations* and *codes*.
Since these applications are fully explained in the text, we simply present solutions here.

Key Definitions and Concepts

differential equation	p.522	6.7	See detailed remarks before Thm 6.31. Here we show the equations for *first* and *second* order. *first-order*: $y' + ay = 0$ (homogeneous, because $= 0$). Solution: $y = e^{-at}$ (see the statement of Thm 6.31) *second-order*: $y'' + ay' + by = 0$ (homogeneous). Solution: $y = c_1 e^{\lambda_1 t} + c_2 e^{\lambda_2 t}$ (see the statement of Thm 6.31) Where: λ_1 and λ_2 are solutions of $\lambda^2 + a\lambda + b = 0$
linear code	p.529	6.7	A *p*-ary *linear code* is a subspace of C of $\mathbb{Z}_p^n$.
Reed-Muller	p.532	6.7	The (first order) *Reed-Muller codes* R_n (see details, p.532).

Theorems

Thm 6.31	p.522	6.7	The set S of all solutions to $y' + ay = 0$ is a subspace of $\Im F$.
Thm 6.32	p.523	6.7	If S is the solution space of $y' + ay = 0$, then: dim $= 1$ and $\{e^{-at}\}$ is a basis for S.
Thm 6.33	p.526	6.7	Let S be the solution space of $y'' + ay' + by = 0$ and let λ_1 and λ_2 be the roots of $\lambda^2 + a\lambda + b = 0$. a. If $\lambda_1 \neq \lambda_2$, then $\{e^{\lambda_1 t}, e^{\lambda_2 t}\}$ is a basis for S. b. If $\lambda_1 = \lambda_2$, then $\{e^{\lambda_1 t}, te^{\lambda_1 t}\}$ is a basis for S.
Thm 6.34	p.531	6.7	Let C be an (n, k) linear code. a. The dual code $C^\perp$ is an $(n, n-k)$ linear code. b. C contains 2^k vectors, and $C^\perp$ contains 2^{n-k} vectors.
Thm 6.35	p.533	6.7	For $n \geq 1$, the *Reed-Muller code* R_n is: a $(2^n, n+1)$ linear code in which every vector (except **0** and **1**) has weight 2^{n-1}.

Solutions to odd-numbered exercises from Section 6.7

1. Since $a = -3$, Theorem 6.32 $\Rightarrow \{e^{3t}\}$ is a basis $\Rightarrow$ The solution is of the form $y(t) = ce^{3t}$.
   So the boundary condition $y(1) = 2 \Rightarrow y(1) = ce^3 = 2 \Rightarrow c = \frac{2}{e^3} \Rightarrow y(t) = \frac{2e^{3t}}{e^3}$.

3. $y'' - 7y' + 12y = 0$ corresponds to the characteristic equation $\lambda^2 - 7\lambda + 12 = 0$.
   Since $\lambda_1 = 3, \lambda_2 = 4$, Theorem 6.33 $\Rightarrow$ solution is of the form $y(t) = c_1 e^{3t} + c_2 e^{4t}$.
   $\begin{array}{l} y(0) = 1 \\ y(1) = 1 \end{array} \Rightarrow \begin{array}{l} c_1 + c_2 = 1 \\ c_1 e^3 + c_2 e^4 = 1 \end{array} \Rightarrow \begin{array}{l} c_1 = \frac{1-e^4}{e^3-e^4} \\ c_2 = \frac{e^3-1}{e^3-e^4} \end{array} \Rightarrow y(t) = \frac{1}{e^3-e^4}((1-e^4)e^{3t} + (e^3-1)e^{4t})$.

5. $f'' - f' - f = 0$ corresponds to the characteristic equation $\lambda^2 + \lambda - 12 = 0$.
   Since $\lambda_1 = \frac{1+\sqrt{5}}{2}, \lambda_2 = \frac{1-\sqrt{5}}{2}$, Theorem 6.33 $\Rightarrow f(t) = c_1 e^{(1+\sqrt{5})t/2} + c_2 e^{(1-\sqrt{5})t/2}$.
   $\begin{array}{l} f(0) = 0 \\ f(1) = 1 \end{array} \Rightarrow \begin{array}{l} c_1 + c_2 = 0 \\ c_1 e^{(1+\sqrt{5})/2} + c_2 e^{(1-\sqrt{5})/2} = 1 \end{array} \Rightarrow \begin{array}{l} c_1 = \frac{e^{(\sqrt{5}-1)/2}}{e^{\sqrt{5}}-1} \\ c_2 = -\frac{e^{(\sqrt{5}-1)/2}}{e^{\sqrt{5}}-1} \end{array} \Rightarrow$
   $x(t) = \frac{e^{(\sqrt{5}-1)/2}}{e^{\sqrt{5}}-1}(e^{(1+\sqrt{5})t/2} - e^{(1-\sqrt{5})t/2})$.

7. $y'' - 2y' + y = 0$ corresponds to the characteristic equation $\lambda^2 - 2\lambda + 1 = 0$.
   Since $\lambda_1 = \lambda_2 = 1$, Theorem 6.33 $\Rightarrow y(t) = c_1 e^t + c_2 t e^t$.
   $\begin{array}{l} y(0) = 1 \\ y(1) = 1 \end{array} \Rightarrow \begin{array}{l} c_1 + c_2 = 1 \\ c_1 e + c_2 e = 1 \end{array} \Rightarrow \begin{array}{l} c_1 = 1 \\ c_2 = -(1 - e^{-1}) \end{array} \Rightarrow y(t) = e^t - (1 - e^{-1})t e^t$.

9. $y'' - k^2 y = 0$ corresponds to the characteristic equation $\lambda^2 - k^2 = 0$.
   Since $\lambda_1 = k, \lambda_2 = -k$, Theorem 6.33 $\Rightarrow y(t) = c_1 e^{kt} + c_2 e^{-kt}$.
   $\begin{array}{l} y(0) = 1 \\ y'(0) = 1 \end{array} \Rightarrow \begin{array}{l} c_1 + c_2 = 1 \\ kc_1 - kc_2 = 1 \end{array} \Rightarrow \begin{array}{l} c_1 = \frac{k+1}{2k} \\ c_2 = \frac{k-1}{2k} \end{array} \Rightarrow y(t) = \frac{1}{2k}((k+1)e^{kt} + (k-1)e^{-kt})$.

11. $f'' - 2f' + 5f = 0$ corresponds to the characteristic equation $\lambda^2 - 2\lambda + 5 = 0$.
    Since $\lambda_1 = 1 + 2i, \lambda_2 = 1 - 2i$, Theorem 6.33 $\Rightarrow f(t) = c_1 e^t \cos(2t) + c_2 e^t \sin(2t)$.
    $\begin{array}{l} f(0) = 1 \\ f(\frac{\pi}{4}) = 0 \end{array} \Rightarrow \begin{array}{l} c_1 = 1 \\ c_2 = 0 \end{array} \Rightarrow f(t) = e^t \cos(2t)$.

13. (a) $p(t) = ce^{kt} \Rightarrow p(0) = ce^{0k} = 100 \Rightarrow c = 100 \Rightarrow p(t) = 100e^{kt}$.
    So, $p(3) = 100e^{3k} = 1600 \Rightarrow e^{3k} = 16 \Rightarrow k = \frac{\ln 16}{3} \Rightarrow p(t) = 100e^{(\ln 16)t/3}$.

    (b) Double $\Rightarrow p(t) = 100e^{(\ln 16)t/3} = 2 \cdot 100 = 200 \Rightarrow e^{(\ln 16)t/3} = 2 \Rightarrow$
    $t = \frac{\ln 2}{(\ln 16)/3} = \frac{3\ln 2}{\ln 16} = \frac{3\ln 2}{\ln 2^4} = \frac{3\ln 2}{4\ln 2} = \frac{3}{4}$ of an hour = 45 minutes.

    (c) One million = $10^6 \Rightarrow p(t) = 100e^{(\ln 16)t/3} = 10^6 \Rightarrow e^{(\ln 16)t/3} = 10^4 \Rightarrow$
    $t = \frac{\ln 10^4}{(\ln 16)/3} = \frac{12\ln 10}{4\ln 2} = \frac{3\ln 10}{\ln 2} \approx 9.968$ hours.

6.7 Applications

15. (a) $m(0) = a e^{-ct} \Rightarrow m(0) = a e^{-0c} = 50 \Rightarrow a = 50 \Rightarrow m(t) = 50 e^{-ct}$.
    Half-life $= 1590$ years $\Rightarrow m(1590) = 50 e^{-1590c} = \frac{1}{2} 50 = 25 \Rightarrow e^{-1590c} = \frac{1}{2} \Rightarrow$
    $c = -\frac{\ln(1/2)}{1590} \Rightarrow m(t) = 50 e^{-ct}$ where $c = -\frac{\ln(1/2)}{1590}$.
    Therefore, $m(1000) = 50 e^{-1000c} \approx 32.33$ mg will be left after 1000 years.
    (b) $m(t) = 50 e^{-ct} = 10 \Rightarrow e^{-ct} = \frac{1}{5} \Rightarrow t = \frac{\ln(1/5)}{-c} = \frac{\ln(1/5)}{\ln(1/2)/1590} = \frac{1590 \ln 5}{\ln 2} \approx 3692$ years.

17. The following analysis comes directly from Section 6.7, Example 6.92.
    Length $= 10$ when $t = 0 \Rightarrow 10 = x(0) = c_1 \cos 0 + c_2 \sin 0 = c_1$.
    Length $= 5$ when $t = 10 \Rightarrow 5 = x(10) = 10 \cos 10\sqrt{K} + c_2 \sin 10\sqrt{K} \Rightarrow$
    $c_2 = \frac{5 - 10 \cos(10\sqrt{K})}{\sin(10\sqrt{K})} \Rightarrow x(t) = \frac{5 - 10 \cos(10\sqrt{K})}{\sin(10\sqrt{K})} \sin \sqrt{K} t + 10 \cos \sqrt{K} t$.

19. The following comes from Example 6.92. Recall, the period of $\sin bt$ and $\cos bt = \frac{2\pi}{b}$.
    (a) Since $\theta'' + \frac{g}{L} \theta = 0$, Example 6.92 $\Rightarrow K = \frac{g}{L} \Rightarrow$
    $\theta(t) = c_1 \cos \sqrt{K} t + c_1 \sin \sqrt{K} t = c_1 \cos \sqrt{\frac{g}{L}} t + c_1 \sin \sqrt{\frac{g}{L}} t \Rightarrow$
    Period, $P = \frac{2\pi}{b} \Rightarrow P = \frac{2\pi}{\sqrt{g/L}} = 2\pi \sqrt{\frac{L}{g}}$.
    In this case, since $L = 1$, the period is $2\pi \sqrt{\frac{1}{g}} = \frac{2\pi}{\sqrt{g}}$.
    (b) θ_1 does not appear in the formula, therefore, the period does not depend on θ_1.
    Furthermore, since g is constant, the period depends only upon the length, L.

21. As stated in the proof of Theorem 6.33a, $\dim S = 2 \Rightarrow$ we need only show $\operatorname{span}(e^{\lambda t}, t e^{\lambda t}) \subseteq S$
    and $e^{\lambda t}, t e^{\lambda t}$ linearly independent $\Rightarrow \operatorname{span}(e^{\lambda t}, t e^{\lambda t}) = S \Rightarrow \{e^{\lambda t}, t e^{\lambda t}\}$ is a basis for S.
    $\operatorname{span}(e^{\lambda t}) \subseteq S$ by proof of Section 6.7, Theorem 6.33b. We will show $\operatorname{span}(t e^{\lambda t}) \subseteq S$.
    Note, λ a solution of $\lambda^2 + a\lambda + b = 0 \Rightarrow \lambda = \frac{-a \pm \sqrt{a^2 - 4b}}{2}$. Now $\lambda_1 = \lambda_2 = \lambda \Rightarrow$
    $a^2 - 4b = 0 \Rightarrow \lambda = -\frac{1}{2} a \Rightarrow 2\lambda + a = 0$. We will use this to show $\operatorname{span}(t e^{\lambda t}) \subseteq S$.
    Let $f(t) = t e^{\lambda t} \Rightarrow f' = e^{\lambda t} + \lambda t e^{\lambda t}$, and $f'' = 2\lambda e^{\lambda t} + \lambda^2 t e^{\lambda t} \Rightarrow$
    $f'' + af' + b = (2\lambda + a)e^{\lambda t} + (\lambda^2 + a\lambda + b) t e^{\lambda t} = 0 + 0 = 0 \Rightarrow \operatorname{span}(t e^{\lambda t}) \subseteq S$.
    $e^{\lambda t}, t e^{\lambda t}$ linearly independent since $c_1 e^{\lambda t} + c_2 t e^{\lambda t} = 0 \Rightarrow$ when $t = 0$, $c_1 = 0 \Rightarrow c_2 = 0$.
    Therefore, $\operatorname{span}(e^{\lambda t}, t e^{\lambda t}) \subseteq S$ and $\dim S = 2 \Rightarrow \{e^{\lambda t}, t e^{\lambda t}\}$ is a basis for S.

23. No, because $\mathbf{c}_1 + \mathbf{c}_2 = \begin{bmatrix} 1 \\ 0 \end{bmatrix} + \begin{bmatrix} 0 \\ 1 \end{bmatrix} = \begin{bmatrix} 0 \\ 0 \end{bmatrix} \notin C$.

25. No, because $\mathbf{c}_1 + \mathbf{c}_3 = \begin{bmatrix} 1 \\ 0 \\ 1 \end{bmatrix} + \begin{bmatrix} 1 \\ 1 \\ 0 \end{bmatrix} = \begin{bmatrix} 0 \\ 1 \\ 1 \end{bmatrix} \notin C$.

27. Yes, since $\mathbf{c}_i + \mathbf{c}_i = \mathbf{0} \in C$, $\mathbf{c}_1 + \mathbf{c}_2 = \mathbf{c}_3 \in C$, and $a \mathbf{c}_i = \mathbf{0}$, $\mathbf{c}_i \in C$.

29. Yes, since $w(\mathbf{x} + \mathbf{y}) = w(\mathbf{x}) + w(\mathbf{y}) - 2\, \text{overlap} \Rightarrow w(\mathbf{x} + \mathbf{y})$ is even.
    $w(\mathbf{x})$ even $\Rightarrow cw(\mathbf{x})$ is even.

31. $c_2 + c_4 = c_3 \Rightarrow \text{span}(c_2, c_4) = \text{span}(c_2, c_3) \Rightarrow \{c_2, c_4\}$ is a basis.

  $c_3 + c_4 = c_2 \Rightarrow \text{span}(c_3, c_4) = \text{span}(c_2, c_3) \Rightarrow \{c_3, c_4\}$ is a basis.

33. The proof of Theorem 6.34a $\Rightarrow \dim(C^\perp)^\perp = \dim C$. Therefore, Section 6.2, Theorem 6.11 $\Rightarrow$ We need only show $C \subseteq (C^\perp)^\perp \Rightarrow C = (C^\perp)^\perp$.

  $c \in C, c' \in C^\perp \Rightarrow c \cdot c' = 0 \Rightarrow c \in (C^\perp)^\perp$ by definition $\Rightarrow C \subseteq (C^\perp)^\perp \Rightarrow C = (C^\perp)^\perp$.

35. We will go through the inductive construction of the bases for R_1, R_2, and R_3 from R_0.

  $R_0 = \{0, 1\} \Rightarrow 1 = 1$ is a basis for $R_0 \Rightarrow$

  Basis for $R_1 = \left\{ 1 = \begin{bmatrix} 1 \\ 1 \end{bmatrix}, u_{01} = \begin{bmatrix} 0 \\ 1 \end{bmatrix} = \begin{bmatrix} 0 \\ 1 \end{bmatrix} \right\} \Rightarrow$

  Basis for $R_2 = \left\{ 1 = \begin{bmatrix} 1 \\ 1 \\ 1 \\ 1 \end{bmatrix}, u_{01|01} = \begin{bmatrix} u_{01} \\ u_{01} \end{bmatrix} = \begin{bmatrix} 0 \\ 1 \\ 0 \\ 1 \end{bmatrix}, u_{00|11} = \begin{bmatrix} u_{00} \\ u_{11} \end{bmatrix} = \begin{bmatrix} 0 \\ 0 \\ 1 \\ 1 \end{bmatrix} \right\} \Rightarrow$

  Basis for $R_3 = \left\{ 1 = \begin{bmatrix} 1 \\ 1 \\ 1 \\ 1 \\ 1 \\ 1 \\ 1 \\ 1 \end{bmatrix}, u_{0101|0101} = \begin{bmatrix} 0 \\ 1 \\ 0 \\ 1 \\ 0 \\ 1 \\ 0 \\ 1 \end{bmatrix}, u_{0011|0011} = \begin{bmatrix} 0 \\ 0 \\ 1 \\ 1 \\ 0 \\ 0 \\ 1 \\ 1 \end{bmatrix}, u_{0000|1111} = \begin{bmatrix} 0 \\ 0 \\ 0 \\ 0 \\ 1 \\ 1 \\ 1 \\ 1 \end{bmatrix} \right\} \Rightarrow$

  $R_3 = \left\{ \begin{bmatrix} 0 \\ 0 \\ 0 \\ 0 \\ 0 \\ 0 \\ 0 \\ 0 \end{bmatrix}, \begin{bmatrix} 0 \\ 0 \\ 0 \\ 0 \\ 1 \\ 1 \\ 1 \\ 1 \end{bmatrix}, \begin{bmatrix} 0 \\ 0 \\ 1 \\ 1 \\ 0 \\ 0 \\ 1 \\ 1 \end{bmatrix}, \begin{bmatrix} 0 \\ 0 \\ 1 \\ 1 \\ 1 \\ 1 \\ 0 \\ 0 \end{bmatrix}, \begin{bmatrix} 0 \\ 1 \\ 0 \\ 1 \\ 0 \\ 1 \\ 0 \\ 1 \end{bmatrix}, \begin{bmatrix} 0 \\ 1 \\ 0 \\ 1 \\ 1 \\ 0 \\ 1 \\ 0 \end{bmatrix}, \begin{bmatrix} 0 \\ 1 \\ 1 \\ 0 \\ 0 \\ 1 \\ 1 \\ 0 \end{bmatrix}, \begin{bmatrix} 0 \\ 1 \\ 1 \\ 0 \\ 1 \\ 0 \\ 0 \\ 1 \end{bmatrix}, \begin{bmatrix} 1 \\ 0 \\ 0 \\ 0 \\ 1 \\ 0 \\ 1 \\ 0 \end{bmatrix}, \begin{bmatrix} 1 \\ 0 \\ 0 \\ 0 \\ 0 \\ 1 \\ 0 \\ 1 \end{bmatrix}, \begin{bmatrix} 1 \\ 0 \\ 0 \\ 1 \\ 1 \\ 1 \\ 0 \\ 0 \end{bmatrix}, \begin{bmatrix} 1 \\ 0 \\ 0 \\ 1 \\ 0 \\ 0 \\ 1 \\ 1 \end{bmatrix}, \begin{bmatrix} 1 \\ 1 \\ 0 \\ 0 \\ 1 \\ 1 \\ 0 \\ 0 \end{bmatrix}, \begin{bmatrix} 1 \\ 1 \\ 0 \\ 0 \\ 0 \\ 0 \\ 1 \\ 1 \end{bmatrix}, \begin{bmatrix} 1 \\ 1 \\ 1 \\ 1 \\ 0 \\ 0 \\ 0 \\ 0 \end{bmatrix}, \begin{bmatrix} 1 \\ 1 \\ 1 \\ 1 \\ 1 \\ 1 \\ 1 \\ 1 \end{bmatrix} \right\}$

6.7 Applications

37. Putting G_2 into standard form $\left[\dfrac{A}{I}\right] \Rightarrow G_2 = \begin{bmatrix} 1 & 0 & 0 \\ 1 & 1 & 0 \\ 1 & 0 & 1 \\ 1 & 1 & 1 \end{bmatrix} \longrightarrow \begin{bmatrix} 1 & 0 & 0 \\ 0 & 1 & 0 \\ 0 & 0 & 1 \\ 1 & 1 & 1 \end{bmatrix} \Rightarrow$

$A = [1\,1\,1] \Rightarrow$ The associated parity check matrix $P = [I\,|\,A] = [1\,|\,1\,1\,1] = [1\,1\,1\,1]$.

39. As suggested, we will show $O' = \{\mathbf{c}_0 + \mathbf{e} : \mathbf{e} \in E\} = O$, vectors in C with odd weight.

$w(\mathbf{c}_0 + \mathbf{e}) = w(\mathbf{c}_0) + w(\mathbf{e}) - 2\,\text{overlap} = \text{odd} \Rightarrow \mathbf{c}_0 + \mathbf{e} \in O \Rightarrow O' \subseteq O$.

Recall, for any vector $\mathbf{c} \in C$, $\mathbf{c} + \mathbf{c} = 2\mathbf{c} = \mathbf{0}$. In particular, $\mathbf{c}_0 + \mathbf{c}_0 = \mathbf{0}$.
Also, note $w(\mathbf{c}_0 + \mathbf{o}) = w(\mathbf{c}_0) + w(\mathbf{o}) - 2\,\text{overlap} = \text{even} \Rightarrow \mathbf{c}_0 + \mathbf{o} \in E$.
We will use these facts to show $O \subseteq O' \Rightarrow O' = O$.

$\mathbf{o} \in O \Rightarrow \mathbf{o} = \mathbf{0} + \mathbf{o} = (\mathbf{c}_0 + \mathbf{c}_0) + \mathbf{o} = \mathbf{c}_0 + (\mathbf{c}_0 + \mathbf{o}) \Rightarrow \mathbf{o} \in O' \Rightarrow O \subseteq O' \Rightarrow O' = O$.

This shows either all the code vectors in C have even weight or exactly half of them do.

Chapter 6 Review

Key Definitions and Concepts

This list includes most but not all of the definitions listed at the end of each section. We begin the list with key symbols and symbol-based definitions.

V	p.433	6.1	V often used to denote a *vector space*
W	p.438	6.1	W often used to denote a *subspace*
span(S)	p.442	6.1	If $S = \{\mathbf{v}_i\}$, then span(S) equals all linear combinations of $\mathbf{v}_i$.
$\{\mathbf{e}_i\}$	p.451	6.2	$\{\mathbf{e}_i\}$, standard basis for $\mathbb{R}^n$ (see Example 6.29)
$\{x^i\}$	p.451	6.2	$\{x^i\}$, standard basis for $\mathscr{P}_n$ (see Example 6.30)
$\{E_{ij}\}$	p.451	6.2	$\mathcal{E} = \{E_{ij}\}$, standard basis for M_{mn} (see Example 6.31)
$[\mathbf{v}]_\mathcal{B}$	p.453	6.2	If $\mathcal{B} = \{\mathbf{u}_i\}$ is a basis and $\mathbf{x} = \sum_{i=1}^{n} c_i \mathbf{u}_i$, then $[\mathbf{v}]_\mathcal{B} = \begin{bmatrix} c_1 \\ \vdots \\ c_n \end{bmatrix}$.
$P_{\mathcal{C} \leftarrow \mathcal{B}}$	p.469	6.3	*change-of-basis* matrix: $P_{\mathcal{C} \leftarrow \mathcal{B}} = \begin{bmatrix} [\mathbf{u}_1]_\mathcal{C} & \cdots & [\mathbf{u}_n]_\mathcal{C} \end{bmatrix}$
$\mathscr{C}[a,b]$	p.477	6.4	The space of continuous functions (See Example 6.52).
$\int_a^b f(x)\,dx$	p.477	6.4	$S: \mathscr{C}[a,b] \to \mathbb{R}$, $S(f) = \int_a^b f(x)\,dx$ (See Example 6.52).
I	p.478	6.4	$I: V \to W$ maps every vector to itself. That is, $I(\mathbf{v}) = \mathbf{v}$ (See Example 6.54).
$S \circ T$	p.481	6.4	If $T: U \to V$ and $S: V \to W$ then the *composition of S with T* is $(S \circ T)(\mathbf{u}) = S(T(\mathbf{u}))$.
T_0	p.478	6.4	$T_0: V \to W$ maps every vector to the zero vector. That is, $T_0(\mathbf{v}) = \mathbf{0}$ (See Example 6.54).
T^{-1}	p.483	6.4	T^{-1}: $T^{-1} \circ T = I_V$, $T \circ T^{-1} = I_W$ (See *invertible*).

Key Definitions and Concepts, symbols, continued

$\ker(T)$	p.486	6.5	$\ker(T) = \{\mathbf{v} \text{ in } V : T(\mathbf{v}) = \mathbf{0}\}$ (See *kernel*.)
nullity(T)	p.488	6.5	nullity$(T) = \dim(\ker(T))$ (See *nullity*.)
range(T)	p.486	6.5	range$(T) = \{\mathbf{w} \text{ in } W : \mathbf{w} = T(\mathbf{v}) \text{ for some } \mathbf{v} \text{ in } V\}$ (See *range*.)
rank(T)	p.488	6.5	rank$(T) = \dim(\text{range}(T))$ (See *rank*.)
$V \cong W$	p.497	6.5	If $T : V \to W$ where T is an *isomorphism*. Then: V and W are *isomorphic*. This is written $V \cong W$.
$[T]_{\mathcal{C} \leftarrow \mathcal{B}}$	p.502	6.6	The matrix of T with respect to bases $\mathcal{B}$ and $\mathcal{C}$. By Thm 6.26 it satisfies: $[T]_{\mathcal{C} \leftarrow \mathcal{B}}[\mathbf{v}]_\mathcal{B} = [T(\mathbf{v})]_\mathcal{C}$ See remarks following Thm 6.26.
$[T]_\mathcal{B}$	p.502	6.6	Special case when $V = W$ and $\mathcal{B} = \mathcal{C}$. By Thm 6.26 it satisfies: $[T]_\mathcal{B}[\mathbf{v}]_\mathcal{B} = [T(\mathbf{v})]_\mathcal{B}$ See remarks following Thm 6.26.
$[T(\mathbf{v})]_\mathcal{C}$	p.502	6.6	$A[\mathbf{v}]_\mathcal{B} = [T(\mathbf{v})]_\mathcal{C}$ We are asked to show this in Exercise 39. See remarks following Thm 6.26.

Chapter 6 Review

Key Definitions and Concepts, continued

basis	p.450	6.2	A subset $\mathcal{B}$ of a vector space V is a *basis* for V if 1. $\mathcal{B}$ spans V, that is span$(\mathcal{B}) = V$ 2. $\mathcal{B}$ is linearly independent
change-of-basis	p.469	6.3	*change-of-basis* matrix: $P_{\mathcal{C}\leftarrow\mathcal{B}} = \begin{bmatrix} [\mathbf{u}_1]_\mathcal{C} & \cdots & [\mathbf{u}_n]_\mathcal{C} \end{bmatrix}$
commutative diagram	p.503	6.6	See detailed remarks following Thm 6.26. The diagram referred to is on p.502.
composition of S with T	p.481	6.4	If $T: U \to V$ and $S: V \to W$ then the *composition of S with T* is the mapping $S \circ T$: $(S \circ T)(\mathbf{u}) = S(T(\mathbf{u}))$.
coordinate vector	p.453	6.2	If $\mathcal{B} = \{\mathbf{u}_i\}$ is a basis and $\mathbf{x} = \sum_{i=1}^{n} c_i \mathbf{u}_i$, then $[\mathbf{v}]_\mathcal{B} = \begin{bmatrix} c_1 \\ \vdots \\ c_n \end{bmatrix}$.
diagonalizable	p.513	6.6	Let V be a finite dimensional vector space. Let $T: V \to V$ be a linear transformation. Then: T is called *diagonalizable* if there is a basis $\mathcal{C}$ such that $[T]_\mathcal{C}$ is diagonal matrix.
differential operator	p.477	6.4	The *differential operator* is defined as follows: $D: \mathscr{D} \to \mathscr{F}$, $D(f) = f'$ (See Example 6.51).
differential equation	p.522	6.7	See detailed remarks before Thm 6.31. Here we show the equations for *first* and *second* order. *first-order*: $y' + ay = 0$ (homogeneous, because $= 0$). Solution: $y = e^{-at}$ (see the statement of Thm 6.31) *second-order*: $y'' + ay' + by = 0$ (homogeneous). Solution: $y = c_1 e^{\lambda_1 t} + c_2 e^{\lambda_2 t}$ (see the statement of Thm 6.31) Where: λ_1 and λ_2 are solutions of $\lambda^2 + a\lambda + b = 0$
dimensional finite-dim infinite-dim zero-dim	p.457	6.2	A vector space is *blank*-dimensional if: A basis has finitely many vectors A basis has infinitely many vectors The dimension of the subspace $\mathbf{0}$ is defined to be 0

Key Definitions and Concepts, continued

identity transformation	p.478	6.4	$I: V \to W$ maps every vector to itself. That is, $I(\mathbf{v}) = \mathbf{v}$ (See Example 6.54).
inverse	p.483	6.4	T^{-1}: $T^{-1} \circ T = I_V$, $T \circ T^{-1} = I_W$ (See *invertible*).
invertible	p.482	6.4	A linear transformation $T: V \to W$ is *invertible* if there is a linear transformation $T': W \to V$ such that: $T' \circ T = I_V$ and $T \circ T' = I_W$ In this case, T' is called the *inverse* for T.
isomorphic	p.497	6.5	If $T: V \to W$ where T is an *isomorphism*. Then: V and W are *isomorphic*. This is written $V \cong W$.
isomorphism	p.497	6.5	$T: V \to W$ is an *isomorphism* if T is *one-to-one* and *onto*.
linear code	p.529	6.7	A *p*-ary *linear code* is a subspace of C of $\mathbb{Z}_p^n$.
linear transformation	p.476	6.4	A *linear transformation* from a vector space V to a vector space W is a mapping $T: V \to W$ such that: 1. $T(\mathbf{u} + \mathbf{v}) = T(\mathbf{u} + \mathbf{v})$ 2. $T(c\mathbf{u}) = cT(\mathbf{u})$ or $T(c_1\mathbf{v}_1 + \cdots + c_k\mathbf{v}_k) = c_1 T(\mathbf{v}_1) + \cdots + c_k T(\mathbf{v}_k)$
linearly dependent	p.447	6.2	vector can be written as linear combination of others $c_1 \mathbf{v}_1 + \cdots + c_n \mathbf{v}_n = \mathbf{0}$ with at least one $c_i \neq 0$
linearly independent	p.447	6.2	no vector can be written as linear combination of others or $c_1 \mathbf{v}_1 + \cdots + c_n \mathbf{v}_n = \mathbf{0} \Leftrightarrow$ all the $c_i = 0$
kernel (of **T**)	p.486	6.5	Let $T: V \to W$ be a linear transformation. The *kernel* of T is the set of all vectors that T maps to zero. That is, $\ker(T) = \{\mathbf{v} \text{ in } V : T(\mathbf{v}) = \mathbf{0}\}$.
nullity (of **T**)	p.488	6.5	Let $T: V \to W$ be a linear transformation. The *nullity* of T is the dimension of the kernel of T. That is, $\text{nullity}(T) = \dim(\ker(T))$.

Chapter 6 Review

Key Definitions and Concepts, continued

one-to-one	p.492	6.5	T maps distinct vectors in V to distinct vectors in W. That is, $\mathbf{v} \neq \mathbf{u}$ implies $T(\mathbf{v}) \neq T(\mathbf{u})$, or $T(\mathbf{v}) = T(\mathbf{u})$ implies $\mathbf{v} = \mathbf{u}$.
onto	p.492	6.5	If range$(T) = W$, then T is called *onto*. For all $\mathbf{w}$, there is a $\mathbf{v}$ such that $\mathbf{w} = T(\mathbf{v})$.
range (of T)	p.486	6.5	Let $T : V \to W$ be a linear transformation. The *range* of T is the set of all vectors that are images under T. That is, range$(T) = \{\mathbf{w}$ in $W : \mathbf{w} = T(\mathbf{v})$ for some $\mathbf{v}$ in $V\}$.
rank (of T)	p.488	6.5	Let $T : V \to W$ be a linear transformation. The *rank* of T is the dimension of the range of T. That is, rank$(T) = \dim(\text{range}(T))$.
Reed-Muller	p.532	6.7	The (first order) *Reed-Muller codes* R_n (see details, p.532).
subspace	p.438	6.1	$W \subseteq V$ and W satisfies the axioms of a vector space
trivial subspaces	p.441	6.1	The subspaces $\{\mathbf{0}\}$ and V are called the *trivial subspaces* of V. Therefore, all other subspaces are nontrivial.
vector space	p.433	6.1	If the following axioms hold, then V is a *vector space*: 1. $\mathbf{u} + \mathbf{v}$ is in V (*Closure under addition*) 2. $\mathbf{u} + \mathbf{v} = \mathbf{v} + \mathbf{u}$ (*Commutativity*) 3. $(\mathbf{u} + \mathbf{v}) + \mathbf{w} = \mathbf{u} + (\mathbf{v} + \mathbf{w})$ (*Associativity*) 4. $\mathbf{u} + \mathbf{0} = \mathbf{u}$ (*Zero vector*) 5. $\mathbf{u} + (-\mathbf{u}) = \mathbf{0}$ 6. $c\mathbf{u}$ is in V (*Closure under scalar multiplication*) 7. $c(\mathbf{u} + \mathbf{v}) = c\mathbf{u} + c\mathbf{v}$ (*Distributivity*) 8. $(c + d)\mathbf{u} = c\mathbf{u} + d\mathbf{u}$ (*Distributivity*) 9. $c(d\mathbf{u}) = (cd)\mathbf{u}$ 10. $1\mathbf{u} = \mathbf{u}$
zero transformation	p.478	6.4	$T_0 : V \to W$ maps every vector to the zero vector. That is, $T_0(\mathbf{v}) = \mathbf{0}$ (See Example 6.54).

Theorems

Thm 6.1   p.437   6.1   Let V be a vector space, $\mathbf{u}$ a vector, c a scalar:
  a. $0\mathbf{u} = \mathbf{0}$
  b. $c\mathbf{0} = \mathbf{0}$
  c. $(-1)\mathbf{u} = -\mathbf{u}$
  d. If $c\mathbf{u} = \mathbf{0}$, then $c = 0$ or $\mathbf{u} = \mathbf{0}$.

Thm 6.2   p.438   6.1   Let $W(\neq \emptyset) \subseteq V$, then W is a subspace if:
  a. $\mathbf{u}, \mathbf{v} \in W \Rightarrow \mathbf{u} + \mathbf{v} \in W$
  b. $\mathbf{u} \in W \Rightarrow c\mathbf{u} \in W$

Thm 6.3   p.445   6.1   Let $\mathbf{v}_1, \mathbf{v}_2, \ldots, \mathbf{v}_k$ be vectors in a vector space V.
  a. $\operatorname{span}(\mathbf{v}_1, \mathbf{v}_2, \ldots, \mathbf{v}_k)$ is a subspace of V
  b. $\operatorname{span}(\mathbf{v}_1, \mathbf{v}_2, \ldots, \mathbf{v}_k)$ is smallest subspace with $\mathbf{v}_1, \mathbf{v}_2, \ldots, \mathbf{v}_k$

Thm 6.4   p.448   6.2   vector can be written as linear combination of others
  $\Leftrightarrow$ *linearly dependent*

Thm 6.5   p.452   6.2   If $\mathcal{B} = \{\mathbf{u}_i\}$ is a basis and $\mathbf{x} = \sum_{i=1}^{n} c_i \mathbf{u}_i$,
  then c_i are unique.

Thm 6.6   p.455   6.2   Let $\mathcal{B} = \{\mathbf{u}_i\}$ be a basis for V then:
  a. $[\mathbf{u} + \mathbf{v}]_\mathcal{B} = [\mathbf{u}]_\mathcal{B} + [\mathbf{v}]_\mathcal{B}$
  b. $[c\mathbf{u}]_\mathcal{B} = c[\mathbf{u}]_\mathcal{B}$

Thm 6.7   p.456   6.2   Let $\mathcal{B} = \{\mathbf{u}_i\}$ be a basis for V then:
  $\{\mathbf{u}_i\}$ are linearly independent in V if and only if
  $[\mathbf{u}_i]_\mathcal{B}$ are linearly independent in $\mathbb{R}^n$

Thm 6.8   p.456   6.2   Let $\mathcal{B} = \{\mathbf{u}_i\}$ be a basis for V then:
  a. Any set of more than n vectors in V must be linearly dependent
  b. Any set of fewer than n vectors cannot span V

Thm 6.9   p.457   6.2   Every basis has exactly the same number of vectors

Thm 6.10   p.458   6.2   Let V be a vector space with $\dim V = n$. Then:
  a. Any linearly independent set in V contains at most n vectors.
  b. Any spanning set for V contains at least n vectors.
  c. Any lin indpt set of exactly n vectors in V is a basis for V.
  d. Any spanning set for V of exactly n vectors is a basis for V.
  e. Any lin indpt set in V can be extended to a basis for V.
  f. Any spanning set for V can be reduced to a basis for V.

Thm 6.11   p.460   6.2   Let W be a subspace of a finite-dimensional vector space V.
  a. W is finite-dimensional and $\dim W \leq \dim V$.
  b. $\dim W = \dim V$ if and only if $W = V$.

Chapter 6 Review

Theorems, continued

Thm 6.12   p.469   6.3   Let $\mathcal{B} = \{\mathbf{u}_1, \ldots, \mathbf{u}_n\}$, $\mathcal{C} = \{\mathbf{v}_1, \ldots, \mathbf{v}_n\}$ be bases for V.
let $P_{\mathcal{C} \leftarrow \mathcal{B}}$ be a *change-of-basis* matrix from $\mathcal{B}$ to $\mathcal{C}$. Then:
a. $P_{\mathcal{C} \leftarrow \mathcal{B}}[\mathbf{x}]_\mathcal{B} = [\mathbf{x}]_\mathcal{C}$ for all $\mathbf{x}$ in V.
b. $P_{\mathcal{C} \leftarrow \mathcal{B}}$ is the unique matrix P with property (a).
c. $P_{\mathcal{C} \leftarrow \mathcal{B}}$ is invertible and $(P_{\mathcal{C} \leftarrow \mathcal{B}})^{-1} = P_{\mathcal{B} \leftarrow \mathcal{C}}$.

Thm 6.13   p.474   6.3   Let $\mathcal{B} = \{\mathbf{u}_1, \ldots, \mathbf{u}_n\}$, $\mathcal{C} = \{\mathbf{v}_1, \ldots, \mathbf{v}_n\}$ be bases for V.
Let $B = [\,[\mathbf{u}_1]_\mathcal{E} \;\cdots\; [\mathbf{u}_n]_\mathcal{E}\,]$, $C = [\,[\mathbf{v}_1]_\mathcal{E} \;\cdots\; [\mathbf{v}_n]_\mathcal{E}\,]$.
Then: $[\,C\,|\,B\,] \longrightarrow [\,I\,|\,P_{\mathcal{C} \leftarrow \mathcal{B}}\,]$.

Thm 6.14   p.479   6.4   Let $T : V \to W$ be a linear transformation. Then:
a. $T(\mathbf{0}) = \mathbf{0}$.
b. $T(-\mathbf{v}) = -T(\mathbf{v})$ for all $\mathbf{v}$ in V.
c. $T(\mathbf{u} - \mathbf{v}) = T(\mathbf{u}) - T(\mathbf{v})$ for all $\mathbf{u}, \mathbf{v}$ in V.

Thm 6.15   p.480   6.4   Let $T : V \to W$ be a linear transformation.
Let $\mathcal{B} = \{\mathbf{v}_1, \ldots, \mathbf{v}_n\}$ be a spanning set for V. Then:
$T(\mathcal{B}) = \{T(\mathbf{v}_1), \ldots, T(\mathbf{v}_n)\}$ is a spanning set for the range of T.

Thm 6.16   p.481   6.4   If $T : U \to V$ and $S : V \to W$ then
$S \circ T : U \to W$, $(S \circ T)(\mathbf{u}) = S(T(\mathbf{u}))$, is a linear transformation.

Thm 6.17   p.483   6.4   If T is an *invertible* linear transformation, then its *inverse* is unique.
If $R \circ T = I_V$, $T \circ R = I_W$ and $S \circ T = I_V$, $T \circ S = I_W$, then $R = S$.

Thm 6.18   p.488   6.5   Let $T : V \to W$ be a linear transformation. Then:
a. The *kernel* of T is a subspace of V
b. The *range* of T is a subspace of W

Thm 6.19   p.490   6.5   Let $T : V \to W$ be a linear transformation. Then:
$\mathrm{rank}(T) + \mathrm{null}(T) = \dim V$

Thm 6.20   p.494   6.5   Let $T : V \to W$ be a linear transformation. Then:
T is *one-to-one* if and only if $\ker(T) = \{\mathbf{0}\}$.

Thm 6.21   p.494   6.5   Let $\dim V = \dim W = n$. Then:
$T : V \to W$ is *one-to-one* if and only if T is *onto*.

Thm 6.22   p.495   6.5   Let $T : V \to W$ be a *one-to-one* linear transformation.
If $S = \{\mathbf{v}_1, \ldots, \mathbf{v}_k\}$ is a linearly independent set in V. Then:
$T(S) = \{T(\mathbf{v}_1), \ldots, T(\mathbf{v}_k)\}$ is a linearly independent set in W.

Cor 6.23   p.495   6.5   Let $\dim V = \dim W = n$. Then:
If T is *one-to-one*, then T maps a basis for V to a basis for W.

Theorems, continued

Thm 6.24   p.495   6.5   Let $T: V \to W$ be a linear transformation. Then:
T is *invertible* if and only if T is *one-to-one* and *onto*.

Thm 6.25   p.498   6.5   Let V and W be two finite dimensional vector spaces. Then:
V is *isomorphic* to W ($V \cong W$) if and only if $\dim V = \dim W$.

Thm 6.26   p.502   6.6   Let V and W be two finite dimensional vector spaces with bases $\mathcal{B}$ and $\mathcal{C}$ respectively, where $\mathcal{B} = \{\mathbf{v}_1, \ldots, \mathbf{v}_n\}$. Let $T: V \to W$ be a linear transformation, then the $n \times n$ matrix $A = \big[\; [T(\mathbf{v}_1)]_\mathcal{C} \,\big|\, [T(\mathbf{v}_2)]_\mathcal{C} \,\big|\, \cdots \,\big|\, [T(\mathbf{v}_n)]_\mathcal{C} \;\big]$ satisfies $A[\mathbf{v}]_\mathcal{B} = [T(\mathbf{v})]_\mathcal{C}$ for every vector $\mathbf{v}$ in V.

Thm 6.27   p.508   6.6   Let U, V, and W be finite dimensional vector spaces with bases $\mathcal{B}$, $\mathcal{C}$, and $\mathcal{D}$ respectively.
Let $T: U \to V$ and $S: V \to W$ be linear transformations. Then:
$[S \circ T]_{\mathcal{D} \leftarrow \mathcal{B}} = [S]_{\mathcal{D} \leftarrow \mathcal{C}} [T]_{\mathcal{C} \leftarrow \mathcal{B}}$

Thm 6.28   p.509   6.6   Let V and W be finite dimensional vector spaces with bases $\mathcal{B}$ and $\mathcal{C}$, respectively.
Let $T: V \to W$ be a linear transformation. Then:
T is invertible if and only if the matrix $[T]_{\mathcal{C} \leftarrow \mathcal{B}}$ is invertible.
In this case, $([T]_{\mathcal{C} \leftarrow \mathcal{B}})^{-1} = [T^{-1}]_{\mathcal{C} \leftarrow \mathcal{B}}$

Thm 6.29   p.512   6.6   Let V be a finite dimensional vector space with bases $\mathcal{B}$ and $\mathcal{C}$.
Let $T: V \to V$ be a linear transformation. Then:
$[T]_\mathcal{C} = P^{-1}[T]_\mathcal{B} P$
where P is the change-of-basis matrix from $\mathcal{C}$ to $\mathcal{B}$.

Thm 6.30   p.516   6.6   *Fundamental Theorem of Invertible Matrices*: Version 4
Let A be an $n \times n$ matrix and
let $T: V \to W$ be a linear transformation
whose matrix $([T]_{\mathcal{C} \leftarrow \mathcal{B}})$ with respect to bases
$\mathcal{B}$ and $\mathcal{C}$ of V and W respectively, is A.
Statements a. through t. are equivalent (see p.516).
p. T is invertible
q. T is one-to-one
r. T is onto
s. $\ker(T) = \mathbf{0}$
t. $\text{range}(T) = W$

Theorems, continued

Thm 6.31	p.522	6.7	The set S of all solutions to $y' + ay = 0$ is a subspace of $\Im F$.
Thm 6.32	p.523	6.7	If S is the solution space of $y' + ay = 0$, then: dim $= 1$ and $\{e^{-at}\}$ is a basis for S.
Thm 6.33	p.526	6.7	Let S be the solution space of $y'' + ay' + by = 0$ and let λ_1 and λ_2 be the roots of $\lambda^2 + a\lambda + b = 0$. a. If $\lambda_1 \neq \lambda_2$, then $\{e^{\lambda_1 t}, e^{\lambda_2 t}\}$ is a basis for S. b. If $\lambda_1 = \lambda_2$, then $\{e^{\lambda_1 t}, te^{\lambda_1 t}\}$ is a basis for S.
Thm 6.34	p.531	6.7	Let C be an (n, k) linear code. a. The dual code $C^\perp$ is an $(n, n-k)$ linear code. b. C contains 2^k vectors, and $C^\perp$ contains 2^{n-k} vectors.
Thm 6.35	p.533	6.7	For $n \geq 1$, the *Reed-Muller code* R_n is: a $(2^n, n+1)$ linear code in which every vector (except **0** and **1**) has weight 2^{n-1}.

Solutions to odd-numbered exercises from Chapter 6 Review

1. We will explain and give counter examples to justify our answers below.

   (a) **False**. See Theorem 6.10 on p.458 of Section 6.2.

   In order for this to be true, we need to know $\dim V = n$ or that a basis for V contains n vectors. What other information would be sufficient?

   Q: In $\mathbb{R}^2$, the set $\left\{ \mathbf{v}_1 = \begin{bmatrix} 1 \\ 0 \end{bmatrix}, \mathbf{v}_2 = \begin{bmatrix} 0 \\ 1 \end{bmatrix}, \mathbf{v}_3 = \begin{bmatrix} 1 \\ 1 \end{bmatrix} \right\}$ is a counterexample.

   That is, $\mathbb{R}^2 = \text{span}(\mathbf{v}_1, \mathbf{v}_2, \mathbf{v}_3)$. Prove this.

   Q: Create a counterexample for M_{22}.

   (b) **True**. See the definition on p.447 in Section 6.2.

   To show $S_a = \{\mathbf{u}+\mathbf{v}, \mathbf{v}+\mathbf{w}, \mathbf{u}+\mathbf{w}\}$ is linearly independent, we need to show if $a(\mathbf{u}+\mathbf{v}) + b(\mathbf{v}+\mathbf{w}) + c(\mathbf{u}+\mathbf{w}) = \mathbf{0}$ then $a = b = c = 0$.

   Regrouping, we have $(a+c)\mathbf{u} + (a+b)\mathbf{v} + (b+c)\mathbf{w} = \mathbf{0}$.

   Since $S = \{\mathbf{u}, \mathbf{v}, \mathbf{w}\}$ is linearly independent, $a+c = a+b = b+c = 0 \Rightarrow b = c \Rightarrow 2b = 0 \Rightarrow b = 0 \Rightarrow c = 0 \Rightarrow a = 0$ as we needed to show.

   Q: What is another way of showing that S_a is linearly independent?
   A: Hint: $\mathbf{u} = \frac{1}{2}[(\mathbf{u}+\mathbf{v}) - (\mathbf{v}+\mathbf{w}) + (\mathbf{u}+\mathbf{w})]$.
   That is, does $\text{span}(S_a) = \text{span}(S)$? Is this sufficient? Is this required?

   (c) **True**. Is there any reason to suspect this is true for all M_{nn}?

   Let $I = \begin{bmatrix} 1 & 0 \\ 0 & 1 \end{bmatrix}$, $J = \begin{bmatrix} 0 & 1 \\ 1 & 0 \end{bmatrix}$, $L = \begin{bmatrix} 1 & 1 \\ 0 & 1 \end{bmatrix}$, and $R = \begin{bmatrix} 1 & 1 \\ 1 & 0 \end{bmatrix}$.

   Let $\mathcal{E} = \{E_{11}, E_{12}, E_{21}, E_{22}\}$ and $\mathcal{B} = \{I, J, L, R\}$.

   Q: Prove $\mathcal{B}$ is a basis by showing $\text{span}(\mathcal{B}) = \text{span}(\mathcal{E})$.
   Why is this sufficient?

   (d) **False**. If $W = \{M \in M_{22} : \text{tr}(A) = 0\}$, then $\dim W = 3$. See Exercise 15.

   Q: Why does $\dim W = 3$ imply M_{22} cannot have a basis of matrices all with trace 0?
   A: Because $\dim M_{22} = 4$. Why is this enough?

   Q: Why does the argument above *not* apply to invertible matrices?
   A: Hint: If A and B are invertible, is $A + B$ necessarily invertible?
   But, if A and B have trace zero, does $A + B$ necessarily have trace zero?

Chapter 6 Review

1. We will explain and give counter examples to justify our answers below. (continued)
   (e) **False**. The Triangle Inequality asserts $\|\mathbf{x}+\mathbf{y}\| \leq \|\mathbf{x}\| + \|\mathbf{y}\|$.
   Q: Which property of a linear transformation does T fails to have?
   A: Hint: To be linear, $T(\mathbf{x}+\mathbf{y}) = T(\mathbf{x}) + T(\mathbf{x})$ and $T(c\mathbf{x}) = cT(\mathbf{x})$.
   Q: Does $T(\mathbf{x}) = \|\mathbf{x}\|$ at least satisfy $T(c\mathbf{x}) = cT(\mathbf{x})$? Why or why not?

   (f) **True**. This is worth proving directly.
   Q: If T is one-to-one, what do we know about the dimension of range(T)?
   A: Hint: If $\{\mathbf{u}_i\}$ is a basis for V, is $\{T(\mathbf{u}_i)\}$ a basis for range(T)?
   Q: If T is onto, what do we know about a basis for W?
   A: Hint: If $\{\mathbf{u}_i\}$ is a basis for V, does $\{T(\mathbf{u}_i)\}$ contain a basis for W?

   (g) **False**. What condition do we need to add to make this statement true?
   Q: If $\ker(T) = V$, what does range(T) equal?
   Q: If $\ker(T) = V$ and T is onto, what does W equal?
   Q: Is $T: \mathbb{R}^2 \to \mathbb{R}^2$ a counterexample to this statement? Why or why not?

   (h) **True**. What Theorem should we cite to prove this?
   Q: What is $\dim M_{33}$?
   Q: If nullity(T) = 4, what is the dimension of range(T)?
   Q: Why do the answers to these two questions imply the statement is true?

   (i) **True**. Prove this by finding a basis for V and thereby the dimension of V.
   Q: Given $V = \{p(x) \in \mathscr{P}_4 : p(1) = 0\}$, find a basis for V.
   A: Hint: is $T: \mathscr{P}_4 \to \mathbb{R}$ defined by $T(p) = p(1)$ a linear transformation?
   Q: Given $T: \mathscr{P}_4 \to \mathbb{R}$ defined by $T(p) = p(1)$, is T onto?
   Q: What is $\ker(T)$?
   Q: What is nullity(T)?
   Q: Is $\dim V = 4$ enough to prove that V is isomorphic to $\mathscr{P}_3$?

   (j) **False**. See Example 6.80 in Section 6.6.
   If bases B and C are different, we get the change-of-basis matrix.

3. As asserted in Theorem 6.2 of Section 6.1, to prove W is a subspace we need only show:
   a. $\mathbf{u}, \mathbf{v} \in W \Rightarrow \mathbf{u} + \mathbf{v} \in W$
   b. $\mathbf{u} \in W \Rightarrow c\mathbf{u} \in W$

In this case, we have $W = \left\{ \begin{bmatrix} a & b \\ c & d \end{bmatrix} : a+b = c+d = a+c = b+d \right\}$.

We proceed as in Exercises 46 and 47 of Section 6.1.

1. $A = \begin{bmatrix} a & b \\ c & d \end{bmatrix}, A' = \begin{bmatrix} a' & b' \\ c' & d' \end{bmatrix} \in W \Rightarrow A + A' = \begin{bmatrix} a+a' & b+b' \\ c+c' & d+d' \end{bmatrix} \in W$
   because $a+b = c+d = a+c = b+d$ and $a'+b' = c'+d' = a'+c' = b'+d'$ imply $(a+a') + (b+b') = (c+c') + (d+d') = (a+a') + (c+c') = (b+b') + (d+d')$.

6. Likewise, $A = \begin{bmatrix} a & b \\ c & d \end{bmatrix} \in W \Rightarrow kA = \begin{bmatrix} ka & kb \\ kc & kd \end{bmatrix} \in W$
   because $a+b = c+d = a+c = b+d$ implies $k(a+b) = k(c+d) = k(a+c) = k(b+d)$ so $ka+kb = kc+kd = ka+kc = kb+kd$.

5. As asserted in Theorem 6.2 of Section 6.1, to prove W is a subspace we need only show:
   a. $\mathbf{u}, \mathbf{v} \in W \Rightarrow \mathbf{u} + \mathbf{v} \in W$
   b. $\mathbf{u} \in W \Rightarrow c\mathbf{u} \in W$

In this case, we have $W = \{f \in \mathscr{F} : f(x + \pi) = f(x) \text{ for all } x\}$.

We follow Example 6.13 and proceed as in Exercises 34 and 37 of Section 6.1.

1. $f(x), g(x) \in W \Rightarrow (f+g)(x) = f(x) + g(x) \in W$
   because $f(x) + g(x) = f(x+\pi) + g(x+\pi) = (f+g)(x+\pi)$.

6. Likewise, $f(x) \in W \Rightarrow kf(x) \in W$ because $kf(x) = kf(x+\pi)$.

7. Given A symmetric ($A^T = A$) and B skew-symmetric ($B^T = -B$), we need to show:
   $S = \{A, B\}$ is a linearly independent set.

For further discussion of *symmetric* and *skew-symmetric* see Section 3.2.
Also see Example 6.26 and Exercises 1 through 4 in Section 6.2.
It is important to note that A and B are *nonzero*. Where do we use these facts below?

To prove $\{A, B\}$ is linearly independent, we need only show $cA + dB = O \Rightarrow c = d = 0$.

Given (1) $cA + dB = O$, we take the transpose of both sides to get (2) $cA^T + dB^T = O$.

Now adding equation (1) to equation (2) yields $(cA + cA^T) + (dB + dB^T) = O$.
   Since A is symmetric $cA + cA^T = cA + cA = 2cA$.
   Likewise, since B is skew-symmetric $cB + cB^T = cB - cB = O$.

So we have $2cA = O \Rightarrow c = 0 \Rightarrow cA + dB = dB = O \Rightarrow d = 0$ as we were required to show.

Q: Why does the fact that $2cA = O$ imply that $c = 0$?
A: Hint: Recall that we are given the fact that A is nonzero.
   Does the implication above necessarily follow if A is the zero matrix?

9. We will find a basis for W using the condition $p(-x) = p(x)$ that defines it.
   For further discussion, see Example 6.34 and Exercises 34, 35, and 36 in Section 6.2.
   Note $W = \{p(x) \in \mathscr{P}_5 : p(-x) = p(x)\}$ implies $W = \{p(x) \in \mathscr{P}_5 : \text{degree of } p(x) \text{ is even }\}$.
   That is, W is all polynomials in $\mathscr{P}_5$ of the form: $W = \{p(x) : p(x) = a_0 + a_2 x^2 + a_4 x^4\}$.
   So, it is obvious that $\mathcal{B} = \{1, x^2, x^4\}$ is a basis for W and $\dim W = 3$.
   Q: Why is it obvious that the degree of $p(x)$ must be even? Prove this.
   A: Hint: Can $p(-x) = p(x)$ if x occurs to an odd power? Like $p(x) = x$, for example?
   Q: Why is it obvious that $\mathcal{B}$ is a basis? Prove this.
   A: Hint: Show that polynomial vectors in $\mathcal{B}$ are linearly independent and span W.

11. **NOTE:** We need $T(\mathbf{x}) = \mathbf{y}\mathbf{x}^T\mathbf{y}$ otherwise it does not go from $\mathbb{R}^2$ to $\mathbb{R}^2$.

    We will show $T(\mathbf{x}) = \mathbf{y}\mathbf{x}^T\mathbf{y}$ is a linear transformation by showing:
    $T(\mathbf{x} + c\mathbf{z}) = T(\mathbf{x}) + cT(\mathbf{z})$.
    Q: What theorem tells us that it is sufficient to show $T(\mathbf{x} + c\mathbf{z}) = T(\mathbf{x}) + cT(\mathbf{z})$?
    A: Hint: See Section 3.5 and Section 6.4.
    Since $T(\mathbf{x} + c\mathbf{z}) = \mathbf{y}(\mathbf{x} + c\mathbf{z})^T \mathbf{y} = \mathbf{y}\mathbf{x}^T\mathbf{y} + c\mathbf{y}\mathbf{z}^T\mathbf{y} = T(\mathbf{x}) + cT(\mathbf{z})$ as required.
    Q: Is $\mathbf{y}\mathbf{x}^T$ a vector, a scalar, or neither?
    Q: Do we need to know the exact value of $\mathbf{y}$ to conclude that T is linear?
    A: No. T is linear for any fixed vector $\mathbf{y}$.
    Q: Does T need to be a mapping from $\mathbb{R}^2$ to $\mathbb{R}^2$ in order to be linear?
    A: No. T can be a mapping from $\mathbb{R}^n$ to $\mathbb{R}^n$ for any n.
    Q: Is T still linear if T is a mapping from M_{nn} to M_{nn}? Why or why not?

13. We will show $T(p(x)) = p(2x - 1)$ is a linear transformation by showing:
    $T(p(x) + cq(x)) = T(p(x)) + cT(q(x))$.
    Also See Exercises 8 and 9 in Section 6.4.
    Recall $p(x) + cq(x) = (p + cq)(x)$.
    So $T(\, p(x) + cq(x) \,) = T(\, (p + cq)(x) \,) = (p + cq)(2x - 1)$
    $\phantom{\text{So } T(\, p(x) + cq(x) \,)} = p(2x - 1) + cq(2x - 1) = T(p(x)) + cT(q(x))$ as required.
    Q: If $2x - 1$ is replaced by $ax + b$ is T still linear?
    A: Yes. T is linear for any such substitution.
    Q: If $2x - 1$ is replaced by $ax^2 + bx + c$ is T still linear? Why or why not?
    Q: If $2x - 1$ is replaced by $a_n x^n + \ldots + a_0$ is T still linear? Why or why not?
    Q: If $2x - 1$ is replaced by $a_n \frac{1}{x^n} + \ldots + a_0$ is T still linear? Why or why not?

15. Given $T(A) = \text{tr}(A)$, we will compute nullity(T) using the Rank Theorem, Theorem 6.19.
    Recall from Exercises 44 through 46 in Section 3.2, we have $\text{tr}(A) = a_{11} + a_{22} + \cdots + a_{nn}$.
    Since $\dim M_{nn} = n^2$ and $\text{tr}(A)$ is a scalar with $\dim \mathbb{R} = 1$, we have nullity(T) $= n^2 - 1$.

    Also see Examples 6.64 through 6.67 and Exercise 6 in Section 6.5.

    Q: In M_{22}, find a basis $\mathcal{B}$ for ker(T). How many basis vectors must $\mathcal{B}$ have?
    A: Since nullity(T) $= n^2 - 1 = 2^2 - 1 = 3$, a basis $\mathcal{B}$ for ker(T) must have 3 vectors.

    Since A is in ker(T) means $a_{11} + a_{22} = 0$, we have $a_{11} = -a_{22}$.

    So $\mathcal{B} = \left\{ \begin{bmatrix} 1 & 0 \\ 0 & -1 \end{bmatrix}, \begin{bmatrix} 0 & 1 \\ 0 & 0 \end{bmatrix}, \begin{bmatrix} 0 & 0 \\ 1 & 0 \end{bmatrix} \right\}$ is a basis for ker(T).

    Q: Can we generalize this result to find a basis for T in M_{nn}?

17. We will find the change-of-basis matrix $[T]_{\mathcal{C} \leftarrow \mathcal{B}}$,
    where $\mathcal{B} = \{1, x, x^2\}$ and $\mathcal{C} = \{E_{11}, E_{12}, E_{21}, E_{22}\}$.

    Also see Examples 6.76 and 6.77 and Exercises 1 through 12 in Section 6.6.

    We compute $T(p(x))$ for the basis vectors in $\mathcal{B}$.
    First we express $\{1, x, x^2\}$ in terms of $\{1, 1+x, 1+x+x^2\}$ from Exercise 14.
    Then we express the result as a linear combination of the basis vectors in $\mathcal{C}$.

    $T(1) = \begin{bmatrix} 1 & 0 \\ 0 & 1 \end{bmatrix} = 1 \cdot E_{11} + 0 \cdot E_{12} + 0 \cdot E_{21} + 1 \cdot E_{22} \Rightarrow [T(1)]_{\mathcal{C}} = \begin{bmatrix} 1 \\ 0 \\ 0 \\ 1 \end{bmatrix}$

    $T(x) = T((1+x) - 1) = T(1+x) - T(1)$

    $= \begin{bmatrix} 1 & 1 \\ 1 & 0 \end{bmatrix} - \begin{bmatrix} 1 & 0 \\ 0 & 1 \end{bmatrix} = \begin{bmatrix} 0 & 1 \\ 0 & 0 \end{bmatrix} = 0 \cdot E_{11} + 1 \cdot E_{12} + 0 \cdot E_{21} + 0 \cdot E_{22} \Rightarrow [T(x)]_{\mathcal{C}} = \begin{bmatrix} 0 \\ 1 \\ 0 \\ 0 \end{bmatrix}$

    $T(x^2) = T((1+x+x^2) - (1+x)) = T(1+x+x^2) - T(1+x)$

    $= \begin{bmatrix} 0 & -1 \\ 1 & 0 \end{bmatrix} - \begin{bmatrix} 1 & 1 \\ 0 & 1 \end{bmatrix} = \begin{bmatrix} -1 & -2 \\ 1 & -1 \end{bmatrix} = 1 \cdot E_{11} - 2 \cdot E_{12} + 1 \cdot E_{21} - 1 \cdot E_{22} \Rightarrow [T(x^2)]_{\mathcal{C}} = \begin{bmatrix} 1 \\ -2 \\ 1 \\ -1 \end{bmatrix}$

    Since these vectors are the columns of $[T]_{\mathcal{C} \leftarrow \mathcal{B}}$, we have: $[T]_{\mathcal{C} \leftarrow \mathcal{B}} = \begin{bmatrix} 1 & 0 & -1 \\ 0 & 1 & -2 \\ 0 & 0 & 1 \\ 1 & 0 & -1 \end{bmatrix}$.

19. Given $S \circ T : U \to W$ and range(T) $\subseteq$ ker(S), we will show $S \circ T$ is the zero map.
    See Figure 6.6 in Section 6.4 and Figure 6.8 in Section 6.5
    for graphical representations of composition, range, and kernel.

    Q: If $\mathbf{v}$ is in ker(S), what is $S(\mathbf{v})$?
    Since range(T) $\subseteq$ ker(S), if $\mathbf{u}$ in U then $T(\mathbf{u})$ is in ker(S). Therefore, $S(T(\mathbf{u})) = \mathbf{0}$.
    So, $S \circ T : U \to W$ is the zero map. That is, $S \circ T$ maps every vector in U to $\mathbf{0}$ in W.

Chapter 7

Distance and Approximation

7.1 Inner Product Spaces

Q: What are our main goals for Section 7.1?
A: To examine *inner product* and *inner product spaces*.
   See Sections 3.5 and 6.1 for comparison to *vector spaces* and *subspaces*.
   Also see Section 5.2 for comparison of orthogonal complements and projections.

Key Definitions and Concepts

$\|\mathbf{v}\|$	p.544	7.1	The *length* (or *norm*) of $\mathbf{v}$ is $\|\mathbf{v}\| = \sqrt{\langle \mathbf{v}, \mathbf{v} \rangle}$
$d(\mathbf{u}, \mathbf{v})$	p.544	7.1	The *distance* between $\mathbf{u}$ and $\mathbf{v}$ is $d(\mathbf{u}, \mathbf{v}) = \|\mathbf{u} - \mathbf{v}\|$
$\langle \mathbf{u}, \mathbf{v} \rangle = 0$	p.544	7.1	$\mathbf{u}$ and $\mathbf{v}$ are orthogonal if $\langle \mathbf{u}, \mathbf{v} \rangle = 0$
$\text{proj}_W(\mathbf{v})$	p.547	7.1	$\text{proj}_W(\mathbf{v}) = \sum_{k=1}^{n} \frac{\langle \mathbf{u}_k, \mathbf{v} \rangle}{\langle \mathbf{u}_k, \mathbf{u}_k \rangle} \mathbf{u}_k$ (*orthogonal projection*)
$\text{perp}_W(\mathbf{v})$	p.547	7.1	$\text{perp}_W(\mathbf{v}) = \mathbf{v} - \text{proj}_W(\mathbf{v})$ (*component of* $\mathbf{v} \perp$ *to* W)
component	p.547	7.1	$\text{perp}_W(\mathbf{v}) = \mathbf{v} - \text{proj}_W(\mathbf{v})$ (*component of* $\mathbf{v} \perp$ *to* W)
distance	p.544	7.1	The *distance* between $\mathbf{u}$ and $\mathbf{v}$ is $d(\mathbf{u}, \mathbf{v}) = \|\mathbf{u} - \mathbf{v}\|$
inner product	p.540	7.1	An *inner product* on a vector space V with $\langle \mathbf{u}, \mathbf{v} \rangle$ has the following properties: 1. $\langle \mathbf{u}, \mathbf{v} \rangle = \langle \mathbf{v}, \mathbf{u} \rangle$ 2. $\langle \mathbf{u}, \mathbf{v} + \mathbf{w} \rangle = \langle \mathbf{u}, \mathbf{v} \rangle + \langle \mathbf{u}, \mathbf{w} \rangle$ 3. $\langle c\mathbf{u}, \mathbf{v} \rangle = c\langle \mathbf{u}, \mathbf{v} \rangle$ 4. $\langle \mathbf{u}, \mathbf{u} \rangle \geq 0$ and $\langle \mathbf{u}, \mathbf{u} \rangle = 0$ if and only if $\mathbf{0} = 0$
inner product space	p.540	7.1	A vector space with an *inner product* is called an *inner product space*
length	p.544	7.1	The *length* (or *norm*) of $\mathbf{v}$ is $\|\mathbf{v}\| = \sqrt{\langle \mathbf{v}, \mathbf{v} \rangle}$
norm	p.544	7.1	The *length* (or *norm*) of $\mathbf{v}$ is $\|\mathbf{v}\| = \sqrt{\langle \mathbf{v}, \mathbf{v} \rangle}$
orthonormal basis	p.546	7.1	A basis that is an orthonormal set
orthonormal set	p.546	7.1	An orthogonal set of unit vectors, that is $\|\mathbf{v}_i\| = 1$
orthogonal	p.544	7.1	$\mathbf{u}$ and $\mathbf{v}$ are orthogonal if $\langle \mathbf{u}, \mathbf{v} \rangle = 0$
orthogonal basis	p.546	7.1	A basis that is an orthogonal set
orthogonal set	p.546	7.1	$\{\mathbf{v}_1, \ldots, \mathbf{v}_k\}$ such that $\langle \mathbf{v}_i, \mathbf{v}_j \rangle = 0$ when $\mathbf{v}_i \neq \mathbf{v}_j$
projection	p.547	7.1	$\text{proj}_W(\mathbf{v}) = \sum_{k=1}^{n} \frac{\langle \mathbf{u}_k, \mathbf{v} \rangle}{\langle \mathbf{u}_k, \mathbf{u}_k \rangle} \mathbf{u}_k$ (*orthogonal projection*)
weighted dot product	p.541	7.1	An *inner product* defined on $\mathbb{R}^n$ such that $\langle \mathbf{u}, \mathbf{v} \rangle = w_1 u_1 v_1 + \ldots + w_n u_n v_n$ with $w_i \geq 0$ (see p.541 for details)

Theorems

Thm 7.1   p.544   7.1   Let $\mathbf{u}$, $\mathbf{v}$, and $\mathbf{w}$ be vectors in an inner product space V and c a scalar:
a. $\langle \mathbf{u} + \mathbf{v}, \mathbf{w} \rangle = \langle \mathbf{u}, \mathbf{v} \rangle + \langle \mathbf{u}, \mathbf{w} \rangle$
b. $\langle \mathbf{u}, c\mathbf{v} \rangle = c\langle \mathbf{u}, \mathbf{v} \rangle$
c. $\langle \mathbf{u}, \mathbf{0} \rangle = \langle \mathbf{0}, \mathbf{v} \rangle = 0$

Thm 7.2   p.546   7.1   Let $\mathbf{u}$ and $\mathbf{v}$ be vectors in an inner product space V. Then $\mathbf{u}$ and $\mathbf{v}$ are orthogonal if and only if $\|\mathbf{u}+\mathbf{v}\|^2 = \|\mathbf{u}\|^2 + \|\mathbf{v}\|^2$

Thm 7.3   p.548   7.1   Let $\mathbf{u}$ and $\mathbf{v}$ be vectors in an inner product space V. Then $|\langle \mathbf{u}, \mathbf{v} \rangle| \leq \|\mathbf{u}\|\|\mathbf{v}\|$ with equality holding if and only if $\mathbf{u}$ and $\mathbf{v}$ are scalar multiples of each other

Thm 7.4   p.549   7.1   Let $\mathbf{u}$ and $\mathbf{v}$ be vectors in an inner product space V. Then $\|\mathbf{u}+\mathbf{v}\| \leq \|\mathbf{u}\| + \|\mathbf{v}\|$

Theorems by name

Thm 7.2   p.546   7.1   **Pythagoras' Theorem**
Let $\mathbf{u}$ and $\mathbf{v}$ be vectors in an inner product space V. Then $\mathbf{u}$ and $\mathbf{v}$ are orthogonal if and only if $\|\mathbf{u}+\mathbf{v}\|^2 = \|\mathbf{u}\|^2 + \|\mathbf{v}\|^2$

Thm 7.3   p.548   7.1   **The Cauchy-Schwarz Inequality**
Let $\mathbf{u}$ and $\mathbf{v}$ be vectors in an inner product space V. Then $|\langle \mathbf{u}, \mathbf{v} \rangle| \leq \|\mathbf{u}\|\|\mathbf{v}\|$ with equality holding if and only if $\mathbf{u}$ and $\mathbf{v}$ are scalar multiples of each other

Thm 7.4   p.549   7.1   **The Triangle Inequality**
Let $\mathbf{u}$ and $\mathbf{v}$ be vectors in an inner product space V. Then $\|\mathbf{u}+\mathbf{v}\| \leq \|\mathbf{u}\| + \|\mathbf{v}\|$

7.1 Inner Product Spaces

Discussion of Key Definitions and Concepts

Computing and defining inner products

Q: How do we determine if $\langle \mathbf{u}, \mathbf{v} \rangle$ is an inner product?
A: By verifying that it possesses all four properties in the definition.
Property 4, $\langle \mathbf{u}, \mathbf{u} \rangle = 0$ if and only if $\mathbf{0} = 0$, is often a good place to start.
Why? Because this property is very restrictive and easily disproven with a counterexample.

Q: How do we determine if W is an inner product space?
A: By verifying that it satisfies all three axioms of Theorem 7.1.

Q: What pattern do we notice when computing inner products, norms, and distance?
A: Regardless of the computation, the inner product, norm, and distance are scalars.
For instance, integration with definite limits results in a scalar. See Exercise 9.

Q: What pattern do we notice when defining inner products, norms, and distance?
A: Regardless of the definition, we are using scalars in the definition.
For instance, with polynomials we use their coefficients (or integrate). See Exercise 5.

Q: What sense should we come away with when defining inner products, norms, and distance?
A: The concept of distance and approximation are limited primarily by our imagination.

Q: What is the point of extending the concept of distance and therefore closeness?
A: To give us a concrete way of determining how objects are related and how closely.

Exploring the meanings of a specific norm

Q: Given the norm, $\int_0^1 f(x)g(x)\,dx$?, when are f and g orthogonal?
A: When the area on $[0,1]$ of $h(x) = f(x)g(x)$ above and below the x-axis is equal.

Q: What does this tell us about the relationship between f and g?
A: Hint: In general the relationship may be very complicated. So consider a simple case:
What happens if f is always zero when g is nonzero and vice-versa?

Q: What relationship must f and g have if the distance between them is zero?
Q: In what context might this definition arise naturally?
Q: What happens if we alter the limits of integration?
Q: Does that change which functions are orthogonal?
Does that change which functions have no distance between them?

Q: What types of transformations preserve orthogonality in this sense?
For instance, if we shift $f(x)$ and $g(x)$ left or right or up or down, what happens?

Solved examples with follow up questions

$\int_0^1 (x^2 - 3x + 2)(a + bx)dx = \frac{b}{4}x^4 + \left(\frac{a-3b}{3}\right)x^3 + \left(\frac{2b-3a}{2}\right)x^2 + 2ax\Big|_0^1 = \frac{b}{4} + \frac{5a}{6} = 0 \Rightarrow 20a = -6b \Rightarrow$
$h(x) = 20x - 6$ is orthogonal to $p(x) = x^2 - 3x + 2$.

Q: Can we find one based on $a + bx^2$?

Noting $(x\sin^2 x)' = \sin^2 x + 2\sin x \cos x \Rightarrow$
$\langle f, h \rangle = \int_0^{2\pi} \sin x(\sin x + 2x \cos x)\, dx = \int_0^{2\pi} (\sin^2 x + 2x \sin x \cos x)\, dx = x\sin^2 x\Big|_0^{2\pi} = 0 \Rightarrow$
$h(x) = \sin x + 2\cos x$ is orthogonal to $f(x) = \sin x$.

Q: Can we find an $h(x)$ by differentiating $x^2 \sin^2 x$? $\sin^2 x \cos x$?
Can we find a simple $h(x)$ that we have not suggested?
The key is we want $\sin x$ to appear in each term of the derivative. Why?

Given $\langle \mathbf{u}, \mathbf{v} \rangle = u_1 v_2 + u_2 v_1$ determine which axioms do not hold and give an example.
Axiom 4 fails: $\mathbf{u} = \begin{bmatrix} u \\ 0 \end{bmatrix} \Rightarrow \langle \mathbf{u}, \mathbf{u} \rangle = u(0) + (0)u = 0$ but $\mathbf{u} \neq \mathbf{0}$.

Q: Can we come up with another form of A that shows axiom 4 fails?

Given $\langle p(x), q(x) \rangle = p(0)q(0)$ determine which axioms do not hold and give an example.
Axiom 4 fails: $p(x) = ax \Rightarrow \langle p(x), p(x) \rangle = (0)(0) = 0$ but $p(x) \neq \mathbf{0}$.

Q: Can we come up with another form of $p(x)$ that shows axiom 4 fails?

Given $\langle A, B \rangle = \det(AB)$ determine which axioms do not hold and give an example.
Axiom 4 fails: $A = \begin{bmatrix} a & a \\ a & a \end{bmatrix} \Rightarrow \langle A, A \rangle = \det A^2 = 4a^4 - 4a^4 = 0$ but $A \neq \mathbf{0}$.

Q: Can we come up with another form of A that shows axiom 4 fails?

Exploration of the meaning of another specific norm

Q: Repeat the exploration of the norm for the norm defined in Example 7.4.
Try to develop intuition about when two polynomials are orthogonal with this norm.

Q: For instance, can two polynomials with all positive coefficients ever be orthogonal?

Q: What types of polynomials are orthogonal to lines through the origin?
A: Hint: All horizontal lines. Why? Consider $p(x) = 1$ and $q(x) = x$.
Then $\langle p(x), q(x) \rangle = 1 \cdot 0 + 0 \cdot 1 + 0 \cdot 0 = 0$.
Based on this, what other types of polynomials are orthogonal to lines through the origin?

Q: What types of transformations preserve orthogonality in this sense?
For instance if we shift $p(x)$ and $q(x)$ left or right the same amount, what happens?

Q: What types of polynomials have no distance between them?

Q: Is there a natural context in which this definition of the norm arises or makes sense?

7.1 Inner Product Spaces

Solutions to odd-numbered exercises from Section 7.1

1. Example 7.2: $\langle \mathbf{u}, \mathbf{v} \rangle = 2u_1 v_1 + 3u_2 v_2 \Rightarrow$

   (a) $\langle \mathbf{u}, \mathbf{v} \rangle = 2(\,(2)(3)\,) + 3(\,(-1)(4)\,) = 0.$
   (b) $\|\mathbf{u}\| = \sqrt{\langle \mathbf{u}, \mathbf{u} \rangle} = \sqrt{2(\,(2)(2)\,) + 3(\,(-1)(-1)\,)} = \sqrt{11}.$
   (c) $d(\mathbf{u}, \mathbf{v}) = \|\mathbf{u} - \mathbf{v}\| = \left\| \begin{bmatrix} -1 \\ -5 \end{bmatrix} \right\| = \sqrt{2(\,(-1)(-1)\,) + 3(\,(-5)(-5)\,)} = \sqrt{77}.$

3. Since $\mathbf{u} = \begin{bmatrix} 2 \\ -1 \end{bmatrix}$ orthogonal means $\langle \mathbf{u}, \mathbf{v} \rangle = 2(2v_1) + 3((-1)v_2) = 0 \Rightarrow$

   $4v_1 - 3v_2 = 0 \Rightarrow 4v_1 = 3v_2 \Rightarrow$ any scalar multiple of $\begin{bmatrix} 3 \\ 4 \end{bmatrix}$ is orthogonal to $\mathbf{u}$.

5. Example 7.4: $\langle p(x), q(x) \rangle = a_0 b_0 + a_1 b_1 + a_2 b_2.$
   It may be useful to note that this is equivalent to $[p(x)]_\mathcal{E}^T \cdot [q(x)]_\mathcal{E}.$

   (a) $\langle p(x), q(x) \rangle = (2)(1) + (-3)(0) + (1)(-3) = -1.$
   (b) $\|p(x)\| = \sqrt{\langle p(x), p(x) \rangle} = \sqrt{(2)(2) + (-3)(-3) + (1)(1)} = \sqrt{14}.$
   (c) $d(p(x), q(x)) = \|p(x) - q(x)\| = \|1 - 3x + 4x^2\| = \sqrt{(1)(1) + (-3)(-3) + (4)(4)} = \sqrt{26}.$

7. Since $p(x) = 2 - 3x + x^2$ orthogonal means $\langle p(x), r(x) \rangle = 2r_0 - 3r_1 + r_2 = 0 \Rightarrow r_2 = 3r_1 - 2r_0.$
   Letting $r_0 = 1, r_1 = 0 \Rightarrow r_2 = -2 \Rightarrow r(x) = 1 - 2x^2$ is orthogonal to $p(x).$
   Note, the many choices and the check: $\langle p(x), r(x) \rangle = (2)(1) + (-3)(0) + (1)(-2) = 0.$

9. Example 7.5: $\langle f, g \rangle = \int f(x) g(x)\, dx$ over $\mathscr{C}[0, 2\pi] \Rightarrow$
   The trigonometric identities $\sin^2 x = \tfrac{1}{2}(1 - \cos 2x)$ and $\cos^2 x = \tfrac{1}{2}(1 + \cos 2x)$ are useful.

   (a) $\langle f, g \rangle = \int_0^{2\pi} \sin x (\sin x + \cos x)\, dx = \int_0^{2\pi} \sin^2 x + \int_0^{2\pi} \sin x \cos x\, dx = \pi.$

   (b) $\|f\| = \sqrt{\langle f, f \rangle} = \sqrt{\int_0^{2\pi} \sin^2 x\, dx} = \sqrt{\pi}.$

   (c) $d(f, g) = \|f - g\| = \|-\cos x\| = \sqrt{\int_0^{2\pi} \cos^2 x\, dx} = \sqrt{\pi}.$

11. Will show $\langle p(x), q(x)\rangle = p(a)q(a) + p(b)q(b) + p(c)q(c)$ satisfies the inner product axioms. Section 6.2, Exercise 62 $\Rightarrow$ a $p(x)$, degree ≤ 2 with 3 distinct zeroes $\Rightarrow p(x) = \mathbf{0}$.

1. $\langle p(x), q(x)\rangle = p(a)q(a) + p(b)q(b) + p(c)q(c) = q(a)p(a) + q(b)p(b) + q(c)p(c)$
   $= \langle q(x), p(x)\rangle$.

2. $\langle p(x), (q+r)(x)\rangle = p(a)(q+r)(a) + p(b)(q+r)(b) + p(c)(q+r)(c)$
   $= (p(a)q(a) + p(b)q(b) + p(c)q(c)) + (p(a)r(a) + p(b)r(b) + p(c)r(c))$
   $= \langle p(x), q(x)\rangle + \langle p(x), r(x)\rangle$.

3. $\langle p(x), (dq)(x)\rangle = p(a)(dq)(a) + p(b)(dq)(b) + p(c)(dq)(c) = d(p(a)q(a) + p(b)q(b) + p(c)q(c))$
   $= d\langle p(x), q(x)\rangle$.

4. $\langle p(x), p(x)\rangle = p(a)p(a) + p(b)p(b) + p(c)p(c) = p(a)^2 + p(b)^2 + p(c)^2 \geq 0 \Rightarrow \langle p(x), p(x)\rangle \geq 0$ and $p(a)^2 + p(b)^2 + p(c)^2 = 0 \Leftrightarrow p(a) = p(b) = p(c) = 0 \Leftrightarrow p(x)$ has three distinct zeroes $\Leftrightarrow p(x)$ is the zero polynomial, that is $\Rightarrow p(x) = \mathbf{0}$.

13. Given $\langle \mathbf{u}, \mathbf{v}\rangle = u_1 v_1$ determine which axioms do not hold and give an example.

    Axiom 4 fails: $\mathbf{u} = \begin{bmatrix} 0 \\ u \end{bmatrix} \Rightarrow \langle \mathbf{u}, \mathbf{u}\rangle = (0)(0) = 0$ but $\mathbf{u} \neq \mathbf{0}$.

15. Given $\langle \mathbf{u}, \mathbf{v}\rangle = u_1 v_2 + u_2 v_1$ determine which axioms do not hold and give an example.

    Axiom 4 fails: $\mathbf{u} = \begin{bmatrix} u \\ 0 \end{bmatrix} \Rightarrow \langle \mathbf{u}, \mathbf{u}\rangle = u(0) + (0)u = 0$ but $\mathbf{u} \neq \mathbf{0}$.

    Can you come up with another form of $\mathbf{u}$ that shows axiom 4 fails?

17. Given $\langle p(x), q(x)\rangle = p(1)q(1)$ determine which axioms do not hold and give an example.

    Axiom 4 fails: $p(x) = a(1 - x) \Rightarrow \langle p(x), p(x)\rangle = (0)(0) = 0$ but $p(x) \neq 0$.

    Can you come up with another form of $p(x)$ that shows axiom 4 fails?

19. Given $\langle \mathbf{u}, \mathbf{v}\rangle = 4u_1 v_1 + u_1 v_2 + u_2 v_1 + 4u_2 v_2$ find a symmetric matrix A such that $\langle \mathbf{u}, \mathbf{v}\rangle = \mathbf{u}^T A \mathbf{v}$.

    $\langle \mathbf{u}, \mathbf{v}\rangle = \mathbf{u}^T A \mathbf{v} = \begin{bmatrix} u_1 & u_2 \end{bmatrix} \begin{bmatrix} a & b \\ b & d \end{bmatrix} \begin{bmatrix} v_1 \\ v_2 \end{bmatrix} = a u_1 v_1 + b u_1 v_2 + b u_2 v_1 + d u_2 v_2 \Rightarrow$

    So, $\langle \mathbf{u}, \mathbf{v}\rangle = 4u_1 v_1 + u_1 v_2 + u_2 v_1 + 4u_2 v_2 \Rightarrow a = 4, b = 1, d = 4 \Rightarrow A = \begin{bmatrix} 4 & 1 \\ 1 & 4 \end{bmatrix}$.

7.1 Inner Product Spaces

21. Given $\langle \mathbf{u}, \mathbf{v} \rangle = u_1 v_1 + \frac{1}{4} u_2 v_2$ sketch $\|\mathbf{u}\| = \sqrt{\langle \mathbf{u}, \mathbf{u} \rangle} = 1 \Rightarrow \langle \mathbf{u}, \mathbf{u} \rangle = 1$. Let $\mathbf{u} = \begin{bmatrix} x \\ y \end{bmatrix} \Rightarrow$

$x^2 + \frac{1}{4} y^2 = 1 \Rightarrow \frac{x^2}{1} + \frac{y^2}{4} = 1$. Section 5.5, Example 5.30 $\Rightarrow$ ellipse intersecting the x-axis at $(\pm 1, 0)$
intersecting the y-axis at $(0, \pm 2)$.

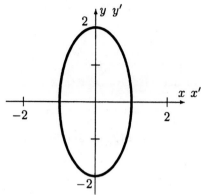
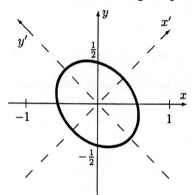

23. Prove Theorem 7.1b. Relevant axioms cited above equal signs below.

$\langle \mathbf{u}, c\mathbf{v} \rangle \stackrel{1}{=} \langle c\mathbf{v}, \mathbf{u} \rangle \stackrel{3}{=} c \langle \mathbf{v}, \mathbf{u} \rangle \stackrel{1}{=} c \langle \mathbf{u}, \mathbf{v} \rangle$.

25. $\langle \mathbf{u} + \mathbf{w}, \mathbf{v} - \mathbf{w} \rangle = \langle \mathbf{u}, \mathbf{v} \rangle - \langle \mathbf{u}, \mathbf{w} \rangle + \langle \mathbf{w}, \mathbf{v} \rangle - \langle \mathbf{w}, \mathbf{w} \rangle = 1 - 5 + 0 - (2)^2 = -8$.

27. $\|\mathbf{u} + \mathbf{v}\| = \sqrt{\langle \mathbf{u} + \mathbf{v}, \mathbf{u} + \mathbf{v} \rangle} = \sqrt{\langle \mathbf{u}, \mathbf{u} \rangle + 2\langle \mathbf{u}, \mathbf{v} \rangle + \langle \mathbf{v}, \mathbf{v} \rangle} = \sqrt{(1)^2 + 2(1) + (\sqrt{3})^2} = \sqrt{6}$.

29. Will show $\mathbf{u} + \mathbf{v} = \mathbf{w}$ by showing $\|\mathbf{u} + \mathbf{v} - \mathbf{w}\| = 0$.

$\|\mathbf{u} + \mathbf{v} - \mathbf{w}\| = \sqrt{\langle \mathbf{u} + \mathbf{v} - \mathbf{w}, \mathbf{u} + \mathbf{v} - \mathbf{w} \rangle}$
$= \sqrt{\langle \mathbf{u}, \mathbf{u} \rangle + 2\langle \mathbf{u}, \mathbf{v} \rangle - 2\langle \mathbf{u}, \mathbf{w} \rangle + \langle \mathbf{v}, \mathbf{v} \rangle - 2\langle \mathbf{v}, \mathbf{w} \rangle + \langle \mathbf{w}, \mathbf{w} \rangle}$
$= \sqrt{(1)^2 + 2(1) - 2(5) + (\sqrt{3})^2 - 2(0) + (2)^2} = 0$.

31. $\langle \mathbf{u} + \mathbf{v}, \mathbf{u} - \mathbf{v} \rangle = \langle \mathbf{u}, \mathbf{u} \rangle + \langle \mathbf{u}, \mathbf{v} \rangle - \langle \mathbf{v}, \mathbf{u} \rangle + \langle \mathbf{v}, \mathbf{v} \rangle = \|\mathbf{u}\|^2 - \|\mathbf{v}\|^2$.

33. $\begin{aligned} \tfrac{1}{2}\|\mathbf{u}+\mathbf{v}\|^2 &= \tfrac{1}{2}\|\mathbf{u}\|^2 + \langle \mathbf{u},\mathbf{v}\rangle + \tfrac{1}{2}\|\mathbf{v}\|^2 \\ +\tfrac{1}{2}\|\mathbf{u}-\mathbf{v}\|^2 &= \tfrac{1}{2}\|\mathbf{u}\|^2 - \langle \mathbf{u},\mathbf{v}\rangle + \tfrac{1}{2}\|\mathbf{v}\|^2 \\ \hline &= \|\mathbf{u}\|^2 + \|\mathbf{v}\|^2 \end{aligned}$

35. $\|\mathbf{u}+\mathbf{v}\| = \|\mathbf{u}-\mathbf{v}\| \Leftrightarrow \|\mathbf{u}+\mathbf{v}\|^2 = \|\mathbf{u}-\mathbf{v}\|^2 \Leftrightarrow \|\mathbf{u}\|^2 + 2\langle\mathbf{u},\mathbf{v}\rangle + \|\mathbf{v}\|^2 = \|\mathbf{u}\|^2 - 2\langle\mathbf{u},\mathbf{v}\rangle + \|\mathbf{v}\|^2 \Leftrightarrow 2\langle\mathbf{u},\mathbf{v}\rangle = -2\langle\mathbf{u},\mathbf{v}\rangle \Leftrightarrow 4\langle\mathbf{u},\mathbf{v}\rangle = 0 \Leftrightarrow \langle\mathbf{u},\mathbf{v}\rangle = 0$ as we needed to show.

Recall, $\mathbf{u}$ and $\mathbf{v}$ are orthogonal means $\langle \mathbf{u}, \mathbf{v} \rangle = 0$.

37. Note, $\langle \mathbf{v}_1, \mathbf{v}_1 \rangle = 2(1)^2 + 3(0)^2 = 2$ and $\langle \mathbf{v}_1, \mathbf{x}_2 \rangle = 2(1) + 3(1)(0) = 2 \Rightarrow \frac{\langle \mathbf{v}_1, \mathbf{x}_2 \rangle}{\langle \mathbf{v}_1, \mathbf{v}_1 \rangle} = \frac{2}{2} = 1$.

$\mathbf{v}_2 = \mathbf{x}_2 - \frac{\langle \mathbf{v}_1, \mathbf{x}_2 \rangle}{\langle \mathbf{v}_1, \mathbf{v}_1 \rangle} \mathbf{v}_1 = \mathbf{x}_2 - \mathbf{v}_1 = \begin{bmatrix} 1 \\ 1 \end{bmatrix} - \begin{bmatrix} 1 \\ 0 \end{bmatrix} = \begin{bmatrix} 0 \\ 1 \end{bmatrix}$.

39. $\mathcal{B} = \{\mathbf{v}_1 = 1, \mathbf{x}_2 = 1 + x, \mathbf{x}_3 = 1 + x + x^2\}$ and $\langle p(x), q(x) \rangle = \sum a_i b_i \Rightarrow$
$\langle \mathbf{v}_1, \mathbf{v}_1 \rangle = 1, \langle \mathbf{v}_1, \mathbf{x}_2 \rangle = 1 \Rightarrow \frac{\langle \mathbf{v}_1, \mathbf{x}_2 \rangle}{\langle \mathbf{v}_1, \mathbf{v}_1 \rangle} = 1$.
$\mathbf{v}_2 = \mathbf{x}_2 - \frac{\langle \mathbf{v}_1, \mathbf{x}_2 \rangle}{\langle \mathbf{v}_1, \mathbf{v}_1 \rangle} \mathbf{v}_1 = \mathbf{x}_2 - \mathbf{v}_1 = (1+x) - 1 = x$.
$\langle \mathbf{v}_1, \mathbf{x}_3 \rangle = 1, \langle \mathbf{v}_2, \mathbf{v}_2 \rangle = 1, \langle \mathbf{v}_2, \mathbf{x}_3 \rangle = 1 \Rightarrow \frac{\langle \mathbf{v}_1, \mathbf{x}_3 \rangle}{\langle \mathbf{v}_1, \mathbf{v}_1 \rangle} = 1$ and $\frac{\langle \mathbf{v}_2, \mathbf{x}_3 \rangle}{\langle \mathbf{v}_2, \mathbf{v}_2 \rangle} = 1 \Rightarrow$
$\mathbf{v}_3 = \mathbf{x}_3 - \frac{\langle \mathbf{v}_1, \mathbf{x}_3 \rangle}{\langle \mathbf{v}_1, \mathbf{v}_1 \rangle} \mathbf{v}_1 - \frac{\langle \mathbf{v}_2, \mathbf{x}_3 \rangle}{\langle \mathbf{v}_2, \mathbf{v}_2 \rangle} \mathbf{v}_2 = \mathbf{x}_3 - \mathbf{v}_1 - \mathbf{v}_2 = (1 + x + x^2) - 1 - x = x^2 \Rightarrow \mathcal{C} = \{1, x, x^2\}$.

41. Note that the limits of integration are -1 and 1. We will find $l_i(x)$ by dividing by the norms.

(a) $\|1\| = \sqrt{\int_{-1}^{1} 1^2 \, dx} = \sqrt{2} \Rightarrow l_0(x) = \frac{1}{\sqrt{2}}$.
$\|x\| = \sqrt{\int_{-1}^{1} x^2 \, dx} = \sqrt{\frac{2}{3}} \Rightarrow l_1(x) = \frac{\sqrt{3} x}{\sqrt{2}}$.
$\|x^2 - \frac{1}{3}\| = \sqrt{\int_{-1}^{1} (x^2 - \frac{1}{3})^2 \, dx} = \sqrt{\frac{8}{45}} = \frac{2\sqrt{2}}{3\sqrt{5}} \Rightarrow l_2(x) = \frac{\sqrt{5}}{2\sqrt{2}}(3x^2 - 1)$.

(b) $\langle \mathbf{v}_1, \mathbf{x}_4 \rangle = \int_{-1}^{1} 1(x^3) \, dx = 0$, $\langle \mathbf{v}_2, \mathbf{v}_4 \rangle = \int_{-1}^{1} x(x^3) \, dx = \frac{2}{5}$, $\langle \mathbf{v}_2, \mathbf{v}_2 \rangle = \int_{-1}^{1} x^2 \, dx = \frac{2}{3}$,
$\langle \mathbf{v}_3, \mathbf{x}_4 \rangle = \int_{-1}^{1} (x^2 - \frac{1}{3}) x^3 \, dx = 0 \Rightarrow$
$\mathbf{v}_4 = \mathbf{x}_3 - \frac{\langle \mathbf{v}_2, \mathbf{x}_4 \rangle}{\langle \mathbf{v}_2, \mathbf{v}_2 \rangle} \mathbf{v}_2 = \mathbf{x}_3 - \frac{2/5}{2/3} \mathbf{v}_2 = \mathbf{x}_3 - \frac{3}{5} \mathbf{v}_2 = x^3 - \frac{3}{5} x \Rightarrow$
$\|x^3 - \frac{3}{5} x\| = \sqrt{\int_{-1}^{1} (x^3 - \frac{3}{5} x)^2 \, dx} = \sqrt{\frac{8}{175}} = \frac{2\sqrt{2}}{5\sqrt{7}} \Rightarrow l_4(x) = \frac{\sqrt{7}}{2\sqrt{2}} (5x^3 - 3x)$.

43. $\mathcal{B} = \{\mathbf{u}_i\}$ orthogonal basis $\Rightarrow \langle \text{proj}_W(\mathbf{v}), \mathbf{u}_j \rangle = \langle \sum \frac{\langle \mathbf{u}_i, \mathbf{v} \rangle}{\langle \mathbf{u}_i, \mathbf{u}_i \rangle} \mathbf{u}_i, \mathbf{u}_j \rangle = \langle \frac{\langle \mathbf{u}_j, \mathbf{v} \rangle}{\langle \mathbf{u}_j, \mathbf{u}_j \rangle} \mathbf{u}_j, \mathbf{u}_j \rangle = \langle \mathbf{u}_j, \mathbf{v} \rangle \Rightarrow$
$\langle \text{perp}_W(\mathbf{v}), \mathbf{u}_j \rangle = \langle \mathbf{v} - \text{proj}_W(\mathbf{v}), \mathbf{u}_j \rangle = \langle \mathbf{v}, \mathbf{u}_j \rangle - \langle \text{proj}_W(\mathbf{v}), \mathbf{u}_j \rangle = \langle \mathbf{v}, \mathbf{u}_j \rangle - \langle \mathbf{u}_j, \mathbf{v} \rangle = 0 \Rightarrow$
$\text{perp}_W(\mathbf{v})$ is orthogonal to every vector in $\mathcal{B} \Rightarrow \text{perp}_W(\mathbf{v})$ is orthogonal to every $\mathbf{w}$ in W.

Exploration: Vectors and Matrices with Complex Entries

Explorations are self-contained, so only odd-numbered solutions will be provided.

Before we begin, it may be useful to review Appendix C, *Complex Numbers*, on p.650.

1. We need to show $\|\mathbf{v}\| = \sqrt{|v_1^2| + |v_2^2| + \cdots + |v_n^2|}$.

   Since $\mathbf{v} = \begin{bmatrix} v_1 \\ v_2 \\ \vdots \\ v_n \end{bmatrix}$ is in $\mathbb{C}^n$, we have the following:

   $\|\mathbf{v}\| = \sqrt{\mathbf{v} \cdot \mathbf{v}} = \sqrt{\overline{v_1}v_1 + \overline{v_2}v_2 + \cdots + \overline{v_n}v_n}$ (as defined in this Exploration)
   $= \sqrt{|v_1^2| + |v_2^2| + \cdots + |v_n^2|}$ as we were to show.

   Q: What basic fact about complex numbers did we use in the last equality?
   A: The fact that $\overline{z}z = |z|^2$. See p.652 in Appendix C, *Complex Numbers*.

   Q: Is it obvious that $|z|^2 = |z^2|$. If so, prove it.
   Q: Do we need this fact to reach the conclusion above?
   A: Yes, so prove that it is true.

3. The key facts are $\mathbf{u} \cdot \mathbf{v} = \sum \overline{u_i} v_i$, $\overline{u_i v_i} = u_i \overline{v_i}$, $\overline{u_i} u_i = \|u_i\|^2$, and $\overline{\overline{u_i}} = u_i$.

   (a) $\mathbf{u} \cdot \mathbf{v} = \sum \overline{u_i} v_i = \sum u_i \overline{v_i} = \overline{\mathbf{v} \cdot \mathbf{u}}$ (because $\overline{\overline{u_i}} = u_i$).
   (b) $\mathbf{u} \cdot (\mathbf{v} + \mathbf{w}) = \sum \overline{u_i}(v_i + w_i) = \sum \overline{u_i} v_i + \sum \overline{u_i} w_i = \mathbf{u} \cdot \mathbf{v} + \mathbf{u} \cdot \mathbf{w}$.
   (c) $(c\mathbf{u}) \cdot \mathbf{v} = \sum \overline{cu_i} v_i = \overline{c} \sum \overline{u_i} v_i = \overline{c}(\mathbf{u} \cdot \mathbf{v})$.
   (d) $\mathbf{u} \cdot \mathbf{u} = \sum \overline{u_i} u_i = \sum \|u_i\|^2 \geq 0$ and equals zero if and only if all $u_i = 0$.

5. All of these facts follow easily from the properties of complex numbers.

   (a) $\overline{\overline{A}} = [\overline{\overline{a_{ij}}}] = [a_{ij}] = A$.
   (b) $\overline{A + B} = [\overline{a_{ij} + b_{ij}}] = [\overline{a_{ij}}] + [\overline{b_{ij}}] = \overline{A} + \overline{B}$.

   The rest of the items are proved similarly and, therefore, left to the reader.

7. We need to show that $\mathbf{u} \cdot \mathbf{v} = \mathbf{u}^* \mathbf{v}$.

   We have: $\mathbf{u} \cdot \mathbf{v} = \sum \overline{u_i} v_i = \begin{bmatrix} \overline{u_1} & \overline{u_2} & \cdots & \overline{u_n} \end{bmatrix} \begin{bmatrix} v_1 \\ v_2 \\ \vdots \\ v_n \end{bmatrix} = \overline{\mathbf{u}}^T \mathbf{v} = \mathbf{u}^* \mathbf{v}$.

9. We will use Exercise 8 and the definition of $A^* = A$ to determine if A is Hermitian.

   (a) Since the diagonal entries of A are not real (2 and i), A is *not* Hermitian.

   (b) Since $A^* = \begin{bmatrix} -1 & 2+3i \\ 2+3i & 5 \end{bmatrix} \neq \begin{bmatrix} -1 & 2-3i \\ 2-3i & 5 \end{bmatrix} = A$, A is *not* Hermitian.

   (c) Since $A^* = \begin{bmatrix} -3 & 1+5i \\ -1-5i & 3 \end{bmatrix} \neq \begin{bmatrix} -3 & -1+5i \\ 1-5i & 3 \end{bmatrix} = A$, A is *not* Hermitian.

   (d) Since $A^* = A$, A *is* Hermitian.

   (e) Since A has all real entries and $A^T \neq A$, A is *not* Hermitian.

   (f) Since A has all real entries and $A^T = A$, A *is* Hermitian.

11. Following the hint, we adapt the proof of Theorem 5.19 on p.399.

    Given $\lambda_1 \neq \lambda_2$ are eigenvalues of an Hermitian matrix A corresponding to eigenvectors $\mathbf{v}_1$ and $\mathbf{v}_2$ respectively, we will show $\mathbf{v}_1$ and $\mathbf{v}_2$ are orthogonal. That is, $\mathbf{v}_1 \cdot \mathbf{v}_2 = 0$.

    So: $\lambda_1(\mathbf{v}_1 \cdot \mathbf{v}_2) = (\lambda_1 \mathbf{v}_1) \cdot \mathbf{v}_2 = (A\mathbf{v}_1) \cdot \mathbf{v}_2 \stackrel{\text{by 7}}{=} (A\mathbf{v}_1)^* \mathbf{v}_2 \stackrel{\text{by 6d.}}{=} (\mathbf{v}_1^* A^*)\mathbf{v}_2$

    Now we use the fact that A is Hermitian to continue:

    $(\mathbf{v}_1^* A^*)\mathbf{v}_2 \stackrel{A^*=A}{=} (\mathbf{v}_1^* A)\mathbf{v}_2 = \mathbf{v}_1^*(A\mathbf{v}_2) = \mathbf{v}_1^*(\lambda_2 \mathbf{v}_2) = \lambda_2(\mathbf{v}_1^* \mathbf{v}_2) = \lambda_2(\mathbf{v}_1 \cdot \mathbf{v}_2)$

    Therefore: $\lambda_1(\mathbf{v}_1 \cdot \mathbf{v}_2) = \lambda_2(\mathbf{v}_1 \cdot \mathbf{v}_2)$ which implies $(\lambda_1 - \lambda_2)(\mathbf{v}_1 \cdot \mathbf{v}_2) = 0$.

    Since $\lambda_1 - \lambda_2 \neq 0$ (because $\lambda_1 \neq \lambda_2$), we have $\mathbf{v}_1 \cdot \mathbf{v}_2 = 0$ as were trying to show.

13. Following the hint, we adapt the proofs of Theorems 5.4 through 5.7 on pp.372 – 373.

    $a \Rightarrow e$ Since $U^*U = I$, $U\mathbf{x} \cdot U\mathbf{y} = (U\mathbf{x})^* U\mathbf{y} = \mathbf{x}^* U^* U\mathbf{y} = \mathbf{x}^* I\mathbf{y} = \mathbf{x}^* \mathbf{y} = \mathbf{x} \cdot \mathbf{y}$.

    $e \Rightarrow d$ Since $U\mathbf{x} \cdot U\mathbf{y} = \mathbf{x} \cdot \mathbf{y}$, $\|U\mathbf{x}\| = \sqrt{U\mathbf{x} \cdot U\mathbf{x}} = \sqrt{\mathbf{x} \cdot \mathbf{x}} = \|\mathbf{x}\|$.

    Since the remaining proofs are equally straightforward adaptations, they are left to the reader.

Exploration: Vectors and Matrices with Complex Entries

15. In each case, we find the eigenvalues, eigenvectors, and normalize to create U.

(a) Let $A = \begin{bmatrix} 2 & i \\ -i & 2 \end{bmatrix}$, then $\lambda_1 = 1$ with $\mathbf{v}_1 = \begin{bmatrix} \frac{1}{\sqrt{2}} \\ \frac{i}{\sqrt{2}} \end{bmatrix}$ and $\lambda_2 = 3$ with $\mathbf{v}_2 = \begin{bmatrix} \frac{1+i}{2} \\ \frac{1-i}{2} \end{bmatrix}$.

So, we have $U = \begin{bmatrix} \frac{1}{\sqrt{2}} & \frac{1+i}{2} \\ \frac{i}{\sqrt{2}} & \frac{1-i}{2} \end{bmatrix}$ and $U^* = \begin{bmatrix} \frac{1}{\sqrt{2}} & -\frac{i}{\sqrt{2}} \\ \frac{1-i}{2} & \frac{1+i}{2} \end{bmatrix}$.

Verify that $A\mathbf{v}_1 = \mathbf{v}_1$, $A\mathbf{v}_2 = 3\mathbf{v}_2$, and $U^*U = I$.

Finally, verify that $U^*AU = D$, where $D = \begin{bmatrix} 1 & 0 \\ 0 & 3 \end{bmatrix}$.

(b) Let $A = \begin{bmatrix} 0 & -1 \\ 1 & 0 \end{bmatrix}$, then $\lambda_1 = i$ with $\mathbf{v}_1 = \begin{bmatrix} \frac{i}{\sqrt{2}} \\ \frac{1}{\sqrt{2}} \end{bmatrix}$ and $\lambda_2 = -i$ with $\mathbf{v}_2 = \begin{bmatrix} \frac{1+i}{2} \\ \frac{-1+i}{2} \end{bmatrix}$.

So, we have $U = \begin{bmatrix} \frac{i}{\sqrt{2}} & \frac{1+i}{2} \\ \frac{1}{\sqrt{2}} & \frac{-1+i}{2} \end{bmatrix}$ and $U^* = \begin{bmatrix} -\frac{i}{\sqrt{2}} & \frac{1}{\sqrt{2}} \\ \frac{1-i}{2} & \frac{-1-i}{2} \end{bmatrix}$.

Verify that $A\mathbf{v}_1 = i\mathbf{v}_1$, $A\mathbf{v}_2 = -i\mathbf{v}_2$, and $U^*U = I$.

Finally, verify that $U^*AU = D$, where $D = \begin{bmatrix} i & 0 \\ 0 & -i \end{bmatrix}$.

Parts (c) and (d) are left to the reader.

17. If a square complex matrix A is *unitarily diagonalizable*, then $A^*A = AA^*$. So A is normal.

Exploration: Geometric Inequalities and Optimization Problems

Explorations are self-contained, so only odd-numbered solutions will be provided.

1. Since $\mathbf{u} = \begin{bmatrix} \sqrt{x} \\ \sqrt{y} \end{bmatrix}$ and $\mathbf{v} = \begin{bmatrix} \sqrt{y} \\ \sqrt{x} \end{bmatrix}$, $\|\mathbf{u}\| = \|\mathbf{v}\| = \sqrt{(\sqrt{x})^2 + (\sqrt{y})^2} = \sqrt{x+y}$.

   Applying Cauchy-Schwarz, $|\langle \mathbf{u}, \mathbf{v} \rangle| \leq \|\mathbf{u}\|\,\|\mathbf{v}\|$, we have: $|\sqrt{xy} + \sqrt{xy}| \leq \sqrt{x+y}\sqrt{x+y}$.
   Recall $x, y \geq 0$, so $\sqrt{x+y}\sqrt{x+y} = |x+y| = x+y$.

   So, we have: $2\sqrt{xy} \leq x+y \Rightarrow \sqrt{xy} \leq \dfrac{x+y}{2}$.

   Now, we show that equality holds $\Leftrightarrow x = y$.

   $x = y \Rightarrow x = \sqrt{xx} = \dfrac{x+x}{2} = x \Rightarrow$ equality holds.

   $\sqrt{xy} = \dfrac{x+y}{2} \Rightarrow 2\sqrt{xy} = x+y \Rightarrow 4xy = (x+y)^2 = x^2 + 2xy + y^2 \Rightarrow$
   $x^2 - 2xy + y^2 = (x-y)^2 = 0 \Rightarrow x - y = 0 \Rightarrow x = y$.

3. Since the area $A = xy = 100$, we have: $10 = \sqrt{xy} \leq \dfrac{x+y}{2} \Rightarrow 40 \leq 2(x+y) = P$

   with equality holding $\Leftrightarrow x = y \Rightarrow 40 = 4x \Rightarrow x = y = 10 \Rightarrow$
   The square is the rectangle with the smallest perimeter, namely 40.

   This follows from the fact that equality holds $\Leftrightarrow x = y$. That is, it is independent of the area.

5. The formula for the volume described is $V = x(10 - 2x)x$. So, we have:
   $\sqrt[3]{V} \leq \dfrac{x + (10 - 2x) + x}{3}$ with equality holding $\Leftrightarrow x = 10 - 2x = x \Rightarrow$
   $3x = 10 \Rightarrow x = \dfrac{10}{3} \Rightarrow 10 - 2x = \dfrac{10}{3} \Rightarrow$

   The dimensions $\dfrac{10}{3} \times \dfrac{10}{3} \times \dfrac{10}{3}$ make the volume of the box as large as possible.

7. We use the substitution $u = x - y \Rightarrow x + \dfrac{8}{y(x-y)} = u + y + \dfrac{8}{uy} \Rightarrow$

   We have an AMGM of the form given in Example 7.10, namely: $\sqrt[3]{xyz} \leq \dfrac{x+y+z}{3} \Rightarrow$

   $\sqrt[3]{uy \cdot \dfrac{8}{uy}} \leq \dfrac{u + y + (8/uy)}{3} \Rightarrow 6 = 3\sqrt[3]{8} \leq u + y + \dfrac{8}{uy} = x + \dfrac{8}{y(x-y)}$

   with equality holding $\Leftrightarrow u = y = \dfrac{8}{uy} \Rightarrow y = \dfrac{8}{y^2} \Rightarrow y = 2 = u \Rightarrow x = u + y = 4$.

   So, the minimum value is 6 and it occurs when $x = 4$ and $y = 2$.

9. Following the method of Example 7.11, we have: $\mathbf{u} = \begin{bmatrix} 1 \\ 1 \\ \sqrt{2} \end{bmatrix}$ and $\mathbf{v} = \begin{bmatrix} x \\ y \\ \frac{1}{\sqrt{2}} z \end{bmatrix}$.

   Then the componentwise form of the Cauchy-Schwarz Inequality gives
   $$(x + y + z)^2 \leq (1^2 + 1^2 + (\sqrt{2})^2)(x^2 + y^2 + \frac{1}{2}z^2)$$
   So, since $x + y + z = 10$, we have:
   $$100 \leq 4(x^2 + y^2 + \frac{1}{2}z^2) \Rightarrow 25 \leq x^2 + y^2 + \frac{1}{2}z^2 \Rightarrow$$
   The minimum value of our function is 25 because the minimum occurs at equality.

11. Following the method of Example 7.11, we have: $\mathbf{u} = \begin{bmatrix} 1 \\ 2 \end{bmatrix}$ and $\mathbf{v} = \begin{bmatrix} x \\ y \end{bmatrix}$

    because we have the constraint $x + 2y = 5$.
    The function we want to minimize is $x^2 + y^2$, the distance from the origin.
    Then the componentwise form of the Cauchy-Schwarz Inequality gives
    $$(x + 2y)^2 \leq (1^2 + 2^2)(x^2 + y^2)$$
    So, since $x + 2y = 5 \Rightarrow y = \frac{1}{2}(5 - x)$, we have:
    $$25 \leq 5(x^2 + y^2) \Rightarrow 5 \leq x^2 + y^2 \Rightarrow 5 \leq x^2 + \frac{1}{4}(5 - x)^2$$
    Now, since the minimum occurs at equality, we solve the equality:
    $$4x^2 + x^2 - 10x + 25 = 20 \Rightarrow 5x^2 - 10x + 5 = 0 \Rightarrow x^2 - 2x + 1 = 0 \Rightarrow (x - 1)^2 = 0 \Rightarrow x = 1 \Rightarrow$$
    The point on the line $x + 2y = 5$ closest to the origin is $(1, 2)$.

13. The area of the rectangle is $A = 2xy$, where $x^2 + y^2 = r^2$.
    Now from Exercise 12, we have:
    $$\sqrt{\frac{x^2 + y^2}{2}} \geq \sqrt{xy} \Rightarrow \sqrt{x^2 + y^2} \geq \sqrt{2xy} \Rightarrow \sqrt{r^2} = r \geq \sqrt{2xy} = \sqrt{A} \Rightarrow A \leq r^2.$$
    So, since the maximum occurs at equality, $A = r^2$ is the maximum area.

15. Squaring the inequality from Exercise 12, we have:
$$\frac{x^2+y^2}{2} \geq \left(\frac{x+y}{2}\right)^2 \geq \left(\frac{2}{1/x+1/y}\right)^2.$$

So, we have:
$$\frac{\left(x+\frac{1}{x}\right)^2 + \left(y+\frac{1}{y}\right)^2}{2} \geq \left(\frac{\left(x+\frac{1}{x}\right)+\left(y+\frac{1}{y}\right)}{2}\right)^2.$$

Recall, $x+y=1 \Rightarrow$
$$\left(\frac{\left(x+\frac{1}{x}\right)+\left(y+\frac{1}{y}\right)}{2}\right)^2 = \left(\frac{1}{2}+\frac{\frac{1}{x}+\frac{1}{y}}{2}\right)^2 \geq \left(\frac{1}{2}+\frac{2}{x+y}\right)^2.$$

Again, use the fact that $x+y=1 \Rightarrow$
$$\left(\frac{1}{2}+\frac{2}{x+y}\right)^2 = \left(\frac{5}{2}\right)^2 = \frac{25}{4} \Rightarrow f(x,y) = \frac{25}{2} \text{ is the minimum when } x+y=1.$$

This equality holds $\Leftrightarrow x=y$. Since $x+y=1$, the minimum occurs when $x=y=\frac{1}{2}$.

7.2 Norms and Distance Functions

Q: What are our main goals for Section 7.2?
A: To see how matrix norms are induced by vector norms.
   This process is subtle. Spend time working through each step of its development.

Key Definitions and Concepts

$\|\mathbf{v}\|_1 = \|\mathbf{v}\|_s$	p.562	7.2	$\|\mathbf{v}\|_1 = \|\mathbf{v}\|_s = |v_1| + \cdots + |v_n|$
$\|\mathbf{v}\|_\infty = \|\mathbf{v}\|_m$	p.562	7.2	$\|\mathbf{v}\|_\infty = \|\mathbf{v}\|_m = \max\{|v_1|, \ldots, |v_n|\}$
$\|\mathbf{v}\|_p$	p.562	7.2	$\|\mathbf{v}\|_p = (|v_1|^p + \cdots + |v_n|^p)^{1/p}$
$\|\mathbf{v}\|_2 = \|\mathbf{v}\|_E$	p.562	7.2	$\|\mathbf{v}\|_2 = \|\mathbf{v}\|_E = \sqrt{|v_1|^2 + \cdots + |v_n|^2}$
$\|\mathbf{v}\|_H$	p.563	7.2	$\|\mathbf{v}\|_H = w(\mathbf{v})$ (weight). See Example 7.15.
compatible	p.565	7.2	A matrix norm $\|A\|$ is compatible with a vector norm $\|\mathbf{x}\|$ if $\|A\mathbf{x}\| \le \|A\|\|\mathbf{x}\|$
condition number	p.571	7.4	The condition number of a square matrix is $\|A^{-1}\|\|A\|$. If A is not invertible, we define $\text{cond}(A) = \infty$.
conditioned	p.570	7.2	For description of *ill-* and *well-conditioned*, see text.
distance	p.563	7.2	See Example 7.16 for $d_s(\mathbf{u},\mathbf{v})$, $d_E(\mathbf{u},\mathbf{v})$, $d_H(\mathbf{u},\mathbf{v})$ ...
matrix norm	p.565	7.2	A *matrix norm* on M_{nn} has the following properties: 1. $\|A\| \ge 0$, and $\|A\| = 0$ if and only if $A = O$ 2. $\|cA\| = |c|\|A\|$ 3. $\|A + B\| \le \|A\| + \|B\|$ 4. $\|AB\| \le \|A\|\|B\|$
norm normed linear space	p.561	7.2	A *norm* on a vector space V has the following properties: 1. $\|\mathbf{v}\| \ge 0$, and $\|\mathbf{v}\| = 0$ if and only if $\mathbf{v} = \mathbf{0}$ 2. $\|c\mathbf{v}\| = |c|\|\mathbf{v}\|$ 3. $\|\mathbf{u} + \mathbf{v}\| \le \|\mathbf{u}\| + \|\mathbf{v}\|$
Euclidean norm	p.562	7.2	$\|\mathbf{v}\|_E = \|\mathbf{v}\|_2 = \sqrt{|v_1|^2 + \cdots + |v_n|^2}$ (same as 2-norm)
Frobenius norm	p.565	7.2	$\|A\|_F = \sqrt{\sum_{i,j=1}^n a_{ij}^2}$. See Example 7.18.
Hamming norm	p.563	7.2	$\|\mathbf{v}\|_H = w(\mathbf{v})$ (weight). See Example 7.15.
max norm	p.562	7.2	$\|\mathbf{v}\|_m = \|\mathbf{v}\|_\infty = \max\{|v_1|, \ldots, |v_n|\}$ (same as ∞-norm)
operator norm	p.568	7.2	$\|A\| = \max\|A\mathbf{x}\|$ where $\|\mathbf{x}\| = 1$. See Theorem 7.6. $\|A\|_1 = \max\|A\mathbf{x}\|_s$, $\|A\|_2 = \max\|A\mathbf{x}\|_E$, $\|A\|_\infty = \max\|A\mathbf{x}\|_m$
p-norm	p.562	7.2	$\|\mathbf{v}\|_p = (|v_1|^p + \cdots + |v_n|^p)^{1/p}$
sum norm	p.561	7.2	$\|\mathbf{v}\|_s = \|\mathbf{v}\|_1 = |v_1| + \cdots + |v_n|$ (same as 1-norm)

Theorems

Thm 7.5 p.564 7.2 Let d be a distance function defined on a normed linear space V. The following properties hold for all vectors $\mathbf{u}$, $\mathbf{v}$, and $\mathbf{w}$ in V.
a. $d(\mathbf{u}, \mathbf{v}) \geq 0$, and $d(\mathbf{u}, \mathbf{v}) = 0$ if and only if $\mathbf{u} = \mathbf{v}$
b. $d(\mathbf{u}, \mathbf{v}) = d(\mathbf{v}, \mathbf{u})$
c. $d(\mathbf{u}, \mathbf{w}) \leq d(\mathbf{u}, \mathbf{v}) + d(\mathbf{v}, \mathbf{w})$

Thm 7.6 p.567 7.2 If $\|\mathbf{x}\|$ is a vector norm on $\mathbb{R}^n$, then $\|A\| = \max \|A\mathbf{x}\|$ where $\|\mathbf{x}\| = 1$ defines a matrix norm on M_{nn} that is compatible with the vector norm that induces it.

Thm 7.7 p.569 7.2 Let A have column vectors $\mathbf{a}_i$ and row vectors $\mathbf{A}_i$ then:
a. $\|A\|_1 = \max\{\|\mathbf{a}_i\|_s\} = \max\left\{\sum_{i=1}^{n} |a_{ij}|\right\}$
b. $\|A\|_\infty = \max\{\|\mathbf{A}_i\|_s\} = \max\left\{\sum_{j=1}^{n} |a_{ij}|\right\}$

7.2 Norms and Distance Functions

Analysis of how the sum norm $\|\mathbf{v}\|_s$ *induces the matrix norm* $\|A\|_1$

Q: What is the definition of the sum norm, $\|\mathbf{v}\|_s = \|\mathbf{v}\|_1$?
A: The sum of the absolute value of the components, $\|\mathbf{v}\|_s = \|\mathbf{v}\|_1 = |v_1| + \cdots + |v_n|$.

Q: How is this definition used to induce the matrix norm $\|A\|_1$?
A: As stated in Theorem 7.6, $\|A\|_1 = \max \|A\mathbf{x}\|_s$ where $\|\mathbf{x}\|_s = 1$.

Q: Is the matrix norm $\|A\|_1$ induced in this way compatible with the vector norm $\|\mathbf{v}\|_s$?
A: Yes. That is precisely the assertion of Theorem 7.6.

Q: What does it mean for $\|A\|_1$ to be compatible with $\|\mathbf{v}\|_s$?
A: It means for a given matrix A, $\|A\mathbf{x}\|_s \leq \|A\|_1 \|\mathbf{x}\|_s$ for all vectors $\mathbf{x}$.

Q: If $\|\mathbf{x}\|_s = 1$, how does the above inequality simplify?
A: If $\|\mathbf{x}\|_s = 1$, $\|A\mathbf{x}\|_s \leq \|A\|_1 \|\mathbf{x}\|_s \Rightarrow \|A\mathbf{x}\|_s \leq \|A\|_1$.

Q: Why is it obvious that $\|A\|_1$ satisfies this simplified inequality?
A: Because $\|A\|_1 = \max \|A\mathbf{x}\|_s$ where $\|\mathbf{x}\|_s = 1$.

Q: How does this ensure that the inequality $\|A\mathbf{x}\|_s \leq \|A\|_1 \|\mathbf{x}\|_s$ holds for all vectors $\mathbf{x}$?
A: Hint: Normalize $\|\mathbf{x}\|_s$ to get $\mathbf{w} = \frac{1}{\|\mathbf{x}\|_s}\mathbf{x}$ so $\|\mathbf{w}\|_s = 1$.
See how the scalar $\frac{1}{\|\mathbf{x}\|_s}$ factors out of both sides of the inequality?

Q: In Theorem 7.7, we take the maximum only for the columns of A.
That is, $\|A\|_1 = \max\{\|\mathbf{a}_i\|_s\}$.
How does this ensure that the inequality $\|A\mathbf{x}\|_s \leq \|A\|_1 \|\mathbf{x}\|_s$ holds for all vectors $\mathbf{x}$?
A: Hint: Note that $\|A\mathbf{e}_i\|_s = \|\mathbf{a}_i\|_s$, so $\|A\|_1 \geq \|\mathbf{a}_i\|_s$ for all s.
Without consulting the proof, explain why $\|A\|_1 \leq \|\mathbf{a}_k\|_s$ for some k.
Hint: Use the fact that $\|A\|_1 = \max \|A\mathbf{x}\|_s$ where $\|\mathbf{x}\|_s = 1$.

Q: Repeat this analysis for $\|A\|_\infty$.

Solutions to odd-numbered exercises from Section 7.2

1. $\|\mathbf{u}\|_E = \sqrt{(-1)^2 + (4)^2 + (-5)^2} = \sqrt{42}$.
   $\|\mathbf{u}\|_s = |-1| + |4| + |-5| = 10$.
   $\|\mathbf{u}\|_m = \max\{|-1|, |4|, |-5|\} = 5$.

3. $d_E\langle\mathbf{u}, \mathbf{v}\rangle = \sqrt{(-3)^2 + (6)^2 + (-5)^2} = \sqrt{70}$.
   $d_s\langle\mathbf{u}, \mathbf{v}\rangle = |-3| + |6| + |-5| = 14$.
   $d_m\langle\mathbf{u}, \mathbf{v}\rangle = \max\{|-3|, |6|, |-5|\} = 6$.

5. $\|\mathbf{u}\|_H = w(\mathbf{u})$ = number of 1s in $\mathbf{u} = 4$.
   $\|\mathbf{v}\|_H = w(\mathbf{v})$ = number of 1s in $\mathbf{v} = 5$.

7. (a) $\|\mathbf{v}\|_E = \sqrt{\sum v_i^2} = \max\{|v_i|\} \Leftrightarrow v_j = 0, j \neq i, |v_i| = \max \Leftrightarrow$ *non*-max components $= 0$.
   (b) $\|\mathbf{v}\|_s = \sum |v_i| = \max\{|v_i|\} \Leftrightarrow v_j = 0, j \neq i, |v_i| = \max \Leftrightarrow$ all *non*-max components $= 0$.
   (c) (a) and (b) $\Rightarrow \|\mathbf{v}\|_E = \|\mathbf{v}\|_s = \|\mathbf{v}\|_m = \max\{|v_i|\} \Leftrightarrow$ all *non*-max components $= 0$.

9. Let $\|\mathbf{v}\|_m = \max\{|v_i|\} = |v_k|$. Then we have: $\|\mathbf{v}\|_E = \sqrt{\sum v_i^2} \geq \sqrt{v_k^2} = |v_k| = \|\mathbf{v}\|_m$.

11. $\|\mathbf{v}\|_s = \sum |v_i| \leq n \max\{|v_i|\} = n\|\mathbf{v}\|_m$.

13. $\|\mathbf{u}\|_s = |x| + |y| = 1$ and $\|\mathbf{u}\|_m = \max\{|x|, |y|\} = 1 \Rightarrow$ four line segments each.

For $\|\mathbf{u}\|_s = |x| + |y| = 1$ we have:
$x + y = 1;\ x \geq 0,\ y \geq 0$
$x - y = 1;\ x \geq 0,\ y \leq 0$
$-x + y = 1;\ x \leq 0,\ y \geq 0$
$-x - y = 1;\ x \leq 0,\ y \leq 0$

And for $\|\mathbf{u}\|_m = \max\{|x|, |y|\} = 1$ we have:
$x = 1,\ |y| \leq 1$
$x = -1,\ |y| \leq 1$
$y = 1,\ |x| \leq 1$
$y = -1,\ |x| \leq 1$

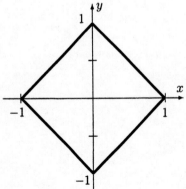
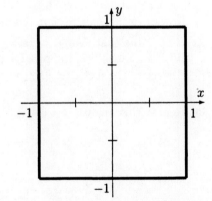

7.2 Norms and Distance Functions

15.
1. Absolute value in definition $\Rightarrow \geq 0$, $\max\{|2a|, |3b|\} = 0 \Rightarrow a = b = 0 \Rightarrow \mathbf{v} = \mathbf{0}$.
2. $\|c\mathbf{v}\| = \max\{|2ca|, |3cb|\} = |c|\max\{|2a|, |3b|\} = |c|\|\mathbf{v}\|$.
3. $\|\mathbf{u}+\mathbf{v}\| = \max\{|2a+2c|, |3b+3d|\} \leq \max\{|2a|+|2c|, |3b|+|3d|\}$
   $\leq \max\{|2a|, |3b|\} + \max\{|2c|, |3d|\} = \|\mathbf{u}\| + \|\mathbf{v}\|$.

17.
1. Absolute value in definition $\Rightarrow \geq 0$, $\int |f(x)|\,dx = 0 \Rightarrow |f(x)| = 0 \Rightarrow f(x) = 0$.
2. $\|cf\| = \int |cf(x)|\,dx = |c| \int |f(x)|\,dx = |c|\|f\|$.
3. $\|f+g\| = \int |f(x)+g(x)|\,dx \leq \int |f(x)|+|g(x)|\,dx = \int |f(x)|\,dx + \int |g(x)|\,dx = \|f\|+\|g\|$.

19. $\|\mathbf{u}-\mathbf{v}\| = |-1|\|\mathbf{u}-\mathbf{v}\| = \|-1(\mathbf{u}-\mathbf{v})\| = \|-\mathbf{u}+\mathbf{v}\| = \|\mathbf{v}-\mathbf{u}\|$.

21. $\|A\|_F = \sqrt{0^2 + (-1)^2 + (-3)^2 + 3^2} = \sqrt{19}$.
$\|A\|_1 = \max\{|0|+|-3|, |-1|+|3|\} = 4$.
$\|A\|_\infty = \max\{|0|+|1|, |-3|+|3|\} = 6$.

23. $\|A\|_F = \sqrt{2(2^2) + 5(1^2) + 2(3^2)} = \sqrt{31}$.
$\|A\|_1 = \max\{4, 5, 6\} = 6$.
$\|A\|_\infty = \max\{4, 6, 5\} = 6$.

25. $\|A\|_F = \sqrt{4^2 + \ldots + 0^2} = 2\sqrt{11}$.
$\|A\|_1 = \max\{7, 6, 3\} = 7$.
$\|A\|_\infty = \max\{7, 3, 6\} = 7$.

27. $\|A\|_1 = \max \|\mathbf{a}_i\|_s = \|\mathbf{a}_2\|_s = \|A\mathbf{x}\|_s$, where $\mathbf{x} = \begin{bmatrix} 0 \\ 1 \end{bmatrix}$ and $\|\mathbf{x}\|_s = 1$ as required.

$\|A\|_\infty = \max \|\mathbf{A}_i\|_s = \|\mathbf{A}_2\|_s = \|A\mathbf{y}\|_m$, where $\mathbf{y} = \begin{bmatrix} -1 \\ 1 \end{bmatrix}$ and $\|\mathbf{y}\|_m = 1$ as required.

29. $\|A\|_1 = \max \|\mathbf{a}_i\|_s = \|\mathbf{a}_3\|_s = \|A\mathbf{x}\|_s$, where $\mathbf{x} = \begin{bmatrix} 0 \\ 0 \\ 1 \end{bmatrix}$ and $\|\mathbf{x}\|_s = 1$ as required.

$\|A\|_\infty = \max \|\mathbf{A}_i\|_s = \|\mathbf{A}_2\|_s = \|A\mathbf{y}\|_m$, where $\mathbf{y} = \begin{bmatrix} 1 \\ 1 \\ 1 \end{bmatrix}$ and $\|\mathbf{y}\|_m = 1$ as required.

31. $\|A\|_1 = \max \|\mathbf{a}_i\|_s = \|\mathbf{a}_1\|_s = \|A\mathbf{x}\|_s$, where $\mathbf{x} = \begin{bmatrix} 1 \\ 0 \\ 0 \end{bmatrix}$ and $\|\mathbf{x}\|_s = 1$ as required.

$\|A\|_\infty = \max \|\mathbf{A}_i\|_s = \|\mathbf{A}_1\|_s = \|A\mathbf{y}\|_m$, where $\mathbf{y} = \begin{bmatrix} 1 \\ -1 \\ -1 \end{bmatrix}$ and $\|\mathbf{y}\|_m = 1$ as required.

33. (a) $\|I\| = \max\|I\mathbf{x}\| = \max\|\mathbf{x}\| = 1$ because the max is taken over $\|\mathbf{x}\| = 1$.

   (b) $\|I\|_F = \sqrt{\sum 1^2} = \sqrt{n} \Rightarrow$ there is no vector norm that induces the Frobenius norm except when $n = 1$ when it is equivalent to absolute value on the real number line.

35. $A^{-1} = \begin{bmatrix} 1 & -\frac{1}{2} \\ -2 & \frac{3}{2} \end{bmatrix} \Rightarrow \begin{array}{l} \text{cond}_1(A) = \|A^{-1}\|_1\|A\|_1 = 7 \cdot 3 = 21 \\ \text{cond}_\infty(A) = \|A^{-1}\|_\infty\|A\|_\infty = 6 \cdot \frac{7}{2} = 21 \end{array} \Rightarrow A$ is well-conditioned.

37. $A^{-1} = \begin{bmatrix} 100 & -99 \\ -100 & 100 \end{bmatrix} \Rightarrow \begin{array}{l} \text{cond}_1(A) = 2 \cdot 200 = 400 \\ \text{cond}_\infty(A) = 2 \cdot 200 = 400 \end{array} \Rightarrow A$ is ill-conditioned.

39. $A^{-1} = \begin{bmatrix} 0 & 0 & 1 \\ 6 & -1 & -1 \\ -5 & 1 & 0 \end{bmatrix} \Rightarrow \begin{array}{l} \text{cond}_1(A) = 7 \cdot 11 = 77 \\ \text{cond}_\infty(A) = 16 \cdot 8 = 128 \end{array} \Rightarrow A$ is moderately ill-conditioned.

41. (a) $A^{-1} = \begin{bmatrix} -\frac{1}{k-1} & \frac{k}{k-1} \\ \frac{1}{k-1} & -\frac{1}{k-1} \end{bmatrix} \Rightarrow \text{cond}_\infty(A) = (\max\{|k|+1, 2\})(\max\{|\frac{k}{k-1}|+|\frac{1}{k-1}|, |\frac{2}{k-1}|\})$.

   (b) $k \to 1 \Rightarrow \text{cond}_\infty(A) \to \infty$ as it should since $A \longrightarrow \begin{bmatrix} 1 & 1 \\ 1 & 1 \end{bmatrix}$ which is not invertible.

43. (a) $A^{-1} = \begin{bmatrix} -\frac{9}{10} & 1 \\ 1 & -1 \end{bmatrix} \Rightarrow \text{cond}_\infty(A) = 20 \cdot 2 = 40$.

   (b) $\Delta A = \begin{bmatrix} 10 & 10 \\ 10 & 11 \end{bmatrix} - \begin{bmatrix} 10 & 10 \\ 10 & 9 \end{bmatrix} = \begin{bmatrix} 0 & 0 \\ 0 & 2 \end{bmatrix} \Rightarrow \|\Delta A\| = 2 \Rightarrow$

   $\frac{\|\Delta \mathbf{x}\|_m}{\|\mathbf{x}'\|_m} \leq \text{cond}_\infty(A) \frac{\|\Delta A\|_\infty}{\|A\|_\infty} = 40 \cdot \frac{2}{20} = 4 \Rightarrow$ can produce at most a 400% relative change.

   (c) $[A|\mathbf{b}] \longrightarrow \begin{bmatrix} 1 & 0 & 9 \\ 0 & 1 & 1 \end{bmatrix} \Rightarrow \mathbf{x} = \begin{bmatrix} 9 \\ 1 \end{bmatrix}$. $[A'|\mathbf{b}] \longrightarrow \begin{bmatrix} 1 & 0 & 11 \\ 0 & 1 & -1 \end{bmatrix} \Rightarrow \mathbf{x}' = \begin{bmatrix} 11 \\ -1 \end{bmatrix} \Rightarrow$

   $\Delta \mathbf{x} = \mathbf{x}' - \mathbf{x} = \begin{bmatrix} 2 \\ 2 \end{bmatrix} \Rightarrow \frac{\|\Delta \mathbf{x}\|_m}{\|\mathbf{x}\|_m} = \frac{2}{9} \approx 22\%$ actual relative error.

   (d) $\Delta \mathbf{b} = \begin{bmatrix} 100 \\ 101 \end{bmatrix} - \begin{bmatrix} 100 \\ 99 \end{bmatrix} = \begin{bmatrix} 0 \\ 2 \end{bmatrix} \Rightarrow \|\Delta \mathbf{b}\|_m = 2 \Rightarrow$

   $\frac{\|\Delta \mathbf{x}\|_m}{\|\mathbf{x}\|_m} \leq \text{cond}_\infty(A) \frac{\|\Delta \mathbf{b}\|_m}{\|\mathbf{b}\|_m} = 40 \cdot \frac{2}{100} = 0.8 \Rightarrow$ at most an 80% relative change.

   (e) $[A|\mathbf{b}] \longrightarrow \begin{bmatrix} 1 & 0 & 9 \\ 0 & 1 & 1 \end{bmatrix} \Rightarrow \mathbf{x} = \begin{bmatrix} 9 \\ 1 \end{bmatrix}$. $[A|\mathbf{b}'] \longrightarrow \begin{bmatrix} 1 & 0 & 11 \\ 0 & 1 & -1 \end{bmatrix} \Rightarrow \mathbf{x}' = \begin{bmatrix} 11 \\ -1 \end{bmatrix} \Rightarrow$

   $\Delta \mathbf{x} = \mathbf{x}' - \mathbf{x} = \begin{bmatrix} 2 \\ 2 \end{bmatrix} \Rightarrow \frac{\|\Delta \mathbf{x}\|_m}{\|\mathbf{x}\|_m} = \frac{2}{9} \approx 22\%$ actual relative error.

45. Exercise 33(a) $\Rightarrow 1 = \|I\| = \|A^{-1}A\| \leq \|A^{-1}\|\|A\| = \text{cond}(A)$.

47. Section 4.3, Theorem 4.18b $\Rightarrow \lambda$ is an eigenvalue of $A \Leftrightarrow \frac{1}{\lambda}$ is an eigenvalue of A^{-1}.

   So, Exercise 34 in Section 4.3 $\Rightarrow \text{cond}(A) = \|A^{-1}\|\|A\| \geq |\frac{1}{\lambda_n}||\lambda_1| = \frac{|\lambda_1|}{|\lambda_n|}$.

7.2   Norms and Distance Functions

49. Section 2.5, Exercise 3 and Section 7.2, Example 7.11 $\Rightarrow M = -D^{-1}(L+U)$

$$M = -\begin{bmatrix} 4.5 & 0 \\ 0 & -3.5 \end{bmatrix}^{-1} \begin{bmatrix} 0 & -0.5 \\ 1 & 0 \end{bmatrix} = \begin{bmatrix} -\frac{2}{9} & 0 \\ 0 & \frac{2}{7} \end{bmatrix} \begin{bmatrix} 0 & -\frac{1}{2} \\ 1 & 0 \end{bmatrix} = \begin{bmatrix} 0 & \frac{1}{9} \\ \frac{2}{7} & 0 \end{bmatrix} \Rightarrow \|M\|_\infty = \frac{2}{7}.$$

$$\mathbf{b} = \begin{bmatrix} 1 \\ -1 \end{bmatrix} \Rightarrow \mathbf{c} = D^{-1}\mathbf{b} = \begin{bmatrix} \frac{2}{9} & 0 \\ 0 & -\frac{2}{7} \end{bmatrix} \begin{bmatrix} 1 \\ -1 \end{bmatrix} = \begin{bmatrix} \frac{2}{9} \\ \frac{2}{7} \end{bmatrix} \Rightarrow \|\mathbf{c}\|_m = \frac{2}{7}.$$

Note, $\mathbf{x}_0 = \mathbf{0} \Rightarrow \mathbf{x}_1 = \mathbf{c} \Rightarrow \|M\|_\infty^k \|\mathbf{x}_1 - \mathbf{x}_0\|_m = \|M\|_\infty^k \|\mathbf{c}\|_m < 0.0005 \Rightarrow k > \frac{4+\log_{10}(\|\mathbf{c}\|_m/5)}{\log_{10}(1/\|M\|_\infty)}$.

So, in this case with $\|M\|_\infty = \frac{2}{7}$ and $\|\mathbf{c}\|_m = \frac{2}{7} \Rightarrow k > \frac{4+\log_{10}(2/35)}{\log_{10}(7/2)} \approx 5.06 \Rightarrow k \geq 6$.

$\mathbf{x}_{k+1} = M\mathbf{x}_k + \mathbf{c} \Rightarrow \mathbf{x}_6 = \begin{bmatrix} 0.2622 \\ 0.3606 \end{bmatrix}$. Compare $\mathbf{x} = \mathbf{x}_{actual} = \begin{bmatrix} 0.2623 \\ 0.3606 \end{bmatrix}$.

51. $M = -\begin{bmatrix} 3 & 0 & 0 \\ 0 & 4 & 0 \\ 0 & 0 & 3 \end{bmatrix}^{-1} \begin{bmatrix} 0 & 1 & 0 \\ 1 & 0 & 1 \\ 0 & 1 & 0 \end{bmatrix} = \begin{bmatrix} 0 & -\frac{1}{3} & 0 \\ -\frac{1}{4} & 0 & -\frac{1}{4} \\ 0 & -\frac{1}{3} & 0 \end{bmatrix} \Rightarrow \|M\|_\infty = \frac{2}{4} = \frac{1}{2}$.

$\mathbf{b} = \begin{bmatrix} 1 \\ 1 \\ 1 \end{bmatrix} \Rightarrow \mathbf{c} = D^{-1}\mathbf{b} = \begin{bmatrix} \frac{1}{3} & 0 & 0 \\ 0 & \frac{1}{4} & 0 \\ 0 & 0 & \frac{1}{3} \end{bmatrix} \begin{bmatrix} 1 \\ 1 \\ 1 \end{bmatrix} = \begin{bmatrix} \frac{1}{3} \\ \frac{1}{4} \\ \frac{1}{3} \end{bmatrix} \Rightarrow \|\mathbf{c}\|_m = \frac{1}{3}$.

In this case with $\|M\|_\infty = \frac{1}{2}$ and $\|\mathbf{c}\|_m = \frac{1}{3} \Rightarrow k > \frac{4+\log_{10}(1/15)}{\log_{10} 2} \approx 9.3 \Rightarrow k \geq 10$.

$\mathbf{x}_{k+1} = M\mathbf{x}_k + \mathbf{c} \Rightarrow \mathbf{x}_{10} = \begin{bmatrix} 0.299 \\ 0.099 \\ 0.299 \end{bmatrix}$. Compare $\mathbf{x} = \mathbf{x}_{actual} = \begin{bmatrix} 0.3 \\ 0.1 \\ 0.3 \end{bmatrix}$.

7.3 Least Squares Approximation

Q: What are our main goals for Section 7.3?
A: To examine the connection between approximations and pseudoinverse matrices, A^+. Review Section 5.2 for the development of orthogonal *complements* and *projections*. Thm 5.16 (QR Factorization Thm) of Section 5.3 is essential to the proof of Thm 7.10.

Key Definitions and Concepts

$\bar{\mathbf{v}}$	p.579	7.3	A vector $\bar{\mathbf{v}}$ is *best approximation* to $\mathbf{v}$ in W if $\|\mathbf{v} - \bar{\mathbf{v}}\| < \|\mathbf{v} - \mathbf{w}\|$ for all $\mathbf{w} \neq \bar{\mathbf{v}}$
$\|\mathbf{e}\|$	p.581	7.3	$\|\mathbf{e}\| = \sqrt{\varepsilon_1^2 + \varepsilon_2^2 + \varepsilon_3^2}$ is the *least squares error* of the approximation
A^+	p.594	7.3	If A is a matrix with linearly independent columns, then $A^+ = (A^T A)^{-1} A^T$ is the *pseudoinverse*.
best approximation	p.579	7.3	A vector $\bar{\mathbf{v}}$ is *best approximation* to $\mathbf{v}$ in W if $\|\mathbf{v} - \bar{\mathbf{v}}\| < \|\mathbf{v} - \mathbf{w}\|$ for all $\mathbf{w} \neq \bar{\mathbf{v}}$
least squares error	p.581	7.3	$\|\mathbf{e}\| = \sqrt{\varepsilon_1^2 + \varepsilon_2^2 + \varepsilon_3^2}$ is the *least squares error* of the approximation
least squares solution	p.583	7.3	A *least squares solution* of $A\mathbf{x} = \mathbf{b}$ is $\bar{\mathbf{x}}$ such that $\|\mathbf{b} - A\bar{\mathbf{x}}\| \leq \|\mathbf{b} - A\mathbf{x}\|$ for all $\mathbf{x}$ in $\mathbb{R}$
normal equations	p.584	7.3	$A^T A \bar{\mathbf{x}} = A^T \mathbf{b}$
Penrose conditions	p.595	7.3	A^+ the pseudoinverse of A satisfies the *Penrose conditions*. a. $AA^+A = A$ b. $A^+AA^+ = A^+$ c. AA^+ and A^+A are symmetric
pseudoinverse	p.594	7.3	If A is a matrix with linearly independent columns, then $A^+ = (A^T A)^{-1} A^T$ is the *pseudoinverse*.

Theorems

Thm 7.8   p.579   7.3   If W is a finite-dimensional subspace of an inner product space V
and if $\mathbf{v}$ is a vector in V,
then $\operatorname{proj}_W(\mathbf{v})$ is the *best approximation* to $\mathbf{v}$ in W.

Thm 7.9   p.584   7.3 $A\mathbf{x} = \mathbf{b}$ has at least one *least squares solution*, $\bar{\mathbf{x}}$.
  a. $\bar{\mathbf{x}}$ is a *least squares solution* of $A\mathbf{x} = \mathbf{b}$ if and only if
     $\bar{\mathbf{x}}$ is a solution of the normal equations $A^T A \bar{\mathbf{x}} = A^T \mathbf{b}$.
  b. A has linearly independent columns $\Leftrightarrow A^T A$ is invertible.
     Then $\bar{\mathbf{x}} = (A^T A)^{-1} A^T \mathbf{b}$ is the *unique* solution.

Thm 7.10   p.591   7.3   Let A have linearly independent columns
with QR factorization $A = QR$ (see Theorem 5.16 below).
Then $\bar{\mathbf{x}} = R^{-1} Q^T \mathbf{b}$ is the *unique* solution of $A\mathbf{x} = \mathbf{b}$.

Thm 7.11   p.592   7.3   Let the columns of A be a basis for W.
Then $P : \mathbb{R}^m \to \mathbb{R}^n$ that projects $\mathbb{R}^m$ onto W
has matrix $A(A^T A)^{-1} A^T$ and
$\operatorname{proj}_W(\mathbf{v}) = A(A^T A)^{-1} A^T \mathbf{v}$.

Thm 7.12   p.595   7.3   Let A have linearly independent columns.
Then A^+ the pseudoinverse of A satisfies the *Penrose conditions*.
  a. $A A^+ A = A$
  b. $A^+ A A^+ = A^+$
  c. $A A^+$ and $A^+ A$ are symmetric

Theorems by name

Thm 7.8   p.579   7.3   **The Best Approximation Theorem**
If W is a finite-dimensional subspace of an inner product space V
and if $\mathbf{v}$ is a vector in V,
then $\operatorname{proj}_W(\mathbf{v})$ is the *best approximation* to $\mathbf{v}$ in W.

Thm 7.9   p.584   7.3   **The Least Squares Theorem**
$A\mathbf{x} = \mathbf{b}$ has at least one *least squares solution*, $\bar{\mathbf{x}}$.
  a. $\bar{\mathbf{x}}$ is a *least squares solution* of $A\mathbf{x} = \mathbf{b}$ if and only if
     $\bar{\mathbf{x}}$ is a solution of the normal equations $A^T A \bar{\mathbf{x}} = A^T \mathbf{b}$.
  b. A has linearly independent columns $\Leftrightarrow A^T A$ is invertible.
     Then $\bar{\mathbf{x}} = (A^T A)^{-1} A^T \mathbf{b}$ is the *unique* solution.

Theorems from Section 5.3

Thm 5.16   p.390   5.3   **The QR Factorization**:
Let A be an $m \times n$ matrix with linearly independent columns.
Then A can be factored as $A = QR$, where
Q is an $m \times n$ matrix with orthonormal columns and
R is an an invertible upper triangular matrix.

7.3 Least Squares Approximation

Examining the relationship between projections and approximations

Q: What is the formula for $\text{proj}_W(\mathbf{v})$?
A: The formula for the *orthogonal projection* given below is on page 379 in Section 5.2:
$$\text{proj}_W(\mathbf{v}) = \sum_{k=1}^{n} \left(\frac{\mathbf{u}_k \cdot \mathbf{v}}{\mathbf{u}_k \cdot \mathbf{u}_k} \right) \mathbf{u}_k = \sum_{k=1}^{n} \text{proj}_{\mathbf{u}_k}(\mathbf{v}) \text{ where } \{\mathbf{u}_k\} \text{ is an orthogonal basis for } W.$$

Q: If $W = \text{span}(\mathbf{w})$, what does the formula for $\text{proj}_W(\mathbf{v})$ simplify to?
A: In this case, the formula for the *orthogonal projection* simplifies to:
$$\text{proj}_W(\mathbf{v}) = \text{proj}_\mathbf{w}(\mathbf{v}) = \left(\frac{\mathbf{w} \cdot \mathbf{v}}{\mathbf{w} \cdot \mathbf{w}} \right) \mathbf{w}$$

Q: What is the relationship between $\text{proj}_\mathbf{w}(\mathbf{v})$ and $\mathbf{w}$?
A: From the formula, we see $\text{proj}_\mathbf{w}(\mathbf{v})$ is a scalar multiple of $\mathbf{w}$.
So $\text{proj}_\mathbf{w}(\mathbf{v})$ lies in same direction as $\mathbf{w}$ but *may* have different length.

Q: What is the scalar multiple k between $\text{proj}_\mathbf{w}(\mathbf{v})$ and $\mathbf{w}$?
A: From the formula, we see the scalar multiple is: $k = \frac{\mathbf{w} \cdot \mathbf{v}}{\mathbf{w} \cdot \mathbf{w}}$.

Q: Prove: if $\mathbf{v} = c\mathbf{w}$, then $\text{proj}_\mathbf{w}(\mathbf{v}) = \mathbf{v}$.
Q: Prove: if $\mathbf{v}$ is in W, then $\text{proj}_W(\mathbf{v}) = \mathbf{v}$.
Q: So, what is the best approximation to $\mathbf{v}$ in W if $\mathbf{v}$ is in W?
A: If $\mathbf{v}$ is in W, the best approximation to $\mathbf{v}$ in W is $\mathbf{v}$ (which makes sense).

Q: Prove: the best approximation to $\mathbf{v}$ in $W = \text{span}(\mathbf{w})$ is $\text{proj}_\mathbf{w}(\mathbf{v})$.
A: In the proof, use the ideas above and then adapt the proof of the text.
What insights do we gain by using these differing methods of proof?

Linear independence and the pseudoinverse

Q: Do we have to know the columns of A are linearly independent
in order to define a pseudoinverse?
A: As mentioned at the end of *this* section, in the *next* section
we will extend the definition of a pseudoinverse to include all matrices.

Q: Currently, where are we using the fact that columns of A are linearly independent?
A: Our current definition is $A^+ = (A^T A)^{-1} A^T$.
As implied in the proof of the Least Squares Theorem (Theorem 7.9),
$A^T A^{-1}$ exists if and only if the columns of A are linearly independent.

Q: Without looking ahead to the next section, how might we extend the pseudoinverse idea?
Can we relax or replace $A^T A^{-1}$ with another matrix that does not require
the columns of A to be linearly independent?

Solutions to odd-numbered exercises from Section 7.3

1. $(1,0) \Rightarrow \varepsilon_1 = 0 - (2 \cdot 1 - 2) = 0$,
   $(2,1) \Rightarrow \varepsilon_2 = 1 - (2 \cdot 2 - 2) = -1$,
   $(3,5) \Rightarrow \varepsilon_3 = 5 - (2 \cdot 3 - 2) = 1 \Rightarrow$
   $\|e\| = \sqrt{0^2 + (-1)^2 + 1^2} = \sqrt{2}$.

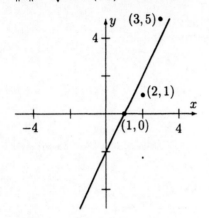

3. $(1,0) \Rightarrow \varepsilon_1 = 0 - (2.5 \cdot 1 - 3) = 0.5$,
   $(2,1) \Rightarrow \varepsilon_2 = 1 - (2.5 \cdot 2 - 3) = -1$,
   $(3,5) \Rightarrow \varepsilon_3 = 5 - (2.5 \cdot 3 - 3) = 0.5 \Rightarrow$
   $\|e\| = \sqrt{0.5^2 + (-1)^2 + 0.5^2} = \sqrt{6}/2$.

5. $(-5,3) \Rightarrow \varepsilon_1 = 3 - 2.5 = 0.5$,
   $(0,3) \Rightarrow \varepsilon_2 = 3 - 2.5 = 0.5$,
   $(5,2) \Rightarrow \varepsilon_3 = 2 - 2.5 = -0.5$,
   $(10,0) \Rightarrow \varepsilon_4 = 0 - 2.5 = 2.5 \Rightarrow$
   $\|e\| = \sqrt{3(0.5)^2 + (2.5)^2} = \sqrt{7}$.

7. $\bar{\mathbf{x}} = (A^T A)^{-1} A^T \mathbf{b} = \begin{bmatrix} \frac{7}{3} & -1 \\ -1 & \frac{1}{2} \end{bmatrix} \begin{bmatrix} 1 & 1 & 1 \\ 1 & 2 & 3 \end{bmatrix} \begin{bmatrix} 0 \\ 1 \\ 5 \end{bmatrix} = \begin{bmatrix} -3 \\ \frac{5}{2} \end{bmatrix} \Rightarrow y = -3 + \frac{5}{2} x.$

$\mathbf{e} = \mathbf{b} - A\bar{\mathbf{x}} = \begin{bmatrix} 0 \\ 1 \\ 5 \end{bmatrix} - \begin{bmatrix} 1 & 1 \\ 1 & 2 \\ 1 & 3 \end{bmatrix} \begin{bmatrix} -3 \\ \frac{5}{2} \end{bmatrix} = \begin{bmatrix} \frac{1}{2} \\ -1 \\ \frac{1}{2} \end{bmatrix} \Rightarrow \|e\| = \sqrt{(\frac{1}{2})^2 + (-1)^2 + (\frac{1}{2})^2} = \frac{\sqrt{6}}{2}.$

7.3 Least Squares Approximation

9. $\bar{x} = (A^TA)^{-1}A^Tb = \begin{bmatrix} \frac{5}{6} & -\frac{1}{2} \\ -\frac{1}{2} & \frac{1}{2} \end{bmatrix} \begin{bmatrix} 0 & 1 & 2 \\ 1 & 1 & 1 \end{bmatrix} \begin{bmatrix} 4 \\ 1 \\ 0 \end{bmatrix} = \begin{bmatrix} \frac{11}{3} \\ -2 \end{bmatrix} \Rightarrow y = \frac{11}{3} - 2x.$

$e = b - A\bar{x} = \begin{bmatrix} 4 \\ 1 \\ 0 \end{bmatrix} - \begin{bmatrix} 1 & 0 \\ 1 & 1 \\ 1 & 2 \end{bmatrix} \begin{bmatrix} \frac{11}{3} \\ -2 \end{bmatrix} = \begin{bmatrix} \frac{1}{3} \\ -\frac{2}{3} \\ \frac{1}{3} \end{bmatrix} \Rightarrow \|e\| = \sqrt{(\frac{1}{3})^2 + (-\frac{2}{3})^2 + (\frac{1}{3})^2} = \frac{\sqrt{6}}{3}.$

11. $\bar{x} = (A^TA)^{-1}A^Tb = \begin{bmatrix} \frac{3}{10} & -\frac{1}{50} \\ -\frac{1}{50} & \frac{1}{125} \end{bmatrix} \begin{bmatrix} 1 & 1 & 1 & 1 \\ -5 & 0 & 5 & 10 \end{bmatrix} \begin{bmatrix} -1 \\ 1 \\ 2 \\ 4 \end{bmatrix} = \begin{bmatrix} \frac{7}{10} \\ \frac{8}{25} \end{bmatrix} \Rightarrow y = \frac{7}{10} + \frac{8}{25}x.$

$e = \begin{bmatrix} -1 \\ 1 \\ 2 \\ 4 \end{bmatrix} - \begin{bmatrix} 1 & -5 \\ 1 & 0 \\ 1 & 5 \\ 1 & 10 \end{bmatrix} \begin{bmatrix} \frac{7}{10} \\ \frac{8}{25} \end{bmatrix} = \begin{bmatrix} -\frac{1}{10} \\ \frac{3}{10} \\ -\frac{3}{10} \\ \frac{1}{10} \end{bmatrix} \Rightarrow \|e\| = \sqrt{(-\frac{1}{10})^2 + (\frac{3}{10})^2 + (-\frac{3}{10})^2 + (\frac{1}{10})^2} = \frac{\sqrt{5}}{5}.$

13. $\bar{x} = (A^TA)^{-1}A^Tb = \begin{bmatrix} \frac{11}{10} & -\frac{3}{10} \\ -\frac{3}{10} & \frac{1}{10} \end{bmatrix} \begin{bmatrix} 1 & 1 & 1 & 1 & 1 \\ 1 & 2 & 3 & 4 & 5 \end{bmatrix} \begin{bmatrix} 1 \\ 3 \\ 4 \\ 5 \\ 7 \end{bmatrix} = \begin{bmatrix} -\frac{1}{5} \\ \frac{7}{5} \end{bmatrix} \Rightarrow y = -\frac{1}{5} + \frac{7}{5}x.$

$e = \begin{bmatrix} 1 \\ 3 \\ 4 \\ 5 \\ 7 \end{bmatrix} - \begin{bmatrix} 1 & 1 \\ 1 & 2 \\ 1 & 3 \\ 1 & 4 \\ 1 & 5 \end{bmatrix} \begin{bmatrix} -\frac{1}{5} \\ \frac{7}{5} \end{bmatrix} = \begin{bmatrix} -\frac{1}{5} \\ \frac{2}{5} \\ 0 \\ -\frac{2}{5} \\ \frac{1}{5} \end{bmatrix} \Rightarrow \|e\| = \sqrt{(-\frac{1}{5})^2 + (\frac{2}{5})^2 + 0^2 + (-\frac{2}{5})^2 + (\frac{1}{5})^2} = \frac{\sqrt{10}}{5}.$

15. $A^TA = \begin{bmatrix} 1 & 1 & 1 & 1 \\ 1 & 2 & 3 & 4 \\ 1 & 4 & 9 & 16 \end{bmatrix} \begin{bmatrix} 1 & 1 & 1 \\ 1 & 2 & 4 \\ 1 & 3 & 9 \\ 1 & 4 & 16 \end{bmatrix} = \begin{bmatrix} 40 & 10 & 30 \\ 10 & 30 & 100 \\ 30 & 100 & 354 \end{bmatrix}, A^Tb = \begin{bmatrix} 1 & 1 & 1 & 1 \\ 1 & 2 & 3 & 4 \\ 1 & 4 & 9 & 16 \end{bmatrix} \begin{bmatrix} 1 \\ -2 \\ 3 \\ 4 \end{bmatrix} = \begin{bmatrix} 6 \\ 22 \\ 84 \end{bmatrix} \Rightarrow$

$[\,A^TA\,|\,A^Tb\,] \longrightarrow [\,I\,|\,\bar{x}\,] \Rightarrow \bar{x} = \begin{bmatrix} 3 \\ -\frac{8}{15} \\ 1 \end{bmatrix} \Rightarrow y = 3 - \frac{18}{5}x + x^2.$

17. $A^T A = \begin{bmatrix} 1 & 1 & 1 & 1 & 1 \\ -2 & -1 & 0 & 1 & 2 \\ 4 & 1 & 0 & 1 & 4 \end{bmatrix} \begin{bmatrix} 1 & -2 & 4 \\ 1 & -1 & 1 \\ 1 & 0 & 0 \\ 1 & 1 & 1 \\ 1 & 2 & 4 \end{bmatrix} = \begin{bmatrix} 5 & 0 & 10 \\ 0 & 10 & 0 \\ 10 & 0 & 34 \end{bmatrix}, A^T \mathbf{b} = \begin{bmatrix} 13 \\ -17 \\ 19 \end{bmatrix} \Rightarrow$

$[\, A^T A \,|\, A^T \mathbf{b} \,] \longrightarrow [\, I \,|\, \bar{\mathbf{x}} \,] \Rightarrow \bar{\mathbf{x}} = \begin{bmatrix} \frac{18}{5} \\ -\frac{17}{10} \\ -\frac{1}{2} \end{bmatrix} \Rightarrow y = \frac{18}{5} - \frac{17}{10} x - \frac{1}{2} x^2.$

19. $[\, A^T A \,|\, A^T \mathbf{b} \,] \longrightarrow [\, I \,|\, \bar{\mathbf{x}} \,] \Rightarrow \begin{bmatrix} 11 & 6 & | & 5 \\ 6 & 6 & | & 4 \end{bmatrix} \longrightarrow \begin{bmatrix} 1 & 0 & | & \frac{1}{5} \\ 0 & 1 & | & \frac{7}{15} \end{bmatrix} \Rightarrow \bar{\mathbf{x}} = \begin{bmatrix} \frac{1}{5} \\ \frac{7}{15} \end{bmatrix}.$

21. $[\, A^T A \,|\, A^T \mathbf{b} \,] \longrightarrow [\, I \,|\, \bar{\mathbf{x}} \,] \Rightarrow \begin{bmatrix} 14 & 8 & | & 12 \\ 8 & 38 & | & -21 \end{bmatrix} \longrightarrow \begin{bmatrix} 1 & 0 & | & \frac{4}{3} \\ 0 & 1 & | & -\frac{5}{6} \end{bmatrix} \Rightarrow \bar{\mathbf{x}} = \begin{bmatrix} \frac{4}{3} \\ -\frac{5}{6} \end{bmatrix}.$

23. $[\, A^T A \,|\, A^T \mathbf{b} \,] \longrightarrow [\, I \,|\, \bar{\mathbf{x}} \,] \Rightarrow \begin{bmatrix} 3 & 0 & 2 & 1 & | & 2 \\ 0 & 3 & -2 & -1 & | & -5 \\ 2 & -2 & 3 & 2 & | & 3 \\ 1 & -1 & 2 & 2 & | & -1 \end{bmatrix} \longrightarrow \begin{bmatrix} 1 & 0 & 0 & -1 & | & 4 \\ 0 & 1 & 0 & 1 & | & -5 \\ 0 & 0 & 1 & 2 & | & -5 \\ 0 & 0 & 0 & 0 & | & 0 \end{bmatrix} \Rightarrow \bar{\mathbf{x}} = \begin{bmatrix} 4+t \\ -5-t \\ -5-2t \\ t \end{bmatrix}.$

25. $[\, A^T A \,|\, A^T \mathbf{b} \,] \longrightarrow [\, I \,|\, \bar{\mathbf{x}} \,] \Rightarrow \begin{bmatrix} 11 & 7 & -5 & | & 35 \\ 7 & 6 & -5 & | & 18 \\ -5 & -5 & 7 & | & -1 \end{bmatrix} \longrightarrow \begin{bmatrix} 1 & 0 & 0 & | & \frac{42}{11} \\ 0 & 1 & 0 & | & \frac{19}{11} \\ 0 & 0 & 1 & | & \frac{42}{11} \end{bmatrix} \Rightarrow \bar{\mathbf{x}} = \begin{bmatrix} \frac{42}{11} \\ \frac{19}{11} \\ \frac{42}{11} \end{bmatrix}.$

27. $\bar{\mathbf{x}} = R^{-1} Q^T \mathbf{b} = \begin{bmatrix} \frac{1}{3} & -\frac{1}{3} \\ 0 & 1 \end{bmatrix} \begin{bmatrix} \frac{2}{3} & \frac{2}{3} & \frac{1}{3} \\ \frac{1}{3} & -\frac{2}{3} & \frac{2}{3} \end{bmatrix} \begin{bmatrix} 2 \\ 3 \\ 1 \end{bmatrix} = \begin{bmatrix} \frac{5}{3} \\ -2 \end{bmatrix}.$

29. $A^T A = \begin{bmatrix} 1 & 1 & 1 & 1 & 1 & 1 \\ 20 & 40 & 48 & 60 & 80 & 100 \end{bmatrix} \begin{bmatrix} 1 & 20 \\ 1 & 40 \\ 1 & 48 \\ 1 & 60 \\ 1 & 80 \\ 1 & 100 \end{bmatrix} = \begin{bmatrix} 6 & 348 \\ 348 & 24304 \end{bmatrix}, A^T \mathbf{b} = \begin{bmatrix} 259.5 \\ 18058 \end{bmatrix} \Rightarrow$

$[\, A^T A \,|\, A^T \mathbf{b} \,] \longrightarrow [\, I \,|\, \bar{\mathbf{x}} \,] \Rightarrow \bar{\mathbf{x}} \approx \begin{bmatrix} 0.92 \\ 0.73 \end{bmatrix} \Rightarrow y = 0.92 + 0.73\, x.$

31. (a) $A^T A = \begin{bmatrix} 8 & 28 \\ 28 & 140 \end{bmatrix}, A^T \mathbf{b} = \begin{bmatrix} 534.5 \\ 1992.9 \end{bmatrix} \Rightarrow [\, A^T A \,|\, A^T \mathbf{b} \,] \longrightarrow [\, I \,|\, \bar{\mathbf{x}} \,] \Rightarrow$

$\bar{\mathbf{x}} \approx \begin{bmatrix} 56.6 \\ 0.29 \end{bmatrix} \Rightarrow s(t) = 56.6 + 0.29\, t \Rightarrow$

The life expectancy of someone born in 2000 is $s(80) = 56.6 + 0.29\,(80) = 79.8$ years.

(b) $\mathbf{e} = \mathbf{b} - A\bar{\mathbf{x}} \Rightarrow \|\mathbf{e}\| \approx 29.2 \Rightarrow$ Error almost half the output $\Rightarrow$ model is not good.

7.3 Least Squares Approximation

33. (a) $p(t) = ce^{kt} \Rightarrow p(0) = 150 \Rightarrow p = 150e^{kt} \Rightarrow \ln p = \ln(150e^{kt}) = \ln 150 + kt \Rightarrow kt = \ln p - \ln 150 \Rightarrow kt = \ln \frac{p}{150}$.

So, Example 7.29 $\Rightarrow A = \begin{bmatrix} t_1 \\ t_2 \\ t_3 \\ t_4 \\ t_5 \end{bmatrix} = \begin{bmatrix} 1 \\ 2 \\ 3 \\ 4 \\ 5 \end{bmatrix}$ and $\mathbf{b} = \begin{bmatrix} kt_1 \\ kt_2 \\ kt_3 \\ kt_4 \\ kt_5 \end{bmatrix} = \begin{bmatrix} \ln \frac{179}{150} \\ \ln \frac{203}{150} \\ \ln \frac{227}{150} \\ \ln \frac{250}{150} \\ \ln \frac{281}{150} \end{bmatrix} \Rightarrow$

$A^T A = [55]$ and $A^T \mathbf{b} \approx [7.21] \Rightarrow \bar{\mathbf{x}} \approx \left[\frac{7.21}{55}\right] = 0.131 = k \Rightarrow p(t) = 150e^{0.131t}$.

(b) The population in 2010 will be $p(6) = 150e^{0.131(6)} \approx 329$ million.

35. Example 7.29 $\Rightarrow A = \begin{bmatrix} t_1 \\ t_2 \\ t_3 \end{bmatrix} = \begin{bmatrix} 30 \\ 60 \\ 90 \end{bmatrix}$ and $\mathbf{b} = \begin{bmatrix} kt_1 \\ kt_2 \\ kt_3 \end{bmatrix} = \begin{bmatrix} \ln \frac{172}{200} \\ \ln \frac{148}{200} \\ \ln \frac{128}{200} \end{bmatrix} \Rightarrow$

$A^T A = [12600]$ and $A^T \mathbf{b} \approx [-62.8] \Rightarrow \bar{\mathbf{x}} \approx \left[-\frac{62.8}{12600}\right] = -0.005 = k \Rightarrow p(t) = ce^{-0.005t}$.

So, Section 6.7, Example 6.88 $\Rightarrow$ we can find the half-life as follows: $\frac{1}{2}c = ce^{-0.005t} \Rightarrow$

$e^{-0.005t} = \frac{1}{2} \Rightarrow -0.005t = \ln(1/2) \Rightarrow t = -\frac{\ln(1/2)}{0.005} \approx 139$ days is the half life.

37. $A = \begin{bmatrix} 1 \\ 1 \end{bmatrix} \Rightarrow A^T A = [2] \Rightarrow (A^T A)^{-1} = [\frac{1}{2}]$. So, with $P =$ the standard matrix we have:

$P = A(A^T A)^{-1} A^T = \begin{bmatrix} 1 \\ 1 \end{bmatrix} [\frac{1}{2}] [1 \ 1] = \begin{bmatrix} \frac{1}{2} & \frac{1}{2} \\ \frac{1}{2} & \frac{1}{2} \end{bmatrix} \Rightarrow \text{proj}_W(\mathbf{v}) = P\mathbf{v} = \begin{bmatrix} \frac{1}{2} & \frac{1}{2} \\ \frac{1}{2} & \frac{1}{2} \end{bmatrix} \begin{bmatrix} 3 \\ 4 \end{bmatrix} = \begin{bmatrix} \frac{7}{2} \\ \frac{7}{2} \end{bmatrix}$.

39. $A = \begin{bmatrix} 1 \\ 1 \\ 1 \end{bmatrix} \Rightarrow A^T A = [3] \Rightarrow (A^T A)^{-1} = [\frac{1}{3}]$. So, $P = A(A^T A)^{-1} A^T \Rightarrow$

$P = \begin{bmatrix} 1 \\ 1 \\ 1 \end{bmatrix} [\frac{1}{3}] [1 \ 1 \ 1] = \begin{bmatrix} \frac{1}{3} & \frac{1}{3} & \frac{1}{3} \\ \frac{1}{3} & \frac{1}{3} & \frac{1}{3} \\ \frac{1}{3} & \frac{1}{3} & \frac{1}{3} \end{bmatrix} \Rightarrow \text{proj}_W(\mathbf{v}) = P\mathbf{v} = \begin{bmatrix} \frac{1}{3} & \frac{1}{3} & \frac{1}{3} \\ \frac{1}{3} & \frac{1}{3} & \frac{1}{3} \\ \frac{1}{3} & \frac{1}{3} & \frac{1}{3} \end{bmatrix} \begin{bmatrix} 1 \\ 2 \\ 3 \end{bmatrix} = \begin{bmatrix} 2 \\ 2 \\ 2 \end{bmatrix}$.

41. $A = \begin{bmatrix} 1 & 1 \\ 0 & 1 \\ -1 & 1 \end{bmatrix} \Rightarrow A^T A = \begin{bmatrix} 2 & 0 \\ 0 & 3 \end{bmatrix} \Rightarrow (A^T A)^{-1} = \begin{bmatrix} \frac{1}{2} & 0 \\ 0 & \frac{1}{3} \end{bmatrix} \Rightarrow$. So, $P = A(A^T A)^{-1} A^T \Rightarrow$

$P = \begin{bmatrix} 1 & 1 \\ 0 & 1 \\ -1 & 1 \end{bmatrix} \begin{bmatrix} \frac{1}{2} & 0 \\ 0 & \frac{1}{3} \end{bmatrix} \begin{bmatrix} 1 & 0 & -1 \\ 1 & 1 & 1 \end{bmatrix} = \begin{bmatrix} \frac{5}{6} & \frac{1}{3} & -\frac{1}{6} \\ \frac{1}{3} & \frac{1}{3} & \frac{1}{3} \\ -\frac{1}{6} & \frac{1}{3} & \frac{5}{6} \end{bmatrix} \Rightarrow \text{proj}_W(\mathbf{v}) = P\mathbf{v} = P \begin{bmatrix} 1 \\ 0 \\ 0 \end{bmatrix} = \begin{bmatrix} \frac{5}{6} \\ \frac{1}{3} \\ -\frac{1}{6} \end{bmatrix}$.

43. $A = \begin{bmatrix} 1 & 1 \\ 1 & 3 \\ 0 & 1 \end{bmatrix} \Rightarrow A^T A = \begin{bmatrix} 2 & 4 \\ 4 & 11 \end{bmatrix} \Rightarrow (A^T A)^{-1} = \begin{bmatrix} \frac{11}{6} & -\frac{2}{3} \\ -\frac{2}{3} & \frac{1}{3} \end{bmatrix} \Rightarrow$. So, $P = A(A^T A)^{-1} A^T \Rightarrow$

$P = \begin{bmatrix} 1 & 1 \\ 1 & 3 \\ 0 & 1 \end{bmatrix} \begin{bmatrix} \frac{11}{6} & -\frac{2}{3} \\ -\frac{2}{3} & \frac{1}{3} \end{bmatrix} \begin{bmatrix} 1 & 1 & 0 \\ 1 & 3 & 1 \end{bmatrix} = \begin{bmatrix} \frac{5}{6} & \frac{1}{6} & -\frac{1}{3} \\ \frac{1}{6} & \frac{5}{6} & \frac{1}{3} \\ -\frac{1}{3} & \frac{1}{3} & \frac{1}{3} \end{bmatrix}$ as found in Example 7.31.

45. $A^T A = \begin{bmatrix} 5 \end{bmatrix} \Rightarrow (A^T A)^{-1} = \begin{bmatrix} \frac{1}{5} \end{bmatrix} \Rightarrow A^+ = (A^T A)^{-1} A^T = \begin{bmatrix} \frac{1}{5} \end{bmatrix} \begin{bmatrix} 1 & 2 \end{bmatrix} = \begin{bmatrix} \frac{1}{5} & \frac{2}{5} \end{bmatrix}$.

47. $A^T A = \begin{bmatrix} 2 & 2 \\ 2 & 14 \end{bmatrix} \Rightarrow (A^T A)^{-1} = \begin{bmatrix} \frac{7}{12} & -\frac{1}{12} \\ -\frac{1}{12} & \frac{1}{12} \end{bmatrix} \Rightarrow A^+ = (A^T A)^{-1} A^T = \begin{bmatrix} \frac{1}{3} & -\frac{2}{3} & -\frac{1}{6} \\ \frac{1}{6} & \frac{1}{6} & \frac{1}{6} \end{bmatrix}$.

49. $A^T A = \begin{bmatrix} 1 & 1 \\ 1 & 2 \end{bmatrix} \Rightarrow (A^T A)^{-1} = \begin{bmatrix} 2 & -1 \\ -1 & 1 \end{bmatrix} \Rightarrow A^+ = (A^T A)^{-1} A^T = \begin{bmatrix} 1 & -1 \\ 0 & 1 \end{bmatrix}$.

51. $A^T A = \begin{bmatrix} 3 & 1 & 2 \\ 1 & 2 & 2 \\ 2 & 2 & 3 \end{bmatrix} \Rightarrow (A^T A)^{-1} = \begin{bmatrix} \frac{2}{3} & \frac{1}{3} & -\frac{2}{3} \\ \frac{1}{3} & \frac{5}{3} & -\frac{4}{3} \\ -\frac{2}{3} & -\frac{4}{3} & \frac{5}{3} \end{bmatrix} \Rightarrow A^+ = (A^T A)^{-1} A^T = \begin{bmatrix} \frac{2}{3} & 0 & -\frac{1}{3} & \frac{1}{3} \\ \frac{1}{3} & -1 & \frac{1}{3} & \frac{2}{3} \\ -\frac{2}{3} & 1 & \frac{1}{3} & -\frac{1}{3} \end{bmatrix}$

53. (a) $A^+ = (A^T A)^{-1} A^T = A^{-1} (A^T)^{-1} A^T = A^{-1}$.

(b) Recall, A has orthonormal columns $\mathbf{q}_i$ means $\mathbf{q}_i \cdot \mathbf{q}_j = \begin{cases} 0 \text{ if } i \neq j \\ 1 \text{ if } i = j \end{cases}$.

So, $A^T A = \begin{bmatrix} \mathbf{q}_1 \\ \vdots \\ \mathbf{q}_n \end{bmatrix} [\mathbf{q}_1 | \ldots | \mathbf{q}_n] = I_n \Rightarrow A^+ = (A^T A)^{-1} A^T = I_n A^T = A^T$.

Note, we should have expected this since if A were an $n \times n$ matrix, $A^{-1} = A^T$.

55. Will show $A^+ A = I$, which is symmetric. $A^+ A = (A^T A)^{-1} A^T A = (A^T A)^{-1} (A^T A) = I$.

57. We will construct A as in Example 7.26, then show the columns of A are linearly independent.

Theorem 7.9 $\Rightarrow A\mathbf{x} = \mathbf{b} = \begin{bmatrix} y_1 \\ \vdots \\ y_n \end{bmatrix} \in \mathbb{R}^n$ has unique least squares solution $\bar{\mathbf{x}} = \begin{bmatrix} a_0 \\ a_1 \end{bmatrix} \in \mathbb{R}^2 \Rightarrow$

$p(x) = a_0 + a_1 x$ is the unique least squares approximating line.

So, let $A = \begin{cases} a_{1i} = 1 & \text{for all } i \\ a_{2i} = x_i & \text{for all } i \end{cases}$, that is $A = \begin{bmatrix} 1 & x_1 \\ \vdots & \vdots \\ 1 & x_n \end{bmatrix}$.

The fact that all the points do not lie on the same vertical line $\Rightarrow$ there exists $x_j \neq x_k$. We will use this fact to show that the columns of A, that is $\mathbf{a}_i$, are linearly independent.

$a_0 \mathbf{a}_1 + a_1 \mathbf{a}_2 = 0 \Leftrightarrow a_0 + a_1 x_i = 0$ for all $i \Rightarrow p(x) = a_0 + a_1 x$ has at least two distinct zeroes (since $x_j \neq x_k$). Therefore, Section 6.2, Exercise 62 $\Rightarrow p(x)$ is the zero polynomial $\Rightarrow a_0 = a_1 = 0 \Rightarrow$ the columns of A, that is $\mathbf{a}_i$, are linearly independent.

7.4 The Singular Value Decomposition

Q: What are our main goals for Section 7.4?
A: To examine and explore *singular value decomposition.*
Also, pay special attention to Theorem 7.19,
The Fundamental Theorem of Invertible Matrices: Final Version.

Key Definitions and Concepts

σ_i	p.599	7.4	If A is an $m \times n$ matrix, the *singular values* of A are the square roots of the eigenvalues of $A^T A$. They are denoted by $\sigma_1, \ldots, \sigma_n$. They are usually arranged $\sigma_1 \geq \sigma_2 \geq \ldots \geq \sigma_n$.
Σ	p.601	7.4	Σ is created from the σ_i (detailed description p.601)
$A = U\Sigma V^T$	p.601	7.4	A *singular value decomposition* (SVD) of $A = U\Sigma V^T$. where U and and V are orthogonal matrices. Σ is constructed using the σ_i.
$A^+ = V\Sigma^+ U^T$	p.611	7.4	The *pseudoinverse* or *Moore-Penrose inverse* is $A^+ = V\Sigma^+ U^T$.
left singular vectors	p.602	7.4	Given a singular value decomposition (SVD) of $A = U\Sigma V^T$. The columns of U (an orthogonal matrix) are called the *left singular values* of A.
pseudoinverse	p.611	7.4	The *pseudoinverse* or *Moore-Penrose inverse* is $A^+ = V\Sigma^+ U^T$.
right singular vectors	p.602	7.4	Given a singular value decomposition (SVD) of $A = U\Sigma V^T$. The columns of V (an orthogonal matrix) are called the *right singular values* of A.
singular values	p.599	7.4	If A is an $m \times n$ matrix, the *singular values* of A are the square roots of the eigenvalues of $A^T A$. They are denoted by $\sigma_1, \ldots, \sigma_n$. They are usually arranged $\sigma_1 \geq \sigma_2 \geq \ldots \geq \sigma_n$.
singular value decomposition	p.601	7.4	A *singular value decomposition* (SVD) of $A = U\Sigma V^T$. where U and and V are orthogonal matrices. Σ is constructed using the σ_i.

Theorems

Thm 7.13   p.602   7.4   **The Singular Value Decomposition Theorem**
Let A be an $m \times n$ matrix with singular values
$\sigma_1 \geq \sigma_1 \geq \ldots \geq \sigma_r > 0$ and $\sigma_{r+1} = \sigma_{r+2} = \ldots = \sigma_n = 0$.
Then there exist an $m \times m$ orthogonal matrix U,
an $n \times n$ orthogonal matrix V,
and an $m \times n$ matrix Σ (see p.601)
such that $A = U\Sigma V^T$.

Thm 7.14   p.605   7.4   Let A be an $m \times n$ matrix with singular values
$\sigma_1 \geq \sigma_1 \geq \ldots \geq \sigma_r > 0$ and $\sigma_{r+1} = \sigma_{r+2} = \ldots = \sigma_n = 0$.
Let $\mathbf{u}_1, \ldots, \mathbf{u}_r$ be left singular values and
let $\mathbf{v}_1, \ldots, \mathbf{v}_r$ be left singular values. Then:
$A = \sigma_1 \mathbf{u}_1 \mathbf{v}_1^T + \cdots + \sigma_r \mathbf{u}_r \mathbf{v}_r^T$.

Thm 7.15   p.606   7.4   Let $A = U\Sigma V^T$ be an *SVD* of an $m \times m$ matrix A
with nonzero singular values $\sigma_1, \ldots, \sigma_r$.
a. The rank of A is r.
b. $\{\mathbf{u}_1, \ldots, \mathbf{u}_r\}$ is an orthonormal basis for $\text{col}(A)$.
c. $\{\mathbf{u}_{r+1}, \ldots, \mathbf{u}_m\}$ is an orthonormal basis for $\text{null}(A^T)$.
d. $\{\mathbf{v}_1, \ldots, \mathbf{v}_r\}$ is an orthonormal basis for $\text{row}(A)$.
e. $\{\mathbf{v}_{r+1}, \ldots, \mathbf{v}_m\}$ is an orthonormal basis for $\text{null}(A)$.

Thm 7.16   p.607   7.4   Let $A = U\Sigma V^T$ be an *SVD* of an $m \times n$ matrix A with rank r.
Then the image of the unit sphere in $\mathbb{R}^n$
under the matrix transformation that maps $\mathbf{x}$ to $A\mathbf{x}$ is
a. The surface of an ellipsoid in $\mathbb{R}^m$ if $r = n$.
b. A solid ellipsoid in $\mathbb{R}^m$ if $r < n$.

Thm 7.17   p.610   7.4   Let A be an $m \times n$ matrix and
let $\sigma_1, \ldots, \sigma_r$ be all the nonzero singular values of A.
Then: $\|A\|_F = \sqrt{\sigma_1^2 + \cdots + \sigma_r^2}$

Thm 7.18   p.613   7.4   The least squares problem $A\mathbf{x} = \mathbf{b}$
has a unique least squares solution $\bar{\mathbf{x}}$ of minimal length
that is given by $\bar{\mathbf{x}} = A^+\mathbf{b}$.

Theorem 5.6 from Section 5.1 and its analogue from Section 7.4

Thm 5.6   p.372   5.1   a. through c. give equivalent conditions for Q to be orthogonal:
a. Q is orthogonal
b. $\|Q\mathbf{x}\| = \|\mathbf{x}\|$ for every $\mathbf{x}$ in $\mathbb{R}^n$
c. $Q\mathbf{x} \cdot Q\mathbf{x} = \mathbf{x} \cdot \mathbf{y}$ for every $\mathbf{x}, \mathbf{y}$ in $\mathbb{R}^n$

Thm 5.6   p.372   5.1   If A is an $m \times n$ matrix and Q is an $m \times m$ orthogonal matrix.
Then: $\|QA\|_F = \|A\|_F$

7.4 The Singular Value Decomposition

The Fundamental Theorem of Invertible Matrices: Final Version

Thm 7.19   p.614   7.4   a. through u. give equivalent conditions for A to be invertible:

  a. A is invertible.
  b. $A\mathbf{x} = \mathbf{b}$ has a unique solution for every $\mathbf{b}$ in $\mathbb{R}^n$
  c. $A\mathbf{x} = \mathbf{0}$ has only the trivial solution.
  d. The reduced row echelon form of A is I_n.
  e. A is a product of elementary matrices.
  f. $\text{rank}(A) = n$
  g. $\text{nullity}(A) = 0$
  h. The column vectors of A are linearly independent.
  i. The column vectors of A span $\mathbb{R}^n$.
  j. The column vectors of A form a basis for $\mathbb{R}^n$.
  k. The row vectors of A are linearly independent.
  l. The row vectors of A span $\mathbb{R}^n$.
  m. The row vectors of A form a basis for $\mathbb{R}^n$.
  n. $\det(A) \neq 0$
  o. 0 is not an eigenvalue of A
  p. T is invertible
  q. T is one-to-one
  r. T is onto
  s. $\ker(T) = \{\mathbf{0}\}$
  t. $\text{range}(T) = W$
  u. 0 is not a singular value of A

Theorems by name

Thm 7.13   p.602   7.4   **The Singular Value Decomposition Theorem**
Let A be an $m \times n$ matrix with singular values
$\sigma_1 \geq \sigma_1 \geq \ldots \geq \sigma_r > 0$ and $\sigma_{r+1} = \sigma_{r+2} = \ldots = \sigma_n = 0$.
Then there exist an $m \times m$ orthogonal matrix U,
an $n \times n$ orthogonal matrix V,
and an $m \times n$ matrix Σ (see p.601)
such that $A = U\Sigma V^T$.

The pseudoinverse and singular value decomposition

Q: In the previous section, we defined the pseudoinverse as $A^+ = (A^T A)^{-1} A^T$. How did we relax the need for $A^T A^{-1}$ in creating the singular value decomposition?
A: By examining $A^T A$ in much more detail as we see in the discussion below.

Q: Why does it make sense to start the process of looking for an inverse with $A^T A$?
A: This is a square matrix naturally associated with A.

Q: Where do the singular values of A, σ_i, come from?
A: They are the square roots of the eigenvalues of $A^T A$.

Q: Why do we take the square roots of the eigenvalues instead of the eigenvalues themselves?
A: Because when we compute $A^T A$ we are in effect *almost* squaring A.

Q: How do we use the singular values in the singular value decomposition of A?
A: They are used to define Σ. Make sure to understand equation (1) on p.601.

Matrices U and V of the singular value decomposition

Q: What desirable property do U and V of $A = U \Sigma V^T$ possess by defining Σ in this way?
A: As stated in Theorem 7.13, **The Singular Value Decomposition Theorem**, U and V are orthogonal.

Q: In particular, how does U relate to the vector spaces naturally associated with A?
A: As stated in Theorem 7.15, the columns of U that correspond to the *nonzero* singular values of A form an orthonormal basis for col(A).

Q: How does U relate to the vector spaces naturally associated with A^T?
A: As stated in Theorem 7.15, the columns of U that correspond to the *zero* singular values of A form an orthonormal basis for null(A^T).

Q: How does V relate to the vector spaces naturally associated with A?

Key property of the pseudoinverse

Q: As defined in this section, what desirable property does $A^+ = U \Sigma^+ V^T$ possess?
A: As stated in Theorem 7.18, the least squares problem $A\mathbf{x} = \mathbf{b}$ has a unique least squares solution $\bar{\mathbf{x}}$ of minimal length given by $\bar{\mathbf{x}} = A^+ \mathbf{b}$.

Q: What underlying property makes this definition of the pseudoinverse so important?
A: When defined in this way, the pseudoinverse can be constructed for *any* matrix A.

7.4 The Singular Value Decomposition

Solutions to odd-numbered exercises from Section 7.4

1. $A = \begin{bmatrix} 2 & 0 \\ 0 & 3 \end{bmatrix} \Rightarrow A^T A = \begin{bmatrix} 4 & 0 \\ 0 & 9 \end{bmatrix} \Rightarrow$ the eigenvalues of $A^T A$ are $\lambda_1 = 9, \lambda_2 = 4 \Rightarrow$ the singular values of A are $\sigma_1 = \sqrt{9} = 3, \sigma_2 = \sqrt{4} = 2$.

3. $A^T A = \begin{bmatrix} 1 & 1 \\ 1 & 1 \end{bmatrix} \Rightarrow$ eigenvalues $2, 0 \Rightarrow$ singular values of A are $\sigma_1 = \sqrt{2}, \sigma_2 = 0$.

5. $A^T A = \begin{bmatrix} 25 \end{bmatrix} \Rightarrow$ eigenvalue $25 \Rightarrow$ singular value of A is $\sigma_1 = 5$.

7. $A^T A = \begin{bmatrix} 4 & 0 \\ 0 & 9 \end{bmatrix} \Rightarrow$ eigenvalues $9, 4 \Rightarrow$ singular values of A are $\sigma_1 = 3, \sigma_2 = 2$.

9. $A^T A = \begin{bmatrix} 4 & 0 & 2 \\ 0 & 4 & 0 \\ 2 & 0 & 1 \end{bmatrix} \Rightarrow$ eigenvalues $5, 4, 0 \Rightarrow$ singular values of A are $\sigma_1 = \sqrt{5}, \sigma_2 = 2, \sigma_3 = 0$.

11. $A^T A = \begin{bmatrix} 1 & 1 \\ 1 & 1 \end{bmatrix} \Rightarrow$ eigenvectors of $A^T A$ are $\begin{bmatrix} \frac{1}{\sqrt{2}} \\ \frac{1}{\sqrt{2}} \end{bmatrix}, \begin{bmatrix} \frac{1}{\sqrt{2}} \\ -\frac{1}{\sqrt{2}} \end{bmatrix} \Rightarrow$

$\mathbf{u}_1 = \frac{1}{\sigma_1} A\mathbf{v}_1 = \frac{1}{\sqrt{2}} \begin{bmatrix} 1 & 1 \\ 0 & 0 \end{bmatrix} \begin{bmatrix} \frac{1}{\sqrt{2}} \\ \frac{1}{\sqrt{2}} \end{bmatrix} = \begin{bmatrix} 1 \\ 0 \end{bmatrix}$. $\sigma_2 = 0 \Rightarrow$ extend with $\mathbf{u}_2 = \begin{bmatrix} 0 \\ 1 \end{bmatrix} \Rightarrow$

$A = U\Sigma V^T = \begin{bmatrix} 1 & 0 \\ 0 & 1 \end{bmatrix} \begin{bmatrix} \sqrt{2} & 0 \\ 0 & 0 \end{bmatrix} \begin{bmatrix} \frac{1}{\sqrt{2}} & \frac{1}{\sqrt{2}} \\ \frac{1}{\sqrt{2}} & -\frac{1}{\sqrt{2}} \end{bmatrix} = \begin{bmatrix} 1 & 1 \\ 0 & 0 \end{bmatrix}$.

13. $A^T A = \begin{bmatrix} 9 & 0 \\ 0 & 4 \end{bmatrix} \Rightarrow$ eigenvectors of $A^T A$ are $\begin{bmatrix} -1 \\ 0 \end{bmatrix}, \begin{bmatrix} 0 \\ -1 \end{bmatrix} \Rightarrow$

$\mathbf{u}_1 = \frac{1}{\sigma_1} A\mathbf{v}_1 = \frac{1}{3} \begin{bmatrix} 0 & -2 \\ -3 & 0 \end{bmatrix} \begin{bmatrix} -1 \\ 0 \end{bmatrix} = \begin{bmatrix} 0 \\ 1 \end{bmatrix}$. $\mathbf{u}_2 = \frac{1}{\sigma_2} A\mathbf{v}_2 = \frac{1}{2} \begin{bmatrix} 0 & -2 \\ -3 & 0 \end{bmatrix} \begin{bmatrix} 0 \\ -1 \end{bmatrix} = \begin{bmatrix} 1 \\ 0 \end{bmatrix}$.

$A = U\Sigma V^T = \begin{bmatrix} 0 & 1 \\ 1 & 0 \end{bmatrix} \begin{bmatrix} 3 & 0 \\ 0 & 2 \end{bmatrix} \begin{bmatrix} -1 & 0 \\ 0 & -1 \end{bmatrix} = \begin{bmatrix} 0 & -2 \\ -3 & 0 \end{bmatrix}$.

15. $A^T A = \begin{bmatrix} 25 \end{bmatrix} \Rightarrow$ eigenvector of $A^T A$ is $\begin{bmatrix} 1 \end{bmatrix} \Rightarrow$

$\mathbf{u}_1 = \frac{1}{\sigma_1} A\mathbf{v}_1 = \frac{1}{5} \begin{bmatrix} 3 \\ 4 \end{bmatrix} \begin{bmatrix} 1 \end{bmatrix} = \begin{bmatrix} \frac{3}{5} \\ \frac{4}{5} \end{bmatrix}$. There is no $\sigma_2 \Rightarrow$ extend with $\mathbf{u}_2 = \begin{bmatrix} -\frac{4}{5} \\ \frac{3}{5} \end{bmatrix} \Rightarrow$

$A = U\Sigma V^T = \begin{bmatrix} \frac{3}{5} & -\frac{4}{5} \\ \frac{4}{5} & \frac{3}{5} \end{bmatrix} \begin{bmatrix} 5 \\ 0 \end{bmatrix} \begin{bmatrix} 1 \end{bmatrix} = \begin{bmatrix} 3 \\ 4 \end{bmatrix}$.

17. $A^T A = \begin{bmatrix} 4 & 0 \\ 0 & 9 \end{bmatrix} \Rightarrow$ eigenvectors of $A^T A$ are $\begin{bmatrix} 0 \\ 1 \end{bmatrix}, \begin{bmatrix} 1 \\ 0 \end{bmatrix} \Rightarrow$

$\mathbf{u}_1 = \frac{1}{\sigma_1} A\mathbf{v}_1 = \frac{1}{3}\begin{bmatrix} 0 & 0 \\ 0 & 3 \\ -2 & 0 \end{bmatrix}\begin{bmatrix} 0 \\ 1 \end{bmatrix} = \begin{bmatrix} 0 \\ 1 \\ 0 \end{bmatrix}$. $\mathbf{u}_2 = \frac{1}{2} A\mathbf{v}_2 = \begin{bmatrix} 0 \\ 0 \\ -1 \end{bmatrix}$.

Extend $\mathbf{u}_3 = \begin{bmatrix} 1 \\ 0 \\ 0 \end{bmatrix} \Rightarrow A = U\Sigma V^T = \begin{bmatrix} 0 & 0 & 1 \\ 1 & 0 & 0 \\ 0 & -1 & 0 \end{bmatrix}\begin{bmatrix} 3 & 0 \\ 0 & 2 \\ 0 & 0 \end{bmatrix}\begin{bmatrix} 0 & 1 \\ 1 & 0 \end{bmatrix} = \begin{bmatrix} 0 & 0 \\ 0 & 3 \\ -2 & 0 \end{bmatrix}$.

19. $A^T A = \begin{bmatrix} 4 & 0 & 2 \\ 0 & 4 & 0 \\ 2 & 0 & 1 \end{bmatrix} \Rightarrow$ eigenvectors of $A^T A$ are $\begin{bmatrix} \frac{2}{\sqrt{5}} \\ 0 \\ \frac{1}{\sqrt{5}} \end{bmatrix}, \begin{bmatrix} 0 \\ 1 \\ 0 \end{bmatrix}, \begin{bmatrix} \frac{1}{\sqrt{5}} \\ 0 \\ -\frac{2}{\sqrt{5}} \end{bmatrix} \Rightarrow$

$\mathbf{u}_1 = \frac{1}{\sigma_1} A\mathbf{v}_1 = \frac{1}{\sqrt{5}}\begin{bmatrix} 2 & 0 & 1 \\ 0 & 2 & 0 \end{bmatrix}\begin{bmatrix} \frac{2}{\sqrt{5}} \\ 0 \\ \frac{1}{\sqrt{5}} \end{bmatrix} = \begin{bmatrix} 1 \\ 0 \end{bmatrix}$. $\mathbf{u}_2 = \frac{1}{2} A\mathbf{v}_2 = \begin{bmatrix} 0 \\ 1 \end{bmatrix}$.

$A = U\Sigma V^T = \begin{bmatrix} 1 & 0 \\ 0 & 1 \end{bmatrix}\begin{bmatrix} \sqrt{5} & 0 & 0 \\ 0 & 2 & 0 \end{bmatrix}\begin{bmatrix} \frac{2}{\sqrt{5}} & 0 & \frac{1}{\sqrt{5}} \\ 0 & 1 & 0 \\ \frac{1}{\sqrt{5}} & 0 & -\frac{2}{\sqrt{5}} \end{bmatrix} = \begin{bmatrix} 2 & 0 & 1 \\ 0 & 2 & 0 \end{bmatrix}$.

21. Exercise 11 $\Rightarrow \sigma_1 = \sqrt{2}, \sigma_2 = 0, \mathbf{v}_1 = \begin{bmatrix} \frac{1}{\sqrt{2}} \\ \frac{1}{\sqrt{2}} \end{bmatrix}, \mathbf{v}_2 = \begin{bmatrix} \frac{1}{\sqrt{2}} \\ -\frac{1}{\sqrt{2}} \end{bmatrix}, \mathbf{u}_1 = \begin{bmatrix} 1 \\ 0 \end{bmatrix}, \mathbf{u}_2 = \begin{bmatrix} 0 \\ 1 \end{bmatrix} \Rightarrow$

outer product form of SVD is $A = \sigma_1 \mathbf{u}_1 \mathbf{v}_1^T + \sigma_2 \mathbf{u}_2 \mathbf{v}_2^T = \sqrt{2}\begin{bmatrix} 1 \\ 0 \end{bmatrix}\begin{bmatrix} \frac{1}{\sqrt{2}} & \frac{1}{\sqrt{2}} \end{bmatrix} + 0\begin{bmatrix} 0 \\ 1 \end{bmatrix}\begin{bmatrix} \frac{1}{\sqrt{2}} & -\frac{1}{\sqrt{2}} \end{bmatrix}$.

23. Exercise 17 $\Rightarrow A = \sigma_1 \mathbf{u}_1 \mathbf{v}_1^T + \sigma_2 \mathbf{u}_2 \mathbf{v}_2^T = 3\begin{bmatrix} 0 \\ 1 \\ 0 \end{bmatrix}\begin{bmatrix} 0 & 1 \end{bmatrix} + 2\begin{bmatrix} 0 \\ 0 \\ -1 \end{bmatrix}\begin{bmatrix} 1 & 0 \end{bmatrix}$.

25. Counter examples: In Exercise 11, we could extend with $\mathbf{u}_2 = \begin{bmatrix} 0 \\ -1 \end{bmatrix} \Rightarrow$

$A = U\Sigma V^T = \begin{bmatrix} 1 & 0 \\ 0 & -1 \end{bmatrix}\begin{bmatrix} \sqrt{2} & 0 \\ 0 & 0 \end{bmatrix}\begin{bmatrix} \frac{1}{\sqrt{2}} & \frac{1}{\sqrt{2}} \\ \frac{1}{\sqrt{2}} & -\frac{1}{\sqrt{2}} \end{bmatrix} = \begin{bmatrix} 1 & 1 \\ 0 & 0 \end{bmatrix}$.

In Exercise 14, we could take eigenvectors $\mathbf{v}_1 = \begin{bmatrix} -1 \\ 0 \end{bmatrix}, \mathbf{v}_2 = \begin{bmatrix} 0 \\ -1 \end{bmatrix} \Rightarrow U = \begin{bmatrix} -\frac{1}{\sqrt{2}} & \frac{1}{\sqrt{2}} \\ -\frac{1}{\sqrt{2}} & -\frac{1}{\sqrt{2}} \end{bmatrix} \Rightarrow$

$A = U\Sigma V^T = \begin{bmatrix} -\frac{1}{\sqrt{2}} & \frac{1}{\sqrt{2}} \\ -\frac{1}{\sqrt{2}} & -\frac{1}{\sqrt{2}} \end{bmatrix}\begin{bmatrix} \sqrt{2} & 0 \\ 0 & \sqrt{2} \end{bmatrix}\begin{bmatrix} -1 & 0 \\ 0 & -1 \end{bmatrix} = \begin{bmatrix} 1 & -1 \\ 1 & 1 \end{bmatrix}$.

7.4   The Singular Value Decomposition

27. (a) Recall U, V orthogonal $\Rightarrow U^T = U^{-1}, V^T = V^{-1}$. So, $A = U\Sigma V^T \Rightarrow U^T A V = \Sigma$, where Σ is diagonal. Therefore, we only need to show $Q = U = V$. From Exercise 27(b), A positive definite $\Rightarrow \sigma_i = \lambda_i$, where λ_i is an eigenvalue of A. Also, recall $\mathbf{v}_i$ is the eigenvector of A corresponding to λ_i that is $A\mathbf{v}_i = \lambda_i \mathbf{v}_i$. So, $\mathbf{u}_i = \frac{1}{\sigma_i} A\mathbf{v}_i = \frac{1}{\lambda_i} A\mathbf{v}_i = \frac{1}{\lambda_i}\lambda_i \mathbf{v}_i = \mathbf{v}_i \Rightarrow Q = U = V$ as we needed to show.

    (b) $A = \sigma_1 \mathbf{u}_1 \mathbf{v}_1^T + \ldots + \sigma_n \mathbf{u}_n \mathbf{v}_n^T = \lambda_1 \mathbf{q}_1 \mathbf{q}_1^T + \ldots + \lambda_n \mathbf{q}_n \mathbf{q}_n^T$ the spectral decomposition because A positive definite $\Rightarrow \sigma_i = \lambda_i$ and part (a) $\Rightarrow Q = U = V$.

29. We need to show the columns of U (the left singular vectors) are the eigenvectors of AA^T. $AA^T = (U\Sigma V^T)(U\Sigma V^T)^T = U\Sigma(V^T V)\Sigma U^T = U\Sigma^2 U^T \Rightarrow U^T AA^T U = \Sigma^2$ (diagonal). Therefore, by Section 4.4, Theorem 4.23, the columns of U are the eigenvectors of AA^T.

31. We need to show the eigenvalues of $A^T A$ are the same as the eigenvalues of $(QA)^T(QA)$. In fact, we will use the fact that Q is orthogonal to show that $(QA)^T(QA) = A^T A$. So, $(QA)^T(QA) = A^T Q^T (QA) = A^T(Q^T Q)A = A^T A$.

    Q: What key property of orthogonal matrices did we use in the equation above?
    A: When Q is orthogonal $Q^{-1} = Q^T$, therefore, $Q^T Q = I$.

33. Exercises 3 and 11 $\Rightarrow \text{rank}(A) = 1 < 2$ (from $\mathbb{R}^2$), so the proof of Theorem 7.16 $\Rightarrow y_1 \leq 1$. Furthermore, since y_1 corresponds to $\mathbf{e}_1 = \begin{bmatrix} 1 \\ 0 \end{bmatrix} \Rightarrow y_1 \leq 1$ is the interval $[-1, 1]$.

35. Exercises 9 and 19 $\Rightarrow \text{rank}(A) = 2 < 3$ (from $\mathbb{R}^3$), so the proof of Theorem 7.16 $\Rightarrow (\frac{y_1}{\sqrt{5}})^2 + (\frac{y_2}{2})^2 \leq 1$ (because $\sigma_1 = \sqrt{5}$ and $\sigma_2 = 2$). After simplifying, we have $\frac{y_1^2}{5} + \frac{y_2^2}{4} \leq 1$.

37. Exercise 3 $\Rightarrow$ (a) $\|A\|_2 = \sigma_1 = \sqrt{2}$.   (b) $\text{cond}_2(A) = \frac{\sigma_1}{\sigma_2} = \frac{\sqrt{2}}{0} = \infty$.

39. $A^T A = \begin{bmatrix} 2 & 1.9 \\ 1.9 & 1.81 \end{bmatrix} \Rightarrow$ the eigenvalues of $A^T A$ are $\lambda_1 \approx 3.807, \lambda_2 \approx 0.0026 \Rightarrow$

    (a) $\|A\|_2 = \sigma_1 = \sqrt{\lambda_1} \approx \sqrt{3.807} \approx 1.95$.   (b) $\text{cond}_2(A) = \frac{\sigma_1}{\sigma_2} \approx \frac{\sqrt{\lambda_1}}{\sqrt{\lambda_0}} \approx \frac{\sqrt{3.807}}{\sqrt{0.0026}} \approx 38.27$.

41. Exercise 11 $\Rightarrow A^+ = V\Sigma^+ U^T = \begin{bmatrix} \frac{1}{\sqrt{2}} & \frac{1}{\sqrt{2}} \\ \frac{1}{\sqrt{2}} & -\frac{1}{\sqrt{2}} \end{bmatrix} \begin{bmatrix} \frac{1}{\sqrt{2}} & 0 \\ 0 & 0 \end{bmatrix} \begin{bmatrix} 1 & 0 \\ 0 & 1 \end{bmatrix} = \begin{bmatrix} \frac{1}{2} & 0 \\ \frac{1}{2} & 0 \end{bmatrix}$.

43. Exercise 19 $\Rightarrow A^+ = V\Sigma^+ U^T = \begin{bmatrix} \frac{2}{\sqrt{5}} & 0 & \frac{1}{\sqrt{5}} \\ 0 & 1 & 0 \\ \frac{1}{\sqrt{5}} & 0 & -\frac{2}{\sqrt{5}} \end{bmatrix} \begin{bmatrix} \frac{1}{\sqrt{5}} & 0 \\ 0 & \frac{1}{2} \\ 0 & 0 \end{bmatrix} \begin{bmatrix} 1 & 0 \\ 0 & 1 \end{bmatrix} = \begin{bmatrix} \frac{2}{5} & 0 \\ 0 & \frac{1}{2} \\ \frac{1}{5} & 0 \end{bmatrix}$.

45. $A^T A = \begin{bmatrix} 5 & 10 \\ 10 & 20 \end{bmatrix} \Rightarrow \sigma_i = 5, 0 \Rightarrow V = \begin{bmatrix} \frac{1}{\sqrt{5}} & \frac{2}{\sqrt{5}} \\ \frac{2}{\sqrt{5}} & -\frac{1}{\sqrt{5}} \end{bmatrix}, \Sigma = \begin{bmatrix} 5 & 0 \\ 0 & 0 \end{bmatrix}, U = \begin{bmatrix} \frac{1}{\sqrt{5}} & \frac{2}{\sqrt{5}} \\ \frac{2}{\sqrt{5}} & -\frac{1}{\sqrt{5}} \end{bmatrix} \Rightarrow$

    $A^+ = \begin{bmatrix} \frac{1}{\sqrt{5}} & \frac{2}{\sqrt{5}} \\ \frac{2}{\sqrt{5}} & -\frac{1}{\sqrt{5}} \end{bmatrix} \begin{bmatrix} \frac{1}{5} & 0 \\ 0 & 0 \end{bmatrix} \begin{bmatrix} \frac{1}{\sqrt{5}} & \frac{2}{\sqrt{5}} \\ \frac{2}{\sqrt{5}} & -\frac{1}{\sqrt{5}} \end{bmatrix} = \begin{bmatrix} \frac{1}{25} & \frac{2}{25} \\ \frac{2}{25} & \frac{4}{25} \end{bmatrix} \Rightarrow \bar{\mathbf{x}} = \begin{bmatrix} \frac{1}{25} & \frac{2}{25} \\ \frac{2}{25} & \frac{4}{25} \end{bmatrix} \begin{bmatrix} 3 \\ 5 \end{bmatrix} = \begin{bmatrix} 0.52 \\ 1.04 \end{bmatrix}$.

47. $A^T A = \begin{bmatrix} 3 & 3 \\ 3 & 3 \end{bmatrix} \Rightarrow \sigma_i = \sqrt{6}, 0 \Rightarrow V = \begin{bmatrix} \frac{1}{\sqrt{2}} & \frac{1}{\sqrt{2}} \\ \frac{1}{\sqrt{2}} & -\frac{1}{\sqrt{2}} \end{bmatrix}$ and $U = \begin{bmatrix} \frac{1}{\sqrt{3}} & \frac{2}{\sqrt{6}} & 0 \\ \frac{1}{\sqrt{3}} & -\frac{1}{\sqrt{6}} & \frac{1}{\sqrt{2}} \\ \frac{1}{\sqrt{3}} & -\frac{1}{\sqrt{6}} & \frac{1}{\sqrt{2}} \end{bmatrix} \Rightarrow$

$A^+ = \begin{bmatrix} \frac{1}{\sqrt{2}} & \frac{1}{\sqrt{2}} \\ \frac{1}{\sqrt{2}} & -\frac{1}{\sqrt{2}} \end{bmatrix} \begin{bmatrix} \frac{1}{\sqrt{6}} & 0 & 0 \\ 0 & 0 & 0 \end{bmatrix} \begin{bmatrix} \frac{1}{\sqrt{3}} & \frac{1}{\sqrt{3}} & \frac{1}{\sqrt{3}} \\ \frac{2}{\sqrt{6}} & -\frac{1}{\sqrt{6}} & -\frac{1}{\sqrt{6}} \\ 0 & \frac{1}{\sqrt{2}} & \frac{1}{\sqrt{2}} \end{bmatrix} = \begin{bmatrix} \frac{1}{6} & \frac{1}{6} & \frac{1}{6} \\ \frac{1}{6} & \frac{1}{6} & \frac{1}{6} \end{bmatrix} \Rightarrow \bar{\mathbf{x}} = \begin{bmatrix} \frac{1}{6} & \frac{1}{6} & \frac{1}{6} \\ \frac{1}{6} & \frac{1}{6} & \frac{1}{6} \end{bmatrix} \begin{bmatrix} 1 \\ 2 \\ 3 \end{bmatrix} = \begin{bmatrix} 1 \\ 1 \end{bmatrix}$.

49. (a) Normal equations: $A^T A \bar{\mathbf{x}} = A^T \mathbf{b}$. So, $A^T A = \begin{bmatrix} 2 & 2 \\ 2 & 2 \end{bmatrix}$ and $A^T \mathbf{b} = \begin{bmatrix} 1 \\ 1 \end{bmatrix}$.

So, since $\begin{bmatrix} 2 & 2 & | & 1 \\ 2 & 2 & | & 1 \end{bmatrix} \longrightarrow \begin{bmatrix} 1 & 1 & | & \frac{1}{2} \\ 0 & 0 & | & 0 \end{bmatrix} \Rightarrow \bar{\mathbf{x}} = \begin{bmatrix} \frac{1}{2} \\ 0 \end{bmatrix} \Rightarrow x + y = \frac{1}{2}$.

(b) Following the method of Section 1.3, Example 1.20, we have:

$\mathbf{x} = \mathbf{p} + t\mathbf{d} \Rightarrow \begin{bmatrix} x \\ y \end{bmatrix} = \begin{bmatrix} \frac{1}{2} \\ 0 \end{bmatrix} + t \begin{bmatrix} 1 \\ -1 \end{bmatrix} = \begin{bmatrix} \frac{1}{2} + t \\ t \end{bmatrix}$.

(c) We need to minimize $\left\| \begin{bmatrix} \frac{1}{2} + t \\ t \end{bmatrix} \right\|^2 = (1 + t)^2 + (-t)^2 = 2t^2 + t + \frac{1}{4}$.

This is an upward opening parabola, so its minimum occurs at $t = -\frac{b}{2a} \Rightarrow$

$t = -\frac{1}{2(2)} = -\frac{1}{4} \Rightarrow \bar{\mathbf{x}}$ of minimum length $= \begin{bmatrix} \frac{1}{2} + (-\frac{1}{4}) \\ -(-\frac{1}{4}) \end{bmatrix} = \begin{bmatrix} \frac{1}{4} \\ \frac{1}{4} \end{bmatrix}$ as found in Example 7.40.

51. Note, $\Sigma_{n \times m}^+ \Sigma_{m \times n} = \begin{bmatrix} I_{r \times r} & 0 \\ 0 & 0 \end{bmatrix}_{n \times n} \Rightarrow \Sigma \Sigma^+ \Sigma = \Sigma (\Sigma^+ \Sigma) = \begin{bmatrix} D_{r \times r} & 0 \\ 0 & 0 \end{bmatrix} \begin{bmatrix} I_{r \times r} & 0 \\ 0 & 0 \end{bmatrix}_{n \times n} = \Sigma$.

(a) Will show $AA^+ A = A$ using the fact that $\Sigma \Sigma^+ \Sigma = \Sigma$ (shown above) in step 4.
$AA^+ A = A(V\Sigma^+ U^T) A = (U\Sigma V^T)(V\Sigma^+ U^T)(U\Sigma V^T) = U\Sigma(V^T V)\Sigma^+(U^T U)\Sigma V^T$
$= U(\Sigma \Sigma^+ \Sigma) V^T = U\Sigma V^T = A$.
We are also using the fact (again) that U, V orthogonal $\Rightarrow V^T V = U^T U = I$.

(b) Will show $A^+ A A^+ = A^+$ using the fact that $\Sigma^+ \Sigma \Sigma^+ = \Sigma^+$ (see note) in step 4.
$A^+ A A^+ = (V\Sigma^+ U^T) A (V\Sigma^+ U^T) = (V\Sigma^+ U^T)(U\Sigma V^T)(V\Sigma^+ U^T)$
$= V\Sigma^+ (U^T U)\Sigma(V^T V)\Sigma^+ V^T = V(\Sigma^+ \Sigma \Sigma^+)U^T = V\Sigma^+ U^T = A^+$.

(c) Will show $(AA^+)^T = AA^+$ using the fact that $(\Sigma^+)^T (\Sigma)^T = \Sigma \Sigma^+$ (see note) in step 5.
$(AA^+)^T = (AV\Sigma^+ U^T)^T = U(\Sigma^+)^T V^T A^T = U(\Sigma^+)^T (V^T V)(\Sigma)^T U^T$
$= U(\Sigma^+)^T (\Sigma)^T U^T = U\Sigma \Sigma^+ U^T = (U\Sigma V^T)(V\Sigma^+ U^T) = AA^+$.
Likewise, $(\Sigma)^T (\Sigma^+)^T = \Sigma^+ \Sigma \Rightarrow (A^+ A)^T = A^+ A$.

7.4 The Singular Value Decomposition

53. We will show A satisfies the Penrose conditions for A^+, so by Exercise 52 $A = (A^+)^+$.
Comparing the Penrose conditions for A^+ and A, we have:

$(a)^+$ $A^+(A^+)^+A^+ = A^+$ (a) $AA^+A = A$
$(b)^+$ $(A^+)^+A^+(A^+)^+ = (A^+)^+$ (b) $A^+AA^+ = A^+$
$(c)^+$ $i.\ (A^+(A^+)^+)^T = A^+(A^+)^+$ (c) $i.\ (AA^+)^T = AA^+$
 $\quad\ \ ii.\ ((A^+)^+A^+)^T = (A^+)^+A^+$ $\quad\ \ ii.\ (A^+A)^T = A^+A$

Since A satisfies conditions $(a), (b), (c)$, will show A satisfies conditions $(a)^+, (b)^+, (c)^+$.

$(a)^+$ A satisfies $(b) \Rightarrow A^+AA^+ = A^+ \Rightarrow A$ satisfies $(a)^+$, that is $A^+(A^+)^+A^+ = A^+$.
$(b)^+$ A satisfies $(a) \Rightarrow AA^+A = A \Rightarrow A$ satisfies $(b)^+$, that is $(A^+)^+A^+(A^+)^+ = (A^+)^+$.
$(c)^+$ A satisfies $(c)\ ii. \Rightarrow (A^+A)^T = A^+A \Rightarrow A$ satisfies $(c)^+\ i.$,
that is $(A^+(A^+)^+)^T = A^+(A^+)^+$. Likewise, for $(c)^+\ ii. \Rightarrow$

A satisfies the Penrose conditions for $A^+ \Rightarrow A = (A^+)^+$.

55. We will show A satisfies the Penrose conditions for A, so by Exercise 52 $A = A^+$.
Comparing Penrose conditions for A and conditions derived from $A = A^T = A^2$, we have:

(a) $AA^+A = A$ $(a)'$ $AAA = A^2A = A^2 = A$
(b) $A^+AA^+ = A^+$ $(b)'$ $AAA = A^2A = A^2 = A$
(c) $i.\ (AA^+)^T = AA^+$ $(c)'$ $i.\ (AA)^T = A^TA^T = AA$
$\quad\ \ ii.\ (A^+A)^T = A^+A$ $\quad\ \ ii.\ (AA)^T = A^TA^T = AA$

Since A satisfies conditions $(a)', (b)', (c)'$ will show A satisfies conditions $(a), (b), (c)$.

(a) A satisfies $(a)' \Rightarrow AAA = A \Rightarrow A$ satisfies (a), that is $AA^+A = A$.
(b) A satisfies $(b)' \Rightarrow AAA = A \Rightarrow A$ satisfies (b), that is $A^+AA^+ = A^+$.
(c) A satisfies $(c)'\ i. \Rightarrow A$ satisfies $(c)\ i.$ Likewise, for $(c)\ ii. \Rightarrow$

A satisfies the Penrose conditions for $A \Rightarrow A = A^+$.

57. From Section 5.5, p.415, we know a positive definite matrix A is *symmetric*. So, $A^T = A$.
Therefore, $A^T = (U\Sigma V^T)^T = V\Sigma U^T = U\Sigma V^T = A$.
Q: Is this enough to prove that $U = V$? Why or why not?
If not, what additional implication of being positive definite do we need to evoke?

59. We follow the hint to prove $\|A\|_2^2 \leq \|A\|_1 \|A\|_\infty$.
Let σ be the largest singular value of A, then as noted in the hint: $\|A\|_2^2 = \sigma^2 = |\lambda|$.
Now since $\|A^T\|_1 = \|A\|_\infty$, $\|A^TA\|_1 = \|A^T\|_1\|A\|_1 = \|A^T\|_1\|A\|_1 = \|A\|_\infty\|A\|_1$.
By Exercise 34 of Section 7.2, $\|A\|_1\|A\|_\infty \geq |\lambda| = \sigma^2 = \|A\|_2^2$ as required.

61. Recall, from Exercises 3 and 11, $U = I_2$, so we have:

$$A = RQ = (U\Sigma U^T)(UV^T) = (I_2\ \Sigma\ I_2^T)(I_2\ V^T) = \Sigma V^T = \begin{bmatrix} \sqrt{2} & 0 \\ 0 & 0 \end{bmatrix} \begin{bmatrix} \frac{1}{\sqrt{2}} & \frac{1}{\sqrt{2}} \\ \frac{1}{\sqrt{2}} & -\frac{1}{\sqrt{2}} \end{bmatrix}.$$

63. Though it is possible to solve this problem exactly, you may wish to solve it using a CAS. However, your answers should be exact as the following analysis shows.

$$A = \begin{bmatrix} 1 & 2 \\ -3 & -1 \end{bmatrix} \Rightarrow A^T A = \begin{bmatrix} 10 & 5 \\ 5 & 5 \end{bmatrix} \Rightarrow \begin{vmatrix} 10-\lambda & 5 \\ 5 & 5-\lambda \end{vmatrix} = 0 \Rightarrow$$

$$\lambda_1 = \frac{15 + 5\sqrt{5}}{2}, \mathbf{v}_1 = \begin{bmatrix} \frac{1+\sqrt{5}}{2} \\ 1 \end{bmatrix} \text{ and } \lambda_2 = \frac{15 - 5\sqrt{5}}{2}, \mathbf{v}_2 = \begin{bmatrix} \frac{1-\sqrt{5}}{2} \\ 1 \end{bmatrix}.$$

Normalizing these vectors, we create V and then use the columns of V to create U:

$$V = \begin{bmatrix} \sqrt{\frac{2}{5-\sqrt{5}}} & -\sqrt{\frac{2}{5+\sqrt{5}}} \\ \sqrt{\frac{2}{5+\sqrt{5}}} & \sqrt{\frac{2}{5-\sqrt{5}}} \end{bmatrix} \text{ so } V \text{ has the form } V = \begin{bmatrix} v_{11} & -v_{21} \\ v_{21} & v_{11} \end{bmatrix} \Rightarrow$$

$$\mathbf{u}_1 = \frac{1}{\sigma_1} A\mathbf{v}_1 = \sqrt{\frac{1}{\lambda_1}} \begin{bmatrix} 1 & 2 \\ -3 & -1 \end{bmatrix} \begin{bmatrix} v_{11} \\ v_{21} \end{bmatrix} = \begin{bmatrix} \sqrt{\frac{1}{\lambda_1}}(v_{11} + 2v_{21}) \\ -\sqrt{\frac{1}{\lambda_1}}(3v_{11} + v_{21}) \end{bmatrix} = \begin{bmatrix} v_{21} \\ -v_{11} \end{bmatrix}.$$

This last equality and ones like it are the reason this solution is exact. We show one below:

$$\sqrt{\frac{1}{\lambda_1}}(v_{11} + 2v_{21}) = v_{21} \Leftrightarrow \sqrt{\frac{1}{\lambda_1}}(\frac{v_{11}}{v_{21}} + 2)v_{21} = v_{21} \Leftrightarrow \frac{v_{11}}{v_{21}} + 2 = \sqrt{\lambda_1} \text{ that is:}$$

$$\frac{\sqrt{\frac{2}{5-\sqrt{5}}}}{\sqrt{\frac{2}{5+\sqrt{5}}}} + 2 = \sqrt{\frac{15 + 5\sqrt{5}}{2}} \Rightarrow \sqrt{\frac{5+\sqrt{5}}{5-\sqrt{5}}} + 2 = \frac{5+\sqrt{5}}{\sqrt{20}} + \frac{2\sqrt{20}}{\sqrt{20}} = \frac{5+5\sqrt{5}}{\sqrt{20}}$$

$$= \sqrt{\frac{(5+5\sqrt{5})^2}{20}} = \sqrt{\frac{150 + 50\sqrt{5}}{20}} = \sqrt{\frac{15 + 5\sqrt{5}}{2}} = \sqrt{\lambda_1}.$$

Likewise, $\mathbf{u}_2 = \frac{1}{\sigma_2} A\mathbf{v}_2 = \begin{bmatrix} v_{11} \\ v_{21} \end{bmatrix} \Rightarrow U = \begin{bmatrix} v_{21} & v_{11} \\ -v_{11} & v_{21} \end{bmatrix}$. Therefore, $Q = UV^T$

$$= \begin{bmatrix} v_{21} & v_{11} \\ -v_{11} & v_{21} \end{bmatrix} \begin{bmatrix} v_{11} & v_{21} \\ -v_{21} & v_{11} \end{bmatrix} = \begin{bmatrix} v_{11}v_{21} - v_{11}v_{21} = 0 & v_{11}^2 + v_{21}^2 = 1 \\ -(v_{11}^2 + v_{21}^2) = -1 & v_{11}v_{21} - v_{11}v_{21} = 0 \end{bmatrix} = \begin{bmatrix} 0 & 1 \\ -1 & 0 \end{bmatrix}.$$

Similarly, we find $R = U\Sigma U^T$

$$= \begin{bmatrix} v_{21} & v_{11} \\ -v_{11} & v_{21} \end{bmatrix} \begin{bmatrix} \sqrt{\lambda_1} & 0 \\ 0 & \sqrt{\lambda_2} \end{bmatrix} \begin{bmatrix} v_{11} & v_{21} \\ -v_{21} & v_{11} \end{bmatrix}$$

$$= \begin{bmatrix} \sqrt{\lambda_2}v_{11}^2 + \sqrt{\lambda_1}v_{21}^2 = 2 & -\sqrt{\lambda_1}v_{11}v_{21} + \sqrt{\lambda_2}v_{11}v_{21} = -1 \\ -\sqrt{\lambda_1}v_{11}v_{21} + \sqrt{\lambda_2}v_{11}v_{21} = -1 & \sqrt{\lambda_1}v_{11}^2 + \sqrt{\lambda_2}v_{21}^2 = 3 \end{bmatrix} = \begin{bmatrix} 2 & -1 \\ -1 & 3 \end{bmatrix}.$$

So the polar decomposition of A is $A = RQ = \begin{bmatrix} 2 & -1 \\ -1 & 3 \end{bmatrix} \begin{bmatrix} 0 & 1 \\ -1 & 0 \end{bmatrix} = \begin{bmatrix} 1 & 2 \\ -3 & -1 \end{bmatrix}.$

7.5 Applications

Q: What are our main goals for Section 7.5?
A: To find best approximations to functions and error-correcting codes.
Since these applications are fully explained in the text, we simply present solutions here.

Key Definitions and Concepts

best approximation	p.620	7.5	If $\{\mathbf{u}_1,\ldots,\mathbf{u}_k\}$ is an orthogonal basis for W then $\text{proj}_W(f) = \frac{\langle \mathbf{u}_1,f \rangle}{\langle \mathbf{u}_1,\mathbf{u}_1 \rangle}\mathbf{u}_1 + \cdots + \frac{\langle \mathbf{u}_k,f \rangle}{\langle \mathbf{u}_k,\mathbf{u}_k \rangle}\mathbf{u}_k$
correction	p.627	7.5	See text for full description of *correcting k errors*.
decoding	p.627	7.5	See text for full description of *nearest neighbor decoding*.
detection	p.627	7.5	See text for full description of *detecting k errors*.
Fourier approximation	p.624	7.5	The best approximation using the *Fourier coefficients*. Given below, used in a *trigonometric polynomial*.
Fourier coefficients	p.624	7.5	These coefficients are used in the *Fourier approximation*: $a_0 = \frac{\langle 1,f \rangle}{\langle 1,1 \rangle} = \frac{1}{2\pi}\int_{-\pi}^{\pi} f(x)\,dx$ $a_k = \frac{\langle \cos kx, f \rangle}{\langle \cos kx, \cos kx \rangle} = \frac{1}{\pi}\int_{-\pi}^{\pi} f(x)\cos kx\,dx$ $b_k = \frac{\langle \sin kx, f \rangle}{\langle \sin kx, \sin kx \rangle} = \frac{1}{\pi}\int_{-\pi}^{\pi} f(x)\sin kx\,dx$
Fourier series	p.626	7.5	Let the *Fourier trigonometric polynomial* become infinite. Then the result is called the *Fourier series*. $f(x) = a_0 + \sum_{k=1}^{\infty}(a_k \cos kx + b_k \sin kx)$ on $[-\pi,\pi]$
minimum distance	p.626	7.5	The smallest distance between any two distinct vectors in C. $d(C) = \min\{d_H(\mathbf{x},\mathbf{y}) : \mathbf{x} \neq \mathbf{y} \text{ in } C\}$
root mean square error	p.621	7.5	This is the *root mean square error*: $\|f - g\| = \sqrt{\int_{-1}^{1}(f(x) - g(x))^2\,dx}$
trigonometric polynomial	p.623	7.5	A function of this form is a *trigonometric polynomial*. $p(x) = a_0 + a_1 \cos x + a_2 \cos 2x + \cdots + a_n \cos nx + b_1 \sin x + b_2 \sin 2x + \cdots + b_n \sin nx.$

Theorems

Thm 7.20	p.628	7.5	Let C be a (binary) code with minimum distance d. a. C detects k errors if and only if $d \geq k+1$. b. C corrects k errors if and only if $d \geq 2k+1$.
Thm 7.21	p.628	7.5	Let C be a (n,k) with parity check matrix P. Then the minimum distance of C is the smallest integer d for which there are d linearly dependent columns.

Solutions to odd-numbered exercises from Section 7.5

1. From Example 7.41, we have $g(x) = \text{proj}_W(x^2) = \frac{\langle 1, x^2 \rangle}{\langle 1, 1 \rangle} 1 + \frac{\langle x, x^2 \rangle}{\langle x, x \rangle} x$, so we compute:

$\langle 1, x^2 \rangle = \int_{-1}^{1} x^2 \, dx = \frac{1}{3} x^3 \big|_{-1}^{1} = \frac{2}{3}$ $\qquad$ $\langle x, x^2 \rangle = \int_{-1}^{1} x^3 \, dx = \frac{1}{4} x^4 \big|_{-1}^{1} = 0$

$\langle 1, 1 \rangle = \int_{-1}^{1} 1 \, dx = x \big|_{-1}^{1} = 2$ $\qquad$ $\langle x, x \rangle = \int_{-1}^{1} x^2 \, dx = \frac{1}{3} x^3 \big|_{-1}^{1} = \frac{2}{3}$

$g(x) = \text{proj}_W(x^2) = \frac{\langle 1, x^2 \rangle}{\langle 1, 1 \rangle} 1 + \frac{\langle x, x^2 \rangle}{\langle x, x \rangle} x = \frac{2/3}{2} 1 + \frac{0}{2/3} x = \frac{1}{3}$ is the best linear approximation.

3. Building on our work in Exercises 1 and 2, we compute:

$\langle 1, x^3 \rangle = \int_{-1}^{1} x^3 \, dx = \langle x, x^2 \rangle = 0$ $\qquad$ $\langle x, x^3 \rangle = \int_{-1}^{1} x^4 \, dx = \frac{1}{5} x^5 \big|_{-1}^{1} = \frac{2}{5}$

$g(x) = \frac{\langle 1, x^3 \rangle}{\langle 1, 1 \rangle} 1 + \frac{\langle x, x^3 \rangle}{\langle x, x \rangle} x = \frac{0}{2} 1 + \frac{2/5}{2/3} x = \frac{3}{5} x$ is the best linear approximation.

5. From Example 7.42, we have the orthogonal basis $\{1, x, x^2 - \frac{1}{3}\} \Rightarrow$

$\langle 1, |x| \rangle = \int_0^1 x \, dx - \int_{-1}^0 x \, dx = 1$ $\qquad$ $\langle x^2 - \frac{1}{3}, |x| \rangle = \int_0^1 (x^3 - \frac{1}{3} x) \, dx - \int_{-1}^0 (x^3 - \frac{1}{3} x) \, dx = \frac{1}{6}$

$\langle x, |x| \rangle = \int_0^1 x^2 \, dx - \int_{-1}^0 x^2 \, dx = 0$ $\qquad$ $\langle x^2 - \frac{1}{3}, x^2 - \frac{1}{3} \rangle = \int_{-1}^1 (x^2 - \frac{1}{3})^2 \, dx = \frac{8}{45} \Rightarrow$

$g(x) = \frac{\langle 1, |x| \rangle}{\langle 1, 1 \rangle} 1 + \frac{\langle x, |x| \rangle}{\langle x, x \rangle} x + \frac{\langle x^2 - \frac{1}{3}, |x| \rangle}{\langle x^2 - \frac{1}{3}, x^2 - \frac{1}{3} \rangle} x = \frac{1}{2} 1 + \frac{0}{2/3} x + \frac{1/6}{8/45} (x^2 - \frac{1}{3}) = \frac{3}{16} + \frac{15}{16} x^2$.

7. We have $\mathbf{x}_2 = x$ and $\mathbf{v}_1 = \mathbf{x}_1 = 1 \Rightarrow$

$\langle 1, 1 \rangle = \int_0^1 1 \, dx = x \big|_0^1 = 1$ and $\langle 1, x \rangle = \int_0^1 x \, dx = \frac{1}{2} x^2 \big|_0^1 = \frac{1}{2} \Rightarrow$

$\mathbf{v}_2 = x - \frac{\langle 1, x \rangle}{\langle 1, 1 \rangle} 1 = x - \frac{1/2}{1} 1 = x - \frac{1}{2} \Rightarrow$ an orthogonal basis for $\mathscr{P}_1[0, 1]$ is $\mathcal{B} = \{1, x - \frac{1}{2}\}$.

9. Building on our work in Exercise 8, we have:

$g(x) = \frac{\langle 1, x^2 \rangle}{\langle 1, 1 \rangle} 1 + \frac{\langle x - \frac{1}{2}, x^2 \rangle}{\langle x - \frac{1}{2}, x - \frac{1}{2} \rangle} (x - \frac{1}{2}) = \frac{1/3}{1} 1 + \frac{1/12}{1/12} (x - \frac{1}{2}) = x - \frac{1}{6}$.

11. $\langle x - \frac{1}{2}, e^x \rangle = \int_0^1 (x e^x - \frac{1}{2} e^x) \, dx = x e^x - \frac{3}{2} e^x \big|_0^1 = -\frac{1}{2} e + \frac{3}{2} \Rightarrow$

$g(x) = \frac{\langle 1, e^x \rangle}{\langle 1, 1 \rangle} 1 + \frac{\langle x - \frac{1}{2}, e^x \rangle}{\langle x - \frac{1}{2}, x - \frac{1}{2} \rangle} (x - \frac{1}{2}) = \frac{e-1}{1} 1 + \frac{-e/2 + 3/2}{1/12} (x - \frac{1}{2}) = (4e - 10) + (18 - 6e) x$.

13. Similar to our work in Exercise 7 and building on our work in Exercise 8, we have:

$\langle 1, x^3 \rangle = \int_0^1 x^3 \, dx = \frac{1}{4} x^4 \big|_0^1 = \frac{1}{4}$ and

$\langle x - \frac{1}{2}, x^3 \rangle = \langle x, x^3 \rangle - \frac{1}{2} \langle 1, x^3 \rangle = \int_0^1 x^4 \, dx - \frac{1}{2} (\frac{1}{4}) = \frac{1}{5} x^5 \big|_0^1 - \frac{1}{8} = \frac{1}{5} - \frac{1}{8} = \frac{3}{40}$.

$\langle x^2 - x + \frac{1}{6}, x^3 \rangle = \langle x^2, x^3 \rangle - \langle x, x^3 \rangle + \frac{1}{6} \langle 1, x^3 \rangle = \int_0^1 x^5 \, dx - 1(\frac{1}{5}) + \frac{1}{6}(\frac{1}{4})$.

$= \frac{1}{6} x^6 \big|_0^1 - \frac{1}{5} + \frac{1}{24} = \frac{1}{6} - \frac{19}{120} = \frac{1}{120}$.

$\langle x^2 - x + \frac{1}{6}, x^2 - x + \frac{1}{6} \rangle = \langle x^2, x^2 \rangle - 2 \langle x, x^2 \rangle + \frac{4}{3} \langle x, x \rangle - \frac{1}{3} \langle 1, x \rangle + \frac{1}{36} \langle 1, 1 \rangle$

$= 1(\frac{1}{5}) - 2(\frac{1}{4}) + \frac{4}{3}(\frac{1}{3}) - \frac{1}{3}(\frac{1}{2}) + \frac{1}{36}(1) = \frac{1}{180}$.

$g(x) = \frac{\langle 1, x^3 \rangle}{\langle 1, 1 \rangle} 1 + \frac{\langle x - \frac{1}{2}, x^3 \rangle}{\langle x - \frac{1}{2}, x - \frac{1}{2} \rangle} (x - \frac{1}{2}) + \frac{\langle x^2 - x + \frac{1}{6}, x^3 \rangle}{\langle x^2 - x + \frac{1}{6}, x^2 - x + \frac{1}{6} \rangle} (x^2 - x + \frac{1}{6})$

$= \frac{1/4}{1} 1 + \frac{3/40}{1/12} (x - \frac{1}{2}) + \frac{1/120}{1/180} (x^2 - x + \frac{1}{6}) = \frac{1}{20} - \frac{3}{5} x + \frac{3}{2} x^2$.

7.5 Applications

15. Building on our work in Exercise 11, we have:
$$\langle x^2 - x + \tfrac{1}{6}, e^x\rangle = \langle x^2, e^x\rangle - \langle x, e^x\rangle + \tfrac{1}{6}\langle 1, e^x\rangle$$
$$= \int_0^1 x^2 e^x\,dx - 1(1) + \tfrac{1}{6}(e-1) = x^2 e^x - 2x\,e^x + 2e^x\Big|_0^1 + \left(\tfrac{1}{6}e - \tfrac{7}{6}\right)$$
$$= (e-2) + \left(\tfrac{1}{6}e - \tfrac{7}{6}\right) = \tfrac{7}{6}e - \tfrac{19}{6}$$
$$g(x) = \frac{\langle 1, e^x\rangle}{\langle 1, 1\rangle}1 + \frac{\langle x - \tfrac{1}{2}, e^x\rangle}{\langle x - \tfrac{1}{2}, x - \tfrac{1}{2}\rangle}(x - \tfrac{1}{2}) + \frac{\langle x^2 - x + \tfrac{1}{6}, e^x\rangle}{\langle x^2 - x + \tfrac{1}{6}, x^2 - x + \tfrac{1}{6}\rangle}(x^2 - x + \tfrac{1}{6})$$
$$= [(4e - 10) + (18 - 6e)x] + \frac{(7e-19)/6}{1/180}(x^2 - x + \tfrac{1}{6})$$
$$= (39e - 105) + (588 - 216e)x + (210e - 570)x^2.$$

17. We need to show $\langle 1, \cos(kx)\rangle = 0$ and $\langle 1, \sin(kx)\rangle = 0$ in $\mathscr{C}[-\pi, \pi]$.
$$\langle 1, \cos(kx)\rangle = \int_{-\pi}^{\pi}\cos(kx)\,dx = \tfrac{1}{k}\sin(kx)\Big|_{-\pi}^{\pi} = \tfrac{2}{k}\sin(k\pi) = \tfrac{2}{k}\cdot 0 = 0.$$
$$\langle 1, \sin kx\rangle = \int_{-\pi}^{\pi}\sin kx\,dx = -\tfrac{1}{k}\cos(kx)\Big|_{-\pi}^{\pi} = -\tfrac{1}{k}\big(\cos(k\pi) - \cos(k\pi)\big) = -\tfrac{1}{k}\cdot 0 = 0.$$

19. We need to show $\langle \sin(jx), \sin(kx)\rangle = 0$ in $\mathscr{C}[-\pi, \pi]$, provided $j \neq k$.
$$\langle \sin(jx), \sin(kx)\rangle = \int_{-\pi}^{\pi}\sin(jx)\sin(kx)\,dx = \tfrac{1}{2}\int_{-\pi}^{\pi}\cos(j-k)x - \cos(j+k)x\,dx$$
$$= \tfrac{1}{2}\left(\frac{\sin(j-k)x}{j-k} - \frac{\sin(j+k)x}{j+k}\right)\Big|_{-\pi}^{\pi} = 0.$$
Note $j \neq k \Rightarrow$ we can divide by $j - k \neq 0$, as we did in the last step.

21. $a_0 = \tfrac{1}{2\pi}\int_{-\pi}^{\pi}|x|\,dx = \tfrac{1}{2\pi}\left(\int_0^{\pi} x\,dx + \int_{-\pi}^0 x\,dx\right) = \tfrac{1}{\pi}\int_0^{\pi} x\,dx = \tfrac{1}{\pi}\tfrac{x^2}{2}\Big|_0^{\pi} = \tfrac{\pi}{2}.$
$a_1 = \tfrac{1}{\pi}\int_{-\pi}^{\pi}|x|\cos x\,dx = \tfrac{1}{\pi}\left(\int_0^{\pi} x\cos x\,dx + \int_{-\pi}^0 x\cos x\,dx\right) = \tfrac{2}{\pi}\int_0^{\pi} x\cos x\,dx$
$= \tfrac{2}{\pi}(x\sin x + \cos x)\Big|_0^{\pi} = -\tfrac{4}{\pi}.$
$a_2 = \tfrac{2}{\pi}\left(\tfrac{1}{2}x\sin(2x) + \tfrac{1}{4}\cos(2x)\right)\Big|_0^{\pi} = \tfrac{2}{\pi}\left(\tfrac{1}{4} - \tfrac{1}{4}\right) = 0.$
$a_3 = \tfrac{2}{\pi}\left(\tfrac{1}{3}x\sin(3x) + \tfrac{1}{9}\cos(3x)\right)\Big|_0^{\pi} = \tfrac{2}{\pi}\left(-\tfrac{1}{9} - \tfrac{1}{9}\right) = -\tfrac{4}{9\pi}.$
$b_k = \tfrac{1}{\pi}\int_{-\pi}^{\pi}|x|\sin(kx)\,dx = \tfrac{1}{\pi}\left(\int_0^{\pi} x\sin(kx)\,dx - \int_0^{\pi} x\sin(kx)\,dx\right) = 0.$
So, the third-order Fourier approximation of $f(x) = |x|$ on $[-\pi, \pi]$ is:
$$\text{proj}_W(|x|) = a_0 + a_1\cos x + a_2\cos(2x) + a_3\cos(3x)$$
$$= \tfrac{\pi}{2} - \tfrac{4}{\pi}\cos x + 0\cos(2x) - \tfrac{4}{9\pi}\cos(3x) = \tfrac{\pi}{2} - \tfrac{4}{\pi}\left(\cos x + \frac{\cos(3x)}{9}\right).$$

23. $a_0 = \tfrac{1}{2\pi}\int_{-\pi}^{\pi}f(x)\,dx = \tfrac{1}{2\pi}\left(\int_0^{\pi} 1\,dx + \int_{-\pi}^0 0\,dx\right) = \tfrac{1}{2\pi}\int_0^{\pi} 1\,dx = \tfrac{1}{2\pi}x\Big|_0^{\pi} = \tfrac{1}{2}.$
$a_k = \tfrac{1}{\pi}\int_{-\pi}^{\pi}f(x)\cos(kx)\,dx = \tfrac{1}{\pi}\int_0^{\pi}\cos(kx)\,dx = \tfrac{1}{\pi}\left(\frac{\sin(kx)}{k}\right)\Big|_0^{\pi} = 0.$
$b_k = \tfrac{1}{\pi}\int_{-\pi}^{\pi}f(x)\sin(kx)\,dx = \tfrac{1}{\pi}\int_0^{\pi}\sin(kx)\,dx = \tfrac{1}{\pi}\left(-\frac{\cos(kx)}{k}\right)\Big|_0^{\pi} = \frac{1-(-1)^k}{\pi k}.$

25. $a_0 = \frac{1}{\pi} \int_{-\pi}^{\pi} (\pi - x) \, dx = \frac{1}{\pi} \left(\pi x - \frac{1}{2} x^2 \right) \Big|_{-\pi}^{\pi} = \pi.$

$a_k = \langle \pi - x, \cos(kx) \rangle = \pi \langle 1, \cos(kx) \rangle - \langle x, \cos(kx) \rangle = \frac{1}{2} \int_{-\pi}^{\pi} \cos(kx) \, dx - \frac{1}{\pi} \int_{-\pi}^{\pi} x \cos(kx) \, dx$

$= \frac{1}{2} \left(\frac{\sin(kx)}{k} \right) \Big|_{-\pi}^{\pi} - \frac{1}{\pi} \left(\frac{x \sin(kx)}{k} + \frac{\cos(kx)}{k^2} \right) \Big|_{-\pi}^{\pi} = \frac{1}{2}(0) - \frac{1}{2\pi} \left(\frac{\cos(\pi k)}{k^2} - \frac{\cos(\pi k)}{k^2} \right) = 0.$

$b_k = \langle \pi - x, \sin(kx) \rangle = \pi \langle 1, \sin(kx) \rangle - \langle x, \sin(kx) \rangle = \frac{1}{2} \int_{-\pi}^{\pi} \sin(kx) \, dx - \frac{1}{\pi} \int_{-\pi}^{\pi} x \sin(kx) \, dx$

$= \frac{1}{2} \left(-\frac{\cos(kx)}{k} \right) \Big|_{-\pi}^{\pi} - \frac{1}{\pi} \left(\frac{-x\cos(kx)}{k} + \frac{\sin(kx)}{k^2} \right) \Big|_{-\pi}^{\pi}$

$= \frac{1}{2} \left(-\frac{\cos(\pi k)}{k} + \frac{\cos(\pi k)}{k} \right) - \frac{1}{\pi} \left(-\frac{\pi \cos(\pi k)}{k} - \frac{\pi \cos(\pi k)}{k} \right) = \frac{1}{2}(0) - \frac{1}{\pi} \left(\frac{-2\pi(-1)^k}{k} \right) = \frac{2(-1)^k}{k}.$

27. (a) $\int_{-\pi}^{\pi} f(x) \, dx = \int_0^{\pi} f(x) \, dx + \int_0^{\pi} f(-x) \, dx = \int_0^{\pi} f(x) \, dx - \int_0^{\pi} f(x) \, dx = 0.$

(b) $a_k = \int_{-\pi}^{\pi} f(x) \cos(kx) \, dx = \int_0^{\pi} f(x) \cos(kx) \, dx + \int_0^{\pi} f(-x) \cos(-kx) \, dx$

$= \int_0^{\pi} f(x) \cos(kx) - \int_0^{\pi} f(x) \cos(kx) \, dx = 0.$

29. $a - b = a + b$ in $\mathbb{Z}_2 \Rightarrow d_H(x, y) = \|x - y\|_H = \|x + y\|_H = w(x + y) \Rightarrow$
$w(0 + \mathbf{v}_1) = w(\mathbf{v}_1) = 1, \; w(0 + \mathbf{v}_2) = w(\mathbf{v}_2) = 2, \; w(\mathbf{v}_1 + \mathbf{v}_2) = w(\mathbf{e}_1) = 1 \Rightarrow$.
Therefore, since $d(C) = \min \{ d_H(x, y) : x \neq y \in C \}$ we have $d(C) = \min \{1, 2, 1\} = 1.$

31. Note, $\mathbf{t} = \begin{bmatrix} 1 \\ 1 \\ 0 \\ \vdots \\ 0 \end{bmatrix} \in E_n \Rightarrow w(0 + \mathbf{t}) = w(\mathbf{t}) = 2 \Rightarrow d(C) \leq 2.$

We will show $\mathbf{x} \neq 0 \in E_n \Rightarrow d_H(x, y) \geq 2$ which will imply $d(C) = 2$.

Let $j \geq k$ ($j \geq 1$ i.e. $\mathbf{x} \neq 0$) with $w(\mathbf{x}) = 2j$, $w(\mathbf{y}) = 2k$, $i = $ overlap (so $i \leq k$) $\Rightarrow$
$d_H(x, y) = w(\mathbf{x}) + w(\mathbf{y}) - 2 \text{ overlap} = 2j + 2k - 2i \geq 2j + 2k - 2k = 2j \geq 2 \Rightarrow d(C) = 2.$

33. Recall in $\mathbb{Z}_2$, $\mathbf{a}_k + \mathbf{a}_j = 0 \Leftrightarrow \mathbf{a}_k - \mathbf{a}_j = 0 \Leftrightarrow \mathbf{a}_k = \mathbf{a}_j.$
Let $A = \begin{bmatrix} \mathbf{a}_1 & \mathbf{a}_2 & \mathbf{a}_3 & \mathbf{a}_4 & \mathbf{a}_5 & \mathbf{a}_6 & \mathbf{a}_7 \end{bmatrix}$. Since none of the $\mathbf{a}_i$ are identical, $d(C) > 2$.
On the other hand, $\mathbf{a}_1 + \mathbf{a}_2 + \mathbf{a}_3 = 0 \Rightarrow d(C) \leq 3 \Rightarrow d(C) = 3.$

35. With $C = \{\mathbf{c}_1, \mathbf{c}_2, \mathbf{c}_3, \mathbf{c}_4\}$, we have:

$d_H(\mathbf{c}_1, \mathbf{c}_2) = 3, \quad d_H(\mathbf{c}_1, \mathbf{c}_3) = 4, \quad d_H(\mathbf{c}_1, \mathbf{c}_4) = 3$
$d_H(\mathbf{c}_2, \mathbf{c}_3) = 3, \quad d_H(\mathbf{c}_2, \mathbf{c}_4) = 4, \quad d_H(\mathbf{c}_3, \mathbf{c}_4) = 3 \Rightarrow d(C) = \min \{3, 4\} = 3.$

$d_H(\mathbf{c}_1, \mathbf{u}) = 2, \quad d_H(\mathbf{c}_2, \mathbf{u}) = 1, \quad d_H(\mathbf{c}_3, \mathbf{u}) = 4, \quad d_H(\mathbf{c}_4, \mathbf{u}) = 3$
Since $\min \{d_H(\mathbf{c}_i, \mathbf{u})\} = 1$ occurs only for $\mathbf{c}_2 \Rightarrow \mathbf{u}$ decodes as $\mathbf{c}_2$.

$d_H(\mathbf{c}_1, \mathbf{v}) = 3, \quad d_H(\mathbf{c}_2, \mathbf{v}) = 2, \quad d_H(\mathbf{c}_3, \mathbf{v}) = 3, \quad d_H(\mathbf{c}_4, \mathbf{v}) = 2$
Since $\min \{d_H(\mathbf{c}_i, \mathbf{v})\} = 2$ occurs for $\mathbf{c}_2$ and $\mathbf{c}_4$, $\mathbf{v}$ cannot be decoded.

$d_H(\mathbf{c}_1, \mathbf{w}) = 3, \quad d_H(\mathbf{c}_2, \mathbf{w}) = 2, \quad d_H(\mathbf{c}_3, \mathbf{w}) = 1, \quad d_H(\mathbf{c}_4, \mathbf{w}) = 4$
Since $\min \{d_H(\mathbf{c}_i, \mathbf{w})\} = 1$ occurs only for $\mathbf{c}_3 \Rightarrow \mathbf{w}$ decodes as $\mathbf{c}_3$.

37. $\text{Rep}_8 = \{0, 1\}$ in $\mathbb{Z}_2^8$ works since we saw in Exercise 32 that $d(\text{Rep}_n) = n$, so $d(\text{Rep}_8) = 8.$

7.5 Applications

39. A parity check matrix P for such a code is $(8-5) \times 8 = 3 \times 8 \Rightarrow$
the number of rows $\leq 3 \Rightarrow$ the number of linearly independent rows $\leq 3 \Rightarrow$
the number of linearly independent columns $\leq 3 \Rightarrow$
the number of linearly dependent columns $\leq 3 + 1 = 4 \Rightarrow$
$d = \text{d}(C) \leq 4 < 5 \Rightarrow$ No linear $(8,5,5)$ code exists.

41.

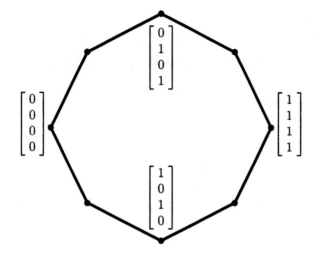

43. A parity check matrix P for such a code is $(n-k) \times n \Rightarrow$ the number of rows $\leq n - k \Rightarrow$
the number of linearly independent rows $\leq n - k \Rightarrow$
the number of linearly independent columns $\leq n - k \Rightarrow$
the number of linearly dependent columns $\leq n - k + 1 \Rightarrow$
$d = $ the minimum number of linearly dependent columns $\leq n - k + 1 \Rightarrow$
$d - 1 \leq n - k$.

Chapter 7 Review

Key Definitions and Concepts

This list includes most but not all of the definitions listed at the end of each section. We begin the list with key symbols and symbol-based definitions.

$\|\mathbf{v}\|$	p.544	7.1	The *length* (or *norm*) of $\mathbf{v}$ is $\|\mathbf{v}\| = \sqrt{\langle \mathbf{v}, \mathbf{v} \rangle}$
$d(\mathbf{u}, \mathbf{v})$	p.544	7.1	The *distance* between $\mathbf{u}$ and $\mathbf{v}$ is $d(\mathbf{u}, \mathbf{v}) = \|\mathbf{u} - \mathbf{v}\|$
$\langle \mathbf{u}, \mathbf{v} \rangle = 0$	p.544	7.1	$\mathbf{u}$ and $\mathbf{v}$ are orthogonal if $\langle \mathbf{u}, \mathbf{v} \rangle = 0$
$\text{proj}_W(\mathbf{v})$	p.547	7.1	$\text{proj}_W(\mathbf{v}) = \sum_{k=1}^{n} \frac{\langle \mathbf{u}_k \cdot \mathbf{v} \rangle}{\langle \mathbf{u}_k \cdot \mathbf{u}_k \rangle} \mathbf{u}_k$ (*orthogonal projection*)
$\text{perp}_W(\mathbf{v})$	p.547	7.1	$\text{perp}_W(\mathbf{v}) = \mathbf{v} - \text{proj}_W(\mathbf{v})$ (*component of* $\mathbf{v} \perp$ *to* W)
$\|\mathbf{v}\|_1 = \|\mathbf{v}\|_s$	p.562	7.2	$\|\mathbf{v}\|_1 = \|\mathbf{v}\|_s = |v_1| + \cdots + |v_n|$
$\|\mathbf{v}\|_\infty = \|\mathbf{v}\|_m$	p.562	7.2	$\|\mathbf{v}\|_\infty = \|\mathbf{v}\|_m = \max\{|v_1|, \ldots, |v_n|\}$
$\|\mathbf{v}\|_p$	p.562	7.2	$\|\mathbf{v}\|_p = (|v_1|^p + \cdots + |v_n|^p)^{1/p}$
$\|\mathbf{v}\|_2 = \|\mathbf{v}\|_E$	p.562	7.2	$\|\mathbf{v}\|_2 = \|\mathbf{v}\|_E = \sqrt{|v_1|^2 + \cdots + |v_n|^2}$
$\|\mathbf{v}\|_H$	p.563	7.2	$\|\mathbf{v}\|_H = w(\mathbf{v})$ (weight). See Example 7.15.
$\overline{\mathbf{v}}$	p.579	7.3	A vector $\overline{\mathbf{v}}$ is *best approximation* to $\mathbf{v}$ in W if $\|\mathbf{v} - \overline{\mathbf{v}}\| < \|\mathbf{v} - \mathbf{w}\|$ for all $\mathbf{w} \neq \overline{\mathbf{v}}$
$\|\mathbf{e}\|$	p.581	7.3	$\|\mathbf{e}\| = \sqrt{\varepsilon_1^2 + \varepsilon_2^2 + \varepsilon_3^2}$ is the *least squares error* of the approximation
A^+	p.594	7.3	If A is a matrix with linearly independent columns, then $A^+ = (A^T A)^{-1} A^T$ is the *pseudoinverse*.
σ_i	p.599	7.4	If A is an $m \times n$ matrix, the *singular values* of A are the square roots of the eigenvalues of $A^T A$. They are denoted by $\sigma_1, \ldots, \sigma_n$. They are usually arranged $\sigma_1 \geq \sigma_1 \geq \ldots \geq \sigma_n$.
Σ	p.601	7.4	Σ is created from the σ_i (detailed description p.601)
$A = U\Sigma V^T$	p.601	7.4	A *singular value decomposition* (SVD) of $A = U\Sigma V^T$. where U and and V are orthogonal matrices. Σ is constructed using the σ_i.
$A^+ = V\Sigma^+ U^T$	p.611	7.4	The *pseudoinverse* or *Moore-Penrose inverse* is $A^+ = V\Sigma^+ U^T$.

Key Definitions and Concepts, symbols, continued

best approximation	p.579	7.3	A vector $\bar{\mathbf{v}}$ is *best approximation* to $\mathbf{v}$ in W if $\|\mathbf{v}-\bar{\mathbf{v}}\| < \|\mathbf{v}-\mathbf{w}\|$ for all $\mathbf{w} \neq \bar{\mathbf{v}}$
best approximation	p.620	7.5	If $\{\mathbf{u}_1,\ldots,\mathbf{u}_k\}$ is an orthogonal basis for W then $\text{proj}_W(f) = \frac{\langle \mathbf{u}_1, f\rangle}{\langle \mathbf{u}_1, \mathbf{u}_1\rangle}\mathbf{u}_1 + \cdots + \frac{\langle \mathbf{u}_k, f\rangle}{\langle \mathbf{u}_k, \mathbf{u}_k\rangle}\mathbf{u}_k$
compatible	p.565	7.2	A matrix norm $\|A\|$ is compatible with a vector norm $\|\mathbf{x}\|$ if $\|A\mathbf{x}\| \leq \|A\|\|\mathbf{x}\|$
condition number	p.571	7.4	The condition number of a square matrix is $\|A^{-1}\|\|A\|$. If A is not invertible, we define $\text{cond}(A) = \infty$.
conditioned	p.570	7.2	For description of *ill-* and *well-conditioned*, see text.
correction	p.627	7.5	See text for full description of *correcting k errors*.
decoding	p.627	7.5	See text for full description of *nearest neighbor decoding*.
detection	p.627	7.5	See text for full description of *detecting k errors*.
distance	p.544	7.1	The *distance* between $\mathbf{u}$ and $\mathbf{v}$ is $d(\mathbf{u},\mathbf{v}) = \|\mathbf{u}-\mathbf{v}\|$
distance	p.563	7.2	See Example 7.16 for $d_s(\mathbf{u},\mathbf{v})$, $d_E(\mathbf{u},\mathbf{v})$, $d_H(\mathbf{u},\mathbf{v})$ ...
Euclidean norm	p.562	7.2	$\|\mathbf{v}\|_E = \|\mathbf{v}\|_2 = \sqrt{|v_1|^2+\cdots+|v_n|^2}$ (same as 2-norm)
Fourier approximation	p.624	7.5	The best approximation using the *Fourier coefficients*. Given below, used in a *trigonometric polynomial*.
Fourier coefficients	p.624	7.5	These coefficients are used in the *Fourier approximation*: $a_0 = \frac{\langle 1,f\rangle}{\langle 1,1\rangle} = \frac{1}{2\pi}\int_{-\pi}^{\pi} f(x)\,dx$ $a_k = \frac{\langle \cos kx, f\rangle}{\langle \cos kx, \cos kx\rangle} = \frac{1}{\pi}\int_{-\pi}^{\pi} f(x)\cos kx\,dx$ $b_k = \frac{\langle \sin kx, f\rangle}{\langle \sin kx, \sin kx\rangle} = \frac{1}{\pi}\int_{-\pi}^{\pi} f(x)\sin kx\,dx$
Fourier series	p.624	7.5	Let the *Fourier trigonometric polynomial* become infinite. Then the result is called the *Fourier series*. $f(x) = a_0 + \sum_{k=1}^{\infty}(a_k\cos kx + b_k\sin kx)$ on $[-\pi,\pi]$
Frobenius norm	p.565	7.2	$\|A\|_F = \sqrt{\sum_{i,j=1}^{n} a_{ij}^2}$. See Example 7.18.

Chapter 7 Review

Key Definitions and Concepts, symbols, continued

Term	Page	Section	Description
Hamming norm	p.563	7.2	$\|\mathbf{v}\|_H = w(\mathbf{v})$ (weight). See Example 7.15.
inner product	p.540	7.1	An *inner product* on a vector space V with $\langle \mathbf{u}, \mathbf{v} \rangle$ has the following properties: 1. $\langle \mathbf{u}, \mathbf{v} \rangle = \langle \mathbf{v}, \mathbf{u} \rangle$ 2. $\langle \mathbf{u}, \mathbf{v} + \mathbf{w} \rangle = \langle \mathbf{u}, \mathbf{v} \rangle + \langle \mathbf{u}, \mathbf{w} \rangle$ 3. $\langle c\mathbf{u}, \mathbf{v} \rangle = c\langle \mathbf{u}, \mathbf{v} \rangle$ 4. $\langle \mathbf{u}, \mathbf{u} \rangle \geq 0$ and $\langle \mathbf{u}, \mathbf{u} \rangle = 0$ if and only if $\mathbf{0} = 0$
inner product space	p.540	7.1	A vector space with an *inner product* is called an *inner product space*
least squares error	p.581	7.3	$\|\mathbf{e}\| = \sqrt{\varepsilon_1^2 + \varepsilon_2^2 + \varepsilon_3^2}$ is the *least squares error* of the approximation
least squares solution	p.583	7.3	A *least squares solution* of $A\mathbf{x} = \mathbf{b}$ is $\bar{\mathbf{x}}$ such that $\|\mathbf{b} - A\bar{\mathbf{x}}\| \leq \|\mathbf{b} - A\mathbf{x}\|$ for all $\mathbf{x}$ in $\mathbb{R}$
length	p.544	7.1	The *length* (or *norm*) of $\mathbf{v}$ is $\|\mathbf{v}\| = \sqrt{\langle \mathbf{v}, \mathbf{v} \rangle}$
left singular vectors	p.602	7.4	Given a singular value decomposition (SVD) of $A = U\Sigma V^T$. The columns of U (an orthogonal matrix) are called the *left singular values* of A.
matrix norm	p.565	7.2	A *matrix norm* on M_{nn} has the following properties: 1. $\|A\| \geq 0$, and $\|A\| = 0$ if and only if $A = O$ 2. $\|cA\| = \|c\|\|A\|$ 3. $\|A + B\| \leq \|A\| + \|B\|$ 4. $\|AB\| \leq \|A\|\|B\|$
max norm	p.562	7.2	$\|\mathbf{v}\|_m = \|\mathbf{v}\|_\infty = \max\{|v_1|, \ldots, |v_n|\}$ (same as ∞-norm)
minimum distance	p.626	7.5	The smallest distance between any two distinct vectors in C. $d(C) = \min\{d_H(\mathbf{x}, \mathbf{y}) : \mathbf{x} \neq \mathbf{y} \text{ in } C\}$

Key Definitions and Concepts, symbols, continued

norm	p.544	7.1	The *length* (or *norm*) of $\mathbf{v}$ is $\|\mathbf{v}\| = \sqrt{\langle \mathbf{v}, \mathbf{v} \rangle}$

norm
 normed
   linear
     space
	p.561	7.2	A *norm* on a vector space V has the following properties:
			 1. $\|\mathbf{v}\| \geq 0$, and $\|\mathbf{v}\| = 0$ if and only if $\mathbf{v} = \mathbf{0}$
			 2. $\|c\mathbf{v}\| = |c|\|\mathbf{v}\|$
			 3. $\|\mathbf{u} + \mathbf{v}\| \leq \|\mathbf{u}\| + \|\mathbf{v}\|$

normal equations	p.584	7.3	$A^T A \overline{\mathbf{x}} = A^T \mathbf{b}$
operator norm	p.568	7.2	$\|A\| = \max\|A\mathbf{x}\|$ where $\|\mathbf{x}\| = 1$. See Theorem 7.6. $\|A\|_1 = \max\|A\mathbf{x}\|_s$, $\|A\|_2 = \max\|A\mathbf{x}\|_E$, $\|A\|_\infty = \max\|A\mathbf{x}\|_*$
orthonormal basis	p.546	7.1	A basis that is an orthonormal set
orthonormal set	p.546	7.1	An orthogonal set of unit vectors, that is $\|\mathbf{v}_i\| = 1$
orthogonal	p.544	7.1	$\mathbf{u}$ and $\mathbf{v}$ are orthogonal if $\langle \mathbf{u}, \mathbf{v} \rangle = 0$
orthogonal basis	p.546	7.1	A basis that is an orthogonal set
orthogonal set	p.546	7.1	$\{\mathbf{v}_1, \ldots, \mathbf{v}_k\}$ such that $\langle \mathbf{v}_i, \mathbf{v}_j \rangle = 0$ when $\mathbf{v}_i \neq \mathbf{v}_j$
p-norm	p.562	7.2	$\|\mathbf{v}\|_p = (|v_1|^p + \cdots + |v_n|^p)^{1/p}$
Penrose conditions	p.595	7.3	A^+ the pseudoinverse of A satisfies the *Penrose conditions*. a. $AA^+A = A$ b. $A^+AA^+ = A^+$ c. AA^+ and A^+A are symmetric
projection	p.547	7.1	$\text{proj}_W(\mathbf{v}) = \sum\limits_{k=1}^{n} \frac{\langle \mathbf{u}_k \cdot \mathbf{v} \rangle}{\langle \mathbf{u}_k \cdot \mathbf{u}_k \rangle} \mathbf{u}_k$ (*orthogonal projection*)
pseudoinverse	p.594	7.3	If A is a matrix with linearly independent columns, then $A^+ = (A^T A)^{-1} A^T$ is the *pseudoinverse*.
pseudoinverse	p.611	7.4	The *pseudoinverse* or *Moore-Penrose inverse* is $A^+ = V\Sigma^+$

Chapter 7 Review

Key Definitions and Concepts, symbols, continued

right singular vectors	p.602	7.4	Given a singular value decomposition (SVD) of $A = U\Sigma V^T$. The columns of V (an orthogonal matrix) are called the *right singular values* of A.
root mean square error	p.621	7.5	This is the root means square error: $\|f - g\| = \sqrt{\int_{-1}^{1}(f(x)-g(x))^2\,dx}$
singular values	p.599	7.4	If A is an $m \times n$ matrix, the *singular values* of A are the square roots of the eigenvalues of $A^T A$. They are denoted by $\sigma_1, \ldots, \sigma_n$. They are usually arranged $\sigma_1 \geq \sigma_1 \geq \ldots \geq \sigma_n$.
singular value decomposition	p.601	7.4	A *singular value decomposition* (SVD) of $A = U\Sigma V^T$. where U and and V are orthogonal matrices. Σ is constructed using the σ_i.
sum norm	p.561	7.2	$\|\mathbf{v}\|_s = \|\mathbf{v}\|_1 = |v_1| + \cdots + |v_n|$ (same as 1-norm)
trigonometric polynomial	p.623	7.5	A function of this form is a *trigonometric polynomial*. $p(x) = a_0 + a_1 \cos x + a_2 \cos 2x + \cdots + a_n \cos nx + b_1 \sin x + b_2 \sin 2x + \cdots + b_n \sin nx.$
weighted dot product	p.541	7.1	An *inner product* defined on $\mathbb{R}^n$ such that $\langle \mathbf{u}, \mathbf{v} \rangle = w_1 u_1 v_1 + \ldots + w_n u_n v_n$ with $w_i \geq 0$ (see p.541 for details)

Theorems

Thm 7.1 p.544   7.1   Let $\mathbf{u}$, $\mathbf{v}$, and $\mathbf{w}$ be vectors
in an inner product space V and c a scalar:
a. $\langle \mathbf{u}+\mathbf{v},\mathbf{w}\rangle = \langle \mathbf{u},\mathbf{v}\rangle + \langle \mathbf{u},\mathbf{w}\rangle$
b. $\langle \mathbf{u},c\mathbf{v}\rangle = c\langle \mathbf{u},\mathbf{v}\rangle$
c. $\langle \mathbf{u},\mathbf{0}\rangle = \langle \mathbf{0},\mathbf{v}\rangle = 0$

Thm 7.2 p.546   7.1   Let $\mathbf{u}$ and $\mathbf{v}$ be vectors in an inner product space V.
Then $\mathbf{u}$ and $\mathbf{v}$ are orthogonal if and only if $\|\mathbf{u}+\mathbf{v}\|^2 = \|\mathbf{u}\|^2 + \|\mathbf{v}\|^2$

Thm 7.3 p.548   7.1   Let $\mathbf{u}$ and $\mathbf{v}$ be vectors in an inner product space V.
Then $|\langle \mathbf{u},\mathbf{v}\rangle| \leq \|\mathbf{u}\|\|\mathbf{v}\|$ with equality holding if and only if
$\mathbf{u}$ and $\mathbf{v}$ are scalar multiples of each other

Thm 7.4 p.549   7.1   Let $\mathbf{u}$ and $\mathbf{v}$ be vectors in an inner product space V.
Then $\|\mathbf{u}+\mathbf{v}\| \leq \|\mathbf{u}\| + \|\mathbf{v}\|$

Thm 7.5 p.544   7.2   Let d be a distance function defined on a normed linear space V.
The following properties hold for all vectors $\mathbf{u}$, $\mathbf{v}$, and $\mathbf{w}$ in V.
a. $d(\mathbf{u},\mathbf{v}) \geq 0$, and $d(\mathbf{u},\mathbf{v}) = 0$ if and only if $\mathbf{u}=\mathbf{v}$
b. $d(\mathbf{u},\mathbf{v}) = d(\mathbf{v},\mathbf{u})$
c. $d(\mathbf{u},\mathbf{w}) \leq d(\mathbf{u},\mathbf{v}) + d(\mathbf{v},\mathbf{w})$

Thm 7.6 p.567   7.2   If $\|\mathbf{x}\|$ is a vector norm on $\mathbb{R}$,
then $\|A\| = \max\|A\mathbf{x}\|$ where $\|\mathbf{x}\| = 1$
defines a matrix norm on M_{nn} that is compatible with
the vector norm that induces it.

Thm 7.7 p.569   7.2   Let A have column vectors $\mathbf{a}_i$ and row vectors $\mathbf{A}_i$ then:
a. $\|A\|_1 = \max\{\|\mathbf{a}_i\|_s\} = \max\left\{\sum_{i=1}^{n}|a_{ij}|\right\}$
b. $\|A\|_\infty = \max\{\|\mathbf{A}_i\|_s\} = \max\left\{\sum_{j=1}^{n}|a_{ij}|\right\}$

Thm 7.8 p.579   7.3   If W is a finite-dimensional subspace of an inner product space V
and if $\mathbf{v}$ is a vector in V,
then $\text{proj}_W(\mathbf{v})$ is the *best approximation* to $\mathbf{v}$ in W.

Thm 7.9 p.584   7.3 $A\mathbf{x} = \mathbf{b}$ has at least one *least squares solution*, $\bar{\mathbf{x}}$.
a. $\bar{\mathbf{x}}$ is a *least squares solution* of $A\mathbf{x} = \mathbf{b}$ if and only if
$\bar{\mathbf{x}}$ is a solution of the normal equations $A^T A \bar{\mathbf{x}} = A^T \mathbf{b}$.
b. A has linearly independent columns if and only if $A^T A$ is invertible
Then $\bar{\mathbf{x}} = (A^T A)^{-1} A^T \mathbf{b}$ is the *unique* solution.

Theorems, continued

Thm 7.10 p.591 7.3 Let A have linearly independent columns with QR factorization $A = QR$ (see Theorem 5.16 below). Then $\bar{\mathbf{x}} = R^{-1}Q^T\mathbf{b}$ is the *unique* solution of $A\mathbf{x} = \mathbf{b}$.

Thm 7.11 p.592 7.3 Let the columns of A be a basis for W. Then $P : \mathbb{R}^m \to \mathbb{R}^n$ that projects $\mathbb{R}^m$ onto W has matrix $A(A^TA)^{-1}A^T$ and $\text{proj}_W(\mathbf{v}) = A(A^TA)^{-1}A^T\mathbf{v}$.

Thm 7.12 p.595 7.3 Let A have linearly independent columns. Then A^+ the pseudoinverse of A satisfies the *Penrose conditions*.
  a. $AA^+A = A$
  b. $A^+AA^+ = A^+$
  c. AA^+ and A^+A are symmetric

Thm 7.13 p.602 7.4 **The Singular Value Decomposition Theorem**
Let A be an $m \times n$ matrix with singular values $\sigma_1 \geq \sigma_1 \geq \ldots \geq \sigma_r > 0$ and $\sigma_{r+1} = \sigma_{r+2} = \ldots = \sigma_n = 0$. Then there exist an $m \times m$ orthogonal matrix U and an $n \times n$ orthogonal matrix V and an $m \times n$ matrix Σ (see p.601) such that $A = U\Sigma V^T$.

Thm 7.14 p.605 7.4 Let A be an $m \times n$ matrix with singular values $\sigma_1 \geq \sigma_1 \geq \ldots \geq \sigma_r > 0$ and $\sigma_{r+1} = \sigma_{r+2} = \ldots = \sigma_n = 0$. Let $\mathbf{u}_1, \ldots, \mathbf{u}_r$ be left singular values and let $\mathbf{v}_1, \ldots, \mathbf{v}_r$ be left singular values. Then:
$A = \sigma_1 \mathbf{u}_1 \mathbf{v}_1^T + \cdots + \sigma_r \mathbf{u}_r \mathbf{v}_r^T$.

Thm 7.15 p.606 7.4 Let $A = U\Sigma V^T$ be an *SVD* of an $m \times m$ matrix A with nonzero singular values $\sigma_1, \ldots, \sigma_r$.
  a. The rank of A is r.
  b. $\{\mathbf{u}_1, \ldots, \mathbf{u}_r\}$ is an orthonormal basis for $\text{col}(A)$.
  c. $\{\mathbf{u}_{r+1}, \ldots, \mathbf{u}_m\}$ is an orthonormal basis for $\text{null}(A^T)$.
  d. $\{\mathbf{v}_1, \ldots, \mathbf{v}_r\}$ is an orthonormal basis for $\text{row}(A)$.
  e. $\{\mathbf{v}_{r+1}, \ldots, \mathbf{v}_m\}$ is an orthonormal basis for $\text{null}(A)$.

Thm 7.16 p.607 7.4 Let $A = U\Sigma V^T$ be an *SVD* of an $m \times n$ matrix A with rank r. Then the image of the unit sphere in $\mathbb{R}^n$ under the matrix transformation that maps $\mathbf{x}$ to $A\mathbf{x}$ is
  a. The surface of an ellipsoid in $\mathbb{R}^m$ if $r = n$.
  b. A solid ellipsoid in $\mathbb{R}^m$ if $r < n$.

Theorems, continued

Thm 7.17 p.610 7.4 Let A be an $m \times n$ matrix and let $\sigma_1, \ldots, \sigma_r$ be all the nonzero singular values of A. Then: $\|A\|_F = \sqrt{\sigma_1^2 + \cdots + \sigma_r^2}$

Thm 7.18 p.613 7.4 The least squares problem $A\mathbf{x} = \mathbf{b}$ has a unique least squares solution $\bar{\mathbf{x}}$ of minimal length that is given by $\bar{\mathbf{x}} = A^+\mathbf{b}$.

Thm 7.20 p.628 7.5 Let C be a (binary) code with minimum distance d.
a. C detects k errors if and only if $d \geq k + 1$.
b. C detects k errors if and only if $d \geq 2k + 1$.

Thm 7.21 p.628 7.5 Let C be a (n, k) with parity check matrix P. Then the minimum distance of C is the smallest integer d for which there are d linearly dependent columns.

The Fundamental Theorem of Invertible Matrices: Final Version

Thm 7.19   p.614   7.4   a. through u. give equivalent conditions for A to be invertible:

    a. A is invertible.
    b. $A\mathbf{x} = \mathbf{b}$ has a unique solution for every $\mathbf{b}$ in $\mathbb{R}^n$
    c. $A\mathbf{x} = \mathbf{0}$ has only the trivial solution.
    d. The reduced row echelon form of A is I_n.
    e. A is the product of elementary matrices.
    f. $\text{rank}(A) = n$
    g. $\text{nullity}(A) = 0$
    h. The column vectors of A are linearly independent.
    i. The column vectors of A span $\mathbb{R}^n$.
    j. The column vectors of A form a basis for $\mathbb{R}^n$.
    k. The row vectors of A are linearly independent.
    l. The row vectors of A span $\mathbb{R}^n$.
    m. The row vectors of A for a basis for $\mathbb{R}^n$.
    n. $\det(A) \neq 0$
    o. 0 is not an eigenvalue of A
    p. T is invertible
    q. T is one-to-one
    r. T is onto
    s. $\ker(T) = \{\mathbf{0}\}$
    t. $\text{range}(T) = W$
    u. 0 is not a singular value of A

Theorems by name

Thm 7.8 p.579 7.3 **The Best Approximation Theorem**
If W is a finite-dimensional subspace of an inner product space V
and if $\mathbf{v}$ is a vector in V,
then $\text{proj}_W(\mathbf{v})$ is the *best approximation* to $\mathbf{v}$ in W.

Thm 7.3 p.548 7.1 **The Cauchy-Schwarz Inequality**
Let $\mathbf{u}$ and $\mathbf{v}$ be vectors in an inner product space V.
Then $|\langle \mathbf{u}, \mathbf{v} \rangle| \leq \|\mathbf{u}\|\|\mathbf{v}\|$ with equality holding if and only if
$\mathbf{u}$ and $\mathbf{v}$ are scalar multiples of each other

Thm 7.9 p.584 7.3 **The Least Squares Theorem**
$A\mathbf{x} = \mathbf{b}$ has at least one *least squares solution*, $\bar{\mathbf{x}}$.
  a. $\bar{\mathbf{x}}$ is a *least squares solution* of $A\mathbf{x} = \mathbf{b}$ if and only if
     $\bar{\mathbf{x}}$ is a solution of the normal equations $A^T A \bar{\mathbf{x}} = A^T \mathbf{b}$.
  b. A has linearly independent columns if and only if $A^T A$ is invertible
     Then $\bar{\mathbf{x}} = (A^T A)^{-1} A^T \mathbf{b}$ is the *unique* solution.

Thm 7.2 p.546 .7.1 **Pythagoras' Theorem**
Let $\mathbf{u}$ and $\mathbf{v}$ be vectors in an inner product space V.
Then $\mathbf{u}$ and $\mathbf{v}$ are orthogonal if and only if $\|\mathbf{u}+\mathbf{v}\|^2 = \|\mathbf{u}\|^2 + \|\mathbf{v}\|^2$

Thm 7.4 p.549 7.1 **The Triangle Inequality**
Let $\mathbf{u}$ and $\mathbf{v}$ be vectors in an inner product space V.
Then $\|\mathbf{u}+\mathbf{v}\| \leq \|\mathbf{u}\| + \|\mathbf{v}\|$

Thm 7.13 p.602 7.4 **The Singular Value Decomposition Theorem**
Let A be an $m \times n$ matrix with singular values
$\sigma_1 \geq \sigma_1 \geq \ldots \geq \sigma_r > 0$ and $\sigma_{r+1} = \sigma_{r+2} = \ldots = \sigma_n = 0$.
Then there exist an $m \times m$ orthogonal matrix U
and an $n \times n$ orthogonal matrix V
and an $m \times n$ matrix Σ (see p.601)
such that $A = U \Sigma V^T$.

Chapter 7 Review

Solutions to odd-numbered exercises from Chapter 7 Review

1. We will explain and give counter examples to justify our answers below.

   (a) **True.** See definition on p.540 of Section 7.1.
   Simply verify that $\langle \mathbf{u}, \mathbf{v} \rangle = u_1 v_1 + \pi u_2 v_2$ satisfies the definition.
   Q: Does $\langle \mathbf{u}, \mathbf{v} \rangle = u_1 v_1 + c u_2 v_2$ work for any scalar c?
   What about $\langle \mathbf{u}, \mathbf{v} \rangle = u_1 v_2 + c u_2 v_1$?

   (b) **True.** Check the properties carefully.

   (c) **False.** Which property or properties should we expect to fail?
   Consider $A = \begin{bmatrix} 0 & 1 \\ 1 & 0 \end{bmatrix}$, $\langle A, A \rangle = \operatorname{tr}(A) + \operatorname{tr}(A) = 0 + 0 = 0$, but $A \neq 0$.
   Q: Which property does this violate? That is, which property fails?

   (d) **True.** Simply perform the calculation implied by the definitions.
   $$\|\mathbf{u} + \mathbf{v}\| = \sqrt{\langle \mathbf{u} + \mathbf{v}, \mathbf{u} + \mathbf{v} \rangle} = \sqrt{\|\mathbf{u}\|^2 + 2\langle \mathbf{u}, \mathbf{v} \rangle + \|\mathbf{v}\|^2}$$
   $$\sqrt{4^2 + 2 \cdot 2 + (\sqrt{5})^2} = \sqrt{16 + 4 + 5} = \sqrt{25} = 5$$

   (e) **True.** Below we apply the definitions when $n = 1$.

sum norm	p.561	7.2	$\|\mathbf{v}\|_s = |v_1|$
max norm	p.562	7.2	$\|\mathbf{v}\|_m = \max\{|v_1|\} = |v_1|$
Euclidean norm	p.562	7.2	$\|\mathbf{v}\|_E = \sqrt{|v_1|^2} = |v_1|$

   (f) **False.** See definition on p.570 in Section 7.2 and the remarks following.
   Since the definition uses an inequality, the converse may not hold.

   (g) **True.** See definition on p.570 in Section 7.2 and the remarks following.

   (h) **False.** It is not necessarily *unique*. See Theorem 7.9 on p.584 of Section 7.3.
   Q: Under what conditions is the least squares solution of a linear system unique?
   A: Hint: See Theorem 7.10 on p.591 of Section 7.3.

   (i) **True.** See Theorem 7.11 on p.592 of Section 7.3.
   Q: What does the fact that A has orthonormal columns tell us about A?
   A: Hint: Consider the effect upon the product $A^T A$.

   (j) **False.** We construct a simple counterexample below.
   Consider $A = \begin{bmatrix} 2 & 0 \\ 0 & 2 \end{bmatrix}$. A is symmetric with eigenvalue 2 and singular value $\sqrt{2}$.

3. We will show that $\langle A, B \rangle = \text{tr}(A^T B)$ for A and B in M_{22} is an inner product by showing that it satisfies all four properties of the definition.

Let $\mathbf{u} = A = \begin{bmatrix} a & b \\ c & d \end{bmatrix}$ and $\mathbf{v} = B = \begin{bmatrix} e & f \\ g & h \end{bmatrix}$.

1. $\langle A, B \rangle = \text{tr}(A^T B) = ae + cg + bf + dh = \text{tr}(B^T A) = \langle B, A \rangle$.
2. $\langle A, B + C \rangle = \text{tr}(A^T(B + C)) = \text{tr}(A^T B) + \text{tr}(A^T C) = \langle A, B \rangle + \langle A, C \rangle$.
3. $\langle cA, B \rangle = \text{tr}(cA^T B) = c\text{tr}(A^T B) = c\langle A, B \rangle$.
4. $\langle A, A \rangle = \text{tr}(A^T A) = a^2 + b^2 + c^2 + d^2 \geq 0$.
   Also, $a^2 + b^2 + c^2 + d^2 = 0$ if and only if $a = b = c = d = 0$ that is, $A = O$.

Therefore, $\langle A, B \rangle = \text{tr}(A^T B)$ for A and B in M_{22} is an inner product.

5. We need only recall that $\|\mathbf{v}\| = \sqrt{\langle \mathbf{v}, \mathbf{v} \rangle}$ on p.544 in Section 7.1.
So $\|1 + x + x^2\| = \sqrt{\langle 1 + x + x^2, 1 + x + x^2 \rangle} = \sqrt{1 \cdot 1 + 1 \cdot 1 + 1 \cdot 1} = \sqrt{3}$.

7. See Example 7.8 and Exercises 37 through 39 in Section 7.1.

$\mathcal{B} = \left\{ \mathbf{v}_1 = \begin{bmatrix} 1 \\ 1 \end{bmatrix}, \mathbf{x}_2 = \begin{bmatrix} 1 \\ 2 \end{bmatrix} \right\}$ and $\langle \mathbf{u}, \mathbf{v} \rangle = \mathbf{u}^T A \mathbf{v}$, $A = \begin{bmatrix} 6 & 4 \\ 4 & 6 \end{bmatrix}$.

$\langle \mathbf{v}_1, \mathbf{v}_1 \rangle = \mathbf{v}_1^T A \mathbf{v}_1 = \begin{bmatrix} 1 & 1 \end{bmatrix} \begin{bmatrix} 6 & 4 \\ 4 & 6 \end{bmatrix} \begin{bmatrix} 1 \\ 1 \end{bmatrix} = 20$,

$\langle \mathbf{v}_1, \mathbf{x}_2 \rangle = \mathbf{v}_1^T A \mathbf{x}_2 = \begin{bmatrix} 1 & 1 \end{bmatrix} \begin{bmatrix} 6 & 4 \\ 4 & 6 \end{bmatrix} \begin{bmatrix} 1 \\ 2 \end{bmatrix} = 30$,

So $\frac{\langle \mathbf{v}_1, \mathbf{x}_2 \rangle}{\langle \mathbf{v}_1, \mathbf{v}_1 \rangle} = \frac{30}{20} = \frac{3}{2}$.

Therefore, $\mathbf{v}_2 = \mathbf{x}_2 - \frac{\langle \mathbf{v}_1, \mathbf{x}_2 \rangle}{\langle \mathbf{v}_1, \mathbf{v}_1 \rangle} \mathbf{v}_1 = \mathbf{x}_2 - \frac{3}{2}\mathbf{v}_1 = \begin{bmatrix} 1 \\ 2 \end{bmatrix} - \frac{3}{2} \begin{bmatrix} 1 \\ 1 \end{bmatrix} = \begin{bmatrix} -\frac{1}{2} \\ \frac{1}{2} \end{bmatrix}$.

Therefore, the orthogonal set is $\mathcal{C} = \left\{ \begin{bmatrix} 1 \\ 1 \end{bmatrix}, \begin{bmatrix} -\frac{1}{2} \\ \frac{1}{2} \end{bmatrix} \right\}$.

Verify that $\langle \mathbf{v}_1, \mathbf{v}_2 \rangle = 0$.

9. See the definition of *norm* and Example 7.12 on p.561 in Section 7.2.

We will show that $\|\mathbf{v}\| = \mathbf{v}^T \mathbf{v}$ is *not* a norm because it fails to satisfy Property 2.

2. $\|c\mathbf{v}\| = c\mathbf{v}^T c\mathbf{v} = c^2 \mathbf{v}^T \mathbf{v} = c^2 \|\mathbf{v}\| \neq c\|\mathbf{v}\|$.

Therefore, $\|\mathbf{v}\| = \mathbf{v}^T \mathbf{v}$ is *not* a norm.

11. See the definition of *ill-conditioned* and Example 7.20 on p.570 in Section 7.2.

We find A^{-1} then compute $\|A\|\|A^{-1}\|$, the condition number(see p.571).

$A^{-1} = \begin{bmatrix} 1 & -11 & 10 \\ -11 & -990 & 1000 \\ 10 & 1000 & -1000 \end{bmatrix} \Rightarrow \begin{array}{l} \text{cond}_1(A) = 1.21 \cdot 2010 = 2432.1 \\ \text{cond}_\infty(A) = 1.21 \cdot 2010 = 2432.1 \end{array} \Rightarrow A$ is ill-conditioned.

13. See Example 7.26 and Exercises 7 through 14 in Section 7.3.

The given points imply $A = \begin{bmatrix} 1 & 1 \\ 1 & 2 \\ 1 & 3 \\ 1 & 4 \end{bmatrix}$.

Therefore, $A^T A = \begin{bmatrix} 4 & 10 \\ 10 & 30 \end{bmatrix}$ which implies $(A^T A)^{-1} = \begin{bmatrix} \frac{3}{2} & -\frac{1}{2} \\ -\frac{1}{2} & \frac{1}{5} \end{bmatrix}$. So:

$$\bar{x} = (A^T A)^{-1} A^T \mathbf{b} = \begin{bmatrix} \frac{3}{2} & -\frac{1}{2} \\ -\frac{1}{2} & \frac{1}{5} \end{bmatrix} \begin{bmatrix} 1 & 1 & 1 & 1 \\ 1 & 2 & 3 & 4 \end{bmatrix} \begin{bmatrix} 2 \\ 3 \\ 5 \\ 7 \end{bmatrix} = \begin{bmatrix} 0 \\ \frac{17}{10} \end{bmatrix} \Rightarrow y = 0 + \frac{17}{10} x.$$

15. See Example 7.31 and Exercises 37 through 42 in Section 7.3.

Since $A = \begin{bmatrix} 1 & 1 \\ 0 & 1 \\ 1 & 0 \end{bmatrix}$, $A^T A = \begin{bmatrix} 1 & 0 & 1 \\ 1 & 1 & 0 \end{bmatrix} \begin{bmatrix} 1 & 1 \\ 0 & 1 \\ 1 & 0 \end{bmatrix} = \begin{bmatrix} 2 & 1 \\ 1 & 2 \end{bmatrix}$. So, $(A^T A)^{-1} = \begin{bmatrix} \frac{2}{3} & -\frac{1}{3} \\ -\frac{1}{3} & \frac{2}{3} \end{bmatrix}$.

Now we take $P = A(A^T A)^{-1} A^T = \begin{bmatrix} 1 & 0 & 1 \\ 1 & 1 & 0 \end{bmatrix} \begin{bmatrix} \frac{2}{3} & -\frac{1}{3} \\ -\frac{1}{3} & \frac{2}{3} \end{bmatrix} \begin{bmatrix} 1 & 1 \\ 0 & 1 \\ 1 & 0 \end{bmatrix} = \begin{bmatrix} \frac{2}{3} & \frac{1}{3} & \frac{1}{3} \\ \frac{1}{3} & \frac{2}{3} & -\frac{1}{3} \\ \frac{1}{3} & -\frac{1}{3} & \frac{2}{3} \end{bmatrix}.$

So the projection is: $\text{proj}_W(\mathbf{v}) = P\mathbf{v} = \begin{bmatrix} \frac{2}{3} & \frac{1}{3} & \frac{1}{3} \\ \frac{1}{3} & \frac{2}{3} & -\frac{1}{3} \\ \frac{1}{3} & -\frac{1}{3} & \frac{2}{3} \end{bmatrix} \begin{bmatrix} 1 \\ 2 \\ 3 \end{bmatrix} = \begin{bmatrix} \frac{7}{3} \\ \frac{2}{3} \\ \frac{5}{3} \end{bmatrix}.$

17. See Examples 7.33, 7.34, and 7.39 in Section 7.4.

(a) We follow Example 7.33 to find the singular values of A.

The matrix $A^T A = \begin{bmatrix} 1 & 0 & 1 \\ 1 & 0 & -1 \end{bmatrix} \begin{bmatrix} 1 & 1 \\ 0 & 0 \\ 1 & -1 \end{bmatrix} = \begin{bmatrix} 2 & 0 \\ 0 & 2 \end{bmatrix}$ has eigenvalues $\lambda_1 = \lambda_2 = 2$.

So, the singular values of A are $\sigma_1 = \sigma_2 = \sqrt{\lambda_1} = \sqrt{\lambda_2} = \sqrt{2}$.

(b) We follow Example 7.34 to find the singular value decomposition of A.

The corresponding eigenvectors of $A^T A$ are $\mathbf{v}_1 = \begin{bmatrix} 1 \\ 0 \end{bmatrix}$ and $\mathbf{v}_2 = \begin{bmatrix} 0 \\ 1 \end{bmatrix}$.

So, $V = \begin{bmatrix} 1 & 0 \\ 0 & 1 \end{bmatrix}$ and $\Sigma = \begin{bmatrix} \sqrt{2} & 0 \\ 0 & \sqrt{2} \\ 0 & 0 \end{bmatrix}$. To find U, we compute $\mathbf{u}_1, \mathbf{u}_2$.

$\mathbf{u}_1 = \frac{1}{\sigma_1} A \mathbf{v}_1 = \frac{1}{\sqrt{2}} \begin{bmatrix} 1 & 1 \\ 0 & 0 \\ 1 & -1 \end{bmatrix} \begin{bmatrix} 1 \\ 0 \end{bmatrix} = \begin{bmatrix} \frac{1}{\sqrt{2}} \\ 0 \\ \frac{1}{\sqrt{2}} \end{bmatrix}$ and $\mathbf{u}_2 = \frac{1}{\sigma_2} A \mathbf{v}_2 = \begin{bmatrix} \frac{1}{\sqrt{2}} \\ 0 \\ -\frac{1}{\sqrt{2}} \end{bmatrix}$.

Verify that $A = U \Sigma V^T = \begin{bmatrix} \frac{1}{\sqrt{2}} & \frac{1}{\sqrt{2}} & 0 \\ 0 & 0 & 1 \\ \frac{1}{\sqrt{2}} & -\frac{1}{\sqrt{2}} & 0 \end{bmatrix} \begin{bmatrix} \sqrt{2} & 0 \\ 0 & \sqrt{2} \\ 0 & 0 \end{bmatrix} \begin{bmatrix} 1 & 0 \\ 0 & 1 \end{bmatrix} = \begin{bmatrix} 1 & 1 \\ 0 & 0 \\ 1 & -1 \end{bmatrix}$.

(c) We follow Example 7.39 to find the pseudoinverse of A, that is A^+.

Given $\Sigma = \begin{bmatrix} D & O \\ O & O \end{bmatrix}$, we have $\Sigma^+ = \begin{bmatrix} D^{-1} & O \\ O & O \end{bmatrix}$.

So in this case, $\Sigma^+ = \begin{bmatrix} \frac{1}{\sqrt{2}} & 0 & 0 \\ 0 & \frac{1}{\sqrt{2}} & 0 \end{bmatrix}$.

Then $A^+ = V \Sigma^+ U^T = \begin{bmatrix} 1 & 0 \\ 0 & 1 \end{bmatrix} \begin{bmatrix} \frac{1}{\sqrt{2}} & 0 & 0 \\ 0 & \frac{1}{\sqrt{2}} & 0 \end{bmatrix} \begin{bmatrix} \frac{1}{\sqrt{2}} & 0 & \frac{1}{\sqrt{2}} \\ \frac{1}{\sqrt{2}} & 0 & -\frac{1}{\sqrt{2}} \\ 0 & 1 & 0 \end{bmatrix} = \begin{bmatrix} \frac{1}{2} & 0 & \frac{1}{2} \\ \frac{1}{2} & 0 & -\frac{1}{2} \end{bmatrix}$.

19. See Exercise 31 in Section 7.4.

We need to show the eigenvalues of $A^T A =$ the eigenvalues of $(PAQ)^T(PAQ)$.
$(PAQ)^T(PAQ) = Q^T A^T(P^T P)AQ = Q^T(A^T A)Q$. Therefore, by definition
$A^T A$ and $(PAQ)^T(PAQ)$ are similar. So, Section 4.4, Theorem 4.22 $\Rightarrow$
eigenvalues of $A^T A =$ eigenvalues of $(PAQ)^T(PAQ) \Rightarrow$
singular values of $A =$ singular values of PAQ.

Appendix I: Key Definitions and Concepts

This list includes most but not all of the definitions listed at the end of each section. We begin the list with key symbols and symbol-based definitions in chapter order.

Chapter 1

$\mathbf{v} = \mathbf{w}$	p.3	1.1	... if and only if *corresponding* components are equal
$\mathbf{u} + \mathbf{v}$	p.5	1.1	$\mathbf{u} + \mathbf{v} = [u_1 + v_1, u_2 + v_2]$ *vector addition*
$\mathbf{u} = \sum c_i \mathbf{v}_i$	p.12	1.1	the c_i are scalars (*linear combination*)
$\mathbf{u} - \mathbf{v}$	p.8	1.1	$\mathbf{u} - \mathbf{v} = \mathbf{u} + (-\mathbf{v})$ *vector subtraction*
$\mathbf{u} \| \mathbf{v}$	p.8	1.1	$\mathbf{u} \| \mathbf{v} \Leftrightarrow \mathbf{v} = c\mathbf{u}$ if and only if scalar multiples (*parallel*)
$\mathbf{v}$	p.3	1.1	vectors denoted by a single, boldface, lowercase letter
$\|$	p.8	1.1	parallel (word is used not symbol)
$\Rightarrow$	p.4	1.1	implies (**if** vectors have the same direction, **then** they are parallel)
$\Leftrightarrow$	p.4	1.1	if and only if (phrase is used on p. 4 of Section 1.1)
$\sum c_i \mathbf{v}_i$	p.12	1.1	shorthand for $c_1 \mathbf{v}_1 + c_2 \mathbf{v}_2 + \cdots + c_n \mathbf{v}_n$
$\overrightarrow{AB}$	p.3	1.1	an overhead arrow (vector $\overrightarrow{AB}$, differs from line segment $\overline{AB}$)

Chapter 2

$R_i \leftrightarrow R_j$	p.70	2.2	interchange two rows
kR_i	p.70	2.2	multiply a row by a nonzero constant
$R_i + kR_j$	p.70	2.2	add a multiple of a row to another row
A_c		2.3	Created by taking vectors $\mathbf{v}_i$ as its columns
A_r		2.3	Created by taking vectors $\mathbf{v}_i$ as its rows
$\mathbf{b} = \sum c_i \mathbf{v}_i$	p.638	A.5	**b** is a linear combination of $\mathbf{v}_i$ (see Appendix A)
$\mathbf{a}_i$	p.90	2.3	rows or columns of A written as vectors
$[A|\mathbf{b}]$	p.91	2.3	augmented matrix of a linear system
$[A|\mathbf{0}]$	p.97	2.3	augmented matrix of a homogenous linear system

Key Definitions and Concepts, continued, Chapter 3 symbols

Chapter 3

A^T	p.149	3.1	A-transpose, created by switching rows and columns
A^{-1}	p.162	3.3	A-inverse: satisfies $AA^{-1} = I$ and is unique
A^{-n}	p.167	3.3	This is defined to be $(A^{-1})^n = (A^n)^{-1}$ and is unique
$\mathbf{e}_i$	p.142	3.1	standard $1 \times m$ unit vector
$\mathbf{e}_j$	p.142	3.1	standard $n \times 1$ unit vector
E_i	p.169	3.3	Matrix created by an elementary row operation on I
L	p.179	3.4	Often used to stand for a unit lower triangular matrix
O	p.152	3.2	O is commonly used to stand for the zero matrix
P	p.185	3.4	Often used to stand for a permutation matrix
U	p.179	3.4	Often used to stand for an upper triangular matrix
LU **factorization**	p.179	3.4	$A = LU$ with L unit lower triangular and and U upper triangular
LDU **factorization**	p.189	3.4	$A = LDU$ same as LU except D is diagonal and U *unit* upper triangular
$P^T LU$ **factorization**	p.184	3.4	Adjustment to the LU factorization when we have to permute the rows of A
$[\mathbf{v}]_\mathcal{B}$	p.198	3.5	coordinate vector with respect to $\mathcal{B}$
$\text{col}(A)$	p.193	3.5	column space of A spanned by the columns of A
$\dim(S)$	p.201	3.5	the number of vectors in a basis called its *dimension*
$\text{null}(A)$	p.195	3.5	null space of A, $\mathbf{x}$ such that $A\mathbf{x} = \mathbf{0}$
$\text{nullity}(A)$	p.203	3.5	the dimension of $\text{null}(A)$
$\text{row}(A)$	p.193	3.5	row space of A spanned by the rows of A
$\det A$	p.163	3.3	Determinant of A for 2×2 matrices $\det A = ad - bc$

Key Definitions and Concepts, continued, Chapter 4 symbols

Chapter 4

λ	p.253	4.1	λ often used to denote an eigenvalue of A
E_λ	p.255	4.1	$E_\lambda = \text{null}(A - \lambda I) = \{\text{eigenvectors of } \lambda\} \cup \{0\}$
$\sum_{i=1}^{n} a_{ij} C_{ij}$	p.263	4.2	$\sum_{j=1}^{n} a_{ij} C_{ij} = a_{i1} C_{i1} + a_{i2} C_{i2} + \cdots + a_{in} C_{in}$
$\prod_{i=1}^{n} A_i$	p.263	4.2	$\prod_{i=1}^{n} A_i = A_1 A_2 \cdots A_n$
			Shorthand for products like $\sum_{j=1}^{n} a_{ij} C_{ij}$ is for sums
A_{ij}	p.263	4.2	submatrix of A obtained by deleting row i and column j
$\det A_{ij}$	p.263	4.2	$\det A_{ij}$, the (i,j)-minor of A
$A_i(\mathbf{b})$	p.273	4.2	$A_i(\mathbf{b})$ obtained by replacing column i of A by $\mathbf{b}$
			That is, $A_i(\mathbf{b}) = [\,\mathbf{a}_1\ \cdots\ \mathbf{b}\ \cdots\ \mathbf{a}_n\,]$ (see Cramer's Rule)
adj A	p.275	4.2	adj $A = [C_{ij}]^T$, the adjoint of A
$\det A$	p.264	4.2	$\det A = \sum_{j=1}^{n} a_{1j}(-1)^{1+j} \det A_{1j}$ or $\det A = \sum_{j=1}^{n} a_{1j} C_{1j}$
C_{ij}	p.265	4.2	$C_{ij} = (-1)^{i+j} \det A_{ij}$, the (i,j)-cofactor of A
$c_A(\lambda)$	p.289	4.3	The *characteristic* polynomial is $c_A(\lambda) = \det(A - \lambda I)$. It is important to note that $c_A(\lambda)$ is a polynomial in λ. That is, λ is *not* a fixed eigenvalue, but a variable like x.
$\prod_{i=1}^{n}(\lambda_i - \lambda)$	p.289	4.3	$c_A(\lambda)$ using the eigenvalues of A, λ_i.
$\prod_{i=1}^{m}(\lambda_i - \lambda)^{k_i}$	p.289	4.3	$c_A(\lambda)$ emphasizing the algebraic multiplicity of λ_i, k_i.
$A \sim B$	p.298	4.4	$A \sim B$ (A is *similar* to B) if $P^{-1}AP = B$, P invertible
λ_1	p.308	4.5	*dominant eigenvalue*, $|\lambda_1| > |\lambda_i|$
$\mathbf{v}_1$	p.308	4.5	*dominant eigenvector*, associated with dominant eigenvalue

Key Definitions and Concepts, continued, Chapter 5 symbols

Chapter 5

$\mathcal{B}$	p.369	5.1	$\mathcal{B}$ often used to denote a set of basis vectors
$\mathbf{q}_i$	p.369	5.1	$\mathbf{q}_i$ often used to denote a vector in an *orthonormal* basis
Q	p.371	5.1	Q often used to denote an *orthogonal* matrix
$\mathbf{w}^\perp$	p.379	5.2	component of $\mathbf{v}$ orthogonal to W, $\mathbf{v} - \text{proj}_W(\mathbf{v})$
$W^\perp$	p.375	5.2	$W^\perp = \{\mathbf{v} \text{ in } \mathbb{R}^n : \mathbf{v} \cdot \mathbf{w} = 0 \text{ for all } \mathbf{w} \text{ in } W\}$
$\text{perp}_W(\mathbf{v})$	p.379	5.2	component of $\mathbf{v}$ orthogonal to W, $\mathbf{v} - \text{proj}_W(\mathbf{v})$
$\text{proj}_W(\mathbf{v})$	p.379	5.2	*orthogonal projection* of $\mathbf{v}$ onto W, $\sum_{k=1}^{n} \left(\frac{\mathbf{u}_k \cdot \mathbf{v}}{\mathbf{u}_k \cdot \mathbf{u}_k} \right) \mathbf{u}_k$
$\mathbf{1}$	p.411	5.5	$\mathbf{1}$ (the vector in $\mathbb{Z}_2^n$ all of whose entries are 1)
C	p.408	5.5	C (a set of code vectors in $\mathbb{Z}_2^n$)
$C^\perp$	p.408	5.5	$C^\perp = \{\mathbf{x} \text{ in } \mathbb{Z}_2^n : \mathbf{c} \cdot \mathbf{x} = 0 \text{ for all } \mathbf{c} \text{ in } C\}$
G	p.405	5.5	$G = \left[\frac{I_m}{A} \right]$ (standard generator matrix for a code)
P	p.405	5.5	$G = [\, B \,|\, I_{n-m} \,]$ (standard parity check matrix for a code)
$w(\mathbf{x})$	p.411	5.5	$w(\mathbf{x})$ (*weight* of $\mathbf{x}$), the number of 1s in $\mathbf{x}$ (in $\mathbb{Z}_2^n$)
$f(\mathbf{x})$	p.412	5.5	$f(\mathbf{x}) = \mathbf{x}^T A \mathbf{x}$ (*quadratic form*), (A, the matrix associated with f, is symmetric)
$\mathbf{x}^T A \mathbf{x}$	p.412	5.5	$\mathbf{x}^T A \mathbf{x} = a_{11} x_1^2 + a_{22} x_2^2 + \cdots + a_{nn} x_n^2 + \sum_{i<j} 2 a_{ij} x_i x_j$

Key Definitions and Concepts, continued, Chapter 6 symbols

Chapter 6

V	p.433	6.1	V often used to denote a *vector space*
W	p.438	6.1	W often used to denote a *subspace*
$\text{span}(S)$	p.442	6.1	If $S = \{\mathbf{v}_i\}$, then $\text{span}(S)$ equals all linear combinations of $\mathbf{v}_i$.
$\{\mathbf{e}_i\}$	p.451	6.2	$\{\mathbf{e}_i\}$, standard basis for $\mathbb{R}^n$ (see Example 6.29)
$\{x^i\}$	p.451	6.2	$\{x^i\}$, standard basis for $\mathscr{P}_n$ (see Example 6.30)
$\{E_{ij}\}$	p.451	6.2	$\mathcal{E} = \{E_{ij}\}$, standard basis for M_{mn} (see Example 6.31)
$[\mathbf{v}]_{\mathcal{B}}$	p.453	6.2	If $\mathcal{B} = \{\mathbf{u}_i\}$ is a basis and $\mathbf{x} = \sum\limits_{i=1}^{n} c_i \mathbf{u}_i$, then $[\mathbf{v}]_{\mathcal{B}} = \begin{bmatrix} c_1 \\ \vdots \\ c_n \end{bmatrix}$.
$P_{\mathcal{C} \leftarrow \mathcal{B}}$	p.469	6.3	*change-of-basis* matrix: $P_{\mathcal{C} \leftarrow \mathcal{B}} = \begin{bmatrix} [\mathbf{u}_1]_{\mathcal{C}} & \cdots & [\mathbf{u}_n]_{\mathcal{C}} \end{bmatrix}$
$\mathscr{C}[a,b]$	p.477	6.4	The space of continuous functions (See Example 6.52).
$\int_a^b f(x)\,dx$	p.477	6.4	$S: \mathscr{C}[a,b] \to \mathbb{R}$, $S(f) = \int_a^b f(x)\,dx$ (See Example 6.52).
I	p.478	6.4	$I: V \to W$ maps every vector to itself. That is, $I(\mathbf{v}) = \mathbf{v}$ (See Example 6.54).
$S \circ T$	p.481	6.4	If $T: U \to V$ and $S: V \to W$ then the composition of S with T is $(S \circ T)(\mathbf{u}) = S(T(\mathbf{u}))$.
T_0	p.478	6.4	$T_0: V \to W$ maps every vector to the zero vector. That is, $T_0(\mathbf{v}) = \mathbf{0}$ (See Example 6.54).
T^{-1}	p.483	6.4	T^{-1}: $T^{-1} \circ T = I_V$, $T \circ T^{-1} = I_W$ (See *invertible*).

Key Definitions and Concepts, Chapter 6 symbols continued

Chapter 6, continued

$\ker(T)$	p.486	6.5	$\ker(T) = \{\mathbf{v} \text{ in } V : T(\mathbf{v}) = \mathbf{0}\}$ (See *kernel*.)
$\text{nullity}(T)$	p.488	6.5	$\text{nullity}(T) = \dim(\ker(T))$
$\text{range}(T)$	p.486	6.5	$\text{range}(T) = \{\mathbf{w} \text{ in } W : \mathbf{w} = T(\mathbf{v}) \text{ for some } \mathbf{v} \text{ in } V\}$ (See *range*.)
$\text{rank}(T)$	p.488	6.5	$\text{rank}(T) = \dim(\text{range}(T))$ (See *rank*.)
$V \cong W$	p.497	6.5	If $T : V \to W$ where T is an *isomorphism*. Then: V and W are *isomorphic*. This is written $V \cong W$.
$[T]_{C \leftarrow B}$	p.502	6.6	The matrix of T with respect to bases B and C. By Thm 6.26 it satisfies: $[T]_{C \leftarrow B}[\mathbf{v}]_B = [T(\mathbf{v})]_C$ See remarks following Thm 6.26.
$[T]_B$	p.502	6.6	Special case when $V = W$ and $B = C$. By Thm 6.26 it satisfies: $[T]_B[\mathbf{v}]_B = [T(\mathbf{v})]_B$ See remarks following Thm 6.26.
$[T(\mathbf{v})]_C$	p.502	6.6	$A[\mathbf{v}]_B = [T(\mathbf{v})]_C$ We are asked to show this in Exercise 39. See remarks following Thm 6.26.

Key Definitions and Concepts, continued, Chapter 7 symbols

Chapter 7

$\|\mathbf{v}\|$	p.544	7.1	The *length* (or *norm*) of $\mathbf{v}$ is $\|\mathbf{v}\| = \sqrt{\langle \mathbf{v}, \mathbf{v} \rangle}$
$d(\mathbf{u}, \mathbf{v})$	p.544	7.1	The *distance* between $\mathbf{u}$ and $\mathbf{v}$ is $d(\mathbf{u},\mathbf{v}) = \|\mathbf{u}-\mathbf{v}\|$
$\langle \mathbf{u}, \mathbf{v} \rangle = 0$	p.544	7.1	$\mathbf{u}$ and $\mathbf{v}$ are *orthogonal* if $\langle \mathbf{u}, \mathbf{v} \rangle = 0$
$\text{proj}_W(\mathbf{v})$	p.547	7.1	$\text{proj}_W(\mathbf{v}) = \sum_{k=1}^{n} \frac{\langle \mathbf{u}_k \cdot \mathbf{v} \rangle}{\langle \mathbf{u}_k \cdot \mathbf{u}_k \rangle} \mathbf{u}_k$ (*orthogonal projection*)
$\text{perp}_W(\mathbf{v})$	p.547	7.1	$\text{perp}_W(\mathbf{v}) = \mathbf{v} - \text{proj}_W(\mathbf{v})$ This is the *component of $\mathbf{v}$ orthogonal to W*)
$\|\mathbf{v}\|_1 = \|\mathbf{v}\|_s$	p.562	7.2	$\|\mathbf{v}\|_1 = \|\mathbf{v}\|_s = |v_1| + \cdots + |v_n|$
$\|\mathbf{v}\|_\infty = \|\mathbf{v}\|_m$	p.562	7.2	$\|\mathbf{v}\|_\infty = \|\mathbf{v}\|_m = \max\{|v_1|, \ldots, |v_n|\}$
$\|\mathbf{v}\|_p$	p.562	7.2	$\|\mathbf{v}\|_p = (|v_1|^p + \cdots + |v_n|^p)^{1/p}$
$\|\mathbf{v}\|_2 = \|\mathbf{v}\|_E$	p.562	7.2	$\|\mathbf{v}\|_2 = \|\mathbf{v}\|_E = \sqrt{|v_1|^2 + \cdots + |v_n|^2}$
$\|\mathbf{v}\|_H$	p.563	7.2	$\|\mathbf{v}\|_H = w(\mathbf{v})$ (weight). See Example 7.15.
$\overline{\mathbf{v}}$	p.579	7.3	A vector $\overline{\mathbf{v}}$ is *best approximation* to $\mathbf{v}$ in W if $\|\mathbf{v}-\overline{\mathbf{v}}\| < \|\mathbf{v}-\mathbf{w}\|$ for all $\mathbf{w} \neq \overline{\mathbf{v}}$
$\|\mathbf{e}\|$	p.581	7.3	$\|\mathbf{e}\| = \sqrt{\varepsilon_1^2 + \varepsilon_2^2 + \varepsilon_3^2}$ is the *least squares error* of the approximation
A^+	p.594	7.3	If A is a matrix with linearly independent columns, then $A^+ = (A^T A)^{-1} A^T$ is the *pseudoinverse*.
σ_i	p.599	7.4	If A is an $m \times n$ matrix, the *singular values* of A are the square roots of the eigenvalues of $A^T A$. They are denoted by $\sigma_1, \ldots, \sigma_n$. They are usually arranged $\sigma_1 \geq \sigma_1 \geq \ldots \geq \sigma_n$.
Σ	p.601	7.4	Σ is created from the σ_i (detailed description p.601)
$A = U\Sigma V^T$	p.601	7.4	A *singular value decomposition* (SVD) of $A = U\Sigma V^T$. where U and and V are orthogonal matrices. Σ is constructed using the σ_i.
$A^+ = V\Sigma^+ U^T$	p.611	7.4	The *pseudoinverse (Moore-Penrose)* is $A^+ = V\Sigma^+ U^T$.

Key Definitions and Concepts, continued, A and B

A

adjacency matrix	p.236	3.7	$A(G)$ where $a_{ij} = 1$ if edge, 0 otherwise
adjoint	p.275	4.2	$\operatorname{adj} A = [C_{ij}]^T$, the adjoint of A
algebraic multiplicity	p.291	4.3	The multiplicity of λ as a root of $\det(A - \lambda I) = 0$. So, if $c_A(\lambda) = \prod_{i=1}^{m}(\lambda_i - \lambda)^{k_i}$ where λ_i are distinct, then the algebraic multiplicity of λ_i is k_i. So, the algebraic multiplicity of an eigenvalue is equal to the exponent of its associated factor in the characteristic equation.
angle	p.21	1.2	$\cos\theta = \frac{\mathbf{u}\cdot\mathbf{v}}{\|\mathbf{u}\|\|\mathbf{v}\|}$
attractor	p.347	4.6	See definition on p.347 and Example 4.48.
augmented matrix	p.62	2.1	coefficient matrix augmented by the constants

B

back substitution	p.62	2.1	procedure used to solve Example 2.5 on p. 62
basis	p.196	3.5	linearly independent set of vectors that spans S
basis	p.450	6.2	A subset $\mathcal{B}$ of a vector space V is a *basis* for V if 1. $\mathcal{B}$ spans V, that is $\operatorname{span}(\mathcal{B}) = V$ 2. $\mathcal{B}$ is linearly independent
best approximation	p.579	7.3	A vector $\bar{\mathbf{v}}$ is *best approximation* to $\mathbf{v}$ in W if $\|\mathbf{v} - \bar{\mathbf{v}}\| < \|\mathbf{v} - \mathbf{w}\|$ for all $\mathbf{w} \neq \bar{\mathbf{v}}$
best approximation	p.620	7.5	If $\{\mathbf{u}_1, \ldots, \mathbf{u}_k\}$ is an orthogonal basis for W then $\operatorname{proj}_W(f) = \frac{\langle \mathbf{u}_1, f\rangle}{\langle \mathbf{u}_1, \mathbf{u}_1\rangle}\mathbf{u}_1 + \cdots + \frac{\langle \mathbf{u}_k, f\rangle}{\langle \mathbf{u}_k, \mathbf{u}_k\rangle}\mathbf{u}_k$
binary	p.47	1.4	see the text discussion of arithmetic in $\mathbb{Z}_2$
branch	p.104	2.4	a directed edge of a network

Key Definitions and Concepts, continued, C

C

change-of-basis	p.469	6.3	*change-of-basis* matrix: $P_{C \leftarrow B} = \begin{bmatrix} [\mathbf{u}_1]_C & \cdots & [\mathbf{u}_n]_C \end{bmatrix}$
characteristic equation	p.289	4.3	The *characteristic* equation is $\det(A - \lambda I) = 0$. Solutions of $\det(A - \lambda I) = 0$ are the *eigenvalues* of A.
characteristic polynomial	p.289	4.3	The *characteristic* polynomial is $c_A(\lambda) = \det(A - \lambda I)$. Solutions of $\det(A - \lambda I) = 0$ are the *eigenvalues* of A.
check digit	p.49	1.4	component added to vector to make *parity* even
circuit	p.237	3.7	a closed *path* (ends and begins at the same *vertex*
code	p.48	1.4	a set of vectors of the same *length* (m-ary)
code	p.240	3.7	See discussion prior to Example 3.69 and Section 1.4
coefficient matrix	p.62	2.1	a matrix of coefficients taken from a linear system
coefficients	p.59	2.1	the a_i in $a_1 x_1 + a_2 x_2 + \cdots + a_n x_n = b$
cofactor	p.265	4.2	$C_{ij} = (-1)^{i+j} \det A_{ij}$, the (i,j)-cofactor of A
column space	p.193	3.5	$\mathrm{col}(A)$ spanned by the columns of A
commutative diagram	p.503	6.6	See detailed remarks following Thm 6.26. The diagram referred to is on p.502.
compatible	p.565	7.2	A matrix norm $\|A\|$ is compatible with a vector norm $\|\mathbf{x}\|$ if $\|A\mathbf{x}\| \leq \|A\| \|\mathbf{x}\|$
components	p.3	1.1	individual coordinates of a vector like 3, 2 of $[3, 2]$
composition of S with T	p.481	6.4	If $T: U \to V$ and $S: V \to W$ then the *composition of S with T* is the mapping $S \circ T$: $(S \circ T)(\mathbf{u}) = S(T(\mathbf{u}))$.

Key Definitions and Concepts, C continued and D

C, continued

conditioned	p.570	7.2	For description of *ill-* and *well-conditioned*, see text.
condition number	p.571	7.4	Condition number of a square matrix is $\|A^{-1}\|\|A\|$. If A is not invertible, we define $\text{cond}(A) = \infty$.
conditions	p.333	4.6	*initial conditions.* See definition p. 333.
conservation of flow	p.104	2.4	at each node, the flow in equals the flow out
consistent	p.61	2.1	system of equations with at least one solution
constant term	p.59	2.1	b in $a_1 x_1 + a_2 x_2 + \cdots + a_n x_n = b$
converges	p.123	2.5	when iterates approach a solution
coordinate vector	p.453	6.2	If $\mathcal{B} = \{\mathbf{u}_i\}$ is a basis and $\mathbf{x} = \sum_{i=1}^{n} c_i \mathbf{u}_i$, then $[\mathbf{v}]_\mathcal{B} = \begin{bmatrix} c_1 \\ \vdots \\ c_n \end{bmatrix}$.
correction	p.627	7.5	See text for description of *correcting k errors*.
Current Law (nodes)	p.106	2.4	sum flowing into a node equals sum out

D

decoding	p.48	1.4	converting code vectors into a message
decoding	p.627	7.5	See text for full description of *nearest neighbor decoding*.
definite	p.415	5.5	A quadratic form $f(\mathbf{x}) = \mathbf{x}^T A \mathbf{x}$ is classified as follows: 1. *positive definite* $f(\mathbf{x}) > 0$ for all $\mathbf{x} \neq \mathbf{0}$ 2. *positive semidefinite* $f(\mathbf{x}) \geq 0$ for all $\mathbf{x}$ 3. *negative definite* $f(\mathbf{x}) < 0$ for all $\mathbf{x} \neq \mathbf{0}$ 4. *negative semidefinite* $f(\mathbf{x}) \leq 0$ for all $\mathbf{x}$ 5. *indefinite* if $f(\mathbf{x})$, both positive and negative. The associated matrix A is classified in the same way.
detection	p.627	7.5	See text for full description of *detecting k errors*.
determinant	p.163	3.3	Determinant of A for 2×2 matrices $\det A = ad - bc$
determinant	p.264	4.2	$\det A = \sum_{j=1}^{n} a_{1j}(-1)^{i+1} \det A_{1j}$ or $\det A = \sum_{j=1}^{n} a_{1j} C_{1j}$

Key Definitions and Concepts, D continued

D, continued

diagonalizable	p.300	4.4	$A \sim D$ if $P^{-1}AP = D$ where D is diagonal
diagonalizable	p.513	6.6	Let V be a finite dimensional vector space. Let $T : V \to V$ be a linear transformation. Then: T is called *diagonalizable* if there is a basis C such that $[T]_C$ is diagonal matrix.
differential equation	p.522	6.7	See detailed remarks before Thm 6.31. Here we show the equations for *first* and *second* order. *first-order*: $y' + ay = 0$ (homogeneous, because $= 0$). Solution: $y = e^{-at}$ (see Thm 6.31) *second-order*: $y'' + ay' + by = 0$ (homogeneous). Solution: $y = c_1 e^{\lambda_1 t} + c_2 e^{\lambda_2 t}$ (see Thm 6.31) Where: λ_1 and λ_2 are solutions of $\lambda^2 + a\lambda + b = 0$
differential operator	p.477	6.4	The *differential operator* is defined as follows: $D : \mathscr{D} \to \mathscr{F}$, $D(f) = f'$ (See Example 6.51).
dimension	p.201	3.5	$\dim(S)$, the number of vectors in a basis
dimensional finite-dim infinite-dim zero-dim	p.457	6.2	A vector space is *blank*-dimensional if: A basis has finitely many vectors A basis has infinitely many vectors The dimension of the subspace **0** is defined to be 0
direction vector	p.32	1.3	**d**, parallel to any vector on line ℓ
distance	p.20	1.2	$d(\mathbf{u}, \mathbf{v}) = \|\mathbf{u} - \mathbf{v}\|$
distance	p.544	7.1	The *distance* between $\mathbf{u}$ and $\mathbf{v}$ is $d(\mathbf{u}, \mathbf{v}) = \|\mathbf{u} - \mathbf{v}\|$
distance	p.563	7.2	See Example 7.16 for $d_s(\mathbf{u}, \mathbf{v})$, $d_E(\mathbf{u}, \mathbf{v})$, $d_H(\mathbf{u}, \mathbf{v})$ ...
divergence	p.125	2.5	when iterates do not approach a solution
dominant eigenvalue	p.308	4.5	$|\lambda_1| > |\lambda_i|$, where λ_i are the *other* eigenvalues of A
dominant eigenvector	p.308	4.5	The eigenvector associated with $|\lambda_1| > |\lambda_i|$, that is, $A\mathbf{x}_1 = \lambda_1 \mathbf{x}_1$, where λ_1 is dominant
dot product	p.15	1.2	$\mathbf{u} \cdot \mathbf{v} = u_1 v_1 + u_2 v_2 + \ldots + u_n v_n$

Key Definitions and Concepts, continued, E and F

E

eigenspace	p.255	4.1	$E_\lambda = \text{null}(A - \lambda I) = \{\text{eigenvectors of } \lambda\} \cup \{\mathbf{0}\}$ All vectors such that $(A - \lambda I)\mathbf{x} = \mathbf{0}$ or $A\mathbf{x} = \lambda\mathbf{x}$
eigenvalue	p.253	4.1	if $A\mathbf{x} = \lambda\mathbf{x}$, then A has eigenvalue λ
eigenvalue	p.289	4.3	The solutions of $\det(A - \lambda I) = 0$.
eigenvector	p.253	4.1	if $A\mathbf{x} = \lambda\mathbf{x}$, then $\mathbf{x}$ is an eigenvector of A
elementary matrix	p.169	3.3	Any matrix that can be obtained by performing an elementary row operation on an identity matrix
elementary row operations	p.70	2.2	EROs: $R_i \leftrightarrow R_j$, kR_i, $R_i + kR_j$
encoding	p.48	1.4	converting a message into code vectors
equivalent	p.61	2.1	linear systems that have the same solution set
Euclidean norm	p.562	7.2	$\|\mathbf{v}\|_E = \|\mathbf{v}\|_2 = \sqrt{|v_1|^2 + \cdots + |v_n|^2}$ (same as 2-norm)
expansion	p.265	4.2	$\det A = \sum_{j=1}^{n} a_{ij} C_{ij}$ along the ith row $\det A = \sum_{i=1}^{n} a_{ij} C_{ij}$ along the jth column

F

Fourier approximation	p.624	7.5	The best approximation using the *Fourier coefficients*. Given below and used in a *trigonometric polynomial*.
Fourier coefficients	p.624	7.5	These coefficients are used in the *Fourier approximation*: $a_0 = \frac{\langle 1, f \rangle}{\langle 1, 1 \rangle} = \frac{1}{2\pi} \int_{-\pi}^{\pi} f(x)\, dx$ $a_k = \frac{\langle \cos kx, f \rangle}{\langle \cos kx, \cos kx \rangle} = \frac{1}{\pi} \int_{-\pi}^{\pi} f(x) \cos kx\, dx$ $b_k = \frac{\langle \sin kx, f \rangle}{\langle \sin kx, \sin kx \rangle} = \frac{1}{\pi} \int_{-\pi}^{\pi} f(x) \sin kx\, dx$
Fourier series	p.624	7.5	When *Fourier trigonometric polynomial* becomes infinite, the result is called the *Fourier series*. $f(x) = a_0 + \sum_{k=1}^{\infty} (a_k \cos kx + b_k \sin kx)$ on $[-\pi, \pi]$

Key Definitions and Concepts, F continued and G

F, continued

free variable	p.75	2.2	a variable free to take on any value
Frobenius norm	p.565	7.2	$\|A\|_F = \sqrt{\sum_{i,j=1}^{n} a_{ij}^2}$. See Example 7.18.
fundamental subspaces	p.377	5.2	There are four *fundamental subspaces* (two pair): $\mathrm{row}(A), \mathrm{null}(A)$ and $\mathrm{col}(A), \mathrm{null}(A^T)$

G

Gauss-Jordan Elimination	p.76	2.2	see procedure described in box on p. 77
Gauss-Seidel method	p.122	2.5	This process is applied in Example 2.35
Gaussian Elimination	p.72	2.2	see procedure described in box on p. 72
general form	p.33	1.3	$ax + by = c$ where $\mathbf{n} = \begin{bmatrix} a \\ b \end{bmatrix}$ is normal for ℓ
generator matrix	p.240	3.7	G: generates a *code*. See definition before Thm 3.34
generator matrix	p.406	5.5	For $n > m$, an $n \times m$ matrix G and an $(n-k) \times n$ matrix P (with entries in $\mathbb{Z}_2$) are a *generator matrix* and a *parity check matrix* respectively for an (n,k) binary code C if: 1. The columns of G are linearly independent. 2. The rows of P are linearly independent. 3. $PG = O$.
geometric multiplicity	p.291	4.3	Dimension of eigenspace associated with λ, $\dim E_\lambda$, where $\dim E_\lambda$ is the number of vectors in a basis. That is, *geometric multiplicity* $= \dim E_\lambda$.
Gerschgorin disk	p.316	4.5	$D_i = \{z \text{ in } \mathbb{C} : |z - a_{ii}| \le r_i\}$ where $r_i = \sum_{j \ne i} |a_{ij}|$
graph	p.236	3.7	finite set of *points* and *edges*

Key Definitions and Concepts, H and I

H

Hamming code	p.242	3.7	See Example 3.70 for construction
Hamming norm	p.563	7.2	$\|\mathbf{v}\|_H = w(\mathbf{v})$ (weight). See Example 7.15.
homogeneous	p.79	2.2	system in which each constant term is 0

I

identity transformation	p.478	6.4	$I: V \to W$ maps every vector to itself. That is, $I(\mathbf{v}) = \mathbf{v}$ (See Example 6.54).
if and only if	p.17	1.2	signals a *double implication*
inconsistent	p.61	2.1	a system of linear equations with no solutions
induction	p.147	3.1	See Appendix B, *Mathematical Induction*
inner product	p.15	1.2	generalized notion of the dot product (Ch. 7)
inner product	p.540	7.1	An *inner product* on a vector space V with $\langle \mathbf{u}, \mathbf{v} \rangle$ has the following properties: 1. $\langle \mathbf{u}, \mathbf{v} \rangle = \langle \mathbf{v}, \mathbf{u} \rangle$ 2. $\langle \mathbf{u}, \mathbf{v} + \mathbf{w} \rangle = \langle \mathbf{u}, \mathbf{v} \rangle + \langle \mathbf{u}, \mathbf{w} \rangle$ 3. $\langle c\mathbf{u}, \mathbf{v} \rangle = c\langle \mathbf{u}, \mathbf{v} \rangle$ 4. $\langle \mathbf{u}, \mathbf{u} \rangle \geq 0$ and $\langle \mathbf{u}, \mathbf{u} \rangle = 0$ if and only if $\mathbf{0} = 0$
inner product space	p.540	7.1	A vector space with an *inner product* is called an *inner product space*
inverse	p.161	3.3	A-inverse: satisfies $AA^{-1} = I$ and is unique
inverse	p.483	6.4	T^{-1}: $T^{-1} \circ T = I_V$, $T \circ T^{-1} = I_W$ (See *invertible*).
inverse transformation	p.219	3.6	$S \circ T = I_n$ and $T \circ S = I_n$ $S(T(\mathbf{v})) = [S][T]\mathbf{v} = \mathbf{v} \Rightarrow [T] = [S]^{-1}$

Key Definitions and Concepts, I continued and J, K

I, continued

invertible	p.161	3.3	If A^{-1} exists, A is called *invertible*.
invertible	p.482	6.4	A linear transformation $T: V \to W$ is *invertible* if there is a linear transformation $T': W \to V$ such that: $T' \circ T = I_V$ and $T \circ T' = I_W$. In this case, T' is called the *inverse* for T.
isomorphic	p.497	6.5	If $T: V \to W$ where T is an *isomorphism*. Then: V and W are *isomorphic*. This is written $V \cong W$.
isomorphism	p.497	6.5	$T: V \to W$ is an *isomorphism* if T is *one-to-one* and *onto*.
iterates	p.123	2.5	vectors found through the iterative process

J

Jacobi's method	p.122	2.5	This process is applied in Example 2.35

K

kernel (**of T**)	p.486	6.5	Let $T: V \to W$ be a linear transformation. *Kernel* of T is the set of all vectors that T maps to zero. That is, $\ker(T) = \{\mathbf{v} \text{ in } V : T(\mathbf{v}) = \mathbf{0}\}$.
Kirchoff's Laws	p.106	2.4	Current Law (nodes), Voltage Law (circuits)

Key Definitions and Concepts, continued, L

L

leading entry	p.68	2.2	the first nonzero entry in a row of a matrix
least squares error	p.581	7.3	$\|\mathbf{e}\| = \sqrt{\varepsilon_1^2 + \varepsilon_2^2 + \varepsilon_3^2}$ is the *least squares error* of the approximation
least squares solution	p.583	7.3	A *least squares solution* of $A\mathbf{x} = \mathbf{b}$ is $\bar{\mathbf{x}}$ such that $\|\mathbf{b} - A\bar{\mathbf{x}}\| \leq \|\mathbf{b} - A\mathbf{x}\|$ for all $\mathbf{x}$ in $\mathbb{R}$
left singular vectors	p.602	7.4	Given a singular value decomposition (SVD) of $A = U\Sigma V^T$. The columns of U (an orthogonal matrix) are called the *left singular values* of A.
length	p.17	1.2	$\|\mathbf{v}\| = \sqrt{\mathbf{v} \cdot \mathbf{v}}$ (Means the same thing as *norm*)
length	p.237	3.7	number of edges a *path* contains
length	p.544	7.1	The *length* (or *norm*) of $\mathbf{v}$ is $\|\mathbf{v}\| = \sqrt{\langle \mathbf{v}, \mathbf{v} \rangle}$
length (m-ary)	p.51	1.4	the number of components in in a vector
Leslie matrix	p.234	3.7	Population matrix L described after Example 3.66
linear code	p.529	6.7	A *p-ary linear code* is a subspace of C of $\mathbb{Z}_p^n$.
linear combination	p.12	1.1	$\mathbf{u} = c_1\mathbf{v}_1 + c_2\mathbf{v}_2 + \cdots + c_n\mathbf{v}_n$ where c_i are scalars
linear combination	p.95	2.3	$\mathbf{v} = c_1\mathbf{v}_1 + \cdots + c_n\mathbf{v}_n$ $\mathbf{v} = \sum c_i\mathbf{v}_i$
linear combination	p.152	3.2	$\mathbf{B} = c_1\mathbf{A}_1 + \cdots + c_n\mathbf{A}_n$ $\mathbf{B} = \sum c_i\mathbf{A}_i$
linearly dependent	p.155	3.2	one *can* be written as linear combination of others $c_1\mathbf{A}_1 + \cdots + c_n\mathbf{A}_n = O$ with at least one $c_i \neq 0$
linearly dependent	p.447	6.2	vector can be written as linear combination of others $c_1\mathbf{v}_1 + \cdots + c_n\mathbf{v}_n = \mathbf{0}$ with at least one $c_i \neq 0$
linearly dependent set of vectors	p.94	2.3	vector can be written as linear combination of others $c_1\mathbf{v}_1 + \cdots + c_n\mathbf{v}_n = 0$ with at least one $c_i \neq 0$ $\sum c_i\mathbf{v}_i = 0$ with at least one $c_i \neq 0$

Key Definitions and Concepts, L continued and M

L, continued

linear equation	p.59	2.1	$a_1x_1 + a_2x_2 + \cdots + a_nx_n = b$
linearly independent	p.155	3.2	matrices are *not* linear combinations of each other $c_1\mathbf{A}_1 + \cdots + c_n\mathbf{A}_n = O \Leftrightarrow$ all the $c_i = 0$
linearly independent	p.447	6.2	no vector can be written as linear combination of others or $c_1\mathbf{v}_1 + \cdots + c_n\mathbf{v}_n = \mathbf{0} \Leftrightarrow$ all the $c_i = 0$
linearly independent set of vectors	p.94	2.3	no vector in the set can be written as a linear combination of the other vectors in the set $c_1\mathbf{v}_1 + \cdots + c_n\mathbf{v}_n = 0 \Leftrightarrow$ all the $c_i = 0$ $\sum c_i\mathbf{v}_i = 0 \Leftrightarrow$ all the $c_i = 0$
linear system	p.59	2.1	a set of linear equations with the same variables
linear transformation	p.211	3.6	$T(c\mathbf{v}) = cT(\mathbf{v})$ and $T(\mathbf{u}+\mathbf{v}) = T(\mathbf{u}+\mathbf{v})$ If T is linear, then $T(\mathbf{v}) = A\mathbf{v}$ for some matrix A.
linear transformation	p.476	6.4	A *linear transformation* from a vector space V to a vector space W is a mapping $T: V \to W$ such that: 1. $T(\mathbf{u}+\mathbf{v}) = T(\mathbf{u}+\mathbf{v})$ 2. $T(c\mathbf{u}) = cT(\mathbf{u})$ or $T(c_1\mathbf{v}_1 + \cdots + c_k\mathbf{v}_k) = c_1T(\mathbf{v}_1) + \cdots + c_kT(\mathbf{v}_k)$

M

m-ary	p.51	1.4	see the text discussion of arithmetic in $\mathbb{Z}_m$
Markov chain	p.228	3.7	See description given prior to Example 3.64
matrix	p.62	2.1	a rectangular array of numbers in rows and columns
matrix norm	p.565	7.2	A *matrix norm* on M_{nn} has the following properties: 1. $\|A\| \geq 0$, and $\|A\| = 0$ if and only if $A = O$ 2. $\|cA\| = \|c\|\|A\|$ 3. $\|A+B\| \leq \|A\| + \|B\|$ 4. $\|AB\| \leq \|A\|\|B\|$

Key Definitions and Concepts, M continued and N

M, continued

matrix product	p.139	3.1	$C = AB \Rightarrow c_{ij} = a_{i1}b_{1j} + a_{i2}b_{2j} + \cdots + a_{in}b_{nj}$
max norm	p.562	7.2	$\|\mathbf{v}\|_m = \|\mathbf{v}\|_\infty = \max\{|v_1|, \ldots, |v_n|\}$ (also ∞-norm)
minimum distance	p.626	7.5	The smallest distance between any two distinct vectors in C: $\mathrm{d}(C) = \min\{\mathrm{d}_H(\mathbf{x},\mathbf{y}) : \mathbf{x} \neq \mathbf{y} \text{ in } C\}$
minor	p.263	4.2	$\det A_{ij}$, the (i,j)-minor of A
modular	p.50	1.4	see the text development of Modular Arithmetic
multiplier	p.181	3.4	the scalar k in $R_i - kR_j$

N

network	p.104	2.4	nodes connected by a series of branches
norm	p.17	1.2	$\|\mathbf{v}\| = \sqrt{\mathbf{v}\cdot\mathbf{v}}$ (using the dot product)
norm	p.544	7.1	The *length* (or *norm*) of $\mathbf{v}$ is $\|\mathbf{v}\| = \sqrt{\langle\mathbf{v},\mathbf{v}\rangle}$
norm	p.561	7.2	A *norm* on a vector space V satisfies:
normal equations	p.584	7.3	$A^T A\bar{\mathbf{x}} = A^T \mathbf{b}$
normal form	p.33	1.3	$\mathbf{n}\cdot(\mathbf{x}-\mathbf{p}) = 0$ or $\mathbf{n}\cdot\mathbf{x} = \mathbf{n}\cdot\mathbf{p}$
normalizing	p.18	1.2	a unit vector in same direction ($\mathbf{u} = \frac{1}{\|\mathbf{v}\|}\mathbf{v}$)
normal vector	p.31	1.3	$\mathbf{n}$, orthogonal to any vector on line ℓ or plane $\mathscr{P}$
normed linear space			1. $\|\mathbf{v}\| \geq 0$, and $\|\mathbf{v}\| = 0$ if and only if $\mathbf{v} = \mathbf{0}$ 2. $\|c\mathbf{v}\| = |c|\|\mathbf{v}\|$ 3. $\|\mathbf{u}+\mathbf{v}\| \leq \|\mathbf{u}\| + \|\mathbf{v}\|$
nullity	p.203	3.5	the dimension of $\mathrm{null}(A)$
nullity (of T)	p.488	6.5	Let $T: V \to W$ be a linear transformation. The *nullity* of T is the dimension of the kernel of T. That is, $\mathrm{nullity}(T) = \dim(\ker(T))$.
null space	p.195	3.5	$\mathrm{null}(A)$, $\mathbf{x}$ such that $A\mathbf{x} = \mathbf{0}$

Key Definitions and Concepts, continued, O

O

Ohm's Law	p.106	2.4	force = resistance × current, $E = RI$
one-to-one	p.492	6.5	T maps distinct vectors in V to distinct vectors in W. That is, $\mathbf{v} \neq \mathbf{u}$ implies $T(\mathbf{v}) \neq T(\mathbf{u})$, or $T(\mathbf{v}) = T(\mathbf{u})$ implies $\mathbf{v} = \mathbf{u}$.
onto	p.492	6.5	If range$(T) = W$, then T is called *onto*. For all $\mathbf{w}$, there is a $\mathbf{v}$ such that $\mathbf{w} = T(\mathbf{v})$.
operator norm	p.568	7.2	$\|A\| = \max\|A\mathbf{x}\|$ where $\|\mathbf{x}\| = 1$. See Theorem 7.6. $\|A\|_1 = \max\|A\mathbf{x}\|_s$, $\|A\|_2 = \max\|A\mathbf{x}\|_E$, $\|A\|_\infty = \max\|A\mathbf{x}\|_m$
ordered	p.3	1.1	$[3,2]$ *vs.* $[2,3]$: these are *not* the same vector
orthogonal	p.23	1.2	$\mathbf{u}$ and $\mathbf{v}$ are orthogonal if $\mathbf{u} \cdot \mathbf{v} = 0$
orthogonal	p.544	7.1	$\mathbf{u}$ and $\mathbf{v}$ are orthogonal if $\langle \mathbf{u}, \mathbf{v} \rangle = 0$
orthogonal set	p.546	7.1	$\{\mathbf{v}_1, \ldots, \mathbf{v}_k\}$ such that $\langle \mathbf{v}_i, \mathbf{v}_j \rangle = 0$ when $\mathbf{v}_i \neq \mathbf{v}_j$
orthogonal		5.1	This adjective is used for sets, bases, and matrices:
set	p.366		$\mathbf{v}_i \cdot \mathbf{v}_j = 0$ when $i \neq j$
basis	p.367		a basis $\mathcal{B}$ that is an orthogonal set *square* matrix
matrix	p.371		whose columns form an orthonormal set. That is, Q where columns $\{\mathbf{q}_i\}$ are orthonormal. Matrix is ortho*gonal* when columns are ortho*normal*
orthogonal		5.2	This adjective applies to *complements* and *projections*:
complement	p.375		$W^\perp = \{\mathbf{v} \text{ in } \mathbb{R}^n : \mathbf{v} \cdot \mathbf{w} = 0 \text{ for all } \mathbf{w} \text{ in } W\}$
component	p.379		$\text{perp}_W(\mathbf{v}) = \mathbf{w}^\perp = \mathbf{v} - \text{proj}_W(\mathbf{v})$
projection	p.379		$\text{proj}_W(\mathbf{v}) = \sum_{k=1}^{n} \left(\frac{\mathbf{u}_k \cdot \mathbf{v}}{\mathbf{u}_k \cdot \mathbf{u}_k} \right) \mathbf{u}_k = \sum_{k=1}^{n} \text{proj}_{\mathbf{u}_k}(\mathbf{v})$
orthogonally diagonalizable	p.397	5.4	If there exists an orthogonal Q and diagonal D such that $Q^T A Q = D$, A is *orthogonally diagonalizable*.
orthonormal	p.369	5.1	This adjective is used for sets and bases: For a set to be orthonormal it must satisfy: $\mathbf{q}_i \cdot \mathbf{q}_j = \begin{cases} 0 & \text{if } i \neq j \quad \text{Property 1} \\ 1 & \text{if } i = j \quad \text{Property 2} \end{cases}$
set	p.369		an orthogonal set of *unit* vectors
basis	p.369		a basis $\mathcal{B}$ that is an orthonormal set
orthonormal set	p.546	7.1	An orthogonal set of unit vectors, that is $\|\mathbf{v}_i\| = 1$
outer product	p.145	3.1	see description of process after Example 3.10

Key Definitions and Concepts, continued, P

P

p-norm	p.562	7.2	$\|\mathbf{v}\|_p = (|v_1|^p + \cdots + |v_n|^p)^{1/p}$
parallel	p.8	1.1	if and only if scalar multiples of each other
parametric equations	p.33	1.3	component equations from $\mathbf{x} = \mathbf{p} + t\mathbf{d}$
parity	p.49	1.4	the number of 1s in a code vector
parity check	p.406	5.5	See *generator matrix* for details: the definitions and conditions are associated.
parity matrix	p.240	3.7	P: checks a *code*. See definition before Thm 3.34
path	p.237	3.7	set of *edges* that connects one *vertex* to another
Penrose conditions	p.595	7.3	A^+ (pseudoinverse) satisfies the *Penrose conditions*. a. $AA^+A = A$ b. $A^+AA^+ = A^+$ c. AA^+ and A^+A are symmetric
permutation matrix	p.185	3.4	$P = P_k \cdots P_2 P_1$ where multiplication by P_i performs a row interchange like $R_i \leftrightarrow R_j$
Power Method	p.312	4.5	This two-step iterative method is described in detail on p. 312 and illustrated in Example 4.31
position vector	p.3	1.1	vector with tail at the origin O, i.e. $\overrightarrow{OA}$
pivoting	p.70	2.2	see explanation in solution to Example 2.9 p. 70
probability vector	p.229	3.7	vector with nonnegative components that add to 1
projection	p.24	1.2	$\operatorname{proj}_\mathbf{u}(\mathbf{v}) = \left(\frac{\mathbf{u}\cdot\mathbf{v}}{\mathbf{u}\cdot\mathbf{u}}\right)\mathbf{u}$
projection	p.547	7.1	$\operatorname{proj}_W(\mathbf{v}) = \sum_{k=1}^{n} \frac{\langle \mathbf{u}_k\cdot\mathbf{v}\rangle}{\langle \mathbf{u}_k\cdot\mathbf{u}_k\rangle}\mathbf{u}_k$ (*orthogonal projection*)
projection form	p.402	5.4	If A is a real symmetric matrix with $A = QDQ^T$, then the *projection form of the Spectral Theorem* is $A = \lambda_1 \mathbf{q}_1 \mathbf{q}_1^T + \cdots + \lambda_n \mathbf{q}_n \mathbf{q}_n^T = \sum_{i=1}^{n} \lambda_i \mathbf{q}_i \mathbf{q}_i^T$.
pseudoinverse	p.594	7.3	If A is a matrix with linearly independent columns, then $A^+ = (A^T A)^{-1} A^T$ is the *pseudoinverse*.
pseudoinverse	p.611	7.4	The *pseudoinverse* (*Moore-Penrose*) is $A^+ = V\Sigma^+ U^T$.

Key Definitions and Concepts, continued, Q and R

Q

quadratic form   p.412   5.5 $f(\mathbf{x}) = \mathbf{x}^T A \mathbf{x}$ (*quadratic form*),
(A, the matrix associated with f, is symmetric)

R

range (of T)   p.486   6.5   Let $T: V \to W$ be a linear transformation. The *range* of T is the set of all vectors that are images under T. So, range$(T) = \{\mathbf{w}$ in $W : \mathbf{w} = T(\mathbf{v})$ for some $\mathbf{v}$ in $V\}$.

rank   p.75   2.2   number of nonzero rows in the REF of a matrix

rank   p.202   3.5   dimension of row(A) or col(A) (they are the same)

rank (of T)   p.488   6.5   Let $T: V \to W$ be a linear transformation. The *rank* of T is the dimension of the range of T. That is, rank$(T) = $ dim(range(T)).

recurrence   p.333   4.6   *linear recurrence.* See definition p. 333.

reduced row echelon form   p.76   2.2   RREF: REF; leading entries, 1; all else, 0s

Reed-Muller   p.532   6.7   The (first order) *Reed-Muller codes* R_n (see p.532).

repeller   p.349   4.6   See definition on p.349 and Example 4.50.

right singular vectors   p.602   7.4   Given a singular value decomposition (SVD) of $A = U\Sigma V^T$. The columns of V (an orthogonal matrix) are called the *right singular values* of A.

root mean square error   p.621   7.5   This is the root means square error:
$\|f - g\| = \sqrt{\int_{-1}^{1}(f(x) - g(x))^2\,dx}$

row echelon form   p.68   2.2   REF: zero rows, bottom; leading entries, left

row equivalent   p.72   2.2 A can be reduced to B using EROs

row reduction   p.70   2.2   applying EROs to bring a matrix into REF

row space   p.193   3.5   row(A) spanned by the rows of A

Key Definitions and Concepts, continued, S

S

saddle point	p.349	4.6	See definition on p.349 and Example 4.50.
scalar	p.8	1.1	a real number c (that is, c is **not** a vector)
scalar product	p.15	1.2	another name for dot product (result is a scalar)
similar	p.298	4.4	$A \sim B$ if $P^{-1}AP = B$ where P is invertible
simple path	p.237	3.7	a *path* that does not contain the same *edge* twice
singular values	p.599	7.4	The square roots of the eigenvalues of $A^T A$. Usually denoted by $\sigma_1, \ldots, \sigma_n$.
singular value decomposition	p.601	7.4	A *singular value decomposition* (SVD) of $A = U\Sigma V^T$. where U and and V are orthogonal matrices. Σ is constructed using the σ_i.
skew-symmetric	p.160	3.2	$A^T = -A$
solution	p.60	2.1	$[s_1, s_2, \ldots, s_n]$ where $a_1 s_1 + a_2 s_2 + \cdots + a_n s_n = b$
span(S)	p.92	2.3	all linear combinations of $S = \{\mathbf{s}_1, \mathbf{s}_2, \ldots, \mathbf{s}_n\}$
spanning set for $\mathbb{R}^n$	p.92	2.3	set S such that span$(S) = \mathbb{R}^n$, that is set S such that $\mathbf{v}$ in $\mathbb{R}^n \Rightarrow \mathbf{v} = \sum c_i \mathbf{s}_i$
spectral decomposition	p.402	5.4	If A is a real symmetric matrix with $A = QDQ^T$, then $A = \lambda_1 \mathbf{q}_1 \mathbf{q}_1^T + \cdots + \lambda_n \mathbf{q}_n \mathbf{q}_n^T = \sum_{i=1}^{n} \lambda_i \mathbf{q}_i \mathbf{q}_i^T$.
spectrum	p.400	5.4	Eigenvalues of an *orthogonally diagonalizable* matrix.
standard basis	p.19	1.2	$\mathbf{e}_i$ with a 1 in the ith component and 0s elsewhere
state vector	p.229	3.7	$\mathbf{x}_k$: $\mathbf{x}_k = P^k \mathbf{x}_0$ in a Markov Chain
steady state	p.231	3.7	$\mathbf{x}$: $\mathbf{x} = P\mathbf{x}$ in a Markov Chain
strictly diagonally dominant	p.126	2.5	$\|a_{11}\| > \|a_{12}\| + \|a_{13}\| + \cdots + \|a_{1n}\|$ $\|a_{22}\| > \|a_{21}\| + \|a_{23}\| + \cdots + \|a_{2n}\|$ and $\|a_{nn}\| > \|a_{n1}\| + \|a_{n2}\| + \cdots + \|a_{n,n-1}\|$
subspace	p.190	3.5	$\mathbf{0}$, $\mathbf{u} + \mathbf{v}$, and $c\mathbf{v}$ are in S
subspace	p.438	6.1	$W \subseteq V$, where W is a vector space
submatrix	p.263	4.2	obtained from A by deleting row i and column j
sum norm	p.561	7.2	$\|\mathbf{v}\|_s = \|\mathbf{v}\|_1 = \|v_1\| + \cdots + \|v_n\|$ (same as 1-norm)
symmetric	p.149	3.1	$A^T = A$
symmetric	p.149	3.1	A matrix A is *symmetric* if and only if $A^T = A$.

Key Definitions and Concepts, continued, T and U

T

ternary	p.50	1.4	see the text discussion of arithmetic in $\mathbb{Z}_3$
trace	p.160	3.2	$\text{tr}(A) = a_{11} + a_{22} + \cdots + a_{nn}$ So, the *trace* is the sum of the diagonal entries
trajectory	p.346	4.6	See definition on p.346 prior to Example 4.48.
transition matrix	p.229	3.7	P: $\mathbf{x}_k = P^k \mathbf{x}_0$ in a Markov Chain
transpose	p.148	3.1	A^T, create by switching rows and columns
transposition	p.52	1.4	the interchange of two adjacent components
trigonometric polynomial	p.623	7.5	A function of this form is a *trigonometric polynomial*. $p(x) = a_0 + a_1 \cos x + a_2 \cos 2x + \cdots + a_n \cos nx +$ $\qquad\qquad b_1 \sin x + b_2 \sin 2x + \cdots + b_n \sin nx.$
trivial subspaces	p.441	6.1	The subspaces $\{\mathbf{0}\}$ and V are called the *trivial subspaces* of V. Therefore, all other subspaces are nontrivial.

U

unit lower diagonal	p.179	3.4	L with all the entries above the diagonal are 0 and all the entries on the diagonal are 1. That is, $i > j \Rightarrow a_{ij} = 0$ and $a_{ii} = 1$.
unit vector	p.18	1.2	vector of length 1
upper diagonal	p.179	3.4	U where all the entries below the diagonal are 0. That is, $i < j \Rightarrow a_{ij} = 0$.
upper triangular	p.160	3.2	all entries below the main diagonal are zero

Key Definitions and Concepts, continued, V, W, and XYZ

V

vector	p.3	1.1	*directed* line segment with *length* and *direction*
vector addition	p.5	1.1	$\mathbf{u} + \mathbf{v} = [u_1 + v_1, u_2 + v_2]$
vector form	p.33	1.3	$\mathbf{x} = \mathbf{p} + t\mathbf{d}$ where $\mathbf{d}$ is a direction vector for ℓ
vector space	p.433	6.1	If the following axioms hold, then V is a *vector space*:

1. $\mathbf{u} + \mathbf{v}$ is in V (*Closure under addition*)
2. $\mathbf{u} + \mathbf{v} = \mathbf{v} + \mathbf{u}$ (*Commutativity*)
3. $(\mathbf{u} + \mathbf{v}) + \mathbf{w} = \mathbf{u} + (\mathbf{v} + \mathbf{w})$ (*Associativity*)
4. $\mathbf{u} + \mathbf{0} = \mathbf{u}$ (*Zero vector*)
5. $\mathbf{u} + (-\mathbf{u}) = \mathbf{0}$
6. $c\mathbf{u}$ is in V (*Closure under scalar multiplication*)
7. $c(\mathbf{u} + \mathbf{v}) = c\mathbf{u} + c\mathbf{v}$ (*Distributivity*)
8. $(c + d)\mathbf{u} = c\mathbf{u} + d\mathbf{u}$ (*Distributivity*)
9. $c(d\mathbf{u}) = (cd)\mathbf{u}$
10. $1\mathbf{u} = \mathbf{u}$

Voltage Law	p.106	2.4	voltage *drops* equal total voltage

W

weight	p.411	5.5	$w(\mathbf{x})$ (*weight* of $\mathbf{x}$), the number of 1s in $\mathbf{x}$ (in $\mathbb{Z}_2^n$)
weighted dot product	p.541	7.1	An *inner product* defined on $\mathbb{R}^n$ such that $\langle \mathbf{u}, \mathbf{v} \rangle = w_1 u_1 v_1 + \ldots + w_n u_n v_n$ with $w_i \geq 0$ (see p.541 for details)

X Y Z

zero transformation	p.478	6.4	$T_0 : V \to W$ maps every vector to the zero vector. That is, $T_0(\mathbf{v}) = \mathbf{0}$ (See Example 6.54).
zero vector	p.4	1.1	$\mathbf{0}$, *all* components are 0, so length is 0
zero vector	p.23	1.2	$\mathbf{0}$. Note: $0 \cdot \mathbf{v} = \mathbf{0}$ for *every* vector $\mathbf{v}$ in $\mathbb{R}^n$

Appendix II: Theorems

Theorems

In the summary, we will occasionally list only the central result of the theorem. For the complete statement of the theorem, refer to the text.

Theorems with names are listed in alphabetical order at the end of this section.

Theorems, Chapter 1

Thm 1.1	p.10	1.1	Algebraic Properties of Vectors in $\mathbb{R}^n$
Thm 1.2	p.16	1.2	Properties of the dot product ($\mathbf{u} \cdot \mathbf{v} = \mathbf{v} \cdot \mathbf{u}$ ...)
Thm 1.3	p.17	1.2	Properties of the norm ($\|\mathbf{v}\| = 0 \Leftrightarrow \mathbf{v} = \mathbf{0}$ ...)
Thm 1.4	p.19	1.2	Cauchy-Schwarz: $\|\mathbf{u} \cdot \mathbf{v}\| \leq \|\mathbf{u}\| \|\mathbf{v}\|$
Thm 1.5	p.19	1.2	Triangle Inequality: $\|\mathbf{u} + \mathbf{v}\| \leq \|\mathbf{u}\| + \|\mathbf{v}\|$
Thm 1.6	p.23	1.2	Pythagoras: $\|\mathbf{u} + \mathbf{v}\|^2 = \|\mathbf{u}\|^2 + \|\mathbf{v}\|^2 \Leftrightarrow \mathbf{u} \cdot \mathbf{v} = 0$

Theorems, Chapter 2

Thm 2.1	p.72	2.2	A and B are row equivalent $\Leftrightarrow$ they reduce to same REF
Thm 2.2	p.74	2.2	**Rank Thm**: number of free variables $= n - \text{rank}(A)$
Thm 2.3	p.80	2.2	$[A|\mathbf{0}]$: m equations $< n$ variables $\Rightarrow$ infinitely many solutions
Thm 2.4	p.91	2.3	$[A|\mathbf{b}]$ is consistent $\Leftrightarrow \mathbf{b} = \sum c_i \mathbf{a}_i$
Thm 2.5	p.95	2.3	linearly dependent set $\Leftrightarrow$ linear combination of the others
Thm 2.6	p.97	2.3	linearly dependent set $\Leftrightarrow [A_c|\mathbf{0}]$ has nontrivial solution
Thm 2.7	p.98	2.3	linearly dependent set $\Leftrightarrow \text{rank}(A_r) < m$ where A_r is $m \times n$
Thm 2.8	p.99	2.3	m vectors in $\mathbb{R}^n$ are linearly dependent if $m > n$
Thm 2.9	p.124	2.5	A strictly diagonally dominant $\Rightarrow$ iterates converge
Thm 2.10	p.124	2.5	methods converge $\Rightarrow$ they converge to the solution

Theorems, Chapter 3

Thm 3.1	p.142	3.1	$e_i A$ is row of A, $A e_j$ is column of A
Thm 3.2	p.152	3.2	Properties of Matrix Addition and Scalar Multiplication
Thm 3.3	p.156	3.2	Properties of Matrix Multiplication
Thm 3.4	p.157	3.2	Properties of the Transpose, A^T
Thm 3.5	p.159	3.2	a. $A + A^T$ is symmetric if A is square AA^T and $A^T A$
			b. AA^T and $A^T A$ are always symmetric
Thm 3.6	p.162	3.3	If A is invertible, then its inverse is unique.
			If $AA^{-1} = I$, then A^{-1} is unique.
Thm 3.7	p.162	3.3	$A\mathbf{x} = \mathbf{b}$ has unique solution $\mathbf{x} = A^{-1}\mathbf{b}$
			The inverse A^{-1} exists if and only if $\det A = ad - bc \neq 0$
Thm 3.8	p.163	3.3	Formula for A^{-1} for 2×2 matrices using $\det A$
Thm 3.9	p.166	3.3	Key Properties of A^{-1}, for example $(A^{-1})^{-1} = I$
Thm 3.10	p.170	3.3	Perform row operation E on A is equivalent to EA
Thm 3.11	p.171	3.3	E^{-1} is created by undoing the operation that created E
Thm 3.12	p.171	3.3	Five conditions that are equivalent to A being invertible
Thm 3.13	p.173	3.3	If $AB = I$ or $BA = I$, then $B = A^{-1}$
Thm 3.14	p.173	3.3	If $E_n E_{n-1} \cdots E_1 A = I$, then $A^{-1} = E_n E_{n-1} \cdots E_1$
Thm 3.15	p.180	3.4	If $A \longrightarrow$ REF without $R_i \leftrightarrow Rj$, then $A = LU$ exists.
Thm 3.16	p.184	3.4	If A is invertible, then we have the following:
			If $A = L_1 U_1$ and $A = L_2 U_2$ then $L_1 = L_2$ and $U_1 = U_2$.
			That is, if an LU factorization exists, it is unique.
Thm 3.17	p.185	3.4	If P is a permutation matrix, then $P^{-1} = P^T$.
Thm 3.18	p.186	3.4	If A is square, then $A = P^T LU$ exists.
Thm 3.19	p.190	3.5	$\text{span}(\mathbf{v}_1, \mathbf{v}_2, \ldots, \mathbf{v}_n)$ is a subspace
Thm 3.20	p.194	3.5	If $A \longrightarrow B$, then $\text{row}(A) = \text{row}(B)$
Thm 3.21	p.194	3.5	$\text{null}(A)$, $\mathbf{x}$ such that $A\mathbf{x} = \mathbf{0}$, is a subspace
Thm 3.22	p.195	3.5	$A\mathbf{x} = \mathbf{b}$ has exactly none, one, or infinitely many solutions
Thm 3.23	p.200	3.5	Any two bases have the same number of vectors (*dimension*)
Thm 3.24	p.202	3.5	$\text{row}(A)$ and $\text{col}(A)$ have the same dimension
Thm 3.25	p.202	3.5	$\text{rank}(A) = \text{rank}(A^T)$
Thm 3.26	p.203	3.5	**Rank Theorem**: $\text{rank}(A) + \text{nullity}(A) = n$
Thm 3.27	p.204	3.5	**Fundamental Theorem: Version 2**
Thm 3.28	p.205	3.5	$\text{rank}(A^T A) = \text{rank}(A)$ and $A^T A$ invertible $\Leftrightarrow \text{rank}(A) = n$
Thm 3.29	p.206	3.5	If $\{\mathbf{v}_1, \ldots, \mathbf{v}_k\}$ is a basis, then $\mathbf{v} = \sum c_i \mathbf{v}_i$ is unique
Thm 3.30	p.212	3.6	Given A, then $T_A(\mathbf{v}) = A\mathbf{v}$ is linear
Thm 3.31	p.214	3.6	$[T] = \begin{bmatrix} T(\mathbf{e}_1) & T(\mathbf{e}_2) & \ldots & T(\mathbf{e}_n) \end{bmatrix}$
Thm 3.32	p.218	3.6	Given $S \circ T$, $S(T(\mathbf{v})) = [S][T]\mathbf{v}$
Thm 3.33	p.220	3.6	$S \circ T = I_n$, $S(T(\mathbf{v})) = [S][T]\mathbf{v} = \mathbf{v} \Rightarrow [T] = [S]^{-1}$
Thm 3.34	p.241	3.7	Gives conditions on G and P for a given *code*

Theorems, Chapter 4

Thm 4.1	p.265	4.2	$\det A = \sum_{j=1}^{n} a_{ij}C_{ij}$ (any row) or $\sum_{i=1}^{n} a_{ij}C_{ij}$ (any column)
Thm 4.2	p.268	4.2	If A is triangular, then $\det A = a_{11}a_{22}\cdots a_{nn}$
Thm 4.3	p.268	4.2	a. through f. detail row (column) operations effects on $\det A$

    a. $\mathbf{A}_i = \mathbf{0} \Rightarrow \det A = 0$
    b. $A \xrightarrow{R_i \leftrightarrow R_j} B \Rightarrow \det B = -\det A$
    c. $\mathbf{A}_i = \mathbf{A}_j \Rightarrow \det A = 0$
    d. $A \xrightarrow{kR_i} B \Rightarrow \det B = k \det A$
    e. $\mathbf{C}_i = \mathbf{A}_i + \mathbf{B}_i \Rightarrow \det C = \det A + \det B$
    f. $A \xrightarrow{R_i + kR_j} B \Rightarrow \det B = \det A$

Thm 4.4	p.270	4.2	a. through c. detail row (column) operations effects on $\det E$

    a. $I \xrightarrow{R_i \leftrightarrow R_j} E \Rightarrow \det E = -1$
    b. $I \xrightarrow{kR_i} E \Rightarrow \det E = k$
    c. $I \xrightarrow{R_i + kR_j} E \Rightarrow \det E = 1$

Lem 4.5	p.271	4.2	E, elementary, then $\det(EB) = (\det E)(\det B)$
Thm 4.6	p.271	4.2	A is invertible if and only if $\det A \neq 0$
Thm 4.7	p.271	4.2	$\det(kA) = k^n \det A$
Thm 4.8	p.272	4.2	$\det(AB) = (\det A)(\det B)$
Thm 4.9	p.273	4.2	$\det(A^{-1}) = \frac{1}{\det A}$
Thm 4.10	p.273	4.2	$\det A = \det(A^T)$
Thm 4.11	p.274	4.2	If $A\mathbf{x} = \mathbf{b}$, then $x_i = \frac{\det(A_i(\mathbf{b}))}{\det A}$ (Cramer's Rule)
Thm 4.12	p.276	4.2	$A^{-1} = \frac{1}{\det A}\operatorname{adj} A$
Lem 4.13	p.276	4.2	(row 1) $\det A = \sum_{j=1}^{n} a_{1j}C_{1j} = \sum_{i=1}^{n} a_{i1}C_{i1}$ (column 1)
Lem 4.14	p.276	4.2	$A \xrightarrow{R_i \leftrightarrow R_j} B \Rightarrow \det B = -\det A$ (see Thm 4.3(f))

Theorems, Chapter 4, continued

Thm 4.15 p.292 4.3 If A is triangular, then its eigenvalues are its diagonal entries.

Thm 4.16 p.292 4.3 A is invertible if and only if all $\lambda \neq 0$.

Thm 4.17 p.293 4.3 a. through o. give equivalent conditions for A to be invertible.

Thm 4.18 p.293 4.3 a. through c. relate eigenvalues, A^n and A^{-1}:
For any integer n if $A\mathbf{x} = \lambda \mathbf{x}$, then $A^n \mathbf{x} = \lambda^n \mathbf{x}$.

Thm 4.19 p.294 4.3 If $\mathbf{x} \in \text{span}(\mathbf{v}_i)$, then $A^k \mathbf{x} = c_1 \lambda_1^k \mathbf{v}_1 + c_2 \lambda_2^k \mathbf{v}_2 + \cdots + c_m \lambda_m^k \mathbf{v}_m$
where $\mathbf{v}_i$ is the eigenvector corresponding to λ_i.

Thm 4.20 p.294 4.3 If λ_i are distinct, then $\{\mathbf{v}_i\}$ are linearly independent.

Thm 4.21 p.291 4.4 a. $A \sim A$, b. $A \sim B \Rightarrow B \sim A$, c. $A \sim B, B \sim C \Rightarrow A \sim C$

Thm 4.22 p.299 4.4 parts a. through e. list implications of $A \sim B$
a. $\det A = \det B$
b. A is invertible if and only if B is invertible
c. $\text{rank}(A) = \text{rank}(B)$
d. $\det(A - \lambda I) = \det(B - \lambda I)$
e. λ is an eigenvalue of A if and only if λ is an eigenvalue of B

Thm 4.23 p.300 4.4 $P^{-1}AP = D \Leftrightarrow [D]_{ii} = \lambda_i$ and $\mathbf{p}_i = \mathbf{v}_i$ (the eigenvectors)

Thm 4.24 p.302 4.4 If λ_i are distinct, their basis vectors are linearly independent.

Thm 4.25 p.303 4.4 If all the λ_i are distinct, then A is diagonalizable.

Lem 4.26 p.303 4.4 geometric multiplicity $\leq$ algebraic multiplicity:
That is, $\dim(E_{\lambda_i}) \leq k_i$ where $c_A(\lambda) = \prod_{i=1}^{m} (\lambda_i - \lambda)^{k_i}$.

Thm 4.27 p.304 4.4 parts a. through c. list equivalences to $A = P^{-1}DP$
a. A is diagonalizable, $A = P^{-1}DP$
b. bases of the eigenvectors contain n vectors
c. geometric multiplicity = algebraic multiplicity, $\dim(E_{\lambda_i}) = k_i$

Thm 4.28 p.309 4.5 If A is diagonalizable with dominant eigenvalue λ_1,
then $\mathbf{x}_1 = A\mathbf{x}_0$, $\mathbf{x}_2 = A\mathbf{x}_1$, $\mathbf{x}_k = A\mathbf{x}_{k-1}$
where the $\mathbf{x}_k$ are approaching the dominant eigenvector of A

Thm 4.29 p.318 4.5 Every eigenvalue of A is contained in a Gerschgorin Disk

Theorems, Chapter 4, continued

Thm 4.30   p.322   4.6   If P is the $n \times n$ transition matrix of a Markov chain, then 1 is an eigenvalue of P

Thm 4.31   p.322   4.6   Let P be an $n \times n$ transition matrix with eigenvalue λ.
a. $|\lambda| \leq 1$
b. If P is regular and $\lambda \neq 1$, then $|\lambda| < 1$.

Lem 4.32   p.324   4.6   Let P be an $n \times n$ transition matrix.
If P is diagonalizable, then the dominant eigenvalue $\lambda_1 = 1$ has algebraic multiplicity 1.

Thm 4.33   p.325   4.6   Let P be an $n \times n$ transition matrix.
Then as $k \longrightarrow \infty$, P^k approaches an $n \times n$ matrix L whose columns are $\mathbf{x}$, the steady state probability vector for P.

Thm 4.34   p.326   4.6   Let P be an $n \times n$ transition matrix ... as in Theorem 4.33. Then for any probability vector $\mathbf{x}_0$, $\mathbf{x}_k$ approaches $\mathbf{x}$.

Thm 4.35   p.328   4.6   Every Leslie matrix has a unique positive eigenvalue and a corresponding eigenvector with positive components.

Thm 4.36   p.330   4.6   Let A be a positive $n \times n$ matrix. A has a real eigenvalue λ_1:
a. $\lambda_1 > 0$
b. λ_1 has a corresponding positive eigenvector.
c. If λ is any other eigenvalue of A, then $|\lambda| \leq \lambda_1$.

Thm 4.37   p.332   4.6   Let A be an irreducible nonnegative $n \times n$ matrix. Then:
a. $\lambda_1 > 0$
b. λ_1 has a corresponding positive eigenvector.
c. If λ is any other eigenvalue of A, then $|\lambda| \leq \lambda_1$.
    If A is primitive, then this inequality is strict.
d. If λ is an eigenvalue of A such that $|\lambda| = \lambda_1$, then λ is a (complex) root of the equation $\lambda^n - \lambda_1^n = 0$.
e. λ_1 has algebraic multiplicity 1.

Thm 4.38   p.336   4.6   See statement of Theorem on p. 336.

Thm 4.39   p.337   4.6   See statement of Theorem on p. 337.

Thm 4.40   p.339   4.6   See statement of Theorem on p. 339.

Thm 4.41   p.344   4.6   See statement of Theorem on p. 344.

Thm 4.42   p.349   4.6   See statement of Theorem on p. 349.

Theorems, Chapter 5

Thm 5.1	p.366	5.1	$\mathcal{B} = \{\mathbf{v}_1, \mathbf{v}_2, \ldots, \mathbf{v}_n\}$ orthogonal $\Rightarrow \mathcal{B}$ is linearly independent
Thm 5.2	p.368	5.1	If $\mathcal{B} = \{\mathbf{v}_1, \mathbf{v}_2, \ldots, \mathbf{v}_n\}$ is orthogonal and $\mathbf{w} = \sum_{i=1}^{n} c_i \mathbf{v}_i$, then $c_i = \frac{\mathbf{w} \cdot \mathbf{v}_i}{\mathbf{v}_i \cdot \mathbf{v}_i}$.
Thm 5.3	p.370	5.1	$\mathbf{w} = \sum_{i=1}^{n} (\mathbf{w} \cdot \mathbf{q}_1) \mathbf{q}_i$ (this representation is unique)
Thm 5.4	p.371	5.1	$Q^T Q = I$ if and only if $\{\mathbf{q}_i\}$ is an orthonormal set.
Thm 5.5	p.371	5.1	Q is orthogonal if and only if $Q^{-1} = Q^T$
Thm 5.6	p.372	5.1	a. through c. give equivalent conditions for Q to be orthogonal: a. Q is orthogonal b. $\|Q\mathbf{x}\| = \|\mathbf{x}\|$ for every $\mathbf{x}$ in $\mathbb{R}^n$ c. $Q\mathbf{x} \cdot Q\mathbf{x} = \mathbf{x} \cdot \mathbf{y}$ for every $\mathbf{x}, \mathbf{y}$ in $\mathbb{R}^n$
Thm 5.7	p.373	5.1	Q is orthogonal, then its *rows* form an orthonormal set.
Thm 5.8	p.372	5.1	a. through d. list implications of Q being orthogonal: a. Q^{-1} is orthogonal b. $\det Q = \pm 1$ c. If λ is an eigenvalue of Q, then $|\lambda| = 1$. d. If Q_1 and Q_2 are orthogonal $n \times n$ matrices, then so is $Q_1 Q_2$.
Thm 5.9	p.376	5.2	a. through d. give properties of $W^\perp$: a. $W^\perp$ is a subspace of $\mathbb{R}^n$ b. $(W^\perp)^\perp = W$ c. $W \cap W^\perp = \{\mathbf{0}\}$ d. If $W = \text{span}(\mathbf{w}_1, \ldots, \mathbf{w}_k)$, then $\mathbf{v}$ is in $W^\perp$ if and only if $\mathbf{v} \cdot \mathbf{w}_i = 0$ for all i
Thm 5.10	p.376	5.2	$(\text{row}(A))^\perp = \text{null}(A)$ and $(\text{col}(A))^\perp = \text{null}(A^T)$
Thm 5.11	p.381	5.2	**The Orthogonal Decomposition Theorem**: $\mathbf{v} = \mathbf{w} + \mathbf{w}^\perp$
Cor 5.12	p.382	5.2	$(W^\perp)^\perp = W$
Thm 5.13	p.383	5.2	$\dim W + \dim W^\perp = n$
Cor 5.14	p.381	5.2	**The Rank Theorem**: Given A is $m \times n$ then ... $\text{rank}(A) + \text{nullity}(A) = n$ and $\text{rank}(A) + \text{nullity}(A^T) = m$

Theorems, Chapter 5, continued

Thm 5.15   p.389   5.3   **The Gram-Schmidt Process**:
Let $\{\mathbf{x}_1, \ldots, \mathbf{w}_k\}$ be a basis for W then ...
$\mathbf{v}_1 = \mathbf{x}_1$, $W_1 = \text{span}(\mathbf{x}_1)$
$\mathbf{v}_2 = \mathbf{x}_2 - \left(\frac{\mathbf{v}_1 \cdot \mathbf{x}_2}{\mathbf{v}_1 \cdot \mathbf{v}_1}\right) \mathbf{v}_1$, $W_2 = \text{span}(\mathbf{x}_1, \mathbf{x}_2)$
$\mathbf{v}_3 = \mathbf{x}_3 - \left(\frac{\mathbf{v}_1 \cdot \mathbf{x}_3}{\mathbf{v}_1 \cdot \mathbf{v}_1}\right) \mathbf{v}_1 - \left(\frac{\mathbf{v}_2 \cdot \mathbf{x}_3}{\mathbf{v}_2 \cdot \mathbf{v}_2}\right) \mathbf{v}_2$, $W_3 = \text{span}(\mathbf{x}_1, \mathbf{x}_2, \mathbf{x}_3)$
...
$\mathbf{v}_k = \mathbf{x}_k - \sum_{i=1}^{k-1} \frac{\mathbf{v}_i \cdot \mathbf{x}_k}{\mathbf{v}_i \cdot \mathbf{v}_i}$, $W_k = \text{span}(\{\mathbf{x}_k\})$
... $\{\mathbf{v}_1, \ldots, \mathbf{v}_k\}$ is an orthogonal basis for W.

Thm 5.16   p.390   5.3   **The QR Factorization**:
Let A be an $m \times n$ matrix with linearly independent columns.
Then A can be factored as $A = QR$, where
Q is an $m \times n$ matrix with orthonormal columns and
R is an invertible upper triangular matrix.

Thm 5.17   p.398   5.4   If A is *orthogonally diagonalizable*,
then A is *symmetric*.

Thm 5.18   p.398   5.4   If A is a *real* symmetric matrix,
then the eigenvalues of A are *real*.

Thm 5.19   p.399   5.4   If A is a *symmetric* matrix, then any two eigenvectors
corresponding to distinct eigenvalues of A are *orthogonal*.

Thm 5.20   p.400   5.4   **The Spectral Theorem**:
Let A be an $n \times n$ *real* matrix. Then A is *symmetric*
if and only if A is *orthogonally diagonalizable*.

Theorems, Chapter 5, continued

Thm 5.21   p.409   5.5   If C is an (n,k) binary code with generator matrix G and parity check matrix P, then $C^\perp$ is an $(n, n-k)$ binary code such that
a. G^T is parity check matrix for $C^\perp$.
b. P^T is generator matrix for $C^\perp$.

Thm 5.22   p.411   5.5   If C is a self dual code, then
a. Every vector in C has even weight.
b. **1** is in C.

Thm 5.23   p.411   5.5   **The Principal Axes Theorem**:
Every quadratic form can be diagonalized:
$\mathbf{x}^T A \mathbf{x} = \mathbf{y}^T D \mathbf{y} = \lambda_1 y_1^2 + \cdots + \lambda_n y_n^2$
(see statement of Theorem in the text for details)

Thm 5.24   p.416   5.5   The quadratic form $f(\mathbf{x}) = \mathbf{x}^T A \mathbf{x}$ is:
a. *positive definite* $\Leftrightarrow \lambda_i > 0$
b. *positive semidefinite* $\Leftrightarrow \lambda_i \geq 0$
c. *negative definite* $\Leftrightarrow \lambda_i < 0$
d. *negative semidefinite* $\Leftrightarrow \lambda_i \leq 0$
e. *indefinite* $\Leftrightarrow \lambda_i$ are both positive and negative

Thm 5.25   p.417   5.5   Given $f(\mathbf{x}) = \mathbf{x}^T A \mathbf{x}$,
$\lambda_1 \geq \lambda_2 \geq \cdots \geq \lambda_n$, and $\|\mathbf{x}\| = 1$:
a. $\lambda_1 \geq f(\mathbf{x}) \geq \lambda_n$
b. $\max f(\mathbf{x}) = \lambda_1$ occurs for a unit eigenvector
c. $\min f(\mathbf{x}) = \lambda_n$ occurs for a unit eigenvector

Theorems, Chapter 6

Thm 6.1	p.437	6.1	Let V be a vector space, **u** a vector, c a scalar: a. $0\mathbf{u} = \mathbf{0}$ b. $c\mathbf{0} = \mathbf{0}$ c. $(-1)\mathbf{u} = -\mathbf{u}$ d. If $c\mathbf{u} = \mathbf{0}$, then $c = 0$ or $\mathbf{u} = \mathbf{0}$.
Thm 6.2	p.438	6.1	Let $W(\neq \emptyset) \subseteq V$, then W is a subspace if: a. $\mathbf{u}, \mathbf{v} \in W \Rightarrow \mathbf{u} + \mathbf{v} \in W$ b. $\mathbf{u} \in W \Rightarrow c\mathbf{u} \in W$
Thm 6.3	p.445	6.1	Let $\mathbf{v}_1, \mathbf{v}_2, \ldots, \mathbf{v}_k$ be vectors in a vector space V. a. $\text{span}(\mathbf{v}_1, \mathbf{v}_2, \ldots, \mathbf{v}_k)$ is a subspace of V b. $\text{span}(\mathbf{v}_1, \mathbf{v}_2, \ldots, \mathbf{v}_k)$, smallest subspace with $\mathbf{v}_1, \mathbf{v}_2, \ldots, \mathbf{v}_k$
Thm 6.4	p.448	6.2	vector can be written as linear combination of others $\Leftrightarrow$ *linearly dependent*
Thm 6.5	p.452	6.2	If $\mathcal{B} = \{\mathbf{u}_i\}$ is a basis and $\mathbf{x} = \sum_{i=1}^{n} c_i \mathbf{u}_i$, then c_i are unique.
Thm 6.6	p.455	6.2	Let $\mathcal{B} = \{\mathbf{u}_i\}$ be a basis for V then: a. $[\mathbf{u} + \mathbf{v}]_\mathcal{B} = [\mathbf{u}]_\mathcal{B} + [\mathbf{v}]_\mathcal{B}$ b. $[c\mathbf{u}]_\mathcal{B} = c[\mathbf{u}]_\mathcal{B}$
Thm 6.7	p.456	6.2	Let $\mathcal{B} = \{\mathbf{u}_i\}$ be a basis for V then: $\{\mathbf{u}_i\}$ are linearly independent in V if and only if $[\mathbf{u}_i]_\mathcal{B}$ are linearly independent in $\mathbb{R}^n$
Thm 6.8	p.456	6.2	Let $\mathcal{B} = \{\mathbf{u}_i\}$ be a basis for V then: a. Any set of more than n vectors in V must be linearly dependent b. Any set of fewer than n vectors cannot span V
Thm 6.9	p.457	6.2	Every basis has exactly the same number of vectors
Thm 6.10	p.458	6.2	Let V be a vector space with $\dim V = n$. Then: a. Any linearly independent set in V contains at most n vectors. b. Any spanning set for V contains at least n vectors. c. Any lin indpt set of exactly n vectors in V is a basis for V. d. Any spanning set for V of exactly n vectors is a basis for V. e. Any lin indpt set in V can be extended to a basis for V. f. Any spanning set for V can be reduced to a basis for V.
Thm 6.11	p.460	6.2	Let W be a subspace of a finite-dimensional vector space V. a. W is finite-dimensional and $\dim W \leq \dim V$. b. $\dim W = \dim V$ if and only if $W = V$.

Theorems, Chapter 6, continued

Thm 6.12 p.469 6.3 Let $\mathcal{B} = \{\mathbf{u}_1, \ldots, \mathbf{u}_n\}$, $\mathcal{C} = \{\mathbf{v}_1, \ldots, \mathbf{v}_n\}$ be bases for V.
let $P_{\mathcal{C} \leftarrow \mathcal{B}}$ be a *change-of-basis* matrix from $\mathcal{B}$ to $\mathcal{C}$. Then:
a. $P_{\mathcal{C} \leftarrow \mathcal{B}} [\mathbf{x}]_\mathcal{B} = [\mathbf{x}]_\mathcal{C}$ for all $\mathbf{x}$ in V.
b. $P_{\mathcal{C} \leftarrow \mathcal{B}}$ is the unique matrix P with property (a).
c. $P_{\mathcal{C} \leftarrow \mathcal{B}}$ is invertible and $(P_{\mathcal{C} \leftarrow \mathcal{B}})^{-1} = P_{\mathcal{B} \leftarrow \mathcal{C}}$.

Thm 6.13 p.474 6.3 Let $\mathcal{B} = \{\mathbf{u}_1, \ldots, \mathbf{u}_n\}$, $\mathcal{C} = \{\mathbf{v}_1, \ldots, \mathbf{v}_n\}$ be bases for V.
Let $B = \begin{bmatrix} [\mathbf{u}_1]_\mathcal{E} & \cdots & [\mathbf{u}_n]_\mathcal{E} \end{bmatrix}$, $C = \begin{bmatrix} [\mathbf{v}_1]_\mathcal{E} & \cdots & [\mathbf{v}_n]_\mathcal{E} \end{bmatrix}$.
Then: $[\, C \,|\, B \,] \longrightarrow [\, I \,|\, P_{\mathcal{C} \leftarrow \mathcal{B}} \,]$.

Thm 6.14 p.479 6.4 Let $T : V \to W$ be a linear transformation. Then:
a. $T(\mathbf{0}) = \mathbf{0}$.
b. $T(-\mathbf{v}) = -T(\mathbf{v})$ for all $\mathbf{v}$ in V.
c. $T(\mathbf{u} - \mathbf{v}) = T(\mathbf{u}) - T(\mathbf{v})$ for all $\mathbf{u}, \mathbf{v}$ in V.

Thm 6.15 p.480 6.4 Let $T : V \to W$ be a linear transformation.
Let $\mathcal{B} = \{\mathbf{v}_1, \ldots, \mathbf{v}_n\}$ be a spanning set for V. Then:
$T(\mathcal{B}) = \{T(\mathbf{v}_1), \ldots, T(\mathbf{v}_n)\}$ is a spanning set for the range of T.

Thm 6.16 p.481 6.4 If $T : U \to V$ and $S : V \to W$ then
$S \circ T : U \to W$, $(S \circ T)(\mathbf{u}) = S(T(\mathbf{u}))$, is a linear transformation.

Thm 6.17 p.483 6.4 If T is an *invertible* linear transformation, then its *inverse* is unique.
If $R \circ T = I_V$, $T \circ R = I_W$ and $S \circ T = I_V$, $T \circ S = I_W$, then $R = S$.

Thm 6.18 p.488 6.5 Let $T : V \to W$ be a linear transformation. Then:
a. The *kernel* of T is a subspace of V
b. The *range* of T is a subspace of W

Thm 6.19 p.490 6.5 Let $T : V \to W$ be a linear transformation. Then:
$\mathrm{rank}(T) + \mathrm{null}(T) = \dim V$

Thm 6.20 p.494 6.5 Let $T : V \to W$ be a linear transformation. Then:
T is *one-to-one* if and only if $\ker(T) = \{\mathbf{0}\}$.

Thm 6.21 p.494 6.5 Let $\dim V = \dim W = n$. Then:
$T : V \to W$ is *one-to-one* if and only if T is *onto*.

Thm 6.22 p.495 6.5 Let $T : V \to W$ be a *one-to-one* linear transformation.
If $S = \{\mathbf{v}_1, \ldots, \mathbf{v}_k\}$ is a linearly independent set in V. Then:
$T(S) = \{T(\mathbf{v}_1), \ldots, T(\mathbf{v}_k)\}$ is a linearly independent set in W.

Cor 6.23 p.495 6.5 Let $\dim V = \dim W = n$. Then:
If T is *one-to-one*, then T maps a basis for V to a basis for W.

Theorems, Chapter 6, continued

Thm 6.24 p.495 6.5 Let $T : V \to W$ be a linear transformation. Then:
T is *invertible* if and only if T is *one-to-one* and *onto*.

Thm 6.25 p.498 6.5 Let V and W be two finite dimensional vector spaces. Then:
V is *isomorphic* to W ($V \cong W$) if and only if $\dim V = \dim W$.

Thm 6.26 p.502 6.6 Let V and W be two finite dimensional vector spaces
with bases $\mathcal{B}$ and $\mathcal{C}$ respectively, where $\mathcal{B} = \{\mathbf{v}_1, \ldots, \mathbf{v}_n\}$.
Let $T : V \to W$ be a linear transformation, then
the $n \times n$ matrix $A = \big[\, [T(\mathbf{v}_1)]_\mathcal{C} \,\big|\, [T(\mathbf{v}_2)]_\mathcal{C} \,\big|\, \cdots \,\big|\, [T(\mathbf{v}_n)]_\mathcal{C} \,\big]$
satisfies $A[\mathbf{v}]_\mathcal{B} = [T(\mathbf{v})]_\mathcal{C}$ for every vector $\mathbf{v}$ in V.

Thm 6.27 p.508 6.6 Let U, V, and W be finite dimensional vector spaces
with bases $\mathcal{B}$, $\mathcal{C}$, and $\mathcal{D}$ respectively.
Let $T : U \to V$ and $S : V \to W$ be linear transformations. Then:
$[S \circ T]_{\mathcal{D} \leftarrow \mathcal{B}} = [S]_{\mathcal{D} \leftarrow \mathcal{C}} [T]_{\mathcal{C} \leftarrow \mathcal{B}}$

Thm 6.28 p.509 6.6 Let V and W be finite dimensional vector spaces
with bases $\mathcal{B}$ and $\mathcal{C}$, respectively.
Let $T : V \to W$ be a linear transformation. Then:
T is invertible if and only if the matrix $[T]_{\mathcal{C} \leftarrow \mathcal{B}}$ is invertible.
In this case, $([T]_{\mathcal{C} \leftarrow \mathcal{B}})^{-1} = [T^{-1}]_{\mathcal{C} \leftarrow \mathcal{B}}$

Thm 6.29 p.512 6.6 Let V be a finite dimensional vector space
with bases $\mathcal{B}$ and $\mathcal{C}$.
Let $T : V \to V$ be a linear transformation. Then:
$[T]_\mathcal{C} = P^{-1} [T]_\mathcal{B} P$
where P is the change-of-basis matrix from $\mathcal{C}$ to $\mathcal{B}$.

Thm 6.30 p.516 6.6 *Fundamental Theorem of Invertible Matrices*: Version 4
Let A be an $n \times n$ matrix and
let $T : V \to W$ be a linear transformation
whose matrix $([T]_{\mathcal{C} \leftarrow \mathcal{B}})$ with respect to bases
$\mathcal{B}$ and $\mathcal{C}$ of V and W respectively, is A.
Statements a. through t. are equivalent (see p.516).
p. T is invertible
q. T is one-to-one
r. T is onto
s. $\ker(T) = \mathbf{0}$
t. $\text{range}(T) = W$

Theorems, Chapter 6, continued

Thm 6.31 p.522 6.7 The set S of all solutions to $y' + ay = 0$ is a subspace of $\mathscr{F}$.

Thm 6.32 p.523 6.7 If S is the solution space of $y' + ay = 0$, then:
dim $= 1$ and $\{e^{-at}\}$ is a basis for S.

Thm 6.33 p.526 6.7 Let S be the solution space of $y'' + ay' + by = 0$ and let λ_1 and λ_2 be the roots of $\lambda^2 + a\lambda + b = 0$.
a. If $\lambda_1 \neq \lambda_2$, then $\{e^{\lambda_1 t}, e^{\lambda_2 t}\}$ is a basis for S.
b. If $\lambda_1 = \lambda_2$, then $\{e^{\lambda_1 t}, te^{\lambda_1 t}\}$ is a basis for S.

Thm 6.34 p.531 6.7 Let C be an (n, k) linear code.
a. The dual code $C^\perp$ is an $(n, n-k)$ linear code.
b. C contains 2^k vectors, and $C^\perp$ contains 2^{n-k} vectors.

Thm 6.35 p.533 6.7 For $n \geq 1$, the *Reed-Muller code* R_n is:
a $(2^n, n+1)$ linear code in which
every vector (except **0** and **1**) has weight 2^{n-1}.

Theorems, Chapter 7

Thm 7.1 p.544 7.1 Let **u**, **v**, and **w** be vectors
in an inner product space V and c a scalar:
a. $\langle \mathbf{u} + \mathbf{v}, \mathbf{w} \rangle = \langle \mathbf{u}, \mathbf{v} \rangle + \langle \mathbf{u}, \mathbf{w} \rangle$
b. $\langle \mathbf{u}, c\mathbf{v} \rangle = c\langle \mathbf{u}, \mathbf{v} \rangle$
c. $\langle \mathbf{u}, \mathbf{0} \rangle = \langle \mathbf{0}, \mathbf{v} \rangle = 0$

Thm 7.2 p.546 7.1 Let **u** and **v** be vectors in an inner product space V.
Then **u** and **v** are orthogonal if and only if $\|\mathbf{u} + \mathbf{v}\|^2 = \|\mathbf{u}\|^2 + \|\mathbf{v}\|^2$

Thm 7.3 p.548 7.1 Let **u** and **v** be vectors in an inner product space V.
Then $|\langle \mathbf{u}, \mathbf{v} \rangle| \le \|\mathbf{u}\| \|\mathbf{v}\|$ with equality holding if and only if
u and **v** are scalar multiples of each other

Thm 7.4 p.549 7.1 Let **u** and **v** be vectors in an inner product space V.
Then $\|\mathbf{u} + \mathbf{v}\| \le \|\mathbf{u}\| + \|\mathbf{v}\|$

Thm 7.5 p.544 7.2 Let d be a distance function defined on a normed linear space V.
The following properties hold for all vectors **u**, **v**, and **w** in V.
a. $d(\mathbf{u}, \mathbf{v}) \ge 0$, and $d(\mathbf{u}, \mathbf{v}) = 0$ if and only if $\mathbf{u} = \mathbf{v}$
b. $d(\mathbf{u}, \mathbf{v}) = d(\mathbf{v}, \mathbf{u})$
c. $d(\mathbf{u}, \mathbf{w}) \le d(\mathbf{u}, \mathbf{v}) + d(\mathbf{v}, \mathbf{w})$

Thm 7.6 p.567 7.2 If $\|\mathbf{x}\|$ is a vector norm on $\mathbb{R}$,
then $\|A\| = \max \|A\mathbf{x}\|$ where $\|\mathbf{x}\| = 1$
defines a matrix norm on M_{nn} that is compatible with
the vector norm that induces it.

Thm 7.7 p.569 7.2 Let A have column vectors $\mathbf{a}_i$ and row vectors $\mathbf{A}_i$ then:
a. $\|A\|_1 = \max\{\|\mathbf{a}_i\|_s\} = \max\left\{\sum_{i=1}^{n} |a_{ij}|\right\}$
b. $\|A\|_\infty = \max\{\|\mathbf{A}_i\|_s\} = \max\left\{\sum_{j=1}^{n} |a_{ij}|\right\}$

Thm 7.8 p.579 7.3 If W is a finite-dimensional subspace of an inner product space V
and if **v** is a vector in V,
then $\text{proj}_W(\mathbf{v})$ is the *best approximation* to **v** in W.

Thm 7.9 p.584 7.3 $A\mathbf{x} = \mathbf{b}$ has at least one *least squares solution*, $\bar{\mathbf{x}}$.
a. $\bar{\mathbf{x}}$ is a *least squares solution* of $A\mathbf{x} = \mathbf{b}$ if and only if
$\bar{\mathbf{x}}$ is a solution of the normal equations $A^T A \bar{\mathbf{x}} = A^T \mathbf{b}$.
b. A has linearly independent columns $\Leftrightarrow A^T A$ is invertible.
Then $\bar{\mathbf{x}} = (A^T A)^{-1} A^T \mathbf{b}$ is the *unique* solution.

Theorems, Chapter 7, continued

Thm 7.10 p.591 7.3 Let A have linearly independent columns
with QR factorization $A = QR$ (see Theorem 5.16 below).
Then $\bar{\mathbf{x}} = R^{-1}Q^T\mathbf{b}$ is the *unique* solution of $A\mathbf{x} = \mathbf{b}$.

Thm 7.11 p.592 7.3 Let the columns of A be a basis for W.
Then $P : \mathbb{R}^m \to \mathbb{R}^n$ that projects $\mathbb{R}^m$ onto W
has matrix $A(A^TA)^{-1}A^T$ and
$\text{proj}_W(\mathbf{v}) = A(A^TA)^{-1}A^T\mathbf{v}$.

Thm 7.12 p.595 7.3 Let A have linearly independent columns.
Then A^+ the pseudoinverse of A satisfies the *Penrose conditions*.
a. $AA^+A = A$
b. $A^+AA^+ = A^+$
c. AA^+ and A^+A are symmetric

Thm 7.13 p.602 7.4 **The Singular Value Decomposition Theorem**
Let A be an $m \times n$ matrix with singular values
$\sigma_1 \geq \sigma_1 \geq \ldots \geq \sigma_r > 0$ and $\sigma_{r+1} = \sigma_{r+2} = \ldots = \sigma_n = 0$.
Then there exist an $m \times m$ orthogonal matrix U
and an $n \times n$ orthogonal matrix V
and an $m \times n$ matrix Σ (see p.601)
such that $A = U\Sigma V^T$.

Thm 7.14 p.605 7.4 Let A be an $m \times n$ matrix with singular values
$\sigma_1 \geq \sigma_1 \geq \ldots \geq \sigma_r > 0$ and $\sigma_{r+1} = \sigma_{r+2} = \ldots = \sigma_n = 0$.
Let $\mathbf{u}_1, \ldots, \mathbf{u}_r$ be left singular values and
let $\mathbf{v}_1, \ldots, \mathbf{v}_r$ be left singular values. Then:
$A = \sigma_1\mathbf{u}_1\mathbf{v}_1^T + \cdots + \sigma_r\mathbf{u}_r\mathbf{v}_r^T$.

Thm 7.15 p.606 7.4 Let $A = U\Sigma V^T$ be an SVD of an $m \times m$ matrix A
with nonzero singular values $\sigma_1, \ldots, \sigma_r$.
a. The rank of A is r.
b. $\{\mathbf{u}_1, \ldots, \mathbf{u}_r\}$ is an orthonormal basis for $\text{col}(A)$.
c. $\{\mathbf{u}_{r+1}, \ldots, \mathbf{u}_m\}$ is an orthonormal basis for $\text{null}(A^T)$.
d. $\{\mathbf{v}_1, \ldots, \mathbf{v}_r\}$ is an orthonormal basis for $\text{row}(A)$.
e. $\{\mathbf{v}_{r+1}, \ldots, \mathbf{v}_m\}$ is an orthonormal basis for $\text{null}(A)$.

Thm 7.16 p.607 7.4 Let $A = U\Sigma V^T$ be an SVD of an $m \times n$ matrix A with rank r.
Then the image of the unit sphere in $\mathbb{R}^n$
under the matrix transformation that maps $\mathbf{x}$ to $A\mathbf{x}$ is
a. The surface of an ellipsoid in $\mathbb{R}^m$ if $r = n$.
b. A solid ellipsoid in $\mathbb{R}^m$ if $r < n$.

Theorems, Chapter 7, continued

Thm 7.17 p.610 7.4 Let A be an $m \times n$ matrix and let $\sigma_1, \ldots, \sigma_r$ be all the nonzero singular values of A. Then: $\|A\|_F = \sqrt{\sigma_1^2 + \cdots + \sigma_r^2}$

Thm 7.18 p.613 7.4 The least squares problem $A\mathbf{x} = \mathbf{b}$ has a unique least squares solution $\bar{\mathbf{x}}$ of minimal length that is given by $\bar{\mathbf{x}} = A^+\mathbf{b}$.

Thm 7.20 p.628 7.5 Let C be a (binary) code with minimum distance d.
a. C detects k errors if and only if $d \geq k + 1$.
b. C detects k errors if and only if $d \geq 2k + 1$.

Thm 7.21 p.628 7.5 Let C be a (n, k) with parity check matrix P. Then the minimum distance of C is the smallest integer d for which there are d linearly dependent columns.

Theorems, Chapter 7, continued

The Fundamental Theorem of Invertible Matrices: Final Version

Thm 7.19 p.614 7.4 a. through u. give equivalent conditions for A to be invertible:

    a. A is invertible.
    b. $A\mathbf{x} = \mathbf{b}$ has a unique solution for every $\mathbf{b}$ in $\mathbb{R}^n$
    c. $A\mathbf{x} = \mathbf{0}$ has only the trivial solution.
    d. The reduced row echelon form of A is I_n.
    e. A is the product of elementary matrices.
    f. $\text{rank}(A) = n$
    g. $\text{nullity}(A) = 0$
    h. The column vectors of A are linearly independent.
    i. The column vectors of A span $\mathbb{R}^n$.
    j. The column vectors of A form a basis for $\mathbb{R}^n$.
    k. The row vectors of A are linearly independent.
    l. The row vectors of A span $\mathbb{R}^n$.
    m. The row vectors of A for a basis for $\mathbb{R}^n$.
    n. $\det(A) \neq 0$
    o. 0 is not an eigenvalue of A
    p. T is invertible
    q. T is one-to-one
    r. T is onto
    s. $\ker(T) = \{\mathbf{0}\}$
    t. $\text{range}(T) = W$
    u. 0 is not a singular value of A

Theorems by name

| Thm 7.8 | p.579 | 7.3 | **The Best Approximation Theorem** |

If W is a finite-dimensional subspace of an inner product space V and if $\mathbf{v}$ is a vector in V,
then $\text{proj}_W(\mathbf{v})$ is the *best approximation* to $\mathbf{v}$ in W.

| Thm 1.4 | p.19 | 1.2 | **The Cauchy-Schwarz Inequality** |

$|\mathbf{u} \cdot \mathbf{v}| \leq \|\mathbf{u}\| \|\mathbf{v}\|$

| Thm 7.3 | p.548 | 7.1 | **The Cauchy-Schwarz Inequality** |

Let $\mathbf{u}$ and $\mathbf{v}$ be vectors in an inner product space V.
Then $|\langle \mathbf{u}, \mathbf{v} \rangle| \leq \|\mathbf{u}\| \|\mathbf{v}\|$ with equality holding if and only if $\mathbf{u}$ and $\mathbf{v}$ are scalar multiples of each other

| Thm 4.11 | p.274 | 4.2 | **Cramer's Rule** |

If $A\mathbf{x} = \mathbf{b}$, then $x_i = \frac{\det(A_i(\mathbf{b}))}{\det A}$

| Thm 5.15 | p.386 | 5.3 | **The Gram-Schmidt Process** |

Let $\{\mathbf{x}_1, \ldots, \mathbf{w}_k\}$ be a basis for W then ...
$\mathbf{v}_1 = \mathbf{x}_1$, $W_1 = \text{span}(\mathbf{x}_1)$
$\mathbf{v}_2 = \mathbf{x}_2 - \left(\frac{\mathbf{v}_1 \cdot \mathbf{x}_2}{\mathbf{v}_1 \cdot \mathbf{v}_1}\right) \mathbf{v}_1$, $W_2 = \text{span}(\mathbf{x}_1, \mathbf{x}_2)$
$\mathbf{v}_3 = \mathbf{x}_3 - \left(\frac{\mathbf{v}_1 \cdot \mathbf{x}_3}{\mathbf{v}_1 \cdot \mathbf{v}_1}\right) \mathbf{v}_1 - \left(\frac{\mathbf{v}_2 \cdot \mathbf{x}_3}{\mathbf{v}_2 \cdot \mathbf{v}_2}\right) \mathbf{v}_2$, $W_3 = \text{span}(\mathbf{x}_1, \mathbf{x}_2, \mathbf{x}_3)$
...
$\mathbf{v}_k = \mathbf{x}_k - \sum_{i=1}^{k-1} \frac{\mathbf{v}_i \cdot \mathbf{x}_k}{\mathbf{v}_i \cdot \mathbf{v}_i}$, $W_k = \text{span}(\{\mathbf{x}_k\})$
... $\{\mathbf{v}_1, \ldots, \mathbf{v}_k\}$ is an orthogonal basis for W.

| Thm 4.1 | p.265 | 4.2 | **The Laplace Expansion Theorem** |

$\det A = \sum_{j=1}^{n} a_{ij} C_{ij}$ (any row) or $\sum_{i=1}^{n} a_{ij} C_{ij}$ (any column)

| Thm 7.9 | p.584 | 7.3 | **The Least Squares Theorem** |

$A\mathbf{x} = \mathbf{b}$ has at least one *least squares solution*, $\bar{\mathbf{x}}$.

a. $\bar{\mathbf{x}}$ is a *least squares solution* of $A\mathbf{x} = \mathbf{b}$ if and only if $\bar{\mathbf{x}}$ is a solution of the normal equations $A^T A \bar{\mathbf{x}} = A^T \mathbf{b}$.
b. A has linearly independent columns $\Leftrightarrow A^T A$ is invertible. Then $\bar{\mathbf{x}} = (A^T A)^{-1} A^T \mathbf{b}$ is the *unique* solution.

| Thm 5.11 | p.381 | 5.2 | **The Orthogonal Decomposition Theorem** |

$\mathbf{v} = \mathbf{w} + \mathbf{w}^\perp$

| Thm 4.36 | p.330 | 4.6 | **Perron's Theorem** |

See Theorem 4.36.

| Thm 4.37 | p.332 | 4.6 | **Perron-Frobenius Theorem** |

See Theorem 4.37.

Theorems by name, continued

Thm 5.23   p.411   5.5   **The Principal Axes Theorem**
Every quadratic form can be diagonalized:
$\mathbf{x}^T A \mathbf{x} = \mathbf{y}^T D \mathbf{y} = \lambda_1 y_1^2 + \cdots + \lambda_n y_n^2$
(see statement of Theorem in the text for details)

Thm 1.6   p.23   1.2   **Pythagoras' Theorem**
$\|\mathbf{u} + \mathbf{v}\|^2 = \|\mathbf{u}\|^2 + \|\mathbf{v}\|^2 \Leftrightarrow \mathbf{u} \cdot \mathbf{v} = 0$

Thm 7.2   p.546   7.1   **Pythagoras' Theorem**
Let $\mathbf{u}$ and $\mathbf{v}$ be vectors in an inner product space V.
Then $\mathbf{u}$ and $\mathbf{v}$ are orthogonal $\Leftrightarrow \|\mathbf{u} + \mathbf{v}\|^2 = \|\mathbf{u}\|^2 + \|\mathbf{v}\|^2$

Thm 5.16   p.390   5.3   **The QR Factorization**
Let A be an $m \times n$ matrix with linearly independent columns.
Then A can be factored as $A = QR$, where
Q is an $m \times n$ matrix with orthonormal columns and
R is an invertible upper triangular matrix.

Thm 2.2   p.75   2.2   **The Rank Theorem**: number of free variables $= n - \text{rank}(A)$

Thm 3.26   p.203   3.5   **The Rank Theorem**: $\text{rank}(A) + \text{nullity}(A) = n$

Cor 5.14   p.381   5.2   **The Rank Theorem**
Given A is $m \times n$ then ...
$\text{rank}(A) + \text{nullity}(A) = n$ and $\text{rank}(A) + \text{nullity}(A^T) = m$

Thm 6.19   p.490   6.5   **The Rank Theorem**
Let $T: V \to W$ be a linear transformation. Then:
$\text{rank}(T) + \text{null}(T) = \dim V$

Thm 5.20   p.400   5.4   **The Spectral Theorem**
Let A be an $n \times n$ *real* matrix. Then A is *symmetric*
if and only if A is *orthogonally diagonalizable*.

Thm 1.5   p.19   1.2   **The Triangle Inequality**: $\|\mathbf{u} + \mathbf{v}\| \leq \|\mathbf{u}\| + \|\mathbf{v}\|$

Thm 7.4   p.549   7.1   **The Triangle Inequality**
Let $\mathbf{u}$ and $\mathbf{v}$ be vectors in an inner product space V.
Then $\|\mathbf{u} + \mathbf{v}\| \leq \|\mathbf{u}\| + \|\mathbf{v}\|$

Thm 7.13   p.602   7.4   **The Singular Value Decomposition Theorem**
Let A be an $m \times n$ matrix with singular values
$\sigma_1 \geq \sigma_1 \geq \ldots \geq \sigma_r > 0$ and $\sigma_{r+1} = \sigma_{r+2} = \ldots = \sigma_n = 0$.
Then there exist an $m \times m$ orthogonal matrix U
and an $n \times n$ orthogonal matrix V
and an $m \times n$ matrix Σ (see p.601)
such that $A = U\Sigma V^T$.